# The Mammalian Cochlear Nuclei
Organization and Function

# The Mammalian Cochlear Nuclei

Organization and Function

Edited by
## Miguel A. Merchán
University of Salamanca
Salamanca, Spain

## José M. Juiz
University of Alicante
Alicante, Spain

## Donald A. Godfrey
Medical College of Ohio
Toledo, Ohio

and

## Enrico Mugnaini
University of Connecticut
Storrs, Connecticut

Springer Science+Business Media, LLC

Proceedings of a NATO Advanced Research Workshop on
The Mammalian Cochlear Nuclei: Organization and Function,
held September 14–17, 1991,
in Salamanca, Spain

**NATO-PCO-DATA BASE**

The electronic index to the NATO ASI Series provides full bibliographical references (with keywords and/or abstracts) to more than 30,000 contributions from international scientists published in all sections of the NATO ASI Series. Access to the NATO-PCO-DATA BASE is possible in two ways:

—via online FILE 128 (NATO-PCO-DATA BASE) hosted by ESRIN, Via Galileo Galilei, I-00044 Frascati, Italy

—via CD-ROM "NATO-PCO-DATA BASE" with user-friendly retrieval software in English, French, and German ( ©WTV GmbH and DATAWARE Technologies, Inc. 1989)

The CD-ROM can be ordered through any member of the Board of Publishers or through NATO-PCO, Overijse, Belgium.

```
      Library of Congress Cataloging-in-Publication Data

The Mammalian cochlear nuclei : organization and function / edited by
  Miguel A. Merchán ... [et al.].
       p.   cm. -- (NATO ASI series. Series A, Life sciences ; 239)
    "Proceedings of a NATO Advanced Research Workshop on the Mammalian
  Cochlear Nuclei: Organization and Function, held September 14-17,
  1991, in Salamanca, Spain"--T.p. verso.
    "Published in cooperation with NATO Scientific Affairs Division."
    Includes bibliographical references and index.

    1. Cochlear nucleus--Congresses.  I. Merchán, Miguel A.
  II. North Atlantic Treaty Organization.  Scientific Affairs
  Division.  III. NATO Advanced Research Workshop on the Mammalian
  Cochlear Nuclei: Organization and Function (1991 : Salamanca, Spain)
  IV. Title.  V. Series.
    [DNLM: 1. Cochlea--anatomy & histology--congresses. 2. Cochlea-
  -physiology--congresses.  3. Cochlear Nerve--anatomy & histology-
  -congresses.  4. Cochlear Nerve--physiology--congresses.  WV 250
  M265 1991]
  QP379.M26   1993
  599'.01825--dc20
  DNLM/DLC
  for Library of Congress                                    92-48455
                                                                 CIP
```

ISBN 978-1-4613-6273-9         ISBN 978-1-4615-2932-3 (eBook)
DOI 10.1007/978-1-4615-2932-3

© Springer Science+Business Media New York 1993
Originally published by Plenum Press, New York 1993
Softcover reprint of the hardcover 1st edition 1993

All rights reserved

No part of this book may be reproduced, stored in a retrieval system, or transmitted in any form or by any means, electronic, mechanical, photocopying, microfilming, recording, or otherwise, without written persmission from the Publisher

# Preface

The presence of sophisticated auditory processing in mammals has permitted perhaps the most significant evolutionary development in humans: that of language. An understanding of the neural basis of hearing is thus a starting point for elucidating the mechanisms that are essential to human communication. The cochlear nucleus is the first region of the brain to receive input from the inner ear and is therefore the earliest stage in the central nervous system at which auditory signals are processed for distribution to higher centers. Clarifying its role in the central auditory pathway is crucial to our knowledge of how the brain deals with complex stimuli such as speech, and is also essential for understanding the central effects of peripheral sensorineural hearing loss caused by, for example, aging, ototoxic drugs, and noise. Ambitious new developments to assist people with total sensorineural deafness, including both cochlear and cochleus nuclear implants, require a detailed knowledge of the neural signals received by the brainstem and how these are processed. Recently, many new data have been obtained on the structure and function of the cochlear nucleus utilizing combinations of anatomical, physiological, pharmacological and molecular biological procedures. Approaches such as intracellular dye-filling of physiologically identified neurons, localization of classical neurotransmitters, peptides, receptors and special proteins, or gene expression have opened the door to novel morphofunctional correlations.

In order to provide an opportunity for leading scientists working on the cochlear nucleus to present and discuss their recent results, a NATO Advanced Research Workshop was held at the University of Salamanca, Spain, in September, 1991. The resulting volume consolidates the large amount of information presented at this meeting into a book that will be available to the scientific community. The studies presented here reemphasize our knowledge on how parallel channels of information processing are established in the cochlear nucleus and indicate means by which particular cell types in the nucleus are modulated by intrinsic and feed-back mechanisms. We hope that these proceedings will not only help advance further neural network strategies, some of which were discussed at the workshop, but also stimulate the design of mammalian neuroethological approaches, which have so far been promoted almost solely in bats.

The workshop as well as the book were dedicated to the memory of the Spanish neuroscientist, Dr. Rafael Lorente de Nó, (1902-1990). Born in Zaragoza and a follower of the famous neurohistologist Ramón y Cajal, Lorente de Nó produced one of the most extensive early descriptions of the morphology of the cochlear nucleus, a body of work that is still widely referred to by auditory scientists today. During the preparation of the present book, Dr. A. Gallego, one of the contributors to the meeting and a colleague of Lorente de Nó, died in Madrid. Because of his relevant role in the continuity of the Spanish scientific tradition during the last decades, we also wish to extend this dedication to the memory of Dr. Gallego.

Many people were involved in the organization of the meeting and the production of this volume. We gratefully acknowledge the invaluable financial support from the NATO Scientific Affairs Division and the contribution from the Spanish Ministry of Science and Education (MEC-DGICYT), Fondo de Investigaciones Sanitarias de la Seguridad Social (FISS), Town Council of Salamanca, and the Institute of Neuroscience of the University of Alicante. We wish to take this opportunity to acknowledge all the contributors and participants, the Scientific Committee, and the team in Salamanca for their support, advice, and hard work in making this conference and volume a reality. We also wish to extend our thanks to Lisbet Sørensen, who processed the entire text, and to Maria E. Rubio and Adelaida Baso for their help with the tedious work of figure placement.

*M. Merchán (Salamanca, Spain)*
*J. Juiz (Alicante, Spain)*
*D. Godfrey (Toledo, USA)*
*E. Mugnaini (Storrs, USA)*

# In Memory of Rafael Lorente de Nó

The National Institute on Deafness and Other Communication Disorders joins in your salute to Rafael Lorente de Nó. An innovator who successfully translated experimental results into new concepts, Lorente de Nó's significant contributions remain as testimony to his creative genius in central nervous system research. His concentualization of the organization of the cochlear nuclei provides the foundation for understanding the auditory system.

A scientist who held positions with the Central Institute for the Deaf, Rockefeller University, and the University of California, Los Angeles, Lorente de Nó's interest in the biochemical mechanisms of nerve conduction continued throughout his career. Honored on both sides of the Atlantic for his work, he received the highest scientific accolade the United States can bestow, membership of the National Academy of Sciences. Our thoughts are with you today as you honor the memory of this distinguished scientist, a man whose insight and discoveries form a legacy of knowledge for all investigators of the auditory system.

*James B. Snow, Jr., M.D.*
*Director of National Institute on Deafness and*
*Other Communication Disorders*
*September, 1991*

# In Memory of Antonia Gallego

While this book was being prepared for publication, one of its contributors, Prof. A. Gallego, died in Madrid in February 1992, shortly after writing that paper. Antonio Gallego was the most prestigious Spanish neurobiologist of the last 50 years and also a symbol of the scientists that struggled to develop a favorable scientific environment in a country devastated by the material and psychological consequences of a Civil War. Antonio Gallego was born in Madrid in 1915. He obtained his M.D. degree at the University of Madrid and initiated there, still as a medical student, his career in physiology. After a forced interruption during the war and post-war years because of his alignment with the republican side, Gallego travelled to the Rockefeller Institute to work with Rafael Lorente de Nó on electrophysiology of the nerve. Back in Spain in 1950, Antonio Gallego resumed his academic activity and finally became Professor and Chairman of the Department of Physiology at the University of Madrid Medical School in 1960. There, he attracted a whole generation of young Spanish scientists that were impressed by his intellectual independence, his modern scientific views, and his courage to defy established doctrines in politics and science. Most of the present professors of Physiology in Spanish medical schools have been, directly or indirectly, pupils of Antonio Gallego.

His scientific work was mainly dedicated to study the functional structure of the retina. Main findings include the description of a new type of horizontal cell devoid of axon in the external plexiform layer, the report with J. Cruz of retinal association ganglion cells, and the first morphological description of the "interplexiform cell," to which he gave this name.

Antonio Gallego was deeply involved in university politics, directing his enormous energy, intelligence, and brilliant personality to the renovation of medical education and the improvement of research resources at the Spanish university. A whole generation of Spanish scientists owes a debt of gratitude to this man, who sacrificed his time and personal success to obtain a better future for science in his country.

*Carlos Belmonte, M.D., Ph.D.*
*Director, Institute of Neuroscience*
*University of Alicante*
*June, 1992*

# Main Authors*

Joe C. Adams
Department of Otolaryngology, Massachusetts
Eye and Ear Infirmary, 243 Charles Street,
Boston, Mass. 02114, USA

Richard A. Altschuler
Kresge Hearing Research Institute, University of
Michigan, Ann Arbor, MI 48109, USA

Albert S. Berrebi
Laboratory of Neuromorphology, Graduate
Degree Program in Biobehavioral Sciences, The
University of Connecticut, Storrs, Connecticut
06269-4154, USA

M. Christian Brown
Harvard Medical School, Eaton-Peabody
Laboratory, Massachusetts Eye & Ear Infirmary,
243 Charles St., Boston, MA 02114, USA

Nell Beatty Cant
Department of Neurobiology Duke University
Medical Center Durham, NC 27710, USA

Donald M. Caspary
Southern Illinois University School of Medicine,
P.O. Box 19230, Springfield, Illinois, 62794-9230, USA

John H. Casseday
Department of Neurobiology, Duke University
Medical Center, Durham, NC 27710, USA

Ellen Covey
Department of Neurobiology, Duke University,
Medical Center, Durham, NC 27710, USA

Edward F. Evans
Department of Communication and Neuroscience,
Keele University, Keele, Staffs. ST5 5BG, U.K.

Alfonso Fairén
Instituto Cajal, CSIC, Madrid, Spain

Eckhard Friauf
Department of Animal Physiology, University of
Tübingen, Auf der Morgenstelle 28, D-7400
Tübingen 1, Federal Republic of Germany

Antonio Gallego
Department of Physiology, Medical School,
Complutense University, Madrid, Spain

Donald A. Godfrey
Department of Otolaryngology - Head & Neck
Surgery, Medical College of Ohio, Toledo, Ohio
43699-0008, USA

Vicente Honrubia
Division of Head and Neck Surgery, UCLA
School of Medicine, University of California, Los
Angeles, USA

Ricardo Insausti
Department of Anatomy, University of Navarra,
Apdo. 273,
31080 Pamplona, Spain

José M. Juiz
Dpt. of Histology and Institute of Neuroscience,
University of Alicante, Apdo. 374, E-03080
Alicante, Spain

Lawrence Kruger
Brain Research Institute, UCLA Medical Center,
Los Angeles, CA 90024, USA

Patricia A. Leake
Epstein and Coleman Laboratories, Department of
Otolaryngology U499, University of California
San Francisco, San Francisco, CA 94143-0732,
USA

Dolores E. López
Departamento de Biología Celular y Patología,
Facultad de Medicina, Universidad de Salamanca,
E-37007 Salamanca, Spain

---

*There is an Abstracts Book with a complete list of participants, including contributors to the poster sessions of the workshop. The Abstracts Book also contains the summaries of these poster presentations, some of which are cited in several chapters of this volume.

Paul B. Manis
Department of Otolaryngology-Head and Neck Surgery, The Johns Hopkins University School of Medicine, Baltimore, MD 21205, USA

Ray Meddis
Speech and Hearing Laboratory, University of Technology, Loughborough, UK

D. Kent Morest
Department of Anatomy, The University of Connecticut Health Center, Farmington, CT 06030, USA

John K. Niparko
Department of Otolaryngology-Head & Neck Surgery, Johns Hopkins University, Baltimore, 21203 Maryland, USA

Donata Oertel
Department of Neurophysiology, University of Wisconsin, Madison, Wisconsin, USA

Douglas L. Oliver
Department of Anatomy, University of Connecticut Health Center, Farmington, CT 06030, USA

Alan R. Palmer
MRC Institute of Hearing Research, University of Nottingham, University Park, Nottingham NG7 2RD, U.K.

Steven J. Potashner
Department of Anatomy, University of Connecticut Health Center, Farmington, CT, 06030, USA

David K. Ruygo
Department of Otolaryngology - Head & Neck Surgery, Johns Hopkins University School of Medicine, Baltimore, MD 21205, USA

Murray B. Sachs
Department of Biomedical Engineering and Center for Hearing Sciences, Johns Hopkins University, School of Medicine, Baltimore, Maryland 21205, USA

Enrique Saldaña
Departamento de Biología Celular y Patología, Facultad de Medicina, Universidad de Salamanca, E- 37007 Salamanca, Spain

Philip H. Smith
Department of Neurophysiology, University of Wisconsin, Madison, Wisconsin, USA

Richard L. Saint Marie
Department of Anatomy, University of Connecticut Health Center, Farmington, 06030 Connecticut, USA

Larry W. Swanson
Department of Biological Sciences, University of Southern California, Los Angeles, California, 90089-2520, USA

Douglas E. Vetter
Laboratory of Neuromorphology, Graduate Degree Program in Biobehavioral Sciences, University of Connecticut, Storrs, CT 06269-4154, USA

Robert J. Wenthold
Laboratory of Neurochemistry, National Institute of Deafness and Other Communication Disorders, NIH, Bethesda, MD 20892, USA

Robert E. Wickesberg
Hearing Research Laboratory, S.U.N.Y. at Buffalo, Buffalo, New York 14214, USA

Frank H. Willard
Department of Anatomy, University of New England, Biddeford, Maine, 04005, USA

Thomas A. Woolsey
James L. O'Leary Division of Experimental Neurology, Department of Neurology and Neurological Surgery, Washington University School of Medicine, St. Louis, Missouri 63110, USA

Eric D. Young
Department of Biomedical Engineering, The Johns Hopkins University, Baltimore MD, 21205, USA

John M. Zook
Dept. Zoological & Biomedical Science, Ohio University, Athens, Ohio, 45701, USA

# Contents

The cellular basis for signal processing in the mammalian cochlear nuclei
    *D. Kent Morest* .................................................................. 1

## I. DEVELOPMENTAL ISSUES

Cell birth, formation of efferent connections, and establishment of tonotopic order in the rat cochlear nucleus
    *Eckhard Friauf and Karl Kandler* ................................ 19

Postnatal development of auditory nerve projections to the cochlear nucleus in *Monodelphis Domestica*
    *Frank H. Willard* ...................................................... 29

## II. PRIMARY INPUTS

Anatomical and physiological studies of type I and type II spiral ganglion neurons
    *M. Christian Brown* ................................................. 43

Topographic organization of inner hair cell synapses and cochlear spiral ganglion projections to the ventral cochlear nucleus
    *Patricia A. Leake, Russell L. Snyder and Gary T. Hradek* ................ 55

Ultrastructural analysis of synaptic endings of auditory nerve fibers in cats: correlations with spontaneous discharge rate
    *David K. Ryugo, Debora D. Wright, and Tan Pongstaporn* ............... 65

## III. INTRINSIC CONNECTIONS

Intrinsic connections in the cochlear nuclear complex studied *in vitro* and *in vivo*
    *Robert E. Wickesberg and Donata Oertel* ...................... 77

The synaptic organization of the ventral cochlear nucleus of the cat: the peripheral cap of small cells
    *Nell Beatty Cant* ...................................................... 91

Alterations in the dorsal cochlear nucleus of cerebellar mutant mice
    *Albert S. Berrebi and Enrico Mugnaini* ......................... 107

## IV. DESCENDING PROJECTIONS

Non-cochlear projections to the ventral cochlear nucleus: are they mainly inhibitory?
    Richard L. Saint Marie, E.-Michael Ostapoff,
    Christina G. Benson, D. Kent Morest and Steven J. Potashner ......... 121

Non-primary inputs to the cochlear nucleus visualized using immunocytochemistry
    Joe C. Adams ...................................................................................... 133

Superior olivary cells with descending projections to the cochlear nucleus
    John M. Zook and Nobuyuki Kuwabara ......................................... 143

Descending projections from the inferior colliculus to the cochlear nuclei in mammals
    Enrique Saldaña ................................................................................. 153

## V.- NEUROTRANSMITTERS OF THE COCHLEAR NUCLEUS

Localizing putative excitatory endings in the cochlear nucleus by quantitative immunocytochemistry
    José M. Juiz, Maria E. Rubio, Robert H. Helfert
    and Richard A. Altschuler ................................................................ 167

Excitatory amino acid receptors in the rat cochlear nucleus
    Robert J. Wenthold, Chyren Hunter and Ronald S. Petralia ................ 179

Glycine and GABA: transmitter candidates of projections descending to the cochlear nucleus
    Steven J. Potashner, Christina G. Benson, E.-Michael Ostapoff,
    Nancy Lindberg and D. Kent Morest .............................................. 195

Inhibitory amino acid synapses and pathways in the ventral cochlear nucleus
    Richard A. Altschuler, José M. Juiz, Susan E. Shore,
    Sanford C. Bledsoe, Robert H. Helfert, and Robert J. Wenthold ......... 211

Glycinergic inhibition in the cochlear nuclei: evidence for tuberculoventral neurons being glycinergic
    Donata Oertel and Robert E. Wickesberg ...................................... 225

GABA and glycine inputs control discharge rate within the excitatory response area of primary-like and phase-locked AVCN neurons
    Donald M. Caspary, Peggy S. Palombi, Patricia M. Backoff,
    Robert H. Helfert and Paul G. Finlayson ...................................... 239

Neuropharmacological and neurophysiological dissection of inhibition in the mammalian dorsal cochlear nucleus
    Edward F. Evans and Wei Zhao ...................................................... 253

Comparison of quantitative and immunohistochemistry for choline acetyltransferase in the rat cochlear nucleus
    *Donald A. Godfrey* .................................................. 267

Choline acetyltransferase in the rat cochlear nuclei: immunolocalization with a monoclonal antibody
    *Douglas E. Vetter, Costantino Cozzari, Boyd K. Hartman and Enrico Mugnaini* .................................................. 279

## VI. PROJECTIONS AND RESPONSE PROPERTIES OF COCHLEAR NUCLEUS NEURONS

The cochlear root neurons in the rat, mouse and gerbil
    *Dolores E. López, Miguel A. Merchán, Victoria M. Bajo and Enrique Saldaña* .................................................. 291

Projections of cochlear nucleus to superior olivary complex in an echolocating bat: relation to function
    *John H. Casseday, John M. Zook and Nobuyuki Kuwabara* ............... 303

The monaural nuclei of the lateral lemniscus: parallel pathways from cochlear nucleus to midbrain
    *Ellen Covey* .................................................. 321

Ascending projections from the cochlear nucleus to the inferior colliculus and their interactions with projections from the superior olivary complex
    *Douglas L. Oliver and Gretchen E. Beckius* .................................................. 335

Responses of cochlear nucleus cells and projections of their axons
    *Philip H. Smith, Philip X. Joris, Matthew I. Banks and Tom C.T. Yin* .................................................. 349

Physiology of the dorsal cochlear nucleus molecular layer
    *Paul B. Manis, John C. Scott, and George A. Spirou* ............... 361

Coding of the fundamental frequency of voiced speech sounds and harmonic complexes in the cochlear nerve and ventral cochlear nucleus
    *Alan R. Palmer and Ian M. Winter* .................................................. 373

## VII. COMPUTER MODELLING OF THE COCHLEAR NUCLEUS

Computer modelling of the cochlear nucleus
    *Ray Meddis and Michael J. Hewitt* .................................................. 385

Regularity of discharge constrains models of ventral cochlear nucleus bushy cells
    *Eric D. Young, Jason S. Rothman, and Paul B. Manis* ............... 395

Cross-correlation analysis and phase-locking in a model of the ventral cochlear nucleus stellate cell
    *Murray B. Sachs, Xiaogin Wang and Scott C. Molitor* ............... 411

## VIII. APPENDIX: COCHLEAR NUCLEUS PROSTHESES

The development and evaluation of cochlear nucleus prostheses
*John K. Niparko, David J. Anderson, Kensall D. Wise
and Josef M. Miller* .................................................................. 421

## IX. SPECIAL CONTRIBUTIONS IN HONOUR OF R. LORENTE DE Nó

Lorente de Nó's scientific life
*Antonio Gallego* ...................................................................... 431

Sensoritopic and topologic organization of the vestibular nerve
*Vicente Honrubia, Larry F. Hoffman, Anita Newman, Eri Naito,
Yasushi Naito and Karl Beykirch* ................................................ 437

Lorente de Nó and the hippocampus: neural modeling in the 1930s
*Larry W. Swanson* ................................................................... 451

The rat entorhinal cortex. Limited cortical input, extended cortical output
*Ricardo Insausti* ..................................................................... 457

Axonal patterns of interneurons in the cerebral cortex: in memory of Rafael Lorente de Nó
*Alfonso Fairén* ....................................................................... 467

Glomérulos, barrels, columns and maps in cortex: an homage to Dr. Rafael Lorente de Nó
*Thomas A. Woolsey* ................................................................ 479

Lorente de Nó: the electrophysiological experiments of the latter years
*Lawrence Kruger* .................................................................... 503

**INDEX** .................................................................................... 515

# THE CELLULAR BASIS FOR SIGNAL PROCESSING IN THE MAMMALIAN COCHLEAR NUCLEI

D. Kent Morest

Department of Anatomy and Center for Neurological Sciences,
The University of Connecticut Health Center, Farmington, CT 06030;
Laboratory of Molecular Neurobiology, The Salk Institute, La Jolla, CA 92037, USA

## The Spanish legacy

Ramón y Cajal taught us that information in the central nervous system is differentiated and transmitted between individual nerve cells. Lorente de Nó taught us that the cells in the CNS are individuals with unique personalities, which depend on the exact arrangement of the synaptic contacts on their dendrites and the precise distributions of their axons, all of their branches and synaptic endings. In fact Lorente de Nó ('43) emphasized the axonal branching pattern as a criterion for distinguishing neurons.

In the mammalian brain we cannot yet study each individual neuron, by name and number, so to speak. We can, however, recognize different families or types of neurons, which share common characteristics. Thus it has been possible to study the structures, connections, and functions of individual types of neurons in the central nervous system. The auditory system, and the cochlear nucleus in particular, have provided perhaps the best opportunity to pursue this approach. There are several reasons for this. First of all, the cochlear nucleus consists of populations of specific types of neurons (Fig. 1). Each type is defined by a specific and different structural pattern of somatic morphology, dendritic architecture, synaptic organization, and axonal projections. Yet cells of each type receive their input from the cochlear ganglion, and often from the very same axons. These axons supply the ascending and descending branches that define tonotopy in the cochlear nucleus (Fig. 2). There is a tendency for cells of different types to segregate into separate groups, hence the rationale for detailed parcellation schemes (Figs. 1, 3, 4).

The Golgi methods, of which Lorente de Nó was one of the last great champions, have provided much of the basis for our study of the cell types in the cochlear nucleus (see review by Morest, '83). The author has attempted to identify in Figures 1 and 2 from the classic papers of Lorente de Nó ('33a,b; reproduced in Lorente de Nó, '81) some of the equivalent cell groups and neuronal types as described in two of the standard references (Osen, '69; Brawer et al., '74; in the cat) and in more recent versions (see, e.g., Hackney et al., '90; Morest et al., '90; in rodents and cat) (Figs. 1-4). A concordance of the classical cytoarchitectonic schemes has been published (Brawer et al., '74). The uppercase abbreviations used in this paper (from Brawer et al., '74) are explained in Figure 3.

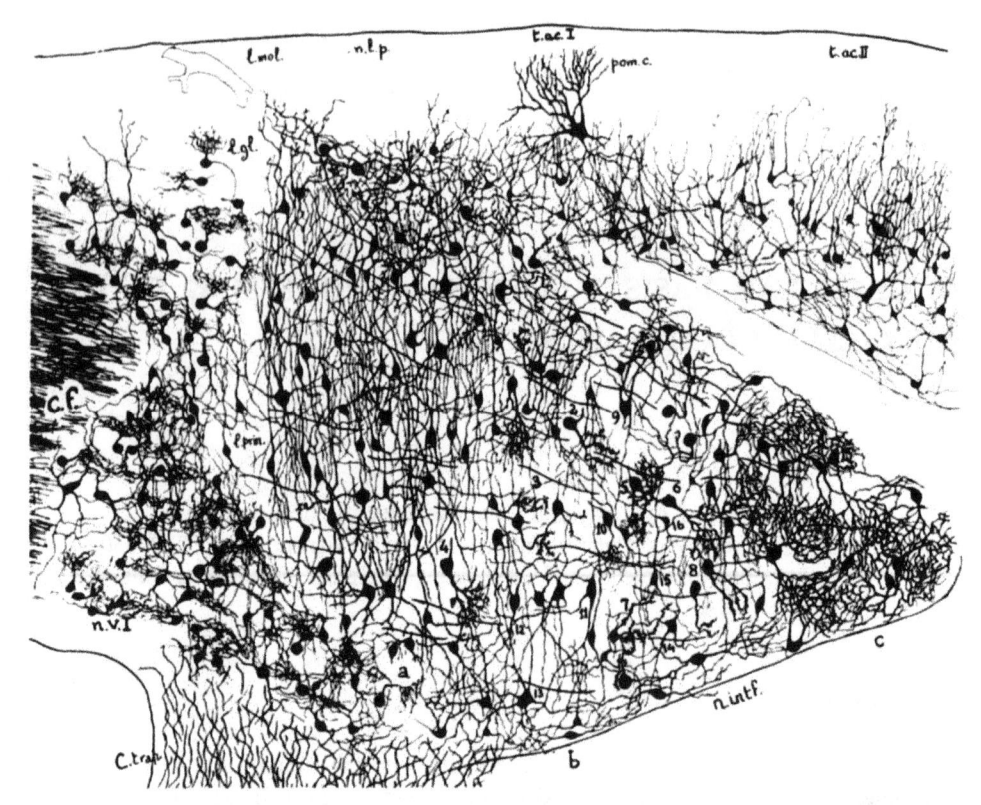

**Figure 1.** An oblique transverse section through the cochlear nucleus of an adult cat (Golgi-Cox method) (from Lorente de Nó, '81). In Figure 1 the section passes through the caudal edge of the cochlear nerve root and zone of bifurcation ("c.f.") and the ventral part (*PV*) of *AVCNp* at the left ("n.v.I"), extends caudally and dorsally at the right, passing through the *PVCN* ("n.int.f.") to the octopus cell area ("c") and to *DCN* ("t.ac. I, t.ac. II") at the right, and extends from the lateral surface of the external granular layer and small cell cap ("n.l.p., l.mol.," and the region to the right of "l.gl.") ventrally and medially to the edge of the trapezoid body at the bottom ("C.trap."). The ventromedial part of *AVCNp* (Morest et al., '90) is represented by the five small neurons (3 globular bushy, 1 stellate, and 1 elongate) nearest the label, "n.v.I." The ventral part (PV) of *AVCNp* (extending between "n.V.I." and "l.gl.") is characterized by larger globular bushy cells, large stellate cells, and elongate cells. The anterodorsal part (*AD*) of *PVCN* (extending from above "a" to the right of "l.prin.") contains large stellate cells, elongate cells, clavate cells, and giant cells. *AD* is gradually replaced laterally by the stellate cells of the cap region (small cell cap of Osen, '69; *E* of Brawer et al., '74). This transition is hard to follow because of the oblique plane, but it occurs in the territory between the labels, "l.prin." and "l.gl.". The anterior part (*A*) of *PVCN* is located above "b" and contains variations of several different neuronal types (#1-16), e.g., bushy cells ("1,5"), large stellate cells ("13,14"), large ("15") and small ("16") elongate cells. In the *DCN*, only the deep layers are impregnated, including the three types of giant cells, vertical ("pom.c."?), horizontal and radiate (caudodorsally), surrounded by many small stellate and "corn" cells.

Figure 2 shows that the subdivisions are most clearly and sharply delineated on the basis of the morphological pattern of the axonal plexus and do not always correspond to the distribution of specific cell types (compare Figs. 1, 3, 4). It also shows the anatomical basis for the tonotopic organization, which is established by the layering of the cochlear nerve branches. In fact there is really only one cochleotopic map, which is sampled unevenly by different cell groups (Figs. 2, 4). The apparent reversals of best frequency sequences are produced by the 3-dimensional geometry of the afferent fiber layers, as Rasmussen et al. ('60) and Rose et al. ('60) demonstrated. It follows that neither the absence nor the presence of

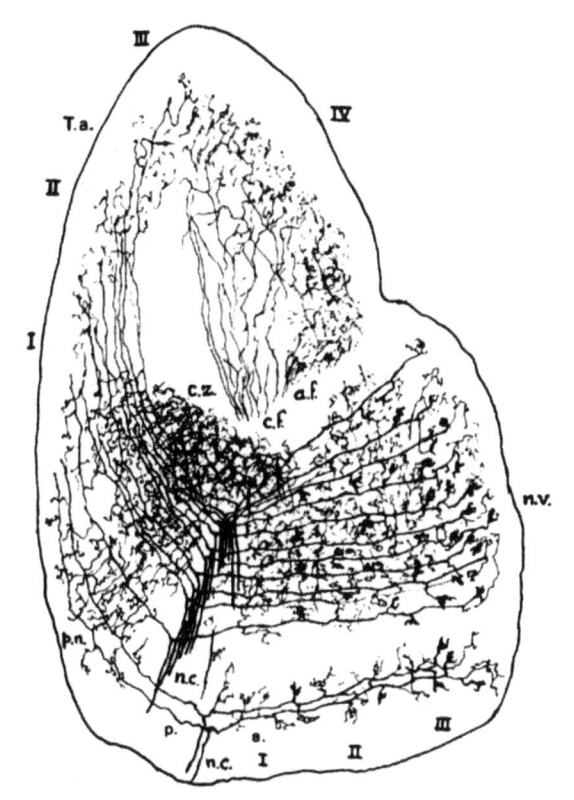

**Figure 2.** Parasagittal section of the cochlear nucleus of a 4 day old cat; rapid Golgi method (from Lorente de Nó, '81) (see Fig. 3 for uppercase abbreviations). In the AVCN ("n.v."), zones "III, II, and I" correspond, respectively, to AA (large spherical cell area of Osen, '69), AP (small spherical cell area of Osen, '69), and AVCNp (anterior region of the multipolar and globular cell areas of Osen '69). "III" marks the caudal boundary of AA, "I" is at the rostral border of AVCNp, which can be further subdivided into PV caudoventrally and PD rostrodorsally by a line extending from the dorsalmost bifurcation of "n.c." to "I." The rostral border of PVCN ("p.n.") corresponds to a line drawn through the bifurcations of the cochlear nerve root ("n.c."). Very little is impregnated in the small cell cap (Osen, '69), which lies in the clear zone containing the labels, "c.z., c.f., a.f." In PVCN AD occupies the dense neuropil ventral to the label "c.z." (confluence zone). A is just above and to the right of the label, "p.n.," while the octopus cell area is in the relatively clear area between the labels "p.n." and "I" caudally. In DCN ("T.a.," tuberculum acusticum) only the deep layers are impregnated; its zones ("I--IV") have not been described by other authors. The anterior or ascending branch ("a") of the cochlear nerve root and the posterior or descending branch ("p") form layers arranged in a dorsomedial to ventrolateral sequence, which represents the tonotopic map (see Fig. 4). At "a.f." are the fibers of the ventro-tubercular and tuberculo-ventral tracts, which interconnect the dorsal and ventral nuclei. At "c.f." the centrifugal fibers from extrinsic sources, including the superior olivary complex and the inferior colliculus, project into DCN (Rasmussen, '67; Saldaña, this volume).

such reversals are necessarily reliable for locating cytoarchitectonic boundaries in the cochlear nucleus, or anywhere else in the auditory system.

## Signal processing

In order to understand the cellular basis of signal processing in the cochlear nucleus, many different kinds of information are needed. One needs to have a detailed cytoarchi-

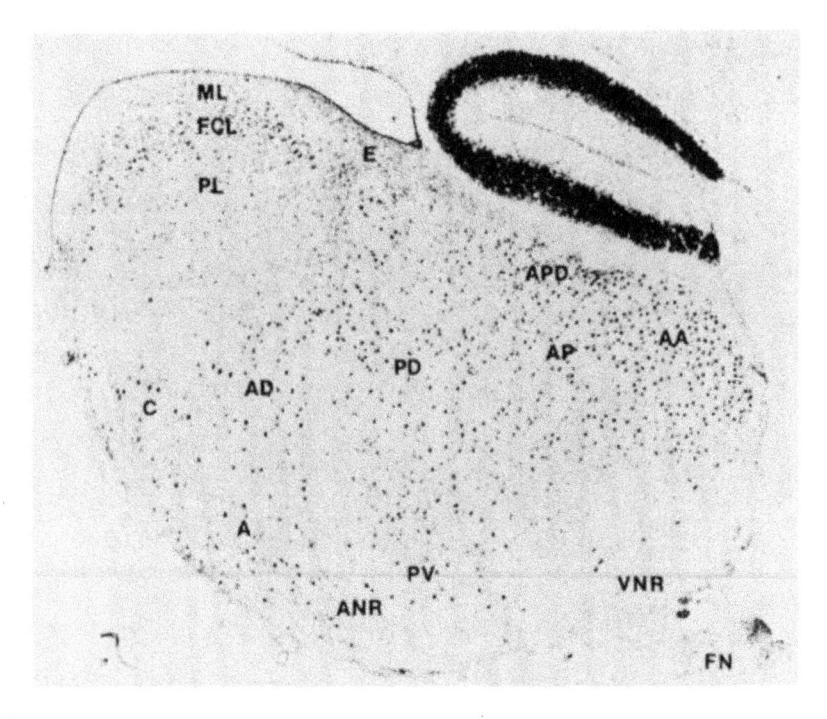

**Figure 3.** Nissl-stained parasagittal section through the approximate center of the cochlear nucleus of an adult cat (from Brawer et al, '74). The pin holes next to the label, "APD," and beneath the label, "PL," were used to determine shrinkage (approx. 17%). Note the tendency of neurons to occur in clusters in PD and PV. (X 25). **Abbreviations:** Anteroventral nucleus. Anterior division (*AVCNa*): *AA*, anterior; *AP*, posterior; and *APD*, posterodorsal parts; Posterior division (*AVCNp*): *PD*, dorsal, and *PV*, ventral parts. Posteroventral nucleus (*PVCN*): *A*, anterior; *AD*, anterodorsal; and *C*, central parts (octopus cell area). Dorsal nucleus (*DCN*): *ML*, molecular; *FCL*, fusiform cell; and *PL*, polymorphic (deep) layers. *E*, (external) granular region: this extends as the cap ventrally into the region between the "AD" and "PL" labels. *ANR*, cochlear; *VNR*, vestibular; and *FN*, facial nerve roots.

tecture and a definition of cell types based on functionally relevant morphological criteria, including characteristics of the cell bodies, dendrites, afferent axons, and their synaptic structure. The synaptic organization and physiological properties of each cell type have to be correlated and analyzed for their functional significance. Information is processed in the circuits formed between individual neurons. In order to establish their function, one needs to know, at the very least, the connections characteristically established by the respective cell types, with each other and with the neuronal types composing the rest of the auditory pathway. We call these connections, the "circuits of cell types." To explain the biological mechanisms that account for these functions also requires information on the biophysics of the electrogenic membranes (membrane properties) for these cell types, on their synaptic transmitters, and on the molecular properties of their synaptic membranes. The dynamic properties of the synaptic organization have to be defined, inasmuch as synapses are subject to long-term changes, based on use, as well as pathological perturbations. Finally, functional models of the cell types and their circuits need to be developed to correlate all of the above data, to rationalize our study of processing, and to provide new insights.

Some of the key features of the cell types in the cochlear nucleus and their circuits are summarized in a series of diagrams (Figs. 5-13). Certainly these are not comprehensive summaries. They are simplifications, which are subject to qualifications when the details are

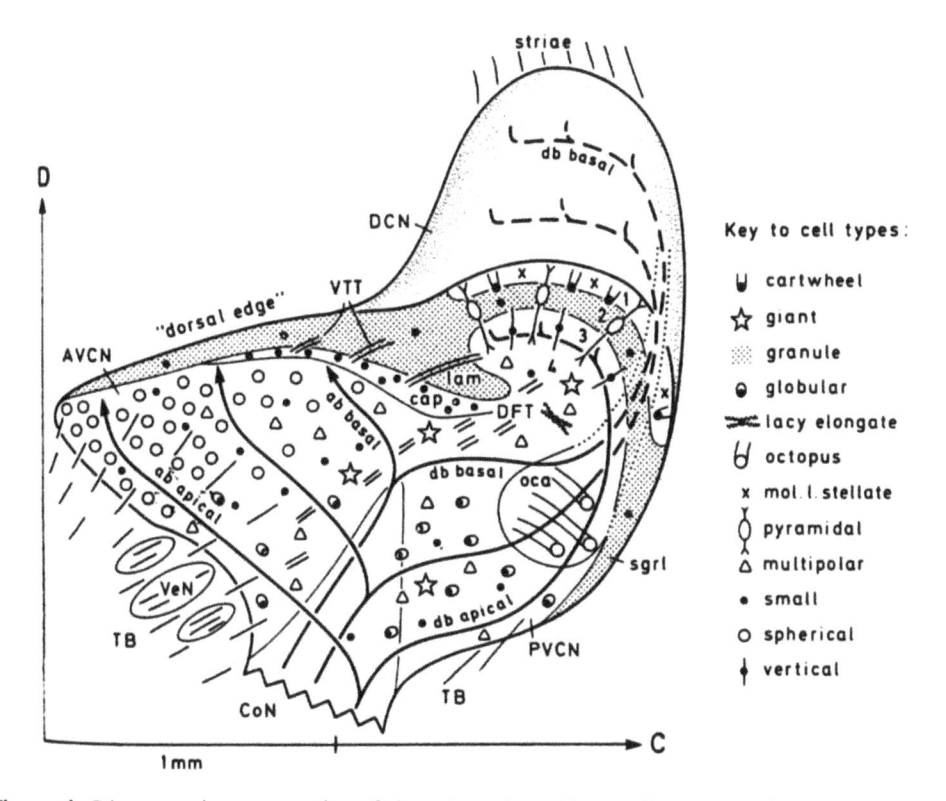

**Figure 4.** Diagrammatic representation of the guinea pig cochlear nucleus on a sagittal plane showing, qualitatively, the distribution of the predominant cell types, the paths of the cochlear afferents, the ventrotubercular tract (VTT), and the descending (centrifugal) fibers (DFT) to DCN. The tonotopic arrangement of the cochlear nerve projection (CoN) is indicated with reference to the ascending (ab) and descending (db) branches from the apical and basal turns of the cochlea. C, caudal; D, dorsal; sgrl, superficial (external) granule cell layer; TB, trapezoid body; VeN, vestibular nerve root. (From Hackney et al, '90). The "pyramidal" cell corresponds to the fusiform cell, the "multipolar" cell to the large stellate cell, and the "vertical" cell to the small elongate ("corn" cell), or special small elongate variety, of Brawer et al ('74). cap, cap region (see Cant, this volume). Certain other rodents contain cochlear root neurons (Morest et al., '90; Lopez et al., this volume).

considered. But they do provide a format for comparing the cell types and their circuits. The legends of the figures should be consulted *before* the corresponding text, since the latter is a commentary, not a review. The literature is not reviewed here; however, most of it is cited in the other chapters of this volume.

## Cochlear input (Fig. 5)

Nearly everything we know on this subject concerns the cochlear nerve fibers innervating the inner hair cells (concerning outer hair cells see Brown, this volume). Six of the most studied physiological properties of such fibers are summarized in Figure 5 (see figure legend). The subsequent figures compare the same six properties for some of the main cell types in the cochlear nucleus. No doubt other properties could be added. For example, considerable attention has been given in recent years to the measure of so-called "spontaneous activity" in the cochlear nerve fibers. It is not yet clear how this measure is related to central processing (see chapter by Leake et al., Ryugo et al., and Young et al., this volume, and presentation by Liberman at this meeting).

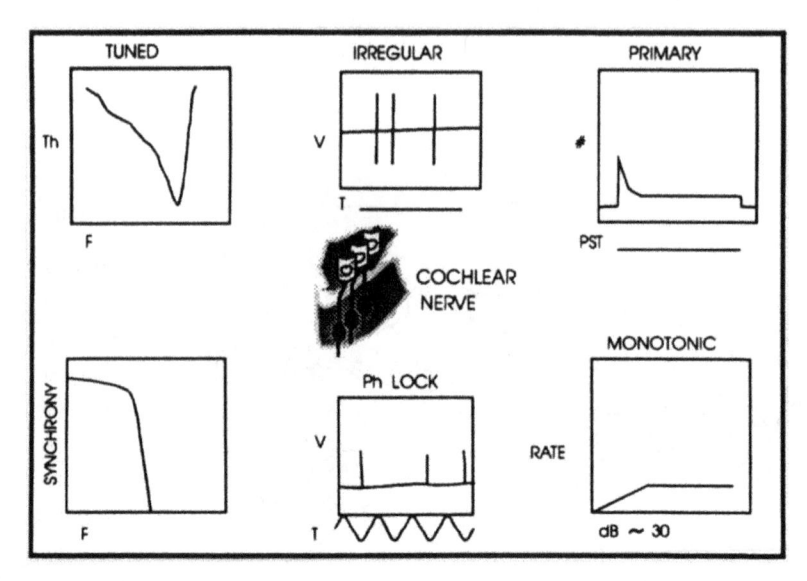

**Figure 5.** Six aspects of cochlear nerve physiology are schematically summarized. As a rule the individual fibers are well tuned to a characteristic frequency. This is shown by a tuning curve (upper left), which is an iso-rate function based on threshold (Th) vs. frequency (F) of a pure tone stimulus. Fibers do not show side band inhibitory zones. Although the first-spike latency can be short, the sustained portion of the response shows irregularly spaced action potentials (upper middle: voltage, V, vs. time, T). The temporal discharge pattern, based on summing the response to many stimulus presentations (upper right: peri-stimulus time histogram, PST) presents a sharp onset in the number of spikes (#), which rapidly decays to a steady state level. Fibers tuned to low frequencies will phase lock to pure tones. The ability to phase-lock begins to drop off rapidly past 1 kHz, as shown in the synchrony vs. frequency function (lower left). The lower middle panel shows that phase-locking for cochlear nerve fibers does not occur on a cycle-by-cycle basis. Rate/level functions (lower right) are monotonic over a limited range, 30-40 dB, more or less, depending on level of spontaneous activity. No single fiber approaches the dynamic range observed in human hearing (>100 dB).

## Binaural timing circuit (Figs. 6, 7)

This is represented by the excitatory, presumably glutamatergic projections of the end-bulbs of the cochlear nerve fibers (see chapters by Juiz et al. and Wenthold et al., and poster presentations by Hackney et al. and Rubio et al. at this meeting), one, occasionally two, very large axonal endings synapsing on each spherical bushy cell body in AVCNa. The synaptic organization of the cochlear input and the associated physiological properties facilitate its role in establishing a secure basis for detection of interaural time differences used for lateralization, and localization, of low frequency sounds by way of the medial superior olivary nucleus and the inferior colliculus (Fig. 7). There is clear evidence that these neurons respond generally with an irregular discharge but phase lock to low frequency tones (e.g., Rhode et al., '83; Rhode and Smith, '86). The synaptic and membrane properties of the spherical bushy cell are such that it produces a non-linear (outwardly rectifying) current-voltage relationship and onset responses in response to depolarization *in vitro* (Wu and Oertel, '84). *In situ* studies have demonstrated that neurons, presumably spherical bushy cells, with a primary-like discharge pattern to tones, can produce onset firing, followed by other, irregularly spaced spikes, in response to depolarizing current (Feng, '89; Feng et al., '88; presentation by Kuwada et al., at this meeting). These membrane properties are well adapted for preserving the temporal information of cochlear nerve fibers used for coincidence detection in the binaural timing circuit (Fig. 7). These properties are very different from those of other cell types, especially the stellate cells. They imply that the cell types of the

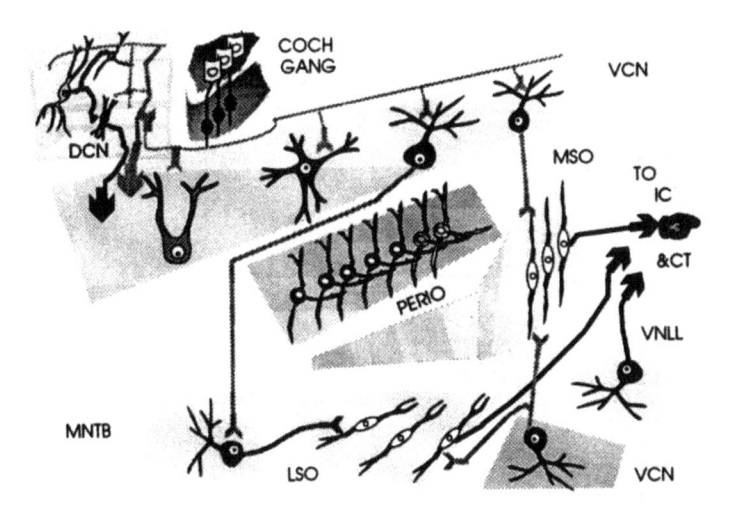

**Figure 6.** A schematic of the distribution of the input from the cochlear ganglion (COCH GANG) and nerve to the cell types of the ventral cochlear nucleus (VCN) (from right to left: spherical bushy cell, globular bushy cell, large stellate cell, octopus cell) and of the dorsal cochlear nucleus (DCN) (from top to bottom: inhibitory interneurons (here represented by a cartwheel cell), fusiform cell at the left, granule cell (an excitatory, possibly glutamatergic interneuron) at the right, and a corn cell (an inhibitory, glycinergic interneuron); (the last two cell types project to the VCN]. Not shown are the giant cell and the fusiform cell projections to the inferior colliculus (IC). All of these neuron types can receive more or less synaptic input from the cochlear nerve fibers (Cohen et al, '72; Kane, '74a, b). The binaural timing circuit is represented by bilateral excitatory projections from the spherical bushy cells to the disc-shaped principal cells of the medial superior olivary nucleus (MSO), which, in turn, project to the ipsilateral IC. The binaural intensity circuit is represented by the excitatory globular bushy cell projection to the principal cells of the contralateral medial trapezoid nucleus (MNTB), which, in turn, provide an inhibitory, glycinergic projection to the principal disc-shaped cells of the ipsilateral lateral superior olivary nucleus (LSO). The LSO receives an excitatory projection from the ipsilateral VCN and projects, in turn, to the IC bilaterally by way of excitatory and inhibitory pathways. Other Abbreviations. PERIO, periolivary nuclei; VNLL, ventral nucleus of the lateral lemniscus.

In this and the following schematics, a "duckbill" (--<) indicates a synaptic connection between specific kinds of neurons in the circuit of cell types, while an arrow (-->) indicates a connection between cell groups where one or both types of neurons involved are not established. Light gray lines of projection indicate excitatory pathways, dark gray lines, inhibitory pathways, and black lines, both excitatory and inhibitory projections or pathways of unknown sign. The gray area beneath the label, "PERIO," represents a gray area in our knowledge of the nuclei of the periolivary region and of the lateral lemniscus, which, like certain empty niches in gothic cathedrals, await future occupants, holy or unholy.

cochlear nucleus differ at the most fundamental level, that even the molecular structure of their membranes differs, and that the number, kinds, and distribution of ion channels on the cell surface are entirely characteristic for each kind of neuron. It is therefore of interest to find that the putative glutamate receptors for these cells may have different molecular properties.

Recent studies in molecular neurobiology have shown that the glutamate receptors in the brain may differ in the composition of their subunit proteins. At least six genes have been cloned to date from the ionotropic non-NMDA receptor family, and their transcripts are differentially expressed in the brain (Bettler et al., '90; Boulter et al., '90; Egeberg et al., '91; Hollmann et al., '89; Keinanen et al., '90; Nakanishi et al., '90). When the genes for these subunits are expressed singly or in combinations within frog oocytes, there are dramatic differences in the electrophysiological and pharmacological properties of the resulting glutamate receptors (Boulter et al., '90). For example, with GluR1 there is a non-linear voltage relationship with agonist-elicited currents, which may be associated with significant $Ca^{++}$ fluxes. By contrast, the combination of GluR1 or GluR3 with GluR2 may produce a

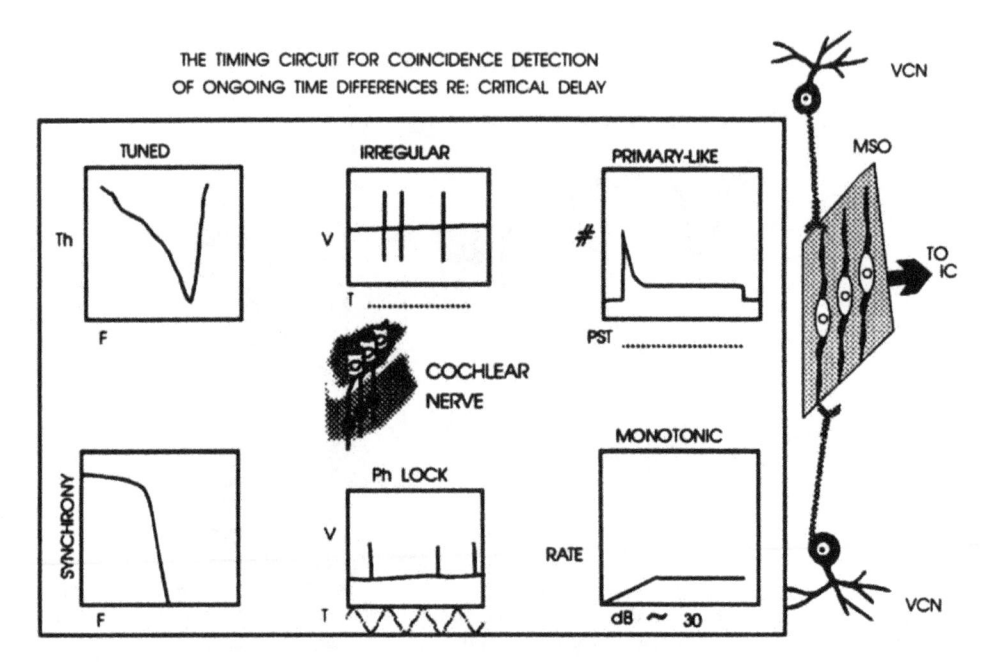

**Figure 7.** In the binaural timing circuit, the cochlear input to the spherical bushy cells of VCN is synaptically secure, thus ensuring a conservation of primary-like physiological properties, which are here compared to those of the cochlear nerve (compare Fig. 5). There are no striking differences between the responses of cochlear nerve fibers and those of spherical bushy cells. The bilateral, excitatory projections to the disc-shaped principal neurons of MSO provide for the possibility of a coincidence detection of ongoing time differences between the two ears. As a result characteristic delays can be defined for the principal cells, which project ipsilaterally to unknown types of IC neurons.

receptor associated with a linear current-voltage curve and no significant $Ca^{++}$ flux (Hollman et al., '91). Our own work (Morest and Hermans-Borgmeyer, unpublished) with *in situ* hybridization, using [a-$^{35}$S]UTP-labeled antisense cRNA polynucleotide probes in the adult mouse cochlear nucleus, suggests that GluR1, GluR2, and GluR3 are differentially expressed in bushy and stellate cells (see also Wenthold et al., this volume). These findings raise the possibility that the functions of bushy and stellate cells may be correlated with glutamate receptors having different subunit structures and synaptic properties. While research on this topic is still in the early stages, a promising new approach is now open for exploration.

Our studies of the adult synaptic organization often assume that the system is hard-wired. This assumption may be safe as long as our measures address populations of neurons, which express some common pattern or mode. However, it can become an issue where one attempts to generalize on the basis of small numbers of observations. It is no secret that the developing nervous system is subject to plastic changes and even structural modifications as the result of normal activity, not to mention traumatic events. The growth and development of the central nervous system are regulated by endogenous as well as exogenous factors (see Friauf et al., and Willard, this volume). Some of these factors could continue to produce plastic change in the adult. To explore the potential for structural variation and new growth at central synapses, Jones et al. ('92) have studied the morphological patterns of the end-bulbs of cochlear nerve axons in adult cats. Growth cones and retraction clubs were present on most, if not all of the end-bulbs in every adult cat studied (Fig. 8). A systematic survey of the end-bulb patterns revealed a continuous gradient of variation, in which each synaptic type forms a distinct mode. These findings lead us to hypothesize that the end-bulbs are in a

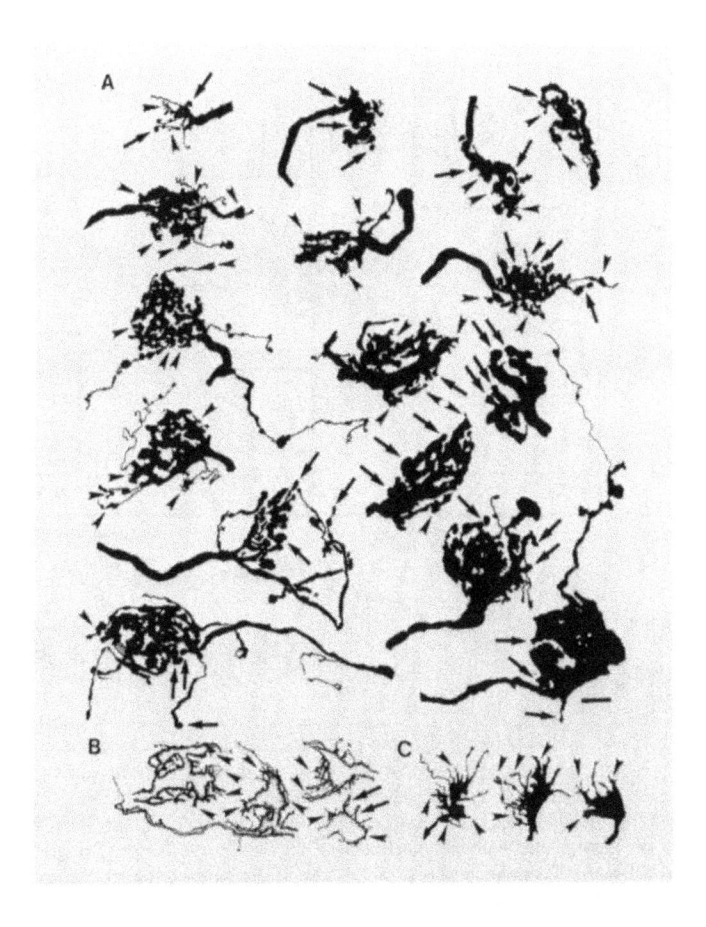

**Figure 8.** Growth cones (arrowheads) and terminal clubs (arrows) on Golgi-impregnated end-bulbs in AVCNa of A. adult cats, B. two-week old cats, and C. two-day old cats (from Jones et al, '92). Growth cones become smaller and less numerous with maturation. Scale = 10 $\mu$m.

continual state of structural and functional flux. These endings should prove useful for studies on the modifiable properties of central synapses.

A corollary of this finding is the discovery that the surviving axonal endings in the cochlear nucleus can sprout following noise damage to the cochlea (Morest et al., '87). Possibly overstimulation of synapses, associated with glutamate toxicity as well as trophic effects, could be implicated in the degenerative changes which precede new growth (see, e.g., Morest and Bohne, '83 ; Pujol, presentation at this meeting).

So far almost all of our research has focused on the excitatory end-bulbs, but in fact it was shown long ago that most of the synaptic endings on the spherical bushy cell are probably inhibitory (see Saint Marie et al., this volume). It comes as no surprise to neuroanatomists therefore to learn that near-center inhibition could be demonstrated in bushy cells (see Caspary et al., this volume).

## Binaural intensity circuit (Figs. 6, 9)

This circuit is represented by pathways to the LSO from spherical bushy cells in the ipsilateral cochlear nucleus and by globular bushy cells in the contralateral cochlear nucleus

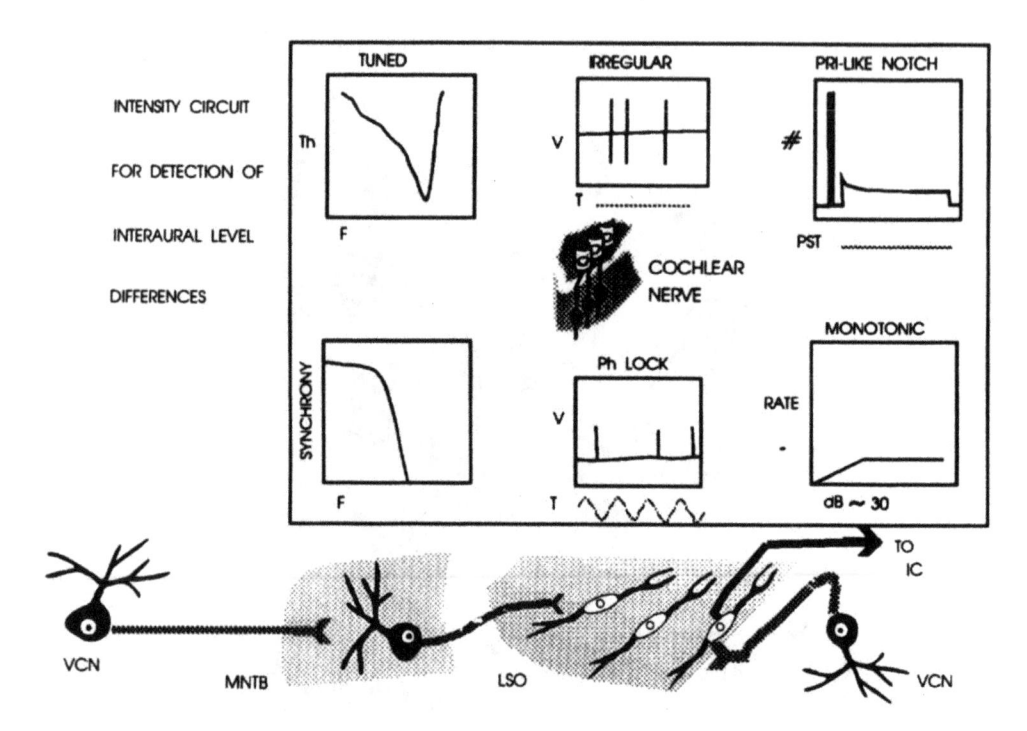

**Figure 9.** In the binaural intensity circuit (see Fig. 6) the globular bushy cell reproduces many of the properties of the cochlear nerve input and compares with the spherical bushy cell in many respects. This applies to the PST histogram, except that the initial onset is immediately followed by a brief silent period, or notch (so-called "primary-like-with-notch" pattern. Neurons that have primary-like-with-notch responses can have about twice the spontaneous rate as those associated with spherical cells or cochlear nerve fibers. The globular bushy cells project to the principal cells of the contralateral MNTB by way of a single excitatory axo-somatic ending, or calyx, to each cell body. While the principal cells resemble globular bushy cells in overall form, they differ in their synaptic organization. They are thought to be glycinergic, inhibitory interneurons projecting to the disc-shaped principal cells of the LSO, which in turn project to undetermined cell types in the IC bilaterally, possibly forming glycinergic, inhibitory synapses as well as excitatory endings (Glendenning et al, '85; Saint Marie et al, '89). Binaural interactions mediated by the contralateral inhibitory pathway and the ipsilateral excitatory pathway from VCN are thought to generate the responses of LSO neurons to interaural level differences, which may be used for lateralization, and localization, of high frequency sounds. Not shown here are the collateral projections of globular bushy cells to the periolivary nuclei, where they may also contact inhibitory interneurons.


(Figs. 6, 9). Globular bushy cells differ from spherical bushy cells in that they are contacted by many more cochlear nerve fibers. The heavy convergence of many cochlear nerve fibers would no doubt produce a less variable latency, compared to the spherical bushy cell, a feature that could explain the notch in the PST histogram (which may reflect the refractory period of the cell) and also facilitate the appearance of PST onset patterns. The heavy excitatory convergence of primary afferents could also produce higher firing rates, even during phase locking (Smith et al., this volume), their high spontaneous rate, and their high driven rate to tones (Rhode and Smith, 1986; Smith et al., this volume). A companion feature to the heavy convergence of primary afferents is the extensive overlap resulting from their numerous axodendritic synapses, especially those on the terminal dendritic tufts, which commonly surround neighboring globular cell bodies (Ostapoff and Morest, '91). Conceivably these features of their synaptic organization would tend to keep bushy cells at a level of on-going activity where inhibitory input could play a more decisive role in their responses. They might also lead to activity levels which have a greater uniformity or coherence in whole clusters of these cells (see Fig. 3, PD, PV). In many respects the globular cell provides much the same high level of synaptic security as the spherical bushy cell (compare Figs. 7 and 9). A heavy

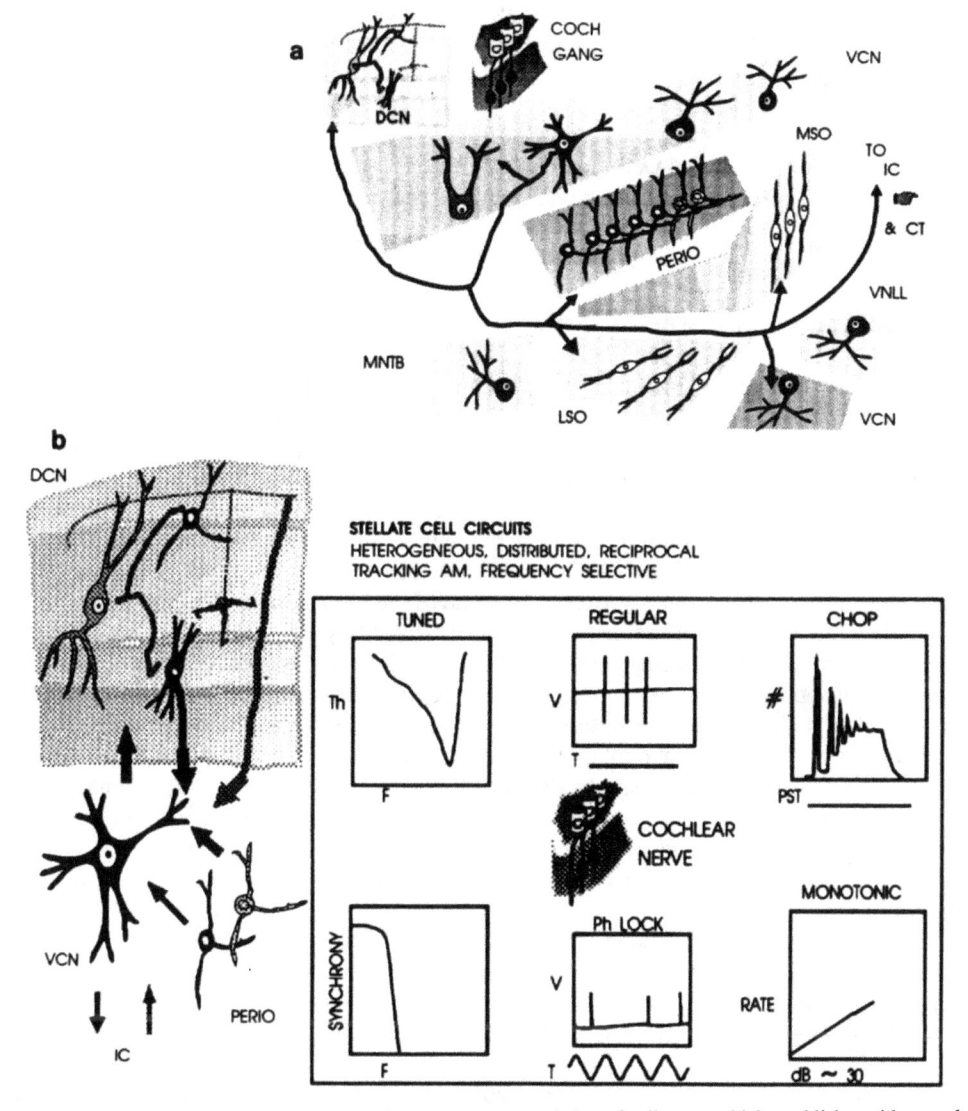

**Figure 10.** Stellate cell circuits. a. This is a heterogeneous population of cell types which establish a widespread, or distributed, series of projections, including the likelihood of reciprocal connections with DCN, the periolivary nuclei, and IC. Although the DCN granule cells and corn cells project to the AVCN, detailed information on the cell types with which the stellate cells are connected or the nature of their transmitters is not yet available. b. Summarized here are some of the physiological properties of large stellate cells in AVCN. These are, for the most part, quite different from those of the cochlear nerve fibers. Although these neurons are usually well tuned, their firing pattern has a regular periodicity not directly related to the stimulating frequency (upper middle panel), corresponding to a chopper pattern in the PST (upper right panel), which can be either sustained or transient. Rate/level functions are monotonic and often have dynamic ranges greater than those of cochlear nerve fibers, so that they may play a role in coding for stimulus level. Although choppers can phase-lock, their high end cut off frequency for phase-locking is lower than that for cochlear nerve fibers (lower left panel). They can follow amplitude modulated signals very well and have been implicated in frequency selectivity.

afferent convergence would seem unnecessary for this purpose, indeed could even be counterproductive (see Young et al., this volume). The answer to this paradox may become clear when we understand better the synaptic organization of the inhibitory endings. These non-cochlear endings constitute the majority of its synaptic input and are very well distributed over the entire surface of the neuron (Fig. 13). They intermingle with the small cochlear endbulbs in such a way that integration and interaction between the different inputs could be facilitated. After years of relative neglect the role of inhibitory interneurons in central processing has become a topic of physiological interest (see e.g., Caspary et al., Oertel and Wickesberg, Wickesberg and Oertel, and others, this volume). Meanwhile it is hard to relate the physiological properties of the globular bushy cell to its role in the binaural intensity circuit.

## The distributed pathway of stellate cells in the VCN (Fig. 10)

One of the most enigmatic groups of neurons in the VCN is comprised of the stellate cells, a heterogeneous population of several distinct varieties. In the AVCN alone we have defined two types (I and II) of large stellate cells, one subgroup of which contains GABA. There are also the small stellate cells of the lateral margins, some of which contain glycine or GABA, not to mention those of the small cell cap (see Saint Marie et al., and Cant, this volume). This leaves open the question of the nature of the elongate cells and all of the varieties of stellate cells of the PVCN and DCN. Moreover, very little is known of the projections of the stellate neurons, except for some of the local inhibitory interneurons of the DCN. Nevertheless, there are some general structural and physiological properties of interest with regard to the large stellate cells of AVCN, which are summarized in Figure 10. Although synaptic profiles, like those provided for globular bushy cells (Fig. 13), are not available for stellate cells, the observations to date suggest that many of these receive numbers of small excitatory and inhibitory afferents, the latter of which may be dominant on the cell bodies of the type II variety (Saint Marie et al., this volume). The synaptic and membrane properties are such that they have a linear current-voltage relationship in response to depolarization and a regular discharge in response to electrical current or acoustic stimulation (Wu and Oertel, '84; Feng, '89; Feng et al., '88; Kuwada, presentation at this meeting).

Because of their interesting coding characteristics the mechanisms producing signal transformation in these cells are attractive for the development of functional models (Arle and Kim, '91a; see Sachs et al., Palmer and Winter, Meddis and Hewitt, this volume). This effort should be greatly abetted by the results of synaptic profile analyses (Fig. 13) as these become available for different types of stellate cells. However, it is unlikely that much progress will be made in understanding their functions until their synaptic organization has been better characterized and their connections in the circuits of cell types defined, especially with respect to the role of inhibitory pathways descending from the DCN, the periolivary nuclei, and the inferior colliculus (see e.g., Adams, Saint Marie et al., Zook et al., Oertel and Wickesberg, and others in this volume).

## The distributed pathways of periolivary cells (Fig. 11)

Like the stellate cells of VCN, the periolivary neurons generate a highly distributed series of projections to practically all of the brain stem nuclei connected with the cochlear nuclei, including the organ of Corti. Some of these connections are probably reciprocal. Except for the olivo-cochlear fibers, which contribute some collaterals to the cochlear nuclei, there is very little direct information on their physiological properties. There is evidence that

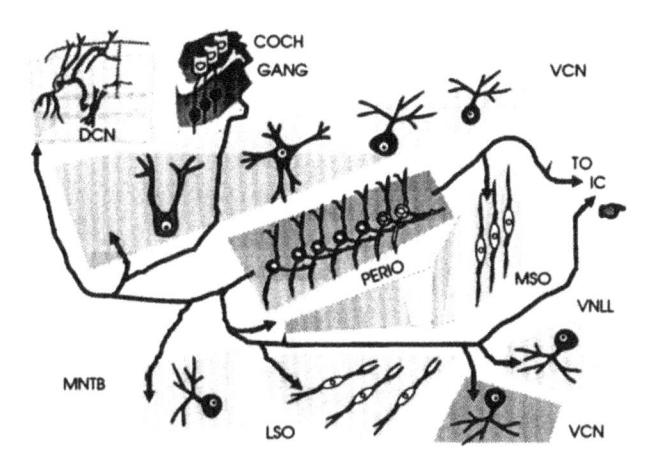

**Figure 11.** The periolivary pathways arise from a heterogeneous population of cell types located in a number of different nuclei surrounding the main olivary segments and project to widespread targets, all of which are connected with the cochlear nuclei. Neither the cell types forming these projections nor the kinds of neurons receiving them are well defined. It is likely that a number of different circuits of cell types compose these pathways. Not shown here are the inputs to the periolivary nuclei from the VCN, some of which project from the globular bushy, octopus, and possibly the stellate cells. While many of the periolivary neurons contain GABA and glycine and are thought to be inhibitory interneurons, the possibility that other inhibitory or excitatory transmitters could be involved needs further study.

many of these cells may use GABA and/or glycine as transmitters, and thus could function in inhibitory feedback circuits (see Potashner et al., Zook et al., Altschuler et al., Adams, Saint Marie et al., this volume). On the other hand, the possibility that some of these neurons provide excitatory feedback projections to the cochlear nucleus remains to be explored. Thus a number of these neurons undoubtedly form cholinergic projections (see Godfrey, this volume). There is a high concentration of receptors in the cochlear nuclei which bind $\alpha$-bungarotoxin and nicotine (Hunt and Schmidt, '78; Wada et al., '89) and which may belong to excitatory synapses. Further progress on the periolivary pathways will depend largely on developing the same kinds of systematic, detailed analyses of cell types, their synaptic organization and circuits, as in the cochlear nucleus.

## Monaural pathways (Fig. 12)

Octopus cells are perhaps the most fascinating cells in the VCN. There is evidence that these neurons project to the periolivary nuclei bilaterally and to VNLL contralaterally, where they may form axo-somatic end-bulbs on the globular-like cells, which project to the IC (see Covey, this volume). Thus the octopus cells may form part of a monaural pathway. If so, their remarkable physiological properties, e.g., unusual capacity to phase-lock and to respond to noise, and to frequency sweeps with directional sensitivity (Morest et al., '73; Godfrey et al., '75a; Godfrey, personal communication), could be related to coding for pitch, timbre, and other acoustic qualities, including vocal transients, which do not depend on binaural function. As shown in Figure 12, the phase-locking response at low frequencies, although having a higher threshold, achieves a higher output than at higher frequencies (hence the term, "heterotonic," in Fig. 12) (see also the by Winter and Palmer concerning dynamic range in other types of onset units). The mechanism responsible for generating the onset response is still open to discussion, although the results from an intracellular study comparing responses to pure tones and current injection are consistent with the hypothesis invoking a

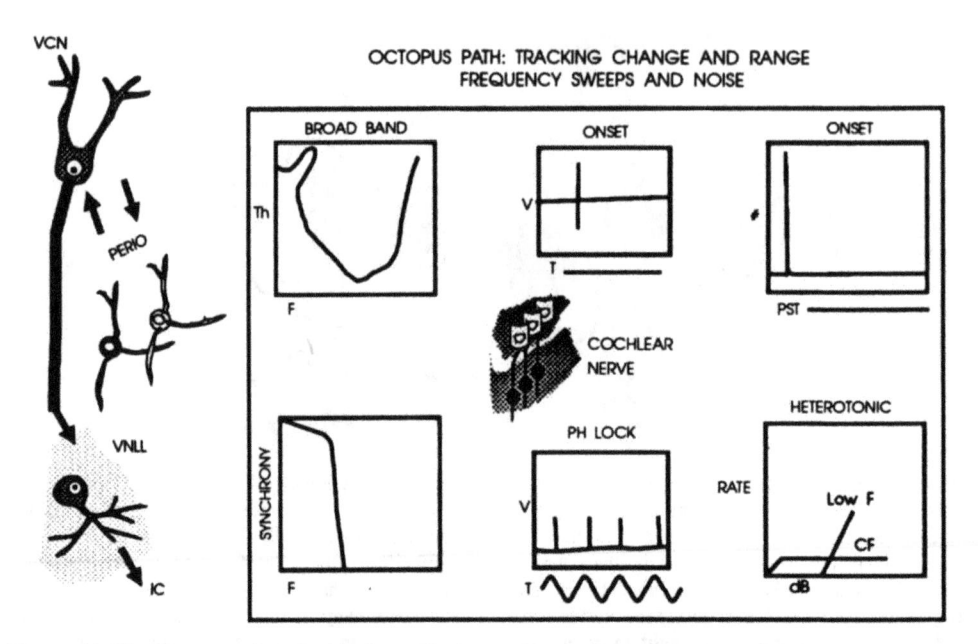

**Figure 12.** The octopus path to the inferior colliculus may be part of a monaural circuit involving reciprocal connections with the periolivary nuclei. The physiological properties of octopus cells differ in many respects from those of the cochlear nerve. They often have broad tuning curves (upper left panel), onset ($O_l$) responses (upper middle and upper right panels) at their characteristic frequency, and cycle-by-cycle phase-locking (lower middle panel) in their low-frequency tails. Their input-output functions are complex (heterotonic). At frequencies which evoke only an onset, individual neurons would not be expected to code for intensity. At frequencies that evoke phase-locking, rate-level functions can be monotonic but have elevated starting levels (lower right panel). Phase locking performance differs from that of the nerve in achieving exceptionally high synchrony and rates. There is evidence for unusual sensitivity to changing stimuli, including sweeping frequencies and especially noise. Varieties of onset responses, having a low-level sustained discharge following the onset (e.g., $O_l$, $O_c$), have been described, with somewhat different properties. Not all of these belong to octopus cells. There is indirect evidence that these cells can achieve a considerable dynamic range under some circumstances, depending partly on frequency.

depolarization blockade (Ritz and Brownell, '82; Feng, '89; Kuwada, presentation at this meeting). It remains to note that different physiological and morphological varieties of octopus cells may be anticipated, as the pioneering studies of Kane ('73) and Godfrey et al. ('75a) suggested (see also, Rhode et al., '83; Oertel et al., '90). Moreover, there are different varieties of onset responses, which are not limited to octopus cells.

The possibility that stellate cells also subserve monaural functions also should be explored (Oliver and Beckius, this volume).

## Dorsal cochlear nucleus (Figs. 6, 10)

One could muster the synaptic and physiological properties of the individual cell types of the DCN in the same way as for the VCN. This information is in itself very interesting from a physiological perspective. Unit recordings and intracellular studies on fusiform, giant, cartwheel, and even granule cells have revealed several details concerning their response properties, including some very striking inhibitory phenomena (e.g., Godfrey et al., '75b; Young and Voigt, '82; Manis, and Evans and Zhao, this volume). Structural and functional models have been forthcoming (e.g., Kane, '74; Wouterlood et al., '84; Pont and Damper,

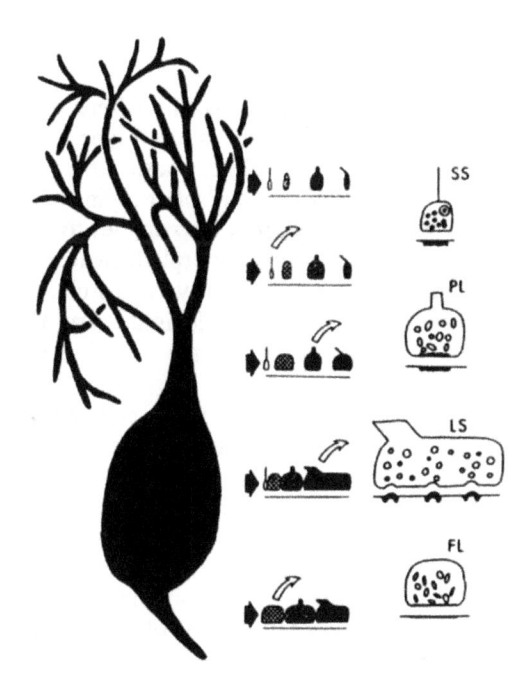

**Figure 13.** Synaptic profile analysis of globular bushy cells (from Ostapoff and Morest, '91). On the left is an idealized globular bushy cell. Solid arrows point from the respective cell regions to icons in the middle column that represent the types of endings contacting the cell surface. The width of the individual icons is proportional to the concentration of that ending type contacting the cell region indicated. The middle column shows that the overall concentration of endings decreases with distance from the cell body and that this decrease is mainly due to the relative paucity of the LS (black) and FL (crosshatched) endings, while the concentration of PL (hatched) endings remains relatively constant and the SS (open) endings increase. Open arrows point from the iconic representation of an ending type to a prototypical drawing (right column), which shows the vesicle and membrane specializations.

'91; Young et al., '88; Arle and Kim, '91b; Meddis and Hewitt, this volume). However, there is still not enough detail concerning the synaptic organization of cell types or a realistic definition of a circuit of cell types. The following discussion suggests that this nucleus has an organization and function which may need another approach than that developed for the VCN.

The DCN in most mammals is a cortical structure, consisting of distinct layers of neuropil, viz., the superficial layer (1), fusiform cell layer (2), and deep layer (subdivided into layers 3 and 4 of Lorente de Nó: Figs. 1-4). The essence of a cortical structure derives from the pattern of the afferent and efferent connections of each neuropil layer and the arrangement by which the dendrites of the neurons cross, and thus differentially sample, the afferent synapses in these layers of neuropil. By virtue of this interesting property, the DCN has some structural features in common with all of the other cortical structures in the nervous system. One thinks especially of the dorsal horn of the spinal cord, the cerebellar cortex, the hippocampus, and the olfactory bulb, all of which contain granule cells, inhibitory interneurons, and projection neurons. It may be possible to gain some insight into the functional significance of the DCN from our knowledge of these other structures. This is not an easy exercise, inasmuch as the analogy with the cerebellar cortex, originally proposed by Lorente de Nó, may derive more from overlap in embryological development and cell lineage than in any functional specificity (see Berrebi and Mugnaini, this volume). However, one cannot help being impressed with the structural analogy in the dorsal horn, where the interneurons, including the excitatory granule cells with parallel fibers and the inhibitory stellate cells, are thought to play a role in gating the sensory processing of the pain pathway. An analogous concept may well apply to the DCN (see, e.g., Wickesberg and Oertel, this volume), for the DCN could very well regulate the input/output transforms of the VCN with respect to the timing and intensity values transmitted a function, at least in principle, not unlike that of the cerebellar cortex as well. The stellate cells of the VCN may play a pivotal role in this process (Fig. 10). Nevertheless, despite some analogies there are many differences. In the end, we will have to study in detail each cortical structure for its own sake in order to understand its particular organization and function.

# Putting it all together: synaptic profile analysis (Fig. 13)

Data from many different kinds of experiments, including approaches as different as electrophysiology and molecular biology, have to be correlated with the same cell type, and the cell types cast in the appropriate circuits. In order to bring all of these kinds of information and data together in correlation, one must have some way of mapping the synaptic endings on an entire cell. We must be able to show that such a map applies to all of the individuals thought to belong to the cell type defined by various morphological and cytochemical methods. In other words, we need to anticipate that different varieties of the principal cell types will be discovered. To meet these demands we have devised a way to map the synaptic organization of cell types in the CNS and to assess the variation of individual neurons in the same category (Fig. 13).

Non-cochlear inputs to globular bushy cells in particular, and to other cell types, potentially form a complex circuitry, whose function is not at all clear at the present time. To study this circuitry one needs a means of determining the contributions of different inputs to the signal transformations occurring in the cochlear nucleus. Synaptic profile analysis provides a means by which inputs can be mapped for cell types and compared quantitatively. Endings originating from any source may be analyzed not only by synaptic morphology, but also according to the region of the postsynaptic cell that they contact and the kind of transmitter molecules that they contain. In addition, the relative contributions of the axonal endings, i.e., the strength of their synaptic inputs to each region of the neuron, and hence the limits of the functional role that it may play, can now be estimated in the synaptic profiles from experimental material.

In effect a synaptic profile provides a map of the synaptic space of a neuron. Such a map allows one to differentiate the entire complement of synaptic inputs impinging on a specific cell type and to compare different neuronal types. Synaptic profile analysis should lead to insights into the functional roles of the different non-cochlear inputs and should prove useful for generating models of the input-output functions of CN neurons.

ACKNOWLEDGEMENTS

The author's work has been supported by grants from the National Institutes of Health. I thank Drs. Charles Stevens and Steve Heinemann for their generous support in the Laboratory of Molecular Neurobiology of The Salk Institute. I thank my colleagues in the hearing research community at The University of Connecticut Health Center and members of The Salk Institute, Laboratory of Molecular Neurobiology, for their help.

REFERENCES

Arle, J.E. and Kim, D.O., 1991a, Neural modeling of intrinsic and spike-discharge properties of cochlear nucleus neurons, *Biol. Cybernetics*, 64:273-283.

Arle, J.E. and Kim, D.O., 1991b, Simulations of cochlear nucleus neural circuitry:excitatory-inhibitory response-area types I-IV, *J. Acoust. Soc. Amer.*, 90:3106-3121.

Bettler, B., Boulter, B., Hermans-Borgmeyer, I., O'Shea-Greenfield, A., Deneris, E., Moll, C., Borgemeyer, U., Hollmann, J., and Heinemann, S., 1990, Cloning of a novel glutamate receptor subunit, GluR5: expression in the nervous system during development, *Neuron*, 5:583-595.

Boulter, J., Hollmann, M., O'Shea-Greenfield, A., Hartley, M., Deneris, E., Marion, C., and Heinemann, S., 1990, Molecular cloning and functional expression of glutamate receptor subunit genes, *Science*, 249:1033-1036.

Brawer, J., Morest, D.K., and Kane, E., 1974, The neuronal architecture of the cochlear nucleus of the cat, *J. Comp. Neurol.*, 155:251-300.

Cohen, E.S., Brawer, J.R., and Morest, D.K., 1972, Projections of the cochlea to the dorsal cochlear nucleus in the cat, *Exp. Neurol.*, 35:470-479.

Egeberg, J., Bettler, B., Hermans-Borgmeyer, I., and Heinemann, S., 1991, Cloning of a cDNA for a glutamate receptor subunit activated by kainate but not AMPA, *Nature*, 361:745-748.

Feng, J., 1989, The Coding of Sound in the Cochlear Nucleus: Mechanisms Underlying the Generation of Response Patterns by Different Morphological Cell Types. Doctoral Dissertation. The University of Connecticut, Farmington.

Feng, J., Ostapoff, E.-M., Kuwada, S., and Morest, D.K., 1988, Synaptic and membrane properties shape response patterns of neurons in the gerbil cochlear nucleus, *Soc. Neurosci. Abstr.*, 14:490.

Glendenning K.K., Hutson, K.A., Nudo, R.J., Masterton, R.B., 1985, Acoustic chiasm II: Anatomical basis of binaurality in lateral superior olive of cat, *J. Comp. Neurol.*, 232:261-285.

Godfrey, D.A., Kiang, N.Y.S., and Norris, B.E., 1975a, Single unit activity in the posteroventral cochlear nucleus of the cat, *J. Comp. Neurol.*, 162:247-268.

Godfrey, D.A., Kiang, N.Y.S., and Norris, B.E. (1975b) Single unit activity in the dorsal cochlear nucleus of the cat. *J. Comp. Neurol.*, 162:269-284.

Hackney, C.M., Osen, K.K., and Kolston, J. 1990, Anatomy of the cochlear nuclear complex of guinea pig, *Anat. Embryol.*, 182:123-149.

Hunt, S. and Schmidt, J., 1978, Some observations on the binding patterns of α-bungarotoxin in the central nervous system of the rat, *Brain Res.*, 157:213-232.

Hollmann, M., Hartley, M., and Heinemann, S., 1991, $Ca^{2+}$ permeability of KA-AMPA-gated glutamate receptor channels depends on subunit composition, *Science*, 252:851-853.

Hollman, M., O'Shea-Greenfield, A., Rogers, S.W., and Heinemann, S., 1989, Cloning by functional expression of a member of the glutamate receptor family, *Nature*, 342:643-648.

Jones, D.R., Hutson, K.A., and Morest, D.K., 1992, Growth cones and structural variation of synaptic endbulbs in the cochlear nucleus of the adult cat brain, *Synapse*, 10:291-309.

Kane, E.S.C., 1973, Octopus cells in the cochlear nucleus of the cat: heterotypic synapses upon homeotypic neurons, *Int. J. Neurosci.*, 5:251-279.

Kane, E.S., 1974a, Synaptic organization in the dorsal cochlear nucleus of the cat: a light and electron microscopic study, *J. Comp. Neurol.*, 155:301-330.

Kane, E.S., 1974b, Patterns of degeneration in the caudal cochlear nucleus of the cat after cochlear ablation, *Anat. Rec.*, 179:67-92.

Keinanen, K., Wisden, W., Sommer, B., Werner, P., Herb, A., Verdoorn, T.A., Sakman, B., and Seeberg, P.H., 1990, A family of AMPA-selective glutamate receptors, *Science*, 249:556-560.

Kiang, N.Y.S., Morest, D.K., Godfrey, D.A., Guinan, J.J. Jr., and Kane, E.C., 1973, Stimulus coding at caudal levels of the cat's auditory nervous system.I. Response characteristics of single units, *in:* "Basic Mechanisms in Hearing", A.Møller, ed., Academic Press, New York, pp. 455-478.

Lorente de Nó, R., 1933a, Anatomy of the eighth nerve. The central projection of the nerve endings of the internal ear, *Laryngoscope*, 43:1-38.

Lorente de Nó, R., 1933b, Anatomy of the eighth nerve. III. General plan of structure of the primary cochlear nuclei, *Laryngoscope*, 43:327-350.

Lorente de Nó, R., 1943, Cerebral cortex: architecture, intracortical connections, motor projections, *in:* "Physiology of the Nervous System. 2nd ed.", J.F. Fulton, Oxford University Press, London, p. 274-301.

Lorente de Nó, R., 1981, "The Primary Acoustic Nuclei", Raven, New York.

Morest, D.K., 1983, Book Review: R. Lorente de Nó, "The Primary Acoustic Nuclei", New York, Raven, 1981, *Ann. Otol. Rhinol. Larngol.*, 92:212-213.

Morest, D.K., and Bohne, B.A., 1983, Noise-induced degeneration in the brain and representation of inner and outer hair cells, *Hearing Res.*, 9:145-151.

Morest, D.K., Hutson, K.A., and Kwok, S. 1990, Cytoarchitectonic atlas of the cochlear nucleus of the chinchilla, *Chinchilla laniger*, *J. Comp. Neurol.*, 300:230-248.

Morest, D., Jones, D.R., Kwok, S., Ard, M.D., Bohne, B., Yurgelun-Todd, D.A., 1987, Response of the brain to acoustic damage of the cochlea, *Assoc. Res. Otolaryngol. Abstr.*, 10:3-4.

Morest, D.K., Kiang, N.Y.S., Kane, E.C., Guinan, J.J. Jr., and Godfrey, D.A., 1973, Stimulus coding at caudal levels of the cat's auditory nervous system. II. Patterns of synaptic organization, *in:* "Basic Mechanisms in Hearing", A. Møller, ed., Academic Press, New York, pp. 479-504.

Nakanishi, N.A., Schneider, R., and Axel, R. 1990, A family of glutamate receptor genes - evidence for the formation of heteromultimeric receptors with distinct channel properties, *Neuron*, 5:569-581.

Oertel, D., Wu, S.H., Garb, M.W., and Dizack, C. 1990, Morphology and physiology of cells in slice preparations of the posteroventral cochlear nucleus of mice, *J. Comp. Neurol.*, 295:136-154.

Osen, K., 1969, Cytoarchitecture of the cochlear nuclei in the cat, *J. Comp. Neurol.*, 136:453-484.

Ostapoff, E.-M., and Morest, D.K., 1991, Synaptic organization of globular bushy cells in the ventral cochlear nucleus of the cat: a quantitative study, *J. Comp. Neurol.*, 314:598-613.

Pont, R.I., and Damper, M., 1991, A computational model of afferent neural circuitry from the cochlear to the dorsal acoustic stria, *J. Acoust. Soc. Amer.*, 89:1213-1228.

Rasmussen, G.L., 1967, Efferent connections of the cochlear nucleus, *in*: "Sensorineural Hearing Processes and Disorders", A.B. Graham, ed., Little, Brown, Boston, p. 61-75.

Rasmussen, G.L., Gacek, R.R., McCrane, E.P., and Baker, C.C., 1960, Model of cochlear nucleus (cat) displaying its afferent and efferent connections, *Anat. Rec.*, 136:344 (cited by Lorente de Nó, '81; model currently housed in the laboratory of D. Kent Morest, Farmington, CT).

Rhode, W.A., Oertel, D., and Smith, P.H., 1983, Physiological response properties of cells labelled intracellularly with horseradish peroxidase in cat ventral cochlear nucleus, *J. Comp. Neurol.*, 213:448-463.

Rhode, W.S., and Smith, P.H., 1986, Encoding timing and intensity in the ventral cochlear nucleus of the cat, *J. Neurophysiol.*, 56:287-307.

Ritz, L.A., and Brownell, W.E., 1982, Single unit analysis of the posteroventral cochlear nucleus of the decerebrate cat, *Neuroscience*, 7:1995-2010.

Rose, J.E., Galambos, R., and Hughes, J.R., 1960, Organization of frequency sensitive neurons in the cochlear nuclear complex of the cat, *in*: "Neural Mechanisms of the Auditory and Vestibular Systems", G.L.Rasmussen and W.F.Windle, eds., Thomas, Springfield, p. 116-136.

Saint Marie, R.L., Ostapoff, E.-M., Morest, D.K., Wenthold, R.J., 1989, Glycine-immunoreactive projection of the cat lateral superior olive: possible role in midbrain ear dominance, *J. Comp. Neurol.*, 279:382-396.

Smith, P.H. and Rhode, W.S., 1989, Structural and functional properties distinguish two types of multipolar cells in the ventral cochlear nucleus, *J. Comp. Neurol.*, 282:595-616.

Wada, E., Wada, K., Boutler, J., Deneris, E., Heineman, S., Patrick, J., and Swanson, L.W., 1989, Distribution of alpha 2, alpha 3, alpha 4, and beta 2 neuronal nicotinic receptor subunit mRNAs in the central nervous system: a hydridization histochemical study in the rat, *J., Comp. Neurol.*, 284:314-335.

Wouterlood, F.G., Mugnaini, E., Osen, K.K., and Dahl, A.L., 1984, Stellate neurons in rat dorsal cochlear nucleus studied with combined Golgi impregnation and electron microscopy: synaptic connections and mutual coupling by gap junctions, *J. Neurocytol.*, 13:639-664.

Wu, S.H., and Oertel, D., 1984, Intracellular injection with horseradish peroxidase of physiologically characterized stellate and bushy cells in slices of mouse anteroventral cochlear nucleus, *J. Neurosci.*, 4:1577-1588.

Young, E.D., Shofner, W.P., White, J.A., Robert, J.M., and Voigt, H.F., 1988, Response properties of cochlear nucleus neurons in relationship to physiological mechanisms, *in:* "Auditory Function: Neurobiological Bases of Hearing", Edelman et al., eds, Wiley, New York, pp. 277-312.

Young, E.D., and Voigt, H.F., 1982, Response properties of type II and type III units in dorsal cochlear nucleus, *Hearing Res.*, 6:153-169.

# I. DEVELOPMENTAL ISSUES

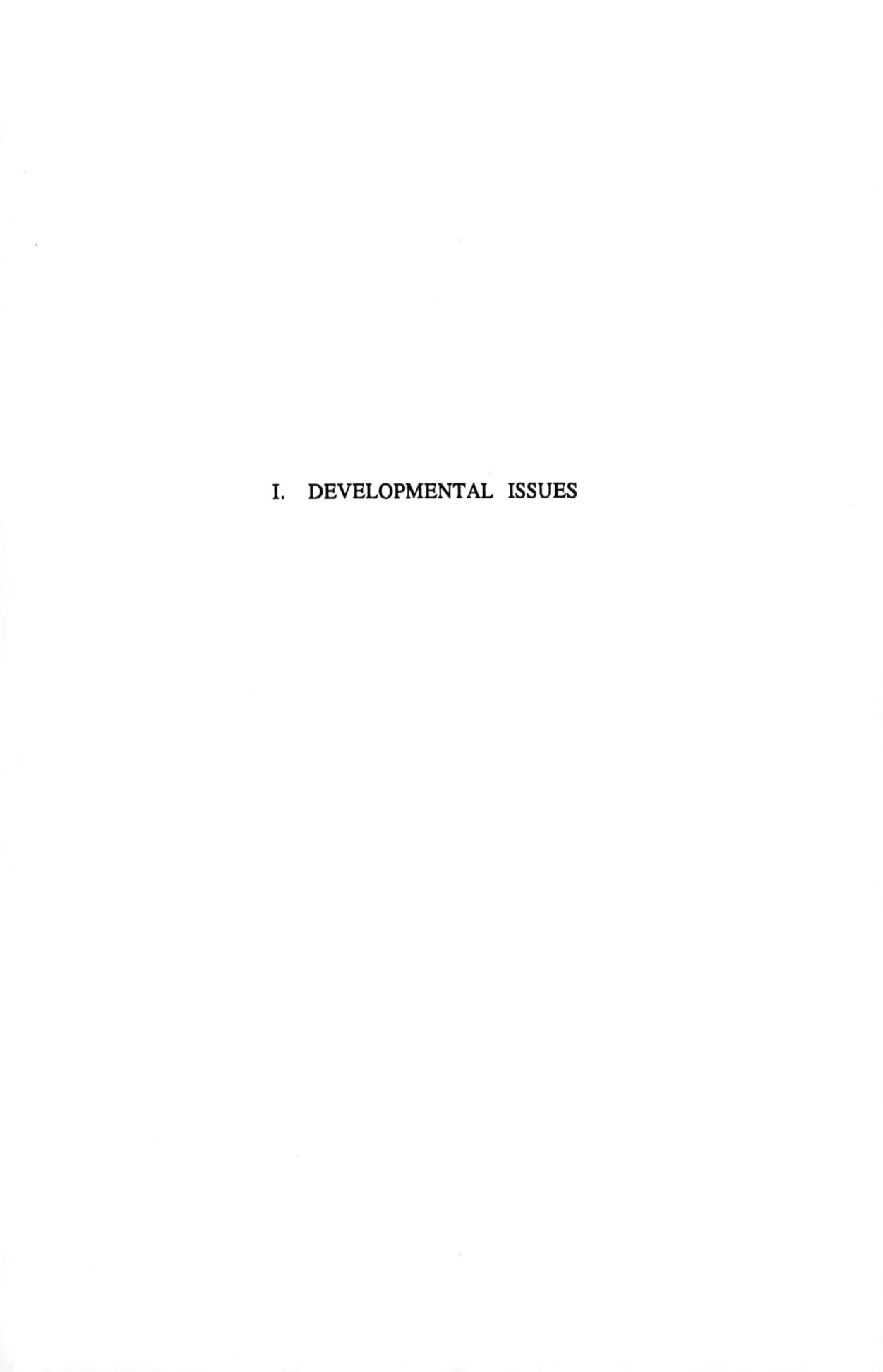

# CELL BIRTH, FORMATION OF EFFERENT CONNECTIONS, AND ESTABLISHMENT OF TONOTOPIC ORDER IN THE RAT COCHLEAR NUCLEUS

Eckhard Friauf and Karl Kandler

Department of Animal Physiology, University of Tübingen, Auf der Morgenstelle 28, D-7400 Tübingen 1, Federal Republic of Germany

The development of the cochlear nucleus (CN) involves proliferation of cells, migration of postmitotic neuroblasts to their final destinations, neurite outgrowth, synapse formation, and cell death. Ultimately, these events lead to the establishment of the high degree of topographic organization observed in the adult. The auditory system of rats is immature at birth which makes this species preferable for developmental studies. Physiological hearing in rats begins about two weeks after birth. The outer ear canals do not open before postnatal day 12 (P12), and first auditory brainstem responses can be reliably recorded at P12-P14 (Jewett and Romano, '72; Tokimoto et al., '77; Blatchley et al., '87). Since gestation in rats lasts 22 days (i.e. birth usually occurs at embryonic day 22, E22 = P0), there is a period of several weeks until the onset of hearing occurs, during which the above mentioned major developmental events must take place.

Until now, not much has been known about the ontogeny of the mammalian central auditory system (see Coleman, '90 and Kitzes, '90 for recent reviews). This is somewhat surprising in light of the fact that the auditory system has unique characteristics which enable us to address specific and challenging questions in development. One of these characteristics is the large number of hierarchically organized nuclei and pathways that make up the auditory brainstem. Do these nuclei develop sequentially or in parallel? Do higher order pathways (e.g. that from the superior olivary complex to the inferior colliculus) develop later than lower order pathways (e.g. that from the cochlear nucleus to the superior olivary complex), or are these pathways formed simultaneously?

In this article, we will consider three fundamental events in the maturation of the cochlear nucleus: neurogenesis, formation of efferent connections to other brainstem nuclei, and establishment of tonotopic order. We will present evidence that these events take place early during development, most of them in fact while the animals are still devoid of normal acoustic input, i.e. virtually deaf.

## Neurogenesis of cochlear nucleus neurons

Neuronal birth is defined by the last mitotic division of precursor cells. Since mitotic division involves previous incorporation of nucleotides into the DNA, during the "S" phase

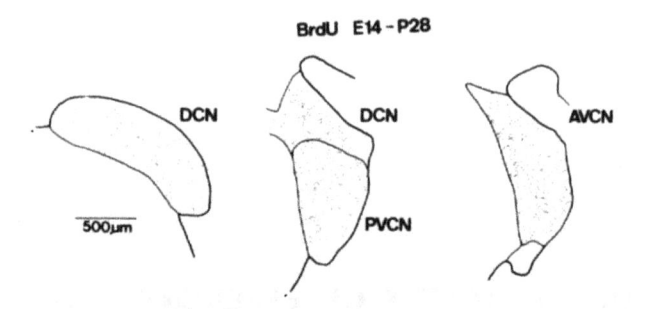

**Figure 1.** Neurogenesis occurs synchronously in all three subdivisions of the CN. This semischematic diagram illustrates the location of BrdU-immunoreactive CN neurons that are produced on E14. On this day, about 50% of the CN neurons are generated. Each dot represents one BrdU-immunoreactive neuron. There is no obvious difference between the three CN subdivisions and also no obvious spatial gradient within any subdivision. Dorsal is up, medial to the left.

of the cycle, dividing cells can be labeled by the administration of analogues of thymidine, a DNA-specific nucleotide. Classically, tritiated thymidine is injected into pregnant females and offsprings are killed later, examining their brains autoradiographically (Sidman, '70). We have applied an alternative technique using the thymidine analogue 5-bromo-2'- deoxyuridine (BrdU), which is also incorporated into the DNA and can be visualized immunohistochemically (Miller and Nowakowski, '88). By injecting a single pulse of BrdU intraperitoneally into pregnant rats between E10-E21 (we consider the day of conception as E0) or into rat pups between P0-P3, we determined the time course of neurogenesis in the cochlear nucleus. These studies showed that most neurons comprising the three subdivisions of the CN are generated prenatally with the exception of neurons in the granule cell layer which are generated perinatally. There is a relatively short period of 2 days (between E14 and E15) during which more than two thirds of the total number of CN neurons are produced. The peak of neurogenesis occurs at E14 in all three subdivisions. The results of a BrdU-experiment performed at E14 are illustrated in Figure 1. From these results it is obvious, that there is significant temporal overlap in the birth dates of the neurons of the cochlear nucleus subdivisions. This overlap is nearly complete in the case of the anteroventral cochlear nucleus (AVCN) and the posteroventral cochlear nucleus (PVCN), but less pronounced in the dorsal cochlear nucleus (DCN). In the latter, neurogenesis lasts over an extended period covering more than 12 days (from E11 to P0). However, as the majority of the DCN neurons are also generated on E14, we conclude that most neurogenesis in the CN occurs in parallel. Consequently, there is no significant temporal gradient in the cell production of the three major CN subdivisions. Our observations confirm earlier birthdating studies using tritiated thymidine in mice and rats (Taber-Pierce, '67; Altman and Bayer, '80). These studies have also shown no statistically significant difference in the birth dates of the neurons of the three CN subdivisions.

One exception to the overlapping neuronal birth are the granule cells which are generated later and whose production lasts into early postnatal life. In contrast to all other CN neurons which are produced in the primary rhombic lip, granule cells are generated in the secondary rhombic lip (Taber-Pierce, '67). They are local circuit neurons (Mugnaini et al., '80a, b) and interneurons have been shown to be invariably generated last in other systems, as well (Jacobson, '78).

Neurons in other nuclei of the rat auditory system are also generated predominantly prenatally, and neurogenesis in the superior olivary complex and the inferior colliculus overlaps with the cell production in the CN (Altman and Bayer, '80, '81; Weber et al., '91). For instance, neurons in the medial nucleus of the trapezoid body and in anterior aspects of the inferior colliculus are generated at E15, a day when neurogenesis is also apparent in the CN. It can therefore be concluded that there is no simple temporal gradient in the auditory brainstem, by which, for example, lower order nuclei (such as the cochlear nucleus) would be formed earlier than higher order nuclei (such as the inferior colliculus).

We speculate that the observed overlap in neurogenesis may be mirrored by an overlap in the formation of internuclear pathways. If this is true, then the connections from the cochlear nucleus to the superior olivary complex would develop synchronously with those from the cochlear nucleus to the inferior colliculus, since neurogenesis in all these regions occurs at E15. We have started to address this question and will provide some results in the following section.

## Formation of efferent connections from the cochlear nucleus

Besides neurogenesis, the differentiation of neuroblasts is another major event during ontogeny. The differentiation of neuroblasts involves the outgrowth of their axons and the formation of appropriate synaptic contacts with postsynaptic neurons. Previous retrograde labeling studies have shown that the projections from the rat CN to the inferior colliculus are present by birth and are therefore established early in development (Friauf and Kandler, '90; Maxwell and Coleman, '89). To further investigate the prenatal formation of internuclear connections, we retrogradely traced the afferents to the inferior colliculus and anterogradely traced the efferents from the cochlear nucleus. Both approaches used the lipophilic carbocyanine dye DiI which was implanted into aldehyde-fixed embryonic brains and allowed to diffuse anterogradely and retrogradely within the neuronal membranes (Godement et al., '87; Honig and Hume, '89). When tiny DiI crystals were implanted into the inferior colliculus, the earliest day that neurons in the subdivisions of the contralateral CN were retrogradely labeled was E17, six days after the generation of the earliest DCN neurons and only two to three days after the generation of neurons in the ventral CN. At this early stage tracer application had to involve rostroventral aspects of the inferior colliculus in order to label CN neurons. Two days later, at E19, retrograde tracing was also successful when the DiI crystals were implanted dorsocaudally into the contralateral inferior colliculus. These results indicate that there is a temporal gradient in the innervation of the inferior colliculus from the CN. This gradient runs from rostral to caudal and is thus in register with the gradient described for the neurogenesis of the inferior colliculus (Altman and Bayer, '81).

Anterograde tracing studies with DiI implants into the CN corroborate the findings of a prenatal formation of the major efferent connections from the rat CN (Kandler and Friauf, '91). These studies not only revealed the projection to the inferior colliculus, but also to other brainstem nuclei such as those in the superior olivary complex (SOC) and the lateral lemniscus (NLL), all of which are known to receive input from the CN in the adult rodent (Harrison and Irving, '66; Friauf and Ostwald, '88; Müller, '90). At E15, efferent fibers from CN neurons had invaded the ventral pontine brainstem, and some had already crossed the midline. Axonal collaterals became apparent in the ipsilateral SOC as early as E18 and appeared two days later in the contralateral SOC and the NLL as well. Although these axonal collaterals are not a clear sign of synaptic innervation, they may be suggestive of synapses, as many of the highly branched arbors terminated within the limits of the nuclei and numerous axonal swellings were present. It will be interesting to see when the synapses in the SOC and other auditory brainstem nuclei begin to function. Studies designed to identify synaptic elements such as the synaptic vesicle-associated proteins synapsin and synaptophysin and/or electrophysiological experiments should clarify when the formation of functional synapses occurs.

The results of our anterograde and retrograde DiI tracing experiments both confirm the hypothesis that connections in the ascending auditory pathway are formed as early as possible to minimize the considerable problems of pathfinding and target selection posed by the progressive growth of the brain. Due to the fact that the main ascending connections of the rat auditory brainstem nuclei are established prenatally and because the peripheral auditory system is immature at least until P12 (e.g. the ear canals do not open before P12), there is a period of more than two weeks during which the auditory neurons, although interconnected, remain devoid of normal acoustic input. One can speculate that the early presence of CN axons in the ventral pontine brainstem might be a prerequisite for the proper maturation of neurons in the superior olivary complex by means of afferent regulation (Hashisaki and Rubel, '89). At the moment, it is unclear whether this regulation is accomplished via classical synaptic transmission or by secretion of a 'trophic' factor from CN axons. The long interval between E18 and the onset of physiological hearing is probably bridged by processes of synaptic maturation. These processes may include an increase in the synaptic efficiency of appropriate connections and the elimination of inappropriate synapses. So far, we have no physiological evidence for this speculation. However, postnatal changes in neurons of the rat auditory brainstem nuclei have been described previously: there is spatial reorientation of the dendritic branches and reabsorption of dendritic spines in inferior colliculus neurons (Dardennes et al., '84), and there is initial growth and successively a reduction of dendritic branches in neurons of the medial superior olive (Rogowski and Feng, '81). It should be mentioned, however, that most of these changes occur after P12. In our tracing studies we observed significant morphological maturation in the SOC early in postnatal life. For instance, in the medial nucleus of the trapezoid body, the shape of the calyces of Held, the characteristic presynaptic terminals from CN globular-bushy cells, changes considerably in the neonate (Morest, '68; Kandler and Friauf, '91). The endbulbs of Held in the rat AVCN (Neises et al., '82) also mature about a week into postnatal life, providing further evidence for the considerable changes that occur between birth and the onset of hearing.

Another example of postnatal maturation is given by the projection neurons in the DCN. We determined their morphological appearance from our photoconverted DiI material (Sandell and Masland, '88) after implanting crystals into the inferior colliculus. Three characteristic DCN neurons at E21 are illustrated in Figure 2. Already at this late fetal age, the gross morphology resembles that of fusiform/bipolar pyramidal neurons which are the main projection neurons of the mammalian DCN, projecting through the dorsal acoustic stria to the inferior colliculus, mainly on the contralateral side (Oliver, '84; Collia et al., '88). Basal and apical dendritic arbors are observed as they have been previously described respectively, in the deep and molecular layer of the DCN (Oertel and Wu, '89; Hackney et al., '90). However, the apical arbors clearly lack the dendritic spines which are abundant in the adult. In contrast, they show large expansions on their distal dendrites which resemble growth cones, thus providing evidence for the immature fine morphology of these neurons. These findings are consistent with an earlier Golgi study in the hamster where large expansions were also frequent on distal dendrites and spines were scarce even at P15 (Schweitzer and Cant, '85). At present, we do not know when the spines appear on the fusiform neurons in the rat DCN, but it is clear that they mature postnatally since we did not observe them in material from animals aged P2 and P4. In summary, we have shown that the major efferent connections from the rat cochlear nucleus form remarkably early during ontogeny, like those of the ferret (Brunso-Bechtold et al., '90) and the opossum (Willard and Martin, '86). Our findings suggest that these connections achieve some of their adult-like appearance early, but are likely to undergo modification and fine tuning in the neonatal rat.

## Establishment of tonotopic order

One of the basic principles of physiological organization in the auditory system is the tonotopic order. We investigated the appearance and development of frequency representation

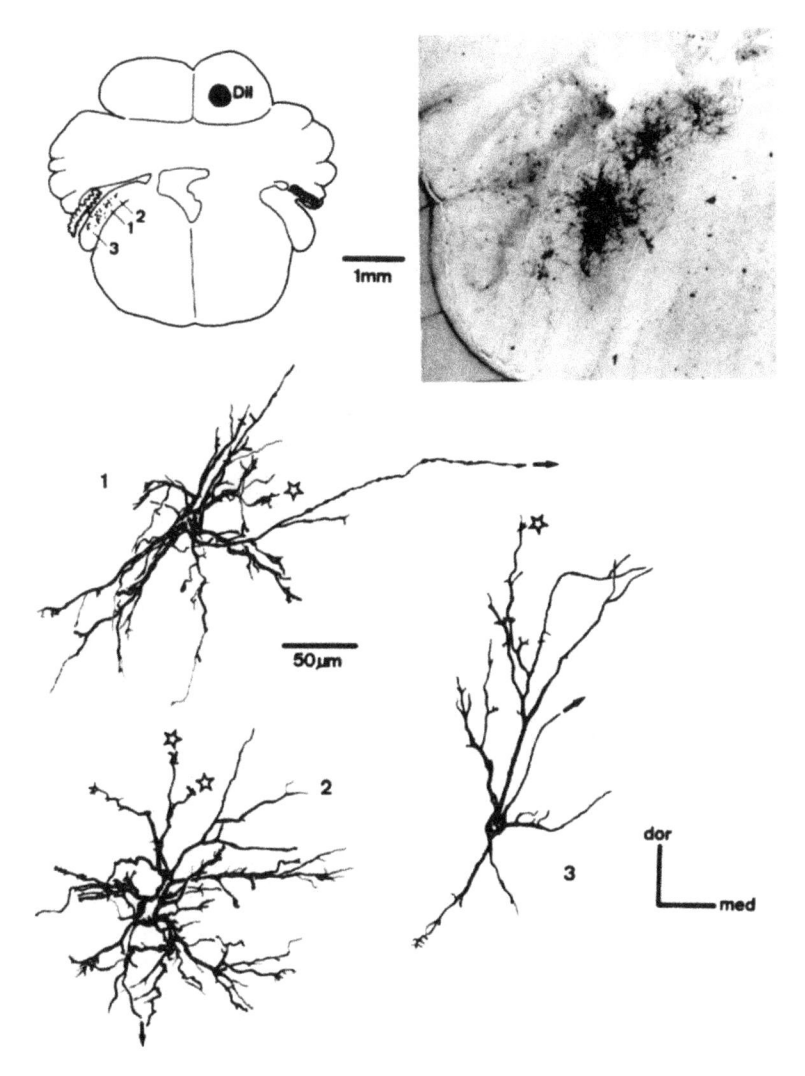

**Figure 2.** Location and morphology of neurons in the DCN that project to the contralateral inferior colliculus at E21. The lipophilic tracer DiI was implanted into fixed fetal brains and allowed to diffuse within the neuronal membranes. Retrogradely labeled cells resemble mature fusiform /bipolar pyramidal neurons in terms of their location and gross morphology, but they have large expansions reminiscent of growth cones on their distal dendrites (stars) and lack dendritic spines on their apical arbors. Axonal course is indicated by arrows.

in the CN subdivisions in intact animals by a time-saving technique that offers the advantage of labeling isofrequency bands at a single cell resolution. We stimulated alert and unrestrained rats at various ages with pure tone pulses and processed their brains immunohistochemically for the nuclear phosphoprotein FOS. FOS can be synthesized in neurons after the expression of the proto-oncogene c-fos in response to synaptic excitation (Sheng and Greenberg, '90; Morgan and Curran, '91). The FOS protein may therefore serve as an excellent marker for mapping neuronal activity. In all three CN subdivisions of adult rats, low frequencies were represented ventrally and high frequencies dorsally, consistent with the known tonotopic order described electrophysiologically (Kaltenbach and Lazor, '91) or by the 2-deoxyglucose method (Ryan et al., '82, '88). An example of the dorsomedial location of the 50 kHz band in the DCN is illustrated in Figure 3. A striking finding was that FOS-immunoreactive

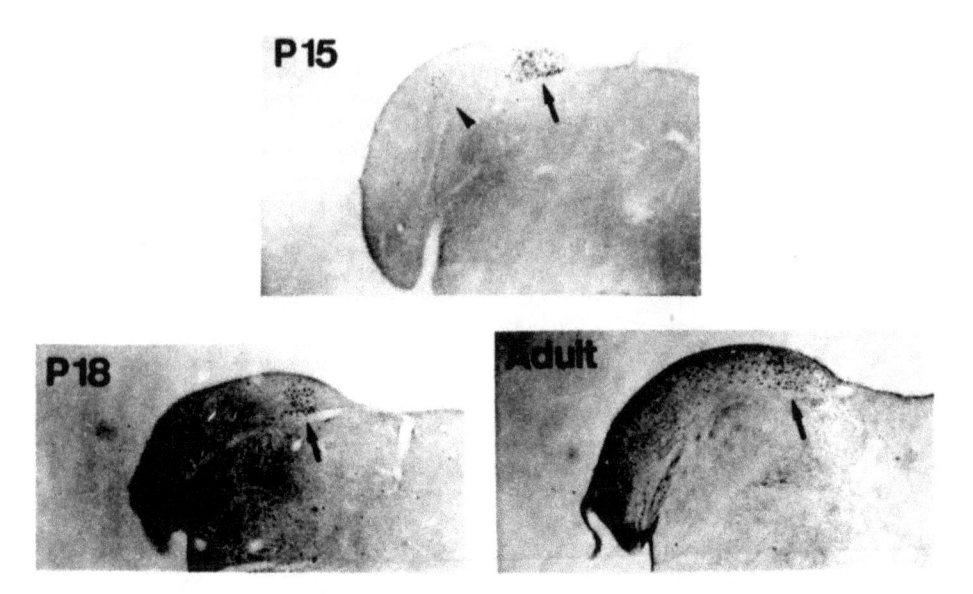

**Figure 3.** Location and distribution of FOS-immunoreactive neurons in the DCN of a P15, a P18 and an adult rat after stimulation with 50 kHz (arrows). 50 kHz tone pulses successfully induced FOS-immunoreactivity as early as P14, suggesting that the rat auditory system is already capable of detecting high frequencies at this age. There are no age-related changes in the intensity, thickness, and center of the labeled band of neurons in the dorsomedial aspect of the DCN. However, at P15, an additional band is obvious further lateral (arrowhead) which was never detected after P17, indicating that some maturation in tonotopic order occurs shortly after the onset of hearing. Dorsal is up, medial to the right in all panels.

neurons are already intensely labeled at P15 (Fig. 3), and we could label them as early as P14 in our experiments. At this age, all frequencies tested were effective in inducing FOS-immunoreactivity in the majority of animals. This indicates that rats are capable of hearing over a large frequency range two weeks after birth and thus only shortly after their peripheral auditory system becomes functional. This result is in contrast to other findings describing a delayed and gradual maturation of the sensitivity to high frequencies (Aitkin and Moore, '75; Woolf and Ryan, '85; Ryan and Woolf, '88; Rübsamen et al., '89; Romand and Ehret, '90). The difference may be attributed to different sensitivities of the various methods employed. Most other researchers used anesthetized animals and microelectrode recording techniques, whereas we performed our experiments in conscious rats. It is, however, unclear why only high frequencies should be impaired by anesthesia.

The locations of the FOS-labeled isofrequency bands in young animals coincided very well with those in adults (Fig. 3). However, when we stimulated P15 rats with 50 kHz, we observed a second band of FOS-immunoreactive neurons at a location that corresponds to significantly lower frequencies in the adult (Fig. 3). This additional band was not seen after P17 and also never occurred when we stimulated with 16 kHz and even lower frequencies. This finding is interesting in light of several reports which described shifts in frequency-place code (Aitkin and Moore, '75; Woolf and Ryan, '85; Ryan and Woolf, '88; Rübsamen et al., '89; Romand and Ehret, '90), thereby confirming Rubel's frequency shift hypothesis (Rubel, '84; Lippe and Rubel, '85). We did not see a shift of the location of the additional band, but it disappeared rather quickly between P17 and P18, after which the adult-like tonotopic organization was present. Whether its disappearance may correspond to gradual shifts of isofrequency bands remains unclear at present.

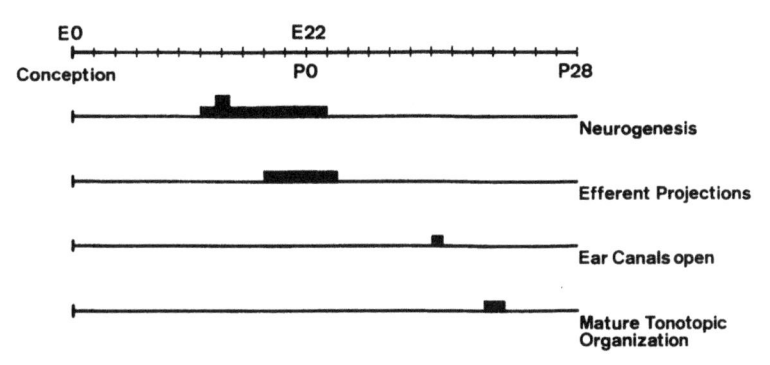

**Figure 4.** Summary of the major events and time periods over which cells in the rat cochlear nucleus develop during ontogeny.

The literature on shifting place code is still controversial. Some authors described shifts between two octaves to four octaves (Ryan and Woolf, '88; Romand and Ehret, '90; Harris and Dallos, '84; Sanes et al., '89), others described shifts of only one octave and even less (Rübsamen et al., '89; Lippe and Rubel, '85; Echteler et al., '89; Rübsamen and Schäfer, '90; Müller et al., '90), and still others found a developmental stability in the place code (Manley et al., '87). Although our data does not clarify the controversy at present, it sheds new light on the topic by describing the transient appearance of an isofrequency band for high frequencies that disappears shortly before the mature organization occurs.

The data presented in this article implies that many events from neuronal birth to differentiation, such as the formation of efferent connections with other auditory brainstem nuclei, occur in the rat CN within approximately one week around birth, and thus strikingly early during ontogeny. Figure 4 summarizes the time course of these major ontogenic events. It makes sense that internuclear connections are formed when the brain is still small to keep at its minimum the considerable problem of pathfinding and target selection posed in a progressively growing brain (Shatz et al., '91). If this assumption is correct, then it is logical that the interval between neurogenesis and synaptogenesis be kept relatively short.

A great deal of development occurs prenatally in the rat's cochlear nucleus, because neurogenesis, cell migration and the formation of major ascending pathways are progressing, if not fully accomplished prior to birth. Another important period occurs after the onset of physiological hearing. There are several reports describing this time as a 'critical period' during which appropriate auditory input is required for the normal development and for the fine tuning of physiological characteristics (e.g. Webster, '83; Sanes and Constantine-Paton, '83, 85; Knudsen et al., '82; Moore, '86). Thus, it is conceivable that after about P14, the auditory system matures under the control of auditory experience. This is in contrast to the preceding neonatal period during which rat pups are virtually deaf. At present, we do not know much about this period and the alterations that occur without proper auditory input. Nevertheless, the fact that it lasts about two weeks may be indicative of its relevance. It is safe to conclude that during this early postnatal interval, neural structures are not simply dormant in the auditory brainstem but undergo significant maturation. What these maturational processes are, when they occur and what they are controlled by (neuronal activity, genetic factors), will be interesting to investigate in the future.

ACKNOWLEDGEMENTS

This research was supported by a grant from the Deutsche Forschungsgemeinschaft (Fr 772/1-1) to E.F. and a Predoctoral Fellowship (Graduiertenkolleg Neurobiologie/Tübingen) to K.K. We wish to thank Drs. H. Herbert, M. Koch and T. Voigt for their generous technical suggestions and H. Zillus and F. Weber for expert histological assistance. Special thanks to K. Lingenhöhl for valuable comments on the manuscript and to Dinnie Goldring for correcting our English.

REFERENCES

Aitkin, L.M. and Moore, D.R., 1975, Inferior colliculus. II. Development of tuning characteristics and tonotopic organisation in central nucleus of the neonatal cat, *J. Neurophysiol.*, 38:1208-1216.

Altman, J. and Bayer, S.A., 1980, Development of the brain stem in the rat. III. Thymidine-radiographic study of the time of origin of neurons of the vestibular and auditory nuclei of the upper medulla, *J. Comp. Neurol.*, 194:877-904.

Altman, J. and Bayer, S.A., 1981, Time of origin of neurons of the rat inferior colliculus and the relations between cytogenesis and tonotopic order in the auditory pathway, *Exp. Brain Res.* 42:411-423.

Blatchley, B.J., Cooper, W.A., and Coleman, J.R., 1987, Development of auditory brainstem response to tone pip stimuli in the rat, *Dev. Brain Res.*, 32:75-84.

Brunso-Bechtold, J.K., Henkel, C.K., and Vinsant, S.L., 1990, Embryonic development of the mammalian hindbrain auditory decussation, *Soc. Neurosci. Abstr.*, 16:298.1.

Coleman, J.R., 1990, Development of auditory system structures, *in*: "Development of Sensory Systems in Mammals", J.R. Coleman, ed, Wiley, New York.

Collia, F., Lopez, D.E., Malmierca, M.S., and Merchan, M., 1988, Study with horseradish peroxidase (HRP) of the connections between the cochlear nuclei and the inferior colliculus of the rat, *in*: "Auditory Pathway. Structure and Function", J. Syka and R.B. Masterton, eds, Plenum Press, New York, London.

Dardennes, R., Jarreau, P.H., and Meininger, V., 1984, A quantitative Golgi analysis of the postnatal maturation of dendrites in the central nucleus of the inferior colliculus of the rat, *Dev. Brain Res.*, 16:159-169.

Echteler, S.M., Arjmand, E., and Dallos, P., 1989, Developmental alterations in the frequency map of the mammalian cochlea, *Nature*, 341:147-149.

Friauf, E. and Kandler, K., 1990, Auditory projections to the inferior colliculus of the rat are present by birth, *Neurosci. Lett.*, 120:58-61.

Friauf, E. and Ostwald, J., 1988, Divergent projections of physiologically characterized rat ventral cochlear nucleus neurons as shown by intra-axonal injection of horseradish peroxidase, *Exp. Brain Res.* 73:263-284.

Godement, P., Vanselow, J., Thanos, S., and Bonhoeffer, F., 1987, A study in developing visual systems with a new method of staining neurones and their processes with carbocyanine dyes in fixed tissue, *Development*, 101:697-713.

Hackney, C.M., Osen, K.K., and Kolston, J., 1990, Anatomy of the cochlear nuclear complex of guinea pig, *Anat. Embryol.*, 182:123-149.

Harris, D.M. and Dallos, P., 1984, Ontogenetic changes in frequency mapping of a mammalian ear, *Science*, 225:741-743.

Harrison, J.M. and Irving, R., 1966, Ascending connections of the anterior ventral cochlear nucleus in the rat, *J. Comp. Neurol.*, 126:51-64.

Hashisaki, G.T. and Rubel, E.W., 1989, Effects of unilateral cochlea removal on anteroventral cochlear nucleus neurons in developing gerbils, *J. Comp. Neurol.*, 283:465-473.

Honig, M.G. and Hume, R.I., 1989, DiI and DiO: Versatile fluorescent dyes for neuronal labelling and pathway tracing, *Trends Neurosci.*, 12:333-341.

Jacobson, M., 1978, Developmental Neurobiology, Plenum Press, New York, London.

Jewett, D.L. and Romano, M.N., 1972, Neonatal development of auditory system potentials averaged from the scalp of rat and cat, *Brain Res.*, 36:101-115.

Kaltenbach, J.A. and Lazor, J., 1991, Tonotopic maps obtained from the surface of the dorsal cochlear nucleus of the hamster and the rat, *Hearing Res.*, 51:149-160.

Kandler, K. and Friauf, E., 1991, Development of efferent connections of the cochlear nucleus in the rat, *Soc. Neurosci. Abstr.*, 17:182.9.

Kitzes, L.M., 1990, Development of auditory system physiology, *in*: "Development of Sensory Systems in Mammals", J. Coleman, ed, Wiley, New York.

Knudsen, E., Knudsen, P., and Esterly, S., 1982, Early auditory experience modifies sound localization in barn owls, *Nature*, 295:238-240.

Lippe, W. and Rubel, E.W., 1985, Ontogeny of tonotopic organization of brain stem auditory nuclei in the chicken: Implications for development of the place principle, *J. Comp. Neurol.*, 237:273-289.

Manley, G.A., Brix, J., and Kaiser, A., 1987, Developmental stability of the tonotopic organization of the chick's basilar papilla, *Science*, 237:655-656.

Maxwell, B. and Coleman, J.R., 1989, Differential timetable of projections into the developing inferior colliculus in rat, *Soc. Neurosci. Abstr.*, 15:745.

Miller, M.W. and Nowakowski, R.S., 1988, Use of bromodeoxyuridine-immunohistochemistry to examine the proliferation, migration and time of origin of cells in the central nervous system, *Brain Res.* 457:44-52.

Moore, D.R., 1986, Critical periods for binaural interaction and spatial representation, *Acta Otolaryngol. (Stockh.) Suppl.*, 429:51-55.

Morest, D.K., 1968, The growth of synaptic endings in the mammalian brain: A study of the calyces of the trapezoid body, *Z. Anat. Entwicklungs-Gesch.*, 127:201-220.

Morgan, J.I. and Curran, T., 1991, Stimulus-transcription coupling in the nervous system: Involvement of the inducible proto-oncogenes fos and jun, *Annu. Rev. Neurosci.*, 14:421-451.

Mugnaini, E., Osen, K.K., Dahl, A.-L., Friedrich jr, V.L., and Korte, G., 1980, Fine structure of granule cells and related interneurons (termed Golgi cells) in the cochlear nucleus complex of cat, rat and mouse, *J. Neurocytol.*, 9:537-570.

Mugnaini, E., Warr, W.B., and Osen, K.K., 1980, Distribution and light microscopic features of granule cells in the cochlear nuclei of cat, rat and mouse, *J. Comp. Neurol.*, 191:58-606.

Müller, M., 1990, Quantitative comparison of frequency representation in the auditory brainstem nuclei of the gerbil, "Pachyuromys duprasi", *Exp. Brain Res.*, 81:140-149.

Müller, M., Roth, B., and Bruns, V., 1990, Postnatal development of the cochlea in the rat: Morphology and tonotopy, in "Brain-Perception-Cognition. Proc. 18th Göttingen Neurobiol. Conf." N. Elsner and G. Roth, eds, Thieme, Stuttgart, New York.

Neises, G.R., Mattox, D.E., and Gulley, R.L., 1982, The maturation of the endbulb of Held in the rat anteroventral cochlear nucleus, *Anat. Rec.*, 204:271-279.

Oertel D. and Wu, S.H., 1989, Morphology and physiology of cells in slice preparations of the dorsal cochlear nucleus of mice, *J. Comp. Neurol.*, 283:228-247.

Oliver, D.L., 1984, Dorsal cochlear nucleus projections to the inferior colliculus in the cat: A light and electron microscopic study, *J. Comp. Neurol.* 224:155-172.

Rogowski., B.A., and Feng, A.S., 1981, Normal postnatal development of medial superior olivary neurons in the albino rat: A Golgi and Nissl study, *J. Comp. Neurol.*, 196:85-97.

Romand, R. and Ehret, G., 1990, Development of tonotopy in the inferior colliculus. I. Electrophysiological mapping in house mice, *Dev. Brain Res.*, 54:221-234.

Rubel, E.W., 1984, Ontogeny of auditory system function, *Annu. Rev. Physiol.*, 46:213-229.

Rübsamen, R. and Schäfer, M., 1990, Ontogenesis of auditory fovea representation in the inferior colliculus of the Sri Lankan rufous bat, "Rhinolophus rouxi", *J. Comp. Physiol. A*, 167:757-769.

Rübsamen, R., Neuweiler, G., and Marimuthu, G., 1989, Ontogenesis of tonotopy in inferior colliculus of a hipposiderid bat reveals postnatal shift in frequency-place code, *J. Comp. Physiol. A*, 165:755-769.

Ryan, A.F. and Woolf, N.K., 1988, Development of tonotopic representation in the central auditory system of the mongolian gerbil: a 2-deoxyglucose study, *Dev. Brain Res.*, 41:61-70.

Ryan, A.F., Furlow, Z., Woolf, N.K., and Keithley, E.M., 1988, The spatial representation of frequency in the rat dorsal cochlear nucleus and inferior colliculus, *Hearing Res.*, 36:181-190.

Ryan, A.F., Woolf, N.K., and Sharp, F.R., 1982, Tonotopic organization in the central auditory pathway of the mongolian gerbil: a 2-deoxyglucose study, *J. Comp. Neurol.*, 207:369-380.

Sandell, J.H., and Masland, R.H., 1988, Photoconversion of some fluorescent markers to a diaminobenzidine product, *J. Histochem. Cytochem.*, 26:555-559.

Sanes, D.H. and Constantine-Paton, M., 1983, Altered activity patterns during development reduce neural tuning, *Science*, 221:1183-1185.

Sanes, D.H. and Constantine-Paton, M., 1985, The sharpening of frequency tuning curves requires patterned activity during development in the mouse, "Mus musculus", *J. Neurosci.*, 5:1152-1166.

Sanes, D.H., Merickel, M., and Rubel, E., 1989, Evidence for an alteration of the tonotopic map in the gerbil cochlea during development, *J. Comp. Neurol.*, 279:436-444.

Schweitzer, L. and Cant, N.B., 1985, Differentiation of the giant and fusiform cells in the dorsal cochlear nucleus of the hamster, *Dev. Brain Res.*, 20:69-82.

Shatz, C.J., Ghosh, A., McConnell, S.K., Allendoerfer, K.L., Friauf, E., and Antonini, A., 1991, Subplate neurons and the development of neocortical connections, in: "Development of the Visual Cortex", D.M. Lam and C.J. Shatz, eds, MIT Press.

Sheng, M. and Greenberg, M.E., 1990, The regulation and function of c-fos and other immediate early genes in the nervous system, *Neuron*, 4:477-485.

Sidman, R.L., 1970, Autoradiographic methods and principles for study of the nervous system with thymidine-H³, *in*: "Contemporary Research Methods in Neuroanatomy", W.J.H. Nauta and S.O.E. Ebbeson, eds, Springer, New York.

Taber-Pierce, E., 1967, Histogenesis of the dorsal and ventral cochlear nuclei in the mouse: an autoradiographic study, *J. Comp. Neurol.*, 131:27-54.

Tokimoto, T., Osako, S., and Matsuura, S., 1977, Development of auditory evoked cortical and brain stem responses during the early postnatal period in the rat, *Osaka City Med. J.*, 23:141-153.

Weber, F., Zillus, H., and Friauf, E., 1991, Neuronal birth in the rat auditory brainstem, *in*: "Synapse-Transmission-Modulation. Proc. 19th Göttingen Neurobiology Conference". N. Elsner and H. Penzlin, eds, Thieme, Stuttgart, New York.

Webster, D.B., 1983, A critical period during postnatal auditory development of mice, *Int. J. Pediatr. Otorhinolaryngol.* 6:107-118.

Willard, F.H. and Martin, G.F., 1986, The development and migration of large multipolar neurons into the cochlear nucleus of the North American opossum, *J. Comp. Neurol.*, 248:119-132.

Woolf, N.G. and Ryan, A.F., 1985, Ontogeny of neural discharge patterns in the ventral cochlear nucleus of the mongolian gerbil, *Dev. Brain Res.*, 17:131-147.

# POSTNATAL DEVELOPMENT OF AUDITORY NERVE PROJECTIONS TO THE COCHLEAR NUCLEUS IN *MONODELPHIS DOMESTICA*

Frank H. Willard

Department of Anatomy, University of New England, Biddeford, Maine, 04005, USA

The external world is represented in the central nervous system by a series of topographic maps or graphs (Changeux, '86; Udin and Fawcett, '88). These maps are composed of precisely ordered sets of neurons whose organization is a function of the distribution of sensory receptors in the skin, eyes, or ears. Understanding the development of these patterned relationships between the receptor organ and its central pathway is critical to the knowledge of sensory system ontogeny.

A model describing the ontogeny of ordered pathways has emerged from studies of the visual system (Shatz, '90). As the axons from retinal ganglion cells grow into the lateral geniculate nucleus, projections from the two eyes overlap extensively; this embryonic state differs significantly from the segregation of input that will characterize the adult retinogeniculate connections. Before stimulus-driven activity begins, functional synapses appear between retinal axons and geniculate neurons. These nascent connections support spontaneous neural activity and a period of activity-dependent remodeling ensues (Shatz, '91). Additional stimulus-dependent remodeling and refinement of the precision in the visual map is accomplished as the receptors begin responding to visual information (Movshon and Kiorpes, '90). Does a similar sequence also occur in the development of the auditory system? This article will focus on the first step of this developmental sequence, the question of uniformity in the early ontogeny of primary afferent connections related to the mammalian cochlear nucleus.

The mammalian cochlear nucleus is organized in a topographic or tonotopic manner (Clopton et al., '74). A description of the complicated spatial relationships between cochlea and cochlear nucleus has been given by Lorente de No ('33). This orderly pattern is based on the distribution of peripheral processes of spiral ganglion cells along the basilar membrane of the cochlea. The central processes of the spiral ganglion cells assemble in an orderly fashion in the auditory nerve, which extends in a spiral to reach the ventral border of the cochlear nucleus. As the nerve enters the nucleus it unwinds, distributing its axons in an

---

ABBREVIATIONS: abr - anterior branch; AN - auditory nerve; D - dorsal; DAS - Dorsal acoustic stria; DCN - dorsal cochlear nucleus; ICP - Inferior cerebellar peduncle; pbr - posterior branch; PND - postnatal day; TB, tb - Trapezoid body; trs - Tract of spinal trigmenial; VCN - ventral cochlear nucleus; VG - .vestibular ganglion; VIIIn - vestibuloacoustic nerve; Vstl - Lateral vestibular nucleus; ANN - Auditory nerve nucleus.

orderly progression. In this manner, a series of stripes, each representing a narrow region of the basilar membrane is superimposed upon the cochlear nucleus (Osen, '70; Fekete et al., '82; Leake and Snyder, '89). Within the cochlear nucleus, as well as at more central auditory structures, the fidelity of the topographic map is maintained by geometric constraints on the distribution of axon terminals and dendritic domains of target neurons (Morest, '73; Scheibel, '74; Oliver, '84; Oliver, '87).

To date, studies of the ontogeny of auditory nerve projections to the mammalian cochlear nucleus have not reported detectable periods of heterogenous (mixed or diffuse) projections similar to those seen in the visual system. Instead, topographic order appears evident in the auditory nerve projections to the cochlear nucleus quite early in the development of the mouse (Lorente de No, '33), hamster (Schweitzer and Cant, '84), and rat (Angulo et al., '90). However, since these studies have used fiber stains, Golgi impregnations, or mass transport of HRP from the cochlea, it is difficult to assess the precision of point-to-point relationships in the auditory primary afferent fibers. The analysis of the ontogeny of ordered projections from the cochlea will require identification of auditory nerve fibers from restricted portions of the basilar membrane very early in development.

The study of growing primary afferent fibers in the cochlear nucleus was attempted in the marsupial *Monodelphis domestica*, because this species is born with an immature auditory system (otocyst has 3/4 turn with no identifiable presumptive hair cells) and its postnatal maturation is quite slow, occurring over a one month period (Willard and Munger, submitted). In addition, these animals experience a postnatal migration of precochlear neurons into dorsal cochlear nucleus (DCN; Willard and Martin, '86). These precochlear neurons are the precursors of the principal (fusiform or pyramidal) and giant cells of the nucleus and are instrumental in the transfer of the cochleotopic map from primary afferent fibers, through DCN, and on to the central nucleus of the inferior colliculus in most mammals (Adams, '79; Ryugo et al., '81; Rhode et al., '83; Oliver, '84; Ryugo and Willard, '85; Smith and Rhode, '85; Willard et al., '91). Thus in *Monodelphis*, it will be possible to study the interactions of primary afferent fibers as they develop orderly relations with a specific group of neurons during their migration into the cochlear nucleus. An understanding of the events occurring in the establishment of these orderly projections from auditory nerve through DCN to midbrain will help explain how patterned connections are organized and what factors influence their development.

It is evident that *Monodelphis* will become an important species in understanding the development of the nervous system in general (Saunders et al., '89) as well as the auditory system specifically (Willard and Munger, submitted) . In this chapter, the structure of the cochlear nucleus in *Monodelphis* will be outlined and a description provided of the cochleotopic pattern of primary afferent innervation in the adult. The ontogeny of this innervation will be examined in the form of two specific questions: 1) "Who arrives at the presumptive DCN first, the primary afferent fibers or the precochlear neurons?" and 2) "Are the narrow bands of auditory nerve fascicles, characteristic of this topographic projection in the adult cochlear nucleus, present during the very early phases of cochlear duct ontogeny?"

## Cytoarchitecture of the *Monodelphis* cochlear nucleus

Three major divisions of the cochlear nucleus are present in *Monodelphis*: dorsal (DCN), anteroventral (AVCN), and posteroventral (PVCN). These divisions are similar to those present in the North American opossum, *Didelphis* (Willard and Martin, '83) as well as eutherian mammals such as cat (Brawer et al., '74; Osen, '69), mouse (Webster and Trune, '82; Willard and Ryugo, '83), rabbit (Disterhoft et al., '80), chinchilla (Morest et al., '90) and bat (Vater and Feng, '90). Although the shape and position of the divisions within the cochlear nucleus are altered slightly in marsupials, strong homologies with eutherian mammals are seen in cell and fiber architecture (Aitkin et al., '86; Willard and Martin, '83). In addition to the three main divisions, there is a highly distinct auditory nerve nucleus, characteristic of *Monodelphis* and *Didelphis*, which is completely separated from the cochlear nucleus by fascicles of the auditory nerve (Figs. 2-4).

### Dorsal cochlear nucleus (DCN)

The DCN is located medial to the inferior cerebellar peduncle, in contrast to its usual lateral position in eutherian mammals. However, metatherian DCN does contain four fibrocellular layers organized around a row of polarized principal cells similar to that reported in other species (Figs. 3-4)(Osen, '69; Brawer et al., '74; Ryugo and Willard, '85).

Although layer I of DCN is only sparsely populated with cell bodies, it does contain a dense mat of apical dendrites from the principal neurons. Layer II, contains many granule cells in addition to the cell bodies of the principal neurons. This layer is continuous along its medial border with the internuclear granule cell lamina. Layer III contains the basal dendrites of the principal cells along with the giant neurons and auditory nerve fibers (Willard et al., '84). Axons of the large neurons gather in layer IV and leave the nucleus in the dorsal acoustic stria, a prominent fiber tract in the dorsolateral portion of the medulla.

### Anteroventral cochlear nucleus (AVCN)

AVCN is the rostralmost division of the nucleus; it can be subdivided into anterolateral and posteromedial regions. The anterolateral portion contains cells homologous to the spherical and stellate cells described in cats (Figs. 1 and 2; Brawer et al., '74; Cant and Morest, '79a; Cant and Morest, '79b; Osen, '69). The posteromedial portion or interstitial nucleus (INT, Figs. 2 and 3) features neurons with homologies to the globular and stellate cells described in cats (Brawer et al., '74; Osen, '69; Tolbert and Morest, '82). These cells are embedded in the fibrolaminae of the primary afferent axons (Fig. 3) forming the root of the auditory nerve. A thin granule cell cap lies along the dorsal apex of AVCN and is continuous caudally with the internuclear granule cell layer separating ventral from dorsal cochlear nucleus. Directly beneath the AVCN and embedded in the fibers of the auditory nerve root, is the prominent auditory nerve nucleus (Figs. 2-4).

### Posteroventral cochlear nucleus (PVCN)

PVCN is located between the inferior cerebellar peduncle and the lateral vestibular nucleus (Figs. 3 and 4). It extends into the caudal tip of the cochlear nucleus, underlying DCN. PVCN contains stellate cells of various sizes arranged in rows separated by fascicles of the auditory nerve. Medially, it has a very prominent octopus cell area containing large neurons with homologies to the octopus cells described in the cat (Fig. 4; Brawer et al., '74; Kane, '73; Osen, '69). The primary afferent fibers course through PVCN in tight fascicles to reach the ventral border of the DCN.

Further partitioning of the *Monodelphis* cochlear nucleus is possible along lines similar to those subdivisions described by Brawer et al. ('74) and Morest et al. ('90). This will be treated along with specific descriptions of its cell types in a separate communication.

## Cochleotopic pattern of primary afferent fibers in the cochlear nucleus of *Monodelphis*

The anterograde transport of horseradish peroxidase (HRP) was used to investigate the trajectory of the auditory nerve fibers in the cochlear nucleus of *Monodelphis*. Thin pledgets

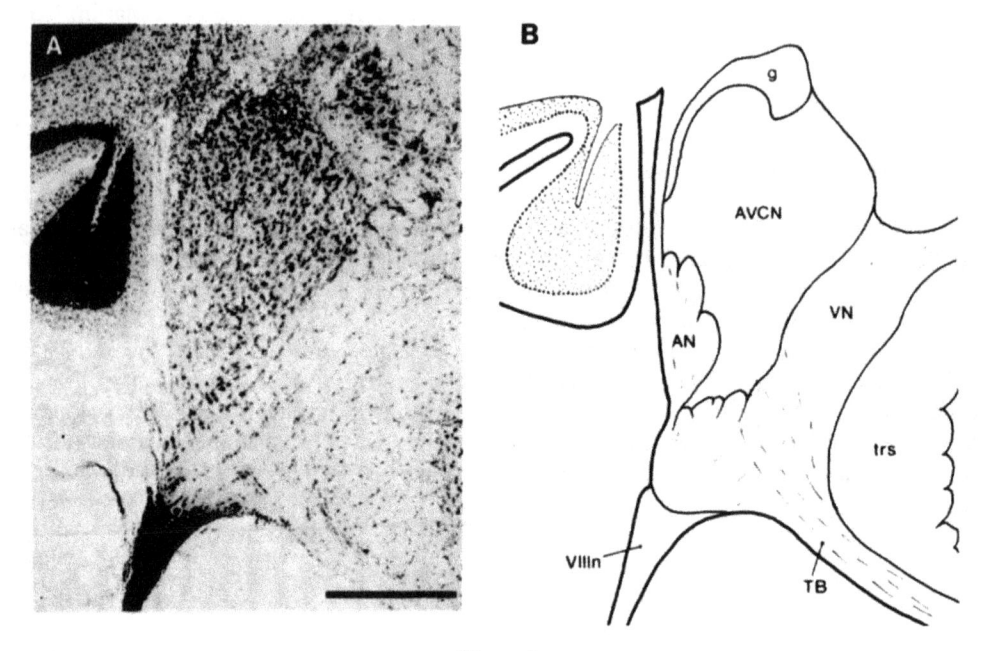

Figure 1

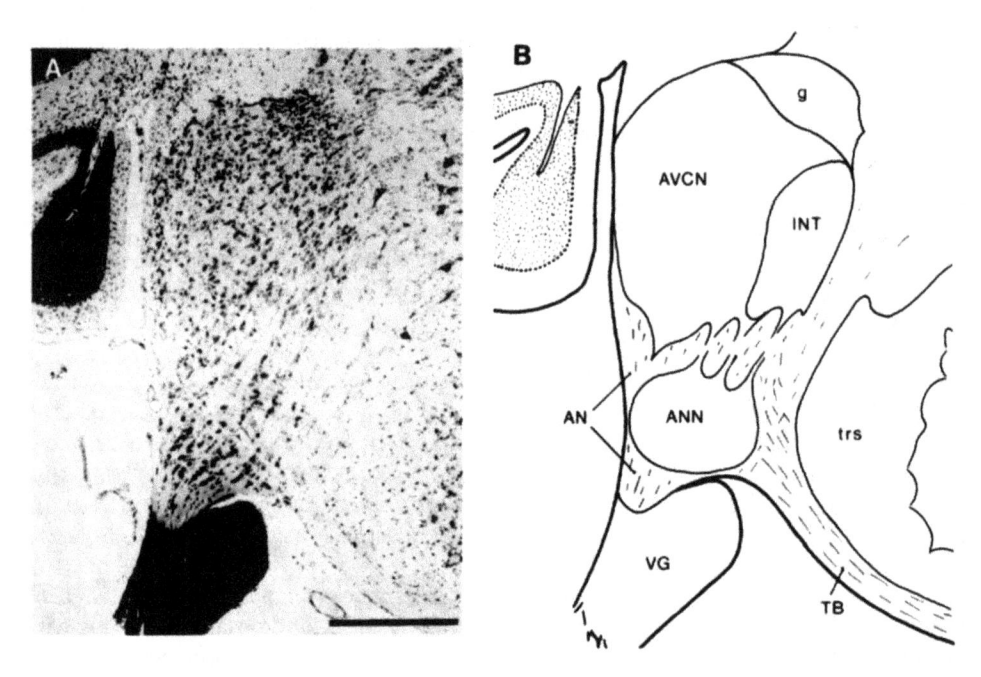

Figure 2

Figures 1-4. The cytoarchitecture of the cochlear nucleus in "Monodelphis". A. Photo-micrographs of Nissl-stained, coronal sections, taken in the middle of each quartile along the rostral to caudal axis of the nucleus. B. Line drawings derived from the accompanying photomicrographs illustrating the major divisions of the cochlear nucleus. The midline is to the right of each plate. The scale bar (lower right of A) on each of the four plates represents 0.5 mm.

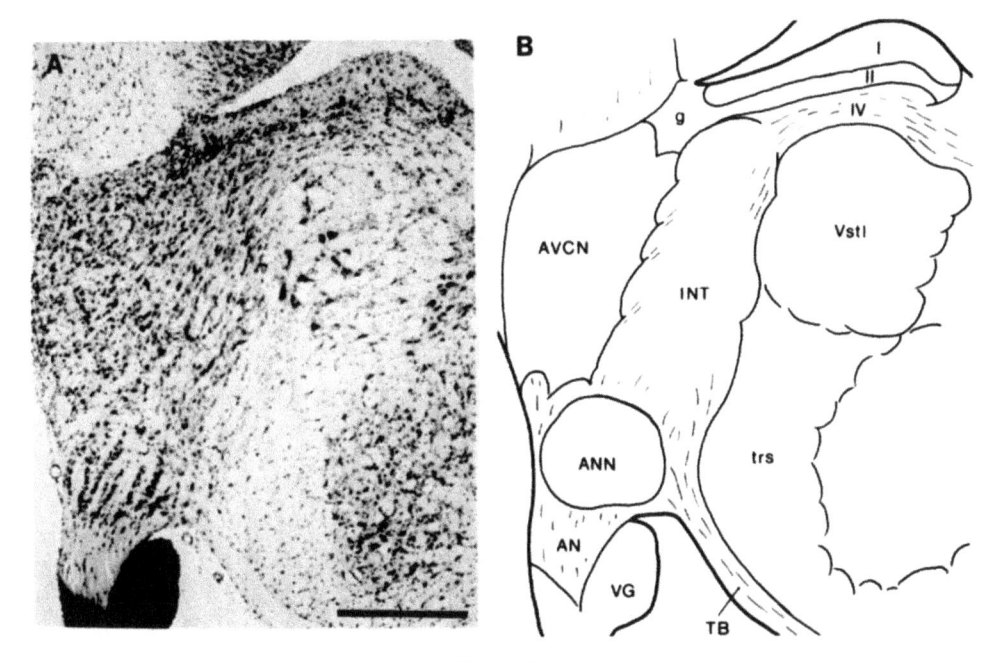

Figure 3

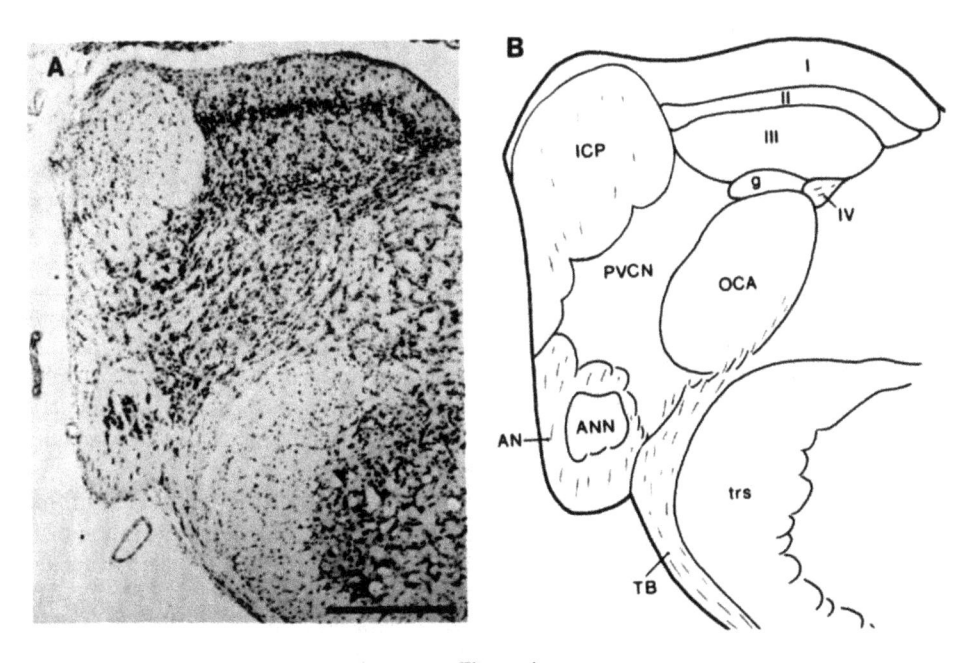

Figure 4

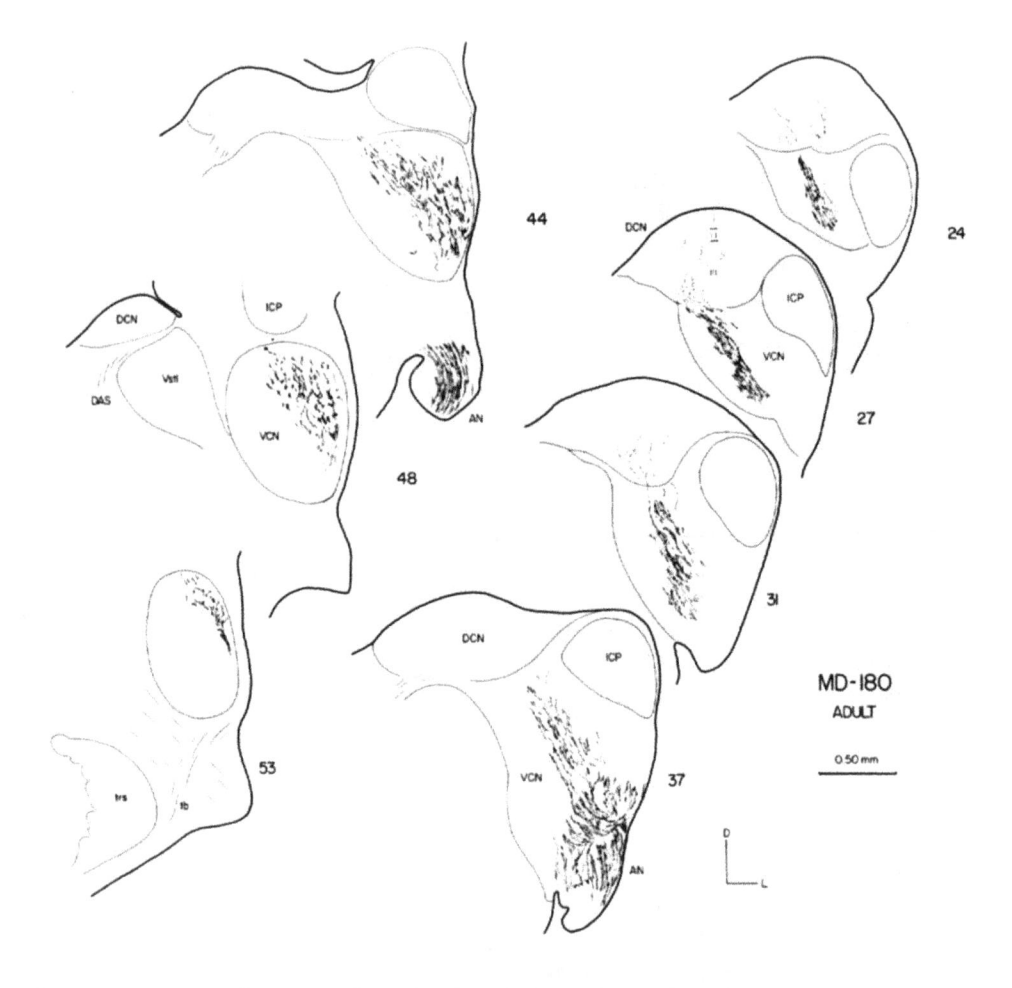

**Figure 5.** A sequence of coronal sections taken through the cochlear nucleus of an adult opossum consequent to a large placement of HRP into the cochlea. The labeled fibers have been drawn with the aid of a camera lucida. Sections are numbered from caudal-to-rostral.

of HRP, solidified on parafilm, were inserted into the base or apex of the cochlea in deeply anesthetized adult animals. The animals were maintained overnight and then perfused transcardially with dilute aldehyde fixatives. The brainstem was removed, stored in phosphate buffer, and sectioned frozen or cut using a vibratome. For visualization of the HRP inauditory nerve fibers, tissue was mounted on glass slides and reacted with tetramethylbendzidine. This method was also employed on postnatal animals.

All three of the major divisions of the cochlear nucleus received labeled primary afferent fibers consequent to HRP placement in the cochlea. (Fig. 5). The bifurcation of the auditory nerve fibers occured in the posteromedial portion of AVCN (interstitial nucleus) shortly after the fibers pass through the auditory nerve nucleus (section 37, Fig. 5). After bifurcation, the distribution of the primary afferent fibers was not uniform. The anterior branch of the auditory nerve fibers passed rostrally through AVCN; distribution of reaction product was quite heavy suggesting many collateral branches and large terminal processes. The posterior branch passed caudally through PVCN in tight fascicles. As these posterior branch fascicles entered DCN, they became much finer in caliber and their density diminished. They were present throughout layers II and III. An endogenous peroxidase reac-

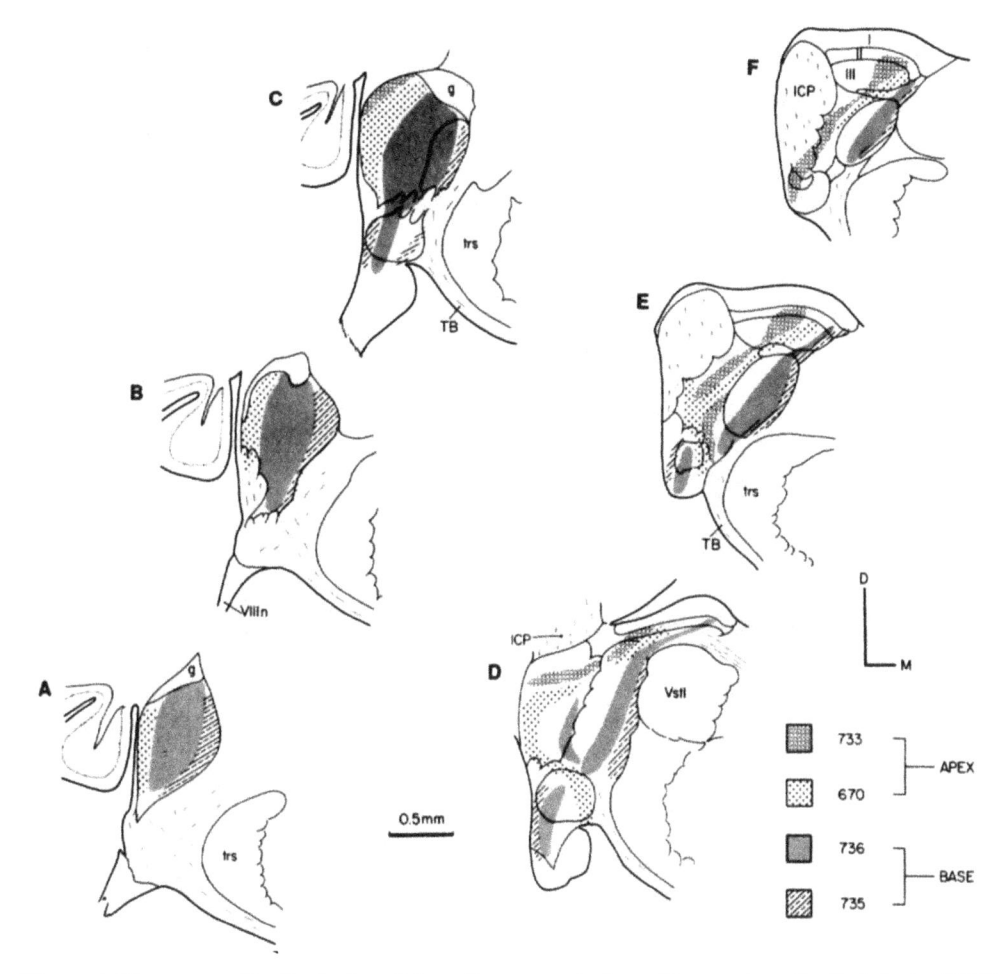

Figure 6. The topographic distribution of the auditory nerve in the cochlear nucleus. Small pledgets of HRP were placed in the cochlea in a series of adult opossums. Plots were constructed of each individual case, and these data were collapsed on to standard sections of the cochlear nucleus to demonstrate the cochleotopic pattern of the auditory nerve. Patterned screening represents the distribution of labeled primary afferent fibers from individual cases. The legend in the figure indicates that cases 733 and 670 involved HRP placements into the apex of the cochlea, while cases 736 and 735 involved placements into the base of the cochlea.

tion was present in layer I making determination of input difficult; however, in no case were labelled primary afferent fibers convincingly traced into this area.

An orderly pattern, reflecting the position of the injection site in the cochlear duct, was present in the medial-to-lateral as well as rostral-to-caudal dimensions of the nucleus (Fig. 6). Primary afferent fibers from the apex of the cochlea extended to the distal tip of DCN, through PVCN, but did not reach the anterior-most portion of AVCN (injection 733 and 670, Fig. 6). Conversely, fibers from the base of the cochlea reached the anterior tip of AVCN, extended through PVCN, but did not reach the distal portion of DCN (injection 736 and 735, Fig. 6).

The projection of primary afferent fibers was on curvilinear planes through the cochlear nucleus. Subsequent to bifurcation, axons in the anterior division of the auditory nerve proceeded rostrally into the anterior portion of AVCN where they curved laterally to end near the rostrolateral border of the nucleus. Axons in the posterior division of the auditory nerve passed caudally through PVCN and entered DCN. All primary afferent fibers

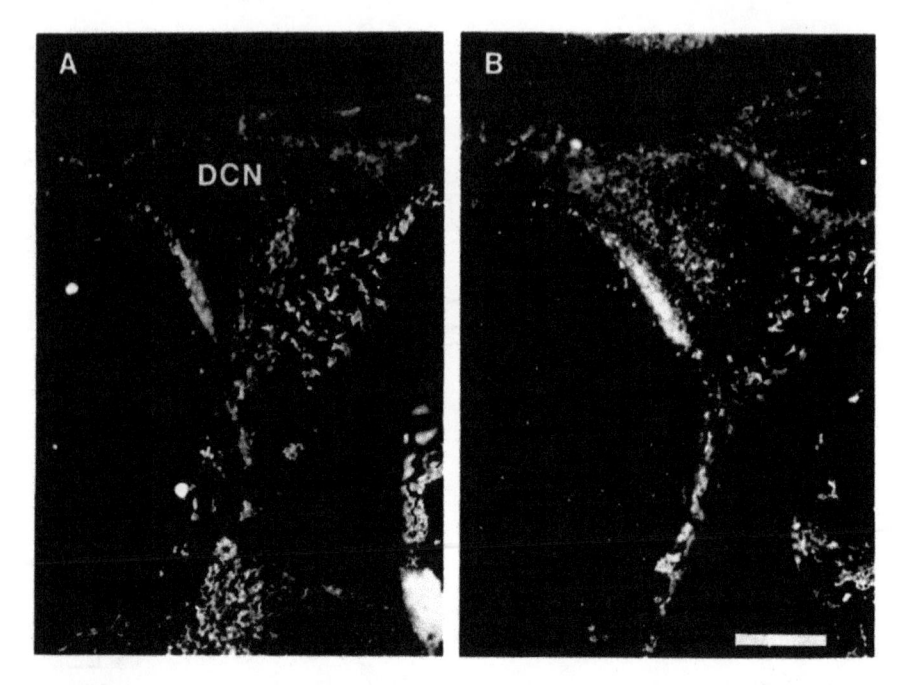

Figure 7. Darkfield photomicrographs of HRP labeled fibers in the presumptive cochlear nucleus of a PND-08 animal after the placement of a small pledget of this enzyme into the cochlear duct. These are coronal sections; A is rostral to B. Dorsal is up and medial is to the left. Note the narrow band of label along the medial (left-hand) boundary of the nucleus. Scale bar in the lower right of B represents 0.2 mm.

to DCN entered near its rostrolateral corner and curved in an arc to terminate along its caudomedial border. Axons from the base of the cochlea had the tightest medial curvature and thus travelled the shortest distance in DCN; while those from the apex had a broader arc, producing a fan of primary afferent fibers in layers II and III of DCN. A similar distribution of auditory nerve axons was reported in DCN of the bat (Osen, '70), mouse (Willard, '81), and hamster (Schweitzer and Cant, '84).

## Postnatal development of the central processes of the spiral ganglion

**Postnatal day 0.** *Monodelphis* pups are born with an otocyst consisting of 3/4 turn and auditory dendrites that have not penetrated the basal lamina of the neuroepithelium (Willard and Munger, submitted). Centrally, the precochlear band of neurons is located on the medial side of the vestibular nerve root. This stage is approximately equivalent to embryonic day 13-14 of the mouse (Lim and Anniko, '85; Sher, '71).

**Postnatal day 6.** This was the youngest age examined for anterograde transport of HRP in primary afferent fibers. As a reference of events, on day 6 the cochlear duct has reached 1.5 turns, the hair cells have just started to differentiate at the base; and centrally, the precochlear band of neurons has just begun to enter DCN (Willard and Munger, submitted). This stage was approximately equivalent to embryonic day 16 in a mouse (Lim and Anniko, '85). Both auditory and vestibular primary afferent fibers were labelled by HRP placements in the cochlea. Individual fascicles of the primary afferent fibers were present throughout presumptive VCN. The area of presumptive DCN contained lightly labelled fibers from the auditory nerve component. Since this date preceded the arrival of the precochlear

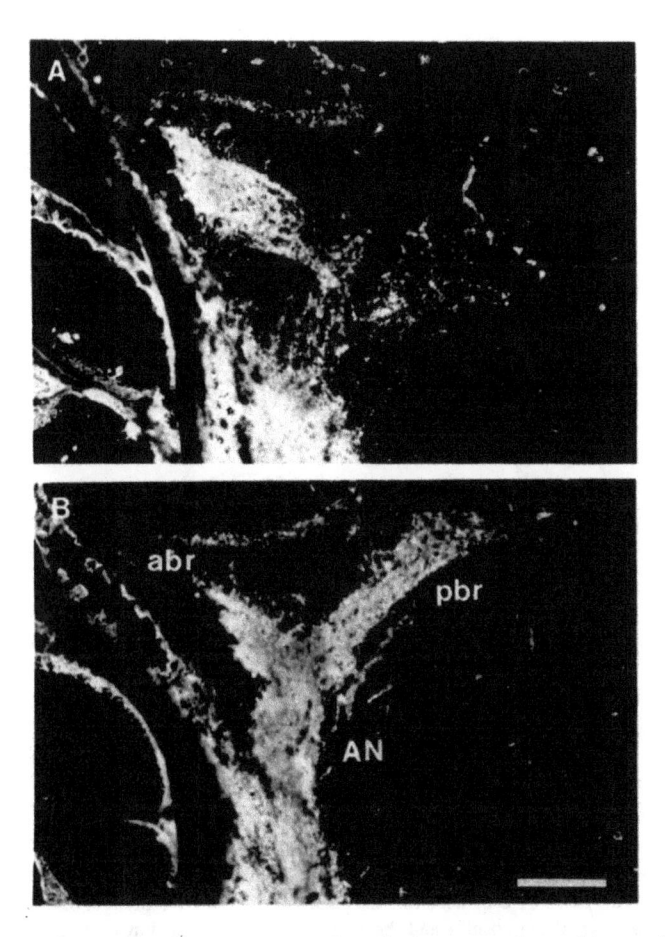

**Figure 8.** Darkfield photomicrographs of HRP labeled fibers in the presumptive cochlear nucleus of a PND-08 animal after the placement of a large pledget of this enzyme into the cochlear duct. These are coronal sections; A is rostral to B. The bifurcation of the auditory nerve (AN) into anterior (abr) and posterior (pbr) branches can be seen; the anterior branch is oriented to the left, the posterior branch to the right in this figure. Note that the posterior branch (in B) extends to the ependymal surface of the fourth ventricle. Scale bar in the lower right of B represents 0.2 mm.

neurons, it suggested that the auditory nerve fibers were the first of the two to reach presumptive DCN in *Monodelphis*.

**Postnatal day 8.** By way of reference, on day 8 the cochlear duct has reached 1.75 turns, the adult value, and the peripheral processes of spiral ganglion cells have reached the bases of the inner, but not outer hair cells. Centrally, the precochlear neurons have begun entering the presumptive DCN and the structure was increasing in size (Willard and Munger, submitted). This stage was approximately equivalent to embryonic day 18 in a mouse (Lim and Anniko, '85). By this date, it was possible to obtain restricted placements of HRP along the basilar membrane (Fig. 7). Injections confined to one coil of the cochlea produced a single, narrow band of labelled primary afferent fibers in the cochlear nucleus (Fig. 7A). Double bands of labelled primary afferent fibers were seen in the VCN when the placement crossed two coils of the cochlea. This double banding was interpreted as demonstrating the early presence of patterned relationships in these connections.

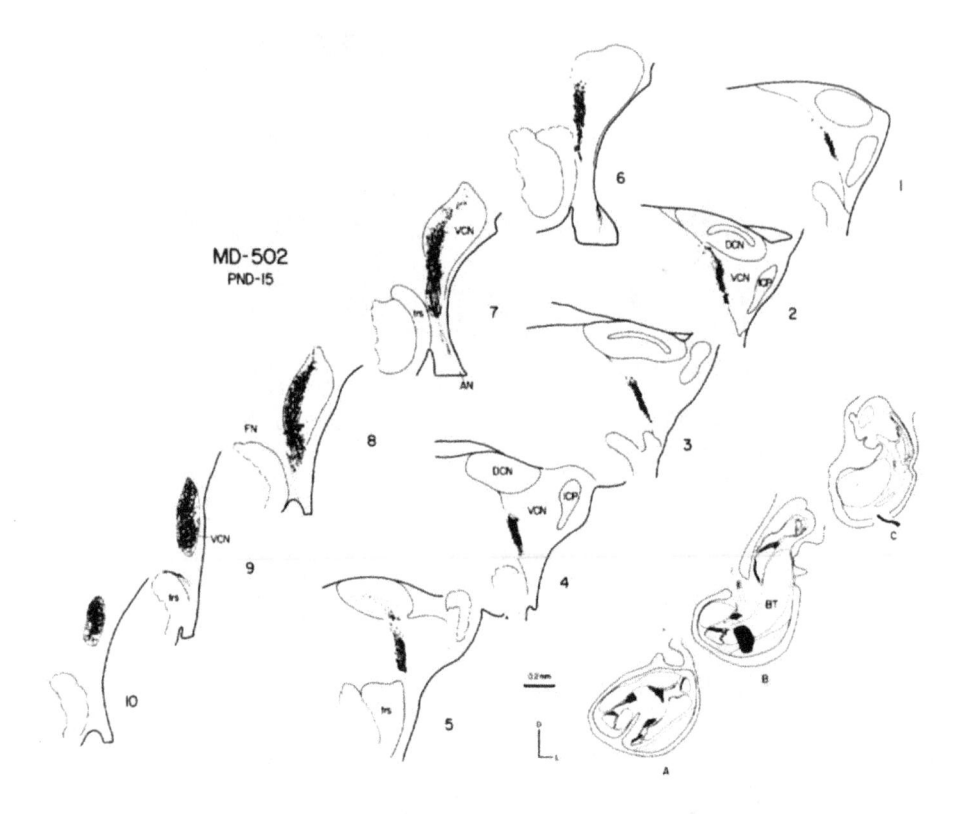

**Figure 9.** The distribution of labeled fibers in a PND-15 animal, consequent to HRP placement in the cochlear duct, has been plotted on sequential coronal sections through the cochlear nucleus. Sections are numbered from caudal (1) to rostral (10). Insets A-C demonstrate the placement of the HRP pledget into the cochlear duct. The placement was in the base of the cochlea and the labeled fibers are distributed along the medial side of the cochlear nucleus.


The posterior division of the auditory nerve root extended caudally through PVCN and into DCN. The bundle was thickest in PVCN where it underlaid the band of migrating precochlear neurons. The labeled fibers coursed along the medial border of DCN with fine caliber fibers (light labeling) extending into the newly forming cell mass (Fig. 7B). Larger placements of HRP verified the general pattern of primary afferent fibers in DCN (Fig. 8B). At their distal extreme, the posterior division fibers could be followed to the ependymal surface of the brain in the vicinity of the presumptive DCN (Fig. 8B). This occurs at a stage in development prior to the formation of layer I.

**Two weeks postnatal.** At this date the cochlear hair cells still have not matured and the tectorial membrane has not developed in all turns; however, the tunnel of Corti has just begun to open (Willard and Munger, submitted). This was approximately equivalent to the mouse at postnatal day 3-4 (Lim and Anniko, '85). Injections of HRP into the cochlea at the end of two weeks postnatal life illustrate the precise mapping the auditory nerve on the cochlear nucleus (Figs. 9 and 10). The base-to-apex, lateral-to-medial topographic relationship has become well defined in the primary afferent fibers. Narrow bands of labelled primary afferent fibers were seen in the cochlear nucleus consequent to HRP placement in the cochlea. When the injection involved two turns of the cochlea, a double band of fibers was present (Fig. 10). Thus the geometry of the restricted projections has been established prior to stimu-

**Figure 10.** The distribution of labeled fibers in a PND-15 animal, consequent to HRP placement in the cochlear duct, has been plotted on sequential coronal sections through the cochlear nucleus. Sections are numbered from caudal (1) to rostral (11). Insets A-C demonstrate the placement of the HRP pledget into the cochlear duct. The placement passed through the basal turn and entered the second turn. Subsequently two bands of labeled fibers are distributed across the cochlear nucleus. These bands are separated by an interband region lacking labelled fibers. This interband interval is interpreted as representing the unlabeled portion of the cochlear duct positioned between the trajectory of the two injection sites.

lus-driven cochlear function which occurs around the end of the first month of postnatal life (Willard and Munger, submitted).

## Summary

The cytoarchitecture of the cochlear nucleus and the distribution of auditory primary afferent fibers has been described in *Monodelphis*. Three divisions of the cochlear nucleus were recognizable, each receiving primary afferent fibers from the auditory nerve. An ordered relationship was present between the origin and distribution of these fibers; the base-to-apex dimension of the cochlea was represented from medial to lateral across the cochlear nucleus. This was quite similar to the arrangement of the cochlear nucleus and distribution of primary afferent fibers in eutherian mammals (Lorente de No, '33; Osen, '70; Fekete et al., '82; Leake and Snyder, '89).

The early development of the central processes of the auditory nerve fibers was confirmed in the marsupial. This pattern of early growth has also been observed in the chick (Whitehead et al., '82; Book and Morest, '90), rat (Angulo et al., '90), mouse (Lorente de No, '33), and human (Cooper, '48).

The presence of an orderly pattern in the central processes of the spiral ganglion cells as they innervate the presumptive cochlear nucleus was also established. Topographic order was demonstrable in the projection of primary afferent fibers in the earliest preparations of *Monodelphis*, at a point in time long before maturation of the cochlea occurs. The early expression of topographic order has also been seen in the chick (Rubel and Parks, '88) and rat (Angulo et al., '90). Yet, in the species examined to date, the response properties (tuning curves) of the immature central projections experience a period of poor precision during early postnatal life (Brugge, '83; Kitzes '90) which gradually resolves into the adult form. This initial period of gradually maturing tuning curves could relate to development of physical properties in the cochlea rather than resolution of heterogeneous projections in the central auditory pathways (Manley, '78; reviewed in Kitzes, '90). The presence of narrow bands of auditory nerve fibers in the PND-8 opossum, reported here, suggests that the geometric precision of the auditory system has been established by this time in the cochlear nucleus. This observation is consistent with the notion that early changes in response characteristics of cells in the central auditory pathways derive, in large part, from shifting physical properties in the receptor organ.

Central processes of the spiral ganglion cells arrived in the DCN prior to the migrating precochlear neurons. A similar pattern has also been observed in the chick, where the central processes of the auditory nerve fibers enter the brainstem prior to the arrival of neurons for the nucleus magnocellularis and nucleus angularis (Book and Morest, '90). However, it is reported that the DCN of the hamster forms on the brainstem prior to invasion by the posterior division of the auditory nerve (Schweitzer and Cant, '84). This discrepancy may represent subtle differences in time of ontogenic events between species; further investigation of the issue is necessary.

In conclusion, it appears that portions of the brainstem auditory pathways form in a rather precise manner, lacking the heterogeneity of projections seen in the early ontogeny of the visual system. This does not mean that remodelling is lacking in the development of the auditory system. Growth-related changes in mammals are seen in the innervation of the outer hair cells (Lenoir et al., '80) as well as in the morphology of axon terminals in the cochlear nucleus (Fekete et al., '82); in addition numerous changes in cell structure have been reported in the maturing chick auditory system (Rubel and Parks, '88). However, the observations reported here do suggest that any such remodelling takes place against a background of well organized primary afferent fibers.

## ACKNOWLEDGEMENTS

A great debt of thanks is owed to Drs. David K. Ryugo and George F. Martin for their guidance as well as Dr. Enrico Mugnaini for his interest in this project. I am thankful to Dr. Nell Cant for her suggestion to begin studying *Monodelphis* and Dr. Barbara Fadem for her kind assistance in establishing the opossum colony at the University of New England. I would like to thank Frank Daly and Melissa Stenner for technical assistance, Liz Havu for assistance with the typing and Holly Haywood with darkroom assistance. This work was supported by NIH grant 1R15NS25978-01.

## REFERENCES

Adams, J.C., 1979, Ascending projections to the inferior colliculus, *J. Comp. Neurol.*, 183:519-538.
Aitkin, L.M., Irvine, D.R.F., Nelson, J.E., Merzenich, M.M., and Clarey, J.C., 1986, Frequency representation in the auditory midbrain and forebrain of a marsupial, the northern native cat (Dasyurus hallucatus), *Brain Behav. Evol.*, 29:17-28.
Angulo, A., Merchan, J.A., and Merchan, M.A., 1990, Morphology of the rat cochlear primary afferents during prenatal development: a Cajal's reduced silver and rapid Golgi study, *J. Anat.*, 168:241-255.
Book, K.J. and Morest, D.K., 1990, Migration of neuroblasts by perikaryal translocation: role of cellular elongation and axonal outgrowth in the acoustic nuclei of the chick embryo medulla, *J. Comp. Neurol.*, 297:55-76.
Brawer, J.R., Morest, D.K., and Kane, E.C., 1974, The neuronal architecture of the cochlear nucleus of the cat, *J. Comp. Neurol.*, 155:251-300.

Brugge, J.F., 1983, Development of the lower brainstem auditory nuclei, in: "Development of Auditory and Vestibular Systems," R. Romand, (eds.), Academic Press, Inc., New York.

Cant, N.B., and Morest, D.K., 1979a, Organization of the neurons in the anterior division of the anteroventral cochlear nucleus of the cat. Light-microscopic observations, Neurosci., 4:1909-1929.

Cant, N.B., and Morest, D.K., 1979b, The bushy cells in the anteroventral cochlear nucleus of the cat. A study with the electron microscope, Neurosci., 4:1925-1945.

Changeux, J., 1986, Neuronal Man, Oxford University Press, New York.

Clopton, B.M., Winfield, J.A., and Flammino, F.J., 1974, Tonotopic organization: Review and Analysis, Brain Res., 76:1-20.

Cooper, E.R.A., 1948, The development of the human auditory pathway from the cochlear ganglion to the medial geniculate body, Acta Anat., 5:99-122.

Disterhoft, J.F., Perkins, R.E., and Evans, S., 1980, Neuronal morphology of the rabbit cochlear nucleus, J. Comp. Neurol., 192:687-702.

Fekete, D.M., Rouiller, E.M., Liberman, M.C., and Ryugo, D.K., 1982, The central projections of intracellularly labeled auditory nerve fibers in cats, J. Comp. Neurol., 229:432-450.

Kane, E.C., 1973, Octopus cells in the cochlear nucleus of the cat: heterotypic synapes upon homeotypic neurons, Intern. J. Neurosci., 5:251-279.

Kitzes, L.M., 1990, Development of auditory system physiology, in: "Development of Sensory Systems in Mammals," J. Coleman, eds., John Wiley & Sons, New York.

Leake, P.A., and Snyder, R.L., 1989, Topographic organization of the central projections of the spiral ganglion in cats, J. Comp. Neurol., 281:612-629.

Lenoir, M., Shnerson, A., and Pujol, R., 1980, Cochlear receptor development in the rat with emphasis on synaptogenesis, Anat. Embryol., 160:253-262.

Lim, D.J. and M. Anniko, 1985, Developmental morphology of the mouse inner ear, Acta Otolaryngol., 422(suppl.):1-69.

Lorente de No, R., 1933, Anatomy of the eighth nerve: The central projections of the nerve endings of the internal ear, The Laryngoscope, 63:1-37.

Manley, G.A., 1978, Cochlear frequency sharpening-A new synthesis, Acta Otolaryngol., 85:167-176.

Morest, D.K., 1973, Auditory neurons of the brain stem, Adv. Oto-Rhino-Laryng., 20:337-356.

Morest, D.K., Hutson, K.A. and Kwok, S., 1990, Cytoarchitectonic atlas of the cochlear nucleus of the chinchilla, chinchilla laniger, J. Comp. Neurol., 300:230-248.

Movshon, J.A. and Kiorpes, L., 1990, The role of experience in visual development, in: "Development of Sensory Systems in Mammals," J. Coleman, (eds.), John Wiley & Sons, New York.

Oliver, D.L., 1984, Neuron types in the central nucleus of the inferior colliculus that project to the medial geniculate body, Neurosci., 11:409-424.

Oliver, D.L., 1987, Projections to the inferior colliculus from the anteroventral cochlear nucleus in the cat: possible substrates for binaural interaction, J. Comp. Neurol., 264:24-46.

Osen, K.K., 1969, The intrinsic organization of the cochlear nuclei, Acta Otolaryngol., 67:352-359.

Osen, K.K., 1970, Course and terminaton of the primary afferents in the cochlear nuclei of the cat, Arch. Ital. Biol., 108:21-51.

Rhode, W.S., Oertel, D., and Smith, P.H., 1983, Physiological response properties of cells labeled intracellularly with horseradish peroxidase in cat ventral cochlear nucleus, J. Comp. Neurol., 213:448-463.

Rubel, E. and Parks, T.N., 1988, Organization and development of the avian brain-stem auditory system, in: "Auditory Function: Neurobiological Bases of Hearing," G.M. Edelman, W.E. Gall, and W.M. Cowan, (eds.), John Wiley & Sons, New York.

Ryugo, D.K., and Willard, F.H., 1985, The dorsal cochlear nucleus of the mouse: a light microscopic analysis of neurons that project to the inferior colliculus, J. Comp. Neurol., 242:381-396.

Ryugo, D.K., Willard, F.H., and Fekete, D.M., 1981, Differential afferent projections to the inferior colliculus from the cochlear nucleus in the albino mouse, Brain Res., 210:342-349.

Saunders, N.R., Adam, E., Reader, M., and Mollgard, K., 1989, *Monodelphis domestica* (grey short-tailed opossum): an accessible model for studies of early neocortical development, Anat. Embryol., 180:227-236.

Scheibel, M.E. and Scheibel, A.B., 1974, Neuropil organization in the superior olive of the cat, Exp. Neurol., 43:339-348.

Schweitzer, L. and Cant, N., 1984, Development of the cochlear innervation of the dorsal cochlear nucleus of the hamster, J. Comp. Neurol., 225:228-243.

Shatz, C.J., 1990, Impulse activity and the patterning of connections during CNS development, Neuron, 5:745-756.

Shatz, C.J., 1991, Role of neural function in fetal visual system development, Dis. Neurosci., 7:95-99.

Sher, A.E., 1971, The embryonic and postnatal development of the inner ear of the mouse, *Acta Otolaryngol.*, Suppl. 285:1-77.

Smith, P.H., and Rhode, W.S., 1985, Electron microscopic features of physiologically characterized, HRP-labeled fusiform cells in the cat dorsal cochlear nucleus, *J. Comp. Neurol.*, 237:127-143.

Tolbert, L.P., and Morest, D.K., 1982, The neuronal architecture of the anteroventral cochlear nucleus of the cat in the region of the cochlear nerve root: electron microscopy, *Neurosci.*, 7:3053-3067.

Udin, S.B. and Fawcett, J.W., 1988, Formation of topographic maps, *Ann. Rev. Neurosci.*, 11:289-327.

Vater, M. and Feng, A.S., 1990, Functional organization of ascending and descending connections of the cochlear nucleus of horseshoe bats, *J. Comp. Neurol.*, 292:373-395.

Webster, D.B., and Trune, D.R., 1982, Cochlear nuclear complex of mice, *Am. J. Anat.*, 163:103-130.

Whitehead, M., Marangos, P., Connolly, S., and Morest, D., 1982, Synapse formation is related to the onset of neuron-specific enolase immunoreactivity in the avian auditory and vestibular systems, *Dev. Neurosci.*, 5:298-307.

Willard, F.H., 1981, Neuroanatomical studies of the auditory brain stem, Doctorial Dissertation, University of Vermont, Collaege of Medicine, Burlington, VT.

Willard, F.H. and Martin, G.F., 1983, The auditory brainstem nuclei and some of their projections to the inferior colliculus in the North American opossum, *Neurosci.*, 10:1203-1232.

Willard, F.H. and Martin, G.F., 1986, The development and migration of large multipolar neurons into the cochlear nucleus of the North American opossum, *J. Comp. Neurol.*, 248:119-132.

Willard, F.H. and Ryugo, D.K., 1983, Anatomy of the central auditory system, "The Auditory Psychobiology of the Mouse" J.F. Willott, (eds.), Charles C. Thomas, Springfield, IL, Springfield, IL.

Willard, F.H., Ho, R.H., and Martin, G.F., 1984, The neuronal types and the distribution of 5-Hydroxytryptamine and enkephalin-like immunoreactive fibers in the dorsal cochlear nucleus of the North American opossum, *Brain Res. Bull.*, 12:253-266.

# II. PRIMARY INPUTS

# ANATOMICAL AND PHYSIOLOGICAL STUDIES OF TYPE I AND TYPE II SPIRAL GANGLION NEURONS

M. Christian Brown

Department of Physiology and Department of Otology and Laryngology, Harvard Medical School, and Eaton-Peabody Laboratory, Massachusetts Eye & Ear Infirmary, 243 Charles St., Boston, MA 02114, USA

## Type I and type II spiral ganglion neurons: anatomical distinctions

Primary auditory neurons transmit information from receptor cells in the cochlea, hair cells, to cochlear-nucleus neurons in the brain. The cell bodies of the primary neurons form the spiral ganglion, which is located in the cochlea. Several early studies demonstrated that there was more than one type of spiral ganglion cell (Suzuki et al., '63; Reinecke, '67; Kellerhals et al., '67; Merck et al., '77). In pioneering work, Spoendlin ('71, '74) showed that most spiral ganglion cells degenerated following section of the auditory nerve. Since the afferent endings on outer hair cells persisted following nerve section, the few remaining "type II" spiral ganglion cells were postulated to innervate the outer hair cells. Using reconstructions of fibers labeled with horseradish peroxidase, Kiang et al. ('82) demonstrated directly that type II neurons sent processes only to outer hair cells whereas type I neurons innervated only the inner hair cells.

Besides their peripheral innervation of hair cells, a number of other anatomical characteristics distinguish the two types of spiral ganglion cells. Type I cell bodies and processes are large whereas type II cell bodies and processes are smaller and unmyelinated (Spoendlin, '71; Kiang et al., '84). There are staining differences between the two types of ganglion cells, observable with toluidine blue, protargol, and with antibodies to phosphorylated neurofilament proteins (Keithley and Feldman, '79; Kiang et al., '84; Berglund and Ryugo, '91). Type I neurons constitute more than 90% of all spiral ganglion cells, with type II neurons making up the remaining 4-10% of the population (Spoendlin, '69; Morrison et al., '75; Kimura, '78; Kiang et al., '84; Brown, '87). In terms of numbers and information-carrying capacity, the type I neurons represent the major pathway for sensory information transfer from the cochlea to the cochlear nucleus.

Two major areas of investigation at the present time are characterizing the differences in central projections and the differences in electrophysiological responses of type I and type II neurons. In this chapter, recent work on the projections of type II neurons into the cochlear nucleus will be summarized, and preliminary results of electrophysiological experiments directed toward recording from type I and type II units in the spiral ganglion will be presented. These results will help to address the functional role of the outer hair cell system in audition.

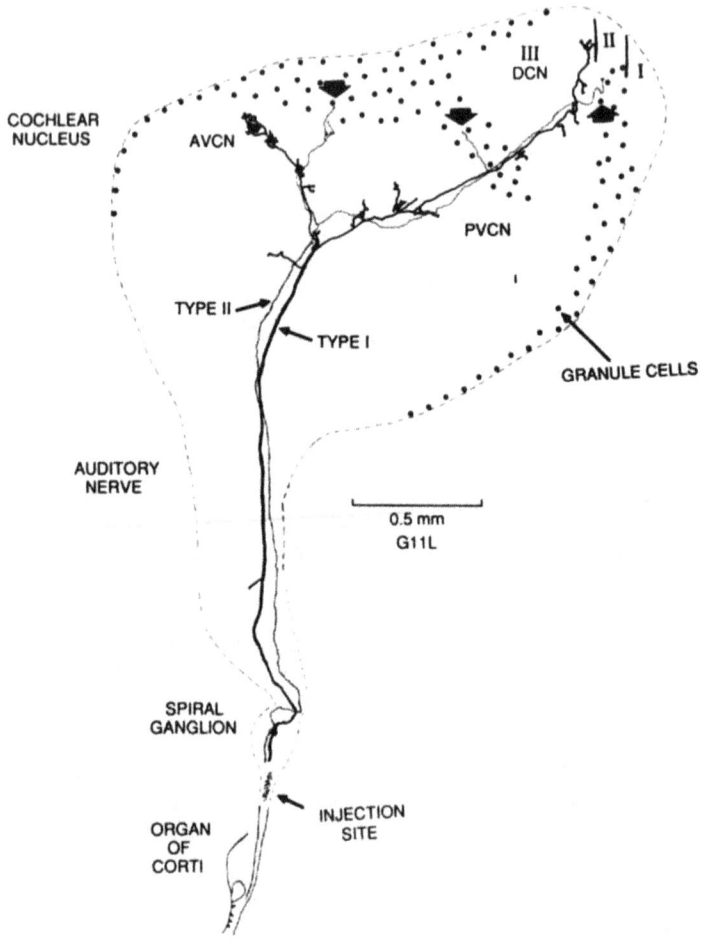

Figure 1. Camera-lucida tracings of labeled axons from type I (thick line) and type II (thin line) spiral ganglion cells in the cochlear nucleus of a gerbil. The HRP injection was made into the spiral ganglion of the basal turn. Dots show granule-cell regions. Thick arrowheads point to the areas in which type II axons terminate that seldom receive terminations from type I axons. These areas are (from left to right): the subpeduncular corner of granule cells, the granule-cell lamina dividing the ventral from the dorsal cochlear nucleus, and the border between layer II and layer III of the dorsal cochlear nucleus. Roman numerals indicate the layers of the dorsal cochlear nucleus. The plane of section is the "modified sagittal" plane (see Fekete et al., '84). Rostral is to the left, and dorsal is uppermost in the figure. AVCN: anteroventral cochlear nucleus, PVCN: posteroventral cochlear nucleus, DCN: dorsal cochlear nucleus. Adapted from Brown et al. ('88a).

## Central projections of type I and type II neurons

Central projections of auditory-nerve fibers have been the subject of a number of studies using Golgi stains (Ramón y Cajal, '09; Lorente de Nó, '81) or horseradish peroxidase (HRP) labeling (Fekete et al., '84; Merchan et al., '85). In order to separate projections from type I and type II neurons, our strategy involves labeling neurons with HRP and tracing individual axons from ganglion cells identified as type I or type II (Brown, '87; Brown et al., '88a). Later studies have relied on the difference in caliber to differentiate between the two types of fibers (Berglund and Brown, '89; Brown and Ledwith, '90; Ryugo et al., '91). Type I and type II axons from neighboring spiral ganglion cells project together

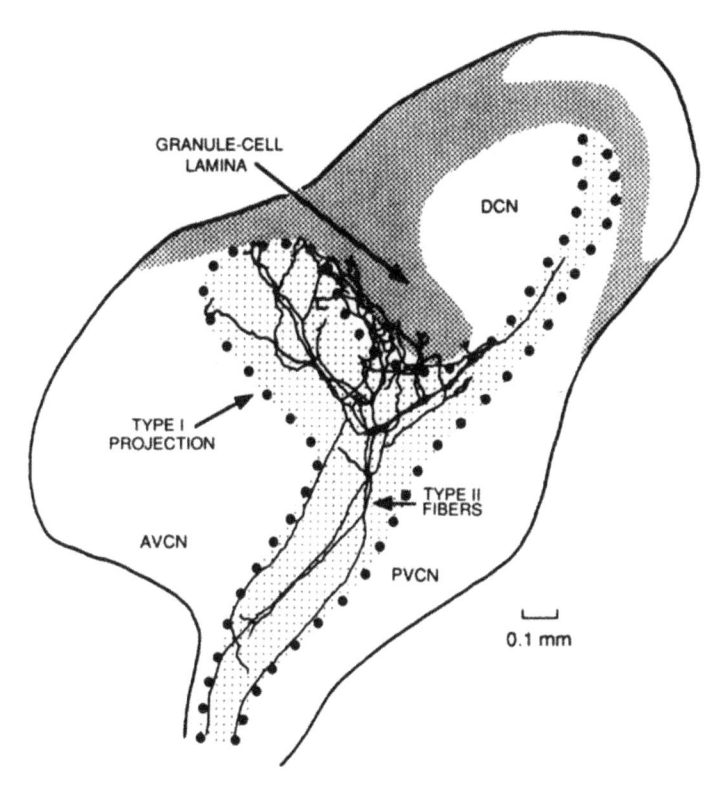

**Figure 2.** Camera-lucida tracings of four type II axons from the cochlear nucleus of a mouse. With the exception of one branch, all branches of these four axons could be reconstructed to their terminal swellings with no evidence of fading of the HRP reaction product. The axons were presumed to be from type II neurons because of their fine caliber. Areas of projection and termination of type I fibers are outlined by large dots. The axons were labeled by an HRP injection made into the spiral ganglion of the basal turn. Shading indicates areas containing high densities of granule cells, such as the granule-cell lamina. Modified sagittal plane (see Fekete et al., '84). Abbreviations as in Figure 1.

in the auditory nerve and nerve root (Fig. 1). The two types of axons bifurcate near one another in the auditory nerve root to form ascending and descending branches. Bifurcation is a fundamental property of most auditory-nerve axons; 60 out of 61 type II axons that have been studied in gerbil and mouse bifurcated (Brown, '87; Brown et al., '88a; Berglund and Brown, '89; Brown and Ledwith, '90).

Throughout much of the cochlear nucleus, type I and type II axons take the same general course. This is true for the ascending branches that course rostrally into the anteroventral cochlear nucleus (AVCN) as well as descending branches that course caudally and dorsally into the posteroventral (PVCN) and dorsal cochlear nucleus (DCN). Thus, many of the *en passant* swellings that are formed by the type II axons are near endings formed by branches of type I fibers. Some of the type II *en passant* swellings, especially those located on the ascending and descending branches, are of angular shape or have short protuberances (Brown and Ledwith, '90). These swellings may represent synaptic sites and their targets could be onto the same cochlear-nucleus cells that receive input from type I neurons.

In contrast, the terminations of type II axons are often near "edge" or "border" regions of the cochlear nucleus that rarely receive projections from type I fibers. In mice, the most consistent of these areas is the granule-cell lamina dividing the ventral from the dorsal cochlear nucleus, especially at its border with the larger cells in the main body of the ventral cochlear nucleus (Fig. 2). This border is especially distinct in mice, and may correspond to

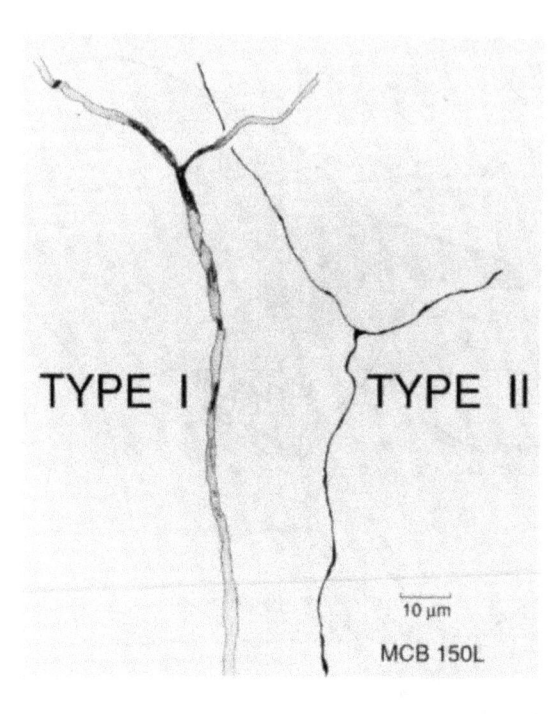

**Figure 3.** Camera-lucida drawing of type I and type II axons at their points of bifurcation in a guinea pig cochlear nucleus. The type II fiber was traced through serial sections of the auditory nerve and nerve root from a type II ganglion cell. The type I fiber is presumed to be from a type I ganglion cell on the basis of its thick diameter. Sagittal section. Dorsal is toward the top of the figure, rostral to the left.

the dorsal part of the "small-cell cap" that is found in other species such as the cat (Osen, '69). Type II axons also sometimes penetrate into the granule-cell lamina to terminate amongst granule cells and small cells. It is within the granule-cell lamina and on its border that type II fibers may end on postsynaptic targets that rarely receive input from type I fibers.

Although initial work on the type II central projections was done with basal-turn fibers, more recent work explored the projections of middle and apical type II fibers (Berglund and Brown, '89). In the main body of the ventral cochlear nucleus, type II and type I fibers from middle and apical turns have similar "tonotopic" projections. Unlike those from basal fibers, type II ascending branches from the middle and apical cochlear turns are not usually directed toward the granule-cell lamina. Their descending branches, however, are almost always directed toward the lamina and its border. Thus, since they receive type II input from a variety of cochlear regions, the lamina and its border may not be tonotopically organized like the main body of the cochlear nucleus.

The *en passant* and terminal swellings of type II fibers are uniformly small in size. In the AVCN, the small, simple type II endings are dwarfed by the large, complicated endbulbs formed by the type I fibers. In Golgi material, Lorente de Nó ('76) classified the ascending branches in AVCN into three general types. The fact that each type of ascending branch gave rise to an endbulb indicates that they are unlikely to emanate from type II neurons, and hence must all be from type I neurons. In general, fine fibers that might correspond to type II fibers appear to be rare in Golgi material (Feldman and Harrison, '69).

A fundamental difference between the central axons of type I and type II neurons is their caliber (Fig. 3). In addition, type I axons have a myelin covering, whereas type II axons are unmyelinated (Ryugo et al., '91). Functionally, these differences should result in vastly different conduction times for impulses in the two different types of axons. Assuming that conduction velocities of auditory-nerve fibers are similar to velocities of somatosensory axons

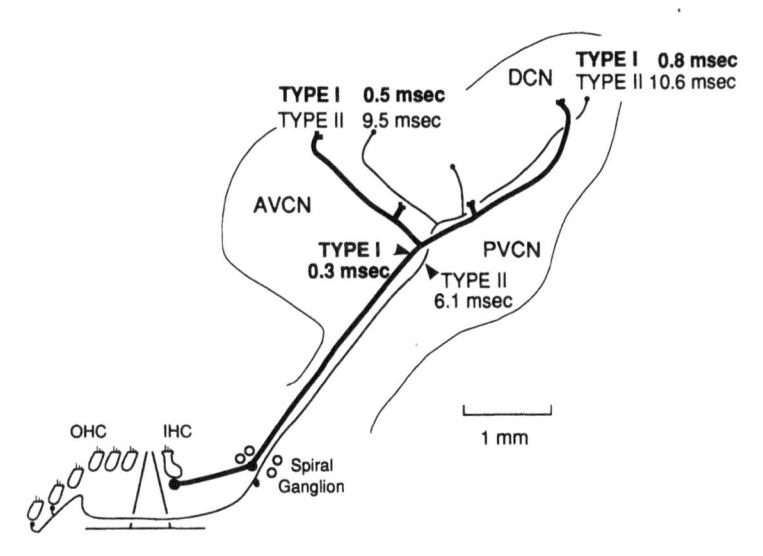

**Figure 4.** Calculated conduction times in guinea pig for impulses originating within terminals near the hair cells of the basal turn and traveling to the cochlear nucleus in type I (bold numbers) or type II (thin numbers) fibers. Conduction times at upper left are to the tips of the ascending branches, those in the middle are to the bifurcation points, and those at upper right are to the tips of the descending branches. Conduction times are calculated based on an assumption that conduction velocities for auditory nerve fibers are similar to myelinated and unmyelinated somatosensory axons of similar diameter (Burgess and Perl, '73). Diameters of type I and type II axons in the cochlea and auditory nerve were taken from previous measurements in guinea pig by Brown ('87). For type I diameters in the cochlear nucleus, average diameters were measured for ten thick central axons that were labeled with HRP. The average diameter of the ascending branches ($\pm$ s.d.) was 1.54 ($\pm$ 0.25) $\mu$m near the bifurcation point and tapered to 1.43 ($\pm$ 0.27) $\mu$m in the DCN. For type II ascending and descending branches, diameters near the bifurcation (0.39 $\mu$m measured by Brown, '87) were assumed constant throughout the CN as indicated by our observations of material from mouse type II fibers. Fiber lengths were measured from serial section and a Pythagorean correction was made for the section thickness. No correction was made for tissue shrinkage. In two cochlear nuclei, the fiber lengths from bifurcation to the tips of the ascending and descending branches of type I axons averaged 1.76 and 2.32 mm, respectively.

of similar diameter, it is possible to make calculations of impulse conduction times for type I and type II fibers (Fig. 4, see Figure Legend for details). The results of such calculations in the guinea pig indicate that impulses initiated beneath the hair cells should reach the bifurcation points in 0.3 msec for type I and 6.1 msec for type II fibers. Impulses should reach the furthest points in AVCN and DCN in less than 1 msec for type I and around 10 msec for type II fibers (Fig. 4). The long conduction time for type II fibers indicates that short-latency auditory reflexes must be initiated by the type I pathway. Whatever processes are initiated or modified by type II fibers must proceed on a much slower time scale.

## Recordings from single units in the spiral ganglion

The nature of the information transmitted to the cochlear nucleus by type II axons is currently unknown. Recordings from type II neurons might enable us to determine their role in information processing in the cochlear nucleus. Recent experiments are aimed at recording from the cell bodies of type I and type II units in the anesthetized guinea pig. Since the processes of type II neurons are so fine, the only conceivable place to make micropipette recordings from these neurons is the cell body. As described by Brown ('89), guinea pigs are anesthetized with urethane (1500 mg/kg i.p.) and Innovar-vet (0.25 ml/kg i.m.) with boosters as needed. Since the appropriate sound stimulus for activation of type II units is unknown,

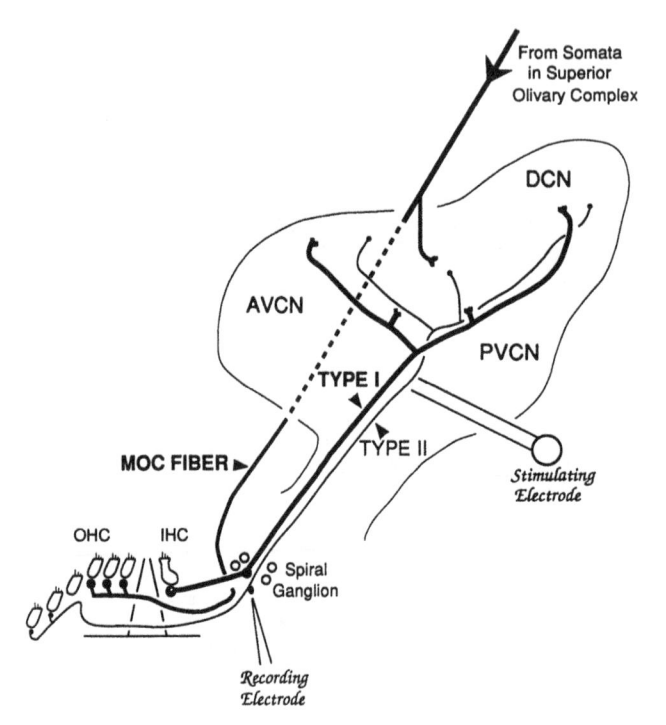

**Figure 5.** Diagram of the experimental setup for recording antidromic responses from spiral ganglion cells in response to shocks of their central axons. Medial olivocochlear (MOC) units can also be recorded from the spiral ganglion; the pathway of one of these fibers is illustrated. Dashed line indicates its course in the vestibular nerve root medial to the cochlear nucleus.

the search stimuli used are shocks to the central axons (Fig. 5). Since their fine axons should have long impulse conduction times, type II units recorded peripherally in the spiral ganglion should have long antidromic latencies to these shocks. The stimulating electrode consisted of a pair of teflon-coated, platinum wires (diameter 0.125 mm). Shocks (0.5-1.0 msec duration) were presented at a rate of 2 per sec. Preliminary experiments indicated that placement of the stimulating electrode in the auditory nerve could cause interruption of the cochlear blood flow (Levine et al., '91), so electrodes were aimed to pass through the cochlear nucleus into the auditory nerve root. Stimulating electrode placement within the auditory nerve root was adjusted while monitoring the antidromic compound action potential recorded from the round window. This short-latency potential is almost certainly due to antidromic activation of type I axons. However, the proximity of the central axons of type I and type II axons (Fig. 1) indicates that correct stimulating electrode positioning for type I fibers should be correct for type II fibers (Brown, '87; Brown et al., '88a; Simmons and Liberman, '88). Glass micropipettes were used to make recordings from the spiral ganglion as described by Brown ('89). Electrodes were aimed at the peripheral half of the ganglion, where type II cell bodies are preferentially distributed (Kellerhals et al., '67; Brown, '87).

Four general classes of units were recorded from the spiral ganglion (Fig. 6). The unit classification scheme is based both on responses to shocks and on responses to tone and noise bursts. The most common type of unit, type I units (n=231 units in 24 guinea pigs), had

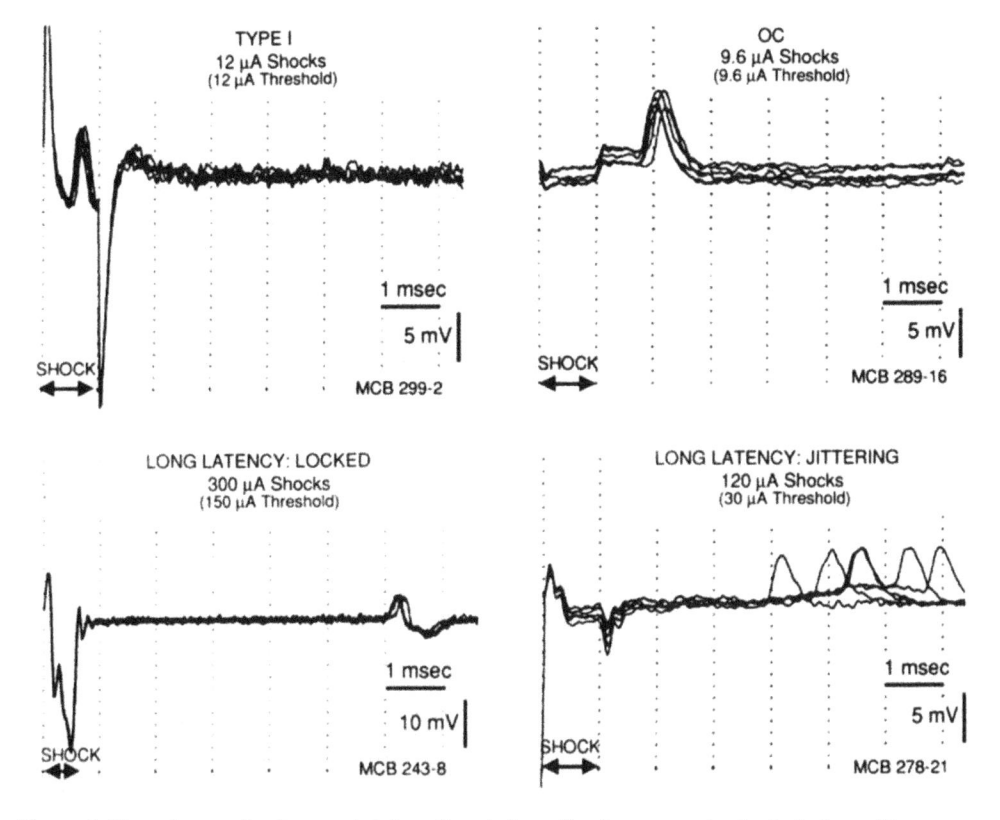

**Figure 6.** Four classes of units recorded from the spiral ganglion in response to shocks to the auditory nerve root. Shock currents used for the figure and threshold for each unit are indicated by numbers at the top of each trace. Shock duration was 1 msec except for unit MCB 243-8, where it was 0.5 msec.

short latencies (0.65-1.55 msec at threshold[1]) and low thresholds to shocks. The unit illustrated (299-2) is shown responding at its threshold (12 $\mu$A). Other type I units in this preparation responded between 10.8 and 108 $\mu$A. These units are most likely to be type I spiral ganglion neurons because of their short antidromic latencies, and because both the spontaneous and sound-evoked interspike intervals are irregular. In addition, these units had short latencies (<2 msec) in response to tone-burst stimuli. Units with irregular interspike intervals and short latencies to sound have been labeled and shown to be type I neurons (Liberman, '82; Robertson, '84; Liberman and Brown, '86; Brown, '89).

A second class of units (n=65) that responded to shocks (Fig. 6, top right) had sound-evoked activity typical of olivocochlear (OC) efferent fibers (Robertson and Gummer, '85; Liberman and Brown, '86; Brown, '89). Approximately one-third of these units responded best to contralateral sound, and the other two-thirds responded best to ipsilateral sound. The sound-evoked discharges of OC units were very regularly spaced in time, so that post-stimulus time histograms show peaks (Fig. 7A). OC units are likely to be from medial OC neurons, since these neurons give rise to thick axons that can be recorded with micropipettes (Liberman and Brown, '86). Typically, OC units had much higher shock thresholds than type I units, on the average 18.5 times higher than the most sensitive type I units in the same

---

[1]*Threshold is defined as the shock current needed to make the unit fire on at least 50% of the shocks. Our steps of shock current were separated by about 1 dB increments and in practice, most units did not respond below threshold and fired on every shock at and above threshold.

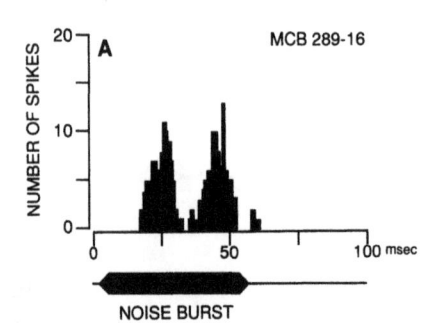

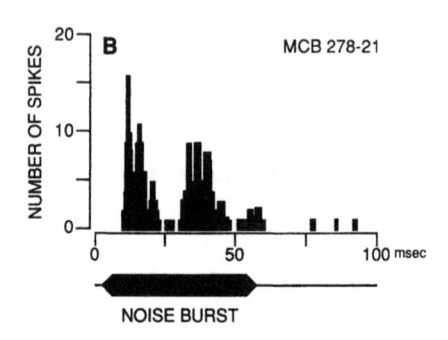

**Figure 7.** Post-stimulus time histograms for an OC unit (A), and a Long-latency:jittering unit (B): Acoustic stimulus was 150 presentations of a broad-band noise burst at about 75 dB SPL, presented to the ipsilateral ear. Neither unit responded to contralateral noise bursts. The response to shocks for the OC unit was shown in Figure 6 (top right) and the response to shocks for the Long-latency:jittering unit was shown in Figure 6 (bottom right).

preparation. Only about half of the OC units responded to shocks (27 of 65 OC units responded) up to shock levels that elicited twitches of the facial musculature. The unit illustrated (289-16) had a very low shock threshold (9.6 $\mu$A). Type-I units in this same preparation had thresholds between 12 and 72 $\mu$A. The stimulating electrode in this case was located at the anterior border of the auditory nerve root, near the vestibular nerve root. This finding might argue that at high shock levels, the stimulating current spreads to the nearby vestibular nerve and vestibular nerve root where olivocochlear fibers run. OC units that do not respond to shocks may be from cases where the stimulating electrode was further from the vestibular nerve and nerve root. Another possibility for activation of OC fibers, however, is via their collaterals to the cochlear nucleus (Brown et al., '88b). In this interpretation, OC units that do not respond to shocks might lack such collaterals. Latencies at threshold for OC units varied between 1.3 and 2.5 msec (avg. 1.96 msec), long enough to be from an OC fiber that is excited through its collateral.

Units (n=9) with much longer latencies in response to shocks were infrequently recorded in the spiral ganglion. Long-latency units consisted of two classes (Fig. 6, lower traces): a class with responses that were tightly locked in time (n=6; Long-latency:locked units) and a class that responded with considerable jitter (n=3; Long-latency:jittering units). The latencies of the Long-latency:locked units ranged from 3.75 to 8 msec. Their shock thresholds were much higher than type I thresholds, ranging between 4 and 33 (average 12) times higher than the most sensitive type I thresholds in the same preparations. The unit illustrated (243-8) is shown responding to shocks of 300 $\mu$A. Its threshold was about 150 $\mu$A. Type-I units in this preparation responded at between 12 and 36 $\mu$A. The recording of this unit terminated in injury discharge, suggesting that the recording was from a cell body. Figure 8 shows the response of this same unit and another Long-latency:locked unit to shocks. The unit at the right (292-5) is shown responding to shocks at threshold (300 $\mu$A). Type I units in this preparation responded to shocks at between 9 and 57 $\mu$A. This unit (292-5) responded to ipsilateral sound but a post-stimulus time histogram was not obtained. Tests of another long-latency:locked unit failed to produce a response to noise bursts up to 85 dB SPL.

Three long-latency units had variable, or "jittering", latencies to shocks (Fig. 6, lower right). Their latencies were between 4 and 15 msec. The shock threshold of only one Long-latency:jittering unit was obtained (30 $\mu$A; Fig. 6); this unit had a threshold close to the thresholds of type I units in the same preparation. A PST histogram for this unit in response to ipsilateral noise bursts is illustrated in Figure 7B. The response pattern is the same general type as seen for the OC unit in Figure 7A. In fact, all three Long-latency:jittering units would be classified as "OC" by their responses to noise bursts.

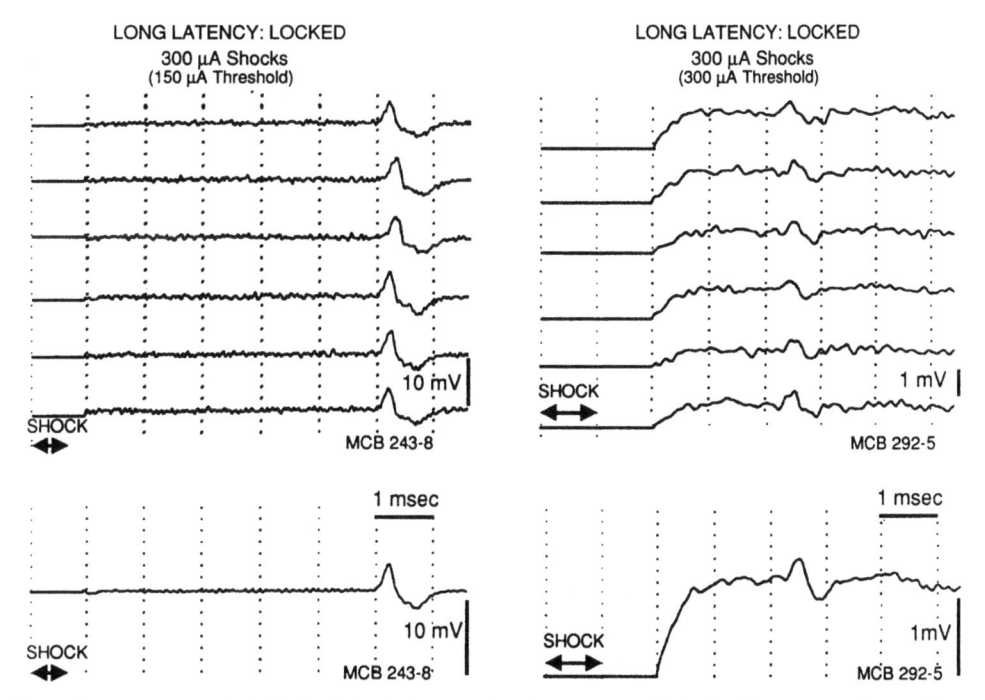

**Figure 8.** Responses to six individual shocks (top traces) and averages to 16 shocks (lower traces) for two Long-latency:locked units. Shock duration was 0.5 msec for MCB 243-8 (left) and 1.0 msec for MCB 292-5 (right). The initial portion of each trace has been blanked in order to eliminate the shock artifact.

Although these data are preliminary, they demonstrate that it is possible to record spiral-ganglion units with long-latency responses to shocks of the auditory nerve root. Two hypotheses that could account for long-latency responses in spiral-ganglion units are diagrammed in Figure 9. The first hypothesis postulates that the Long-latency:locked units correspond to cell bodies of type II neurons. Their responses reach the spiral ganglion much later than the type I responses because of the long conduction times for the thin, unmyelinated axons of the type II neurons. Consistent with this hypothesis, the antidromic responses do not have much shock-to-shock variability in latency. Also consistent with this explanation are the high shock thresholds for these units, an average of twelve times higher than type I thresholds in the same preparations. In cutaneous nerves, unmyelinated C-fiber responses are typically about ten times higher in electrical thresholds than are those of the thicker, myelinated A-fibers (Hallin and Torebjörk, '73; Fitzgerald and Woolf, '81). Robertson ('84) demonstrated that it was possible to record from type II neurons by filling one with an HRP injection, and long-latency units have been previously recorded from the auditory nerve with the use of metal electrodes (Kiang et al., '83). Present results, and those of Robertson ('84), suggest that such recordings are infrequent, as expected from the low percentage of type II neurons and the small size of their cell bodies relative to the cell bodies of type I neurons.

A second hypothesis postulates that Long-latency:jittering units are OC fibers (Fig. 9). In this hypothesis, the shocks to the auditory nerve fibers activate the afferent auditory pathway emanating from the cochlear nucleus. A significant latency would be required for information to travel along the axons and across the synapses of the part of this pathway that leads to the OC neurons. If the synaptic delays in this multi-synaptic pathway differ somewhat from shock to shock, then this hypothesis might account for the Long-latency:jittering units. Latencies of OC fiber responses to intense tone bursts are minimally about five msec. (Fex, '62; Liberman and Brown, '86), comparable to the shock latencies of the Long-latency:jit-

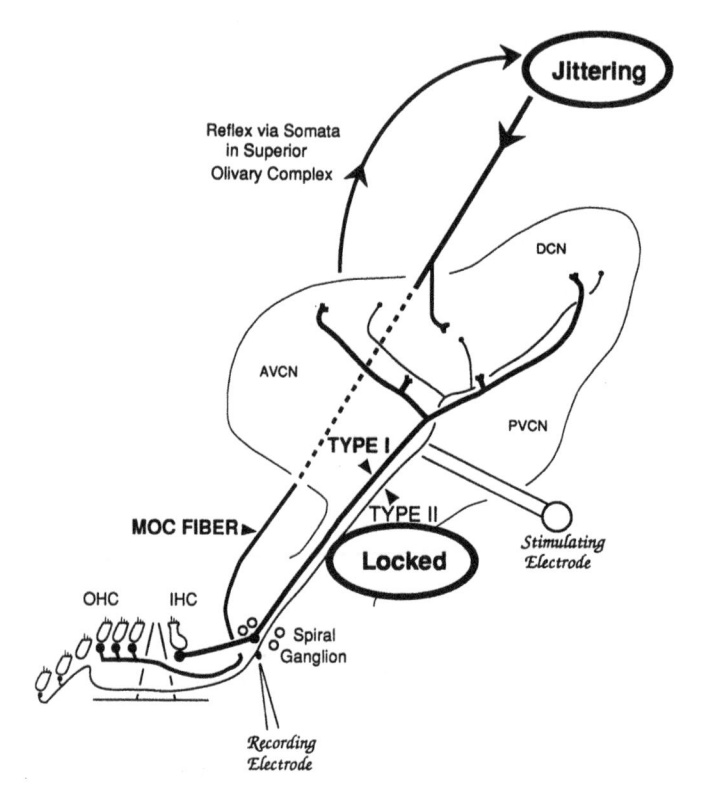

**Figure 9.** Schematic illustration of two possible hypotheses for long-latency units recorded in the spiral ganglion. As indicated by "Locked" on the figure, Long-latency:locked units could be type II neurons responding antidromically to shocks of their central axons. Such responses would be expected to have long latency because of the slow conduction times of their axons. As indicated by "Jittering" on the figure, Long-latency:jittering units may be MOC fibers that are responding to shock activation of the auditory pathway through the MOC reflex. These neurons would be expected to have long latencies because of the synaptic delays in this relatively long pathway.

tering units recorded in the present study. Most OC units recorded in this study were not Long-latency:jittering units, that is, they did not respond to brief shocks at long latency. Correspondingly, most OC units do not respond to brief acoustic stimuli, instead requiring tone bursts of substantial duration (>25 msec) to generate a response (Fex, '62; Liberman and Brown, '86). Future experiments are needed to provide a full description of why a few OC units do respond at long latency. Shock thresholds were obtained for only one of the Long-latency:jittering units. Significantly, the threshold was near the threshold of type I fibers in that preparation, and much lower than Long-latency:locked units. This point further demonstrates that the locked and jittering groups are truly different types of long-latency units. Future experiments will be necessary to confirm the anatomical identities of these subgroups of long-latency units.

ACKNOWLEDGEMENTS

It is a pleasure to acknowledge the assistance of A.M. Berglund, J.V. Ledwith, III, and D.K. Ryugo in the anatomical studies. Many thanks also to J.J. Guinan, Jr., M.C. Liberman, and N.Y.S. Kiang for helpful discussions concerning this research and to K.D. Whitley for technical assistance. (Supported by NIH grants DC 01089 and DC 00119).

# REFERENCES

Berglund, A.M. and Brown, M.C., 1989, Axonal trajectories of type-II spiral ganglion cells from various cochlear regions in mice. *Soc. Neurosci. Abstr.*, 15:742.

Berglund, A.M. and Ryugo, D.K., 1991, Neurofilament antibodies and spiral ganglion neurons of the mammalian cochlea, *J. Comp. Neurol.* 306:393-408.

Brown, M.C., 1987, Morphology of labeled afferent fibers in the guinea pig cochlea, *J. Comp. Neurol.* 260:591-604.

Brown, M.C., 1989, Morphology and response properties of single olivocochlear fibers in the guinea pig, *Hearing Res.*, 40:93-110.

Brown, M.C., Berglund, A.M, Kiang, N.Y.S., 1988a, Central trajectories of type II spiral ganglion neurons, *J. Comp. Neurol.*, 278:581-590.

Brown, M.C., Liberman, M.C., Benson, T.E. and Ryugo, D.K., 1988b, Brainstem branches from olivocochlear axons in cats and rodents. *J. Comp. Neurol.* 278:591-603.

Brown, M.C. and Ledwith, J.V., III, 1990, Projections of thin (type-II) and thick (type-I) auditory-nerve fibers in the cochlear nucleus of the mouse, *Hearing Res.*, 49:105-118.

Burgess, P.R. and Perl, E.R., 1973, Cutaneous mechanoreceptors and nociceptors, in "Handbook of Sensory Physiology, vol. 2, the Somatosensory System", I. Iggo, ed., pp. 29-78, Springer, New York.

Fekete, D.M., Rouiller, E.M., Liberman, M.C. and Ryugo, D.K., 1984, The central projections of intracellularly labeled auditory nerve fibers in cats., *J. Comp. Neurol.*, 229:432-450.

Feldman, M.L. and Harrison, J.M., 1969, The projection of the acoustic nerve to the ventral cochlear nucleus of the rat. A Golgi study, *J. Comp. Neurol.*, 137:267-294.

Fex, J., 1962, Auditory activity in centrifugal and centripetal cochlear fibers in cat, *Acta Physiol. Scan.*, 55 (S189):1-68.

Fitzgerald, M. and Woolf, C.J., 1981, Electrically induced A and C fibre responses in intact human skin nerves, *Exp. Brain Res.*, 16:309-320.

Keithley, E.M. and Feldman, M.L., 1979, Spiral ganglion cell counts in an age-graded series of rat cochleas, *J. Comp. Neurol.*, 188:429-442.

Kellerhals, B., Engstrom, H. and Ades, H.W., 1969, Die Morphologie des Ganglion Spirale cochlea, *Acta Otolaryngol. Suppl.*, 226:1-78.

Kiang, N.Y.S., Rho, J.M., Northrop, C.C., Liberman, M.C. and Ryugo, D.K., 1982, Hair-cell innervation by spiral ganglion cells in adult cats, *Science*, 217:175-177.

Kiang, N.Y.S. Keithley, E.M. and Liberman, M.C., 1983, The impact of auditory nerve experiments on cochlear implant design, *Ann. N.Y. Acad. Sci.*, pp. 114-121.

Kiang, N.Y.S., Liberman, M.C., Gage, J.S., Northrop, C.C., Dodds, L.W. and Oliver, M.E., 1984, Afferent innervation of the mammalian cochlea, in: "Comparative Physiology of Sensory Systems", L.Bolis, R.D.Keynes and S.H.P.Maddrell, eds., pp. 143-161, Cambridge University Press, Cambridge.

Kimura, R., 1978, Differences in innervation of cochlear inner and outer hair cells, Abstracts of the Association for Research in Otolaryngology.

Levine, R.A., Bu-Saba, N. and Brown, M.C., 1991, Laser Doppler flowmetry and electrocochleography during ischemia of the guinea pig cochlea: Implications for acoustic neuroma surgery. Presented at the Association for Research in Otolaryngology 14th Midwinter Research Meeting, February 1991.

Liberman, M.C., 1982, Single-neuron labeling in the cat auditory nerve, *Science*, 216:1239-1241.

Liberman, M.C. and Brown, M.C., 1986, Physiology and anatomy of single olivocochlear neurons in the cat, *Hearing Res.*, 24:17-36.

Lorente de Nó, R., 1976, Some unresolved problems concerning the cochlear nerve, *Ann. Otol., Rhinol., and Laryngol.*, 85 (S34):1-28.

Lorente de Nó, R., 1981, "The Primary Acoustic Nuclei", Raven Press, New York.

Merchan, M.A., Collia, F.P., Merchan, J.A. and Saldana, E., 1985, Distribution of primary afferent fibers in the cochlear nuclei. A silver and horseradish peroxidase (HRP) study, *J. Anat.*, 141:121-130.

Merck, W., Riede, U.N., Löhle, E. and Leupe, M., 1977, Einfluss verschiedener Pufferlösungen auf die Structurerhaltung des Ganglion spirale cochleae, *Archs. Oto-Rhino-Laryngol.*, 215:283-292.

Morrison, D., Schindler, R.A. and Wersall, J., 1975, A quantitative analysis of the afferent innervation of the Organ of Corti in guinea pig, *Acta Otolaryngol.*, 79:11-23.

Osen, K.K., 1969, Cytoarchitecture of the cochlear nuclei in the cat, *J. Comp. Neurol.*, 136:453-484.

Ramón y Cajal, S., 1909, "Histologie du Système Nerveux de l'Homme et des Vertébrés. Vol. I", pp.754-838, Instituto Ramón y Cajal, Madrid, (1952 reprint).

Reinecke, M., 1967, Elektronenmikroskopische Untersuchungen am Ganglion spirale des Meerschweinchens, *Arch. klin. exp. Ohren-Nasen-und Kehlhopgheilk*, 189:158-167.

Robertson, D., 1984, Horseradish peroxidase injection of physiologically characterized afferent and efferent neurons in the guinea pig spiral ganglion, *Hearing Res.*, 15:113-121.

Robertson, D. and Gummer, M., 1985, Physiological and morphological characterization of efferent neurons in the guinea pig cochlea, *Hearing Res.*, 20:63-77.

Ryugo, D.K., Dodds, L.W., Benson, T.E. and Kiang, N.Y.S., 1991, Unmyelinated axons of the auditory nerve in cats, *J. Comp. Neurol.*, 308:209-223.

Simmons, D.D. and Liberman, M.C., 1988, Afferent innervation of outer hair cells in adult cats, I: Light microscopic analysis of fibers labeled with horseradish peroxidase, *J. Comp. Neurol.*, 270:132-144.

Spoendlin, H., 1969, Innervation patterns in the organ of Corti of the cat, *Acta Otolaryngol.*, 67:239-254.

Spoendlin, H., 1971, Degeneration behaviour of the cochlear nerve, *Arch. klin. exp. Ohr.-Nas.-u. Kehlk. Heilk.*, 200:275-291.

Spoendlin, H., 1974, Neuroanatomy of the cochlea, in: "Facts and Models in Hearing", E.Zwicker and E.Terhardt, eds. pp. 18-32, Springer-Verlag, Berlin.

Suzuki, Y., Watanabe, A. and Osada, M., 1963, Cytological and electron microscopic studies on the spiral ganglion cells of the adult guinea pigs and rabbits, *Archiv. Histol. Jap.*, 24:9-33.

# TOPOGRAPHIC ORGANIZATION OF INNER HAIR CELL SYNAPSES AND COCHLEAR SPIRAL GANGLION PROJECTIONS TO THE VENTRAL COCHLEAR NUCLEUS

Patricia A. Leake, Russell L. Snyder and Gary T. Hradek

Epstein and Coleman Laboratories, Department of Otolaryngology U499, University of California San Francisco, San Francisco, CA 94143-0732, USA

In the mammalian cochlea, the somata of the primary afferent auditory neurons are arranged in an irregular spiral channel (Rosenthal's canal) within the bone of the modiolus, and comprise the spiral ganglion. The spiral organization of the ganglion parallels that of the basilar membrane. Approximately 95% of the cells are myelinated type I ganglion cells which exclusively innervate the inner hair cells of the organ of Corti (Spoendlin, '71; Spoendlin, '83). Each inner hair cell receives up to 30 afferent terminals, with each terminal connecting via an unbranched radial nerve fiber to a single spiral ganglion cell (Spoendlin, '71; Spoendlin, '73; Kiang et al., '82; Kiang et al., '84; Liberman et al., '90). The central axons of the bipolar type I spiral ganglion cells come together within the core of the modiolus to form the auditory nerve, which projects in a highly organized manner to each subdivision of the cochlear nuclear complex (Lorente de Nó, '81; Powell and Cowan, '62; Arnesen and Osen, '78). Within the ventral cochlear nucleus (VCN), auditory nerve fibers from the base of the cochlea project dorsally, and progressively more apical cochlear sectors project more ventrally, thus providing the morphological basis for the tonotopic organization of the VCN which is well-known from electrophysiological recording experiments. That is, neurons in the dorsal part of the anteroventral cochlear nucleus (AVCN) respond best to high-frequency tonal stimuli, while those in progressively more ventral areas respond maximally to progressively lower frequencies (Rose et al., '60; Bourk et al., '81). A detailed map of the AVCN frequency organization was provided by Bourk et al. '81, and their data suggested that the AVCN is organized in "isofrequency laminae." Each isofrequency lamina comprises a thin sheet that extends across the AVCN to its boundaries and presumably receives selective input from a spatially restricted cochlear sector. A number of morphological studies have supported this concept of a laminar organization of the VCN, demonstrating that auditory nerve fibers arising from restricted cochlear sectors project in laminae in the VCN in an orderly and sequential arrangement based on cochlear location (i.e., frequency), (Osen, '70; Webster, '71; Webster et al., '78; Ryan et al., '82; Nudo and Masterton, '84).

More recent studies utilizing the technique of intracellular labeling with horseradish peroxidase (HRP) have provided a detailed description of the central trajectories of individual, physiologically characterized auditory nerve fibers. This work has demonstrated that certain morphological features of auditory nerve fibers are correlated with the spontaneous discharge rate (SR) of individual neurons. Specifically, auditory neurons with high SR show less extensive axonal branching, shorter collaterals and larger but less complex terminals than do neurons with low and medium SRs (Fekete et al., '84; Rouiller et al., '86; Ryugo and Rouiller, '88). In addition, in the cochlea it has been reported that synapses of auditory nerve fibers of different spontaneous discharge rates are spatially segregated on the inner hair cells (IHC), such that the pillar face of the IHC is contacted exclusively by terminals of high SR fibers, while the terminals of low and medium SR fibers are found only on the modiolar side of the IHC along with a subpopulation of high SR terminals (Liberman, '82a; Liberman and Oliver, '84). Further, SR has also been correlated with a number of other physiological response properties, such as threshold, gain and dynamic range (Liberman, '78). Documentation of the correlations between the detailed morphology of auditory nerve fibers and their physiological characteristics comprises an important first step toward constructing appropriate models of information processing in the cochlear nucleus. However, it is apparent that it will also be important to better define the overall organization and topography of the spiral ganglion projections to the cochlear nuclei, which will be difficult to reconstruct from this type of individual fiber analysis due to the complexity of the projections and the anatomical variation among individual animals.

The authors' recent work is focused on this mid-level analysis, with the specific goal of providing a better understanding of the structural basis of the isofrequency laminae and their role in stimulus coding within the VCN. Using the technique of extracellular microinjections of HRP directly into the spiral ganglion, the CN projection patterns from small clusters of spiral ganglion neurons were described (Leake and Snyder, '88; Leake and Snyder, '89). Such injections intensely labeled discrete sectors of the spiral ganglion that projected into thin "isofrequency laminae" extending across the entire lateral-to-medial dimension of the VCN. Consistent with the known organization of the VCN as discussed previously, high frequency laminae were situated dorsally in the VCN and lower frequencies were located progressively more ventrally. A distinct projection to the small cell cap area was often observed to be offset from the main projection lamina, depending on the frequency representation of a given isofrequency lamina. In addition, these HRP experiments demonstrated that the VCN laminae are composite structures. That is, smaller, more restricted injections demonstrated a spatial segregation of spiral ganglion projections from a given cochlear location across the lateral-to-medial dimension of the isofrequency laminae. Specifically, spiral ganglion cells within the lower part of the ganglion near the scala tympani projected to the lateral part of the VCN isofrequency laminae, while cells in the upper or scala vestibuli aspect of the ganglion projected to the medial aspect of the isofrequency planes. As summarized in Figure 1, the data indicated a topographic organization of the spiral ganglion and isofrequency laminae orthogonal to the frequency domain. That is, in addition to the spiral dimension of the ganglion (corresponding to the dorsal-to-ventral frequency map in the VCN), there is also an orderly, sequential topographic projection of inputs from the scala tympani-to-scala vestibuli dimension of the spiral ganglion across the lateral-to-medial dimension of the VCN. As mentioned previously, synapses of auditory nerve fibers of different spontaneous discharge rates have been reported to be spatially segregated on the IHC. This led us to hypothesize that this functional segregation of primary afferents expressed across the bases of IHCs may be preserved within the ganglion as a scala vestibuli-to-scala tympani progression, and that this segregation may be carried centrally by the orderly projections in the medial-to-lateral dimension (i.e., across the isofrequency laminae) of the VCN (Leake and Snyder, '89).

In a second series of experiments, restricted lesions of the anteroventral cochlear nucleus (AVCN) were made to selectively ablate either the lateral or the medial aspect of isofrequency laminae (Leake et al., 92; Leake and Snyder, in press). These experiments take advantage of the well-documented phenomenon that spiral ganglion neurons undergo secondary degeneration following central axotomy (Spoendlin, '71; Spoendlin, '73; Spoendlin, '75). Lesions in the cochlear nucleus were created under aseptic conditions, and

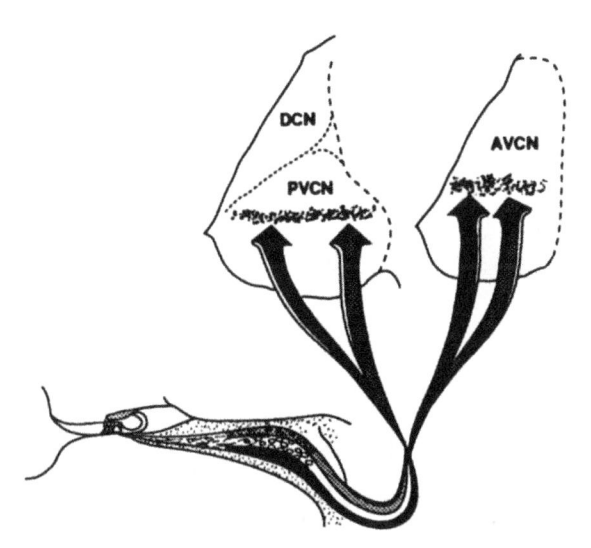

**Figure 1.** Diagram illustrating the topographic organization of projections from the cochlear spiral ganglion to the ventral cochlear nucleus (VCN) suggested by HRP studies. The projection from the entire cell group at a given location comprises a sheet of fibers and terminals that extends across the entire lateral to medial extent of the VCN. The cells in the upper or scala vestibuli aspect of the ganglion contribute fibers in the medial portion of this band, while cells located closer to the scala tympani within the ganglion contribute the lateral portion of the isofrequency laminae.

the cochleas were evaluated after post-surgical intervals of 2 to 3 months to allow for the retrograde degeneration of the spiral ganglion. The hypothesis was that these restricted lesions would induce selective degeneration of spiral ganglion cells. Based on the earlier-described HRP results, ablation of the lateral part of AVCN isofrequency laminae should result in degeneration of cells in the scala tympani portion of the ganglion; medial lesions should induce degeneration of the scala vestibuli portion of the ganglion for the frequency laminae damaged by the lesion.

Results from these lesion experiments support the concept of a lateral-to-medial segregation of inputs to VCN isofrequency laminae suggested by the earlier HRP experiments. Cochlear nucleus lesions that ablated the lateral part of AVCN isofrequency sheets resulted in highly selective losses of spiral ganglion cells in the scala tympani part of Rosenthal's canal. Lesions of the medial VCN induced degeneration of cells in the scala vestibuli portion of the ganglion in the cochlear sector appropriate for the frequencies represented across the damaged laminae (Leake et al., 92; Leake and Snyder, in press). (See below, Figs. 2a,b and 3a,b).

Following these selective spiral ganglion lesions, peripheral and central axons also degenerated, and this axonal loss both within the osseous spiral lamina and within the cochlear modiolus displayed a selective topography that paralleled the cell loss within the ganglion. These data, especially the observation of the selective pattern of peripheral fiber loss, reinforced the suggestion that the spatial segregation of high and low SR terminals on the IHC may be maintained into the spiral ganglion and may be reflected in the demonstrated segregation of projections across the lateral to medial dimension of VCN isofrequency laminae.

To further address this hypothesis, in some of the lesion cases, serial electron microscopic sections of synaptic regions of the inner hair cells (IHC) were cut in a horizontal plane parallel to the basilar membrane to determine the distribution of surviving terminals. Preliminary synaptic data from these lesion cases are presented here, and suggest differential but not completely selective losses of IHC synapses depending on the site of the lesion.

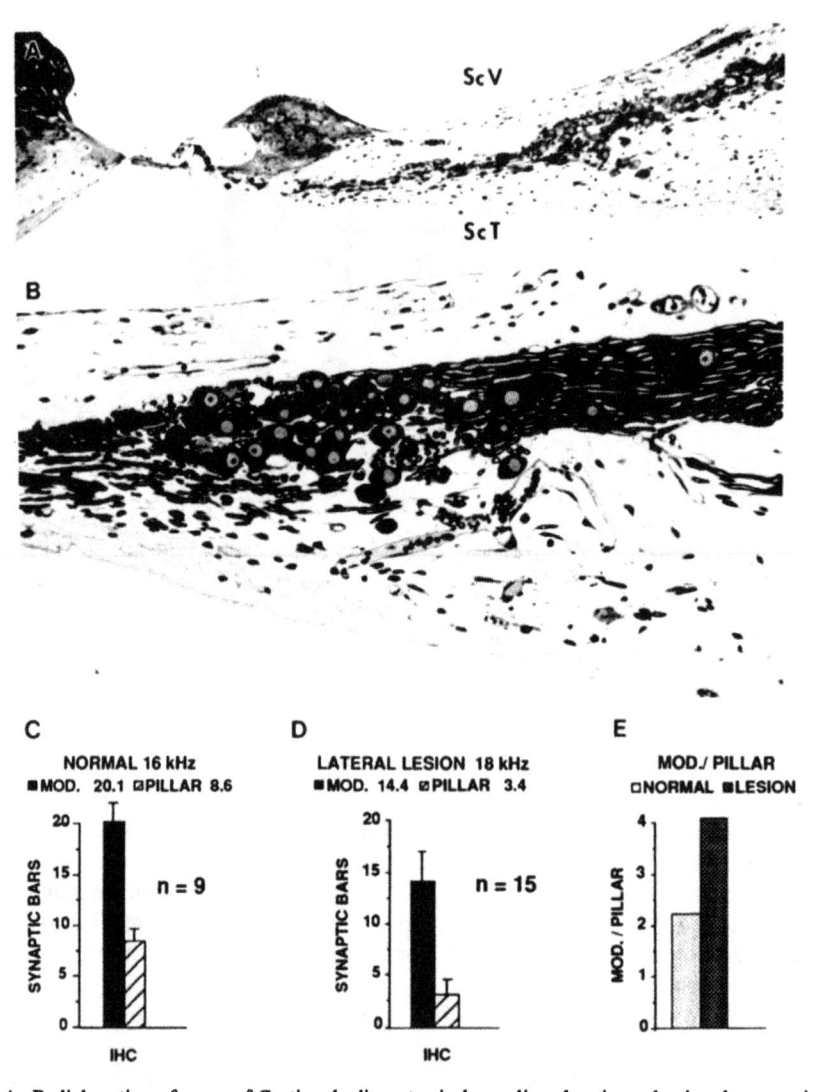

**Figure 2.** A. Radial section of organ of Corti and adjacent spiral ganglion showing selective degeneration in the scala tympani (lower) part of the ganglion caused by a lesion in the dorsolateral aspect of the AVCN. Note the parallel loss of myelinated axons in the tympanic portion of the osseous spiral lamina. ScT, scala tympani; ScV, scala vestibuli. B. Higher magnification of the spiral ganglion in a nearby region of the same cochlea. Faint profiles of satellite cell remnants indicate the outlines of many of the degenerated cells in scala tympani part of the ganglion. C. Data from serial E.M. sections of 9 IHC at the 16.3 kHz region in a normal cat cochlea. Synapses were classified as contacting either the modiolar or the pillar face of the hair cell. Means and standard deviations are shown. D. Synaptic data from IHCs at the 18 kHz region of the lesion case shown above in A and B, showing means and standard deviations for 15 IHC. About 28% of modiolar synapses and 60% of the pillar synapses have degenerated. E. The ratios of modiolar to pillar synapses are compared for normal IHC and lesion data from C and D. In normal IHC this ratio is 2.3:1; in cells from the lesioned cochlea the ratio increases to 4.2:1 indicating that there is proportionately greater degeneration of pillar synapses on IHC in this lesioned cochlea.

Figure 2 illustrates results from a lesion in the lateral aspect of the AVCN which caused highly selective degeneration in the scala tympani part of the spiral ganglion. A shallow ablation damaging the most superficial or dorsolateral portion of the AVCN was made under aseptic conditions, and the cat was euthanized after a post-surgical period of eight weeks to allow retrograde degeneration of the cochlear spiral ganglion cells. Cell loss was observed at the extreme basal end of the cochlea, and extended as far apical as the basilar membrane location corresponding to approximately 8 kHz, as calculated using Greenwood's frequency-position function (Greenwood, '74) with Liberman's revised constants for the cat (Liberman, '82b). The micrographs in Figure 2A and 2B show representative examples of sections taken across this region of cell loss. They illustrate the selective loss of spiral ganglion cells from the scala tympani aspect of the ganglion. At the 27 kHz location shown in A, the lesion is at its maximum development, and more than half of the spiral ganglion cells have degenerated. There is also selective thinning of the myelinated radial nerve fibers within the scala tympani aspect of the osseous spiral lamina, and selective loss of central axons in fascicles of fibers passing into the modiolus from the lower part of the ganglion. The higher magnification micrograph shown in B is taken from the 22 kHz location, and shows the virtually total cell loss within the scala tympani part of the ganglion.

Data from serial electron microscopic sections through the neural poles of 15 inner hair cells in this lesioned sector of the cochlea are shown in the graph in Figure 2D and may be compared with the data from 9 inner hair cells from the 16.3 kHz region of a normal cat cochlea (Fig. 2C). The position of synapses on each IHC was evaluated, and each terminal was classified as contacting either the modiolar or the pillar face of the hair cell. In this high frequency region of the normal cat cochlea there were an average of 28.7 synapses on the IHCs. Of these, 20.1 synapses were located on the modiolar face of the IHC and 8.6 contacted the pillar side of the hair cell. For the 15 IHCs examined in the lesioned case, the mean total number of synapses was 17.8, which is a reduction of about 38% from normal. The mean distribution of the surviving terminals on the IHC was 14.4 modiolar synapses and 3.4 pillar synapses. Compared to normal cells, this represents a significant decrease ($p < 0.001$, Student's T test) in both synaptic populations. When calculated as a percent of the normal number of synapses, these data indicate a loss of about 28% of the modiolar synapses, and a loss of 60% of the pillar synapses. Thus, with this lesion of the lateral AVCN and degeneration of ganglion cells in the scala tympani portion of the ganglion, there appears to be preferential degeneration of pillar (high SR) synapses.

In Figure 2E the proportionate IHC synaptic distributions are compared for these normal and lesion data. Normal IHC at this frequency location have an average of 2.3 times more modiolar synapses than pillar synapses. For inner hair cells from the lesioned cochlea, the ratio of modiolar to pillar synapses increased almost twofold, to 4.2:1, again indicating a disproportionate loss of pillar synapses. These data suggest that most or at least a greater proportion of the high spontaneous rate IHC terminals on the pillar face have cell somata located in the scala tympani part of the spiral ganglion and project to the lateral aspect of the corresponding VCN isofrequency laminae.

It should be noted that in absolute numbers these data represent roughly equal losses of 5.7 modiolar synapses and 5.2 pillar synapses. However, since less than 30% of the terminals are located on the pillar side of the IHC and since 38% of all the terminals had degenerated, it is not surprising that some loss of modiolar synapses occurred. Because the modiolar population contains a mixture of high and low SR terminals, it is not clear whether or not the neuronal losses in this case were completely selective for high SR neurons.

Figure 3 illustrates a second lesion case, in which damage to the medial VCN resulted in selective degeneration of the upper or scala vestibuli part of the spiral ganglion cell cluster. The lesion was located more ventrally in the VCN than in the preceding example, and spiral ganglion cell degeneration in the cochlea was observed over a lower frequency range of 7 to 1.5 kHz. The micrographs in Figure 3A and B illustrate the cell loss in the middle of this lesioned sector. Note that while neuronal degeneration is quite selective to the upper (scala vestibuli) part of the ganglion, there is some sparing of ganglion cells that appear to be scattered widely across this region. Inner hair cells were studied at the basilar membrane location corresponding to about 5.4 kHz. Data were collected from serial E.M. sections for 8 inner hair cells and are shown in the graph in Figure 3D. For comparison in this lower frequency region, data from 9 inner hair cells from the 4.5 kHz location of a normal cat are

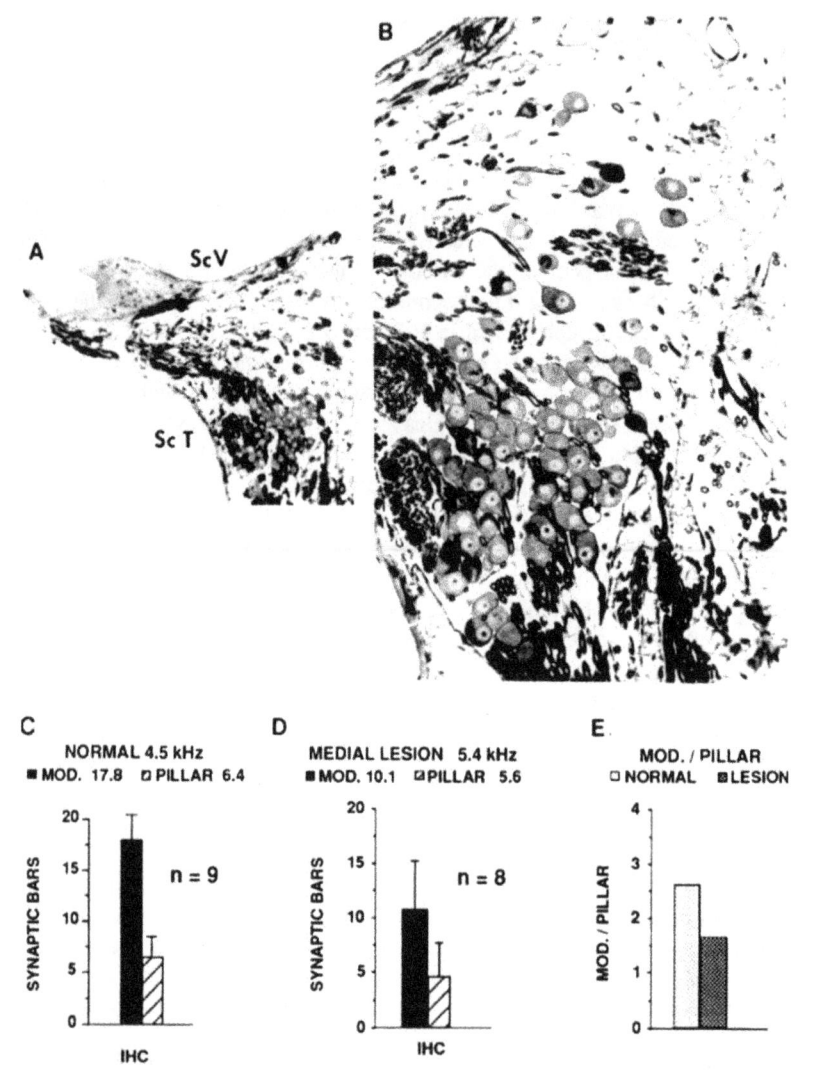

**Figure 3.** A. Low magnification radial section illustrating the selective degeneration in the scala vestibuli (upper) part of the ganglion resulting from a lesion of the medial VCN. ScT, scala tympani; ScV, scala vestibuli. B. Higher magnification of the spiral ganglion in this same region. At the 6 kHz basilar membrane location illustrated here, approximately half of the spiral ganglion cells have degenerated. C. In serial sections through IHC at the 4.5 kHz region of a normal cat cochlea, the snynaptic profiles were classified as modiolar or pillar. Means and standard deviations are shown (n=9). D. Serial section data for synapses on IHC at the 5.4 kHz region of the lesion case shown above (n=8). About 43% of modiolar synapses have degenerated, but only about 14% of pillar synapses are missing. E. The proportionate distributions of synapses are compared for normal IHC and lesion data from C and D. In normal IHC at this frequency there are 2.8 modiolar synapses for each pillar synapse, while in hair cells from the lesioned cochlea this ratio falls to 1.8:1. These data suggest that preferential loss of modiolar synapses is associated with selective cell loss in the scala vestibuli part of the ganglion.

also presented (Figure 3C). The mean number of afferent synapses on normal IHCs at this location was 24.2, with 17.8 terminals contacting the modiolar face and 6.4 on the pillar side of the hair cell. The inner hair cells from the medial lesion case had a mean of 15.7 synapses, which comprises a loss of about 35% when compared to the normal data for the 4.5 kHz location. Of these surviving synaptic profiles, a mean of 10.1 contacted the modiolar side of

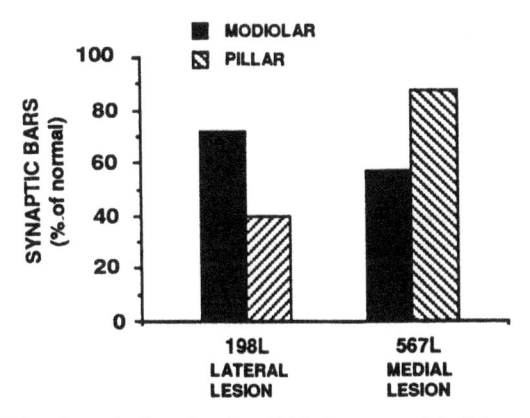

**Figure 4.** Comparison of data from the lateral and medial lesion cases. Modiolar and pillar IHC synapses are shown as percent of normal number for the appropriate frequency region. The lateral lesion with cell loss in the scala tympani part of the ganglion resulted in greater loss of pillar synapses, while the medial lesion caused ganglion cell loss in the scala vestibuli part of the ganglion and selective loss of modiolar synapses.

the inner hair cell and 5.6 were classified as pillar synapses. Calculated as a percentage of normal, these data represent a mean loss of 43% of modiolar synapses, which is significantly different from the normal ($p < 0.001$). The mean number of pillar synapses on these cells was reduced by 14%, and was not significantly different from the normal IHC. The proportionate distribution of synapses for the cells from this lesion case and for the corresponding normal IHC are shown in Figure 3E. For normal IHC in this lower frequency cochlear region, the ratio of modiolar to pillar synapses was 2.8 to 1, while in the medial lesion case, this ratio fell to 1.8 to 1. These data suggest that cell loss in the scala vestibuli aspect of the spiral ganglion caused highly selective degeneration of modiolar synapses on inner hair cells in the lesioned area.

Thus, a lesion which damaged the lateral AVCN and scala tympani portion of the ganglion caused degeneration of a greater percentage of synapses on the pillar face of the IHC; while a medial lesion causing cell loss in the scala vestibuli part of the ganglion resulted in preferential loss of modiolar synapses (Figure 4). These results suggest that the segregation of IHC terminals of different SR groups is maintained in a topographic organization across the vertical (scala tympani to scala vestibuli) dimension of the spiral ganglion and is carried into VCN. This infers that, in addition to the spiral frequency organization represented by the dorsal-to-ventral frequency map in the VCN, there may also be a segregation of inputs from high and low SR fibers across VCN isofrequency laminae. It is interesting to note that recent preliminary reports of single fiber HRP labeling studies support this hypothesis (Kawase and Liberman, '91; Liberman, '91). These authors reported that labeled fibers of high SR had cell somata that were distributed throughout the ganglion but were more common in the scala tympani part, while fibers of low and medium SR had somata limited to the scala vestibuli part of the ganglion. Taken together, the available data suggest that the lateral aspect of VCN isofrequency laminae receives a more predominant input from high SR terminals, especially from the pillar face of the IHC, while medial isofrequency laminae receive preferential input from modiolar terminals, including the low and medium SR populations.

ACKNOWLEDGEMENTS

This work was supported by NIH grant RO1 DC00160.

# REFERENCES

Arnesen, A.R. and Osen, K.K., 1978, The cochlear nerve in the cat, Topography, cochleotopy, and fiber spectrum, *J. Comp. Neurol.*, 178: 661-678.

Bourk, T.R., Mielcarz, J.M.,and Norris, B.E., 1981, Tonotopic organization of the anteroventral cochlear nucleus of the cat, *Hearing Res.*, 4:215-241.

Fekete, D.M., Rouiller, E.M., Liberman, M.C. and Ryugo, D.K., 1984, The central projections of intracellularly labeled auditory nerve fibers in cats, *J. Comp. Neurol.*, 229:432-450.

Greenwood, D.D., 1974, Critical bandwidth in man and some other species in relation to the traveling wave envelope, *in*: "Sensation and Measurement," H.R. Moskowitz and J.C. Stevens, eds., Reidel, Boston, pp. 231-239.

Kawase, T. and Liberman, M.C., 1991, Spatial organization of the spiral ganglion according to spontaneous discharge rate, *Abstr. Assoc. Res. Otolaryngol.*, ed., 14:17.

Kiang, N.Y.S., J.M. Rho, Northrop, C.C., Liberman, M.C., Ryugo, D.K., 1982, Hair cell innervation by spiral ganglion cells in adult cats, *Science*, 217:129-177.

Kiang, N.Y.S., Liberman, M.C., Gage, J.S., Northrop, C.C., Dodds, L.W., Oliver, M.E., 1984, Afferent innervation of the mammalian cochlea, *in*: "Comparative Physiology of Sensory Systems," L. Bolis, R.D. Deynes and S.H.P. Maddrel,eds., pp.143-161.

Leake, P.A. and Snyder, R.L., 1988, An additional dimension in the topographic organization of the spiral ganglion projection to the ventral cochlear nucleus, *Soc. Neurosci. Abstr.* 14:322.

Leake, P.A. and Snyder, R.L., 1989, Topographic organization of the central projections of the spiral ganglion in cats, *J. Comp. Neurol.*, 281:612-629.

Leake, P.A., Snyder, R.L. and Marzeneck, M.M., 1992, Topographic organization of the cochlear spiral ganglion demonstrated by restricted lesions of the anteroventral cochlear nucleus, *J. Comp. Neurol.*, 320:468-478.

Leake, P.A. and Snyder, R.L., (In press), Topographic organization of the projections of the spiral ganglion to the ventral cochlear nucleus, *in*: "Cochlear Nucleus: Structure and Function in Relation to Modelling", W. A. Ainsworth, ed., Adv. Speech, Hear. Lang. Proc., 3.

Liberman, M.C., 1978, Auditory nerve responses from cats raised in a low-noise chamber, *J. Acoust. Soc. Am.*, 63:442-455.

Liberman, M.C., 1982b, The cochlear frequency map for the cat: Labeling auditory-nerve fibers of known characteristic frequency, *J. Acoust. Soc. Am.* 72:1441-1449.

Liberman, M.C. and Oliver, M.E., 1984, Morphometry of intracellularly labeled neurons of the auditory nerve: Correlations with functional properties, *J. Comp. Neurol.*, 223:163-176.

Liberman, M.C., 1984a, Single-neuron labeling in the cat auditory nerve, *Science*, 216:1239-1241.

Liberman, M.C., Dodds, L.W. and Pierce, S., 1990, Afferent and efferent innervation of the cat cochlea: Quantitative analysis using light and electron microscopy, *J. Comp. Neurol.*, 301:443-460.

Liberman, M.C., 1991, Spatial segregation of auditory nerve projections in the cohlear nucleus according to spontaneous discharge rates, *Abstr. Assoc. Res. Otolaryngol.*, 14:42.

Lorente de Nó, 1981, "The Primary Acoustic Nuclei," Raven Press, New York.

Nudo, R.J. and Masterton, R.B., 1984, 2-Deoxyglucose studies of stimulus coding in the brainstem auditory system of the cat, *in*: "Contributions to Sensory Physiology," W.D. Neff, ed., Academic Press, Orlando, Fla:, pp. 77-97.

Osen, K.K., 1970, Course and termination of the primary afferents in the cochlear nuclei of the cat: An experimental anatomical study, *Archives Ital. Biol.*, 108:21-51.

Powell, T.P.S. and Cowan, W.W., 1962, An experimental study of the projection of the cochlea, *J. Anat.* (London), 96:269-284.

Rose, J.E., Galambos, R. and Hughes, J., 1960, Organization of frequency sensitive neurons in the cochlear nuclear complex of the cat, *in*: "Neuronal Mechanisms of the Auditory and Vestibular Systems," G.L. Rassmussen and W.F. Windle, eds. Thomas, Springfield.

Rouiller, E.M., Cronin-Schreiber, R., Fekete, D.M. and Ryugo, D.K., 1986, The central projections of intracellularly labeled auditory nerve fibers in cats: An analysis of terminal morphology, *J. Comp. Neurol.*, 249:261-278.

Ryan, A.F., Woolf, N.K. and Sharp, F.R, 1982, Tonotopic organization in the central auditory pathway of the mongolian gerbil: A 2-deoxyglucose study, *J. Comp. Neurol.*, 207:369-380.

Ryugo, D.K. and Rouiller, E.M., 1988, Central projections of intracellularly labeled auditory nerve fibers in cats: Morphometric correlations with physiological properties, *J. Comp. Neurol.*, 271:130-142.

Spoendlin, H., 1971, Degeneration of cochlear afferent neurons, *Acta Otolaryngol.*, 91:451-456.

Spoendlin, H., 1973, The innervation of the cochlear receptor, *in*: "Basic Mechanisms in Hearing," A.R. Moller, ed., pp. 185-234, Academic Press, New York.

Spoendlin, H., 1975, Retrograde degeneration of the cochlear nerve, *Acta Otolaryngol.*, 79:266-275.

Webster, D.B., 1971, Projection of the cochlea to cochlear nuclei in Meriam's kangaroo rat, *J. Comp. Neurol.*, 143: 323-340.

Webster, D.B., Serviere, J., Batini, C.,and Laplante, S., 1978, Autoradiographic demonstration with 2-[$^{14}$C] deoxyglucose of frequency selectivity in the auditory system of cats under conditions of functional activity, *Neuroscience Lett.*, 10:43-48.

# ULTRASTRUCTURAL ANALYSIS OF SYNAPTIC ENDINGS OF AUDITORY NERVE FIBERS IN CATS: CORRELATIONS WITH SPONTANEOUS DISCHARGE RATE

David K. Ryugo, Debora D. Wright, and Tan Pongstaporn

Center for Hearing Sciences, Departments of Otolaryngology-Head and Neck Surgery and Neuroscience, Johns Hopkins University School of Medicine, Baltimore, MD 21205, USA

In mammals, all known auditory information enters the brain by way of the auditory nerve. The auditory nerve is a bundle of axons whose cell bodies are located in the spiral ganglion within the cochlea. The ganglion cells send peripheral processes out to the organ of Corti to contact acoustic receptor cells and send central processes by way of the auditory nerve to terminate in the cochlear nucleus. In this way, the ganglion cells convey the output of the receptors to neurons of the brain. In turn, cells of the cochlear nucleus give rise to the ascending auditory pathways. The role of the cochlear nucleus is to receive incoming auditory nerve discharges, to preserve or transform the signals, and to distribute outgoing activity to higher brain centers. In order to understand mechanisms of stimulus coding in these early stages of the auditory system, we need to know the nature of the signals conveyed by auditory nerve fibers and structural details of their destination in the cochlear nucleus.

There are at least two different types of ganglion neurons based on cell body size, myelination, and cytoplasmic features (Kellerhals et al., '67; Spoendlin, '73; Berglund and Ryugo, '91), peripheral innervation (Kiang et al., '82; Berglund and Ryugo, '87; Brown, '87; Liberman et al., '91), and central projections (Fekete el al., '84; Brown et al., '88; Ryugo and Rouiller, '88; Ryugo et al., '91). Type I neurons constitute 95% of the ganglion population, have myelinated processes and a high somatic ribosomal content, and contact one or occasionally two inner hair cells (IHCs). In contrast, type II neurons represent 5% of the population, have unmyelinated processes, exhibit high numbers of somatic neurofilaments, and contact many outer hair cells (OHCs).

When a recording microelectrode is inserted into the auditory nerve, the physiological responses of individual fibers can be monitored (Kiang et al., '65; Liberman, '78). It is generally accepted that all single unit recordings in the auditory nerve have come from type I auditory nerve fibers and that virtually nothing is known about the response properties of type II fibers (Liberman, '82; Robertson, '84). Consequently, this report focuses on type I fibers.

It has been suggested that under a limited range of stimulus conditions, the pattern of discharges in any single auditory nerve fiber is qualitatively similar throughout the population (Pfeiffer, 1966). Quantitatively, however, single fiber responses differ along two principal

Table I. TYPE I AUDITORY NERVE FIBERS IN CATS. QUANTITATIVE DATA SUMMARY

|  | Low SR Fibers | High SR Fibers |
|---|---|---|
| Spontaneous Discharge Rate | < 18 s/s | > 18 s/s |
| Representation in Nerve | 40% | 60% |
| Threshold at CF (dB SPL) | 0–40 dB | ≈ 0 dB |
| Maximum Discharge Rate | 50–200 s/s | 150–250 s/s |
| Diameter of Peripheral Ending | < 0.5 μm | ≈ 1.5 μm |
| Number of Central Endings/Fiber | ≈ 100 | ≈ 55 |
| Endbulb Form Factor (Area/Perimeter) | < 0.52 | > 0.52 |

Data from Liberman, 1978, 1982; Fekete et al. 1984; Sento and Ryugo, 1989.

dimensions: frequency selectivity and spontaneous discharge rate. Frequency selectivity or characteristic frequency (CF, that frequency to which the neuron is most sensitive) refers to a fiber's tendency to be most responsive to a single frequency, and is best represented by a threshold tuning curve. A tuning curve describes the envelope of the coordinates of a neuron's response area to tonal stimuli in terms of sound level (dB SPL) and frequency, and specifies both CF and threshold at CF. Threshold at CF is correlated to spontaneous discharge rate.

In the absence of acoustic stimulation, auditory nerve fibers manifest spontaneous, randomly-occurring spike discharges. Among units having similar CFs, SR can range from near zero to greater than 100 spikes per second (s/s). Across the entire range of CF values, SR has a bimodal distribution, and for our purposes, fibers exhibiting < 18 s/s are defined as low SR fibers, whereas fibers exhibiting > 18 s/s are defined as high SR fibers. Many aspects of an auditory nerve fiber response to simple stimuli can be calculated or predicted once a tuning curve and the spontaneous discharge rate (SR) are known (Kiang et al., '65; Anderson et al., '71; Sachs and Abbas, '74; Liberman, '78; Evans and Palmer, '80). For example, fibers of the different SR groups display systematic differences in their threshold at CF, dynamic range, and discharge rate properties in response to sound.

There are morphological correlates to fiber SR. In the peripheral auditory system, SR is proportional to the caliber of the unmyelinated terminal and is correlated with the location of the peripheral synapse on the IHC (Liberman, '82). Centrally, SR-related differences in fiber morphology in the anteroventral cochlear nucleus (AVCN) include the number, size, and distribution of terminal swellings (Fekete et al., '84; Rouiller et al., '86; Ryugo and Rouiller, '88; Ryugo and Sento, '91). These systematic differences in anatomical and physiological features of fibers belonging to separate SR groups (summarized in Table I) suggest that each group may represent functionally distinct components for the processing of acoustic information. The most sensitive fibers might ultimately be involved in threshold detection, whereas the least sensitive fibers might be related to the perception of loudness or contribute to the elicitation of the middle ear muscle reflex. Such a distinction between fibers of the different SR groups might be expressed at the cellular level, whereby members of each group would exhibit unique features in terms of their synaptic connections with different populations of neurons in the cochlear nucleus. We began a direct test of this hypothesis by studying synaptic relationships of intracellularly labeled auditory nerve fibers whose physiological response properties were recorded prior to staining.

In the present paper, we report on our examination of the synaptic connections formed by the ascending branches of two type I auditory nerve fibers having CFs of 1 kHz. One fiber belonged to the low SR group (0.4 spikes/sec), whereas the other belonged to the high SR group (44.8 spikes/sec). Each fiber came from the opposite nerve of the same cat, a circumstance that serves to minimize variation. The methods for surgery, sound stimulation,

CAT 20-1
Left Cochlear Nucleus
CF = 1.0 kHz
SR = 0.4 s/s

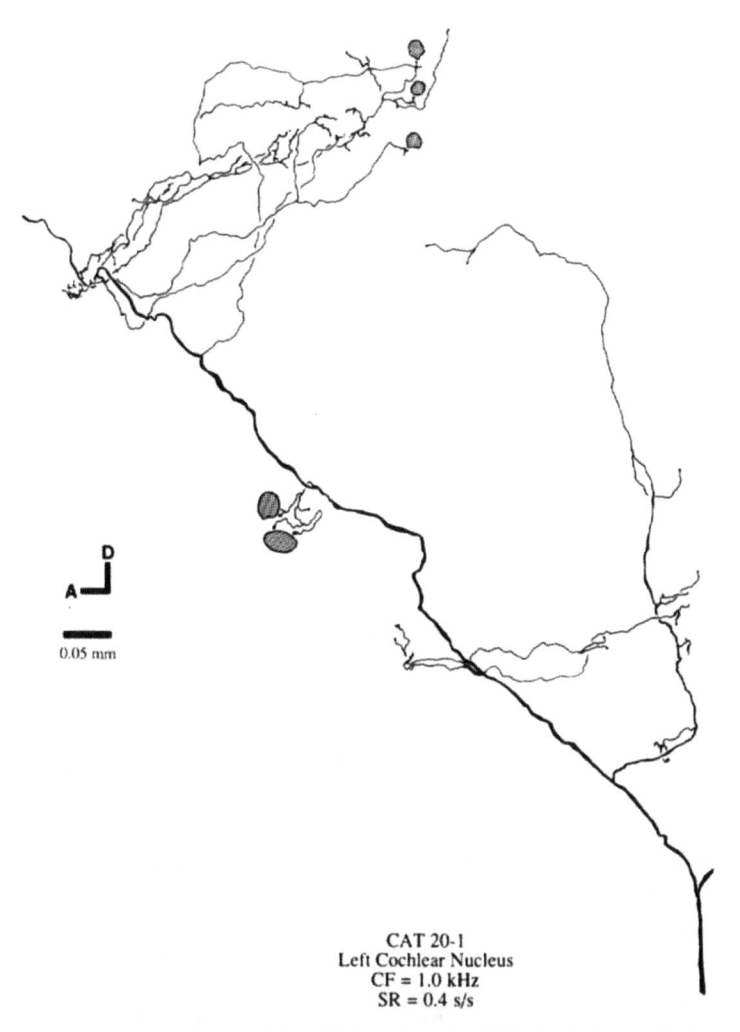

**Figure 1.** Drawing tube reconstruction of the initial portion of an HPR-labeled ascending branch of a low SR fiber. Stippled symbols represent cell bodies in close apposition to labeled terminals which were examined with an electron microscope. The extensive branching is typical for auditory nerve fibers having low SR.

intracellular recording and horseradish peroxidase (HRP) injections, and histological procedures have been published previously (Ryugo and Sento, '91).

The light microscopic appearance of each fiber was entirely consistent with previously published descriptions (Fekete et al., '84; Ryugo and Rouiller, '88). The fiber entered the nucleus by crossing the Schwann-glia border and travelled a few hundred micrometers before bifurcation. The bifurcation gave rise to a prominent ascending branch and descending branch. In the region of the bifurcation, each ascending branch produced collaterals that ramified further, distributing swellings against cell bodies and in the neuropil (Fig. 1,2). This region, called the interstitial nucleus (Lorente de Nó, '33) or globular cell area (Osen, '69), is comprised mostly of so-called globular bushy and stellate (multipolar) cells (Osen, '69; Tolbert and Morest, '82). The low SR fiber emitted additional collaterals that ramified rather extensively and distributed themselves dorsolaterally (Fig. 1), within a region called the small cell cap (Osen, '69).

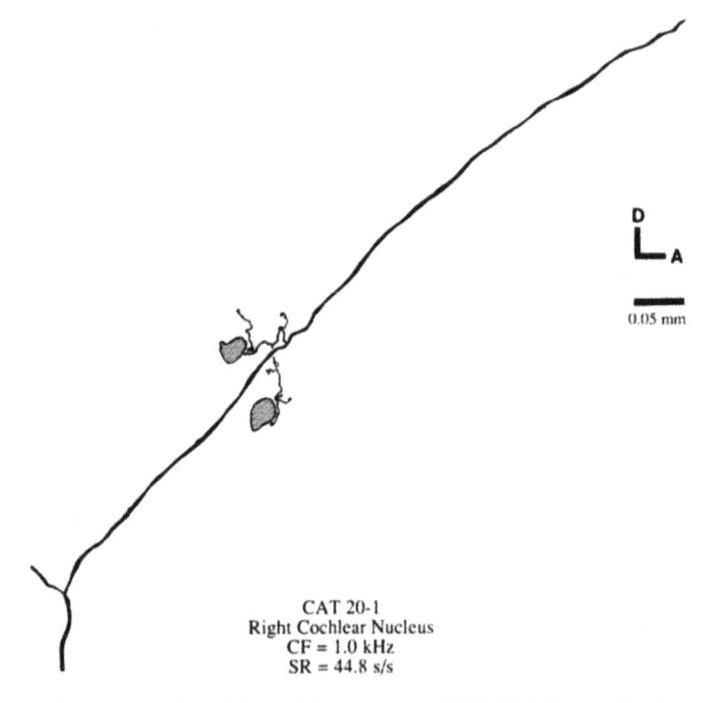

CAT 20-1
Right Cochlear Nucleus
CF = 1.0 kHz
SR = 44.8 s/s

**Figure 2.** Drawing tube reconstruction of the initial portion of an HRP-labeled ascending branch for a high SR fiber. Stippled symbols represent cell bodies in close apposition to labeled terminals which were examined with an electron microscope. The infrequent branching is typical for auditory nerve fibers having high SR.

The first issue addressed by our serial section, electron microscopic investigation was which neurons received labeled endings (Figs. 3,4). Our objective was to determine whether low and high SR fibers contacted different sets of neurons. In fact, however, we discovered that each fiber type formed axosomatic synapses onto both globular bushy cells and stellate cells. Globular bushy cells were identified by an abundance of free cytoplasmic polysomes, an occasional small stack of rough endoplasmic reticulum, and a smooth nuclear membrane. Each globular bushy cell had more than 85% of its somatic surface apposed by synaptic endings, as determined by analyzing 3 representative sections through the cell body which included the nucleus. In contrast, stellate cells were characterized by multiple medium-to-large stacks of rough endoplasmic reticulum and an irregular nuclear membrane. Our cell type distinctions essentially confirm previously published criteria (Tolbert and Morest, '82). The stellate cell postsynaptic to the labeled high SR fiber had an average of 49.4 ± 11.2% of its somatic surface apposed to synaptic endings and that postsynaptic to the labeled low SR fiber had 41.1 ± 7.6%. These stellate cells with their well-innervated cell bodies more resembled the onset-chopper neurons than the sparsely innervated sustained-chopper neurons described in this same region of the nucleus (Smith and Rhode, '89).

We also reconstructed a number of postsynaptic targets of labeled terminals from the auditory nerve fiber having low SR in the small cell cap (see Fig. 12C of Osen, '69), but only some of these targets could be traced back to an identifiable cell body. This region contained three principal cell populations: Golgi, granule, and small cells. These cell types were distinguishable from each other by virtue of differences in their size, shape, and cytological characteristics. A thorough description of cochlear Golgi and granule cells is available in the literature (Mugnaini et al., '80). Briefly, granule cells have the smallest somata (8-10 μm in diameter) with a relatively pale-staining, thin rim of cytoplasm and a

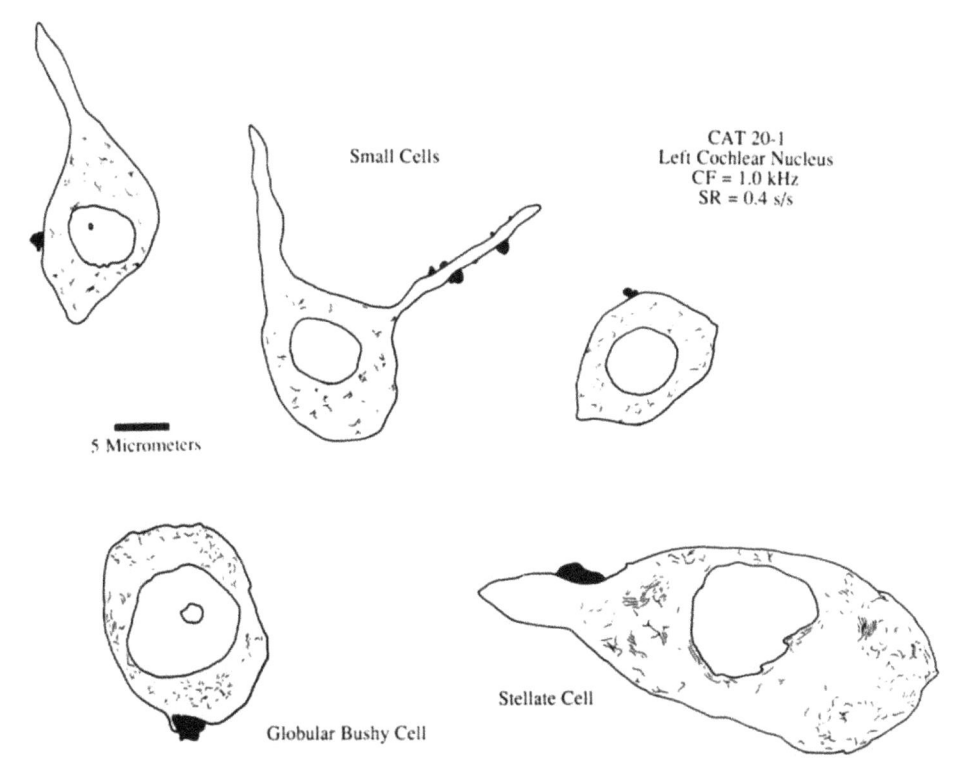

**Figure 3.** Tracing from low magnification (x6,000 total) electron micrograph of cells postsynaptic to labeled terminals of the low SR auditory nerve fiber. The drawings illustrate somatic shape, position and shape of nucleus, cytoplasmic pattern of endoplasmic reticulum, and locations of labeled terminals (indicated in black). Stellate cells were distinguishable from globular bushy cells by their large arrays of endoplasmic reticulum and irregular nuclear membrane.

sharply invaginated nuclear envelope. The granule cell somata and proximal dendrites are largely covered by glial processes. More distally, granule cell dendrites typically have a beaded appearance. Golgi cells have slightly larger somata (10-15 $\mu$m in diameter) with thick (5-8 $\mu$m), gnarled dendrites and short branches. Both the cell body and proximal dendrites are covered with numerous appendages. The somata of small cells are 15-20 $\mu$m in diameter, exhibit dispersed fragments of rough endoplasmic reticulum with no large organized arrays, and contain a centrally placed nucleus with a relatively smooth nuclear membrane. Each small cell gives rise to 4-5 primary dendrites which are characteristically slender and exhibit gentle undulations in thickness. There are few if any somatic or dendritic appendages, and 20-24% of their somata are apposed by small endings (<2 $\mu$m in diameter).

Collaterals of the low SR auditory nerve fiber formed many terminal swellings in this cap region (Fig. 1). Some of these were traceable to the somata of two small cells, whereas others made contact with the dendritic shaft of a third small cell (Fig. 3). This latter cell received seven synaptic swellings within a 12 $\mu$m length along its dendrite. There were labeled endings contacting other dendritic profiles in the neuropil but the cell body of origin could not be determined. The morphologically distinct Golgi cells and their dendrites were never in close apposition to labeled terminals. Likewise, identifiable portions of granule cells were not associated with labeled terminals. Although we could not distinguish between cross-sections of the distal dendrites of granule and small cells, the lack of labeled terminal interactions with glomerular complexes argues that small cells are the major, if not exclusive, recipient of the low SR auditory nerve fiber.

**Table II.** SUMMARY OF SYNAPTIC MORPHOLOGY IN THE INTERSTITIAL NUCLEUS AND SMALL CELL CAP

|  | Low SR Fibers | High SR Fibers | P Value |
|---|---|---|---|
| Number of Endings | 13 | 4 | - |
| Number of Synapses | 17 | 18 | - |
| Vesicle Diameter (nm) | 55.9 ± 11.2 | 54.6 ± 8.9 | n.s |
| Vesicles per $\mu m^2$ | 44.8 ± 15.3 | 46.5 ± 23.0 | n.s |
| Maximum PSD Length ($\mu$m) | 0.248 ± 0.12 | 0.246 ± 0.06 | n.s |
| Mitochondrial Fraction (% ending volume) | 15.4 ± 7.4 | 24.6 ± 7.5 | $p < 0.05$ |

PSD: Postsynaptic density

Within the limits of the methods, we can conclude that both low and high SR fibers make synaptic contact with the cell bodies of globular bushy cells and stellate cells in the region of the auditory nerve root. Several labeled endings made synaptic contact with dendrites of unknown origin in the neuropil, but these profiles are presumably of bushy or stellate cell origin because they are the principal cell types in the region. Since the low SR fiber made additional synaptic contact with small cells, it would appear that small cells have a synaptic relation with low SR fibers which is not shared with high SR fibers.

Our observations that primary endings terminate upon the cell bodies of globular bushy and stellate cells are directly relevant to a model of rate-place representations in the auditory nerve (Winslow et al., '87). The general idea is that at low stimulus levels, the high SR, low threshold fibers are important carriers of information into the brain, whereas at high stimulus levels or in the presence of masking noise, low SR, high threshold fibers would play a major role in sound perception. In this context, a simple neural circuit was hypothesized to account for level-dependent weighting of auditory nerve fiber rate responses. Essentially, the model assumes that high SR fibers with CFs near stimulus frequency form excitatory synapses at distal locations of the dendritic tree of a second order neuron, and that low SR fibers with similar CFs synapse closer to the soma. At low stimulus levels, the rate response of the second order neuron would be qualitatively similar to those of high SR fibers because the low SR fibers would be below threshold. At high stimulus levels, the synaptic current from the more proximal endings of low SR fibers would drive the cell. Chopper units of the ventral cochlear nucleus, known to be the physiological equivalent of stellate neurons (Rhode et al., '83; Rouiller and Ryugo, '84), are the proposed candidates for this selective weighting function (Winslow et al., '87).

On the basis of our data, however, it seems that a differential distribution of primary endings onto separate compartments of the same cell does not occur with respect to fiber SR. That is, endings from both high and low SR fibers synapse upon the cell body of the stellate cell. On the other hand, because we have examined only one low SR fiber and one high SR fiber from a single cat, it is premature to draw strict conclusions. If one considers the differential distribution of SR related inputs onto second order neurons to be a probability function rather than an absolute function, then the model of Winslow et al. ('87) may still be valid. Certainly, the cells postsynaptic to our labeled endings also received input from unlabeled endings exhibiting the morphologic features of primary fibers. Unfortunately, it could not be determined whether such endings arose from high or low SR fibers.

The ability to identify primary endings with respect to fiber SR group, without having to first intracellularly record from and stain them, would be very useful in determining the complement of primary endings onto postsynaptic cell classes, especially with respect to the issues posed above. Such information would also be useful in answering questions such as whether some cell classes in the cochlear nucleus receive little or no input from endings of a particular SR group, and whether some other cell classes receive a relatively equal amount

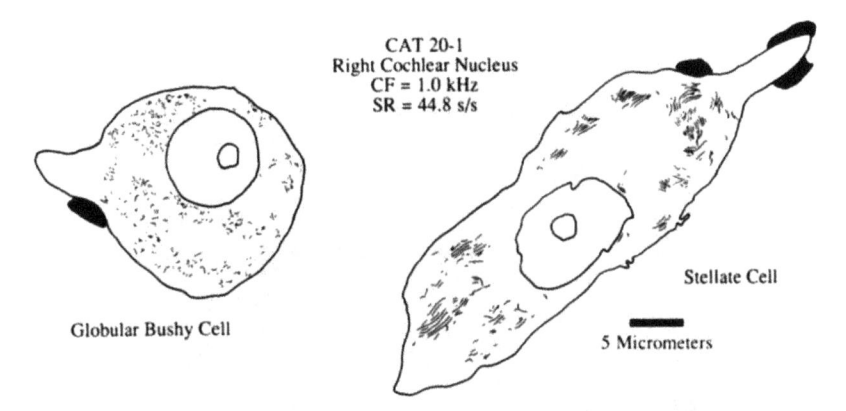

**Figure 4.** Tracing from low magnification(x6,000 total) electron micrograph of cells postsynaptic to labeled terminals of the high SR auditory nerve fiber. The drawings illustrate the locations of labeled endings (indicated in black) and structural differences between the somata of stellate and globular bushy cells.

of input from both SR groups. Because answers to these kinds of questions are of interest to us, we sought distinctive morphologic markers for the endings of different SR fibers by examining the fine structure of labeled endings.

Assessing the fine structure of endings for the separate SR groups using electron microscopy is not trivial because of the often large distances between swellings and because within any single swelling the density of synapses, synaptic vesicles, and mitochondria is not uniform. For example, some swellings were filled with synaptic vesicles, but there were no mitochondria and no apposing postsynaptic densities. Others, however, contained mitochondria and synaptic vesicles, and apposed one or several postsynaptic densities. Morphometric analysis was conducted on coded micrographs and was restricted to endings containing at least one synapse.

The synaptic endings of primary fibers are qualitatively similar to each other, irrespective of fiber SR or location (Table II). Endings contained numerous clear, round synaptic vesicles and mitochondria, and formed asymmetric synapses which were characterized by a prominent postsynaptic density and one or more synaptic vesicles located within a distance of their diameters to the membrane specialization (Fig. 5). The endings of high SR fibers contained vesicles having an average ($\pm$ SD) diameter of 54.6 $\pm$ 8.9 nm (n=505), compared to 55.9 $\pm$ 11.2 nm (n=420) for endings of low SR fibers. These vesicles had a similar packing density where the average number of synaptic vesicles per $\mu m^2$ was 46.5 $\pm$ 23.0 for the high SR fiber and 44.8 $\pm$ 15.3 for the low SR fiber. There was no statistical difference between the two groups of endings with respect to the length of the postsynaptic densities: 0.246 $\pm$ 0.06 $\mu$m for the high SR fiber and 0.248 $\pm$ 0.12 $\mu$m for the low SR fiber.

There were, however, some potentially important differences between the endings of the separate SR groups. On average, the mitochondria within terminals of the high SR fiber occupied a greater proportion of the terminal (24.6 $\pm$ 7.5%) than did those of the low SR fiber (15.4 $\pm$ 7.4%; $p<0.05$). There was also a difference in the number of synapses formed by individual endings: 4.5 synapses per ending for the high SR fiber and 1.3 per ending for the low SR fiber. These morphologic distinctions would suggest that the increased mitochondria provide energy for the higher levels of activity in high SR fibers, and that a key feature of the endings is in the number and not the appearance of synapses.

More data are needed in order to confirm these morphometric values, and there are certain limitations inherent to the methods that serve to temper the conclusions. For example,

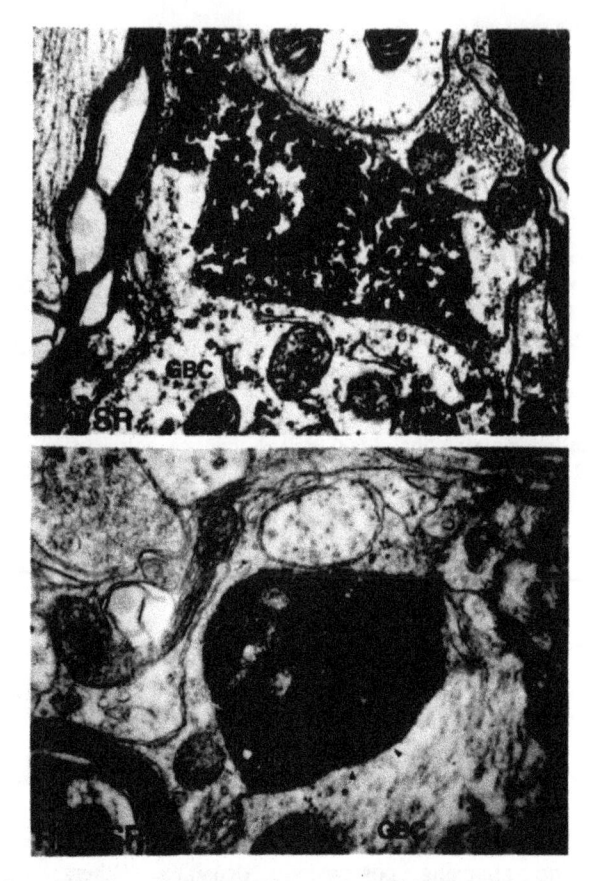

**Figure 5.** Electron micrographs through representative sections of labeled endings contacting the somata of globular bushy cells (GBC). The ending in the upper panel is from the low SR fiber which is synapsing on a somatic spine (arrowheads). The ending in the lower panel is from the high SR fiber which forms a synapse at the junction between the cell body and primary dendrite (arrowheads). Mitochondria, postsynaptic densities, and the lumina of synaptic vesicles are visible. Note the uniform diameter of the synaptic vesicles in both endings and the higher density of mitochondria in the ending of the high SR fiber.

due to the difficult nature of intracellular recording and staining techniques coupled to the labor-intensive serial section electron microscopy, the sample size is relatively small. Furthermore, this sample represents only a fraction of the total complement of endings generated by each fiber. Second, the HRP method can itself introduce a problem when on occasion, the reaction product is so dense that the entire ending is opaque and organelles are obscured. Finally, due to the small size of some of the endings, curvature of the ending membranes within the tissue section often smears the postsynaptic densities, thereby diminishing their visibility. Despite these challenges, our data nevertheless represent an important beginning for morphometric inquiry into the synaptic organization of auditory nerve projections to the cochlear nucleus. These kinds of quantitative studies are necessary if we are to understand further the intricate relationship between structure and function in the biology of hearing.

ACKNOWLEDGEMENTS

This work was supported in part by NIH grant R01 DC00232. Special thanks are due to Dr. Paul B. Manis for presenting the data in this report to the participants of the NATO Workshop, substituting for the primary author who was home helping his wife, Karen, deliver their 3rd son, Nicholas Max. Thanks also to our colleagues of the Center for Hearing Sciences for their helpful discussions of the data.

REFERENCES

Anderson, D.J., Rose, J.E. and Brugge, J.F., 1971, Temporal position of discharges in single auditory nerve fibers within the cycle of a sine-wave stimulus: frequency and intensity effects, *J. Acoust. Soc. Am.*, 49:1131-1139.

Berglund, A.M. and Ryugo, D.F., 1987, Hair cell innervation by spiral ganglion neurons in the mouse, *J. Comp. Neurol.*, 255:560-570.

Berglund, A.M. and Ryugo, D.K., 1991, Neurofilament antibodies and spiral ganglion neurons of the mammalian cochlea, *J. Comp. Neurol.*, 308:209-223.

Brown, M.C., 1987, Morphology of labeled afferent fibers in the guinea pig cochlea, *J. Comp. Neurol.*, 260:591-604.

Brown, M.C., Berglund, A.M., Kiang, N.Y.S. and Ryugo, D.K., 1988, Central trajectories of type II spiral ganglion neurons, *J. Comp. Neurol.*, 278:581-590.

Evans, E.F., and Palmer, A.R., 1980, Relationship between the dynamic range of cochlear nerve fibers and their spontaneous activity, *Exp. Brain Res.*, 40:115-118.

Fekete, D.M., Rouiller, E.M., Liberman M.C. and Ryugo, D.K., 1984, The central projections of intracellularly labeled auditory nerve fibers in cats, *J. Comp. Neurol.*, 229:432-450.

Kellerhals, B., Engström, H. and Ades, H.W., 1967, Die Morphologie des Ganglion spirale Cochleae, *Acta Otolaryngol. Suppl.*, 226:6-33.

Kiang, N.Y.S., Watanabe, T., Thomas, L.C. and Clark, L.F., 1965, Discharge Patterns of Single Fibers in the Cat's Auditory Nerve, MIT Press, Cambridge.

Kiang, N.Y.S., Rho, J.M., Northrup, C.C., Liberman, M.C. and Ryugo, D.K., 1982, Hair-cell innervation by spiral ganglion cells in adult cats, *Science*, 217:175-177.

Liberman, M.C., 1978, Auditorynerve response from cats raised in a low-noise chamber, *J. Acoust. Soc. Am.*, 53:442-455.

Liberman, M.C., 1982, Single-neuron labeling in the cat auditory nerve, *Science*, 216:1239-1241.

Liberman, M.C., 1991, Spatial segregation of auditory-nerve projections in the cochlear nucleus according to spontaneous discharge rates, *Abstr. Assoc. Res. Otolaryngol.*, 14:42.

Liberman, M.C., Dodds, L.W. and Pierce, S., 1991, Afferent and efferent innervation of the cat cochlea: Quantitative analysis with light and electron microscopy, *J. Comp. Neurol.*, 301:443-460.

Lorente de Nó, R., 1933, Anatomy of the eighth nerve. III. General plan of structure of the primary cochlear nuclei, *Laryngoscope*, 43:327-350.

Mugnaini, E., Osen, K.K., Dahl, A., Friedrich Jr., V.L. and Korte, G., 1980, Fine structure of granule cells and related interneurons (termed Golgi cells) in the cochlear nuclear complex of cat, rat and mouse, *J. Neurocytol.*, 9:537-570.

Osen, K.K., 1969, Cytoarchitecture of the cochlear nuclei in the cat, *J. Comp, Neurol.*, 136:453-484.

Pfeiffer, R.R., 1966, Classification of response patterns of spike discharges for units in the cochlear nucleus: tone-burst stimulation, *Exp. Brain Res.*, 1:220-235.

Rhode, W.S., Oertel, D. and Smith, P.H., 1983, Physiological response properties of cells labeled intracellularly with horseradish peroxidase in cat ventral cochlear nucleus, *J. Comp. Neurol.*, 213:448-463.

Robertson, D., 1984, Horseradish peroxidase injection of physiologically characterized afferent and efferent neurones in the guinea pig spiral ganglion, *Hearing Res.*, 15:113-121.

Rouiller, E.M., Cronin-Schreiber, R., Fekete, D.M. and Ryugo, D.K., 1986, The central projections of intracellularly labeled auditory nerve fibers in cats: An analysis of terminal morphology, *J. Comp. Neurol.*, 249:261-278.

Rouiller, E.M. and Ryugo, D.K., 1984, Intracellular marking of physiologically characterized neurons in the ventral cochlear nucleus of the cat, *J. Comp. Neurol.*, 225:167-186.

Ryugo, D.K. and Rouiller, E.M., 1988, The central projections of intracellularly labeled auditory nerve fibers in cats: Morphometric correlations with physiological properties, *J. Comp. Neurol.*, 271:130-142.

Ryugo, D.K., Dodds, L.W., Benson, T. and Kiang, N.Y.S., 1991, Unmyelinated axons of the auditory nerve in cats, *J. Comp. Neurol.*, 308:209-223.

Ryugo, D.K. and Sento, S., 1991, Synaptic connections of the auditory nerve in cats: Relationship between endbulbs of Held and spherical bushy cells, *J. Comp. Neurol.*, 305:35-48.

Sachs, M.B. and Abbas, P.J., 1974, Rate versus level functions for auditory nerve fibers in cats: Tone-burst stimulation, *J. Acoust. Soc. Am.*, 56:1835-1847.

Sento, S. and Ryugo, D.K., 1989, Endbulbs of Held and spherical bushy cells in cats: Morphological correlates with physiological properties, *J. Comp. Neurol.*, 280:553-562.

Smith, P.H. and Rhode, W.S., 1989, Structural and functional properties distinguish two types of multipolar cells in the ventral cochlear nucleus, *J. Comp. Neurol.*, 282:595-616.

Spoendlin, H., 1973, The innervation of the cochlear receptor, *in*: Mechanisms in Hearing, A.R.Møller, ed., Academic Press, New York, pp. 185-229.

Tolbert, L.P. and Morest, D.K., 1982, The neuronal architecture of the anteroventral cochlear nucleus of the cat in the region of the cochlear nerve root: Electron microscopy, *Neuroscience*, 7:3053-3067.

Winslow, R.L., Barta, P.E. and Sachs, M.B., 1987, Rate coding in the auditory-nerve, *in*: Auditory Processing of Complex Signals, W.A.Yost and C.S.Watson, eds., Lawrence Erlbaum Associates, Publishers, Hillsdale, pp. 212-224.

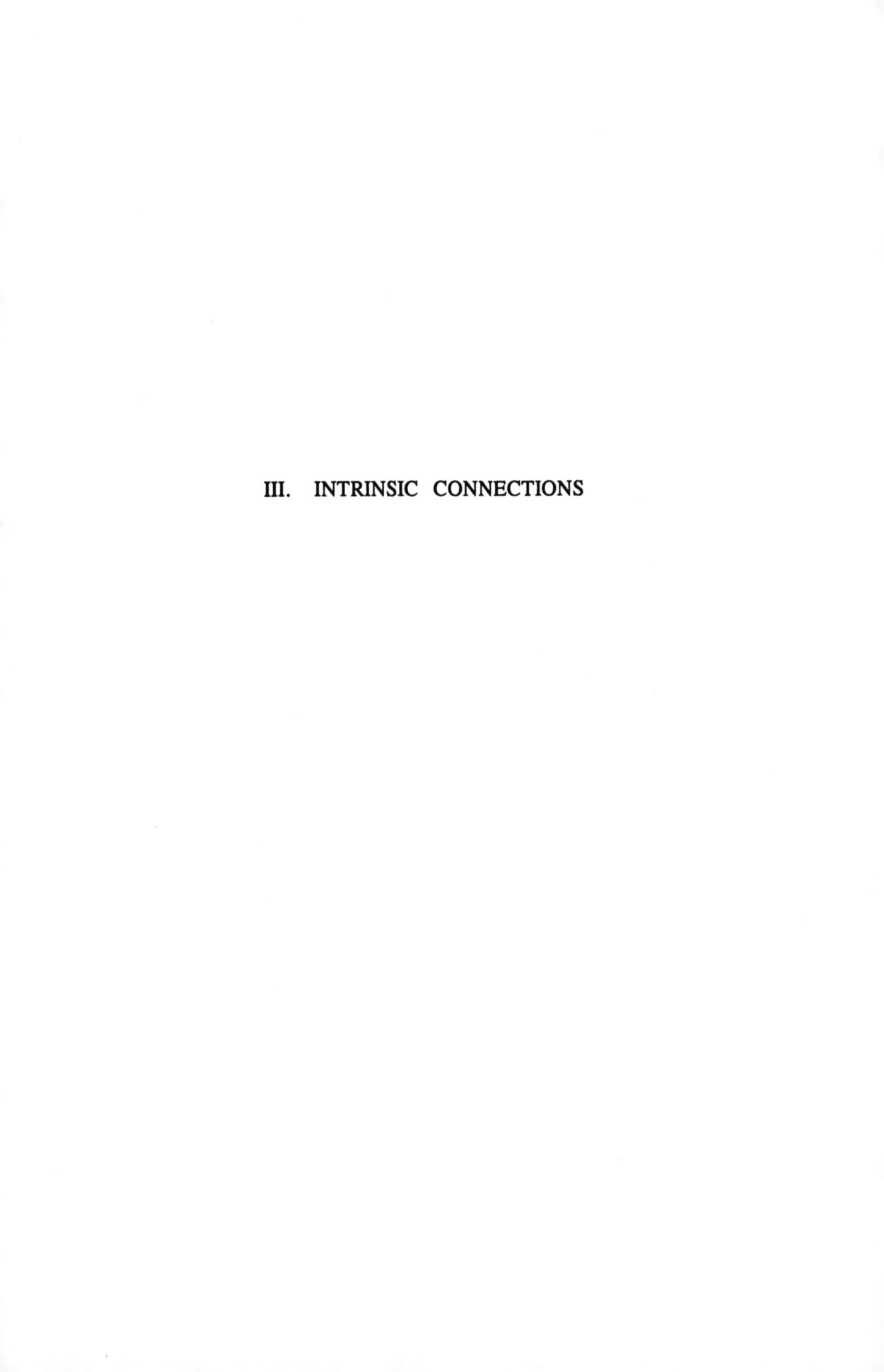

# III. INTRINSIC CONNECTIONS

# INTRINSIC CONNECTIONS IN THE COCHLEAR NUCLEAR COMPLEX STUDIED *IN VITRO* AND *IN VIVO*

Robert E. Wickesberg[1,2] and Donata Oertel[1]

[1]Department of Neurophysiology, University of Wisconsin - Madison
Madison, Wisconsin 53706, USA
[2]Hearing Research Laboratory, S.U.N.Y. at Buffalo
Buffalo, New York 14214, USA

The lateral ventrotubercular tract was described first by Lorente de Nó ('33, '81) in Golgi studies of cat and mouse cochlear nuclei. This tract connects the dorsal (DCN) and the ventral (VCN) cochlear nucleus. The projection from neurons in the deep layer of the DCN to the VCN has been found in bats (Feng and Vater, '85), mice (Wickesberg and Oertel, '88; Wickesberg et al., '91), cats (Snyder and Leake, '88), gerbils (Müller, '90) and guinea pigs (Saint-Marie et al., '91). Because these neurons project from the *tuberculum acusticum* (Cajal, '09) to the VCN, they have been called tuberculoventral cells (Oertel and Wu, '89). This paper reviews the findings of our *in vitro* studies on the anatomy and physiology of this projection, examines the hypothesis that this intrinsic circuit contributes to the monaural suppression of echoes, and presents preliminary results from *in vivo* experiments that begin to examine this hypothesis.

## A topographic projection from the DCN to the VCN

The first indication that the projection from the DCN to the VCN is topographic came from the results of extracellular horseradish peroxidase (HRP) injections in bats (Feng and Vater, '85). In mice, a series of extracellular injections of HRP in the anteroventral cochlear nucleus (AVCN; Fig. 1A) and the rostroventral portion of the posteroventral cochlear nucleus (PVCN; Fig. 1C) labeled cell bodies in the deep layer of the DCN (Wickesberg and Oertel, '88; Wickesberg et al., '91). These injections also labeled a group of auditory nerve fibers that passed through the injection site. Most of the labeled tuberculoventral neurons were located across the rostrocaudal extent of the deep DCN in the dorso-ventral center of the labeled band of auditory nerve fiber terminals. A few labeled neurons were found ventral to this band. As the site of injection was moved from dorsal to ventral in the VCN, the band of labeled auditory nerve fibers and tuberculoventral neurons marched ventrally in the DCN following the topographic organization of the nucleus.

Within the resolution of the anatomical experiments, the projection from the DCN to the VCN is frequency-specific. The pattern of HRP labeling in the DCN indicates that labeled tuberculoventral neurons found within the band of labeled auditory nerve fibers

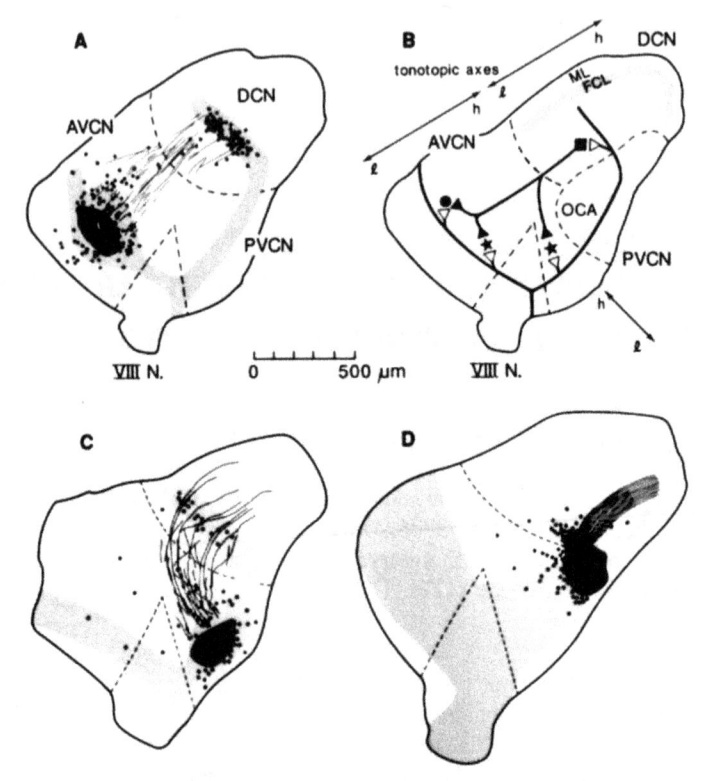

**Figure 1.** *Camera lucida* reconstructions of slices with extracellular HRP injections in (A) AVCN, (C) the multipolar cell area in the rostro-ventral PVCN, and (D) the octopus cell area in PVCN. Each sketch shows the locations of the injection site (black), labeled auditory nerve fibers (stippling), labeled cell bodies (dots), and labeled axons (lines). (B) Summary diagram of the projection from the DCN to the VCN. Tuberculoventral neurons inhibit bushy (dot) and stellate cells (stars) in the AVCN and the rostro-ventral PVCN, but they do not innervate the octopus cell area.

terminate within the injection site. Tuberculoventral neurons and their targets in the VCN are, therefore, innervated by the same auditory nerve fibers and presumably respond to the same frequencies *in vivo* (Wickesberg and Oertel, '88).

Tuberculoventral neurons do not innervate neurons in the octopus cell area (Wickesberg et al., '91). Following extracellular injections of HRP in the octopus cell area (Fig. 1D), labeled neurons are not found in the band of labeled auditory nerve fiber terminals in the deep DCN. A few labeled neurons are observed in the DCN near the border with the VCN close to the injection site. Since the HRP injection spread into the DCN, it is uncertain whether these neurons project to the octopus cell area. Large HRP injections that included both the octopus cell and the multipolar cell areas, but did not spread into the DCN, did not label cells outside the band of labeled auditory nerve fibers (Wickesberg et al., '91).

The structure of the dendritic trees and axonal arbors of tuberculoventral neurons (Fig. 2) is revealed by intracellular labeling with HRP (Oertel and Wu, '89). In mice, the dendritic trees extend horizontally along an approximate isofrequency contour in the DCN, as well as vertically (Oertel and Wu, '89; Wickesberg et al., '91). These neurons were named vertical or corn cells by Lorente de Nó ('33, '81) because of the characteristic vertical orientation of their dendrites, which in the cat rarely have horizontal branches. Each tuberculoventral neuron innervates a long stretch of an isofrequency contour in the VCN (Wickesberg and Oertel, '88; Oertel and Wu, '89). This stretch of isofrequency contour can be in the AVCN, the multipolar cell area of the PVCN, or in both. The axon of a tuberculoventral neuron also

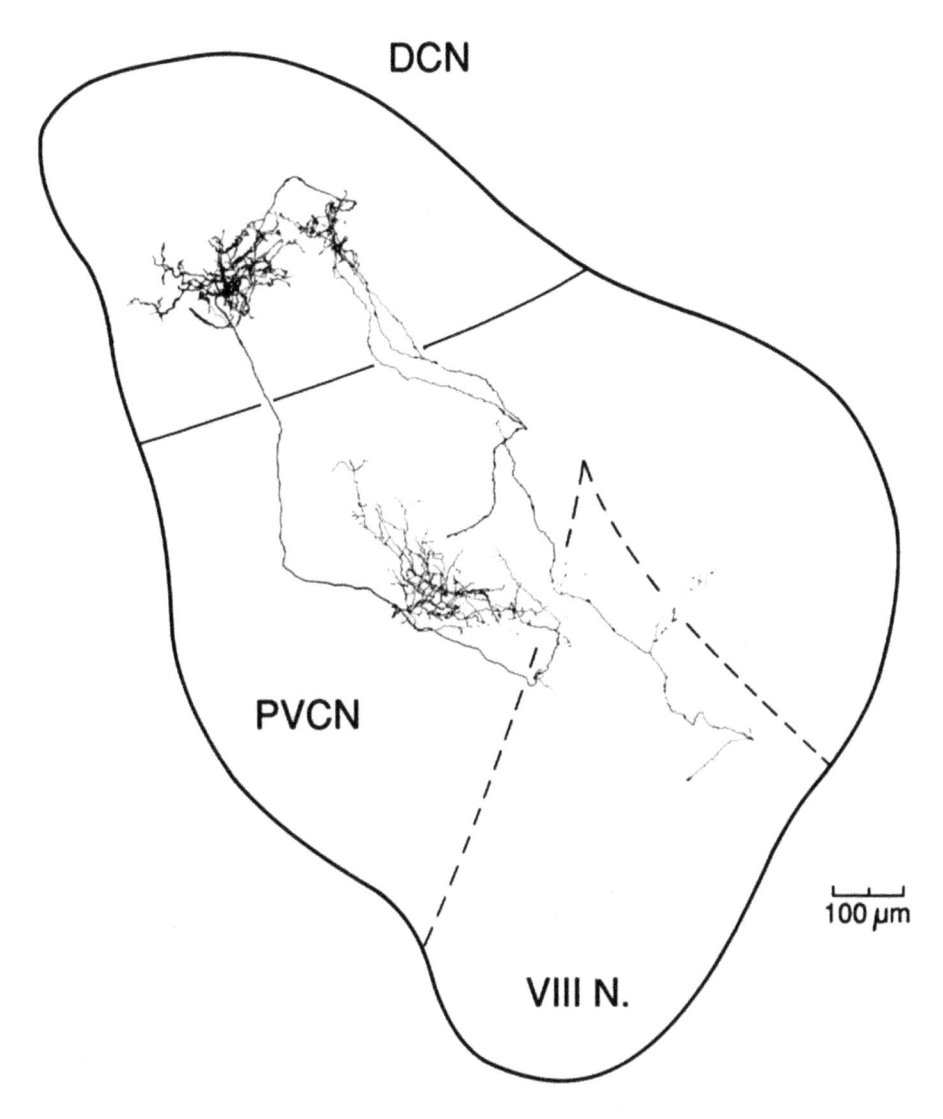

**Figure 2.** A tuberculoventral neuron in the deep layer of the dorsal cochlear nucleus that is labeled following an intracellular injection of HRP. The dendrites of this neuron lie within a 70 μm band in the DCN. The axon branches and innervates approximately the same isofrequency band in the DCN, a long stretch of the PVCN, and a bit of the AVCN (from Oertel and Wu, '89).

has local collaterals that terminate in the DCN approximately within the isofrequency lamina that contains the cell body and dendrites. A summary diagram of the proposed circuitry is shown in Figure 1B.

## The projection from DCN to VCN is inhibitory

Tuberculoventral neurons inhibit their targets in the VCN (Wickesberg and Oertel, '90). Figure 3 shows the results of activating tuberculoventral neurons while recording intracellularly from a target neuron in the VCN. A micropipette filled with 20 mM glutamate

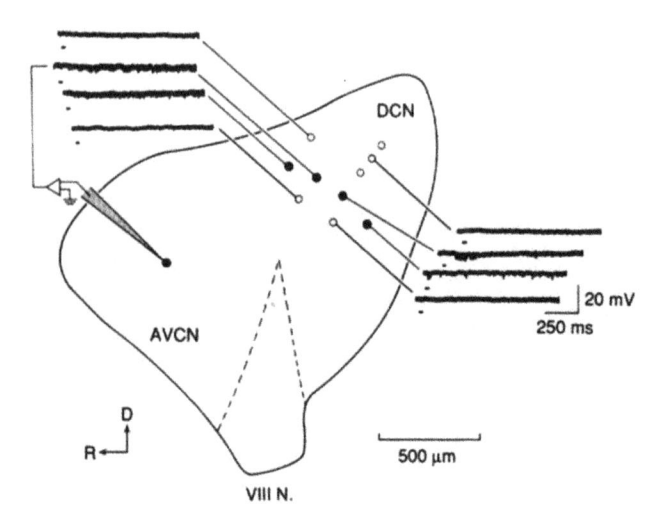

**Figure 3.** Afferent field in the DCN for a stellate cell in AVCN. The traces show the intracellular records from a stellate cell located in AVCN (*filled circle*) following microinjections of 4 mM kainate at different locations in the DCN. The bar below each trace shows the time of the microinjection of kainate. At the locations marked with the *filled circles*, the microinjection of kainate resulted in trains of IPSPs. These locations are in a rostrocaudal band across the DCN. Dorsal and ventral to this band, at the locations shown with open circles, microinjections of kainate produced no intracellular responses. This neuron was classified as a stellate cell on the basis of its regular firing in response to depolarizing current, and had a resting potential of -62 mV.

or 4 mM kainate was inserted into the DCN. Microinjections of an excitatory amino acid excited cell bodies and dendrites of neurons without activating neighboring axons. In each VCN neuron tested, microinjections along a rostrocaudal band in the DCN produced a barrage of inhibitory postsynaptic potentials (IPSPs). The finding that IPSPs sum to produce a steady hyperpolarization indicates that many inhibitory neurons converge on a target cell in the VCN. Dorsal and ventral to this band, microinjections produced no response. Analogous to a "receptive field" for a sensory neuron, this area is called an "afferent field". An afferent field closely resembles the band of DCN neurons labeled by an HRP injection in AVCN. These afferent fields follow the tonotopic organization of the cochlear nuclear complex (Wickesberg and Oertel, '90). The simplest interpretation of these results is that tuberculoventral neurons inhibit cells in the VCN and that each target neuron in the VCN receives inputs from multiple tuberculoventral neurons along an isofrequency lamina in the DCN.

The timing of the inhibition delivered to the AVCN from the DCN relative to the excitation from the auditory nerve can be measured by stimulating the ascending and descending branches of auditory nerve fibers independently while recording intracellularly from neurons in the AVCN (Wickesberg and Oertel, '90). The excitatory synaptic response to activation of the ascending branch has a latency of 0.6 msec. Stimulating the descending branch produces an IPSP in the AVCN neuron that begins about 2.5 msec. after the shock and lasts for about 4 msec. Following a shock to the auditory nerve, the median latency to the start of the excitatory postsynaptic potential for neurons in the DCN is 1.2 msec. (Hirsch and Oertel, '88) and the peak of an action potential occurs 0.3 msec. later. These latencies leave about 1 msec. for the action potential of a tuberculoventral neuron to reach the AVCN and produce an IPSP. This timing indicates that the projection from tuberculoventral cells to targets in the VCN is likely to be monosynaptic.

## Targets of tuberculoventral neurons

Bushy cells in the AVCN are innervated by tuberculoventral neurons (Wickesberg and Oertel, '90). There are two subclasses of bushy cells: spherical and globular (Osen, '69). In brain slices from mice, the targets of tuberculoventral neurons identified physiologically as bushy cells are probably globular bushy cells. Recent evidence of Winter and Palmer (Winter and Palmer, '90) in guinea pigs indicates that tuberculoventral neurons might also terminate on spherical bushy cells, as they reported that units with prepotentials have narrowly-tuned on-frequency inhibition. Bushy cells are specialized to preserve the temporal firing patterns of their inputs from auditory nerve fibers (Kiang et al., '65a; Pfeiffer, '66; Bourk, '76; Rhode et al., '83a; Wu and Oertel, '84; Oertel, '85). Spherical bushy cells respond to tones with "primary-like" peristimulus time histogram patterns while globular bushy cells respond with "primary-like-with-notch" patterns (Pfeiffer, '66; Kiang et al., 73; Bourk, '76; Rhode et al., '83a; Smith and Rhode, '89; Spirou et al., '90). The targets of bushy cells are neurons in the superior olivary complex where interaural timing and intensity comparisons are made (Harrison and Warr, '62; Tolbert et al., '82; Warr, '82; Cant and Casseday, '86).

Stellate cells, a heterogeneous group of neurons (Brawer et al., '74; Cant, '81; Smith and Rhode, '89), are also targets of tuberculoventral neurons (Wickesberg and Oertel, '90). Subclasses of stellate cells have been distinguished anatomically (Cant, '81; Smith and Rhode, '89; Oertel et al., '90) and physiologically (Bourk, '76; Young et al., '88; Blackburn and Sachs, '89). Tuberculoventral neurons inhibit VCN neurons with the same intrinsic electrical characteristics as the "T-stellate" cells that send an axon collateral out the ventral acoustic stria (Wickesberg and Oertel, '90; Oertel et al., '90) to innervate the inferior colliculi (Adams, '79; Ryugo et al., '81; Cant, '82; Oliver, '87), as well as the DCN (Adams, '83; Oertel et al., '90). Stellate cells integrate information from auditory nerve fibers spatially and temporally (Molnar and Pfeiffer, '68; Rhode et al., '83a; Wu and Oertel, '84; Oertel, '85). T-stellate cells respond to tones with "chopper" response patterns (Rhode et al., '83a; Rouiller and Ryugo, '84; Smith and Rhode, '89; Oertel et al., '90). Whether tuberculoventral neurons innervate all physiological or anatomical subclasses of stellate cells is not known.

Fusiform cells in the DCN, which project to the contralateral inferior colliculus (Adams, '79; Ryugo et al., '81; Oliver, '84), are likely to be the targets of the local collaterals of axons from tuberculoventral neurons (Oertel and Wu, '89; Caspary, '72; Godfrey et al., '75b;). In response to tones, fusiform cells have "pauser/buildup" histograms (Rhode et al., '83b) and correspond to the type IV neurons of Young and Brownell (Young and Brownell, '76; Young, '80). Type IV neurons are inhibited by type II neurons with similar characteristic frequencies (Voigt and Young, '80, '90). Many tuberculoventral neurons have type II characteristics (Young, '80) suggesting that the interaction between type IV and type II cells reflects the inhibition of fusiform cells by tuberculoventral neurons. There is a possibility, however, that not all type II cells are tuberculoventral neurons. Only about 20% of type II neurons could be activated antidromically from the VCN (Young, '80), Lorente de Nó reported that not all of the vertical cells in the deep DCN of the cat project to the VCN (Lorente de Nó, '81). Because about 10% of type III neurons could be antidromically driven from the VCN, not all tuberculoventral neurons are of type II (Young, '80).

Two pieces of evidence indicate that octopus cells are not targets of tuberculoventral neurons. First, extracellular HRP injections in the octopus cell area (Fig. 1D) do not label tuberculoventral neurons. Second, immunolabeling of the cochlear nuclear complex of the mouse with an antibody to glycine (Wenthold et al., '87), the putative neurotransmitter of tuberculoventral neurons (Wickesberg and Oertel, '90; Oertel and Wickesberg, this volume), resulted in immunoreactive puncta throughout the VCN except in the octopus cell area (Wickesberg et al., '91). The paucity of glycine-like immunolabeling in the octopus cell area is consistent with the low level of strychnine binding observed in the octopus cell area of the mouse (Frostholm and Rotter, '85).

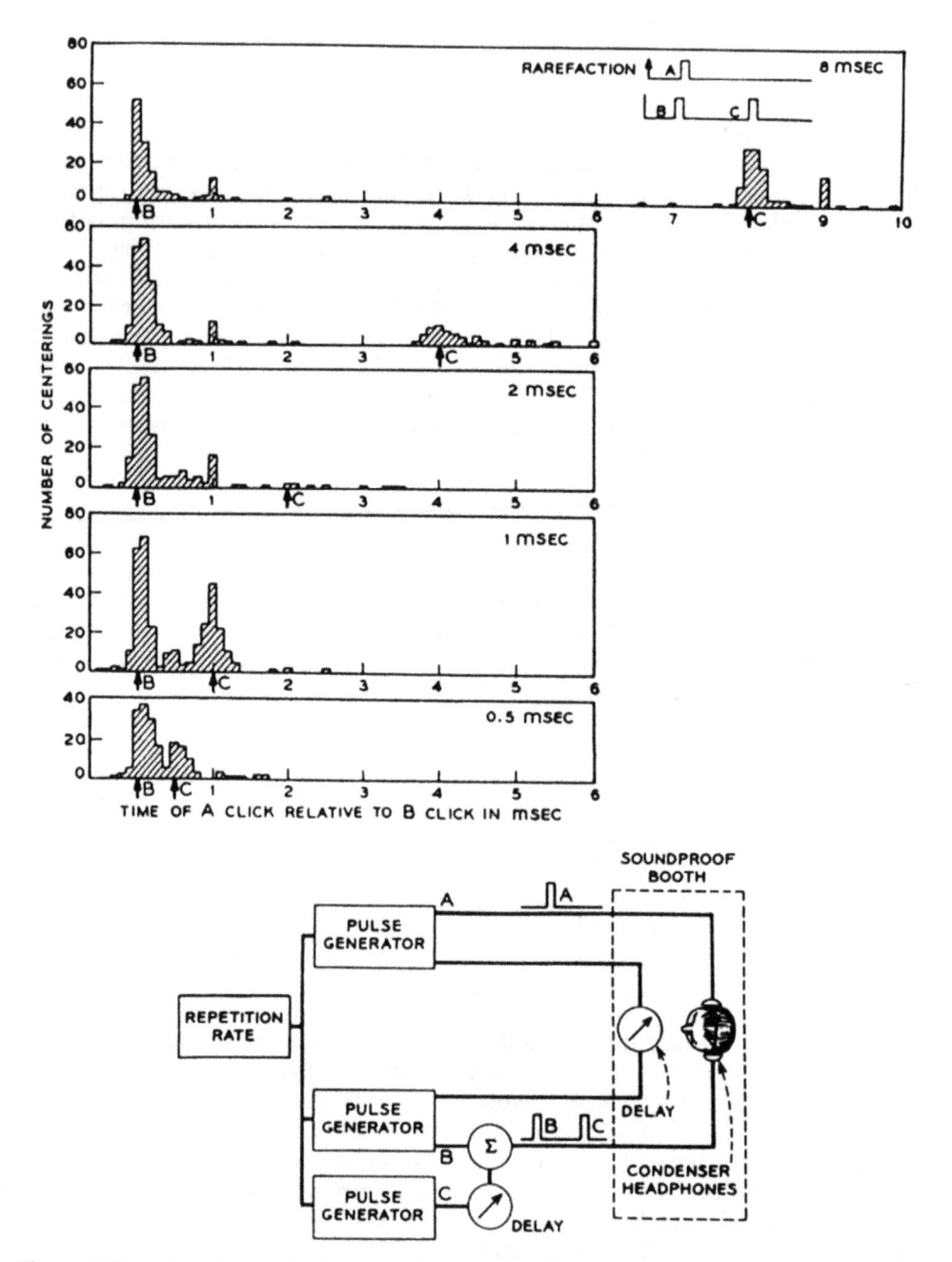

Figure 4. The timing of monaural echo suppression as measured in a psychoacoustic experiment on laterization in humans. The experimental setup is shown below the data. Two clicks, B and C, are presented to one ear (at times indicated by arrows) while a roving click is presented to the opposite ear. All clicks are 70 $\mu$sec, rarefaction impulses, presented 40 dB above threshold. The subject was asked to adjust the time of occurrence of click A to center the image. The delay between clicks B and C is varied between 0.5 and 8 msec. The histograms at the left show the distributions of the timing of A relative to B for five different separations between clicks B and C. Zero time corresponds to arrow B. (Figure is a combination of figures 1 and 2 in the article by Drs. G.G. Harris, J.L. Flanagan, and B.J. Watson that appeared in 1963 in J.A.S.A. 35:672-678 and it is reproduced with the permission of Dr. J.L. Flanagan).

## How might tuberculoventral neurons contribute to the processing of information in the VCN?

The results from *in vitro* experiments indicate that tuberculoventral neurons provide a delayed, frequency-specific inhibition to neurons in the VCN. The latency of the inhibition indicates that maximal suppression should occur in the targets of tuberculoventral neurons for the signals encoding sounds arriving 2 msec. after an acoustic event that results in the activation of tuberculoventral neurons. For intervals less than approximately 2 msec. the signals produced by a trailing sound should not be suppressed. The duration of the inhibition depends on the temporal firing patterns of the tuberculoventral neurons produced by the acoustic event. For stimuli that activate tuberculoventral cells within an isofrequency lamina synchronously, the inhibition lasts about 4 msec. Only those trailing sounds with frequency spectra similar to that initial acoustic event would elicit signals that are subject to suppression by the circuit through the DCN.

Echoes are one class of trailing sounds with the same frequency components as an initial acoustic event. The mammalian auditory system appears to ignore echoes resulting from reflections of the sound from nearby objects in localization tasks. Echoes arriving between 1 and 30 msec. after the primary sound have little influence on the perceived location of the source (Haas, '51; Wallach et al., '49), although these echoes do contribute to the "spaciousness" of the sound (Blauert, '83). The dominance of the directional cues for the primary sounds over cues from echoes in sound localization is known as the precedence effect. The precedence effect requires the suppression of the information carried by echoes for the localization of a sound source (Zurek, '87). The suppression is both monaural and binaural (McFadden, '73; Yost and Hafter, '87).

The time course of monaural echo suppression has been measured in several human psychoacoustic experiments. Harris *et al.* ('63) presented pairs of clicks to one ear and asked listeners to adjust the timing of a roving click to the opposite ear so that the image of the sound source was centered, as shown in Figure 4. When the clicks in the pair were 0.5, 1, 4 or 8 msec. apart the roving click was adjusted by the listeners to occur simultaneously with either click, but when the click interval was 2 msec., the roving click was adjusted to occur simultaneously with only the first click, not with the second. Thus with a 2 msec. interval between clicks, the trailing click appeared to be suppressed. Zurek ('80) found that interaural time and intensity sensitivity was degraded for a period from approximately 0.5 to 10 msec. after the onset of a 1 msec. noise burst, with the lowest sensitivity at delays of 2 to 3 msec., while Gaskell ('83) found the lowest sensitivity occurred about 1 msec. after a 20 $\mu$sec click. Hafter and coworkers (Hafter and Dye, '83; Hafter et al., '88) measured the detection thresholds of interaural time differences for trains of rectangular clicks as a function of the number of clicks and the interclick interval. Thresholds were greatest with interclick intervals of 1 and 2 msec., reflecting an apparent suppression of later clicks in a train at these intervals. Thresholds were lowest with interclick intervals of 5 msec. or longer. The timing of monaural echo suppression measured in these psychoacoustic experiments is almost identical to the timing of the inhibition from the DCN to the VCN measured *in vitro* (Wickesberg and Oertel, '90).

Echo suppression is frequency-specific; it occurs within, but not across, frequency bands. Hafter and his colleagues (Hafter and Wenzel, '83; Hafter et al., '88) measured detection thresholds for interaural time differences using trains of band-pass filtered clicks. If the same center frequency was used for the filtering of all clicks in a train, then with a 2.5 msec. interval between clicks the detection thresholds were higher than with a 5 msec. interclick interval. However, if two different center frequencies were used for alternate clicks in the train, then the detection thresholds obtained with 2.5 msec. and 5 msec. intervals were the same, as if no suppression occurred. Similarly, Divenyi and Blauert ('87) found that echo suppression occurred only when the primary sound and the echo had overlapping spectra.

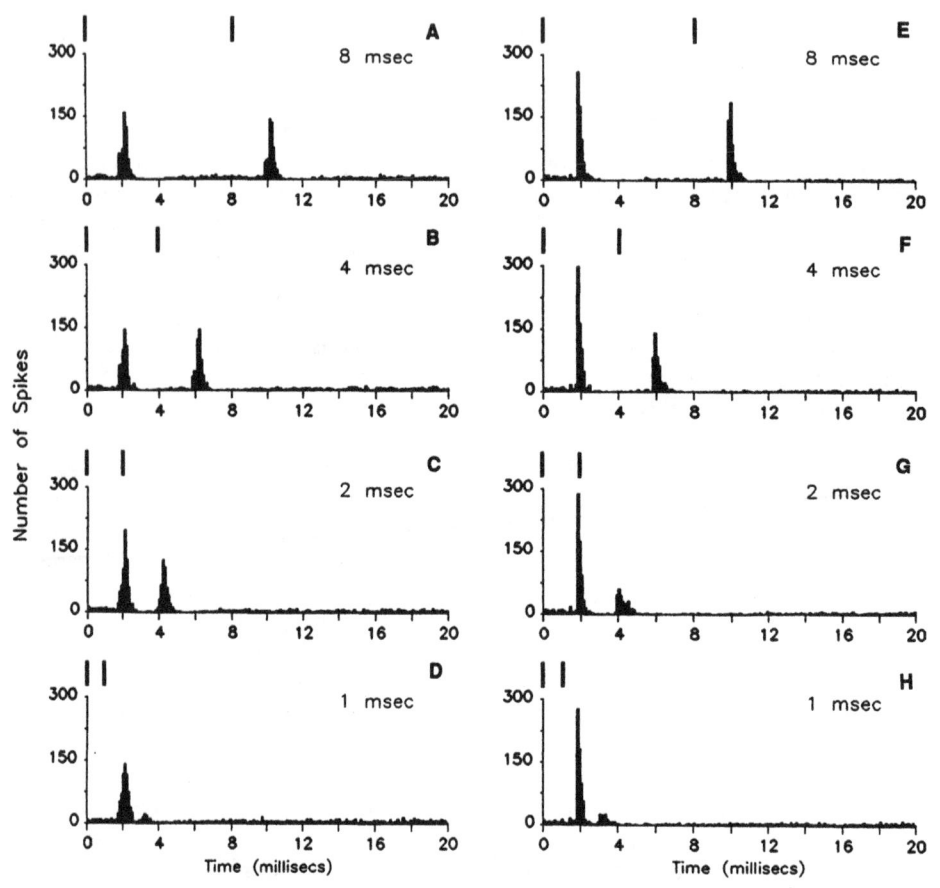

**Figure 5.** Post-stimulus time (PST) histograms calculated from the responses of a "primary-like with notch" unit to pairs of clicks presented near threshold (A-D) and 10 dB higher (E-H). Each click pair was presented 1000 times. Above each histogram, the bars indicate the times of occurence of the clicks with the interclick interval indicated on the right.

Divenyi and Blauert also ruled out contributions from monaural forward masking, since "echo suppression is a situation in which the *presence* of the probe is clearly detectable but its spatial position cannot be recognized."

Whether this short-latency, frequency-specific suppression is a component of the precedence effect or not (see Hafter et al., '88), these psychoacoustic findings have been used to postulate the existence of circuitry within the cochlear nuclear complex that mediates the suppression. Harris et al. ('63) argue that the suppression of the second click of a pair does not occur along the cochlear partition and that it does not result from neural refractory periods. They hypothesize that the suppression is the result of a neural gating mechanism which "...would have to occur before the place of binaural interaction. It seems probable that its position must be at or previous to the superior-olivary complex. In particular, it could occur at the level of the cochlear nucleus." Hafter and coworkers (Hafter et al., '88) also conclude that the frequency-specific suppression, which they term binaural adaptation, "...is the result of a prebinaural process, it seems that the most likely place to look for it is in the activity of individual units or populations of units in the cochlear nucleus, prior to the first stages of binaural interaction."

There is a good correspondence between the characteristics of monaural echo suppression determined in human psychoacoustic experiments and the properties of the circuit involving tuberculoventral neurons determined from anatomical and physiological experiments *in vitro*. Because any circuit that provides a delayed, frequency-specific inhibition can suppress signals encoding echoes, two questions have to be answered to determine whether tuberculoventral neurons can contribute to the monaural suppression of echoes described in the psychoacoustic experiments. The first question is whether the monaural suppression of signals encoding echoes can be observed in the cochlear nuclear complex. The second question is whether tuberculoventral neurons produce a delayed, frequency-specific inhibition *in vivo* that suppresses the responses of their targets in the VCN to the stimuli used in psychoacoustic experiments on sound localization.

## Suppression of the trailing click *in vivo*

The responses of cochlear nucleus units to pairs of clicks have been recorded *in vivo* to examine whether the monaural suppression of echoes can be observed. In rats anesthetized with ketamine and acepromazine, ablation of the overlying cerebellum yields a posterior approach to the cochlear nuclear complex. Stable, extracellular recordings made with glass micropipettes filled with 3 M KCl provide the preliminary results from 22 units in 7 animals. On the basis of responses to short tone bursts and latencies of response, these units were categorized as 7 choppers, 3 primary-like, 3 onset, 3 pauser, 5 VIIIth nerve fibers, and 1 unknown. The pauser units were also determined to have the characteristics of type IV units (Young and Brownell, '76). The locations of these units have not been verified histologically, but given the surgical approach, they were most likely in the PVCN or DCN. Except for the auditory nerve fibers, all of these units show suppression in the responses to the second click of a pair, when the clicks were presented at a level greater than 10 dB above threshold. Near threshold, suppression is not always observed.

The responses of a "primary-like with notch" unit to pairs of clicks presented 8, 4, 2 and 1 msec. apart and at two intensities are shown in Figure 5. Near threshold this unit responds about equally to both clicks for interclick intervals of 8 msec. (A), 4 msec. (B) and 2 msec. (C), but at a 1 msec. interval (D) there is only a small response to the second click. At a level 10 dB higher (E-H), the unit responds equally only to the clicks presented 8 msec. apart (E). For the other intervals (F-H), the height of the peak in the PST histogram for the trailing click is less than half that for the initial click. The total number of spikes recorded in response to the trailing click with either a 4 msec. (B,F) or an 8 msec. (A,E) interclick interval is about the same at both levels, but with the 2 msec. interval the response to the trailing click at the higher intensity (G) is about half the total near threshold (C). Above threshold and at an interclick interval of 2 msec., the response to the trailing click appears, therefore, to be suppressed.

The refractory properties of either the unit or its auditory nerve inputs are probably not responsible for the absence of a peak. At the lower sound pressure, the interspike interval histograms showed that this unit often responded to both clicks during a single presentation. The number of times that this unit responded to both clicks was about the same with interclick intervals of 8, 4, and 2 msec. This result indicates that the refractory properties of this unit do not prevent it from responding to two clicks presented 2 msec. apart and that the auditory nerve delivers inputs for both clicks to this neuron.

Figure 6 shows the responses of a "chopper" unit to clicks. This unit had a best frequency of 4350 Hz, a spontaneous rate of 46 spikes/sec and no apparent high frequency, sideband inhibition in its responses to tone pips. For all click pairs (6A-D), the response to the second click contained fewer spikes. The response to a single click (6E) can be used to predict the time of occurrence of the main peak in response to each click of the pair. At 8

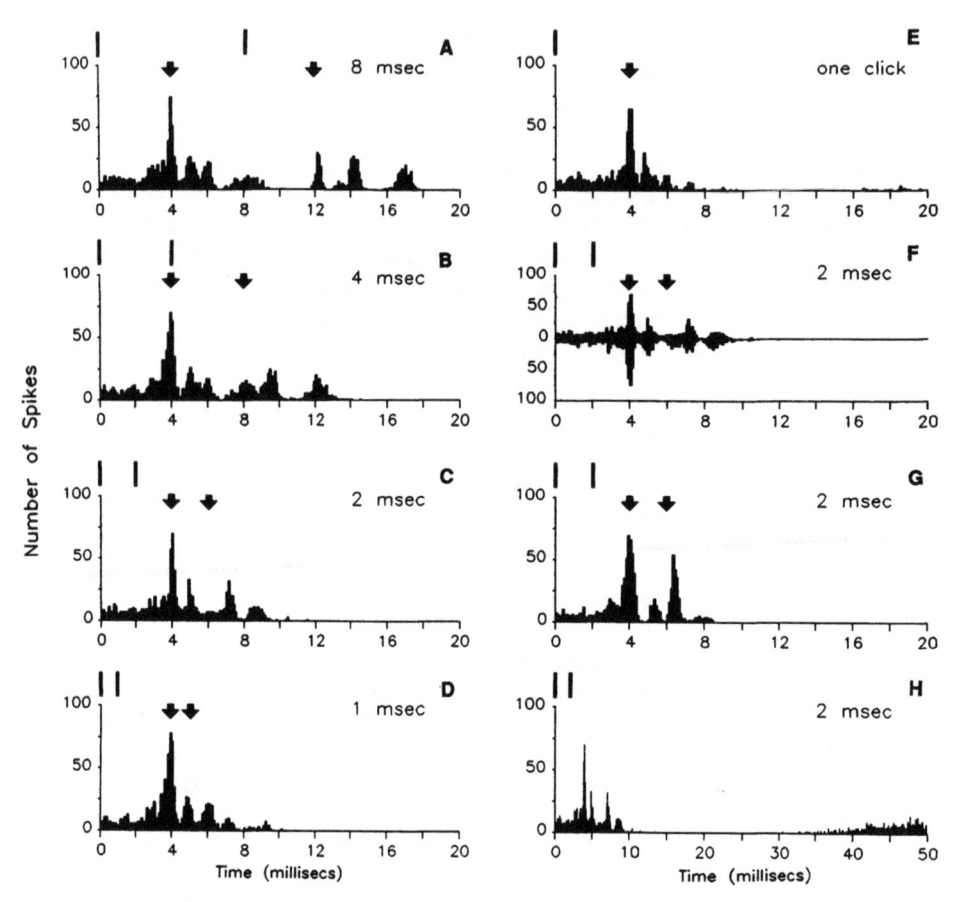

**Figure 6.** PST histograms calculated from the responses of a "chopper" unit to 1000 repetitions of two, same polarity clicks presented (A) 8 msec., (B) 4 msec., (C) 2 msec., or (D) 1 msec. apart. Bars above the histograms show the times of click presentation and the interclick interval is also indicated. (E) The response of this neuron to a single click. The latency of the main peak in single click histogram was used to predict when major peaks should be present in the PST histograms for responses to click pairs, and the times of these predicted responses are indicated by the arrows (A-D,F). (F) "Compound" PST histogram made from the PST histogram for a pair of same polarity clicks presented 2 msec. apart (top, same as C), and inverted below it is the PST histogram for a pair of clicks 2 msec. apart with the second click having polarity opposite to the first. (G) PST histogram of the responses of the same neuron to a pair of clicks presented 2 msec. apart and at a level 20 dB below the level used in A-F. (H) Same response as C, but on a longer time scale to show the duration of suppression of the spontaneous activity.

msec. (6A) and 4 msec. (6B) separations, this peak in the responses to the trailing click is present, but not as large as the responses to a single click. At a 2 msec. separation (6C), the predicted main peak is absent, but the small following peaks are present. With a 1 msec. separation (6D), the response is virtually identical to the response to a single click.

The predicted peak that is missing for a 2 msec. separation of the click pair (Fig. 6C) is not the result of a cancellation along the cochlear partition (Harris et al., '63; Goblick and Pfeiffer, '69; Møller, '70). Figure 6F shows a "compound" PST histogram with the response to the two clicks with the same polarities in the upper portion (the same histogram in Fig. 6C), and inverted beneath this is the PST histogram for the response to pairs of the click 2 msec. apart with the polarity of the second click reversed. Any cancellation of stimulus waveforms along the cochlear partition in the response to one pair of clicks would yield a

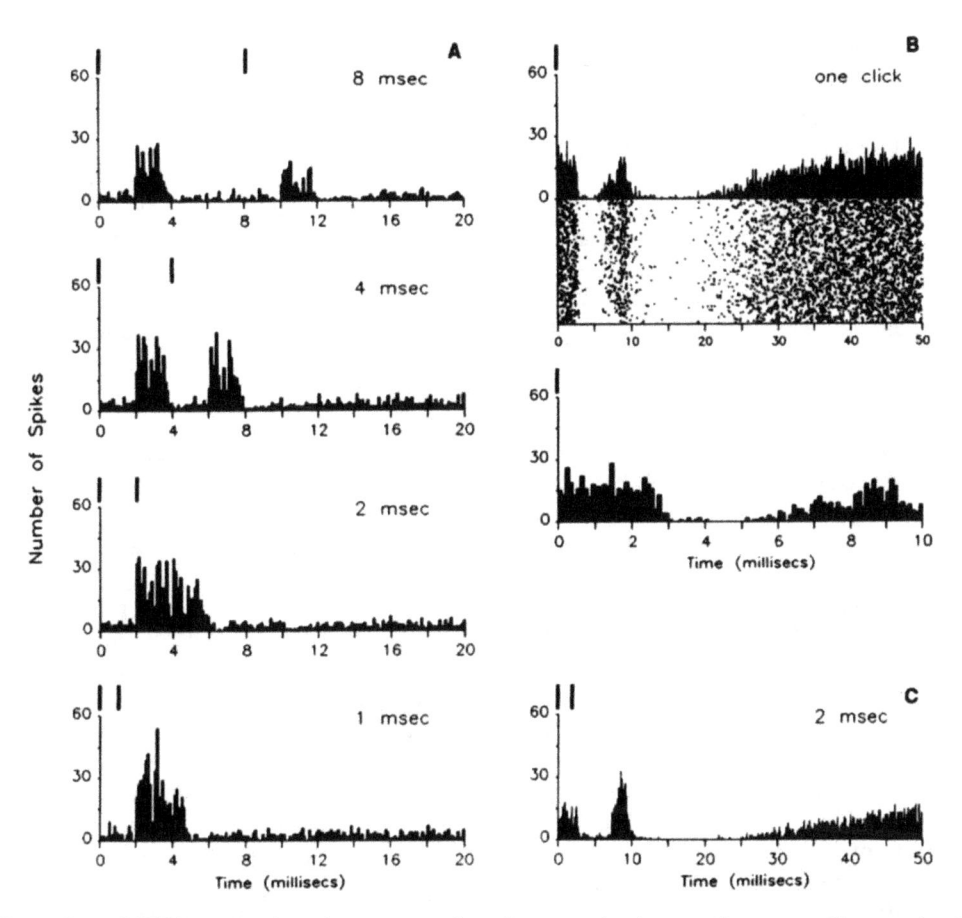

**Figure 7.** A. PST histograms from the responses of a unit presumed to be an auditory nerve fiber to pairs of clicks 8, 4, 2 and 1 msec. apart. Each pair of clicks was presented 1000 times at a level 20 dB above threshold. B. (top) PST histogram of the responses to 1000 repetitions of a single click for a type IV unit encountered shortly after the unit shown in figure 7A. This unit had the same best frequency as the unit in 7A. The level of the click is also the same as that used for the responses in 7A. The latency dot rasters for these responses are shown just beneath the histogram. An early and a late inhibition are apparent in the PST histogram. The initial segment of the PST histogram is shown in the bottom part of 7B. The latency of the early inhibitory component is about 3 msec. C. The PST histogram of the responses of the type IV unit to a pair of clicks 2 msec. apart. The clicks were presented 1000 times at the same level used in 7A and 7B.

summation and an enhanced response for the other click pair; if a cancellation was eliminating the predicted second peak in the upper histogram, it would produce in the inverted histogram a peak at least as large as the initial one. However, the two responses are almost mirror images of one another. The differences between the two responses are similar to those observed between data sets (compare the responses to single clicks in 6A and 6E).

The "chopper" unit also responded to both clicks near threshold. Figure 6G shows that with a 20 dB reduction in click level from that used for Figs. 6A-F, the response to the initial click was as large as that to the louder click (e.g. Fig. 6E), while the response to the trailing click was only slightly reduced. However, this unit did not respond to both clicks during a single presentation. The interspike interval histogram showed only a small peak at 4 msec., the preferred period for chopping of this unit, and not a peak at 2 msec. At this lower intensity, the latency of the trailing click is slightly longer than predicted.

The suppression that is reducing the responses to the trailing clicks is also eliminating the spontaneous activity following the response to both a single click (Fig. 6E) and pairs of clicks. The response of the chopper to two clicks, 2 msec. apart, is plotted on a longer time scale in figure 6H. The spontaneous activity is eliminated for almost 25 msec. following the small response to the second click. Auditory nerve fibers show little or no suppression of spontaneous activity following the response to the clicks (Fig. 7A; Kiang et al., '65b). The suppression of spontaneous activity and its long duration indicates the presence of a neurally mediated inhibition.

For comparison, the responses of an auditory nerve fiber to pairs of clicks are shown in Figure 7A. This unit had a best frequency of about 26 kHz; it was encountered shortly before the type IV unit described below and was, therefore, probably located in the DCN. This unit is presumed to be a nerve fiber because of its unipolar action potential different from those recorded from cell bodies in the DCN, and "primary-like" response to tone pips. The latency of the response of this unit is about 0.3 msec. longer than that of auditory nerve fibers encountered in the VCN, but the responses of this unit to clicks are characteristic of the auditory nerve fibers recorded to date. At a level 20 dB above threshold, at all interclick intervals, this fiber often responded to both clicks in the pair during the same stimulus presentation. Even with a 1 msec. separation, the response was 1 msec. longer than the response to a single click, as observed in the responses to the clicks either 8 or 4 msec. apart.

Another finding of these preliminary experiments was the presence of short latency inhibition on a type IV neuron in the DCN. This neuron had a best frequency of 26 kHz with a threshold of 20 dB SPL and it was encountered shortly after the responses of the auditory nerve fiber were recorded. Figure 7B shows the response of this neuron to a single click presented at a level well above threshold, but the same level used in Figure 7A to collect the responses of the nerve fiber to clicks. There are at least two inhibitory components to the response. The initial inhibition has a latency of less than 3 msec. and with an increase in intensity of 10 dB, the latency of the initial inhibition drops to about 2.6 msec. The later inhibitory component begins at about 10 msec. and reduces the spontaneous activity from about 10 msec. to 35 msec. after the initial click. The response of this type IV neuron to a pair of clicks 2 msec. apart is shown in figure 7C. While a single click did not cause this unit to fire at a rate above the spontaneous rate, the pair of clicks produces a single peak in the response between the inhibition. The histograms of the responses for this unit to pairs of clicks 4 and 8 msec. apart appeared similar making it difficult to judge whether suppression of the trailing click occurs.

This inhibition observed in the type IV unit in Figure 7B indicates that there are inhibitory interneurons within the cochlear nuclear complex that have a short latency response to clicks and terminate on type IV neurons. Given acoustic delays, travel time in the cochlea and the latency of neurons in the DCN to auditory nerve stimulation (Hirsch and Oertel, '88), if one assumes all auditory nerve synapses to be excitatory, there are probably only three synapses interposed between earphone and the type IV neurons to produce an inhibition with a latency of 2.6 msec.: hair cell → auditory nerve, auditory nerve → interneuron, interneuron → type IV cell. Because type IV neurons are probably targets of tuberculoventral neurons, tuberculoventral neurons are candidates for providing this short latency inhibition.

These preliminary results support the proposals by Harris et al., ('63) and Hafter et al. ('88) that inhibition intrinsic to the cochlear nuclear complex suppresses responses to the second click. The absence of a large peak in the histogram for the responses with the polarity of the second click reversed indicates that suppression of the trailing click does not occur along the cochlear partition. This finding was confirmed by the ability of auditory nerve fibers to respond to both clicks at all interclick intervals tested. The responsiveness of auditory nerve fibers to both clicks indicates that the refractory periods of the fibers do not prevent the encoding of the trailing click. The refractory periods of the cochlear nucleus units also do not appear to account for the suppression of the trailing click with an interclick interval of 2 msec., as the primary-like-with notch unit responded to both clicks near threshold. With a 1 msec. interclick interval, the refractory period of a unit in the cochlear nuclear complex may contribute to the suppression, because units rarely responded to two clicks 1 msec. apart. Suppression of the trailing click with a 1 msec. interclick interval was not predicted from the *in vitro* latencies (Wickesberg and Oertel, '90), but in psychoacoustic experiments, Gaskell ('83) observed the most suppression with a 1 msec. interclick interval. The elimination of spontaneous activity following responses to a click found in the responses of units in the cochlear nuclei but not auditory nerve units supports the existence of inhibitory inputs.

## Echoes are heard

While echoes appear to be suppressed for localization tasks, they do contribute to the perception of a sound. In the studies by Harris et al. ('63) and Guttman ('63), subjects described a pair of clicks 2 msec. apart as having a different "timbre and loudness" than a single click, even though the second click was psychoacoustically suppressed in localization tasks. Echoes contribute to the "spaciousness" (Blauert, '83) and "extent" (Perrott, '84) of a sound. One explanation of these findings is that in those pathways from the cochlear nuclear complex concerned with localization, echoes are suppressed, but that other parallel ascending pathways convey information concerning echoes. A consequence of the hypothesis that tuberculoventral neurons contribute to monaural echo suppression is that neurons which are not targets of tuberculoventral neurons can encode echoes and thereby contribute to the perceptions created by echoes. One pathway, therefore, that could encode echoes is through the octopus cell area, since the octopus cell area does not receive a projection from tuberculoventral neurons (Wickesberg et al., '91).

If information about echoes, which contributes to "timbre and loudness" judgments, is transmitted by neurons in the octopus cell area to higher auditory centers, then octopus cells could participate in the encoding of loudness. Psychoacoustic measurements show that mammals perceive loudness changes over a dynamic range of about 120 dB (Hellman and Zwislocki, '61). The dynamic range over which firing rate changes in individual auditory nerve fibers is 30 to 40 dB, although the dynamic range of initial transients can be up to 70 dB (Smith and Brachman, '80; Smith, '88). Some additional mechanism must, therefore, contribute to the encoding of intensity. Two possibilities have been proposed (Smith, '88; Viemeister, '79, '88). The first is that thresholds are staggered so that auditory nerve fibers with high spontaneous firing rates and low thresholds together with those with low spontaneous firing rates and high thresholds cover the dynamic range (Sachs and Abbas, '74; Liberman, '78; Winter et al., '90). The second is that spread of excitation along the cochlea with increasing intensity could be detected in the nervous system (Veimeister, '83). Octopus cells, with their convergent auditory nerve input, sensitivity to transients and broad tuning (Harrison and Warr, '62; Osen, '69; Godfrey et al., '75a; Rhode et al., '83a; Kim et al., '86), may be uniquely positioned and shaped to extract information on intensity.

Perhaps octopus cells, which project to the ventral nucleus of the lateral lemniscus (Warr, '82), are part of a pathway that is similar to the intensity pathway in the barn owl. Separate pathways for encoding intensity and timing were identified in barn owls with a combination of physiology (Sullivan and Konishi, '84; Manley et al., '88), anatomy (Takahashi and Konishi, '88a,b) and reversible injections of local anesthetic (Takahashi et al., '84). Timing information travels from nucleus magnocellularis to nucleus laminaris, then to the dorsal nucleus of the lateral lemnsicus, and the inferior colliculus (Takahashi and Konishi, '88b; Konishi et al., '88). The intensity pathway proceeds from *nucleus angularis* to the ventral nucleus of the lateral lemniscus and to the inferior colliculus.

# REFERENCES

Adams, J.C., 1979, Ascending projections to the inferior colliculus, *J. Comp. Neurol.*, 183:519-538.

Adams, J.C., 1983, Multipolar cells in the ventral cochlear nucleus project to the dorsal cochlear nucleus and the inferior colliculus, *Neurosci. Lett.*, 37:205-208.

Blackburn, C.C. and Sachs, M.B., 1989, Classification of unit types in the anteroventral cochlear nucleus: PST histograms and regularity analysis, *J. Neurophysiol.*, 62:1303-1329.

Blauert, J., 1983, "Spatial Hearing", MIT Press, Cambridge.

Bourk, T.R., 1976, "Electrical responses of neural units in the anteroventral cochlear nucleus of the cat", Doctoral disstertation, Massachusetts Institute of Technology, Cambridge, Massachusetts.

Brawer, J.R., Morest, D.K. and Kane, E.C., 1974, The neuronal architecture of the cochlear nucleus of the cat, *J. Comp. Neurol.*, 155:251-300.

Cajal, S.R., 1909, "Histologie du System Nerveux de l'Homme et des Vertebres", Maloine, Paris.

Cant, N.B., 1981, The fine structure of two types of stellate cells in the anterior division of the anteroventral cochlear nucleus of the cat, *Neurosci.*, 6:2643-2655.

Cant, N.B., 1982, Identification of cell types in the anteroventral cochlear nucleus that project to the inferior colliculus, *Neurosci. Lett.* 32:241-246.

Cant, N.B. and Casseday, J.H., 1986, Projections from the anteroventral cochlear nucleus to the lateral and medial superior olivary nuclei, *J. Comp. Neurol.*, 247:457-476.

Caspary, D.M., 1972, Classification of subpopulations of neurons in the cochlear nuclei of the kangaroo rat, *Exp. Neurol.*, 37: 131-151.

Divenyi, P.L. and Blauert, J., 1987, On creating a precedent for binaural patterns: When is an echo an echo? in: "Auditory Processing of Complex Sounds", W.A. Yost and C.S. Watson, eds., pp. 147-156, Lawrence Erlbaum, Hillsdale, N.J.

Feng, A.S. and Vater, M., 1985, Functional organization of the cochlear nucleus of rufous horseshoe bats (*Rhinolophus rouxi*),: Frequencies and internal connections are arranged in slabs, *J. Comp. Neurol.*, 225:529-553.

Frostholm, A. and Rotter, A., 1985, Glycine receptor distribution in mouse CNS: Autoradiographic localization of $^3$H-strychnine binding sites, *Brain.Res. Bull.*, 15:473-486.

Gaskel, H., 1983, The precedence effect, *Hearing Res.* 11:277-303.

Goblick, T.J. and Pfeiffer, R.R., 1969, Time-domain measurements of cochlear nonlinearities using combination click stimuli, *J. Acoust. Soc. Amer.*, 46:924-946.

Godfrey, D.A., Kiang, N.Y.S. and Norris, B.E., 1975a, Single unit activity in the posteroventral cochlear nucleus of the cat, *J. Comp. Neurol.*, 162:247-268.

Godfrey, D.A. Kiang, N.Y.S. and Norris, B.E., 1975b, Single unit activity in the dorsal cochlear nucleus cat, *J. Comp. Neurol.*, 162: 269-284.

Guttman, N., 1963, Binaural interaction of three clicks, *J. Acoust. Soc. Amer.*, 37:145-150.

Haas, H., 1951, On the influence of a single echo on the intelligibility of speech, *Acustica*, 1:49-58.

Hafter, E.R., Buell, T.N. and Richards, V.M., 1988, Onset-coding in lateralization: Its form, site, and function, *in*: "Auditory Function", G.M. Edelman, W.E. Gall and W.M. Cowan, eds., pp. 647-676, Wiley, New York.

Hafter, E.R. and Dye, Jr., R.H., 1983, Detection of interaural differences of time in trains of high-frequency clicks as a function of interclick interval and number, *J. Acoust. Soc. Amer.*, 73:644-651.

Hafter, E.R. and Wenzel, E.M., 1983, Lateralization of transients presented at high rates: Site of the saturation effect, *in*: "Hearing - Physiological Basis and Psychophysics", R. Klinke and R. Hartmann, eds., pp. 202-208, Springer-Verlag, Berlin.

Harris, G.G., Flanagan, J.L. and Watson, B.J., 1963, Binaural interaction of a click with a click pair, *J. Acoust. Soc. Amer.*, 35:672-678.

Harrison, J.M. and Warr, R.B., 1962, A study of the cochlear nuclei and the ascending auditory pathways of the medulla, *J. Comp. Neurol.*, 119:341-380.

Hellman, R.P. and Zwislocki, J., 1961, Some factors affecting the estimation of loudness, *J. Acoust. Soc. Amer.*, 33:687-694.

Hirsch, J.A. and Oertel, D., 1988, Synaptic connections in the dorsal cochlear nucleus of mice, in vitro, *J. Physiol.*, 396:549-562.

Kiang, N.Y.S., Pfeiffer, R.R., Warr W.B. and Backus, A.S.N., 1965a, Stimulus coding in the cochlear nucleus, *Ann. Otol. Rhino. Laryngol.*, 74:463-486.

Kiang, N.Y.S., Watanabe, T., Thomas E.C. and Clark, L.F., 1965b, "Discharge Patterns of Single Fibers in the Cat's Auditory Nerve", MIT Press, Cambridge, MA.

Kiang, N.Y.S., Morest, D.K., Godfrey, D.A., Guinan, J.J. Jr. and Kane, E.S., 1973, Stimulus coding at caudal levels of th cat's auditory nervous system: I. Response characteristics of single units, *in*: "Basic Mechanisms in hearing", A.Møller, ed., Academic Press, New York, pp. 455-478.

Kim, D.O., Rhode, W.S. and Greenberg, S.R., 1986, Responses of cochlear nucleus neurons to speech signals: neural encoding of pitch, intensity and other parameters, in: "Auditory Frequency Selectivity", B.C.J. Moore and R.D. Patterson, eds., pp. 281-288, Plenum, New York.

Konishi, M., T.T. Takahashi, Wagner, H., Sullivan, W.E. and Carr, C.E., 1988, Neurophysiological and anatomical substrates of sound localization in the owl, in: "Auditory Function" G.M. Edelman, W.E. Gall and W.M. Cowan, eds., Wiley, New York, pp.721-745.

Liberman, M.C., 1978, Auditory-nerve response from cats raised in a low-noise chamber, J. Acoust. Soc. Amer., 63:442-455.

Lorente de Nó, R., 1933, Anatomy of the eighth nerve III. General plans of structure of the primary cochlear nuclei, Laryngoscope, 43:327-350.

Lorente de Nó, R., 1981, "The Primary Acoustic Nuclei", Raven, New York.

Manley, G.A., Köppl, C. and Konishi, M., 1988, A neural map of intensity differences in the brain stem of the barn owl, J. Neurosci., 8:2665-2676.

McFadden, D., 1973, Precedence effects and auditory cells with long characteristic delays, J. Acoust. Soc. Amer., 54:528-530.

Møller, A.R., 1970, Studies of the damped oscillatory response of the auditory frequency anaylzer, Acta Physiol. Scand., 78:299-314.

Molnar, C.D. and Pfeiffer, R.R., 1968, Interpretation of spontaneous spike discharge patterns of neurons in the cochlear nucleus, Proc. IEEE, 56:993-1004.

Müller, M., 1990, Quantitative comparison of frequency representation in the auditory brainstem nuclei of the gerbil, Pachyuromys duprasi, Exp. Brain.Res., 81:140-149.

Oertel, D., 1983, Synaptic responses and electrical properties of cells in brain slices of the mouse anteroventral cochlear nucleus. J. Neurosci., 3:2004-2053.

Oertel, D., 1985, Use of brain slices in the study of the auditory system; Spatial and temporal summation of synaptic inputs in cells in the anteroventral cochlear nucleus of the mouse, J. Acoust. Soc. Amer., 78:328-333.

Oertel, D. and Wu, S.H., 1989, Morphology and physiology of cells in slice preparations of the dorsal cochlear nucleus of mice, J. Comp. Neurol., 283:228-247.

Oertel, D., Wu, S.H., Garb, M.Wand Dizack, C.,1990, Morphology and physiology of cells in slice preparations o the posteroventral cochlear nucleus of mice, J. Comp. Neurol., 295:136-154.

Oliver, D.L., 1984, Dorsal cochlear nucleus projections to the inferior colliculus in the cat. A light and electron microscopic study, J. Comp. Neurol., 224:155-172.

Oliver, D.L., 1987, Projections to the inferior colliculus from the anteroventral cochlear nucleus in the cat: Possible substrates for binaural interaction, J. Comp. Neurol., 264:24-46.

Osen, K.K., 1969, Cytoacrchitechture of the cochlear nuclei in the cat, J. Comp. Neurol., 136:453-484.

Perrott, D.R., 1984, Binaural resolution of the size of an acoustic array: Some experiments with stereophonic arrays, J. Acoust. Soc. Amer., 76:1704-1712.

Pfeiffer, R.R., 1966, Classification of response patterns of spike discharges for units in the cochlear nucleus: tone-burst stimulation, Exp. Brain Res., 1:220-235.

Rhode, W.S., Oertel, D. and Smith, P.H., 1983a, Physiological response properties of cells labeled intracellularly with horseradish peroxidase in cat ventral cochlear nucleus, J. Comp. Neurol., 213:448-463.

Rhode, W.S., Smith, P.H. and Oertel, D., 1983b, Physiological response properties of cells labeled intracellularly with horseradish peroxidase in cat dorsal cochlear nucleus, J. Comp. Neurol., 13:426-447.

Rouiller, E.M. and Ryugo, D.K., 1984, Intracellular marking of physiologically characterized cells in the ventral cochlear nucleus, J. Comp. Neurol., 225:167-186.

Ryugo, D.K., Willard, F.H. and Fekete, D.M.,1981, Differential afferent projections to the inferior colliculus from the cochlear nucleus in the albino mouse, Brain Res., 210:342-349.

Saint-Marie, R.L., Benson, C.G., Ostapoff, E.-M. and Morest, D.K., 1991, Glycine immunoreactive projections from the dorsal to the anteroventral cochlear nucleus, Hearing Res., 51:11-28.

Sachs, M.B. and Abbas, P.J., 1974, Rate versus level functions for auditory-nerve fibers in cats: tone-burst stimuli, J. Acoust. Soc. Amer., 56:1835-1847.

Smith, P.H. and Rhode, W.S., 1989, Structural and functional properties distinguish two types of multipolar cells in the ventral cochlear nucleus, J. Comp. Neurol., 282:595-616.

Smith, R.L., 1988, Encoding of sound intensity by auditory neurons, in: "Auditory Function", G.M. Edelman, W.E. Gall and W.M. Cowan, eds., Wiley, New York. pp. 243-274.

Smith, R.L. and Brachman, M.L., 1980, Operating range and maximum response of single auditory nerve fibers. Brain Res., 184:499-505.

Snyder, R.L. and Leake, P.A., 1988, Intrinsic connections within and between cochlear nucleus subdivisions in cat, J. Comp. Neurol., 278:209-225.

Spirou, G., Brownell, W.E. and Zidanic, M., 1990, Recordings from cat trapezoid body and HRP labeling of globular bushy cell axons, *J. Neurophysiol.*, 63:1169-1190.

Sullivan, W.E. and Konishi, M., 1984, Segregation of stimulus phase and intensity in the cochlear nuclei of the barn owl, *J. Neurosci.*, 4:1787-1799.

Takahashi, T.T. and Konishi, M., 1988, Projections of the cochlear nuclei and nucleus laminaris to the inferior colliculus of the barn owl, *J. Comp. Neurol.*, 274:190-211.

Takahashi, T.T. and Konishi, M., 1988, Projections of nucleus angularis and nucleus laminaris to the lateral lemniscal nuclear complex of the barn owl, *J. Comp. Neurol.*, 274:212-238.

Takahashi, T.T., Moiseff, A. and Konishi, M., 1984, Time and intensity cues are processed independently in the auditory system of the owl, *J. Neurosci.*, 4:1781-1786.

Tolbert, L.P., Morest, D.K. and Yurgelun-Todd, D.A., 1982, The neuronal architecture of the anteroventral cochlear nucleus of the cat in the region of the cochlear nerve root: horseradish peroxidase labeling of identified cell types, *Neurosci.* 7:3031-3052.

Viemeister, N.F., 1979, Temporal modulation transfer functions based upon modulation thresholds, *J. Acoust. Soc. Amer.*, 66:1364-1380.

Viemeister, N.F., 1983, Auditory intensity discrimination at high frequencies in the presence of noise, *Science*, 221:1206-1208.

Viemeister, N.F., 1988, Psychophysical aspects of auditory intensity coding, *in*: "Auditory Function", G.M. Edelman, W.E. Gall and W.M. Cowan, eds., Wiley, New York, pp. 213-241.

Voigt, H.F. and Young, E.D., 1980, Evidence of inhibitory interactions between neurons in dorsal cochlear nucleus, *J. Neurophysiol.*,44:76-96.

Voigt, H.F. and Young, E.D., 1990, Cross-correlation analysis of inhibitory interactions in dorsal cochlear nucleus. *J. Neurophysiol.*, 64:1590-1610.

Wallach, H., E.B. Newman, E.B. and M.R. Rosenzweig, M.R., 1949, The precedence effect in sound localization, *Am. J. Psychol.*, 52:315-336.

Warr, W.B., 1982, Parallel ascending pathways from the cochlear nucleus: Neuroanatomical evidence of functional specialization, *in*: "Contrib. Sens. Physiol.", W.D. Neff, ed., 7:1-38.

Wickesberg, R.E. and Oertel, D., 1988, Tonotopic projection from the dorsal to the anteroventral cochlear nucleus of mice, *J. Comp. Neurol.*, 268:389-399.

Wickesberg, R.E. and Oertel, D., 1990, Delayed, frequency-specific inhibition in the cochlear nuclei of mice: A mechanism for monaural echo suppression, *J. Neurosci.*, 10:1762-1768.

Wickesberg, R.E., Whitlon, D. and Oertel, D., 1991, Tuberculoventral neurons project to the multipolar cell area but not to the octopus cell area of the posteroventral cochlear nucleus, *J. Comp. Neurol.*, 313:457-468.

Winter, I.M. and A.R. Palmer, 1990, Responses of single units in the anteroventral cochlear nucleus of the guinea pig, *Hearing Res.*, 44:1577-1588.

Winter, I.M., Robertson, D. and Yates, G.K., 1990, Diversity of characteristic frequency rate-intensity functions in guinea pig auditory nerve fibers, *Hearing Res.*, 45:191-202.

Wu, S.H. and Oertel, D., 1984, Intracellular injection whith horseradish peroxidase of physiologically characterized stellate and bushy cells in slices of mouse anteroventral cochlear nucleus, *J. Neurosci.*, 4:1577-1588.

Yost, W.A. and Hafter, E.R., 1987, Lateralization, *in*: "Directional Hearing" W.A. Yost & G. Gourevitch, eds., Springer Verlag, New York, pp. 48-84.

Young, E.D., 1980, Identification of response properties of ascending axons from dorsal cochlear nucleus, *Brain Res.*, 200:23-37.

Young, E.D. and Brownell, W.E., 1976, Responses to tones and noise of single cells in dorsal cochlear nucleus of unanesthetized cats, *J. Neurophysiol.*, 39:282-300.

Young, E.D., Robert, J.M. and Shofner, W.P., 1988, Regularity and latency of units in the ventral cochlear nucleus: Implications for unit classification and generation of response properties, *J. Neurophysiol.*, 60:1-29.

Zurek, P.M., 1980, The precedence effect and its possible role in the avoidance of interaural ambiguities, *J. Acoust. Soc. Amer.*, 67:952-964.

Zurek, P.M., 1987, The precedence effect, *in*: "Directional Hearing", W.A. Yost and G. Gourevitch, eds., pp.85-105, Springer Verlag, New York, pp. 85-105.

# THE SYNAPTIC ORGANIZATION OF THE VENTRAL COCHLEAR NUCLEUS OF THE CAT: THE PERIPHERAL CAP OF SMALL CELLS

Nell Beatty Cant

Department of Neurobiology, Duke University Medical Center
Durham, NC 27710, USA

The ventral cochlear nucleus of the cat comprises a number of morphologically distinct subdivisions, each of which appears to have a characteristic pattern of synaptic organization (Lorente de Nó, '33, '81; Osen, '69; Kane, '73; Brawer et al., '74; Cant and Morest, '79b; Mugnaini et al., '80; Cant, '81; Tolbert and Morest, '82b; Adams and Mugnaini, '87; Smith and Rhode, '87, '89; Saint Marie et al., '89). One important gap in our knowledge of the synaptic organization of the ventral cochlear nucleus is that one of its subdivisions, the peripheral cap of small cells, has not been described at the level of the electron microscope.

The purpose of this report is to describe the neuronal organization and the fine structure of the neurons and neuropil of the peripheral cap of small cells and to compare its synaptic organization to that of the adjacent subdivisions of the ventral cochlear nucleus. Electron microscopic examination of the cap reveals several major differences from the adjacent areas of large cells, adding to the accumulating evidence that this subdivision of the ventral cochlear nucleus is functionally distinct.

## The boundaries of the small cell cap

It has been recognized for a long time that marginal areas of the ventral cochlear nucleus differ from the main body of the nucleus (Fuse, '13; Lorente de Nó, '33, '81; van Noort, '69; Brawer et al., '74). Lorente de Nó described at least three marginal areas which he called the anterior and posterior lateral nuclei and the internal marginal layer. When Osen ('69) subdivided the ventral cochlear nucleus on the basis of the morphology of the neurons in different regions, she called these marginal areas collectively the peripheral cap of small cells. By her definition, this cap surrounds the entire ventral nucleus and includes the small cells that lie along both the medial and lateral margins of the ventral nucleus. In the cat, the cap expands dorsolaterally, anterior to the dorsal cochlear nucleus and subjacent to the granular cell layer that covers the lateral surface of the ventral cochlear nucleus.

The boundaries of the peripheral cap of small cells, including its dorsolateral part, are indistinct in Nissl-stained material (Osen, '69). They are more clearly delineated in sections of the nucleus stained for myelin (Osen, '87) or in toluidine blue-stained sections (Fig. 1).

**Figure 1.** Transverse section approximately 2 $\mu$m thick through the ventral cochlear nucleus of a cat; stained with toluidine blue. Three subdivisions in the ventral cochlear nucleus are present in this section. At the top is the granule cell layer (GCL). At the bottom is the posterior division of the AVCN (PD), which contains many large cells. Between these two subdivisions lies the small cell cap (CAP). Bundles of myelinated axons in both the cap and the adjacent large cell areas are indicated by arrows. Scale bar = 150 $\mu$m.

In such preparations, the granule cell domains that form the lateral boundary of much of the cap are characterized both by the high concentration of granule cells and also by a relative scarcity of bundles of myelinated axons. The areas of large cells on the other boundary contain the main trajectory of the ascending and descending branches of the auditory nerve fibers, and, in consequence, contain many bundles of large, myelinated axons. The cap receives branches from auditory nerve fibers, but it does not lie in the main path of the ascending or descending axons (Osen, '70, '87; Cant and Morest, '84). As illustrated in Figure 1, the myelinated axons in the small cell cap appear to be of considerably smaller diameter on average than do those in the main body of the nucleus. The difference in the diameters of the axons in these two regions provides the easiest criterion for defining the boundary of the cap with the adjacent areas of larger cells. In myelin-stained material, the cap is also distinguished from the rest of the ventral cochlear nucleus by the fact that intrinsic axons travelling through it run in a direction different from that followed by the heavily

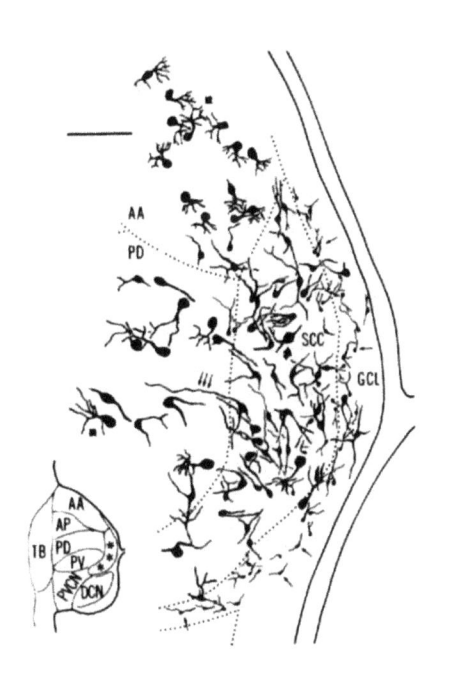

Figure 2. Cells in the small cell cap in the cat impregnated by the Golgi-Kopsch method (Braitenberg et al., '67). The large drawing is a composite of cells from five consecutive horizontal sections through the right cochlear nucleus. The small inset drawing is of the middle section of the series and illustrates the subdivisions of the nucleus as defined by Brawer et al. ('74). Rostral is toward the top of the figure; lateral is to the right. The location of the small cell cap is indicated on the inset by asterisks. In the enlarged drawing of this area, neurons typical of the cap area are illustrated. Scale bar is equal to 250 μm. Abbreviations: AA, anterior part of the anterior division of the anteroventral cochlear nucleus (AVCN); AP, posterior part of the anterior division of the AVCN; PD, dorsal part of the posterior division of the AVCN; PV, ventral part of the posterior division of the AVCN; SCC, small cell cap; GCL, granule cell layer; PVCN, posteroventral cochlear nucleus; DCN, dorsal cochlear nucleus; TB, trapezoid body. Thin single arrows indicate granule cells in the GCL. Triple thin arrows point to the dendrite of a cell in the cap that protrudes into PD. The short thick arrows indicate bushy cells in AA and PD. The curved thick arrow indicates a bushy cell that lies within the boundaries of the cap.

myelinated axons in the rest of the anteroventral cochlear nucleus (e.g., Osen, '87).

The boundaries drawn on the basis of myelination patterns closely resemble in location those drawn in Nissl sections on the basis of cytoarchitecture (Osen, '69; Cant and Morest, '84). The main cytoarchitectonic criteria are size and type of the constituent cells. The appearance of the neurons of the cap in Nissl-stained preparations has been described previously (Osen, '69; Cant and Morest, '84). Most of the neurons are small to medium sized multipolar cells that resemble the multipolar neurons in the anterior division of the anteroventral cochlear nucleus (AVCN-A). Although as the name implies, many of the neurons in the cap are relatively small, the average size of the multipolar cells in the part of the cap that lies along the dorsolateral aspect of the ventral cochlear nucleus is the same as that of the multipolar or stellate cells in AVCN-A (Cant and Morest, '84).

None of the boundaries described above is sharp. Figure 1 illustrates that the granule cell domain and the neuropil of the cap often appear to interdigitate along their border. On the medial side, large cells more typical of the large cell areas (e.g. spherical bushy cells) may invade the cap. As noted above, a thin marginal layer of small cells, bordering almost the entire ventral cochlear nucleus, is also a part of the peripheral cap of small cells as originally defined (Osen, '69). Defining a definite anatomical boundary for this part of the nucleus is not possible, since in many places, it may be only one cell thick. All of the descriptions that follow are based on studies of the dorsolaterally enlarged region of the small cell cap.

## Organization of the neurons of the small cell cap: Golgi impregnations

Boundaries of the cap in Golgi-impregnated cochlear nuclei were estimated using the dimensions established in Nissl and myelin-stained sections and on the basis of cell type and

**Figure 3.** Cells in the small cell cap impregnated by the Golgi-Kopsch method. This is a composite drawing of cells from four horizontal sections 100 µm thick through the right cochlear nucleus. A. Typical neurons of the small cell cap. a and b, cells with relatively smooth dendritic profiles; x and y, cells with spine laden dendrites; g, granule cells. Arrow C indicates the region shown at higher magnification in panel C. Arrow D illustrates the region shown at higher magnification in Panel D. The long arrow indicates a bushy cell that has invaded the SCC. B. Drawing at low magnification to indicate the region of the cochlear nucleus from which the drawings in A were taken (arrow). Rostral is to the top; lateral is to the left. C, D. Enlargement of the two regions indicated in A. Arrows indicate spines and other dendritic appendages. Upper scale bar indicates 100 µm for A and approximately 1.5 µm for B. Lower scale bar indicates 20 µm for C, D. Abbreviations: P, posteroventral cochlear nucleus; g, granule cell layer; other abbreviations as in Figure 2.

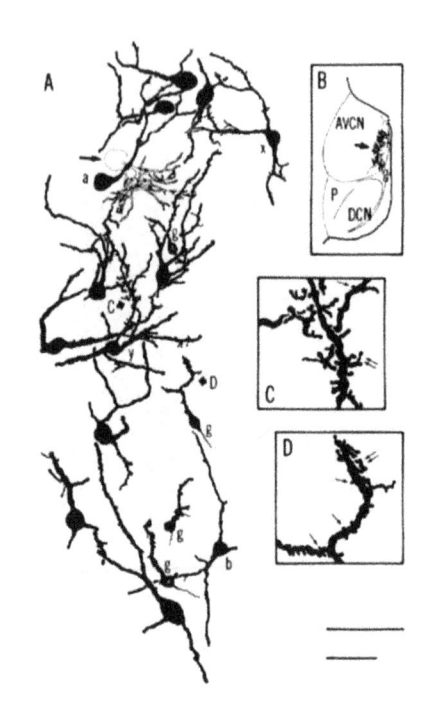

size as outlined above. The general organization of the neurons in the cap as revealed in Golgi impregnations is illustrated in Figures 2-4. Most of the neurons that are impregnated are similar to the stellate and small cells in the anteroventral cochlear nucleus described by Brawer et al. ('74). The dendrites of the neurons radiate in all directions, although there seems to be some tendency for them to follow the rostrocaudal contour of the cap anterodorsally (Figs. 2, 3) and to extend in a lateral to medial direction more posteroventrally (Figs. 2, 4). The distal dendrites of some of the multipolar neurons cross the boundary of the cap to enter the anterior or posterior divisions of the AVCN (e.g., Fig. 2, triple arrows; Fig. 4). Likewise, dendrites of neurons in these divisions may sometimes enter the cap. In addition, a few large spherical or globular bushy cells, characteristic of the adjacent divisions may sometimes be found within the cap (e.g., Fig. 2, curved arrows; Fig. 3A, long arrow), although this is relatively rare. This invasion of the cap by a few spherical or globular cells was also observed in Nissl-stained sections (Cant and Morest, '84). Granule cells in the granule cell layer are sometimes impregnated, along with a few small cells (Fig. 2). Granule cells are also present within the cap, where they are intermingled with the small and medium-sized multipolar neurons (Figs. 2-4).

Details of the dendritic morphology of the multipolar neurons in the cap are illustrated in Figures 3 and 4. The cell bodies give rise to several long dendrites that branch one or a few times. Some of the neurons have smooth dendrites with few appendages (e.g., cells a and b in Figure 2A; Fig. 4), but many of them have dendrites that are densely covered with spines and more complex appendages (e.g., cells x and y in Figure 2A, Fig. 2C). The multipolar cells with very spiny dendrites seem to be characteristic of the small cell cap; the dendrites of neurons in the medially adjacent areas of the AVCN have generally been described as smooth (Brawer et al., '74; Cant and Morest, '79a; Tolbert and Morest, '82a). Granule cells in the cap, like those in the granule cell layer, also give rise to spiny dentrites (Figs. 3A and 4, cells labelled g; Fig. 3D).

Most of the impregnated neurons in the cap have a morphology similar to the neurons illustrated in Figures 2-4, but we have seen at least one case of a Golgi cell as first described

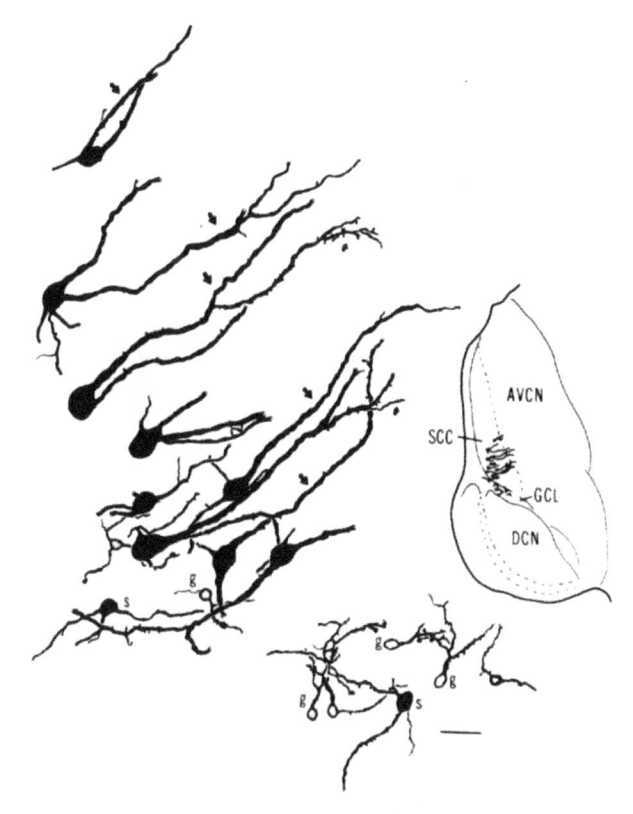

**Figure 4.** Cells in the small cell cap impregnated by the Golgi-Kopsch method. This is a composite drawing of cells from five horizontal sections 100 μm thick through the left cochlear nucleus. Straight arrows indicate dendrites of the cells of the cap that tend to run in a lateral to medial direction. Curved arrows indicate spiny dendritic tufts sometimes seen at the ends of these dendrites. s, small cells; g, granule cells. Inset illustrates the portion of the cochlear nucleus from which these drawings are taken. Abbreviations as in Figure 2. Scale bar is equal to approximately 20 μm for the large drawing and to approximately 0.5 μm for the inset drawing.

by Mugnaini et al. ('80) (Fig. 5). The dendrites of this cell are relatively thick and are heavily encrusted with spines and more elaborate appendages. The cell body also gives rise to spiny protrusions. Although the Golgi cell is difficult to impregnate with the Golgi methods (as also noted by Mugnaini et al., '80), it is seen commonly in electron micrographs.

## The synaptic organization of the small cell cap: Electron microscopic observations

The fine structure of the dorsolateral portion of the small cell caps of two adult cats was examined in the electron microscope. Preparation of the tissue for electron microscopy was as described previously (e.g., Cant and Morest, '79b). The boundaries of the cap were determined in toluidine- blue-stained sections such as that shown in Figure 1. Adjacent thin sections were cut, and boundary lines were transferred to maps drawn of the thin sections. All descriptions presented here are of areas well within the boundaries unless otherwise noted. Sections were cut large enough to include the granule cell layer, the cap and adjacent large cell areas (e.g., Fig. 1) so that comparisons of neuronal and synaptic morphology could be made in the same sections.

**Figure 5.** A Golgi cell (filled) and a granule cell (clear) in the small cell cap of an adult cat. These cells are from the same cat illustrated in Figure 2. Thin arrows indicate dendritic and somatic spines. Thick arrows indicate elaborate dendritic specializations typical of these cells. Scale bar = 10 μm.

Based on their appearance in the electron microscope and correlations with the cell types seen in Nissl-stained and Golgi impregnations, it is possible to identify at least three groups of neurons in the cap. These are cells with smooth somatic outlines, Golgi cells and granule cells. Most of the small and medium-sized cells have smooth somatic outlines (Fig. 6). Except for differences in their size, there are no obvious somatic features that would allow one to distinguish between different groups of these cells. Since the sizes of the somata of cells in the small cell cap form a continuous distribution (Cant and Morest, '84), any distinction based solely on size would have to be arbitrary. In the electron microscope, all of the smooth cells resemble the type I stellate cells of the anterior division of the AVCN (Cant, '81) in that the somata receive few, if any, synaptic contacts. The Golgi cells and granule cells appear to have the same fine structure as those described by Mugnaini et al. ('80) in the granule cell domains. The Golgi cells are the only cell type in the small cell cap that receives extensive somatic input. Otherwise, most of the synaptic contacts in the cap appear to be made with dendrites or dendritic appendages.

It has proven useful in previous electron microscopic studies of the ventral cochlear nucleus to classify synaptic terminals according to the size and shape of their synaptic vesicles. When such criteria are used to classify terminals in the small cell cap, the most common type is a terminal that contains small, round, densely packed synaptic vesicles (type SR terminals) (Figs. 7A, 9, 10, 11). Most of the other terminals in the cap contain flattened or pleomorphic vesicles, similar to those found in other parts of the cochlear nucleus (Figs. 9, 10). In the large cell areas of the AVCN, the most common type of synaptic terminal contains large, round synaptic vesicles (type LR terminals; e.g., Cant and Morest, '79b; Tolbert and Morest, '82b); this type of terminal is found only rarely in the small cell cap (Fig. 7B).

To confirm the impression that there is a difference in size in the round synaptic vesicles in terminals in the cap and those in the adjacent large cell areas, measurements of the cross-sectional areas of synaptic vesicles in randomly chosen terminals in each part of the nucleus were made. The results are given in Figure 8. Although our observations indicate that there are some SR terminals in the large cell areas and some LR terminals in the cap, the proportions are quite different so that, on average, the area of the synaptic vesicles is significantly smaller in the cap. It is also the case that the average size of round vesicles found in terminals in the granule cell layer is less than that in the cap. This difference is small but significant (Fig. 8) and can probably be accounted for by our observation that some

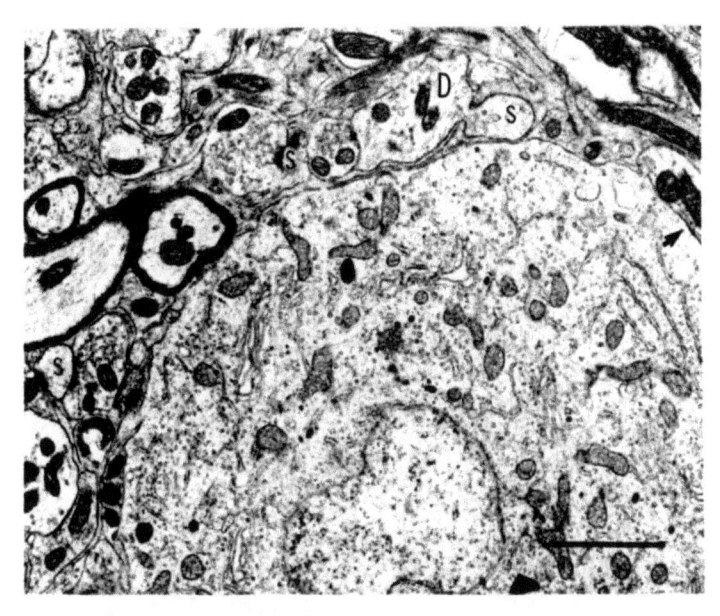

**Figure 6.** Neuron in the small cell cap of the cat. Like most of the larger cell bodies in the cap, this one receives only a few synaptic contacts on its somatic surface; one of these is indicated by the arrow. Most synaptic contacts are formed with dendrites (D) or dendritic spines (s) in the neuropil. Scale bar = 2 µm.

mossy fibers in the granule cell layer are filled with round synaptic vesicles that are very small. Although most of the terminals with round vesicles in the granule cell layer, including many of the mossy fiber terminals, look the same as the SR terminals in the cap, the mossy fibers with very small synaptic vesicles are not uncommon (cf. Mugnaini et al., '80).

The type SR terminals form many of their synaptic contacts with the spinous processes of the neurons of the small cell cap (Figs. 6, 7, 9), although they also contact dendritic shafts (Figs. 10, 11). As might be expected from the Golgi studies, the spinous contacts are very common in the neuropil of the cap. Since all sizes of neurons in the cap may have spiny dendrites, it is not possible to know which particular cell types receive these synapses without labelling the cells in some way. The terminals with flattened or pleomorphic vesicles almost always form synaptic contacts with the dendritic shafts of neurons rather than with the spines (Figs. 9, 10). Since the Golgi cells and granule cells seem to have a synaptic organization very similar to that described by Mugnaini et al. ('80), they will not be discussed further here.

To summarize the electron microscopic observations, there are a number of differences between the cap and the adjacent parts of the ventral cochlear nucleus that contain large cells (in particular, the anterior and posterior divisions of the AVCN). Except for the small Golgi cell, almost none of the neurons in the cap receives extensive somatic input. Although they are by far the most common type of terminal with round vesicles in the large cell areas, type LR terminals are rare in the cap. Terminals with round vesicles are common, but the vesicles are of smaller size (type SR terminals). It is very common to observe SR terminals forming synapses on spines in the cap; synaptic contacts with spines seem much rarer in the neuropil of the adjacent AVCN. In parts of the cap, the neuropil is very similar to that of the granule cell layer described in detail by Mugnaini et al. ('80). Mossy fibers and associated structures similar to those in the granule cell domains are found in the cap.

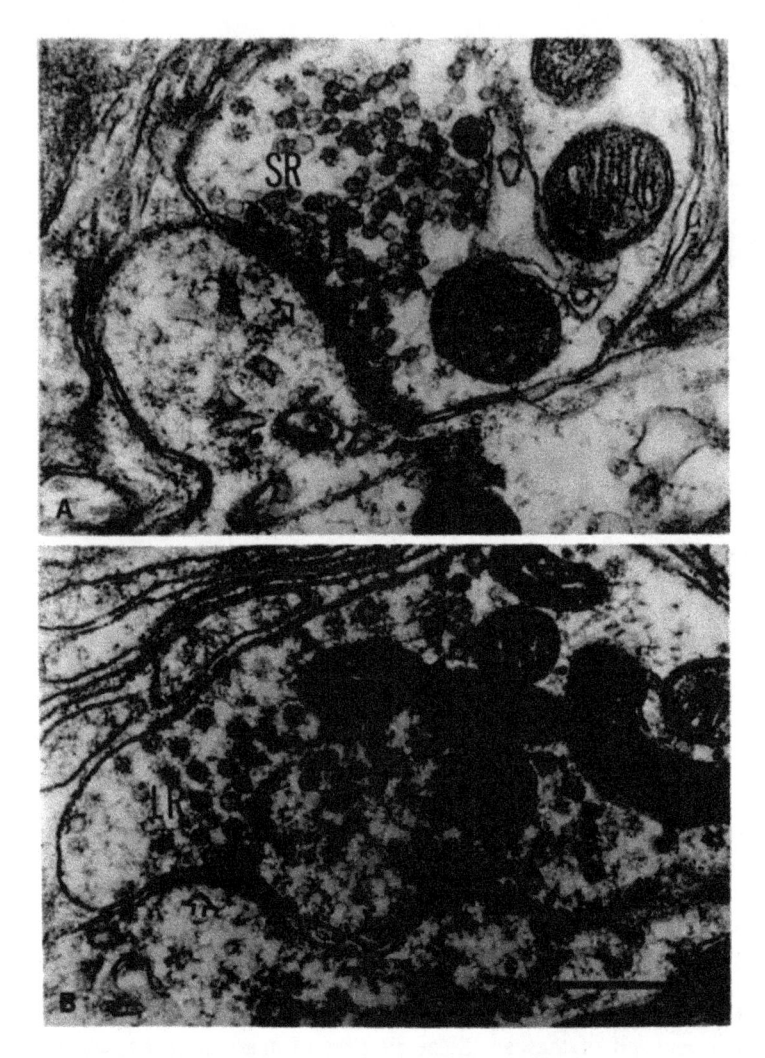

**Figure 7.** Two types of synaptic terminals with round synaptic vesicles in the ventral cochlear nucleus of the cat. Arrows indicate the synaptic contacts in both panels. A. Terminal with small, round (SR) synaptic vesicles. This is the most common type of terminal in the small cell cap; such terminals are relatively rare in the areas of large cells. B. Terminal with large, round (LR) synaptic vesicles. Although this terminal was located in the small cell cap, such terminals were rare in the parts of the cap examined in this study. Scale bar = 0.5 $\mu$m for both panels.

There are several limitations of these studies that prevent more than preliminary conclusions about the overall organization of the cap. Most importantly, only a very limited part of the cap was examined. Almost all of the synaptic contacts in the part of the cap that was examined were in the neuropil and not on the somatic surface of the neurons. This means that it is not possible to ascribe a particular pattern of synaptic organization to a particular cell type, since the origins of both the synaptic terminals and the dendrites in the neuropil are unknown. This makes it impossible to infer synaptic relationships among particular sources of inputs and cell types that give rise to particular output pathways.

Since there is no sharp border either between the cap and the GCL or between the cap and the large cell areas, it is not always possible to know whether some of the features seen in the cap are more properly considered attributes of these other areas. For example, it may be that mossy fiber glomeruli are a feature only of granule cell domains and that those that

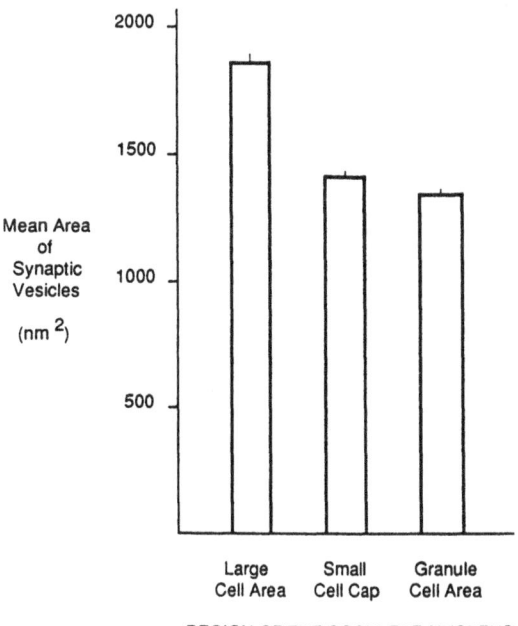

**Figure 8.** Histograms that summarize the results of measurements of the cross-sectional areas of round synaptic vesicles in three parts of the cochlear nucleus. All measurements were made on terminals in the same thin section. The first 15 synaptic terminals with round vesicles encountered in the granule cell layer, the small cell cap and the adjacent large cell area (the posterior division of the ventral cochlear nucleus in this case) were photographed. The areas of synaptic vesicles in each terminal were measured using a computerized measurement system (SigmaScan, Jandel Scientific). The areas of the vesicles from all terminals were averaged to obtain these histograms. The values (with the standard error of the mean) were $1859 \pm 31$ nm² for the posterior division, $1405 \pm 16$ nm² for the small cell cap, and $1340 \pm 18$ nm² for the granule cell layer. Independent t tests indicated that the average area of the synaptic vesicles in the cap is significantly different both from that in the large cell area ($P < 0.001$) and also from that in the granule cell area ($P < 0.01$).

appear to lie within the small cell cap are actually more appropriately considered a part of the granule cell domain that interdigitates with the cap. Likewise, it is possible that the few LR terminals found in the cap actually form synapses with dendrites of cells that lie in the large cell areas or with the few large cells that invade the cap. It would be very difficult to settle such issues at the electron microscopic level without marking systems of neurons in some way.

## Relation of the findings to previous studies of the small cell cap

Although few studies of the peripheral cap of small cells have been reported, some authors mention differences between the organization of marginal layers and that of more interior parts of the ventral cochlear nucleus of the cat. These include differences in the distribution of transmitter substances and transmitter-related enzymes (Osen and Roth, '69; Godfrey et al., '77a, b; Adams and Mugnaini, '87), differences in the distribution of cochlear and non-cochlear afferent axons (Osen, '69; Cant and Morest, '78; Fekete et al., '84; Leake and Snyder, '89), and differences in physiological response properties of single neurons (Bourk, '76). Similar medial vs. lateral differences have been reported in rodents (e.g.,

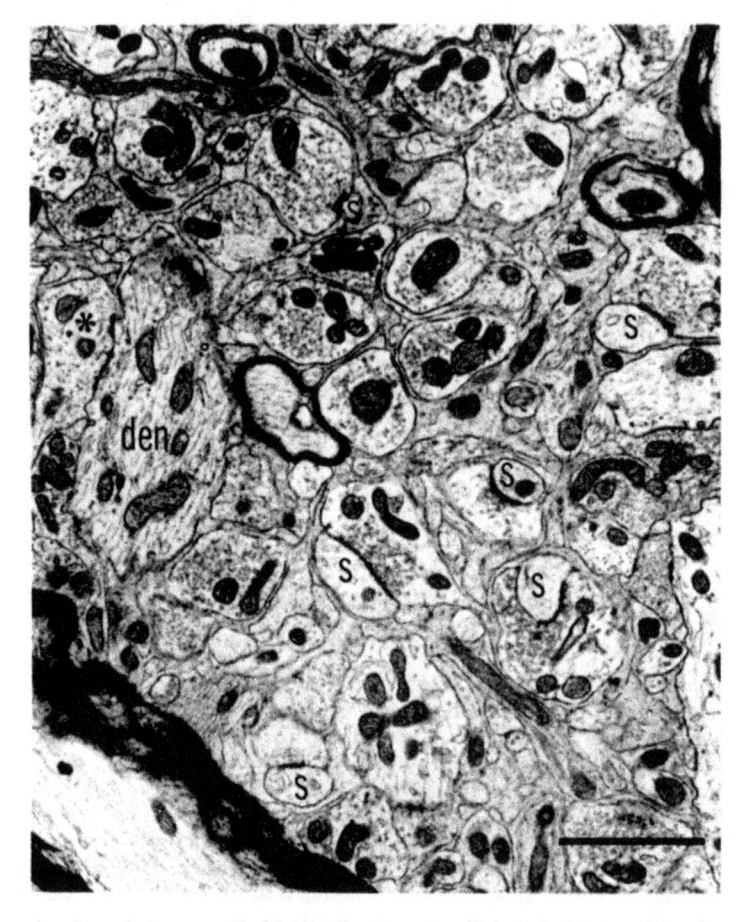

Figure 9. A section through the neuropil of the small cell cap that illustrates the common observation that many of the synapses formed in the cap are on dendritic spines (s). It could be determined at higher magnification that most of the synapses in this figure are made by terminals with small, round vesicles. Some of the synaptic contacts with dendritic shafts (den), however, are made by terminals with flattened or pleomorphic vesicles. One of these is indicated by an asterisk. Scale = 1.0 µm.

Godfrey et al., '78; Merchán et al., '85; Brown and Ledwith, '90), although a distinct peripheral cap of small cells has not been recognized in these animals (see below). Which of these distinctions between medial and more lateral regions represent differential organization *within* the large cell areas (i.e., from medial to lateral within an isofrequency lamina) and which might reflect the different organization of the cap and the large cell areas remains to be determined.

Before hypotheses can be formed about the functional role of the peripheral cap of small cells, it is important to know both the sources of input to its neurons and also their projection targets. Afferent input to the cap appears to arise both in the cochlea and from other sources. The cap receives its cochlear input from collaterals that arise from the main branches of the cochlear nerve (Osen, '70; '87). Recent results (Liberman, presentation at this meeting) indicate that most of this input is from type I auditory nerve fibers with low rates of spontaneous activity (low SR fibers). There is only a moderate input from medium SR fibers and almost no input from high SR fibers. Ryugo et al. (this volume) show that terminals of low SR fibers contain large, round synaptic vesicles (LR terminals). In the present study, it was found that there are few LR terminals in the cap. It is difficult to recon-

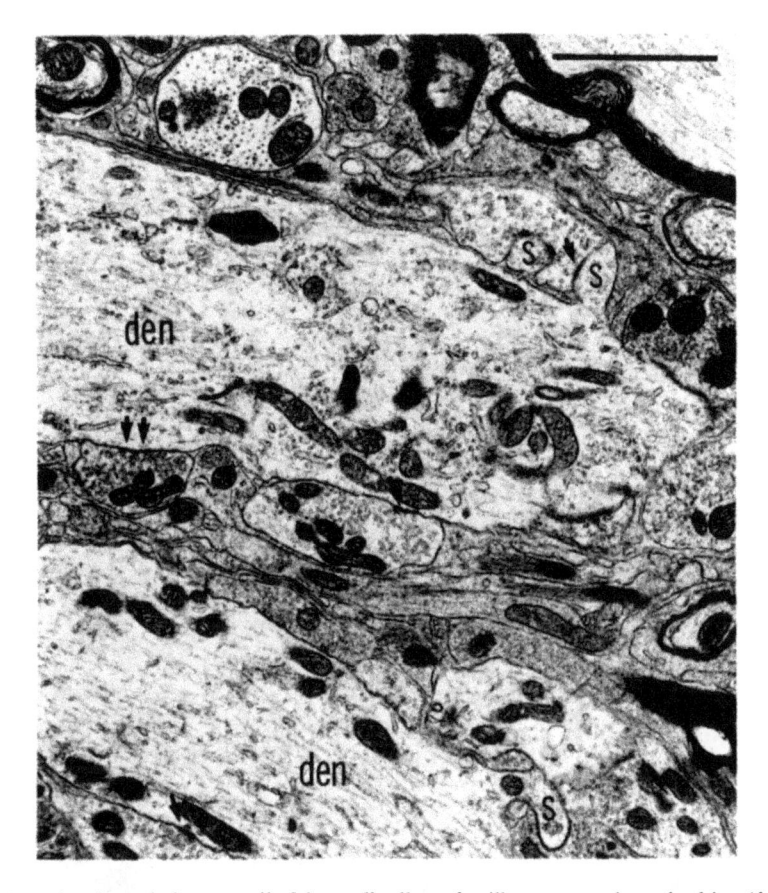

**Figure 10.** A section through the neuropil of the small cell cap that illustrates two large dendrites (den) that form synaptic contacts with terminals containing small, round vesicles (single arrows) or pleomorphic vesicles (double arrows). Some of the contacts are made with dendritic spines (s). Scale = 2 μm.

cile these observations. One possible explanation is that the low SR fibers are distributed unevenly in the cap and are sparse in the regions described here. Another is that low SR terminals may have more than one morphology.

Studies in rodents demonstrate that type II auditory nerve fibers give rise to collaterals that terminate in the neuropil that forms the border between the ventral cochlear nucleus and the granule cell layer (Brown et al., '88a; Brown and Ledwith, '90). The cells along this border resemble those of the small cell cap of cats, although there is no enlargement of the dorsolateral aspect of the border. Whether this thin cell layer is analogous to the cap of cats (which is itself only one cell thick throughout much of its extent) remains to be determined. In the cat, the type II fibers give rise to synapses with small, round vesicles (Ryugo et al., '91). It has not been shown in the cat that the type II axons project into the cap, but the data from rodents suggest this possibility. If so, the type II axons would be a potential source of the many terminals in the cap with small round synaptic vesicles.

In addition to its primary input, the cap receives inputs from other sources. A major source appears to be the superior olivary complex, which gives rise to cholinergic endings that appear to terminate in the cap (Osen and Roth, '69; Godfrey et al., '90). In the mouse, collaterals of the medial olivocochlear system (a cholinergic system) terminate beneath the granule cell layer and form synapses there with the dendrites (and sometimes with spines) of multipolar cells that lie at the border (Brown et al., '88b; Benson and Brown, '90). The

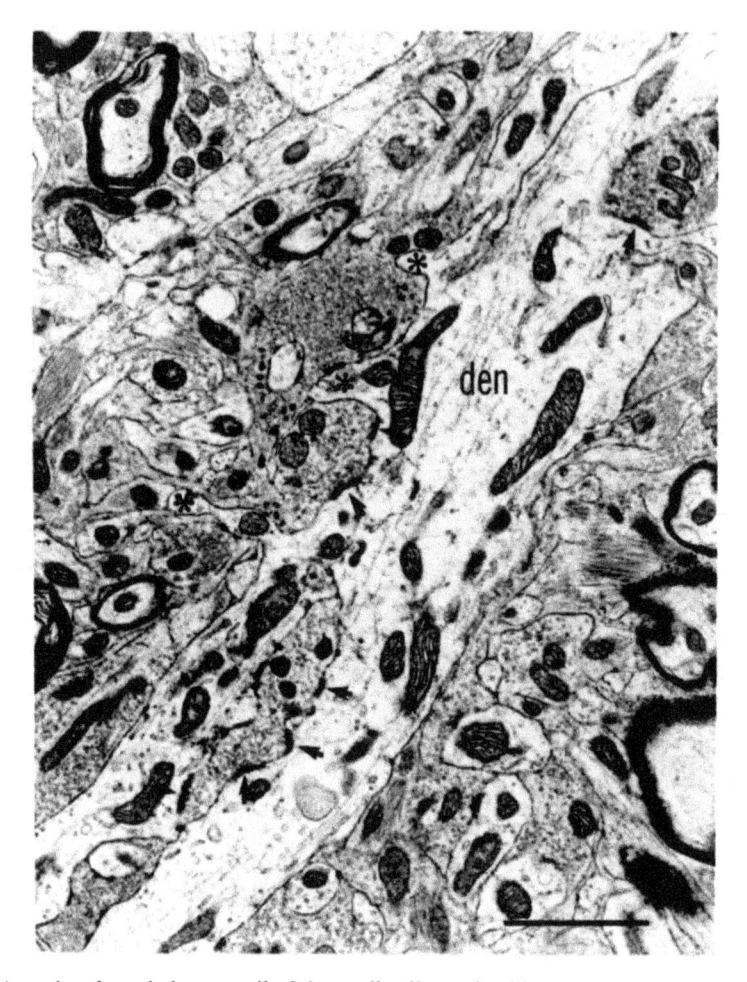

Figure 11. A section through the neuropil of the small cell cap that illustrates a dendrite (den) that receives synaptic contacts from terminals with small, round (SR) vesicles and also from terminals with flattened or pleomorphic vesicles. Asterisks indicate spines or other appendages that arise from the dendrite. Arrows indicate some of the synaptic contacts with SR terminals. The cell type to which this dendrite belongs is not known. It resembles dendrites of Golgi cells in the granule cell domains (Mugnaini et al., '80) or it could be one of the dendrites of a larger multipolar cell with spiny dendrites (cf. Fig. 3). Scale = 2.0 μm.

synaptic terminals contain small, round synaptic vesicles (Benson and Brown, '90). As noted above, the neurons in this border region could be analogous to neurons in the cap of cats. There are substantial projections of the collaterals of the medial olivocochlear axons to other border areas of the ventral cochlear nucleus as well (Brown et al., '91). Axons connecting the anteroventral and dorsal cochlear nuclei pass through the cap, some perhaps forming collaterals (Cant and Morest, '78; Lorente de Nó, '81). This could be a source of the GAD-positive terminals reported to be present in the cap (Adams and Mugnaini, '87). Some of these may correspond to the terminals with flattened or pleomorphic synaptic vesicles.

The efferent connections of the cap have not been studied in any detail, although there is evidence that at least some of the neurons project to the inferior colliculus (Adams, '79). This finding implies that the cells of the cap form part of a relatively short pathway for the higher processing of auditory information. Adams and Mugnaini ('87) report a relatively high number of GAD-positive cell bodies in some parts of the cap. Whether these cells participate in the projection to the inferior colliculus is not known; the possibility that some of these projections might be inhibitory is an interesting idea that awaits further study. Cells of the cap do not appear to project to the superior olivary complex (SOC). The small cell cap is the only part of the ventral cochlear nucleus that does not contain labelled cells after large injections of horseradish peroxidase in the SOC (Cant and Casseday, '86).

As pointed out by Osen ('69), there appears to be an inverse relationship in the prominence of the peripheral small cell cap and the dorsal cochlear nucleus. As the small cell cap becomes more prominent (e.g., as in humans, Moore and Osen, '79), the dorsal cochlear nucleus becomes apparently less well-organized. Both structures contain neurons that project to the inferior colliculus. The difference in relative prominence of the two structures in higher mammals may indicate a shift in importance from the dorsal cochlear nucleus to the small cell cap as a source of projections to the inferior colliculus.

A few studies of the physiology of the ventral cochlear nucleus of the cat have included recordings from single units in the small cell cap (Rose et al., '59; Rose, '60; Bourk, '76). The units in the cap appear to be arranged differently from those in the rest of the nucleus in terms of their characteristic frequencies (CF), a fact that led Bourk ('76) to suggest that this area might form part of a "CF system" separate from the main part of the anteroventral cochlear nucleus. More recent studies of the projection patterns of auditory nerve fibers offer an anatomical correlate of this finding. After small injections of horseradish peroxidase in the cochlea or cochlear nerve, labelled axons in the cap are "offset" from labelled isofrequency laminae in the main body of the nucleus (Fekete et al., '84; Leake and Snyder, '89). Bourk ('76) found that units located in the region of the small cell cap had physiological response properties quite different from those of units in the main subdivisions of the nucleus. He called these units unusual, since they did not fit well into any of the major response types. Although many parts of the nucleus have been described in some detail physiologically, no further studies of the cells of the small cell cap have been reported since these initial observations by Bourk.

The peripheral cap of small cells of the ventral cochlear nucleus, first described by Osen in '69, appears to be substantially different from other subdivisions of the nucleus in terms of its morphology, physiology and connections. What the function of the peripheral cap of small cells might be is completely unknown. Even the limited findings discussed above, however, suggest that it forms a functionally unique part of the ventral cochlear nucleus.

ACKNOWLEDGEMENTS

Research reported here was supported by grants from the NIH (RO1 DC00135) and the Deafness Research Foundation. I wish to thank Ms. Ava Krol for excellent and patient technical assistance.

REFERENCES

Adams, J.C., 1979, Ascending projections to the inferior colliculus, *J. Comp. Neurol.*, 183:519-538.
Adams, J.C., and Mugnaini, E., 1987, Patterns of glutamate decarboxylase immunostaining in the feline cochlear nuclear complex studied with silver enhancement and electron microscopy, *J. Comp. Neurol.*, 262:375-401.
Benson, T.E. and Brown, M.C., 1990, Synapses formed by olivocochlear axon branches in the mouse cochlear nucleus, *J. Comp. Neurol.*, 295:52-70.
Bourk, T.R., 1976, Electrical response of neural units in the anteroventral cochlear nucleus, Massachusetts Institute of Technology, Ph.D. Dissertation, Cambridge, MA.
Brawer, J.R., Morest, D.K., and Kane, E.C., 1974, The neuronal architecture of the cochlear nucleus in the cat, *J. Comp. Neurol.*, 160:491-506.
Braitenberg, V., Guglielmotti, V., and Sada, E., 1967, Correlation of crystal growth with the staining action of axons by the Golgi procedure, *Stain Tech.*, 42:277-283.
Brown, M.C., Berglund, A.M., Kiang, N.Y.S., and Ryugo, D.K., 1988a, Central trajectories of type II spiral ganglion neurons, *J. Comp. Neurol.*, 278:581-590.

Brown, M.C., Liberman, M.C., Benson, T.E., and Ryugo, D.K., 1988b, Brainstem branches from olivocochlear axons in cats and rodents, *J. Comp. Neurol.*, 278:591-603.

Brown, M.C. and Ledwith, J.V., 1990, Projections of thin (type-II) and thick (type-I) auditory-nerve fibers into the cochlear nucleus of the mouse, *Hearing Res.*, 49:105-118.

Brown, M.C., Pierce, S. and Berglund, A.M., 1991, Cochlear-nucleus branches of thick (medial) olivocochlear fibers in the mouse: A cochleotopic projection, *J. Comp. Neurol.*, 303:300-315.

Cant, N.B., 1981, The fine structure of two types of stellate cells in the anterior division of the anteroventral cochlear nucleus of the cat, *Neuroscience*, 6:2643-2655.

Cant, N.B. and Casseday, J.H., 1986, Projections from the anteroventral cochlear nucleus to the lateral and medial superior olivary nuclei, *J. Comp. Neurol.*, 247:457-476.

Cant, N.B., and Morest, D.K., 1978, Axons from non-cochlear sources in the anteroventral cochlear nucleus of the cat. A study with the rapid Golgi method, *Neuroscience*, 3:1003-1029.

Cant, N.B., and Morest, D.K., 1979a, Organization of the neurons in the anterior division of the anteroventral cochlear nucleus of the cat. Light microscopic observations, *Neuroscience*, 4:1909-1923.

Cant, N.B., and Morest, D.K., 1979b, The bushy cells in the anteroventral cochlear nucleus of the cat. A study with the electron microscope, *Neuroscience*, 4:1925-1945.

Cant, N.B., and Morest, D.K., 1984, The structural basis for stimulus coding in the cochlear nucleus of the cat, *in*: "Hearing Science:Recent Advances," C.I. Berlin, ed., College-Hill Press, San Diego.

Fekete, D.M., Rouiller, E.M., Liberman, M.C., and Ryugo, D.K., 1984, The central projections of intracellularly labeled auditory nerve fibers in cats, *J. Comp. Neurol.*, 229:432-450.

Fuse, G., 1913, Das Ganglion ventrale und das Tuberculum acusticum bei einigen Säugern und beim Menschen, *Arb. Hirnanat. Inst. Zürich*, 7:1-210.

Godfrey, D.A., Carter, J.A., Berger, S.J., Lowry, O.H., and Matschinsky, F.M., 1977a, Quantitative histochemical mapping of candidate transmitter amino acids in cat cochlear nucleus, *J. Histochem. Cytochem.*, 25:417-431.

Godfrey, D.A., Williams, A.D., and Matschinsky, F.M., 1977b, Quantitative histochemical mapping of enzymes of the cholinergic system in cat cochlear nucleus, *J. Histochem. Cytochem.*, 25:397-416.

Godfrey, D.A., Carter, J.A., Lowry, O.H., Matschinsky, F.M., 1978, Distribution of gamma-aminobutyric acid, glycine, glutamate and aspartate in the cochlear nucleus of the rat, *J. Histochem. Cytochem.*, 26:118-126.

Godfrey, D.A., Beranek, K.A., Carlson, L., Parli, J.A., Dunn, J.D., Ross, D.C., 1990, Contribution of centrifugal innervation to choline acetyltransferase activity in the cat cochlear nucleus, *Hearing Res.*, 49:259-280.

Kane, E.S.C., 1973, Octopus cells in the cochlear nucleus of the cat: Heterotypic synapses upon homeotypic neurons, *Int. J. Neurosci.*, 5:251-279.

Leake, P.A., and Snyder, R.L., 1989, Topographic organization of the central projections of the spiral ganglion in cats, *J. Comp. Neurol.*, 281:612-629.

Lorente de Nó, R., 1933, Anatomy of the eighth nerve--III. General plans of structure of the primary cochlear nuclei, *Laryngoscope*, 43:327-350.

Lorente de Nó, R., 1981, "The Primary Acoustic Nuclei," Raven Press, New York.

Merchán, M.A., Collia, F.P., Merchán, J.A., and Saldaña, E., 1985, Distribution of primary afferent fibres in the cochlear nuclei. A silver and horseradish peroxidase (HRP) study, *J. Anat.*, 141:121-130.

Moore, J.K., and Osen, K.K., 1979, The cochlear nuclei in man, *J. Comp. Neurol.*, 154:393-418.

Mugnaini, E., Osen, K.K., Dahl, A.-L., Friedrich, V.L., and Korte, G., 1980, Fine structure of granule cells and related interneurons (termed Golgi cells) in the cochlear nuclear complex of cat, rat and mouse, *J. Neurocytol.*, 9:537-570.

Osen, K.K., 1969, Cytoarchitecture of the cochlear nuclei in the cat, *J. Comp. Neurol.*, 136:453-484.

Osen, K.K., 1970, Course and termination of the primary afferents in the cochlear nuclei of the cat. An experimental anatomical study, *Arch. Ital. Biol.*, 108:21-51.

Osen, K.K., 1987, Anatomy of the mammalian cochlear nuclei: A review, *in*: "Auditory Pathway. Structure and Function," J. Syka and R.B. Masterton, eds., Plenum Press, New York.

Osen, K.K., and Roth, K., 1969, Histochemical localization of cholinesterases in the cochlear nuclei of the cat, with notes on the origin of acetylcholinesterase-positive afferents and the superior olive, *Brain Res.*, 16:165-184.

Rose, J.E., Galambos, R., and Hughes, J.R., 1959, Microelectrode studies of the cochlear nuclei of the cat, *Bull. Johns Hopkins Hosp.*, 104:211-251.

Rose, J.E., 1960, Organization of frequency sensitive neurons in the cochlear nuclear complex of the cat, *in*: "Neural Mechanisms of the Auditory and Vestibular Systems," G. L. Rasmussen and W. Windle, eds., Charles C Thomas, Springfield, IL.

Ryugo, D.K., Dodds, L.W., Benson, T.E., and Kiang, N.Y.S., 1991, Unmyelinated axons of the auditory nerve in cats, *J. Comp. Neurol.*, 308:209-223.

Saint Marie, R.L., Morest, D.K., and Brandon, C.J., 1989, The form and distribution of GABAergic synapses on the principal cell types of the ventral cochlear nucleus of the cat, *Hearing Res.*, 42:97-112.

Smith, P.H., and Rhode, W.S., 1987, Characterization of HRP-labeled globular bushy cells in the cat anteroventral cochlear nucleus, *J. Comp. Neurol.*, 266:360-375.

Smith, P.H., and Rhode, W.S., 1989, Structural and functional properties distinguish two types of multipolar cells in the ventral cochlear nucleus, *J. Comp. Neurol.*, 282:595-616.

Tolbert, L.P., and Morest, D.K., 1982a, The neuronal architecture of the anteroventral cochlear nucleus of the cat in the region of the cochlear nerve root: Golgi and Nissl methods, *Neuroscience*, 7:3013-3030.

Tolbert, L.P., and Morest, D.K., 1982b, The neuronal architecture of the anteroventral cochlear nucleus of the cat in the region of the cochlear nerve root: Electron microscopy, *Neuroscience*, 7:3053-3068.

van Noort, J., 1969, The Structure and Connections of the Inferior Colliculus. An Investigation of the Lower Auditory System, Van Gorcum, Assen.

# ALTERATIONS IN THE DORSAL COCHLEAR NUCLEUS OF CEREBELLAR MUTANT MICE

Albert S. Berrebi and Enrico Mugnaini

Laboratory of Neuromorphology, Graduate Degree Program in
Biobehavioral Sciences, The University of Connecticut, Storrs, Connecticut
06269-4154, USA

The idea that the dorsal cochlear nucleus (DCoN) contains a cerebellar-like microcircuit in its superficial layers was introduced several years ago (Lorente de Nó, '81; Osen and Mugnaini, '81). Support for this hypothesis comes from the fact that certain neuronal cell types residing in the superficial DCoN layers resemble those in the cerebellar cortex (Mugnaini et al., '80a,b; Wouterlood et al., '84; Wouterlood and Mugnaini, '84; Mugnaini, '85). Within this framework, our laboratory has been particularly interested in the so-called cartwheel neurons, originally termed globular or Type C cells by Lorente de Nó ('33, '79, '81). These neurons, with rounded cell bodies and extremely spiny dendrites that branch extensively in the molecular layer, are presumed to represent the equivalent of cerebellar Purkinje cells. The two cell populations derive from neighboring regions of the neuroepithelium at the rhombic lip, are generated during the same embryonic period (Miale and Sidman, '61; Altman and Das, '66; Taber-Pierce, '67; Martin and Rickets, '81; Willard and Martin, '86), and share several ultrastructural (Wouterlood and Mugnaini, '84) and neurochemical phenotypes (Mugnaini, '85; Mugnaini and Morgan, '87; Mugnaini et al., '87; Saito et al., '88; Mignery et al., '89; Osen et al., '90; and our unpublished observations using antisera to calbindin-28kD (Christakos et al., '87), the cyclic GMP-dependent protein kinase (Lohmann et al., '81), and the G-substrate (Schlichter et al., '78) (see Table I). Additional evidence militating for a direct lineage relationship between cartwheel and Purkinje cells would be the demonstration that both cell types are equally affected by different genetic mutations mapping to single loci of diverse chromosomes.

We explored this possibility utilizing murine models of cerebellar neuropathology (Berrebi and Mugnaini, '88; Berrebi et al., '90). Our studies, therefore, made use of a set of mutations different from those that affect primarily the inner and middle ear (Green, '81; Steel et al., '83; Crenshaw et al., '91). Several mutations have been reported that produce cerebellar abnormalities (Breakefield, '79; Green, '81), and strains of mice with these gene defects are commercially available from the Jackson Laboratory, Bar Harbor, Maine. To date, however, none of the genes that determine these mutant phenotypes have been fully characterized. In a series of studies, we have examined the DCoN and the cerebellum of four Purkinje cell mutants (Table II) to determine whether cartwheel neurons are affected by neurological mutations which alter Purkinje cells directly, i.e. not through a cascade of transneuronal degeneration.

Table I. Similarities between cerebellar Purkinje cells and DCoN cartwheel neurons

| | |
|---|---|
| Embryological: | -originate from the rhombic lip<br>-"born" on days E12-E13 in mouse |
| Structural: | -high density of dendritic spines<br>-innervated by parallel fibers (glutamatergic)<br>-receive only symmetric synapses on their cell bodies<br>-possess subsurface cistern-mitochondrion complexes<br>-have myelinated axons |
| Chemical: | -express GAD, GABA<br>-express cerebellin<br>-express PEP-19<br>-express protein kinase C<br>-express Inositol 1,4,5-Trisphosphate receptor<br>-express calbindin<br>-express cGMP-dependent protein kinase<br>-express G-substrate |

Table II. Murine mutations which affect cerebellar Purkinje cells

| Mutation | Gene Symbol | Chromosome Number | Mode of Inheritance | Background Strain |
|---|---|---|---|---|
| Lurcher | Lc | 6 | semi-dominant | B6CBA |
| Purkinje Cell Degeneration | pcd | 13 | recessive | C57BR/cdJ |
| Staggerer | sg | 9 | recessive | B6C3Fe |
| Nervous | nr | 8 | recessive | BALB/cGR |
| Reeler | rl | 5 | recessive | 'snowy-bellied' |

Electron microscopy and recently discovered immunocytochemical probes were employed to address this problem. Antiserum to PEP-19, which reveals the entire populations of both cell types (Mugnaini et al., '87) or to L7 protein, which is only expressed by Purkinje cells in brain (Oberdick et al., '88; Berrebi and Mugnaini, '92), were applied to tissue sections according to protocols standardized in our laboratory (Mugnaini and Dahl, '83; Mugnaini et al., '87).

Figure 1 demonstrates PEP-19 immunostaining in the cerebellar cortex and the DCoN of a normal C57BL/10 mouse. Dense staining of the molecular and Purkinje cell layers throughout the cerebellar cortex results from immunolocalization of the antigen in the entire Purkinje neuron: the cell body, dendritic arbor, and axon including the recurrent collaterals of the infra- and supraganglionic plexuses, and the terminals in the deep cerebellar nuclei (Fig. 1A).

In the normal mouse DCoN, PEP-19 antiserum reveals a population of densely immunostained medium-sized cells, identified as cartwheel neurons (Fig. 1B). Their cell bodies and dendrites are situated in layers 1 and 2. Cartwheel cell axons, not visible at low magnification (but see Fig. 9 of Mugnaini et al., '87), ramify throughout layers 1 and 2 and immediately beneath the pyramidal neurons in the upper part of the deep DCoN region.

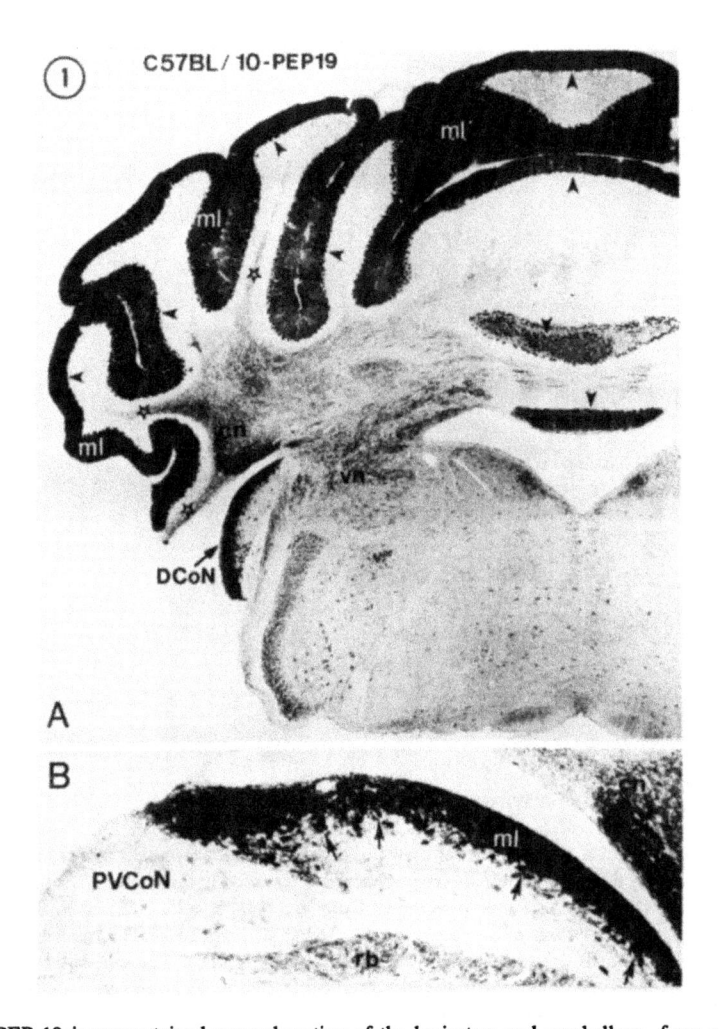

**Figure 1.** A) PEP-19 immunostained coronal section of the brainstem and cerebellum from a normal adult mouse. Throughout the cerebellar cortex, the molecular layer (ml) and the Purkinje cell layer (arrowheads) are densely immunoreactive. Purkinje axons in the folial white matter (stars) and their terminals in the cerebellar nuclei (cn) and vestibular nuclei (vn) are also stained. In this wet-mounted tissue section, the high contrast observed in the cross-cut fiber bundles of the spinal trigeminal tract, the restiform body and the pyramidal tract is due to diffraction and not to immunoreactivity. Moderately labeled neurons are scattered in the brainstem, but the only other region containing a high concentration of densely PEP-19-immunoreactive cells is the DCoN. x20. B) PEP-19 immunostained coronal section of a normal mouse DCoN, rotated 90° clockwise. A portion of the posterior division of the ventral cochlear nucleus (PVCoN) is included to the left. Densely immunoreactive cartwheel cell bodies (arrows) are evident in layers 1 and 2, and their dendritic trees occupy layer 1, the molecular layer (ml). Note the sharp border between the DCoN superficial layers and the PVCoN. cn, cerebellar nuclei; rb, restiform body. x60.

**Mutant Mice:** In the Lurcher (Lc) mutant mouse (Fig. 2), nearly the entire population of cerebellar Purkinje cells undergoes degeneration between the first and third postnatal weeks (P7 - P21) (Caddy and Biscoe, '76, '79). As a result, PEP-19 immunostaining of the adult Lc cerebellum revealed only rare remaining Purkinje cells (Fig. 2A). These had stunted dendritic trees oriented towards the molecular layer, which was much reduced in thickness. In the Lc DCoN, PEP-19 immunostaining revealed only a few immunostained cartwheel neurons (Fig. 2B), and the molecular layer was nearly free of immunoreaction product.

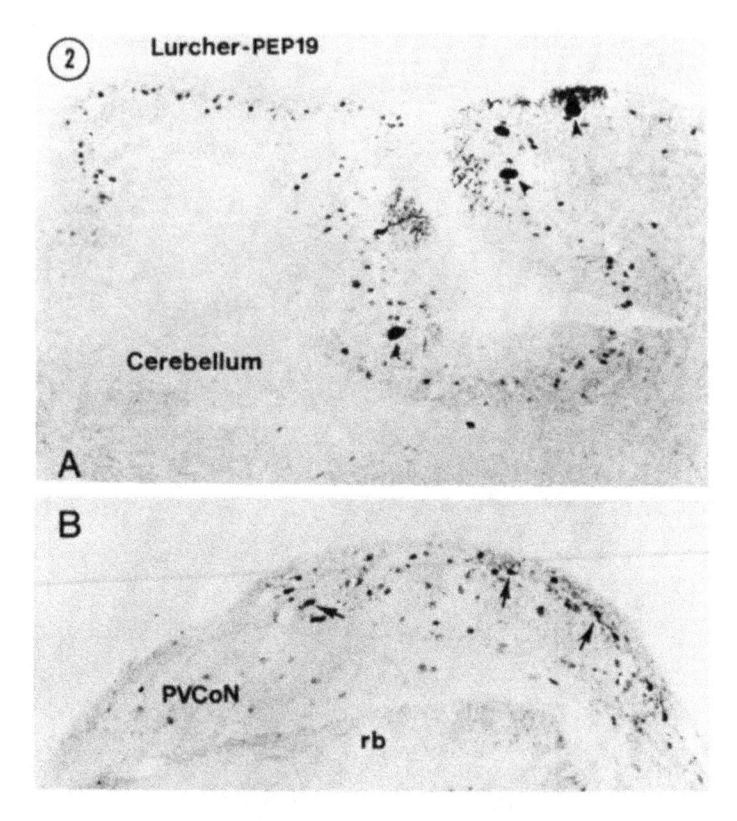

Figure 2. A) Coronal section of the cerebellar cortex of an adult 'Lurcher' mutant mouse, immunostained with PEP-19 antiserum. Only a few Purkinje cells (arrowheads) are scattered in the cortex. The small, faintly stained cell bodies located in the cortex presumably represent Golgi and basket cells. x85. B) PEP-19 immunostained coronal section of the DCoN from a 'Lurcher' mutant, rotated 90° clockwise. The number of immunostained cartwheel neurons (arrows) is drastically reduced (compare with Fig. 1B). The molecular layer is nearly free of immunoreactive dendrites. The actual degeneration of cartwheel neurons was confirmed by electron microscopy (see text). rb, restiform body; PVCoN, posteroventral cochlear nucleus. x60.

This light microscopic observation suggested that the vast majority of cartwheel neurons, like the Purkinje cells, had undergone degeneration as a result of the Lc mutation. However, an alternative explanation for these findings could be that affected cartwheel cells were unable to synthesize PEP-19, and therefore were not revealed by the antiserum. To test this possibility, we undertook examination of the mutant DCoN by electron microscopy, using established ultrastructural criteria to classify murine cartwheel cells (Berrebi and Mugnaini, '88). While normal cartwheel neurons were rarely recognizable, a number of cartwheel neurons in various stages of degeneration were observed. Furthermore, an extreme paucity of dendritic processes in the molecular layer was noted. Other neuronal populations in the cochlear nuclei seemed unaffected by the mutation. We concluded, therefore, that a specific and massive loss of cartwheel neurons takes place in the Lc mutant.

The basic findings of the Lc experiments were replicated in studies of other mutant mice, namely Purkinje Cell Degeneration (pcd) and staggerer (sg) (not shown). In the pcd mouse, the entire population of Purkinje cells degenerates, beginning at P15-P18 and ending about two weeks later (Mullen at al., '76). The sg mutation results in the failure of Purkinje cells to develop branchlet spines during synaptogenesis (Sotelo and Changeux, '74). This leads to the degeneration of a substantial proportion of Purkinje cells, and secondarily, to the loss of the vast majority of granule cells (Sidman et al., '62; Sotelo and Changeux, '74; Herrup and Mullen, '79). The few Purkinje cells that survive into adulthood are delayed in

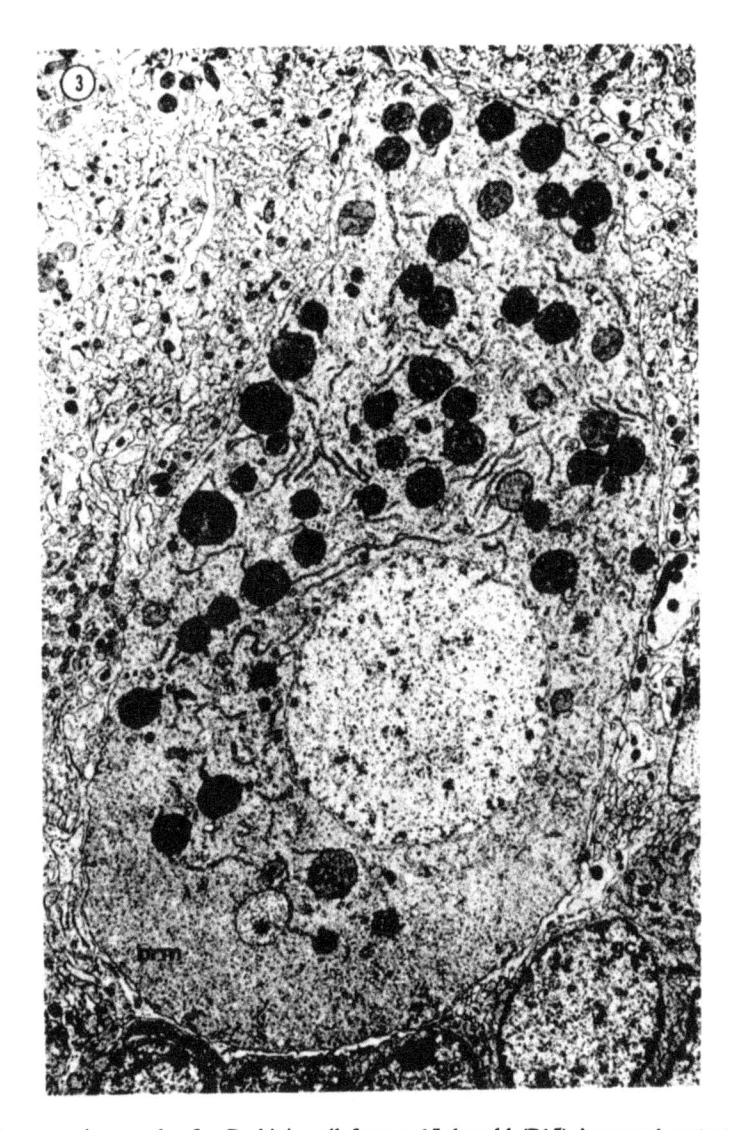

**Figure 3.** Electron micrograph of a Purkinje cell from a 15-day-old (P15) 'nervous' mutant mouse. Most mitochondria are larger, rounder and more electron dense than normal. Asterisk indicates a degenerating, electron-lucent mitochondrion. Granule cells (gc), located immediately beneath the Purkinje cell, have normal mitochondria. brm, basal ribosomal mass. x5,150.

their maturation, and possess short, stub-like dendrites that display few spines (Hirano and Dembitzer, '75; Yoon, '76). In both pcd and sg, PEP-19 immunostaining in the adult cerebellum and DCoN was virtually abolished. Moreover, by electron microscopy we observed degenerating cartwheel neurons in young animals and the delayed formation of dendritic spines by the few remaining cartwheel neurons in adult sg mutants (Berrebi and Mugnaini, '90).

The nervous (nr) mutation results in the loss of 90% of cerebellar Purkinje cells between P23-P50 (Sidman and Green, '70). However, by P15 the entire Purkinje cell population displays a set of characteristic cytological alterations (Landis, '73; Berrebi and Mugnaini, '88). Most conspicuously, mutant Purkinje cells possess a varying number of enlarged, rounded mitochondria with a higher than normal density of cristae (Fig. 3). A

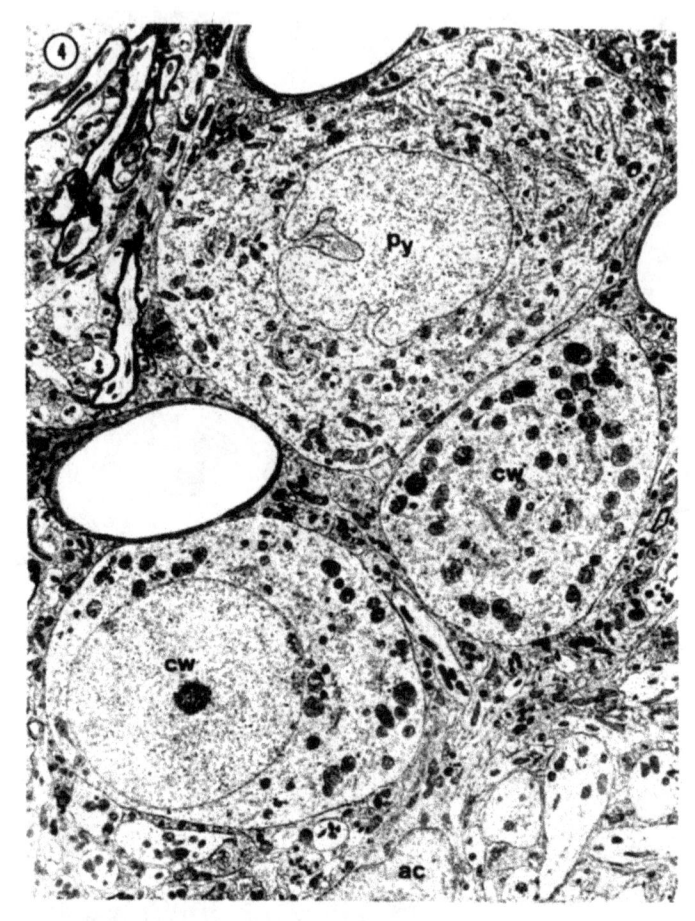

**Figure 4.** Electron micrograph from the DCoN of a P15 'nervous' mutant mouse. Two cartwheel neurons (cw) display enlargement, rounding and increased electron density of mitochondria, while the neighboring pyramidal neuron (py) and an astrocyte (ac) appear unaffected. By this age, all cartwheel neurons are similarly affected by the mutation. x4,950.

similar mitochondrial abnormality was observed at P15 in all nr cartwheel neurons, but not in other DCoN cell classes (Fig. 4).

Work in progress with another mutation, reeler (rl) (see Table II), suggests that the DCoN may contain signal molecules that enhance Purkinje cell growth. In the adult rl, only about 50% of the normal number of Purkinje cells remain (Heckroth et al., '89). However, due to a still unexplained defect acting throughout the entire brain, only a small percentage of Purkinje cells are found in their normal position in the cortex (see Caviness and Rakic, '78 and Caviness et al., '88 for reviews). We used antiserum to L7 protein, a Purkinje cell marker that is not present in cartwheel neurons, to immunostain the rl cerebellum. We observed that the vast majority of mutant Purkinje cells were clustered in the deep cerebellar mass (Fig. 5), as previously shown with other methods (Mariani et al., '77; Heckroth et al., '89). Remarkably, we discovered that misplaced Purkinje cells, located laterally at the base of the cerebellum, had extended their dendrites into the molecular layer of the DCoN, which was otherwise free of L7-immunoreaction (Berrebi and Mugnaini, '89). The axons of these Purkinje cells, therefore, must have remained within the cerebellar nuclei. Immunoelectron microscopy confirmed that the foreign dendrites in DCoN possess numerous spines and form *en passant* synaptic contacts with parallel fiber varicosities that contain

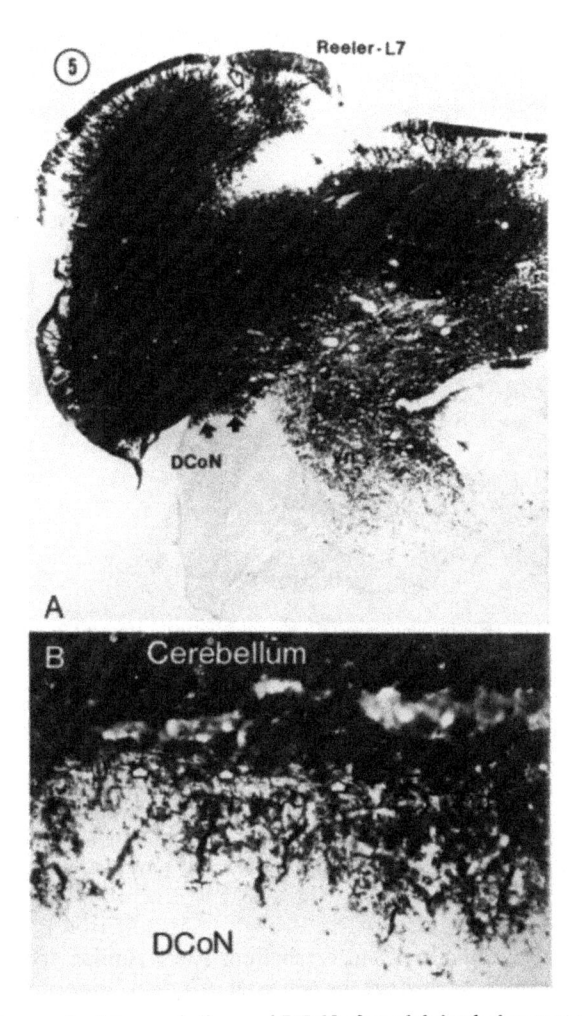

**Figure 5.** A) Light micrograph of the cerebellum and DCoN of an adult 'reeler' mutant mouse, immunostained with antiserum to L7. This antiserum reveals only cerebellar Purkinje cells, most of which form clusters in the deep cerebellar mass. The few Purkinje cells that have reached the cortex are evident (open arrowheads), while those which have remained in the central cerebellum are masked by the thick web of immunoreactive axons and dendrites. The DCoN, which is normally free of L7, contains the immunostained dendrites (arrows) of misplaced Purkinje cells located at the base of the cerebellum. vn, vestibular nucleus. x33. B) High magnification of a region of apposition between the cerebellum and DCoN. The misplaced Purkinje cell bodies remain within the cerebellum, but their dendritic processes have grown into the DCoN. The border between the two structures is demarcated by bundles of thin immunoreactive Purkinje cell axons (thick arrowheads). x165.

round synaptic vesicles. The ultrastructural features of the contacts appear normal, suggesting that the synapses are functional (unpublished observations). Presumably, cochlear nuclear granule cells are sufficiently similar to their counterparts in the cerebellum to induce an aberrant ingrowth of Purkinje cell dendrites, which they innervate with their axons.

These experiments demonstrated that four independent mutations, Lc, pcd, sg, and nr, cause similar defects in cerebellar Purkinje cells and DCoN cartwheel neurons. The results provided compelling evidence that the repertoire of genes expressed in these cell types is very similar. Thus, we proposed that cartwheel and Purkinje cells are derived from closely related precursors that assume different positions at the rhombic lip and are influenced by different developmental cues very early in embryogenesis. Furthermore, the rl data suggest that

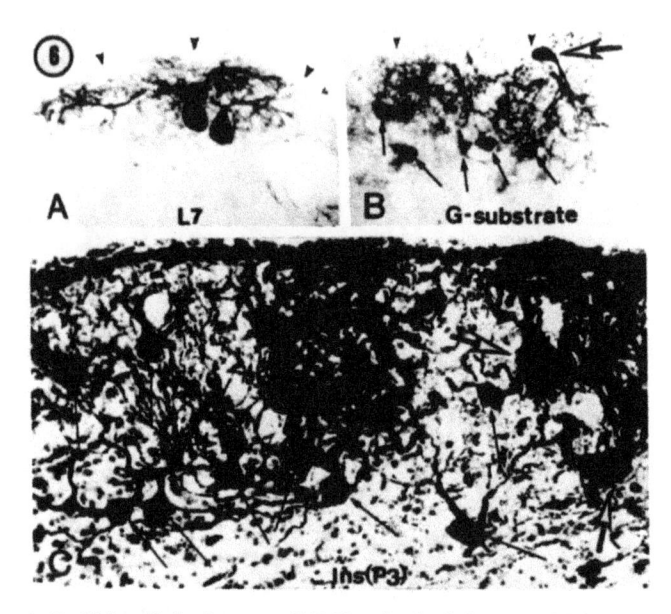

Figure 6. A) Ectopic Purkinje cells in the mouse DCoN maintain their strong L7-immunopositivity. Note the extensive branching of their dendritic trees, which are oriented towards the ependymal surface of the DCoN (marked by small arrowheads). x185. B) Ectopic Purkinje neuron (large arrow), located subependymally in the mouse DCoN, is revealed by G-substrate antiserum. The dendritic field of this cell is inverted, and it arborizes in the molecular layer. Also visible, on the left, are the inverted dendrites of a second Purkinje neuron, whose cell body is not included in this section. DCoN cartwheel neurons (thin arrows) are more lightly stained by this antiserum. Small arrowheads mark the ependymal layer. x185. C) Antiserum to the Ins(P3) receptor stains two ectopic Purkinje cells in the guinea pig DCoN (large arrows). The dendritic arbor of another Purkinje cell, whose soma is not visible, is present in the center of the field. Cartwheel neurons (thin arrows) are smaller and less densely immunoreactive than Purkinje cells. (Photograph kindly provided by Dr. J.C. Adams). x300.

granule cells of the cochlear nuclei and cerebellum exert similar effects on Purkinje cell dendritic growth. Therefore, we believe the neuronal cell populations of the cerebellar cortex and the superficial layers of the DCoN form remarkably similar neuronal circuits.

## Ectopic Purkinje and Cartwheel Neurons

Both Purkinje cells and cartwheel neurons may be found in ectopic positions in the mature brain, as indicated by immunocytochemical studies and Golgi impregnations. Ectopic Purkinje cells have been demonstrated not only within the cerebellar nuclei, in the medullary vela, and along the course of the cerebellar peduncles, but also in the molecular layer of the DCoN (De Camilli et al., '84; Mugnaini, '85; Osen, '85; Mugnaini et al., '87; Mugnaini and Morgan, '87; Mignery et al., '89). Ectopic cartwheel neurons have been observed in all subdivisions of the granule cell domain, including the DCoN, the lamina, the medial sheet, the superficial granular layer and the subpeduncular corner (Osen, '85; Berrebi and Mugnaini, '91). The dendritic arbors of the ectopic cartwheel cells appear somewhat stunted or distorted and their geometry is altered compared to that of normally situated cells. Among the ectopic Purkinje cells, those found in the DCoN have the most developed dendritic fields. Purkinje cells situated in the DCoN can easily be distinguished from cartwheel neurons, in spite of their situation in the same tissue compartment (Fig. 6). Their cell bodies are larger than those of cartwheel neurons, their dendrites are more extensively branched, they appear L7-positive while cartwheel neurons are L7-immunonegative (Fig. 6A) and they stain more

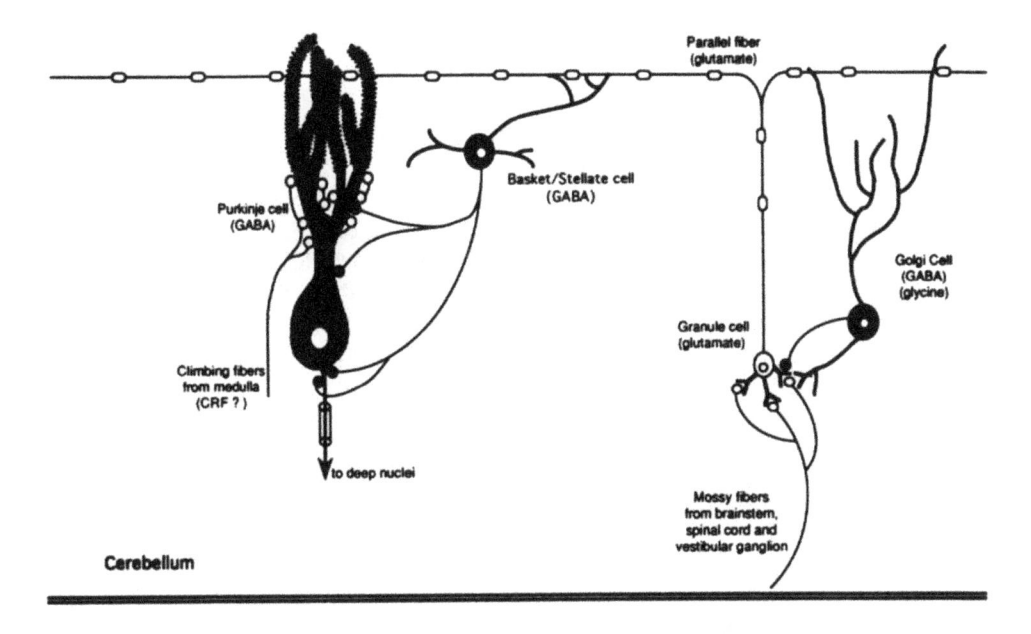

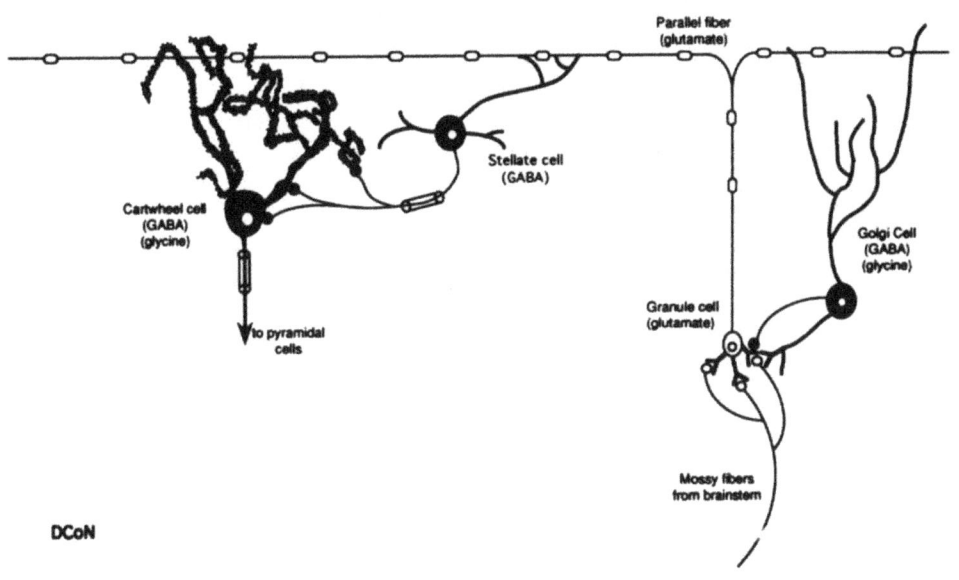

**Figure 7.** This diagram highlights the differences between Purkinje cells of the cerebellum (top) and cartwheel neurons of the DCoN (bottom). See text for detailed description.

densely with antisera to shared markers such as the cyclic GMP-dependent protein kinase substrate (Fig. 6B), and the Ins(P3) receptor (Fig. 6C). The dendritic orientation and branching pattern of ectopic Purkinje cells situated in the DCoN is highly variable, and like that of cartwheel neurons, depends on the position of the cell body. Subependymal Purkinje cell bodies, for example, have inverted dendritic fields (Fig. 6B). These data indicate that certain aspects of the genetic specification of Purkinje and cartwheel neurons, whether determined by cell-intrinsic or cell-extrinsic factors, may be triggered prior to cell migration, and furthermore, that the commitment to one or the other phenotype is irreversible. However, dendritic orientation and the precise geometry of the branching are determined after cells have migrated to their final position.

## Differences Between Purkinje and Cartwheel Neurons

Although we have repeatedly pointed out their similarities, phenotypical differences between Purkinje and cartwheel cells should not be overlooked (Fig. 7). Unlike Purkinje cells, cartwheel neurons do not receive a climbing fiber input onto their dendritic trunks, nor do they receive an inhibitory pinceau formation around their axon initial segments. Furthermore, the cell bodies of cartwheel neurons are not arranged in a monolayer, their dendritic fields are not monoplanar and their axons are relatively short, although these features vary somewhat among species. For example, in the guinea pig, where lamination of the DCoN is most clear, cartwheel neurons are located in a fairly regular band just above layer 2 and show some dendritic field anisotropy (Hackney et al., '90; Berrebi and Mugnaini, '91). A notable dissimilarity in chemical phenotypes is that cartwheel cells contain immunocytochemically detectable amounts of GABA and glycine (Mugnaini, '85; Wenthold et al., '86; Wenthold et al., '87; Osen et al., '90), while Purkinje cells are glycine-negative (Ottersen and Storm-Mathisen, '84; our unpublished observations).

Recently, it has been shown that some degree of chemical heterogeneity is present even within the rather stereotyped Purkinje cell population (Hawkes and LeClerc, '87). Positional variations in certain phenotypes are evident at very early morphogenetic stages, presumably before synaptic afferentation of the cerebellum (Greenberg et al., '90, Smeyne et al., '91). The DCoN is situated beneath the floor of the fourth ventricle, covered by ependyma, and does not have Golgi-Bergmann glia. Therefore, topical cues that may be involved in determining the differences between Purkinje and cartwheel neurons abound.

Our studies demonstrate that the murine mutations determining Purkinje cell abnormalities produce concomitant effects in cartwheel neurons of the DCoN. These observations support the notion that cartwheel neurons represent the homologues of Purkinje cells in the cerebellar-like circuit of the DCoN superficial layers. We speculate that cartwheel neurons are members of the Purkinje cell family that have adapted to carry out cerebellar-like processing of auditory inputs, or perhaps play a role in the regulation of acoustic reflexes. Furthermore, our studies suggest that the cerebellum and DCoN may be useful models for investigations into genetic (cell-intrinsic) and environmental (cell-extrinsic) factors guiding the differentiation of CNS neurons.

Using PEP-19, we have recently demonstrated that cartwheel neurons provide a major inhibitory input onto the soma and apical and basal dendrites of pyramidal neurons, the large efferent neurons of DCoN layer 2 (Berrebi and Mugnaini, '91a). Therefore, cartwheel neurons are capable of profound modulation not only of parallel fiber inputs, but also of primary auditory nerve input to pyramidal neurons. The synaptic rearrangements taking place in the cochlear nuclei as a consequence of the loss of cartwheel neurons remain to be investigated. It may be interesting to examine the changes which result from the degeneration of cartwheel cells from a physiological point of view. Such studies may provide insight into the role of this cell class in normal auditory processing. Recently, a somatosensory input to cochlear granule cells has been demonstrated, suggesting that the DCoN may play a role in polysensory integration (Weinberg and Rustioni, '87, '89; Itoh et al., '87). Therefore, the loss of cartwheel neurons in murine mutants may engender compensatory changes involving not only the auditory system.

ACKNOWLEDGEMENTS

We wish to thank Dr. James I. Morgan of the Roche Institute of Molecular Biology for the generous gifts of PEP-19, L7 and G-substrate antisera and Dr. Pietro DeCamilli, Yale University School of Medicine, for antiserum to the Ins(P3) receptor. This work was supported by US-PHS grant DC 01805-01.

# REFERENCES

Altman, J. and Das, G., 1966, Autoradiographic and histological studies of postnatal histogenesis. II. A longitudinal investigation of kinetics, migration, and transformation of cells incorporating thymidine in infant rats with special reference to postnatal neurogenesis in some brain regions, *J. Comp. Neurol.*, 126:337-390.

Berrebi, A.S., Morgan, J.I. and Mugnaini, E., 1990, The Purkinje cell class may extend beyond the cerebellum, *J. Neurocytol.*, 19:643-654.

Berrebi, A.S. and Mugnaini, E., 1988, Effects of the murine mutation 'nervous' on neurons in cerebellum and dorsal cochlear nucleus, *J. Neurocytol.*, 17:465-484.

Berrebi, A.S. and Mugnaini, E., 1989, Dendrites of Purkinje cells in reeler mutant cerebellum invade the molecular layer of the dorsal cochlear nucleus, *Soc. Neurosci. Abstr.*, 15:125.

Berrebi, A.S. and Mugnaini, E., 1991, Distribution and targets of the cartwheel cell axon in the guinea pig dorsal cochlear nucleus, *Anat. Embryol.*, 183:427-454.

Berrebi, A.S. and Mugnaini, E., 1992, Characteristics of labeling of the cerebellar Purkinje neurons by L7 antiserum, *J. Chem. Neuroanat.*, 5:235-243.

Breakefield, X.O., 1979, "Neurogenetics: Genetic Approaches to the Nervous System", Elsevier, New York.

Caddy, K.W.T. and Biscoe, T.J., 1976, The number of Purkinje cells and olive neurones in the normal and Lurcher mutant mouse, *Brain Res.*, 111:396-398.

Caddy, K.W.T. and Biscoe, T.J., 1979, Structural and quantitative studies on the normal C3H and Lurcher mutant mouse, *Phil. Trans. R. Soc. Lond.*, 287:167-201.

Caviness Jr., V.S., Crandall, J.E. and Edwards, M.A., 1988, The Reeler Malformation: Implications for Neocortical Histogenesis, *in*: Cerebral Cortex, Volume VI, Development and Maturation of Cerebral Cortex, A. Peters and E. G. Jones, eds., Plenum Press, New York.

Caviness Jr., V.S. and Rakic, P., 1978, Mechanisms of cortical development: A view from mutations in mice, *Ann. Rev. Neurosci.*, 1:297-326.

Christakos, S., Rhoten, W.B. and Feldman, S.C., 1987, Rat calbindin-D28k purification, quantitation, immunocytochemical localization and comparative aspects, *Methods Enzymol.*, 139: 534-551.

Crenshaw III, E.B., Ryan, A., Dillon, S.R., Kalla, K. and Rosenfeld, M.G., 1991, Wocko, a neurological mutant generated in a transgenic mouse pedigree, *J. Neurosci.*, 11:1524-1530.

DeCamilli, P., Miller, P.E., Levitt, P., Walter, U. and Greengard, P., 1984, Anatomy of cerebellar Purkinje cells in the rat determined by a specific immunohistochemical marker, *Neurosci.*, 11:761-817.

Green, M.C., 1981, "Genetic Variants and Strains of the Laboratory Mouse", Gustav Fischer Verlag, New York.

Greenberg, J.M., Boehm, T., Sofroniew, M.V., Keynes, R.J., Barton, S.C., Norris, M.L., Surani, M.A., Spillantini, M. G. and Rabbitts, T.H., 1990, Segmental and developmental regulation of a presumptive T-cell oncogene in the central nervous system, *Nature*, 344:158-160.

Hackney, C.M., Osen, K.K. and Kolston, J., 1990, Anatomy of the cochlear nuclear complex of guinea pig, *Anat. Embryol.*, 182:123-149.

Hawkes, R. and LeClerc, N., 1987, An antigenic map of the rat cerebellar cortex: The distribution of parasagittal bands as revealed by a monoclonal anti-Purkinje cell antibody mabQ113, *J. Comp. Neurol.*, 256:29-41.

Heckroth, J.A., Goldowitz, D. and Eisenman, L.M., 1989, Purkinje cell reduction in the reeler mutant mouse: A quantitative immunohistochemical study, *J. Comp. Neurol.*, 279: 546-555.

Herrup, K. and Mullen, R., 1979, Regional variation and absence of large neurons in the cerebellum of the staggerer mouse, *Brain Res.*, 172:1-12.

Hirano, A. and Dembitzer, H.M., 1975, The fine structure of staggerer cerebellum, *J. Neuropathol. Exper. Neurol.*, 34:1-11.

Itoh, K., Kamiya, H., Mitani, A., Yasui, Y., Takada, M. and Mizuni, N., 1987, Direct projections from the dorsal column nuclei and the spinal trigeminal nuclei to the cochlear nuclei in the cat, *Brain Res.*, 400:145-150.

Landis, S.C., 1973, Ultrastructural changes in the mitochondria of cerebellar Purkinje cells of nervous mutant mice, *J. Cell Biol.*, 57:782-797.

Lohmann, S.M., Walter, U., Miller, P.E., Greengard, P. and DeCamilli, P., 1981, Immunohistochemical localization of cyclic GMP-dependent protein kinase in mammalian brain, *Proc. Natl. Acad. Sci.(USA)*, 78:653-657.

Lorente de Nó, R., 1933, Anatomy of the eighth nerve. III. General plan of structure of the primary cochlear nuclei, *Laryngoscope*, 43:327-350.

Lorente de Nó, R., 1979, Central representation of the eighth nerve, *in*: "Ear Diseases, Deafness and Dizziness", V. Goodhill, ed., Harper and Row, Hagerstown, Maryland Lorente de Nó, R., 1981, The Primary Acoustic Nuclei, Raven Press, New York.

Mariani, J., Crepel, F., Mikoshiba, K., Changeux, J.P. and Sotelo, C., 1977, Anatomical, physiological and biochemical studies of the cerebellum from reeler mutant mouse, *Philos. Trans. R. Soc. Lond.*, 281:1-28.

Martin, M.R. and Rickets, C., 1981, Histogenesis of the cochlear nucleus of the mouse, *J. Comp. Neurol.*, 197:169-184.

Miale, I.L. and Sidman, R.L., 1961, An autoradiographic analysis of histogenesis in the mouse cerebellum, *Exp. Neurol.*, 4:277-296.

Mignery, G.A., Sudhof, T.C., Takei, K. and DeCamilli, P., 1989, Putative receptor for inositol 1,4,5-trisphosphate similar to ryanodine receptor, *Nature*, 342:192-195.

Mugnaini, E., 1985, GABA neurons in the superficial layers of the rat dorsal cochlear nucleus: light and electron microscopic immunocytochemistry, *J. Comp. Neurol.*, 235:61-81.

Mugnaini, E., Berrebi, A.S., Dahl, A.-L. and Morgan, J.I., 1987, The polypeptide PEP-19 is a marker for Purkinje neurons in cerebellar cortex and cartwheel neurons in the dorsal cochlear nucleus, *Arch. Ital. de Biol.*, 126:41-67.

Mugnaini, E. and Dahl, A.-L., 1983, Zinc-aldehyde fixation for light microscopic immunocytochemistry of nervous tissues, *J. Histochem. Cytochem.*, 31:1435-1438.

Mugnaini, E. and Mörgan, J.I., 1987, The neuropeptide cerebellin is a marker for two similar neuronal circuits in rat brain, *Proc. Natl. Acad. Sci. (USA)*, 84:8692-8696.

Mugnaini, E., Osen, K.K., Dahl, A.-L., Friedrich Jr., V.L. and Korte, G., 1980(a), Fine structure of granule cells and related interneurons (termed Golgi cells) in the cochlear nuclear complex of the cat, rat and mouse, *J. Neurocytol.*, 9: 537-570.

Mugnaini, E., Warr, W.B. and Osen, K.K., 1980(b), Distribution and light microscopic features of granule cells in the cochlear nuclei of the cat, rat and mouse, *J. Comp. Neurol.*, 191:581-606.

Mullen, R.J., Eicher, E.M. and Sidman, R.L., 1976, Purkinje cell degeneration, a new neurological mutation in the mouse, *Proc. Natl. Acad. Sci. (USA)*, 73:208-212.

Oberdick, J., Levinthal, F. and Levinthal, C., 1988, A Purkinje cell differentiation marker shows a partial DNA sequence homology to the cellular sis/PDGF2 gene, *Neuron*, 1: 367-376 [Erratum 3:385].

Osen, K.K., 1985, Ectopic neurons of the cochlear nuclei, *Neurosci. Lett.*, [Suppl.], 22:S167.

Osen, K.K. and Mugnaini, E., 1981, Neural circuits in the dorsal cochlear nucleus, in: "Neuronal Mechanisms of Hearing", J. Syka and L. Aitkin, eds., Plenum Press, New York.

Osen, K.K., Ottersen, O.P. and Storm-Mathisen, J., 1990, Colocalisation of glycine-like and GABA-like immuno- reactivities. A semi-quantitative study of individual neurons in the dorsal cochlear nucleus of cat, in: "Glycine Neurotransmission", O.P. Ottersen and J. Storm-Mathisen, eds., John Wiley and Sons, Chicester.

Ottersen, O.P. and Storm-Mathisen, J., 1984, Neurons containing or accumulating transmitter amino acids, in: "Handbook of Chemical Neuroanatomy", Vol 3., Classical Transmitters and Transmitter Receptors in the CNS, A. Bjürklund, T. Hükfelt and M. J. Kuhar, eds., Elsevier, Amsterdam.

Saito, N., Kikkawa, U., Nishizuka, Y. and Tanaka, C., 1988, Distribution of protein kinase C-like immunoreactive neurons in rat brain, *J. Neurosci.*, 8:369-382.

Schlichter, D.J., Casnellie, J.E. and Greengard, P., 1978, An endogenous substrate for cGMP-dependent protein kinase in mammalian cerebellum, *Nature*, 273:61-62.

Sidman, R.L. and Green, M.C., 1970, Nervous, a new mutant mouse with cerebellar disease, in: "Les Mutants Pathologiques chez L'Animal", M. Sabourdy, ed., Editions du Centre National de la Recherche Scientifique, Paris.

Sidman, R.L., Lane, P. W. and Dickie, M.M., 1962, Staggerer, a new mutation in the mouse affecting the cerebellum, *Science*, 137:610-612.

Smeyne, R.J., Oberdick, J., Schilling, K., Berrebi, A.S., Mugnaini, E. and Morgan, J.I., 1991, Dynamic organization of developing Purkinje cells revealed by transgene expression, *Science*, 254:719-721.

Sotelo, C. and Changeux, J.-P., 1974, Transsynaptic degeneration 'en cascade' in the cerebellar cortex of staggerer mutant mice, *Brain Res.*, 67:519-526.

Steel, K., Niaussat, M.M. and Bock, G.R., 1983, The genetics of hearing, in: "The Auditory Psychobiology of the Mouse", J. F. Willott, ed., Thomas, Springfield, IL.

Taber-Pierce, E., 1967, Histogenesis of the dorsal and ventral cochlear nuclei of the mouse. An autoradiographic study, *J. Comp. Neurol.*, 131:27-54.

Weinberg, R.J. and Rustioni, A., 1987, A cuneocochlear pathway in the rat, *Neurosicence*, 20:209-219.

Weinberg, R.J. and Rustioni, A., 1989, Brainstem projections to the rat cuneate nucleus, *J. Comp. Neurol.*, 282:142-156.

Wenthold, R.J., Huie, D., Altschuler, R.A. and Reeks, K.A., 1987, Glycine immunoreactivity localized in the cochlear nucleus and superior olivary complex, *Neurosci.*, 22:897-912.

Wenthold, R.J., Zempel, J.M., Parakkal, M.H., Reeks, K.A. and Altschuler, R.A., 1986, Immunocytochemical localization of GABA in the cochlear nucleus of the guinea pig, *Brain Res.*, 380:7-18.

Willard, F.H. and Martin, G.F., 1986, The development and migration of large multipolar neurons into the cochlear nucleus of the North American Opossum, *J. Comp. Neurol.*, 248: 119-132.

Wouterlood, F.G. and Mugnaini, E., 1984, Cartwheel neurons of the dorsal cochlear nucleus. A Golgi-electron microscopic study in the rat, *J. Comp. Neurol.*, 227:136-157.

Wouterlood, F.G., Mugnaini, E., Osen, K.K. and Dahl, A.-L., 1984, Stellate neurons in the dorsal cochlear nucleus of the rat studied with Golgi-impregnation electron microscopy; synaptic connections and mutual coupling by gap junctions, *J. Neurocytol.*, 13:639-664.

Yoon, C.H., 1976, Pleiotropic effect of the staggerer gene, *Brain Res.*, 109:205-215.

# IV. DESCENDING PROJECTIONS

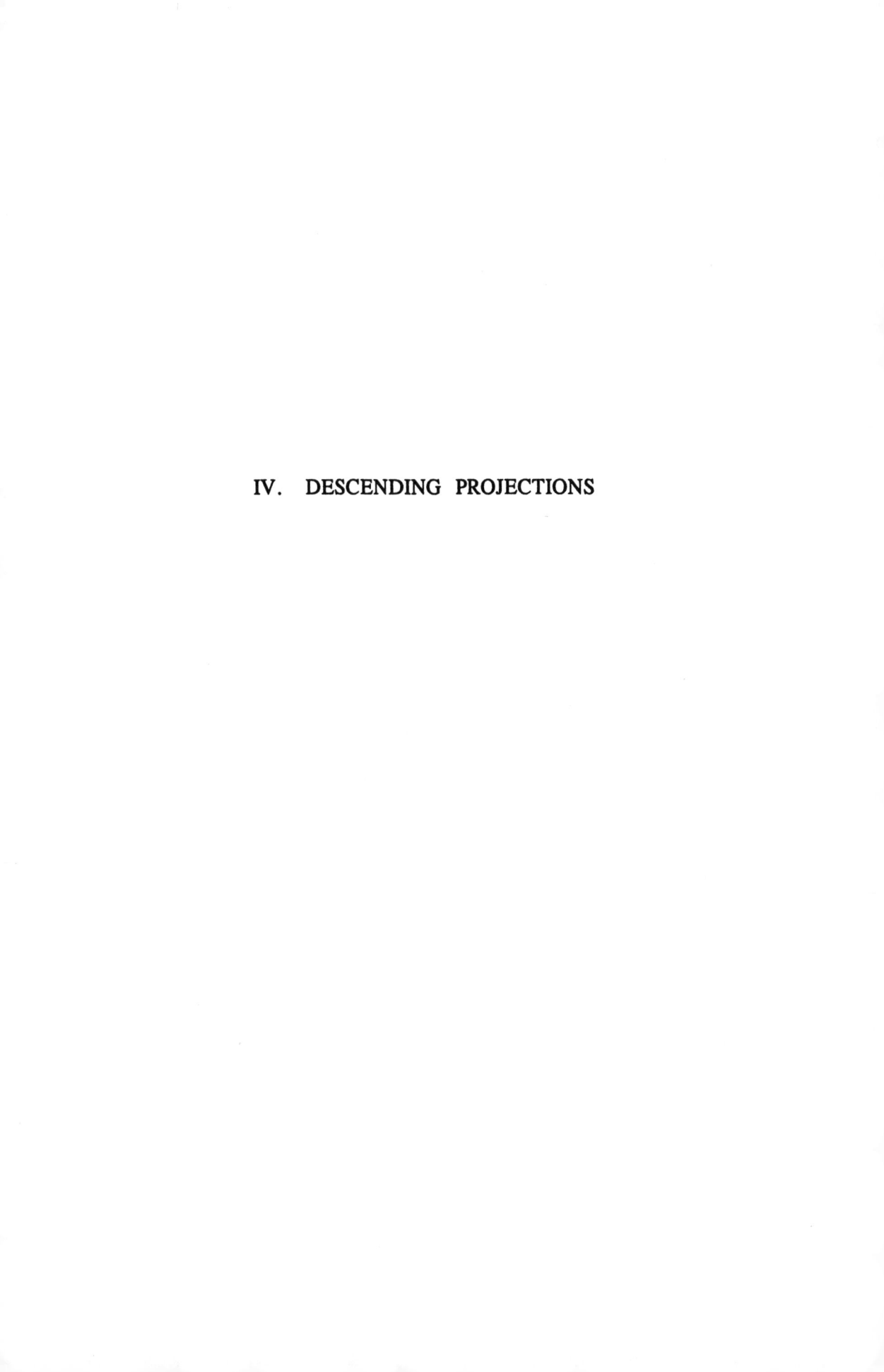

# NON-COCHLEAR PROJECTIONS TO THE VENTRAL COCHLEAR NUCLEUS: ARE THEY MAINLY INHIBITORY?

Richard L. Saint Marie, E.-Michael Ostapoff,
Christina G. Benson, D. Kent Morest and Steven J. Potashner

Department of Anatomy and Center for Neurological Sciences, University
of Connecticut Health Center, Farmington, 06030 Connecticut, USA

Non-cochlear synapses in the ventral cochlear nucleus (VCN), originate from a variety of intrinsic and extrinsic sources. It has been suggested by some studies that these synapses may actually outnumber those of the primary afferents in some regions of the VCN. In this chapter we will review the evidence that most non-cochlear synapses may be inhibitory, examine their distribution on different cell types in the VCN, and attempt to identify their origins.

## Morphology of non-cochlear synapses

Four basic kinds of synapses are found in the VCN in aldehyde-fixed tissue (Kane, '73, '74, '77; Cant and Morest, '79; Cant, '81; Tolbert and Morest, '82; Smith and Rhode, '87, '89; Ostapoff and Morest, '91). Two kinds have been described with morphologies usually associated with excitatory transmission (Gray's Type I, Gray, '69). These have asymmetrical synaptic thickenings and contain either large round or small round synaptic vesicles. Those with large round vesicles are the most numerous and include the large endbulbs of Held and the smaller collateral endings of the primary afferents. Two additional kinds of synapses have also been described with morphologies usually associated with inhibitory transmission (Gray's Type II, Gray, '69). These have symmetrical synaptic thickenings and contain either pleomorphic or mostly flattened synaptic vesicles (see also Ibata and Pappas, '76; Schwartz and Gulley, '78).

After complete cochlear ablation, synapses with asymmetrical thickenings and large round vesicles do not survive in the anterior part of the VCN (AVCN)[1] (Cant and Morest, '79, Cant, '81, Schwartz and Gulley, '78; Tolbert and Morest, '82). The other three kinds of synapses survive and appear normal. Of the surviving synapses, most

---

[1]In the octopus cell area of PVCN the morphological distinction between cochlear and non-cochlear synaptic endings is less clear than in AVCN. Kane ('73, '74) reports that nearly all endings with round vesicles degenerate after cochlear ablation, but that many endings with pleomorphic vesicles also degenerate suggesting that in this regions of the VCN they too may be of cochlear origin.

contain pleomorphic or flattened synaptic vesicles - those with small round vesicles account for only a small fraction of the synapses in the large-cell areas of the AVCN and posterior part of the VCN (PVCN) (Cant and Morest, '79, Cant, '81, Schwartz and Gulley, '78; Smith and Rhode, '87, '89; Ostapoff and Morest, '91; see Cant, this volume) - suggesting that a large majority of non-cochlear synapses in AVCN proper may have an inhibitory function. In fact, these ultrastructural studies were the first indication that non-cochlear inputs to the VCN were probably largely inhibitory.

## Inhibitory transmitters and non-cochlear synapses

Additional evidence that synapses with pleomorphic or flattened vesicles are probably inhibitory comes from transmitter localization studies. Ultrastructural studies using immunocytochemistry have shown that synaptic endings with pleomorphic vesicles in the VCN probably use GABA as a transmitter because they contain elevated levels of GABA and glutamate decarboxylase (GAD), the enzyme which synthesizes GABA (Altschuler et al., '86b; Wenthold et al., '86; Adams and Mugnaini, '87; Oberdorfer et al., '88; Saint Marie et al., '89). Some of these endings also appose $GABA_A$/ benzodiazepine postsynaptic receptors (Juiz et al., '89). Flattened vesicle endings in the VCN, on the other hand, are thought to use glycine as a transmitter, since they always appose postsynaptic thickenings that stain for the glycine receptor (Altschuler et al., '86a; Wenthold et al., '88) and ultrastructurally similar endings elsewhere have been shown to contain elevated levels of glycine (van den Pol and Gorcs, '88) or to accumulate [$^3$H] glycine from the extracellular media (Matus and Dennison, '71; Ljungdahl and Hökfelt, '73). Schwartz ('83, '85) has shown that endings with non-cochlear morphologies have selective uptake mechanisms for [$^3$H] GABA and [$^3$H] glycine, but did not distinguish morphological subtypes of endings. Moreover, Altschuler and colleagues (see chapter, this volume) conclude that synaptic endings in the VCN which contain either pleomorphic or flattened vesicles are immunoreactive for GABA or glycine, respectively, and that some with pleomorphic vesicles are immunoreactive for both transmitter substances.

The considerable evidence that GABA and glycine are major inhibitory transmitters in the VCN has been summarized recently (Saint Marie et al., '89, '91) and is discussed in other chapters of this volume (e.g., see Altschuler et al., Caspary et al., Oertel and Wickesberg, Potashner et al., Wickesberg and Oertel). On the assumption that GABA and glycine are inhibitory transmitters at non-cochlear endings with pleomorphic or flattened synaptic vesicles, it is important to know how prevalent these endings are and where they come from.

## Distribution of non-cochlear synapses on identified cell types

The conclusion that synapses with flattened or pleomorphic vesicles are inhibitory is significant because in many regions of the VCN synapses with these morphologies probably outnumber those of the primary afferents. For example, using conventional electron microscopy it has been estimated that about half of the axosomatic synapses on spherical bushy cells are non-cochlear (Cant and Morest, '79; Cant, '81). Detailed studies of Golgi-impregnated or intracellularly HRP-filled globular bushy cells have shown that more than half of all the synaptic endings (axosomatic and axodendritic) on this cell type have pleomorphic or flattened vesicles (Smith and Rhode, '87; Ostapoff and Morest, '91) (Fig. 1). In all regions but the cell body, where the primary afferents account for about 60% of the endings, putative inhibitory endings clearly outnumber those of the cochlear nerve. Endings with pleomorphic or flattened vesicles presumably predominate on the

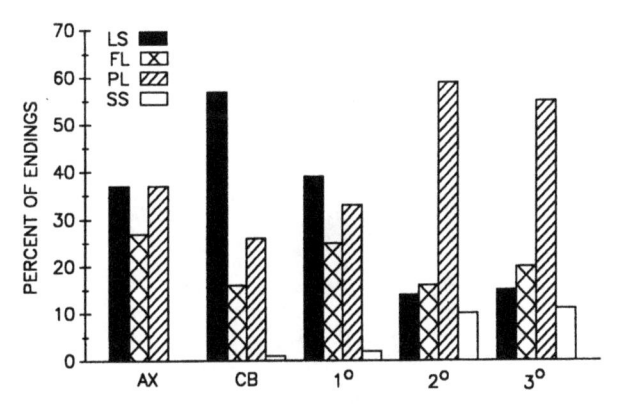

**Figure 1.** Bar graph showing the proportion of 4 kinds of synaptic ending on each of 5 regions of globular bushy cells in the cat AVCN. Endings with four kinds of vesicles are illustrated: large spherical (LS), flattened (FL), pleomorphic (PL), and small spherical (SS). Proportion of endings for five areas of the cell are illustrated: axon hillock (AX), cell body (CB), primary (1°), secondary (2°), and tertiary dendrites (3°). Except on the cell body, putative inhibitory non-cochlear endings (FL and PL) outnumber those of the cochlear nerve (LS). Endings with small spherical vesicles are a minor population on all regions of the cell and are not found at all on the axon hillock. Reprinted with permission from Ostapoff and Morest ('91).

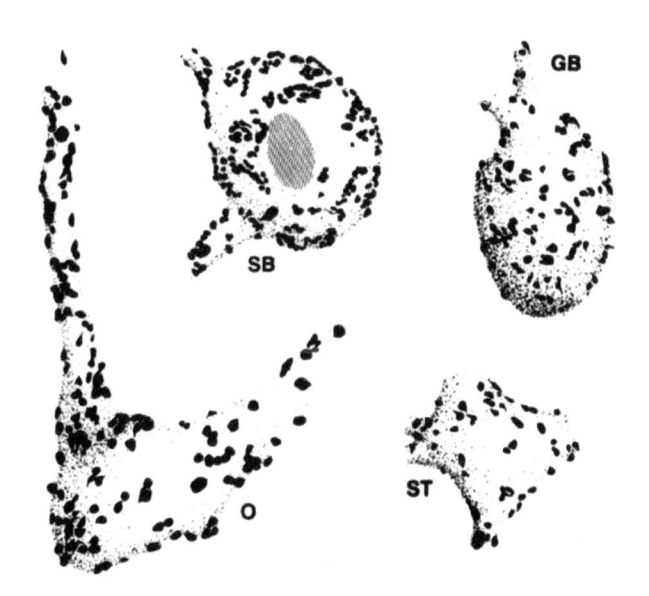

**Figure 2.** GAD-immunoreactive endings on graphical examples of the large cell types found in the cat VCN: SB - spherical bushy; GB - globular bushy; O - octopus; ST - stellate cells. On cells in AVCN (SB, GB and ST) the endings are mostly small. On SB cells they occur in tight clusters separated by large empty patches. On O cells in PVCN the endings are larger and frequently aggregate into small clusters or rows along the somata and dendrites. Modified and reprinted with permission from Saint Marie et al. ('89).

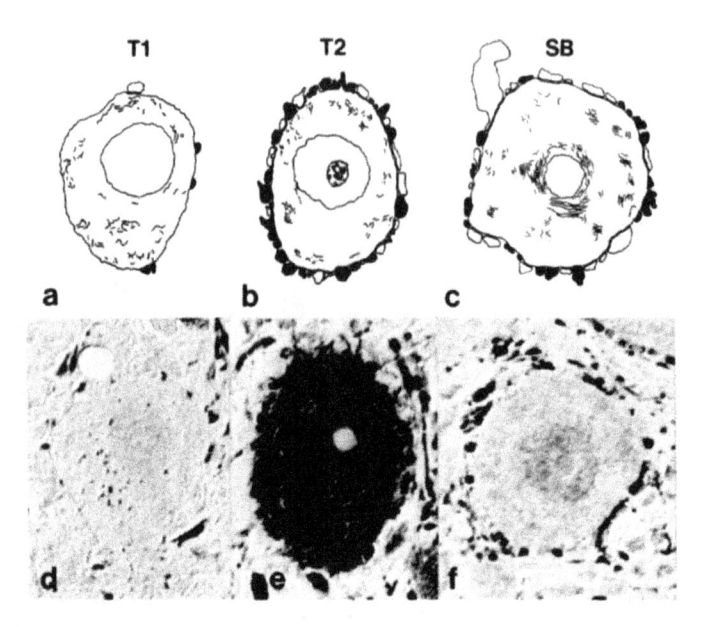

**Figure 3.** Distribution of non-cochlear synaptic endings on Type 1 (T1) and Type 2 (T2) stellate and spherical bushy (SB) cells in the cat AVCN, identified electron microscopically (top) and immunocytochemically (bottom). a-c.- Illustrations are reprinted with permission from Cant ('81) and show the distribution of primary afferent (clear) and non-cochlear (filled) endings on the 3 cell types. d-f.- Antibodies to both GABA and glycine were used to stain putative inhibitory non-cochlear axosomatic endings on the 3 cell types in semi-thin plastic sections. See text for additional details.

large stellate cells in AVCN and PVCN as well (Cant, '81; Tolbert and Morest, '82; Smith and Rhode, '89). An apparent exception is the octopus cell body, on which only a minority of the synaptic endings have pleomorphic or flattened vesicles (Kane, '73). The prevalence of putative inhibitory endings on the different cell types in VCN is also apparent in tissue immunostained for GABA, GAD, or glycine (Shiraishi et al., '85; Thompson et al., '85; Peyret et al., '86; Wenthold et al., '86, '87; Adams and Mugnaini, '87; Moore and Moore, '87; Roberts and Ribak, '87; Adams, '88; Aoki et al. '88; Oberdorfer et al., '88; Saint Marie et al., '89). Figure 2 shows the distribution of GAD-immunoreactive endings on typical examples of the large principal cells in the cat VCN, including spherical bushy (SB), globular bushy (GB), octopus (O), and stellate cells (ST). Each cell type exhibits a distinctive pattern of labeled endings on its surface, with respect to ending sizes, clustering, and overall distribution. In each case the pattern corresponds to that of synaptic endings with pleomorphic vesicles described electron microscopically (Kane, '73; Cant and Morest, '79; Tolbert and Morest, '82; Wenthold et al., '86; Adams and Mugnaini, '87; Saint Marie et al., '89).

One of the difficulties of this and other immunocytochemical studies of the VCN is that Type 1 stellate cells cannot be identified easily because they have so few synaptic endings of any type on their somata (Cant, '81). To get around this problem we chose to examine cells in semi-thin plastic sections and to stain each section with antibodies for both GABA and glycine. The rationale was to distinguish between Type 1 and Type 2 stellate cells by their relative complement of immunoreactive axosomatic endings without having to use electron microscopy. We knew from studies mentioned above that nearly all of the non-cochlear endings on these cells contained either pleomorphic or flattened synaptic vesicles. These endings, therefore, should be immunoreactive for GABA and/or glycine and should constitute about half of the axosomatic endings on these cell types.

When examined in this manner we found that immunoreactive endings formed characteristic patterns on the somata of the different cell types (Fig. 3). Cell types were identified in adjacent plastic sections stained with toluidine blue (not illustrated). When stained with antibodies for both GABA and glycine, spherical bushy cells are contacted by many immunoreactive profiles which form large clusters that alternate with large unstained areas (Fig. 3f). This pattern exactly resembles that of non-cochlear synaptic endings on this cell type, originally illustrated by Cant ('81) based on electron microscopy (Fig. 3c). Cant ('81) also described two types of stellate neuron in the AVCN, depending on the number of endings contacting their cell body. Type 1 stellate cells are contacted by only a few endings, whereas Type 2 stellate cells can be completely surrounded by synaptic endings, many of which are non-cochlear (Fig. 3a,b). The same two types of stellate cells are found in our immunoreacted sections of the cat VCN (Fig. 3d,e).

As illustrated in Figure 3e, there is a special kind of large stellate neuron in the VCN that appears to be inhibitory. These neurons are immunoreactive for GABA, glycine, or both and are almost always contacted by many immunoreactive profiles, identifying them as a kind of Type 2 stellate cell (not all Type 2 stellate cells in the cat VCN are immunoreactive). This observation is comparable to that of Smith and Rhode ('89), who showed that endings made by the axon of a similar type of stellate cell in PVCN contained pleomorphic synaptic vesicles, suggesting that it was probably an inhibitory neuron. This stellate neuron had a dense local axonal plexus, suggesting that cells of its kind may be one possible source of the immunoreactive profiles that we find on cells in the VCN.

More important than the large numbers of putative inhibitory synapses on the principal cells of the VCN may be the strategic positioning of these endings. Many are found on the cell bodies and axon hillocks of bushy (Cant and Morest, '79; Smith and Rhode, '87; Ostapoff and Morest, '91) and stellate cells (Cant, '81; Smith and Rhode, '89), where they would be ideally situated near the spike generator to affect the output of the principal efferent projections of the VCN.

## Origins of non-cochlear synapses

Our goal in the following studies was to identify the origins of non-cochlear synapses in the VCN and to identify which of these may be using GABA or glycine as a transmitter. To do this we combined immunocytochemistry with conventional retrograde HRP labeling of projection neurons in guinea pigs.

**The dorsal cochlear nucleus.** In this study (Saint Marie et al., '91), we examined the projection from the deep layer of the dorsal cochlear nucleus (DCN) to AVCN. This pathway was first described by Lorente de Nó ('33, '81) from Golgi-impregnated material. There were indications from other studies (Wenthold et al., '87; Wickesberg and Oertel, '90) that this projection was inhibitory and probably glycinergic. Following injections of HRP into AVCN, we found that, aside from a few granule cells, all of the labeled neurons in DCN with projections to AVCN were located in the deep layer of the DCN. Most resembled the "corn cells" described by Lorente de Nó ('33) and later referred to as small elongate cells (Brawer et al., '74), vertical cells (Lorente de Nó, '81), or fan cells (Moore, '86). Some of these cells are illustrated in Figure 4. When the retrogradely labeled cells were stained immunocytochemically, nearly all (96%) were immunoreactive for glycine and only a few (3%) were immunoreactive for GABA (also see Adams and Wenthold, '87 for cats). Hence, we concluded that this projection is almost entirely glycine immunoreactive and, therefore, probably inhibitory (see Wickesberg and Oertel, this volume). Not investigated in this study was a possible excitatory pathway to the AVCN by way of granule cell projections (Oliver et al., '83).

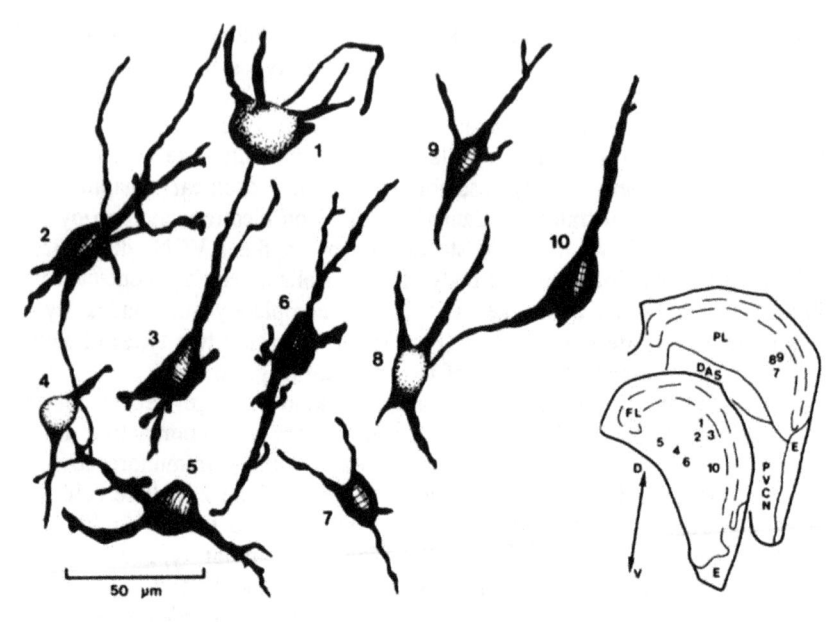

**Figure 4.** Examples of neurons in the deep DCN retrogradely labeled from the guinea pig AVCN. Different morphological types are illustrated in approximately the same proportions and orientations as they appeared in the tissue sections (lower right). Most are "corn" cells elongated toward the surface of the DCN (2-4,6,8-10), but a few are elongated parallel to the surface (5,7). Fewer still are not elongated at all and resemble stellate cells (1). Reprinted with permission from Saint Marie et al. ('91).

**The periolivary nuclei.** Studies in several species have shown that the VCN gets large centrifugal projections bilaterally from the inferior colliculus and periolivary regions of the superior olive and a commissural projection from the contralateral cochlear nucleus (Adams, '76, '83; Kane, '76, '77; Elverland, '77; Kane and Conlee, '79; Cant and Gaston, '82; Covey et al., '84; Spangler et al., '87; Wenthold, '87; Winter et al., '89). We have found that many of these projection neurons, particularly those in the periolivary nuclei and contralateral cochlear nucleus, are immunoreactive for GABA or glycine. Examples of periolivary neurons that are both retrogradely labeled with HRP from the VCN and immunoreactive are shown in Figure 5.

Approximately 90% of the periolivary neurons with descending projections to the VCN stain with antibodies to either GABA or glycine. In Figure 6, it is apparent that most glycine-immunoreactive periolivary projections are from ipsilateral regions, primarily from the lateral and ventral nuclei of the trapezoid body and from the dorsal and dorsomedial periolivary nuclei. GABA-immunoreactive projection neurons, on the other hand, are more equally divided between the ipsilateral and contralateral superior olive, with most found in the ventral nuclei of the trapezoid body on both sides. This difference in the origin of glycine- and GABA-immunoreactive periolivary projections is the same regardless of whether the projection is to AVCN, PVCN, or DCN (Ostapoff, Benson, Saint Marie - in preparation). A brief report by Adams and Wenthold ('87) suggests that periolivary projections to the VCN in cats may be similar but not identical in immunoreactivity to those that we find in the guinea pig. Not illustrated in Figure 6 is the fact that about 28% of the immunoreactive projection neurons in this case stain for both GABA and glycine. This is consistent with light and electron microscopic observations which show that some synaptic endings in the VCN contain both GABA and glycine (Wenthold et al., '87; see Altschuler et al., this volume).

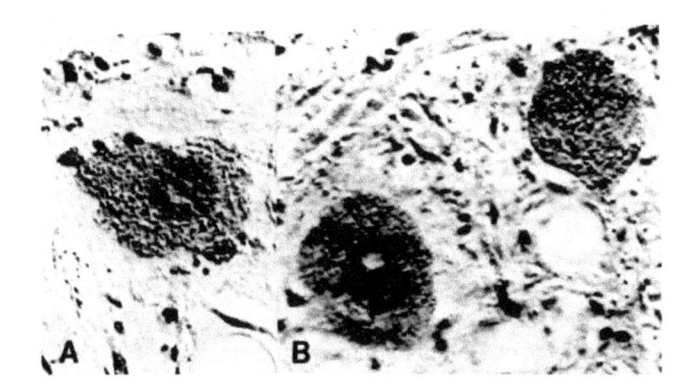

Figure 5. Examples of immunoreactive periolivary neurons with projections to the ipsilateral PVCN of the guinea pig, photographed with Nomarski optics. A. Neuron in the lateral nucleus of the trapezoid body is lightly stained with an antibody to glycine. The coarse appearance of the perikaryon indicates the presence of many small HRP granules retrogradely transported from the PVCN injection. B. Two neurons in the dorsomedial periolivary nucleus are stained for GABA. One of these immunoreactive neurons (upper right) also contains many retrogradely transported HRP granules.

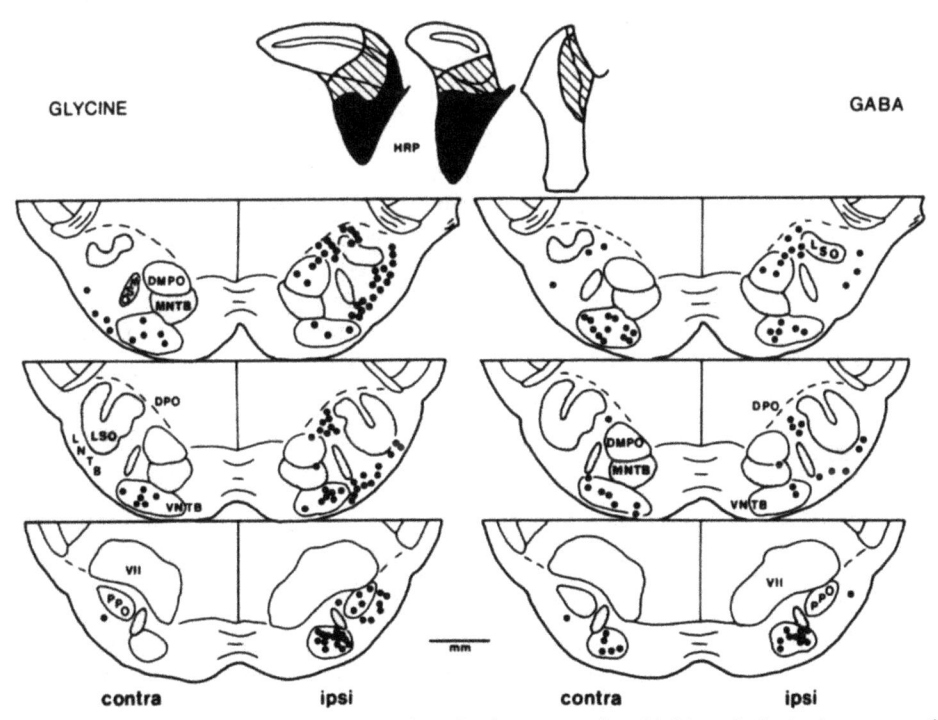

Figure 6. Plot illustrating the distribution of GABA-immunoreactive (right) and glycine-immunoreactive (left) periolivary neurons labeled from an HRP injection in the PVCN (top - center) of a guinea pig. The right PVCN was injected. LSO and MSO - lateral and medial superior olive; DMPO, DPO, and PPO - dorsomedial, dorsal, and posterior periolivary nuclei; LNTB, MNTB, and VNTB - lateral, medial, and ventral nuclei of the trapezoid body; VII - facial nucleus. See text for additional details.

About 10% of the retrogradely labeled neurons in the superior olive do not stain with either antibody. Some of these may represent other inhibitory as well as excitatory projections to the cochlear nucleus, e.g., by way of either muscarinic or nicotinic synapses (Morley et al., '77; Hunt and Schmidt, '78; Wamsley et al., '81; Glendenning and Baker, '88), and use acetylcholine as a transmitter (Caspary et al., '83; Godfrey et al., '87, '90; see Godfrey, this volume).

We conclude from this study that most of the projections to the VCN from periolivary regions are either GABAergic and/or glycinergic and, thus, probably inhibitory. This conclusion is consistent with results obtained using retrograde transport of [$^3$H]GABA or [$^3$H]glycine from the cochlear nucleus (Benson and Potashner, '90; Ostapoff et al., '90), with studies of synaptic release and uptake of [$^3$H]GABA or [$^3$H]glycine following ablation of centrifugal projections (Potashner et al., '85a; Staatz-Benson and Potashner, '88; see Potashner et al., this volume), and with an EM degeneration study of olivary projections to the VCN (Kane, '77).

**The contralateral cochlear nucleus.** As originally reported by Wenthold ('87), we have found that projections from the contralateral cochlear nucleus are glycine immunoreactive. The number of cells in this projection is much smaller than in projections from the DCN, periolivary nuclei, and inferior colliculi. Nevertheless, like the projection from the ipsilateral DCN, this commissural projection appears to be almost entirely glycine immunoreactive. In our sample of 39 cells labeled in the contralateral cochlear nucleus, all but one were immunoreactive for glycine. Presynaptic uptake and retrograde transport of [$^3$H]glycine (Benson and Potashner, '90), but not [$^3$H]GABA (Ostapoff et al., '90) or [$^3$H]D-aspartate (Oliver et al., '83; Jones et al., '84; Potashner et al., '85b), by neurons of this commissural projection further supports the hypothesis that this is largely a glycinergic projection (see Potashner et al., this volume).

**The inferior colliculus.** Nearly one quarter of all neurons in the inferior colliculus stain with antibodies to GABA or GAD (Oliver, Winer, Beckius, Saint Marie - in preparation). Nevertheless, we find that projections to the VCN from the inferior colliculus are neither GABA nor glycine immunoreactive and, therefore, may use an excitatory or other inhibitory transmitter. This projection is primarily to the external granular layer of the VCN, however, and probably does not contribute significantly to the non-cochlear endings in the central large-cell areas (see Saldaña, this volume).

## Conclusions

Ultrastructural studies suggest that a very large proportion of the non-cochlear synapses in the large-cell areas of the VCN have either pleomorphic or flattened vesicles, a morphology usually associated with synaptic inhibition. Immuno-electron-microscopic studies have shown that most if not all synapses in the VCN that have either pleomorphic or flattened vesicles probably use GABA, glycine, or both of these inhibitory substances as transmitters. Available evidence suggests that projections to the VCN from the DCN and contralateral cochlear nucleus are glycine immunoreactive and that projections from periolivary nuclei are immunoreactive for GABA, glycine, or both. This is not to imply that all of the non-cochlear synapses in the VCN are GABAergic or glycinergic. Clearly there is a lot more to learn about other extrinsic projections, especially those using acetylcholine. Also, little has been mentioned of the intrinsic connections made by VCN neurons. Nevertheless, it is evident that a large proportion of non-cochlear synapses in the VCN are probably inhibitory. The fact that these projections arise from several different sources suggests that inhibition must have several important roles in signal processing in the VCN.

ACKNOWLEDGEMENTS

The authors wish to acknowledge Lisa Seman Tobin for her excellent technical and photographic assistance. Antisera for GABA and glycine immunocytochemistry were kindly provided by Dr. Robert J. Wenthold. These investigations were supported by NIH-NIDCD Grants DC00726 (RLSM), DC00127 (DKM), and DC00199 (SJP).

REFERENCES

Adams, J.C., 1976, Central projections to the cochlear nucleus, *Soc. Neurosci. Abst.*, 2:12.
Adams, J.C., 1983, Cytology of periolivary cells and the organization of their projections in the cat, *J. Comp. Neurol.*, 215:275-289.
Adams, J.C., 1988, Glutamate decarboxylase immunostaining in the human cochlear nucleus, in: "Auditory Pathway: Structure and Function," J. Syka and R.B. Masterton, eds., Plenum, New York, pp. 133-139.
Adams, J.C. and Mugnaini, E., 1987, Patterns of glutamate decarboxylase immunostaining in the feline cochlear nuclear complex studied with silver enhancement and electron microscopy, *J. Comp. Neurol.*, 262:375-401.
Adams, J.C. and Wenthold, R.J., 1987, Immunostaining of ascending auditory pathways with glycine antiserum, *Assoc. Res. Otolaryngol. Abst.*, 10:63.
Altschuler, R.A., Betz, H., Parakkal, M.H., Reeks, K.A., and Wenthold, R.J., 1986a, Identification of glycinergic synapses in the cochlear nucleus through immunocytochemical localization of the postsynaptic receptor, *Brain Res.*, 369:316-320.
Altschuler, R.A., Hoffman, D.W., and Wenthold, R.J., 1986b, Neurotransmitters of the cochlea and cochlear nucleus: Immunocytochemical evidence, *Am. J. Otolaryngol.*, 7:100-106.
Aoki, E., Semba, R., Keino, H., Kato, K., and Kashiwamata, S., 1988, Glycine-like immunoreactivity in the rat auditory pathway, *Brain Res.*, 442:63-71.
Benson, C.G. and Potashner, S.J., 1990, Retrograde transport of [$^3$H]glycine from the cochlear nucleus to the superior olive in the guinea pig, *J. Comp. Neurol.*, 296:415-426.
Brawer, J.R., Morest, D.K. and Kane, E.C., 1974, The neuronal architecture of the cochlear nucleus of the cat, *J. Comp. Neurol.*, 155:251-300.
Cant, N.B., 1981, The fine structure of two types of stellate cells in the anterior division of the anteroventral cochlear nucleus of the cat, *Neurosci.*, 6:2643-2655.
Cant, N.B. and Gaston, K.C., 1982, Pathways connecting the right and left cochlear nuclei, *J. Comp. Neurol.* 212:313-326.
Cant, N.B. and Morest, D.K., 1979, The bushy cells in the anteroventral cochlear nucleus of the cat. A study with the electron microscope, *Neurosci.*, 4:1925-1945.
Caspary, D.M. Havey, D.C. and Faingold, C.L., 1983, Effects of acetylcholine on cochlear nucleus neurons, *Exp. Neurol.*, 82:491-498.
Covey, E., Jones, D.R., and Casseday, J.H., 1984, Projections from the superior olivary complex to the cochlear nucleus in the tree shrew, *J. Comp. Neurol.*, 226:289-305.
Elverland, H.H., 1977, Descending connections between the superior olivary and cochlear nuclear complexes in the cat studied by autoradiographic and horseradish peroxidase methods, *Exp. Brain Res.*, 27:397-412.
Glendenning, K.K. and Baker, B.N., 1988, Neuroanatomical distribution of receptors for three potential inhibitory neurotransmitters in the brainstem auditory nuclei of the cat, *J. Comp. Neurol.*, 275:288-308.
Godfrey, D.A., Park-Hellendall, J.L., Dunn, J.D., and Ross, C.D., 1987, Effect of olivocochlear bundle transection on choline acetyltransferase activity in the rat cochlear nucleus, *Hearing Res.*, 28:237-251.
Godfrey, D.A., Beranek, K.L., Carlson, L., Parli, J.A., Dunn, J.D., and Ross, C.D., 1990, Contribution of centrifugal innervation to choline acetyltransferase activity in the cat cochlear nucleus, *Hearing Res.*, 49:259-280.
Gray, E.G., 1969, Electron microscopy of excitatory and inhibitory synapses: A brief review, in: "Mechanisms of synaptic transmission," Progress in Brain Research, Vol. 31, K. Akert and P.G. Waser, eds., Elsevier, Amsterdam, pp. 141-155.

Hunt, S. and Schmidt, J., 1978, Some observations on the binding patterns of α-bungarotoxin in the central nervous system of the rat, *Brain Res.*, 157:213-232.

Ibata, Y. and Pappas, G., 1976, The fine structure of synapses in relation to the large spherical neurons in the anterior ventral cochlear nucleus of the cat, *J. Neurocytol.*, 5:395-406.

Jones, D.R., Morest, D.K., Oliver, D.L., and Potashner, S.J., 1984, Transganglionic transport of D-aspartate from cochlear nucleus to cochlea - a quantitative autoradiographic study, *Hearing Res.*, 15:179-213.

Juiz, J.M., Helfert, R.H., Wenthold, R.J., DeBlas, A.L., and Altschuler, R.A., 1989, Immunocytochemical localization of the GABA$_A$/benzodiazepine receptor in the guinea pig cochlear nucleus: Evidence for receptor localization heterogeneity, *Brain Res.*, 504:173-179.

Kane, E.C., 1973, Octopus cells in the cochlear nucleus of the cat: Heterotypic synapses upon homeotypic neurons, *Intern. J. Neurosci.*, 5:251-279.

Kane, E.C., 1974, Patterns of degeneration in the caudal cochlear nucleus of the cat after cochlear ablation, *Anat. Rec.*, 179:67-92.

Kane, E.S., 1976, Descending inputs to caudal cochlear nucleus in cat: A horseradish peroxidase, HRP, study, *Am. J. Anat.*, 146:433-441.

Kane, E.S., 1977, Descending inputs to the octopus cell area of the cat cochlear nucleus: An electron microscopic study, *J. Comp. Neurol.*, 173:337-354.

Kane, E.S. and Conlee, J.W., 1979, Descending inputs to the caudal cochlear nucleus of the cat: Degeneration and autoradiographic studies, *J. Comp. Neurol.*, 187:759-784.

Ljungdahl, A. and Hökfelt, T., 1973, Autoradiographic uptake patterns of [$^3$H]glycine in central tissues with special reference to the cat spinal cord, *Brain Res.*, 62:587-595.

Lorente de Nó, R., 1933, Anatomy of the eighth nerve. III. General plan of structure of the primary cochlear nuclei, *Laryngoscope*, 43:327-350.

Lorente de Nó, R., 1981, "The Primary Acoustic Nuclei," Raven, New York.

Matus, A.I. and Dennison, M.E., 1971, Autoradiographic localization of tritiated glycine at 'flat vesicle' synapses in spinal cord, *Brain Res.*, 32:195-197.

Moore, J.K., 1986, Cochlear nuclei: Relationship to the auditory nerve, in: "Neurobiology of Hearing: The Cochlea," R. A. Altschuler, D.W. Hoffman, and R.P. Bobbin, eds., Raven, New York, pp. 283-301.

Moore, J.K. and Moore, R.Y., 1987, Glutamic acid decarboxylase-like immunoreactivity in brainstem auditory nuclei of the rat, *J. Comp. Neurol.*, 260:157-174.

Morley, B.J., Lorden, J.F., Brown, G.B., Kemp, G.E., and Bradley, R.J., 1977, Regional distribution of nicotinic acetylcholine receptor in rat, *Brain Res.*, 134:161-166.

Oberdorfer, M.D., Parakkal, M.H., Altschuler, R.A., and Wenthold, R.J., 1988, Ultrastructural localization of GABA-immunoreactive terminals in the anteroventral cochlear nucleus of the guinea pig, *Hearing Res.*, 33:229-238.

Oliver, D.L., Potashner, S.J., Jones, D.R., and Morest, D.K., 1983, Selective labeling of spiral ganglion and granule cells with D-aspartate in the auditory system of cat and guinea pig, *J. Neurosci.*, 3:455-472.

Ostapoff, E.-M. and Morest, D.K., 1991, Synaptic organization of globular bushy cells in the ventral cochlear nucleus of the cat, *J. Comp. Neurol.*, 314:598-613.

Ostapoff, E.-M., Morest, D.K., and Potashner, S.J., 1990, Uptake and retrograde transport of [$^3$H]GABA from the cochlear nucleus to the superior olive in the guinea pig, *J. Chem. Neuroanat.*, 3:285-295.

Peyret, D., Geffard, M., and Aran, J.-M., 1986, GABA immunoreactivity in the primary nuclei of the auditory central nervous system, *Hearing Res.*, 23:115-121.

Potashner, S.J., Lindberg, N., and Morest, D.K., 1985a, Uptake and release of τ-aminobutyric acid in the guinea pig cochlear nucleus after axotomy of cochlear and centrifugal fibers, *J. Neurochem.*, 45:1558-1566.

Potashner, S.J., Morest, D.K., Oliver, D.L., and Jones, D.R., 1985b, Identification of glutamatergic and aspartatergic pathways in the auditory system, in: "Auditory Biochemistry," D.G. Drescher, ed., C.C. Thomas, Springfield, IL, pp. 141-162.

Roberts, R.C. and Ribak, C.E., 1987, GABAergic neurons and axon terminals in the brainstem auditory nuclei of the gerbil, *J. Comp. Neurol.*, 258:267-280.

Saint Marie, R.L., Morest, D.K., and Brandon, C.J., 1989, The form and distribution of GABAergic synapses on the principal cell types of the ventral cochlear nucleus of the cat, *Hearing Res.*, 42:97-112.

Saint Marie, R.L., Benson, C.G., Ostapoff, E.-M., and Morest, D.K., 1991, Glycine immunoreactive projections from the dorsal to the anteroventral cochlear nucleus, *Hearing Res.*, 51:11-28.

Schwartz, A.M. and Gulley, R.L., 1978, Nonprimary afferents to the principal cells of the anteroventral cochlear nucleus of the guinea pig, *Am. J. Anat.*, 153:489-508.

Schwartz, I.R., 1983, Differential uptake of ³H-amino acids in the cat cochlear nucleus, *Am. J. Otol.*, 4:300-304.

Schwartz, I.R., 1985, Autoradiographic studies of amino acid labeling of neural elements in the auditory brainstem, *in*: "Auditory Biochemistry," D.G. Drescher, ed., C.C. Thomas, Springfield, IL, pp. 258-277.

Shiraishi, T., Senba, E., Tohyama, M., Wu, J.-Y., Kubo, T., and Matsunaga, T., 1985, Distribution and fine structure of neuronal elements containing glutamate decarboxylase in the rat cochlear nucleus, *Brain Res.*, 347:183-187.

Smith, P.H. and Rhode, W.S., 1987, Characterization of HRP-labeled globular bushy cells in the cat anteroventral cochlear nucleus, *J. Comp. Neurol.*, 266:360-375.

Smith, P.H. and Rhode, W.S., 1989, Structural and functional properties distinguish two types of multipolar cells in the ventral cochlear nucleus, *J. Comp. Neurol.*, 282:595-616.

Spangler, K.M., Cant, N.B., Henkel, C.K., Farley, G.R., and Warr, W.B., 1987, Descending projections from the superior olivary complex to the cochlear nucleus of the cat, *J. Comp. Neurol.*, 259:452-465.

Staatz-Benson, C. and Potashner, S.J., 1988, Uptake and release of glycine in the guinea pig cochlear nucleus after axotomy of afferent or centrifugal fibers, *J. Neurochem.*, 51:370-379.

Thompson, G.C., Cortez, A.M., and Lam, D.M., 1985, Localization of GABA immunoreactivity in the auditory brainstem of guinea pigs, *Brain Res.*, 339:119-122.

Tolbert, L.P. and Morest, D.K., 1982, The neuronal architecture of the anteroventral cochlear nucleus of the cat in the region of the cochlear nerve root: Electron microscopy, *Neurosci.*, 7:3053-3067.

van den Pol, A.N. and Gorcs, T., 1988, Glycine and glycine receptor immunoreactivity in brain and spinal cord, *J. Neurosci.*, 8:472-492.

Wamsley, J.K., Lewis, M.S., Young, W.S.,III, and Kuhar, M.J., 1981, Autoradiographic localization of muscarinic cholinergic receptors in rat brainstem, *J. Neurosci.*, 1:176-191.

Wenthold, R.J., 1987, Evidence for a glycinergic pathway connecting the two cochlear nuclei: An immunocytochemical and retrograde transport study, *Brain Res.*, 415:183-187.

Wenthold, R.J., Zempel, J.M., Parakkal, M.H., Reeks, K.A., and Altschuler, R.A., 1986, Immunocytochemical localization of GABA in the cochlear nucleus of the guinea pig, *Brain Res.*, 380:7-18.

Wenthold, R.J., Huie, D., Altschuler, R.A., and Reeks, K.A., 1987, Glycine immunoreactivity localized in the cochlear nucleus and superior olivary complex, *Neurosci.*, 22:897-912.

Wenthold, R.J., Parakkal, M.H., Oberdorfer, M.D., and Altschuler, R.A., 1988, Glycine receptor immunoreactivity in the ventral cochlear nucleus of the guinea pig, *J. Comp. Neurol.*, 276:423-435.

Wickesberg, R.E. and Oertel, D., 1990, Delayed, frequency-specific inhibition in the cochlear nuclei of mice: A mechanism for monaural echo suppression, *J. Neurosci.*, 10:1762-1768.

Winter, I.M., Robertson, D., and Cole, K.S., 1989, Descending projections from auditory brainstem nuclei to the cochlea and cochlear nucleus of the guinea pig, *J. Comp. Neurol.*, 280:143-157.

# NON-PRIMARY INPUTS TO THE COCHLEAR NUCLEUS VISUALIZED USING IMMUNOCYTOCHEMISTRY

Joe C. Adams

Department of Otolaryngology, Massachusetts Eye and Ear Infirmary
243 Charles Street, Boston, Mass. 02114, USA

It has been recognized since the time of Held ('93) that there are a number of non-primary inputs to cochlear nucleus (CN) neurons. With passing years, knowledge of a variety of different sources of these inputs has increased (e.g., Elverland,'77; Adams,'83; Spangler et al.,'87) but we are still largely ignorant concerning the details of which cells in a given location give rise to CN inputs and also which CN cells receive inputs from given sources. One tool that can aid in sorting out such details is immunocytochemistry. The availability of antibodies that permit visualization of chemically distinct cell classes and their axonal endings in the CN offers a means of studying origins of given inputs and learning about their terminations within the CN. Unfortunately, usually there is ambiguity regarding the origins of immunostained terminals because there are multiple possible sources of terminals which are immunopositive for most antigens. Because of the complexity of the problems involved, there has been relatively little progress in sorting out specific pathways even with the aid of immunocytochemistry. It is therefore premature to attempt a review of progress of this work. Instead, this report will give a brief status report of what some promising antibodies have revealed about non-primary CN inputs. Results reported here were obtained in cats using previously described procedures (Adams and Mugnaini,'87). Selected examples of immunostained terminals in the ventral CN (VCN) are shown in Figure 1 and the dorsal CN (DCN) in Figure 2, which includes all four DCN layers for each antibody. Neurons that may be the sources of these and other CN terminals are illustrated in Figures 3 and 4.

In terms of their sheer numbers, perhaps the most impressive non-primary CN inputs are the presumed inhibitory inputs that immunostain for glutamate decarboxylase (GAD), the synthesizing enzyme for the inhibitory neurotransmitter GABA (Adams and Mugnaini,'87; Moore and Moore,'87; Saint Marie et al.,'89). GAD immunostained terminals are present on virtually every CN cell and these terminals have characteristic sizes and distributions that are different for various CN cell types. It may be that distinctive GAD-positive endings which characterize a given CN cell class have unique origins that are reflected in the morphology of the endings (Saint Marie et al.,'89). It is also possible that given CN cell classes have GAD-positive inputs from a variety of sources and that the terminals from these different sources cannot be distinguished morphologically. There is presently no means of readily determining which of these suggested possibilities is correct. It may well be that various combinations of the two possibilites hold for different cell classes. One factor that makes the issue difficult to resolve is the number of possible sources of GAD-positive terminals. These

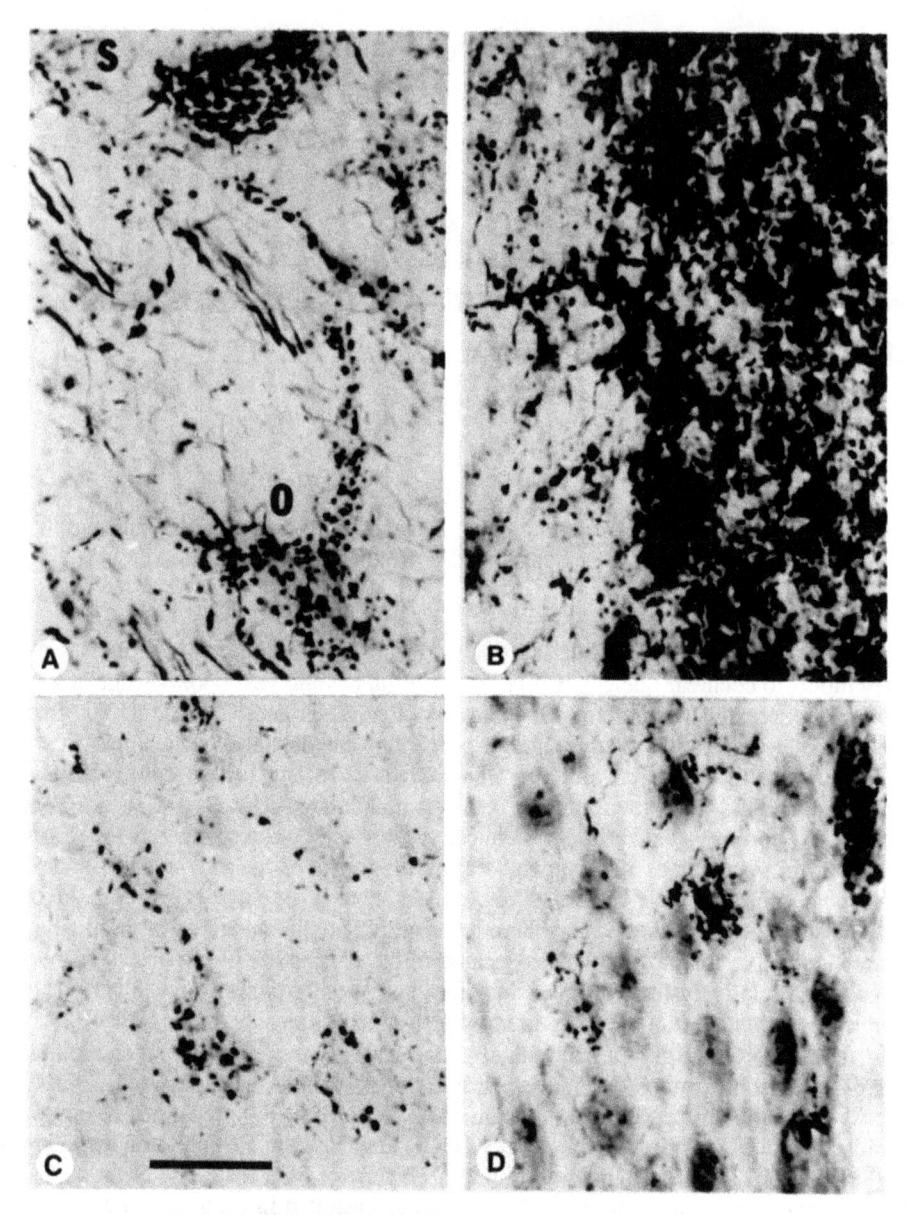

**Figure 1.** VCN. Calibration 40 μm
A. Anti-GAD in PVCN. O, octopus cell. S, Type 2 stellate cell.
B. Anti-ChAT in Granule cell layer (right) of AVCN.
C. Anti-enkephalin in PVCN.
D. Anti-neurotensin in AVCN, Azure

include cells intrinsic to both VCN and DCN, periolivary cells of both sides, and cells in the inferior colliculus (IC)(see e.g., Mugnaini and Oertel, '85).

One approach towards determining the origins of GABAergic inputs has been to inject horseradish peroxidase into the anteroventral CN (AVCN) to retrogradely label inputs, then immunostain for GABA to show which retrogradely labeled cells are also positive for GABA. Such experiments in cat have shown that there are GABA-positive cells in medial and lateral

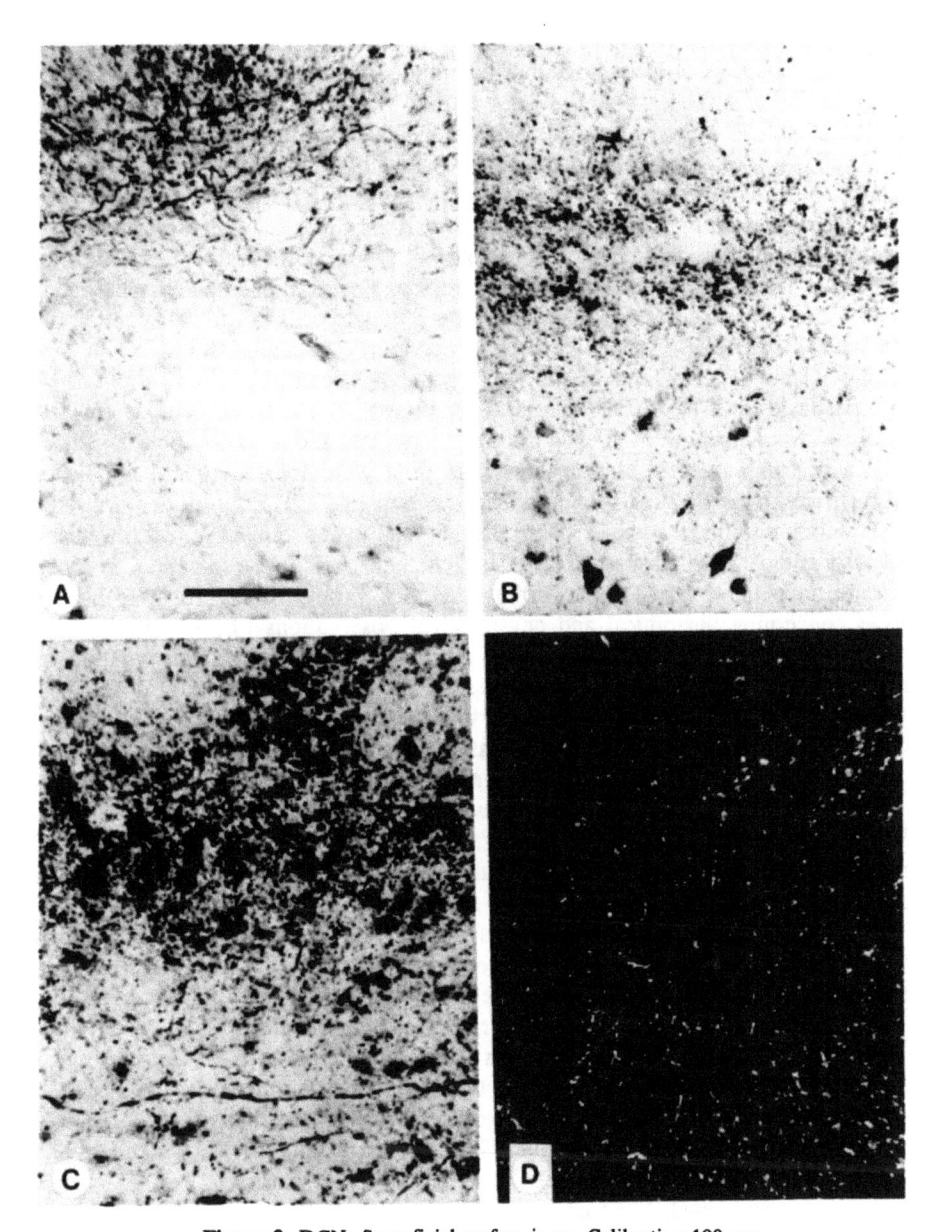

**Figure 2.** DCN. Superficial surface is up. Calibration 100 μm
A. Anti-neurotensin.
B. Anti-enkephalin.
C. Anti-substance P.
D. Anti-CCK.

periolivary cell groups and in the deep DCN that project to the AVCN (Adams and Wenthold,'87.) Similar results have also been found in guinea pig (Ostapoff et al.,'88; Saint and Marie et al.,'91, and this volume). As fruitful as this approach has been, it still does not show whether all GABA-positive endings on given AVCN cell classes arise from one or more particular cell classes. More refined techniques are needed to address this and similar problems. The same strengths and limitations apply equally to the study of glycine-positive inputs which, in fact, have been studied in the same above-mentioned experiments which investigated GABA-positive inputs. The complexity of the problems of determining the origins and terminations of CN glycine-positive inputs is nearly the same as that of GABA glycine-positive systems because there are a variety of known sources of glycine-positive inputs. These include periolivary cells, DCN cells, the contralateral VCN, and probably ipsilateral collaterals of contralaterally projecting CN cells (Adams and Wenthold,'87; Oertel, this volume; Saint Marie et al.,'91, this volume; Wenthold,'87; Wickesberg et al.,'91).

In addition to aiding in the identification of various CN inputs, markers of GABAergic processes are useful for identifying various CN cell types by means of their characteristic complement of immunostained GABA-positive inputs. For example, the characteristic high density of GAD immunostained terminals on type 2 stellate cells permits distinction of type 1 from type 2 cells, a distinction that was previously appreciated only with electron microscopy (Cant,'81). The use of the GAD antibody permitted identification of type 2 stellate cells throughout the VCN (Adams, '91). Figure 1A shows an octopus cell with its characteristic complement of GAD-positive axon terminals and a nearby type 2 stellate cell with a denser complement of terminals. These cells were located in the middle of the "octopus cell area", as the area was described by Osen ('69). The figure demonstrates that the GAD immunostaining permits distinguishing cell types far more readily than was previously possible. The presence of type 2 stellate cells within the so-called "octopus cell area" is not uncommon, and is noteworthy because a variety of investigators have made inferences concerning anatomical and physiological observations of cells presumed to be octopus cells based on the cells' presence in the "octopus cell area". Examination of this region in numerous GAD immunostained cats shows that there is variability in the locations of octopus cells and type 2 stellate cells and that location alone is not a reliable means of identifying cells in this area. These observations emphasize the hazards of assumptions concerning the identities of given cells which are based solely upon the cells' location, even when such assumptions are based on careful, well founded observations. Striking differences between cell types such as shown in Fig. 1A emphasize the great utility of GAD immunostaining for identifying cell classes within the CN. Distinctions made this way can be used for a wide variety of applications in the study of the nucleus.

Other markers that are useful for visualizing selected populations of non-primary inputs are those for cholinergic fibers. Since the work of Rasmussen ('64), a variety of investigators have used various assays for acetylcholinesterase and choline acetyltransferase (ChAT) to investigate presumed olivocochlear (OC) inputs to the CN. The use of antibodies to ChAT offers the advantage that valid results are more compelling than results with markers for acetylcholinesterase because esterases can be present in sites other than cholinergic axons and because the resolution of fine processes can be much better than is possible with histochemical methods for esterases. Figure 1B demonstrates the density of terminal boutons in the granule cell layer of the cat AVCN that immunostain for ChAT. Possible sources of these terminals include medial and lateral OC cells (e.g. Osen et al.,'84; Brown et al., '88; Godfrey et al.,'90;), cells in the monoaminergic cell groups A6 and A7, (Kromer and Moore,'76; personal unpublished observations in cat), (these cells can be positive for ChAT and a variety of peptides), ChAT-positive cells located in the rostral part of the lateral periolivary region (Adams, '89), and/or intrinsic ChAT-positive CN cells (Godfrey and Heaney,'91; Vetter, this volume). Reports of ChAT-positive CN cells have been in rat. Personal preliminary observations in cat have shown similar cells, although there appear to be fewer of these cells in cat than in rat.

The difficulty presented by the presence of multiple possible sources of ChAT-positive terminals can be approached in some cases by comparing results obtained with ChAT with results using antibodies to other antigens that given ChAT-positive populations are known to contain. For example, it is known that cholinergic lateral OC cells are also immunopositive for enkephalins (Altschuler et al.,'84). The CN granule cell domain contains enkephalin-positive terminals (not shown) but the density of these terminals is not nearly as great as those that are ChAT-positive. This indicates that only a minority of the ChAT-positive terminals in the granule cell domain may originate from lateral OC cells. Enkephalin-positive fibers can be seen leaving the olivocochlear bundle and entering the AVCN (not shown). The fact that some enkephalin-positive terminals may be those of olivocochlear fibers does not rule out the possibility that given enkephalin-positive terminals may originate from other sources. There are numerous enkephalin-positive endings in the VCN which are outside the granule cell domain. In the AVCN these are present largely in the neuropil (not shown), where they presumably terminate on dendrites. In more caudal regions they are commonly found on stellate cells (Fig. 1C). It is not yet clear whether the recipients of these terminals are type 1 or type 2 stellate cells. Another region that is rich in enkephalin-positive terminals is layer 2 of the DCN (Fig. 2B). The most obvious candidates for origins of these large, deeply immunoreactive terminals are cells within the medial, ventral periolivary cell group (MVPO or VNTB)(Fig. 3B). This is suggested by the presence of large enkephalin-positive axons that course within the trapezoid body and presumably terminate in the CN. Another possible source of enkephalin-positive terminals in the CN is the external cortex of the IC (Fig. 4C). There are also a few enkephalin-positive cells located in lateral periolivary regions and in the VCN (not shown) that may contribute projections to the CN. Perhaps the greatest number of enkephalin-positive cells known to project to the VCN are the cells in the deep DCN (bottom of Fig. 2B) that give rise to the tuberculoventral tract. These cells are readily immunostained for enkephalins in kittens, but in adult cats they are seen only if the animals have been treated with colchicine to prevent axoplasmic transport. Axons of the tuberculoventral tract are not obviously enkephalin-positive, and it remains to be demonstrated whether there is sufficient enkephalin in the axonal endings of these fibers to be reliably detected by immunocytochemistry. The multitude of sites which contain cells that project to the CN and also contain cells that are enkephalin-positive complicates the problem of determining the origins of all enkephalin-positive terminals within the CN. Clearly, more intricate techniques than simple immunostaining will be required to sort out these CN inputs.

Another peptide, neurotensin, appears in non-primary CN inputs that are different from those which are positive for enkephalin. Figure 1D shows neurotensin immunostaining pericellular plexuses in the rostral AVCN, a low frequency portion of the nucleus (Bourk et al.,'81). In the DCN the densest immunostaining of terminals is in layer 1 (Fig. 2A) and this dense plexus of neurotensin-positive terminals is found exclusively in the ventral (low frequency) portion of the DCN. Other scattered fibers are present within the DCN but these are sparse and their locations are much less consistent between animals than those shown in Fig. 2A. In dorsal portions of the rostral PVCN there are also neurotensin-positive pericellular terminals. There are two candidate sources of the origins of neurotensin-positive terminals in the CN. These include cells in the MVPO (Fig. 3A) and cells in the dorsal cortex of the IC (Fig. 4D). It may be relevant that there are also neurotensin-positive terminals in the lateral (low frequency) portion of the medial nucleus of the trapezoid body (MNTB) and the dorsal (low frequency) cap of the medial superior olive. That is, there may be one source of all the terminals found in low frequency portions of the CN and superior olivary nuclei. The presence of neurotensin-positive axons between the MNTB and MVPO suggests that the inputs to the MNTB come from the MVPO but this and other possible connections remain to be experimentally verified. The neurotensin-positive cells in the IC are few in number and confined to the dorsal perimeter of nucleus. They remain the other candidate source of CN inputs that are neurotensin-positive.

A somewhat simpler situation exists with regard to CN terminals that immunostain for substance P. Figure 2C shows that there are abundant substance P-positive terminals in layer 2 of the DCN. The cluster of positive terminals at the top of the figure is in layer 1 and is within a cluster of granule cells. Not all granule cell clusters in layer 1 receive substance P-positive inputs. There are also substance P-positive terminals generally throughout the granule cell domain of the VCN and along the medial, dorsal and lateral margins of the AVCN. Their density in these regions is less than shown in Figure 2C within layer 2 of the DCN. There is one prime candidate for the origins of the substance P-positive terminals in the CN and that

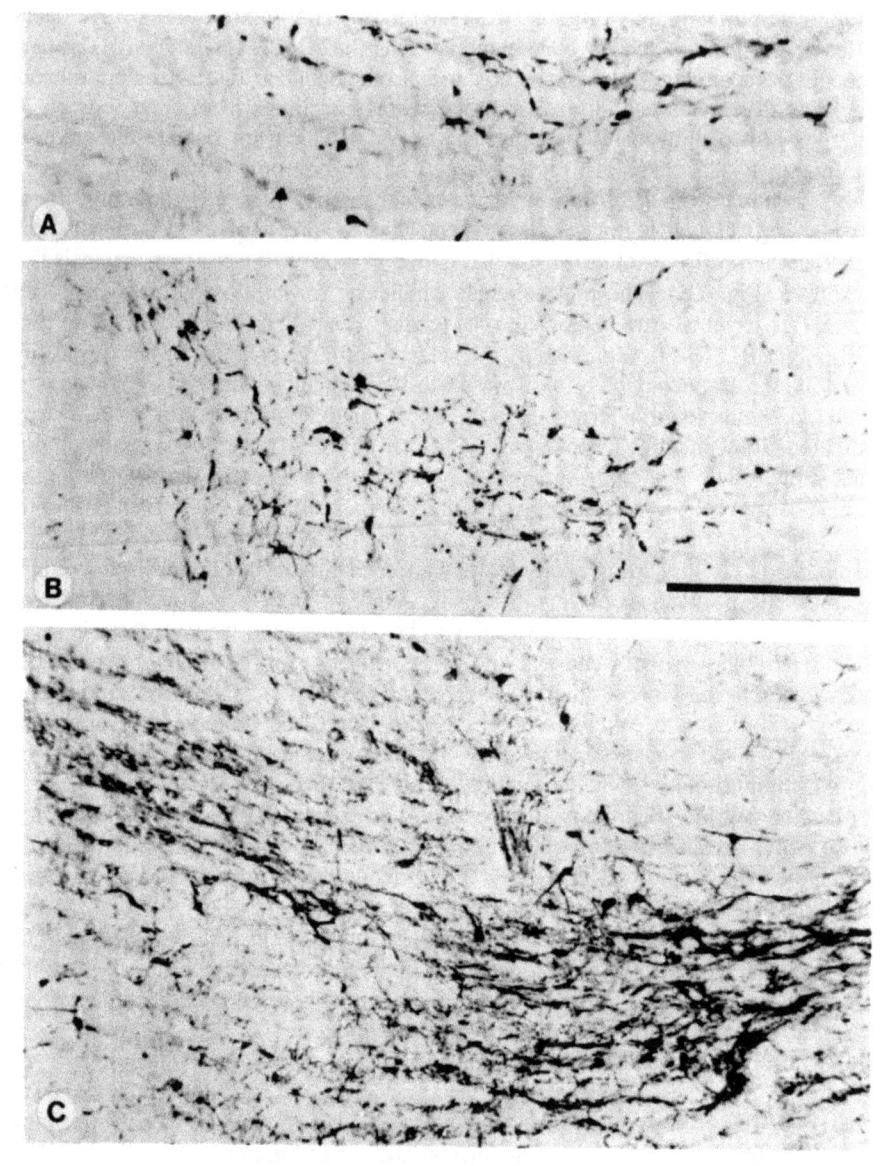

**Figure 3.** MVPO. Calibration 100 μm
A. Anti-neurotensin.
B. Anti-enkephalin.
C. Anti-GAD.

is the IC. There are large numbers of substance P-positive cells within the IC (Fig. 4B). It may be relevant that the other subcollicular target of IC projections, the VMPO (e.g. Rasmussen,'64), also has a rich plexus of substance P-positive terminal boutons (Adams and Mugnaini,'85a).

Another peptide found in terminals in the CN is cholecystokinin (CCK) (Adams and Mugnaini,'85b). Figure 2D shows CCK-positive fibers and terminals in the DCN. These processes do not strongly immunostain and they are small so it is necessary to demonstrate their presence with darkfield microscopy, as illustrated in the figure. In addition to the DCN terminals there are CCK-positive terminals in the VCN. These are most conspicuous in the

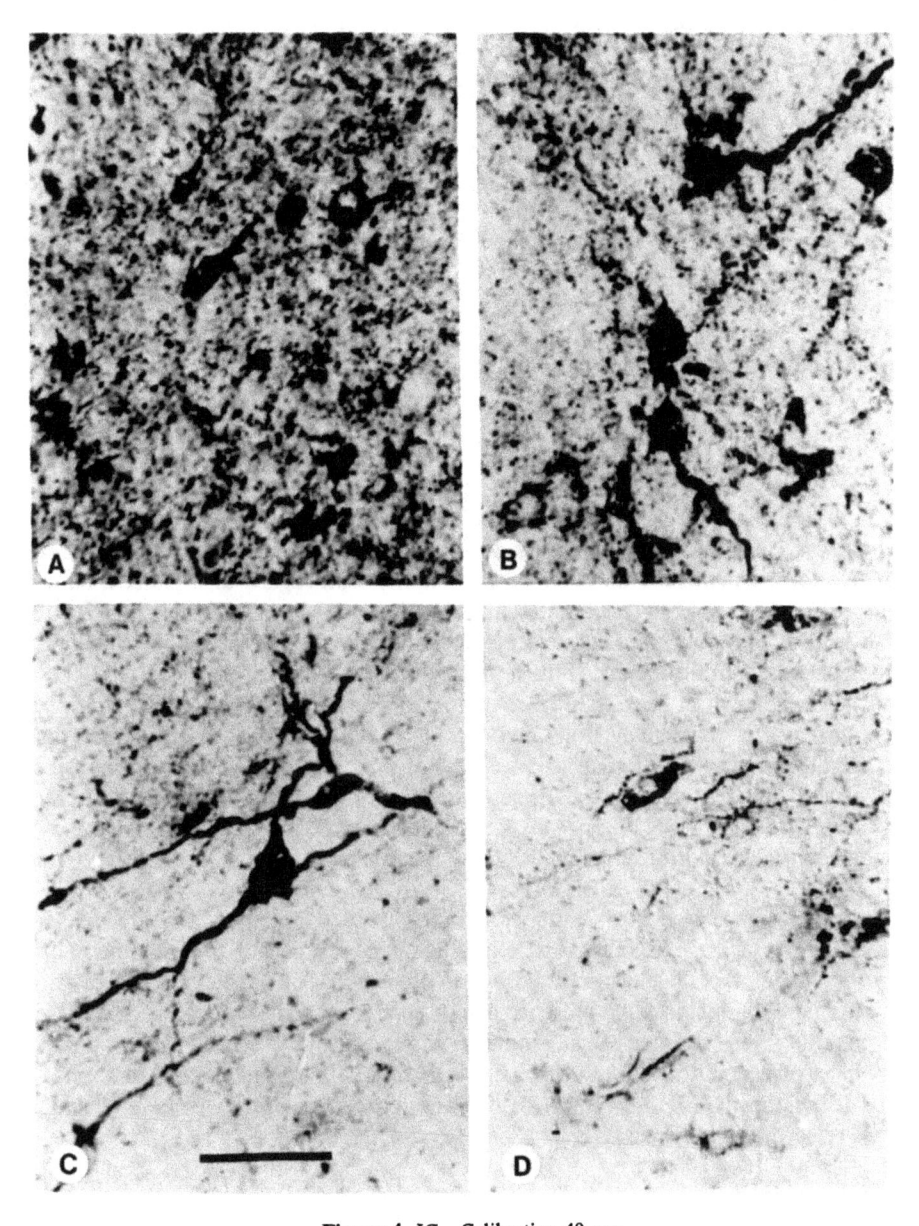

**Figure 4.** IC. Calibration 40 μm
A. Anti-CCK.
B. Anti-substance P.
C. Anti-enkephalin.
D. Anti-neurotensin.

region near the nerve root (Adams and Mugnaini,'85b). There are two probable sources of CCK-positive endings in the CN. These are CCK-positive stellate cells found within and caudal to the nerve root and cells within the IC. Stellate cells that send projections to the IC have been shown to send collateral projections to the DCN (Adams,'83a). It remains to be demonstrated directly that the CCK-positive CN cells project to the IC, but they are strong candidates for being the cells that project both to the IC and the DCN. The IC cells that are CCK-positive are generally smaller (see Fig. 4A) than the larger IC cells that are probably the projection neurons of the IC. Whether the IC CCK-positive cells are interneurons remains to be established.

Figure 3 illustrates the distribution of immunostained cells in the MVPO, a region from which all or most cells send projections to the CN. One difficulty with determining the projections of cytochemically distinct cells in this region is that it is necessary to inject colchicine on the day previous to perfusing the animal in order to obtain intense immunostaining for GAD, enkephalin, or neurotensin in these cells. This makes it difficult to retrogradely label these cells and subsequently immunostain them for these antigens. Given the overlapping distribution of the cells in MVPO that are positive for these three antigens, it is appropriate to ask which of the various antigens may be found within the same cells. One can determine from simply counting labeled cells that there cannot be a complete correspondence of these three antigens within given cells because there are far more GAD-positive cells than enkephalin-positive cells and more enkephalin-positive cells than neurotensin-positive cells. To examine the issue of co-containment of GAD and enkephalin within cells, preliminary experiments have been done in which the MVPO of colchicine-treated animals was first immunostained for enkephalin, and the positive cells' locations mapped; the sections were then immunostained for GAD, and GAD-positive cells' locations mapped. It was clear that most cells were not positive for both antigens. A small minority was positive for both antigens and there was ambiguity with regard to whether a number of remaining cells were positive for both. Further studies will be necessary to achieve more quantitative results on this issue, but it seems clear there is not substantial co-containment of GAD and enkephalin in these cells. If the terminals in the CN that are positive for enkephalin and neurotensin (shown in Figs. 1C, 1D,2A and 2B) arise from MVPO cells, then the two peptides must be in different neurons because the distributions of the terminals that are positive for the two are so different. It is well established that MVPO cells send projections to the CN and that the contralateral projection is more abundant (Adams and Warr,'76; Adams,'83b; Spangler et al.,'87), but it has not been established whether the more pronounced contralateral projection applies to each cytochemically distinct cell type found in MVPO.

In conclusion, there are a number of non-primary inputs to the CN that can be visualized by a variety of antibodies. Space limitations do not permit covering many of them here. In most cases the advantage offered by selectively marking cells and processes using a given antibody is limited by the presence of multiple cell classes which are candidates for the origins of given immunostained endings within the CN. There are a number of experiments which could resolve some of the present ambiguities, but it appears that more advanced techniques will be necessary for determining details of connections of major CN cell classes and their non-primary inputs.

ACKNOWLEDGEMENTS

This work was supported by NIH grant DC 00269-08.

REFERENCES

Adams, J.C., 1983a, Multipolar cells in the ventral cochlear nucleus project to the dorsal cochlear nucleus and the inferior colliculus, *Neurosci. Lett.*, 37:205-208.

Adams, J.C., 1983b, Cytology of periolivary cells and the organization of their projections in the cat, *J. Comp. Neurol.*, 215:275-289.

Adams, J.C., 1989, Non-olivocochlear cholinergic periolivary cells, *Soc. Neurosci. Abstr.*, 15:1114.

Adams, J.C., 1991, Distribution of some cytochemically distinct cell classes in the ventral cochlear nucleus of cat and human with emphasis on octopus cells and their projections, *in*: "Advances in Speech, Hearing and Language Processing" Vol.3, W.A.Ainsworth (Ed.) JAI Press, London. In press.

Adams, J.C. and Mugnaini, E., 1985a, Patterns of immunostaining with antisera to peptides in the auditory brainstem of cat, *Soc. Neurosci. Abstr.*, 11:32.

Adams, J.C. and Mugnaini, E., 1985b, Distribution of cholecystokinin-like immunoreactivity in the brainstem auditory system, *Soc. N.Y. Acad. Sci.*, 448: 563-565.

Adams, J.C. and Mugnaini, E., 1987, Patterns of glutamate decarboxylase immunostaining in the feline cochlear complex using silver enhancement and electron microscopy, *J. Comp. Neurol.*, 375-401.

Adams, J.C. and Warr, W.B., 1976, Origins of axons in the cat's acoustic striae determined by injection of horseradish peroxidase into severed tracts, *J. Comp. Neurol.*, 170:107-122.

Adams, J.C. and Wenthold, R.J., 1987, Immunostaining of GABA-ergic and glycinergic inputs to the cochlear nucleus, *Soc. Neurosci. Abstr.*, 13:1259.

Altschuler, R.A., Fex, J., Parakkal, M.H. and Eckenstein, F., 1984, Colocalization of enkephalin-like and choline acetyltransferase-like immunoreactivities in olivocochlear neurons of the guinea pig, *J. Histochem. Cytochem.*, 32:839-843.

Bourk, T.R., Mielcarz, J.P. and Norris, B.E., 1981, Tonotopic organization of the anteroventral cochlear nucleus of the cat, *Hearing Res.*, 4:215-241.

Brown, M.C., Liberman, M.C. and Ryugo, D.K., 1988, Brainstem branches from olivocochlear axons in cats and rodents, *J. Comp. Neurol.*, 278:591-603.

Cant, N.B., 1981, The fine structure of two types of stellate cells in the anterior division of the anteroventral cochlear nucleus of the cat, *Neuroscience*, 6:2643-2655.

Elverland, H.H., 1977, Descending connections between the superior olivary and cochlear nuclear complexes in the cat studied by autoradiographic and horseradishperoxidase methods, *Exp. Brain Res.*, 17:428-412.

Godfrey, D.A. and Heaney, M.L., 1991, Immunoreactivity for choline acetyltransferase in the rat cochlear nucleus and superior olive, *Abstr. Assoc. Res. Otolaryngol.*, 14:10.

Godfrey, D.A., Beranek, K.L., Carlson, L., Parli, J.A., Dunn, J.D. and Ross, C.D., 1990, Contribution of centrifugal innervation to choline acetyltransferase activity in the cat cochlear nucleus, *Hearing Res.*, 49: 259-280.

Held, H., 1893, Die centrale Gehörleitung, *Arch. Anat. Physiol., Anat. Abt.*, 201-248.

Kromer, L.F. and Moore, R.Y., 1976, Cochlear nucleus innervation by central norepinephrine neurons in the rat, *Brain Res.*, 118: 531-537.

Mugnaini, E. and Oertel, W.H., 1985, An atlas of the distribution of GABAergic neurons and terminals in the rat CNS as revealed by GAD immunohistochemistry, in: Handbook of Chemical Neuroanatomy, Vol 4: GABA and Neuropeptides in the CNS, Part 1. A. Björklund and T. Hokfelt (Eds.) Elsevier, New York, pp. 436-608.

Osen, K.K., 1969, Cytoarchitecture of the cochlear nuclei in the cat, *J. Comp. Neurol.*, 136:453-484.

Osen, K.K., Mugnaini, E., Dahl, A.-L. and Cristiansen, A.H., 1984, Histochemical localization of acetylcholinesterase in the cochlear and superior olivary nuclei. A reappraisal with emphasis on the cochlear granule cell system, *Archiv. Ital. Biol.*, 122: 169-212.

Ostapoff, E.-M., Staatz-Benson, C., Morest, D.K., Potashner, S.J. and Saint Marie, R.L., 1988, GABA and glycine immunoreactivity of descending and commissural inputs to the cochlear nucleus in guinea pig, *Soc. Neurosci. Abstr.*, 14:489.

Rasmussen, G.L., 1964, Anatomic relationships of the ascending and descending auditory systems, in: "Neurological Aspects of Auditory and Vestibular Disorders" W.S. Fields and B.R. Alford (Eds.) Thomas, Springfield, pp. 5-19.

Saint Marie, R.L., Morest, D.K. and Brandon, C.J., 1989, The form and distribution of GABAergic synapses on the principal cell types of the ventral cochlear nucleus of the cat, *Hearing Res.*, 42:97-112.

Saint Marie, R.L., Benson, C.G., Ostapoff, E.-M. and Morest, D.K., 1991, Glycine immunoreactive projections from the dorsal to the anteroventral cochlear nucleus, *Hearing Res.*, 51:11-28.

Spangler, K.M., Cant, N.B., Henkel, C.K., Farley, G.R. and Warr, W.B., 1987, Descending projections from the superior olivary complex to the cochlear nucleus of the cat, *J. Comp. Neurol.*, 259: 452-465.

Wenthold, R.J., 1987, Evidence for a glycinergic pathway connecting the two cochlear nuclei: an immunocytochemical and retrograde transport study, *Brain Res.*, 415:183-187.

Wickesberg, R., Whitlon, D. and Oertel, D., 1991, Axonal transport accounts for some of the dynamic properties of glycine-like immunoreactivity, *Ass. Res. Otolaryngol.*, p.87.

# SUPERIOR OLIVARY CELLS WITH DESCENDING PROJECTIONS TO THE COCHLEAR NUCLEUS

John M. Zook and Nobuyuki Kuwabara

Dept. Zoological & Biomedical Science
Ohio University, Athens, Ohio, 45701, USA

The cochlear nucleus is the target of a greater concentration of descending projections than any other auditory structure, including the cochlea (Rasmussen, '64; Cant and Morest, '78; Spangler and Warr, '91). In contrast to the rest of the central auditory system, parts of the cochlear nucleus may receive more descending afferents than ascending afferents (Kane and Conlee, '79; Conlee and Kane, '82; Spangler et al., '87). Despite their unusual numbers and potential importance, the descending inputs to the cochlear nucleus have received limited attention. Consequently, we have an equally limited understanding of the contribution of descending projections to the functional integrity of the cochlear nucleus. General stimulation of the descending fibers has broad inhibitory effects upon unit activity, although some facilitory effects have also been reported (Pfalz, '62; Comis and Whitfield, '68; Starr and Wernick, '68; Comis and Davies, '69; Comis, '70; Mast, '70; Bourk, '76; Brown and Buchwald, '76; Caspary, '86). Pickles and Comis ('73) and Pickles ('76) have argued that descending input may enhance signal detection in noise by influencing critical bandwidth, but this concept has not been developed. Others have suggested an influence of descending input on the effective encoding of complex signals (Frisina et al., '90; Shore et al., '91). Still other proposed roles of the descending auditory pathways, such as loudness protection, might be expected to function at the level of the cochlea rather than at the level of the cochlear nucleus (Cody and Johnstone, '82; Liberman, '88).

Although there has been a renewed focus on descending inputs to the cochlear nucleus, the fundamental design of the descending auditory system presents a major deterrent to progress. The cells and projections of the descending auditory system are diffusely distributed at all levels of the auditory pathway, in contrast to the more numerous ascending auditory fibers and concentrated ascending auditory nuclei. The cochlear nucleus, for example, receives most of its descending projections from the superior olivary complex. Almost all cells which originate this projection are periolivary, loosely scattered around the more condensed cell groups and concentrated fiber plexi of the principal superior olivary nuclei (Rasmussen, '64; Adams and Warr, '76; Adams, '83; Glendenning and Masterton, '83; Covey et al., '84; Spangler et al., '87; Farley and Warr, '81; Winter et al., '89; Spangler and Warr, '91). Periolivary cells with descending projections form two general populations, distinguished by the course and targets of their axons.

The best known cell population forms the olivocochlear bundle (OCB) which contributes collateral input to the cochlear nucleus as well as providing all of the descending

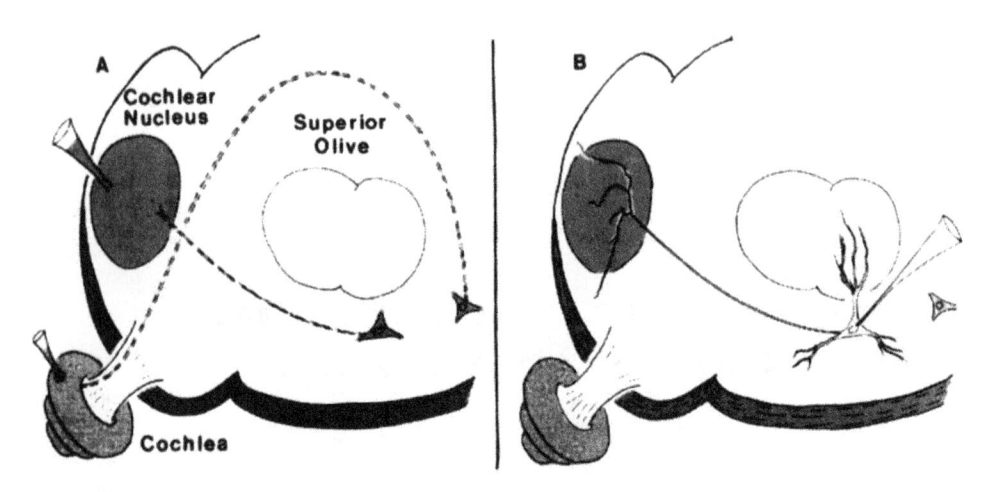

**Figure 1.** Diagram of cell "pre-tagging" and intracellular labeling protocol. A. In the intact subject, a fluorescent dye is deposited in the cochlear nucleus. Following retrograde transport, the somata of olivary cells with projections to the cochlear nucleus are pre-tagged with the fluorescent dye, usually Fluoro-gold, Diamidino Yellow or Fast Blue. In some cases, to distinguish olivocochlear nucleus (OCN) from olivocochlear (OCB) projections, a different dye is deposited respectively into the cochlear nucleus and cochlea. B. Following the preparation of transverse slices of this tissue, the retrogradely pre-tagged somata are identified under a dissecting or compound microscope and a chosen cell is secondarily labeled with a visually guided, intracellular electrode filled with either Lucifer Yellow or biocytin.

input to the cochlea (Warr, '75; Warr and Guinan, '79; Thompson and Thompson, '86; '91; Brown et al., '88; Winter et al., '89; Benson and Brown, '90; Ryan et al., '90; Brown et al., '91). The second, largely unexplored population consists of periolivary cells whose axons descend via the trapezoid body or the acoustic striae and terminate exclusively in the cochlear nucleus (Rasmussen, '64; Adams, '83; Zook and Kuwabara, '91). We will focus on this latter cell population and its projections to the cochlear nuclei. To distinguish this projection from the OCB, we will refer to it as the "olivocochlear nucleus" projection (OCN).

Less is known of OCN cells and their axon projections compared to the OCB largely because OCN fibers are neither as collected nor as distinct as the OCB. OCB fibers form a discrete bundle which crosses the brainstem as a separate tract from the ascending auditory system. The axons which make up the OCN projection, on the other hand, never coalesce into a discrete bundle. Each separate fiber remains in intimate contact with the ascending system and weaves its own path through the trapezoid body or acoustic striae from superior olive to cochlear nucleus.

Despite the difficulty in isolating and focusing upon OCN cells and fibers, we have descriptions of both the general distribution of OCN cell soma in the superior olive (Adams and Warr, '76; Elverland, '77; Adams, '83; Glendenning and Masterton, '83; Winter et al.'89) and the gross distribution of OCN axons within the cochlear nucleus (Rasmussen, '64; Cant and Morest, '78; Spangler et al., '87; Elverland, '77; Covey et al., '84; Shore et al., '91). However, fundamentally important details of this projection have not been determined, such as the complete dendritic morphology of identified OCN cells or the specific terminal pattern of individual OCN axons within the cochlear nucleus. Such details can be revealed with great clarity by the discrete dye-labeling of individual cells with intracellular electrodes.

We will describe here our initial findings on the axonal and dendritic morphology of OCN cells obtained by intracellularly labeling cells in a tissue slice preparation of the auditory brainstem. To selectively locate and focus on this scattered population of cells in a tissue slice, we began each experiment by "pre-tagging" OCN cell soma with retrogradely transported fluorescent dyes from extracellular deposits in the cochlear nucleus *in vivo* (Fig. 1A). This pre-tagged population of cells is then re-labeled intracellularly using visually guided electrodes in an *in vitro* tissue slice (Fig. 1B). Our best examples of OCN cells to date are from tissue slices of the mustache bat, *Pteronotus parnellii*, and the gerbil. However, the general patterns shown here reflect additional data from brainstem tissue slices of the brown bat, *Eptesicus fuscus*, and mouse as well. The specific pre-tagging and tissue slice techniques used in these species have been described elsewhere (Zook and Kuwabara, '91; Kuwabara et al., '91; Kuwabara and Zook, '91).

We have chosen these four species because each shows a different pattern of cell groups within the superior olive. We are particularly interested in the convergence between the ascending and descending systems at this level (Thompson and Thompson, '91; Zook and Kuwabara, '91). As we will show, some OCN cells do show close relationships with the principal nuclei of the superior olive. It is useful to examine species with contrasting patterns of olivary nuclei in order to assess the cross-species consistency or variability of the OCN system (Shore et al., '91). For example, the mustache bat has a very prominent medial superior olive (MSO, see Casseday et al., this volume), lateral superior olive (LSO), and medial nucleus of the trapezoid body (MNTB), but virtually no superior paraolivary nucleus (SPN). The gerbil has a somewhat smaller representation of all three principal olivary cell groups, but has a prominent SPN (Nordeen et al, '83). In both mouse and big brown bat, the MNTB, LSO and SPN are large, while the MSO is particularly small (Willard and Ryugo, '83). Some gross differences in the positions of cells associated with the descending projection to the cochlea have already been noted in these species. In the mustache bat, the lateral OCB cell group lies outside of the LSO neuropil (Bishop and Henson, '87). In the big brown bat, this group lies mainly inside the LSO. In both rodents, the lateral OCB cell group lies completely within the LSO (White and Warr, '83; Aschoff and Ostwald, '87; Campbell and Henson, '88).

There is no equivalent difference between rodent and non-rodent species in the distribution of OCN somata. In the four species used, the large deposits of retrograde tracers in the cochlear nucleus used to "pre-tag" the OCN population revealed a fairly consistent pattern of labeled somata in the superior olive. Virtually all labeled somata were periolivary, with the majority concentrated bilaterally in the ventral nucleus of the trapezoid body (VNTB) and ipsilaterally in the lateral nucleus of the trapezoid body, as seen in the cat, tree shrew and guinea pig (Elverland, '77; Adams, '83; Glendenning and Masterton, '83; Covey et al., '84; Spangler et al., '87; Winter et al., '89, Shore et al., '91). Lesser OCN cell populations were found bilaterally in the ventral periolivary nucleus (VPO) and ipsilaterally in the dorsal periolivary nucleus (DPO) and posterior periolivary nucleus (PPO). A few small species variations were observed. For example, the mustache bat has a concentration of OCN cells in the ventromedial periolivary area below and medial to the MSO, and both bat species have unusual numbers of OCN cells in both the LNTB and the VPO.

From our initial intracellular labeling of fluorescently pre-tagged OCN cells we can suggest several general classes of OCN cells. This classification scheme is based on the relative location and morphology of labeled dendritic fields (Fig. 2). This must be considered a tentative division, as our sample of intracellularly labeled OCN cells is still small (N=27). Examples of each cell category have been found in all four species and in all parts of the superior olivary complex.

Three classes of OCN cells have been identified, one general type and two specific types. The general class contained labeled OCN cells with dendritic fields confined to the periolivary zones of the superior olive (Fig. 2, cell 1). These cells were mainly medium to large multipolar cells and were usually found at some distance from the borders of the principal olivary nuclei. Although these cells represent the majority of OCN cells labeled intracellularly, as yet we have not identified any distinguishable patterns in their dendritic morphology. These pure periolivary OCN cells will be discussed in greater detail in the future as our sample increases.

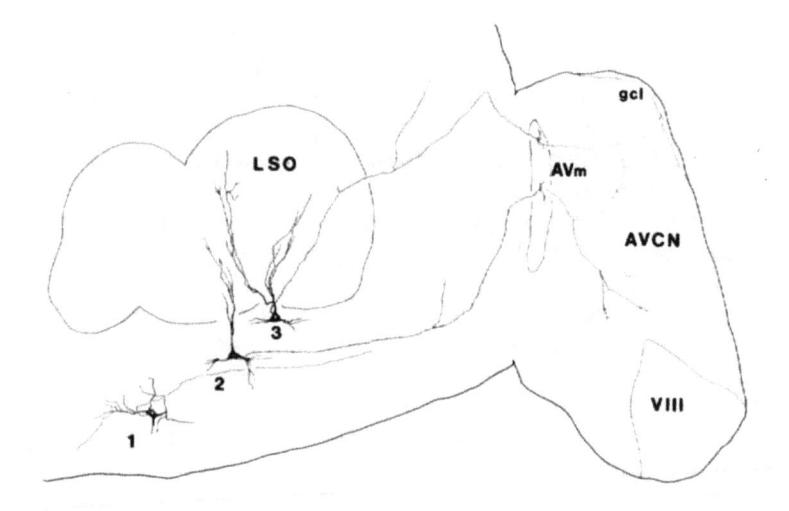

**Figure 2.** Camera lucida drawings of three representative intracellularly labeled cells with projections to the ipsilateral anteroventral cochlear nucleus (AVCN) in the mustache bat. 1.- An OCN cell with dendrites which remain periolivary. The labeled axon of this cell left the slice plane before reaching the cochlear nucleus. 2.- An OCN cell with a narrow band of dendrites invading the overlying lateral superior olive (LSO). 3.- An OCN cell with a widely distributed band of dendrites invading the LSO. Note that all three OCN cell types have at least some dendrites which remain periolivary. Axons were traced from the latter two cells into the AVCN. These axons gave off collaterals upon entering the AVCN to the marginal zone (AVm) and to the vicinity of the granule cell layer (gcl). Eighth nerve (VIII).

The remaining classes of OCN cells are characterized by dendrites which invade the fiber plexus of an adjacent principal olivary cell group. The two cells illustrated (Fig. 2, cells 2-3) both sent their dendrites dorsally into the neuropil of the overlying LSO. Similar OCN cells were labeled near the borders of the MSO and the MNTB (Fig. 3).

The latter two OCN cell types share several common features. Although all showed some dendrites associated with a principal cell group neuropil, these cells also had additional dendrites which remained outside of the principal neuropil. In the few cases where the soma was found within the borders of the principal cell group, all of these cells sent some of their dendrites out of the principal neuropil and into the surrounding periolivary zone.

These principal neuropil-associated cells can be further subdivided into two groups based on their dendritic patterns. The more common cell type showed dendrites with a limited spread in the invaded cell group (Fig. 2, cell 2). As illustrated in the transverse plane, the invading dendrites were confined to a narrow band which extended across the width of the cell group and parallel to the main isofrequency orientation of the fiber plexus. We do not yet have a clear three-dimensional sense of this dendritic pattern as most of our tissue slices have been sectioned in the transverse plane illustrated. In particularly thick slices (600-900 $\mu$m), the labeled dendrites of these cells could be followed rostrally and caudally to the limits of the slice, suggesting a wide rostrocaudal dendritic spread. The least common type of OCN cell labeled (Fig. 2, cell 3) was characterized by dendrites which spread widely across the invaded neuropil viewed in transverse slices.

The two cell types with principal neuropil-associated dendrites probably represent a minority in the overall population of OCN cells. While future work will explore the range of OCN cells, we have focused our initial efforts on these cell types because of their distinctive dendritic patterns (Zook and Kuwabara, '91). OCN cells with principal neuropil - associated dendrites were most commonly located ventral to the LSO, MSO or MNTB. A

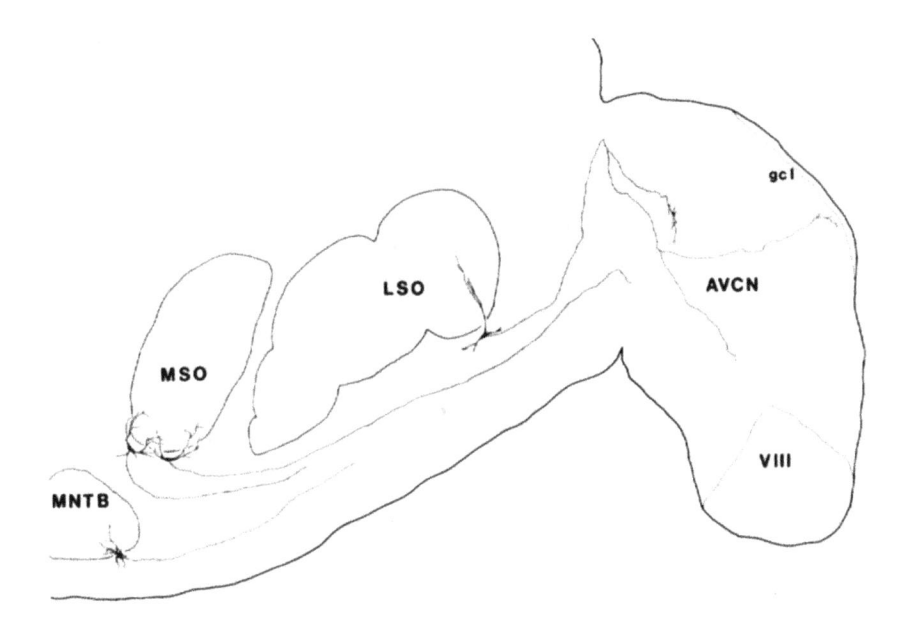

**Figure 3.** Camera lucida drawings showing the dendritic patterns of four intracellularly labeled cells with projections to the ipsilateral cochlear nucleus in the mustache bat. From left to right, dendrites of these cells invaded respectively the medial nucleus of the trapezoid body (MNTB), the medial superior olive (MSO) and the lateral superior olive (LSO). Axons which could be traced as far as the anteroventral cochlear nucleus (AVCN) showed the typical pattern of collateral branches to the medial marginal and the granule cell layer (gcl) of the AVCN.

few additional cells of these types were located either medial, lateral or dorsal to one of these associated cell groups. It is unlikely that this distributional bias is significant, since the majority of these cells have long dendrites spanning the width of the invaded neuropil. Besides the principal cell groups, a few OCN cells were also intracellularly labeled which had dendrites associated with the large SPN that is a prominent part of the superior olive in the gerbil, mouse and big brown bat.

It is our working hypothesis that these principal neuropil-invading dendrites may be positioned to sample the concentrated ascending afferent input that is focused upon each of the associated principal olivary cell groups. Consistent with this hypothesis, spines and spicules were rarely found on the initial dendritic segment outside the principal neuropil but were commonly observed on these same dendrites beginning at the point where they entered the neuropil of the associated principal cell group (Zook and Kuwabara, '91). Of course, to directly test this sampling hypothesis, it is necessary to examine individually identified OCN cells at the ultrastructural level and show the proposed axodendritic contacts formed by ascending afferent axons.

Under the assumption that these axodendritic contacts exist, different sampling strategies can be suggested for the OCN cell types identified so far. We would expect that the output of each cell type might reflect its specific pattern of axodendritic input. The first, general OCN cell type, with dendrites which remain periolivary, might be expected to receive input from axons ascending or descending to this periolivary zone. The output of this class of OCN neurons would be expected to reflect a broad range of (as yet unspecified) axodendritic and axosomatic contacts. The second OCN cell type, with its narrow dendritic pattern, might be expected to sample the ascending input encountered in the narrow span of its dendrites. This class of OCN neurons would be expected to show the response characteristics of the specific ascending input and the narrow frequency tuning associated with the particular band of invaded principal neuropil. In contrast, the third class of OCN neurons, with their widespread dendrites, would be expected to show the characteristics of the ascending input to the associated principal neuropil as well as a correspondingly broader frequency tuning. As the two principal neuropil-sampling cell types generally have both somata and other dendrites which remain periolivary, the output of these cells may reflect a combination of both principal neuropil and periolivary inputs.

These OCN dendritic patterns are in contrast to what we know of OCB dendritic patterns and the distribution of OCB somata. Although there is no published description of the complete dendritic field of a lateral OCB cell, we have intracellularly labeled a few of these cells in the mustache bat and gerbil. In both cases, all of the labeled dendrites were contained within the neuropil of the LSO. This dendritic pattern was especially striking in the mustache bat, where all the OCB somata are located outside the LSO neuropil. Thus, one fundamental difference between LSO-associated OCN cells and lateral OCB cells may be whether or not their dendrites are predominantly or only partially contained within the neuropil of the LSO.

The axonal projections of OCN cells are likely to offer even more ways to characterize the cell population. Although axons could be traced from most intracellularly labeled OCN cells, we have not found many complete terminal fields. This problem is largely due to the fact that viable transverse slices cannot be made thick enough to encompass the entire cochlear nucleus. The limited axon projections illustrated were taken from slices which included a portion of the anteroventral cochlear nucleus (AVCN). Some general projection patterns have been observed which may characterize OCN cells.

The majority of labeled OCN axons gave off a number of thin collateral branches which followed similar courses within the cochlear nucleus. The first collateral or collaterals appeared soon after the main axon entered the cochlear nucleus (Fig. 2-3). In the mustache bat, these initial axon branches arborize extensively along the medial border of the AVCN in a cytoarchitectonically distinct region named the marginal zone (Zook and Casseday, '82). In all species, the main axon usually gave off a second branch which ramified near the lateral edge of the AVCN in or near the granule cell cap or the small cell cap (Osen, '69).

We are particularly interested in the axonal projections of OCN cells with dendrites associated with the principal olivary nuclei. In a few cases, the axons of OCN cells associated with the MSO neuropil could be traced to their terminal arborizations within the rostral AVCN. Figure 4 shows both a photomicrograph and a reconstruction of part of one of these terminal arbors as seen in the gerbil. As shown in the reconstruction, the *en passant* and terminal swellings of this axon were specifically associated with the somata of counterstained AVCN spherical cells, the predominant cell type in this part of the AVCN (Osen, '69; '70; Brawer et al., '72). The terminal pattern of this specific axon arbor bears a close resemblance to the Group I pattern of descending axon terminals in the rostral AVCN as described by Cant and Morest ('78) in cat Golgi material.

Such axonal patterns raise the possibility of a close correspondence between specific OCN cells and AVCN cells which project to the superior olivary complex. The spherical cells of the rostral AVCN are recognized as the main source of the ascending projections to the MSO (Osen, '70; Warr, '82; Cant and Morest, '84; Cant and Casseday, '86; see also Casseday et al., this volume). The axon shown in Figure 4 is direct evidence that MSO-sampling OCN cells may project back to the same spherical cells which supply the ascending input to the MSO.

The significance of the collateral axon patterns seen is less clear. The marginal zone in the mustache bat is an almost homogeneous concentration of small and medium multipolar cells. These cells are similar to the multipolar cells found scattered throughout the posterior two-thirds of the AVCN and the anterior PVCN in the mustache bat and in other mammals (Osen, '69; Brawer et al., '72; Zook and Casseday, '82). The marginal zone is characterized

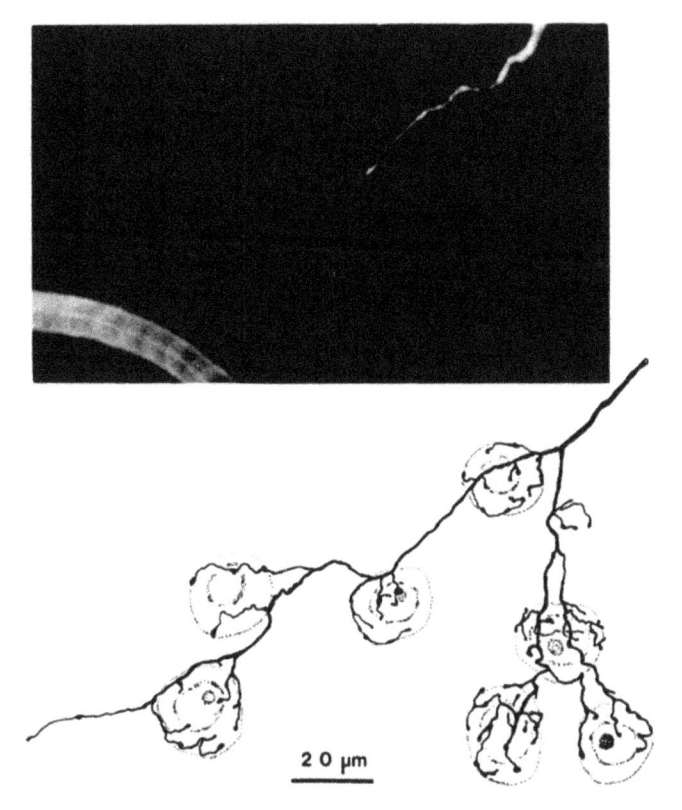

**Figure 4.** "Top": A photomicrograph of a Lucifer Yellow labeled axon in a tissue slice from the gerbil. This axon descended to the rostral AVCN from an OCN cell with dendrites which invaded the medial superior olive. "Bottom": A camera lucida drawing of the same labeled axon in relation to several ethidium bromide counterstained somata. Note that the axon branched and formed varicosities specifically in relation to the somata of these spherical cells.

by an unusually rich plexus of acetylcholinesterase-positive (AChE-positive) fibers. The OCN axons which ramify in this area may represent one source of this AChE-positive fiber plexus (personal observation). AChE-positive fibers are also associated with the peripheral granule cell layer in this bat and other mammals.

Future work will focus upon the relationship between the descending projections of OCB and OCN axons to the cochlear nucleus. The observed collateral projections of OCN axons to the vicinity of the granule cell layer and small cell cap are particularly intriguing as a subclass of multipolar cells found near these regions is the target of descending collateral branches of medial OCB axons (Brown et al., '88; Benson and Brown, '90; Brown et al., '91). While these OCB and OCN projections may overlap at the periphery of the AVCN, the remaining OCN projections may be unique to this system.

Although we have offered some preliminary descriptions of the dendritic morphology and axonal projections of OCN cells, the bulk of this system has yet to be described. Clearly the OCN system contains a rich, diverse population of cells which maintain both subtle and complex relationships with the cell groups of the superior olivary complex. Our initial exploration suggests that there are several different OCN subpopulations, some of which may be closely related to specific ascending cell populations and may narrowly or broadly sample the ascending input to the principal olivary nuclei. Hopefully, by using a similar approach in combination with anterograde labeling techniques, we will be able to characterize these

OCN cells more completely and begin to establish the relationship of these and other OCN cells to the descending or intrinsic fiber populations of the superior olivary complex.

ACKNOWLEDGEMENTS

We wish to acknowledge Dr. S.D. Comis who planted the seeds of this work some eighteen years ago when he introduced a new graduate student to the descending auditory system. The authors thank Dr. L.S. Ross, L. Owen and G. Tuck for their critical reading of the manuscript and to K. Reisig for technical assistance. This work was supported by NIH grants DC00503, DC00038, DC01303 and the Ohio University College of Osteopathic Medicine.

REFERENCES

Adams, J.C., 1983, Cytology of periolivary cells and the organization of their projections in cat, *J. Comp. Neurol.*, 215:275-289.
Adams, J.C. and Warr, W.B., 1976, Origins of axons in the cat's acoustic striae determined by injection of horseradish peroxidase into severed tracts, *J. Comp. Neurol.*, 170:107-122.
Aschoff, A. and Ostwald, J., 1987, Different origins of cochlear efferents in some bat species, rats, and guinea pigs, *J. Comp. Neurol.*, 264:56-72.
Benson, T.E. and Brown, M.C., 1990, Synapses formed by olivocochlear axon branches in the mouse cochlear nucleus, *J. Comp. Neurol.*, 295:52-70.
Bishop, A.L. and Henson, O.W.Jr., 1987, The efferent cochlear projections of the superior olivary complex in the mustache bat, *Hearing Res.*, 31:175-182.
Bourk, T.R., 1976, Electrical responses of neural units in the anteroventral cochlear nucleus of the cat, Doctoral Dissertation, M.I.T., Cambridge.
Brawer, J.R., Morest, D.K., and Kane, E., 1972, The neuronal architecture of the cochlear nucleus of the cat, *J. Comp. Neurol.*, 155:251-300.
Brown, K.A. and Buchwald, J.S., 1976, Response decrements during repetitive tone stimulation in the surgically isolated cochlear nucleus, *Exp. Neurol.*, 53:663-669.
Brown, M.C., Liberman, M.C., Benson, T.E., and Ryugo, D.K., 1988, Brainstem branches from olivocochlear axons in cats and rodents, *J. Comp. Neurol.*, 278:591-603.
Brown, M.C., Pierce, S., and Berglund, A.M., 1991, Cochlear-nucleus branches of thick (medial) olivocochlear fibers in the mouse: a cochleotopic projection, *J. Comp. Neurol.*, 303:300-315.
Campbell, J.P. and Henson, M.M., 1988, Olivocochlear neurons in the brainstem of the mouse, *Hearing Res.*, 35:271-274.
Cant, N.B. and Casseday, J.H., 1986, Projections from the anteroventral cochlear nucleus to the lateral and medial superior olivary nuclei, *J. Comp. Neurol.*, 247:457-476.
Cant, N.B. and Morest, D.K., 1978, Axons from non-cochlear sources in the anteroventral cochlear nucleus of the cat. A study with the rapid Golgi method, *Neurosci.*, 3:1003-1029.
Cant, N.B. and Morest, D.K., The structural basis for stimulus coding in the cochlear nucleus, *in:* "Hearing Sciences: Recent Advances" C. Berlin, ed., College-Hill, San Diego (1984).
Caspary, D.M., 1986, Cochlear nuclei: Functional neuropharmacology of principal cell types, *in:* "Neurobiology of Hearing: The Cochlea," Raven, New York.
Cody, A.R. and Johnstone, B.M., 1982, Temporary threshold shift modified by binaural acoustic stimulation, *Hearing Res.*, 6:199-206.
Comis, S.D. and Whitfield, I.C., 1968, Influence of centrifugal pathways on unit activity in the cochlear nucleus, *J. Neurophysiol.*, 31:62-68.
Comis, S.D. and Davies, W.E., 1969, Acetylcholine as a transmitter in the cat auditory system, *J. Neurochem.*, 16:423-429.
Comis, S.D., 1970, Centrifugal inhibitory processes affecting neurons in the cat cochlear nuclei, *J. Physiol. (London)*, 210:751-760.
Conlee, J.W., and Kane, E.S., 1982, Descending projections from the inferior colliculus to the dorsal cochlear nucleus in the cat: an autoradiographic study, *Neurosci.*, 7:161-178.
Covey, E., Jones, D.R., and Casseday, J.H., 1984, Projections from the superior olivary complex to the cochlear nucleus in the tree shrew, *J. Comp. Neurol.*, 226:289-305.
Elverland, H.H., 1977, Descending connections between superior olivary and cochlear nuclear complexes in the cat studied by autoradiographic and horseradish peroxidase methods, *Exp. Brain. Res.*, 27:397-412.
Farley, G.R., and Warr, W.B., 1981, Some recurrent projections of the superior olive to anteroventral and dorsal cochlear nuclei in cat, *Soc. Neurosci. Abstr.*, 7:56.

Frisina, R.D., Smith, R.L., and Chamberlain, S.C., 1990, Encoding of amplitude modulation in the gerbil cochlear nucleus: II. Possible neural mechanisms, *Hearing Res.*, 44:123-142.

Glendenning, K.K. and Masterton, R.B., 1983, Acoustic chiasm: efferent projections of the lateral superior olive, *J. Neurosci.*, 3:1521-1537.

Kane, E.S. and Conlee, J.W., 1979, Descending inputs to the caudal cochlear nucleus of the cat: degeneration and autoradiographic studies, *J. Comp. Neurol.*, 187:759-784.

Kuwabara, N., DiCaprio, R.A., and Zook, J.M., 1991, Afferents to the medial nucleus of the trapezoid body and their collateral projections, *J. Comp. Neurol.*, 314:684-706.

Kuwabara N. and Zook, J.M. 1991, Classification of the principal cells of the medial nucleus of the trapezoid body, *J. Comp. Neurol.*, 314:707-720.

Liberman, M.C. 1988, Response properties of cochlear efferent neurons: monaural vs. binaural stimulation and the effects of noise, *J. Neurophysiol.*, 60:1779-1798.

Mast, T.R., 1970, Binaural interaction and contralateral inhibition in dorsal cochlear nucleus of the chinchilla, *J. Neurophysiol.*, 33:108-115.

Nordeen, K.W., Killackey, H.P., and Kitzes, L.M., 1983, Ascending auditory projections to the inferior colliculus in the adult gerbil, *J. Comp. Neurol.*, 214:131-143.

Osen, K.K., 1969, Cytoarchitecture of the cochlear nuclei in the cat, *J. Comp. Neurol.*, 136:453-484.

Osen, K.K., 1970, Afferent and efferent connections of three well-defined cell types of the cat cochlear nuclei, *in:* "Excitatory Synaptic Mechanisms," P. Andersen and J.K.S. Jansen, eds., Universitetforlaget, Oslo.

Pfalz, K.R.J., 1962, Centrifugal inhibition of afferent secondary neurons in the cochlear nucleus by sound, *J. Acoust. Soc. Am.*, 34:1472-1477.

Pickles, J.O., 1976, Role of centrifugal pathways to cochlear nucleus in determination of critical bandwidth, *J. Neurophysiol.*, 39:394-400.

Pickles, J.O. and Comis, S.D., 1973, Role of centrifugal pathways to cochlear nucleus in detection of signals in noise, *J. Neurophysiol.*, 36:1131-1137.

Rasmussen, G.L., 1964, Anatomical relationships of the ascending and descending Auditory systems, *in:* "Neurological Aspects of Auditory and Vestibular Disorders," W.S. Fields, and B.R. Alford, eds., Thomas, Springfield.

Ryan, A.F., Keithley, E.M., Wang, Z-X., and Schwartz, I.R., 1990, Collaterals from lateral and medial olivocochlear efferent neurons innervate different regions of the cochlear nucleus and adjacent brainstem, *J. Comp. Neurol.*, 300:572-582.

Shore, S., Helfert, R.H., Bledsoe, S.C., Altschuler, R.A. and Godfrey, D.A., 1991, Descending projections to the dorsal and ventral divisions of the cochlear nucleus in guinea pig, *Hearing Res.*, 52:255-268.

Spangler, K.M., Cant, N.B., Henkel, C.K., Farley, G.R., and Warr, W.B., 1987, Descending projections from the superior olivary complex to the cochlear nucleus of the cat, *J. Comp. Neurol.*, 259:452-465.

Spangler, K.M. and Warr, W.B., 1991, The descending auditory system, *in:* "Neurobiology of Hearing: The Central Auditory System," R.A. Altschuler, ed, Raven, New York.

Starr, A. and Wernick, J.S., 1968, Olivocochlear bundle: Effects on spontaneous and tone-evoked activities of single units in cat cochlear nucleus, *J. Neurophysiol.*, 31:549-564.

Thompson, A.M. and Thompson, G.C., 1991, Posteroventral cochlear nucleus projections to olivocochlear neurons, *J. Comp. Neurol.*, 303:267-285.

Thompson G.C. and Thompson, A.M., 1986, Olivocochlear neurons in the squirrel monkey brainstem, *J. Comp. Neurol.*, 254:246-258.

Warr, W.B., 1975, Olivocochlear and vestibular efferent neurons of the feline brain stem: Their location, morphology and number determined by retrograde axonal transport and acetylcholinesterase histochemistry, *J. Comp. Neurol.*, 161:159-181.

Warr, W.B., 1982, Parallel ascending pathways from the cochlear nucleus: Neuroanatomical evidence of functional specialization, *Contrib. Sens. Physiol.*, 7:1-38.

Warr, W.B. and Guinan, J.J., 1979, Efferent innervation of the organ of Corti: Two separate systems, *Brain Res.*, 173:152-155.

White, J.S. and Warr, W.B., 1983, The dual origins of the olivocochlear bundle in the albino rat, *J. Comp. Neurol.*, 219:203-214.

Willard, F.H. and Ryugo, D.K., 1983, Anatomy of the central auditory system, *in:* "The Auditory Psychobiology of the Mouse," J.F. Willott, ed., Thomas, Springfield.

Winter, I.M., Robertson, D. and Cole, K.S., 1989, Descending projections from auditory brainstem nuclei to the cochlea and cochlear nucleus of the guinea pig, *J. Comp. Neurol.*, 280:143-157.

Zook, J.M. and Casseday, J.H., 1982, Cytoarchitecture of auditory system in lower brainstem of the mustache bat, *Pteronotus parnellii*, *J. Comp. Neurol.*, 207:1-13.

Zook, J.M. and Kuwabara, N., 1991, A possible interface between the ascending and descending auditory systems within the lateral superior olive of rodents and bats, *Abstr. Assoc. Res. Otolaryngel.*, 14:89.

# DESCENDING PROJECTIONS FROM THE INFERIOR COLLICULUS TO THE COCHLEAR NUCLEI IN MAMMALS

Enrique Saldaña

Departamento de Biología Celular y Patología, Facultad de Medicina,
Universidad de Salamanca, 37007-Salamanca, Spain, and
Laboratory of Neuromorphology, Graduate Program in Biobehavioral
Sciences, The University of Connecticut, Storrs, CT 06269-4154, USA

Along phylogenetic evolution, animals have developed efferent neural pathways able to modulate their sensory systems, presumably to enhance or attenuate sensory stimuli with relative biological significance. In the case of mammalian hearing, acoustic information can be modulated at two levels, the periphery and the central nervous system. The peripheral control is carried out by movement of the pinnae, muscles of the middle ear, and by efferent neurons of the olivocochlear systems that innervate the organ of Corti. The central control is carried out by efferent or descending auditory pathways that parallel the ascending auditory pathways. In general, the descending pathways have received little attention compared to the ascending pathways. Although for many years the descending auditory pathways were considered far more simple than the ascending pathways, recent studies are beginning to unravel the complexity of the descending systems and their functions, and the unexpected intricacy of their topographic organization (reviewed by Huffman and Henson,'90; Spangler and Warr, '91; Warr, '92).

Like other primary sensory nuclei, the cochlear nuclei (CoN) receive descending or centrifugal innervation from various higher centers of the acoustic system, namely the superior olivary complex (SOC), the nuclei of the lateral lemniscus (NLL) and the inferior colliculus (IC). The CoN project to the IC either directly or via synapses in the SOC and/or the NLL (for a review, see Oliver and Shneiderman, '91). Thus, nearly all the ascending auditory information converges in the IC before reaching the auditory thalamus and from there the auditory cortex. Since much processing of information takes place within the IC, a collicular projection to the CoN could convey more elaborate information than that from the NLL, the SOC or the contralateral CoN. Therefore, a direct projection from IC to the CoN could inform the CoN of the functional state of higher auditory centers.

This chapter provides a brief review of the relevant literature and summarizes the current knowledge of the direct descending projection from the IC to the CoN. Following a few historical notes and general remarks, the cells that originate the projection, the trajectories of their axons, and their targets in the CoN are successively described.

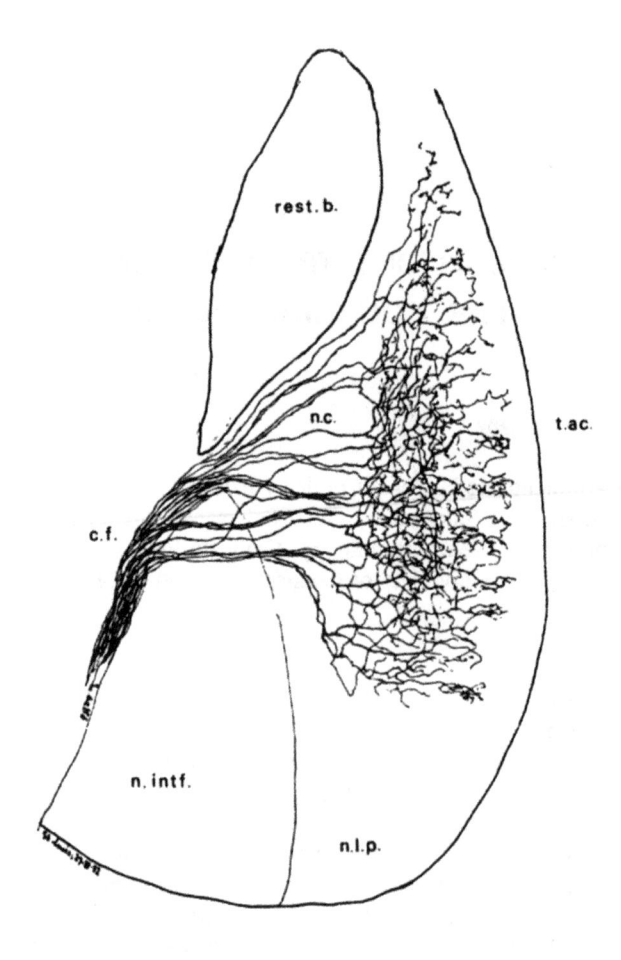

**Figure 1.** Centrifugal fibers reaching the DCoN as drawn by Lorente the Nó from a coronal section impregnated with the rapid Golgi method. Cat, 12 days old. Reproduced from Lorente de Nó: Laryngoscope, 43:327-350, 1933.

## History

It was Held ('93) who first described centrifugal (descending) fibers that reach the CoN through the trapezoid body and the dorsal and intermediate acoustic striae. His findings were partially confirmed by Kölliker ('96). Cajal ('09) accepted the theoretical existence of centrifugal fibers for the CoN, but failed to impregnate them with the Golgi method. The existence of centrifugal fibers was definitely confirmed with the Golgi method by Lorente de Nó ('33, '81), who described different systems of descending fibers that innervate specific regions in the ventral (VCoN) and dorsal cochlear nucleus (DCoN). Lorente de Nó did not observe fibers reaching the CoN through either the intermediate or the dorsal acoustic striae, which led him to believe that all the descending fibers arrive through the trapezoid body. Although this seems to be the case for the collicular fibers (see below), other projections have been demonstrated which reach the CoN via the intermediate and the dorsal acoustic striae (Adams and Warr, '76).

In 1933, Lorente de Nó described in the mouse and the cat the "centrifugal bundle" of fibers that enter the CoN through the trapezoid body and innervate layers II and III of the in the rat, the collicular fibers twist around the NLL as they descend dorsoventrally: at the level of the dorsal nucleus of the LL (DNLL) the collicular fibers occupy a caudal and lateral position within the periphery of the LL; as the fibers run ventrally, they are situated laterally and rostrally; at the ventral end of the LL the collicular fibers occupy a clearly rostral position, so that when the LL bends caudally to enter the SOC, the collicular fibers are found ventrally.

The collicular fibers heavily innervate the nucleus sagulum (Faye-Lund, '86; Saldaña et al., '93a). With respect to more ventral stations, however, there is much discrepancy as to whether the collicular fibers innervate the NLL and the pontine nuclei (for a review, see Faye-Lund, '86; Saldaña et al., '93a). It is very possible that interspecies variations may account for some of the discordant results reported by different authors.

In all species, few, if any, of the descending collicular fibers from one side cross to the contralateral LL via the commissure of the IC or the commissure of the LL (commissure of Probst).

**The descending collicular fibers in the SOC:** When the collicular fibers reach the ipsilateral SOC, they accumulate in periolivary regions known to contain neurons of the medial olivocochlear system (Rasmussen, '55). These regions have received different names: medial preolivary (or periolivary) nucleus (Rasmussen, '64; '67, -cat- ; Moore and Goldberg, '63, -monkey-; van Noort, '69, -cat-; Borg, '73, -rabbit-; Casseday et al., '76, --tree shrew-; Conlee and Kane, '82, -cat-), rostral and medioventral periolivary regions (Osen et al., '84, -rat-; Faye-Lund, '86, -rat-), or ventral nucleus of the trapezoid body (Syka et al., '88, -guinea pig-; Thompson and Thompson, '89, -bush baby-; Vetter and Saldaña, '90, -rat-; Saldaña et al., '93b, -rat-). In the rat, the collicular fibers run rostrocaudally along the ventral half of the ventral nucleus of the trapezoid body (VNTB) in parallel to the main axis of this nucleus and perpendicularly to the fascicles of the trapezoid body (Faye-Lund, '86; Saldaña et al., '93b). The projection from the CNIC to the VNTB, perhaps the most dense of all the descending collicular projections, is topographically (tonotopically) organized, with successively more dorsolateral (lower frequency) zones of CNIC projecting to progressively more lateral zones of VNTB (Saldaña et al., '93b). Within the VNTB, the collicular fibers contact the large neurons of the medial olivocochlear system, mostly on their distal dendrites (Vetter and Saldaña, '90).

There is considerable disagreement as to whether the IC innervates other nuclei of the ipsi- or contralateral SOC (for a review, see Faye-Lund, '86; Saldaña et al., '93b).

**The descending collicular fibers in the trapezoid body:** With very few exceptions[4], the vast majority of the studies with degeneration and tract tracing methods agree that the descending collicular fibers reach the CoN on both sides almost exclusively by way of the trapezoid body (Rasmussen, '60, '64, '67; van Noort, '69; Conlee and Kane, '82; Faye-Lund, '86; Saldaña et al., '93c). This idea is supported by experiments with PHA-L, where only few collicular fibers were seen entering the CoN via the dorsal or intermediate acoustic striae, even in cases in which numerous collicular fibers were found in the trapezoid body (Saldaña et al., '93c). Furthermore, deposits of HRP into the severed intermediate and dorsal acoustic striae failed to retrogradely label neurons in the IC, whereas HRP injections into the DCoN did produce bilateral retrograde labeling in the IC, suggesting that the collicular fibers enter the CoN through the trapezoid body, but not through the dorsal and intermediate acoustic striae (Adams and Warr, '76).

---

[4]In the kangaroo rat, according to Carey and Webster ('71), collicular fibers reach the CoN via the trapezoid body, and "fibers also enter the CoN directly from the LL through the dorsal acoustic stria". Kane and Conlee ('79) observed that after small lesions into the cat dorsal IC "degeneration of fibers of passage was rarely found in the dorsal acoustic stria (DAS) of either side. However, when more of the ventral IC was destroyed there were more degenerating fibers in the DAS".

(Fig. 2) and guinea pig (Ostapoff et al., '90 ; Shore et. al., '91). The distribution of these projecting neurons is roughly similar on the two sides[3], with more cells found caudally than rostrally (Fig. 2). Within CNIC and ECIC, more cells appear to be concentrated ventrally (Hashikawa and Kawamura, '83; Faye-Lund, '86), thus suggesting the idea that a gradient may exist within the IC.

In the cat, rat and guinea pig, colliculo-cochleonuclear cells vary in shape and size, with a predominance of medium-sized (15-20 $\mu$m), round, oval or multipolar somata (Kane and Finn, '77; Faye-Lund, '86; Shore et. al., '91). Neurons larger than 20 $\mu$m are primarily found in the rat ECIC (Faye-Lund, '86) (Fig. 2) and the lateral region of the cat CNIC (Kane and Finn, '77). According to Kane and Finn ('77), the larger feline neurons provide input only to the contralateral CoN, whereas the medium-sized neurons send afferents to the CoN on both sides. This finding, however, has not been confirmed in later studies.

Whether there are differences between the collicular neurons projecting to the ipsilateral and contralateral CoN has not been studied, nor is it known whether individual IC neurons send axon collaterals to the CoN on both sides.

## Trajectories of the collicular axons that innervate the CoN

Most authors agree that the collicular fibers that innervate the CoN descend in the ipsilateral lateral lemniscus (LL) until they reach the SOC. From the ipsilateral SOC, some fibers run laterally, to innervate the ipsilateral CoN, whereas other fibers run medially, cross the midline, and innervate the contralateral CoN. On both sides, the collicular fibers reach the CoN via the trapezoid body.

**The descending collicular fibers in the LL:** The descending axons of the IC neurons run laterally and caudally, to reach the fibrocellular capsule that surrounds the IC. Then the fibers course ventrally to leave the IC, and enter the LL, where they occupy mostly the lateral margin of the tract. In his early studies, Rasmussen ('60) stated that in the cat the collicular fibers "descend for the most part in the medial division of the LL, often referred to as Monakow's bundle". He later refined this concept, reporting that destruction of the CNIC, including the fibrocellular capsule enveloping the IC, produces two streams of degeneration in the LL: a lateral one, more prominent and originating from the fibrocellular capsule and the lateral nucleus of the IC, and a medial one, with fewer fibers, originating from the CNIC; the lateral fiber group is the one that chiefly supplies the CoN with efferents (Rasmussen, '64). However, subsequent studies using a variety of methods and species have failed to confirm these two systems of descending collicular fibers. Most authors concur that the collicular fibers descend mostly in the lateral part of the ipsilateral LL, mainly close to its periphery. This has been shown for the cat (van Noort, '69; Conlee and Kane, '82), the kangaroo rat (Carey and Webster, '71), the albino rat (Lugo and Cooper, '82; Faye-Lund, '86; Saldaña et al., '93a) and the rabbit (Borg, '73). Saldaña et al. ('93a) have shown that,

---

[2]The subdivisions and nomenclature proposed for the IC of different mammals vary considerably. The homologies and equivalences between different cytoarchitectural models have been reviewed and discussed by Huffman and Henson ('90). See also Saldaña and Merchán ('92) for a recent redefinition of the subdivisions of the rat IC.

[3]Kane and Finn ('77) found that after small HRP injections into the DCoN considerably more neurons were retrogradely labeled in the contralateral than in the ipsilateral IC. However, later work from the same laboratory confirms the bilateral nature of the colliculo-cochleonuclear projection (Conlee and Kane, '82).

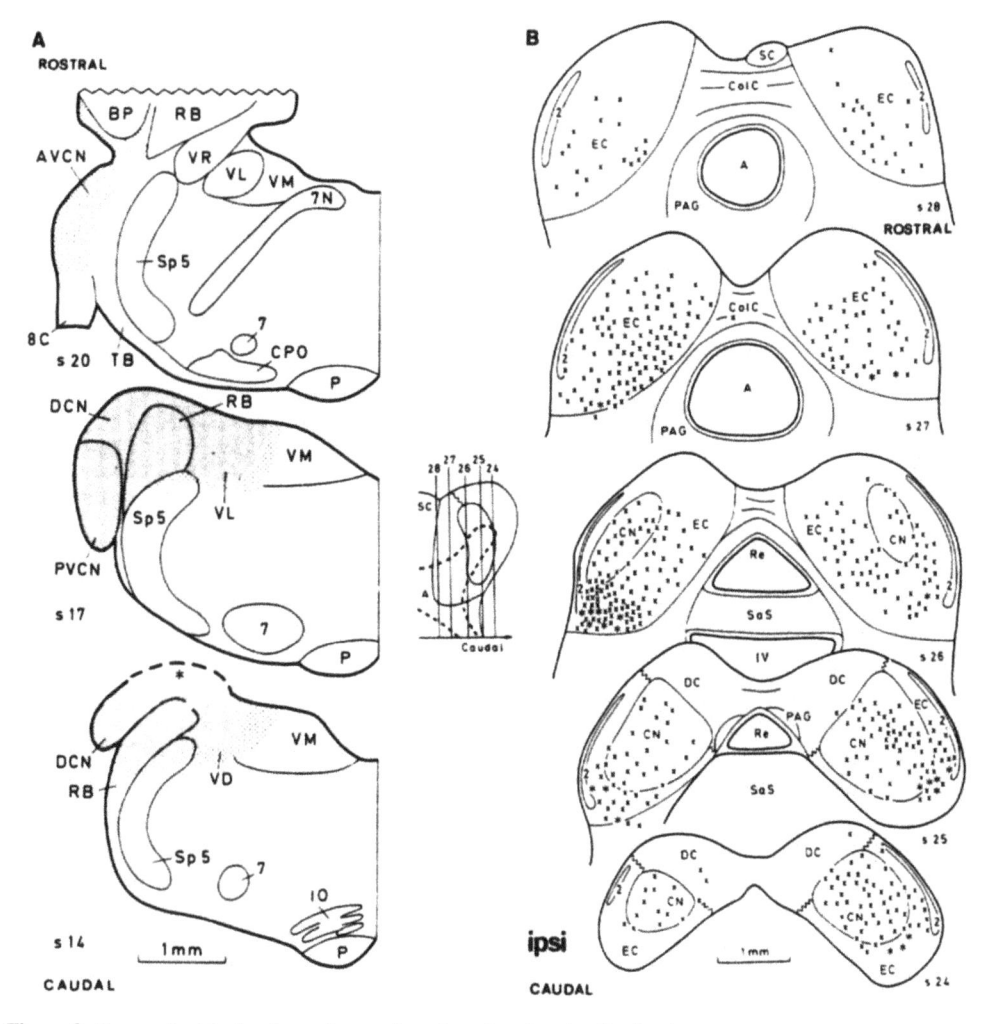

**Figure 2.** Camera lucida drawings of coronal sections showing the distribution of IC neurons that project to the CoN. The cells were retrogradely labeled after an injection of HRP-compounds into the left CoN. Due to space limitations, the crosses and asterisks represent only the relative number of cells. Large cells (>20μm) are indicated by asterisks, smaller cells by crosses. Labeled cells are found bilaterally throughout the larger parts of the central nucleus (CN) and the deep external cortex (EC) of the IC. More cells are found caudally than rostrally, and only occasional cells are found in the dorsal cortex (DC) or the commissure of the IC (CoIC). The majority of large cells are situated in the external cortex. AVCN = AVCoN, DCN = DCoN, PVCN = PVCoN. Reproduced from Faye-Lund, Anat. Embryol., 175:35-52, 1986.

## Cells of origin of the colliculo-cochleonuclear projection

Rasmussen ('64, '67) concluded that the descending collicular fibers originate from spindle-shaped neurons located in the fibrocellular capsule and lateral nucleus of the IC. More recently, injections of retrograde tracers into the CoN have demonstrated that within the IC, the neurons of origin of the IC-CoN projection are located bilaterally in the central nucleus (CNIC) and deep external cortex (ECIC), with very few projecting cells situated in the dorsal cortex (DCIC)[2]. This seems to be the case for the IC of the cat (Adams, '76; Kane and Finn, '77; Hashikawa, '83; Hashikawa and Kawamura, '83), albino rat (Faye-Lund, '86)

DCoN (Fig. 1). Conclusive evidence for the descending nature of the centrifugal bundle was obtained with degeneration methods by Rasmussen ('55, '60), who ascertained that in the cat "the cells of origin are located in the nucleus of the inferior colliculus and dorsal nucleus of the lateral lemniscus[1]". Thus, the existence of a direct projection from the IC to the DCoN was established over thirty years ago.

In the sixties, several degeneration studies by Rasmussen ('60, '64, '67) and especially by van Noort ('69) furnished a large body of evidence for the collicular origin of the centrifugal bundle, as well as details of the course and termination of the descending collicular fibers in the cat (see also Konigsmark, '73). Van Noort ('69) described in detail the "colliculocochlear bundle", and distinguished it from the "tectopontine tract", which originates mostly from the superior colliculus.

In the seventies and eighties, the advent of new retrograde neuroanatomical tracers, such as horseradish peroxidase (HRP) and fluorescent dyes, enabled the analysis of the locations and morphologies of the collicular neurons projecting to the CoN (Adams, '76; Kane and Finn, '77; Hashikawa, '83; Hashikawa and Kawamura, '85 ; Faye-Lund, '86). New experiments with degeneration and autoradiographic techniques and with anterograde transport of HRP-compounds complemented the knowledge of the course and distribution of the colliculo-cochleonuclear projection in the cat and other species (Carey and Webster, '71; Kane and Conlee, '79; Andersen et al., '80; Conlee and Kane, '82; Lugo and Cooper, '82; Druga et al., '85; Faye-Lund, '86; Syka et al., '88). The fine structure of the collicular fibers in the DCoN was also described (Kane, '77b).

Recently, the use of high resolution anterograde tracers such as *Phaseolus vulgaris-leucoagglutinin* (PHA-L) has clarified the trajectory, topography, morphology and possible targets of the descending collicular fibers (Thompson and Thompson, '89; Saldaña et al., '93a,b,c), and has shown a considerable collicular projection to the DCoN and the VCoN, particularly to its anterior division (AVCoN) and the granule cell regions surrounding it (Saldaña et al., '93c). Furthermore, a strict tonotopic arrangement for the IC-CoN projection has been described (Saldaña et al., '93c). Attempts are also being made to elucidate the excitatory or inhibitory nature of the collicular descending projections by means of neurochemical studies after collicular ablations (Bergman et al., '89) and retrograde transport of radiolabeled amino acids (Staatz-Benson and Potashner, '88; Ostapoff et al., '90; see also Potashner et al., and Saint-Marie et al., this volume), although definite conclusions are yet to be reached on this matter.

## Generalities

The IC-CoN projection has been described in a number of mammals, including albino rat, kangaroo rat, guinea pig, chinchilla, rabbit, cat, bat and bush baby. The basic features of this projection are very similar across species. Although there is not complete agreement, most studies show that the colliculo-cochleonuclear projection is bilateral: in most species the main target of the collicular fibers appears to be the DCoN on both sides. However, Saldaña et al., ('93c) have recently uncovered a considerable collicular projection to the rat AVCoN that may also be present in other mammals (see also Thompson and Thompson, '89, -bush baby-). The collicular projections to the ipsi- and contralateral CoN appear to be identical in density, distribution and fiber morphology.

---

[1]In Rasmussen's experiments the lesions included the IC and part of the subjacent dorsal nucleus of the lateral lemniscus. Later studies have shown that the dorsal nucleus of the lateral lemniscus sends very few, if any fibers to the CoN (Adams and Warr, '76; Zook, '86). Therefore, the vast majority of the degenerating fibers seen by Rasmussen in the CoN must have actually originated in the IC.

**Possible neuronal targets of the collicular fibers in DCoN:** Since the collicular fibers innervate all layers of the DCoN, all neuron types within this nucleus are potential targets of the projection. Electron microscopic studies of the DCoN following lesions of the IC (Kane, '77b) have shown that the collicular fibers contact dendrites of granule cells and other neurons in the ipsilateral DCoN, and dendrites and somata of fusiform cells in the contralateral DCoN. However, the criteria used by Kane ('77b) to identify degenerating boutons are questionable and, therefore, her results should be looked upon cautiously. New electron microscopic studies are needed to draw definite conclusions.

## AVCoN

A collicular projection to the AVCoN has been reported in the albino rat, the bat and the bush baby. In the only detailed account thus far of this projection, Saldaña et al. ('93c) have shown that the collicular fibers bilaterally innervate the granule cell regions surrounding the AVCoN, the cap of small cells situated underneath them, and the magnocellular regions of the AVCoN. The granule cell regions that receive most collicular fibers are the medial sheet, the superficial layer, the subpeduncular corner and the lamina (nomenclature of Mugnaini et al., '80). These four regions are nearly continuous, forming an envelope that covers the lateral, dorsal and medial aspects of the the AVCoN. As previously mentioned, the collicular fibers reach the CoN via the trapezoid body, and run dorsally in the medial sheet on their way to the subpeduncular corner and the DCoN. Some of the fibers ramify and give off long collaterals that run laterally spanning the magnocellular regions of AVCoN in parallel to the known isofrequency planes of the nucleus (Bourk et al., '81) to reach the superficial layer of granule cells. As they traverse the AVCoN, these fibers present "en passant" varicosities and short, thin collaterals with one or more terminal boutons, indicating a collicular input to the magnocellular regions of AVCoN. When the fibers reach the cap of small cells and the superficial granular layer, they branch into thinner collaterals that form terminal fields in which "en passant" boutons predominate. The dorsoventral level at which these fibers cross the AVCoN to terminate in the superficial layer appears to be related to the collicular regions where the fibers originate: successively more dorsolateral (lower frequency) areas of the CNIC innervate progressively more ventral areas of AVCoN and its superficial granular layer, indicating that the CNIC-AVCoN projection is tonotopically organized. Multipolar neurons of AVCoN project, also in tonotopic order (Oliver, '87), to the contralateral IC. Therefore, the descending projection to AVCoN appears to be reciprocal to the ascending AVCoN-IC projection. However, while the ascending projection is mostly contralateral, the descending projection is bilateral.

AVCoN and DCoN may be innervated by collaterals of the same axons since multiple ramifications of the collicular fibers on their way to the DCoN are seen at the medial border of AVCoN.

Ultrastructural studies with anterograde tracers or degeneration are needed to identify the neuron types of the AVCoN that are innervated by the collicular fibers, and to determine if the IC is a source of mossy fibers that innervate the granule cell regions (Dunn et al., '92).

In the bush baby, PHA-L experiments have revealed that some collicular fibers "travel in the trapezoid body laterally and then dorsally to terminate within the AVCoN" (Thompson and Thompson, '89). Interestingly, in this study no collicular projection to the DCoN was reported.

In the mustache bat, studies with retrograde transport of HRP show that the marginal zone (a region along the medial and posterior borders of the AVCoN that contains an almost completely homogeneous population of large multipolar neurons) could be the main target of the descending projections from the SOC, NLL and IC (Zook, '85), thus indicating that notable species variations may exist in the descending collicular projections.

## PVCoN

In most studies, no collicular projection to PVCoN has been reported. Kane ('76) and Kane and Finn ('77) failed to retrogradely label cells in either IC after HRP injections confined to the PVCoN. In an early study, Kane ('77a) did not observe preterminal or terminal degeneration in the octopus cell area at the light and electron microscopic level after collicular ablations. However, a later report by Kane and Conlee ('79) described with degeneration and autoradiographic methods that, in the cat, the dorsal IC sends a sparse-to-moderate projection to the caudal and, in particular, the rostral zones of the PVCoN on both sides, and that the projection from the ventral IC to these same areas is moderate-to-dense. According to these authors, the collicular fibers contact the proximal dendrites of octopus cells, and the somata and dendrites of elongate and multipolar neurons. The illustrations of Bergman et al. ('89) also suggest that ablation of the guinea pig IC may result in sparse preterminal degeneration in the PVCoN, although they do not describe it specifically (see their Fig. 2E, F). In the rat, Saldaña et al. ('93c) did not observe a significant collicular projection to the PVCoN; in cases in which many collicular fibers were seen innervating the AVCoN and the DCoN, only occasional fibers were seen in the PVCoN, and they did not seem to innervate the octopus cell area.

## Functional considerations

The excitatory or inhibitory nature of the colliculo-cochleonuclear fibers has not been determined with certainty. Studies of amino acid uptake and release in the guinea pig CoN after IC ablation indicate that most IC neurons that project to the CoN probably do not use glutamate, aspartate, GABA, or glycine as a transmitter (Bergman et al., '89). In these studies, the collicular ablation decreased slightly the uptake and release of GABA and glycine in the AVCoN (although the difference from the control group was not statistically significant), suggesting that only a small proportion of the IC neurons projecting to AVCoN might use GABA or glycine as a neurotransmitter. However, injections of tritiated GABA or glycine into the CoN have failed to retrogradely label cells in either IC, suggesting that the colliculo-cochleonuclear projections are not GABAergic, nor glycinergic (Staatz-Benson and Potashner, '88; Ostapoff et al., '90; see also Potashner et al., and Saint-Marie et al., this volume).

The possible roles of the descending collicular projections have not been exhaustively explored electrophysiologically. No information is available about the specific roles of the colliculo-cochleonuclear projections. With respect to colliculo-olivary projections, Rajan ('90) has shown that electrical stimulation of the IC significantly reduces the temporary threshold shifts in cochlear sensitivity caused by a loud sound exposure, and that this effect is probably mediated by neurons of the medial olivocochlear system. While the stimulation of the IC alone does not appear to directly excite the medial olivocochlear neurons, the collicular input exerts a facilitatory action over the medial olivocochlear neurons that allows them to be more easily activated by the subsequent sound exposure. This facilitatory effect would be the consequence of excitatory synapses formed by the collicular fibers on the distal dendrites of the medial olivocochlear neurons (Vetter and Saldaña, '90).

Clearly, electrophysiological studies are needed to determine the mechanism of action of the collicular fibers in the CoN. It would be interesting to know, for instance, the physiological effect of electrical stimulation of the IC on fusiform cells of the DCoN on both sides. But in addition to the influence exerted by the IC on the CoN by means of the direct projection, one must consider that the IC could influence the activity of the CoN indirectly via the SOC, as many of the olivary neurons that project to CoN could receive collicular inputs (Faye-Lund, '86); moreover, the medial olivocochlear neurons send axonal collaterals to the CoN (Brown et al., '88)

## ACKNOWLEDGEMENTS

This work was supported in part by a grant from the Spanish Research Council to Miguel A. Merchán (DGICYT, PB88/0372), and in part by U.S.A. PHS grant NS-09904 to Enrico Mugnaini.

## REFERENCES

Adams, J.C., 1976, Central projections to the cochlear nucleus, *Soc. Neurosci. Abstr.*, 2:12.
Adams, J.C. and Warr, B.W., 1976, Origins of axons in the cat's acoustic striae determined by injection of horseradish peroxidase into severed tracts, *J. Comp. Neurol.*, 170:107-122.
Andersen, R.A., Roth, G.L., Aitkin, L.M. and Merzenich, M.M., 1980, The efferent projections of the central nucleus and the pericentral nucleus of the inferior colliculus in the cat, *J. Comp. Neurol.*, 194:649-662.
Bergman, M., Staatz-Benson, C. and Potashner, S.J., 1989, Amino acid uptake and release in the guinea pig cochlear nucleus after inferior colliculus ablation, *Hearing Res.*, 42:283-292.
Borg, E., 1973, A neuroanatomical study of the brainstem auditory system of the rabbit. Part II. Descending connections, *Acta Morphol. Neerl.-Scand.*, 11:49-62.
Bourk, T.R., Mielcarz, J.P. and Norris, B.E., 1981, Tonotopic organization of the anteroventral cochlear nucleus of the cat, *Hearing Res.*, 4:215-241.
Brown, M.C., Liberman, M.C., Benson, T.E. and Ryugo, D.K., 1988, Brainstem branches from olivocochlear axons in cats and rodents, *J. Comp. Neurol.*, 278:591-603.
Cajal, S. Ramón y, 1909, "Histologie du Système Nerveux de l'Homme et des Vertébrés, vol. I," (1955 reprint), Consejo Superior de Investigaciones Científicas, Madrid, p. 774-831.
Carey, C.L. and Webster, D.B., 1971, Ascending and descending projections of the inferior colliculus in the kangaroo rat (Dipodomys merriami), *Brain Behav. Evol.*, 4:401-412.
Casseday, J.H., Diamond, I.T. and Harting, J.K., 1976, Auditory pathways to the cortex in Tupaia glis, *J. Comp. Neurol.*, 166:303-340.
Conlee, J.W. and Kane, E.S., 1982, Descending projections from the inferior colliculus to the dorsal cochlear nucleus in the cat: an autoradiographic study, *Neurosci.*, 7:161-178.
Druga, R., Syka, J. and Markow, G., 1985, Descending part of the auditory pathway. Projections from the inferior colliculus in the rat, *Physiol. Bohemosl.*, 34:414.
Dunn, M. E., Vetter, D. E., Berrebi, A. S., Krider, H. M. and Mugnaini, E., 1992, The mossy fiber-granule cell-cartwheel cell system in the mammalian cochlear nuclear complex, *in*: "Advances in Speech Hearing and Language Processing, vol. III," JAI Press, (in press).
Faye-Lund, H., 1986, Projection from the inferior colliculus to the superior olivary complex in the albino rat, *Anat. Embryol.*, 175:35-52.
Hashikawa, T., 1983, The inferior colliculopontine neurons of the cat in relation to other collicular descending neurons, *J. Comp. Neurol.*, 219:241-249.
Hashikawa, T. and Kawamura, K., 1983, Retrograde labeling of ascending and descending neurons in the inferior colliculus. A fluorescent double labeling study in the cat, *Exp. Brain Res.*, 49:457-461.
Held, H, 1893, Die centralen Gehörleitung, *Arch. f. Anat. u. Physiol., Anat. Abtell.*, 201-248.
Huffman, R.F. and Henson, O.W., 1990, The descending auditory pathway and acousticomotor systems: connections with the inferior colliculus, *Brain Res. Rev.*, 15:295-323.
Kane, E.S., 1976, Descending inputs to caudal cochlear nucleus in cats: a horseradish peroxidase (HRP) study, *Am. J. Anat.*, 146:433-441.
Kane, E.S., 1977a, Descending inputs to the octopus cell area of the cat cochlear nucleus: an electron microscopic study, *J. Comp. Neurol.*, 173:337-354.
Kane, E.S., 1977b, Descending inputs to the cat dorsal cochlear nucleus: an electron microscopic study, *J. Neurocytol.*, 6:583-605.
Kane, E.S. and Conlee, J.W., 1979, Descending inputs to the caudal cochlear nucleus of the cat: Degeneration and autoradiographic studies, *J. Comp. Neurol.*, 187:759-784.
Kane, E.S. and Finn, R.C., 1977, Descending and intrinsic inputs to dorsal cochlear nucleus of cats: A horseradish peroxidase study, *Neurosci.*, 2:897-912.
Kölliker, A. von, 1896, "Handbuch der Gewebelehre des Menschen", vol. 2, Engelman, Leipzig.
Konigsmark, B.W., 1973, Neuroanatomy of the auditory system, *Arch. Otolaryngol.*, 98:397-413.
Lorente de Nó, R, 1933, Anatomy of the eighth nerve. III. General plan of structure of the primary cochlear nuclei, *Laryngoscope*, 43:327-350.

Lorente de Nó, R, 1981, "The primary acoustic nuclei," Raven Press, New York, p. 1-177.

Lugo, D.I. and Cooper, M.H., 1982, Descending efferent projections of the inferior colliculus: An autoradiographic study, *Soc. Neurosci. Abstr.*, 8:349.

Moore, R.Y,. and Goldberg J.M., 1963, Ascending projections of the inferior colliculus in the cat, *J. Comp. Neurol.*, 121:109-136.

Mugnaini, E., Warr, W.B. and Osen, K.K., 1980, Distribution and light microscopic features of granule cells in the cochlear nucleus of cat, rat, and mouse, *J. Comp. Neurol.*, 191:581-606.

Noort, J. van, 1969, "The structure and connections of the inferior colliculus. An investigation of the lower auditory system," van Gorcum and Comp. N.V., Leiden.

Oliver, D.L., 1984, Dorsal cochlear nucleus projections to the inferior colliculus in the cat: a light and electron microscopic study, *J. Comp. Neurol.*, 224:155-172.

Oliver, D.L., 1987, Projections to the inferior colliculus from the anteroventral cochlear nucleus in the cat: possible substrate for binaural interaction, *J. Comp. Neurol.*, 264:24-46.

Oliver, D.L. and Shneiderman, A., 1991, The anatomy of the inferior colliculus. A cellular basis for integration of monaural and binaural information, *in*: "Neurobiology of Hearing, vol. II: The central auditory system," R.A. Altschuler, D.W. Hoffman, R.P. Bobbin and B.M. Clopton, eds., Raven Press, New York, p.195-222.

Osen, K.K., Mugnaini, E., Dahl, A.-L. and Christiansen, A.H., 1984, Histochemical localization of acetylcholinesterase in the cochlear and superior olivary nuclei. A reappraisal with emphasis on the cochlear granule cell system, *Arch. Ital. Biol.*, 122:169-212.

Ostapoff, E.-M., Morest, D.K. and S.J. Potashner, 1990, Uptake and retrograde transport of [$^3$H]GABA from the cochlear nucleus to the superior olive in the guinea pig, *J. Chem. Neuroanat.*, 3:285-295.

Rajan, R., 1990, Electrical stimulation of the inferior colliculus at low rates protects the cochlea from auditory desensitization, *Brain Res.*, 506:192-204.

Rasmussen, G.L., 1955, Descending or 'feed-back' connections of auditory system of the cat, *Am. J. Physiol.*, 183:653.

Rasmussen, G.L., 1960, Efferent fibers of the cochlear nerve and cochlear nucleus, *in*: "Neuronal mechanisms of the auditory and vestibular systems," G.L. Rasmussen and W. Windle, eds., Charles C. Thomas, Springfield, Illinois, p. 105-115.

Rasmussen, G.L., 1964, Anatomical relationships of the ascending and descending auditory systems, *in*: "Neurological aspects of auditory and vestibular disorders," W.S. Fields and B.R. Alford, eds., Charles C. Thomas, Springfield, Illinois, p. 1-15.

Rasmussen, G.L., 1967, Efferent connections of the cochlear nucleus, *in*: "Sensorineural hearing processes and disorders," A.B. Graham, ed., Little Brown, Boston, Massachusetts, p. 61-75.

Ryan, A.F., Furlow, Z., Woolf, N.D. and Keithley, E.M. ,1988, The spatial representation of frequency in the rat dorsal cochlear nucleus and inferior colliculus, *Hearing Res.*, 36:181-190.

Saldaña, E. and Merchán, M.A., 1992, Intrinsic and commissural connections of the rat inferior colliculus, *J. Comp. Neurol.*, 319:417-437.

Saldaña, E., Bajo, V, Malmierca, M.S. and Merchán, M.A., 1993a, Descending projections of the central nucleus of the inferior colliculus. I. Nucleus sagulum and nuclei of the lateral lemniscus, (in preparation).

Saldaña, E., López, M.D.E., Bajo, V, and Merchán, M.A., 1993b, Descending projections of the central nucleus of the inferior colliculus. II. Superior olivary complex, (in preparation).

Saldaña, E., Malmierca, M.S., López, M.D.E. and Merchán, M.A., 1993c, Descending projections of the central nucleus of the inferior colliculus. III. Cochlear nuclei, (in preparation).

Shore, S.E., Helfert, R.H., Bledsoe, S.C., Altschuler, R.A. and Godfrey, D.A., 1991, Descending projections to the dorsal and ventral divisions of the cochlear nucleus in guinea pig, *Hearing Res.*, 52:255-268.

Spangler, K.M. and Warr, W.B., 1991, The descending auditory system, *in*: "Neurobiology of hearing, vol. II: The central auditory system," R.A. Altschuler, D.W. Hoffman, R.P. Bobbin and B.M. Clopton, eds., Raven Press, New York, p.27-45.

Staatz-Benson, C. and Potashner, S.J., 1988, Transport of $^3$H-glycine to the superior olivary complex from the cochlear nucleus in the guinea pig, *Soc. Neurosci. Abstr.*, 14:489.

Syka, J., Robertson, D. and Johnstone, B.M., 1988, Efferent descending projections from the inferior colliculus in guinea pig, *in*: "Auditory Pathway: Structure and Function," J. Syka and R.B. Masterton, eds., Plenum Press, New York, p. 299-303.

Thompson, G.C. and Thompson, A.M., 1989, Descendent targets of inferior colliculus efferents in bush baby (Galago crassocaidatus), *Soc. Neurosci. Abstr.*, 15:748.

Warr, W.B., 1992, Organization of olivocochlear efferent systems in mammals, *in*: "Springer Series in Auditory Research, vol. I: The anatomy of mammalian auditory pathways," R.R. Fay, A.N. Popper and D.B. Webster, eds., Springer Verlag, Heidelberg.

Vetter, D.E. and Saldaña, E., 1990, Descending input from the central nucleus of the inferior colliculus to the medial olivocochlear system in rat. A combined PHA-L and CT-HRP study, *Soc. Neurosci. Abstr.*, 16:716.

Zook, J.M., 1985, The descending auditory pathway and the cochlear nucleus in the mustache bat, *Soc. Neurosci. Abstr.*, 11:734.

Zook, J.M., 1986, The contribution of the lateral lemniscus to the descending auditory pathway, *Soc. Neurosci. Abstr.*, 12:1273.

# V. NEUROTRANSMITTERS OF THE COCHLEAR NUCLEUS

# LOCALIZING PUTATIVE EXCITATORY ENDINGS IN THE COCHLEAR NUCLEUS BY QUANTITATIVE IMMUNOCYTOCHEMISTRY

José M. Juiz[1], Maria E. Rubio[1], Robert H. Helfert[2] and Richard A. Altschuler[3]

[1]Dpt. of Histology and Institute of Neuroscience, University of Alicante,
  Alicante, Spain
[2]Dpts. of Surgery and Pharmacology, Southern Illinois University,
  Springfield, IL, USA
[3]Kresge Hearing Research Institute, University of Michigan,
  Ann Arbor, MI, USA

A major characteristic of CN neurons is their rich and complex patterns of synaptic inputs. The input carried by the auditory nerve (AN) is relayed to CN neurons through very secure excitatory synapses (reviewed by Wenthold and Martin, '84 and Caspary, '86, see also chapter by Morest in this volume). In addition, an intricate array of intrinsic and extrinsic (descending) connections from a variety of sources (see chapter by Saint-Marie et al. in this volume), make inhibitory as well as excitatory synapses on CN neurons. This pattern of synaptic inputs is specific for each cell type and determines in part the different processing abilities of CN neurons. Therefore, analyzing the synaptology of CN neurons is essential for an adequate understanding of the roles of CN in auditory processing. Several questions relevant to understand the functional synaptology of the CN are addressed in several chapters of this volume. Here we report on some of our recent findings regarding the chemical anatomy of putative excitatory endings in the CN.

The first attempts to assign functional roles to morphologically identified synapses on CN neurons, stemmed from the pioneer works of Gray ('59, '69) and Uchizono ('65, '67) on the ultrastructure of central nervous system synapses. These authors classified synapses according to the electron microscopic characteristics of the active zones or "synaptic densities" (Gray, '59) and to vesicle shape and size (Uchizono, '65, '67). Correlating these morphological criteria with available physiological and biochemical data, a relatively simple "morpho-functional" classification of synapses was designed (Eccles, '64; for review see Pappas and Waxman, '72), which in broad terms is still widely accepted. Thus, in the CN as in many other synaptic areas of the central nervous system, it is typical to assign a potential excitatory role to synapses containing "large", uniformly round synaptic vesicles and more or less asymmetric densities (Ibata and Pappas, '76; Schwartz and Gulley, '78; Cant and Morest, '79). Synapses with smaller, pleomorphic or flattened vesicles and symmetric pre- and postsynaptic densities (Ibata and Pappas, '76; Kane, '77; Schwartz and Gulley, '78; Cant and Morest, '79) are considered inhibitory. Oversimplification notwithstanding, these concepts have proven very useful at least as general guidelines to help establish basic aspects of synaptic organization in the CN (Ibata and Pappas, '76; Cant and Morest, '79; Tolbert and Morest, '82; Kane, '77). More recently, however, immunocytochemical techniques have allowed to localize at the cellular level putative neurotransmitters, enzymes or synaptic receptors, involved in specific excitatory or inhibitory synaptic mechanisms. The localization and distribution of such "functional" anatomical markers in correlation with biochemical, pharmacological and physiological data adds important evidence to support specific roles for neurotransmission-related molecules in given neuronal or synaptic populations, greatly expanding our ability to analyze functional patterns of synaptic organization in relation to auditory processing in the CN (see chapter by Morest in this volume).

## Immunocytochemical localization of inhibitory versus excitatory amino acids

The role of neurotransmitter amino acids in CN function has attracted considerable interest in recent years (reviewed by Wenthold and Martin, '84; Altschuler et al., '86 and Caspary, '86). In fact, it is almost a basic tenet that a large part of synaptic excitation and inhibition in the central nervous system are mediated by a few short-chain amino acids (Mc Geer et al., '87a; Mc Geer et al., '87b). Glutamic acid (GLU) (or related molecules, called generically excitatory amino acids, EAAs) have been implicated in neurotransmission at most excitatory synapses (Fonnum, '84, McGeer et al., '87a). GABA and glycine (GLY) in turn, serve inhibitory roles (Mc Geer et al., '87b). As in other regions, a wealth of evidence supports glycine and GABA as inhibitory neurotransmitters in descending and intrinsic inputs to the CN (reviewed in Caspary, '86). Putative inhibitory synapses have been the first to be anatomically localized in the CN using several different immunocytochemical approaches (reviewed in the chapters by Altschuler et al., and Saint-Marie et al. in this volume, see other related chapters in this volume). From all these methods, the immunocytochemical localization of GLY and GABA with antibodies against immunogenic conjugates of these amino acids (Somogyi et al., '85a; Somogyi et al., '85b; Wenthold, et al., 86; Ottersen and Storm-Mathissen, '87; Wenthold et al., '87; Ottersen, '89; Ottersen et al., '89) has allowed one to study the distribution of putative GABAergic or glycinergic synapses in the CN with great resolution (see chapter by Altschuler et al. in this volume). One of the reasons for this success has been the fact that under normal conditions both GLY and GABA levels are almost undetectable by immunocytochemical methods outside labeled synaptic endings, axons and cell bodies of origin. This results in unequivocal, almost "all or none" labeling, which greatly facilitates anatomical studies of presumably GABAergic or glycinergic synapses and their correlation with pharmacological, physiological, biochemical and anatomical tract-tracing data (see chapters by Altschuler et al. and Saint-Marie et al. in this volume). However, anatomical identification of putative excitatory endings in the CN, where GLU is probably involved in neurotransmission seems to be a more complicated task for reasons detailed in the next section.

## Some problems associated with the immunocytochemical localization of excitatory endings

Considerable evidence supports an EAA, probably GLU, as a neurotransmitter in AN synapses (Godfrey et al., '77, '78; Altschuler et al., 81; Caspary et al., '81; Oliver et al., '83; Potashner, '83; Altschuler et al., '84; Wenthold and Martin, '84; Martin, '85; Wenthold, '85; Altschuler et al., '86; Wickesberg and Oertel, '89; Juiz et al., '90; see also chapter by Wenthold et al. in this volume). Recent data suggest that in some non-primary synapses like those likely originating from the intrinsic CN granule cells (Mugnaini et al., '80), EAAs may also be involved in neurotransmission (Manis, '89 and chapter in this volume). Therefore, as for GLY and GABA, it should be possible to tag and localize putative excitatory synapses in the CN using GLU immunolabeling. In fact, antibodies against GLU conjugates have been developed for the immunocytochemical detection of GLU in tissue sections (Storm-Mathissen et al., '83; Wanaka et al., '87; Hepler et al., '88; Liu et al., '89; Montero and Wenthold, '89). However, the potential power of this approach clashes with the extraordinary abundance of non-neurotransmitter related pools of GLU in the central nervous system (Fonnum, '84; McGeer et al., '87a), which in some cases has also interfered with the interpretation of biochemical, pharmacological and physiological data on the neurotransmitter role of GLU. In fact, GLU is the single most abundant amino acid in brain, with levels six times higher than GLY and nine times higher than GABA (Mc Geer et al., '87a). The ubiquity of GLU reflects somehow the many different functions of this amino acid in the central nervous system. Besides its suspected neurotransmitter role, GLU is an important building block for many proteins. GLU is also the precursor for GABA, and participates in metabolic steps crucial for brain function, such as handling of ammonium groups (Fonnum, '84; Mc Geer et al., '87a). The consequence of this exceptional abundance is that GLU may be detectable by immunocytochemistry in areas and structures where it is not directly linked to a putative neurotransmitter function. This makes highly unlikely that the simple presence of GLU immunoreactivity in synaptic endings can be used as a reliable anatomical marker for excitatory endings in the CN, because many "non-glutamatergic" endings will probably immunolabel for GLU, as discussed below.

## Quantitative immunocytochemistry of glutamate in the CN

Despite the ubiquity of GLU in the central nervous system including the CN, it is believed that levels of this amino acid are enriched in synaptic endings where GLU may serve a specialized role in neurotransmission (Fonnum, '84; Maycox et al., '90). In fact, it has been shown by immunocytochemistry (Altschuler et al., '81; Altschuler et al., '84) and enzyme cytochemistry (Rubio et al., poster presentation at this meeting) that there are synaptic endings in the CN containing very high levels of aspartate aminotransferase and glutaminase, two enzymes involved in metabolic pathways leading to glutamate. Therefore, from an anatomical perspective demonstration by immunocytochemistry of increased levels of GLU immunoreactivity in morphologically identified synaptic endings is likely to give a more faithful representation of putative excitatory endings than simply considering presence of GLU immunoreactivity.

Immunocytochemistry with antibodies to which colloidal gold particles are attached (Horisberger and Rosset, '77; De Mey, '83; van den Pol, '89) provides a reliable and reproducible way to quantify relative intensities of immunolabeling (Somogyi et al., '86; Ottersen, '89, Ottersen et al., '89; Montero and Wenthold, '89; Maxwell et al., '90). This is currently one of our approaches to study the distribution of putative excitatory endings in the CN. For immunolocalization of GLU in the CN, we use a modified version of the immunogold procedure of Somogyi and Hodgson ('85). Briefly, guinea pigs and rats are perfused through the ascending aorta with a mixture of paraformaldehyde and glutaraldehyde. The brainstems are dissected, sliced, dehydrated and embedded in plastic for electron microscopy. Ultrathin (90 nm) sections of the cochlear nucleus are incubated with an antibody against GLU conjugated to bovine serum albumin with glutaraldehyde (Montero and Wenthold, '89). Antibodies directed against GLU complexes are believed to bind to GLU-glutaraldehyde-protein complexes formed in the tissue during fixation (Ottersen and Storm-Mathissen, '87; Ottersen et al., '89). In a further step, a species-specific "secondary antibody" which recognizes the "primary" anti-GLU antibody and which has colloidal gold particles (average size, 15 nm) attached to it, is used to visualize sites of immunoreaction.

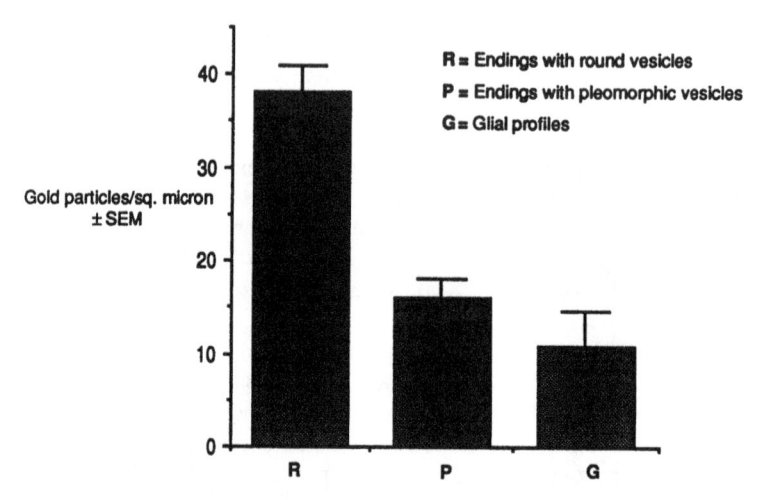

**Figure 1.** Representative experiment showing levels of GLU immunoreactivity in endings and glial profiles in AVCN and DCN. Endings with round vesicles and asymmetric synaptic densities (R) contain higher levels of GLU immunoreactivity than endings with pleomorphic vesicles and symmetric densities (P) or glial profiles (G). It is interesting to note that differences in GLU immunolabeling density between P and G profiles are not statistically significant.

In electron micrographs, colloidal gold particles appear as discrete electron dense particles which is possible to count. It has been shown that the amount of gold particles bears a close to linear relation with the amount of antigen in the tissue sample (Ottersen, '84, '90). It is therefore possible to detect levels of *fixed* GLU present in the section. However, one of the limitations of the technique is that it is difficult to relate levels of fixed GLU with the amount of *free* GLU present in the tissue (Ottersen, '89). Therefore, the technique is essentially semi-quantitative, allowing to determine relative amounts of fixed GLU in different neuronal and glial compartments.

The quantitative analysis shown in this chapter, involved previous identification of glial and synaptic profiles in electron micrographs of sections immunolabeled for GLU. For our purposes, we scored synaptic endings as containing "round" vesicles and apposed by asymmetric synaptic densities ("R" endings) or containing "pleomorphic" (oval/flattened) vesicles and apposed by symmetric synaptic densities ("P" endings). Numbers of gold particles over identified synaptic and glial profiles were manually counted and the perimeter of each profile traced with the aid of a digitizing tablet attached to a microcomputer equipped with an interactive morphometry software. After entering the number of gold particles, the program calculated the density of labeling, expressing the results as gold particles per square micron of tissue. So far, our sample includes 1,200 profiles from the anteroventral and dorsal divisions of the CN (AVCN and DCN).

## Synaptic endings containing increased levels of glutamate immunoreactivity in the CN

Figure 1 shows the results of a representative experiment in which the density of GLU immunolabeling was analyzed in synaptic and glial profiles in AVCN and the fusiform and molecular layers of DCN. In this experiment, the population of "R" endings (examples shown in Figs. 2, 3 and 4) had an average density of gold particles 2.3 times higher than that in "P" endings (examples shown in Figs. 2 and 4) or glial profiles. Although Figure 1 is highly

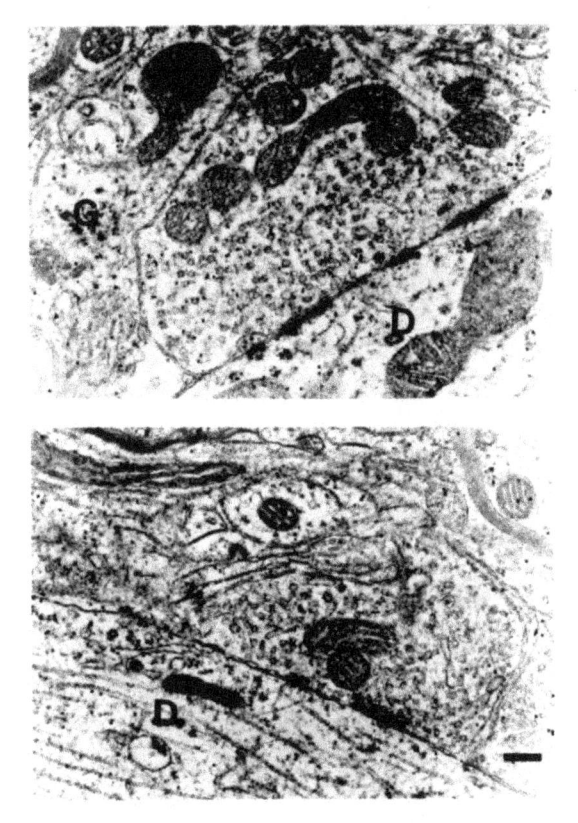

**Figure 2.** Two synaptic endings contacting the same primary dendritic trunk (D) from a spherical bushy cell in AVCN. The one containing round vesicles (top) contains substantially more gold particles than the other (bottom), which has pleomorphic vesicles. Note also the comparatively lower level of GLU immunolabeling in glial profiles (G). (Bar, 200 nm).

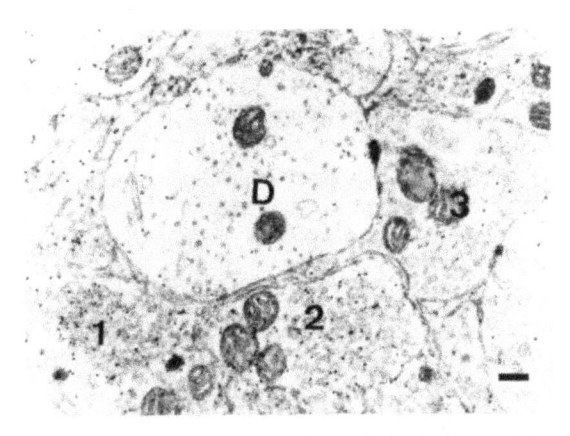

**Figure 3.** Synaptic endings surrounding a dendritic profile (D) in the fusiform cell layer of the DCN. Note the strong GLU immunolabeling in endings labeled 1 and 2. In contrast, GLU immunoreactivity in ending 3 is much weaker. (Bar, 200 nm)

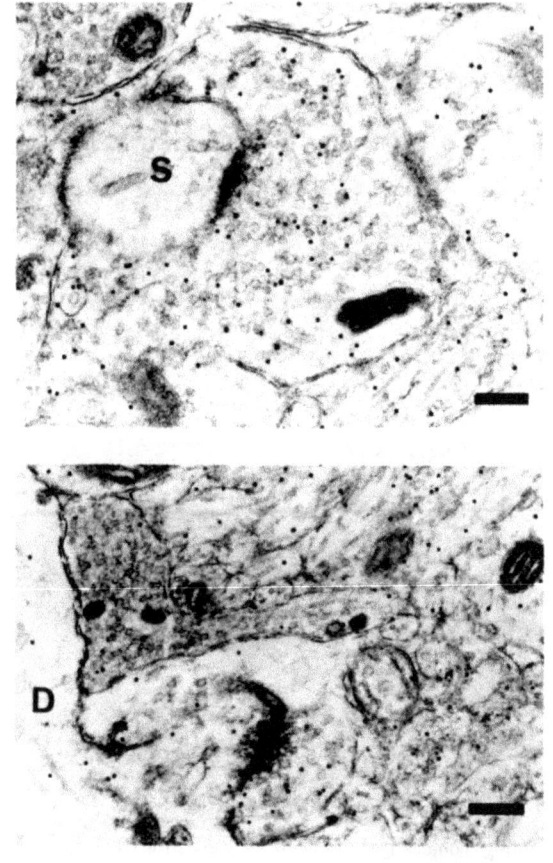

**Figure 4.** GLU immunoreactivity in synaptic endings in the molecular layer of the DCN. An ending containing round vesicles and contacting what appears to be a dendritic spine (S) shows intense GLU immunolabeling. A nearby ending which contains pleomorphic and dense-core vesicles (right side) and contacts a dendritic shaft shows lower intensity of immunolabeling. (Bars, 200 nm).

representative of findings across experiments, some variation was found in *absolute* density values of gold particles obtained from experiments separated from each other weeks or months. This could be due to uncontrollable experimental variables like differences among batches of immunocolloidal gold or subtle variations in the fixation procedure. What is remarkable, however, is that density *ratios* were constant across experiments. Levels of GLU immunoreactivity in "R" endings were always in the range of 2.3 to 2.6 higher than those found in "P" endings and glial profiles, regardless absolute density values.

Many "R" endings in our sample from AVCN (Fig. 2) and fusiform cell layer of DCN (Fig. 3) had morphological characteristics usually associated with AN endings. Similar synaptic endings have been shown to degenerate after sectioning of the acoustic nerve or cochlear ablation (Gentschev and Sotelo, '73; Ibata and Pappas, '76; Tolbert and Morest, '82). Therefore, our findings strongly suggest that AN endings contain increased levels of GLU immunoreactivity. However, not all the strongly GLU immunoreactive (potentially excitatory) "R" endings found on neurons in magnocellular regions of the AVCN and fusiform cell layer of the DCN necessarily originate from the AN. As discussed below, it will be necessary to combine anterograde tract-tracing techniques and quantitative GLU immunocytochemistry to address all the possible origins of excitatory endings in the CN.

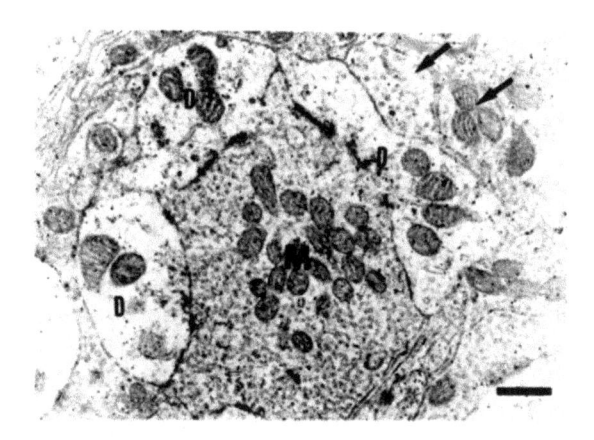

**Figure 5.** Example of a large "mossy-like" ending (M) in the granule cell domain of the DCN forming a glomerular array. Although no quantitative data are still available for this type of endings, note the high intensity of immunolabeling in this particular ending relative to the surrounding dendrites (D), and to endings with pleomorphic vesicles (arrow) which contact the side of the dendritic profiles opposite to that facing the "mossy-like" ending. (Bar, 500 nm).

A separate category of "R" endings included in the experiment shown in Fig. 1, and for which some hints are available regarding its possible origin, was found in the molecular layer of the DCN (Fig. 4). These were synaptic boutons which always contacted dendritic spines in this layer through asymmetrical synaptic densities (Fig 4). This type of endings also contained increased GLU immunoreactivity, which stood out more clearly when the density of gold particles in "R" endings, "P" endings and glial profiles in the molecular layer of the DCN was analyzed separately from the general sample (data not shown). There are reasons to believe that "R" endings with increased levels of GLU immunoreactivity in the molecular layer of the DCN probably originate from granule cells located in this division of the CN (Mugnaini, '80). Endings in the granule cell cap of the VCN (Mugnaini, '80) were not included in our sample, so we still do not know if all endings from granule cells in the CN contain increased levels of GLU immunoreactivity.

We have presented some evidence indicating that there are populations of synaptic endings in the CN that can be recognized by their enrichment in GLU immunoreactivity. It seems reasonable to assume that many of them originate from spiral ganglion neurons and CN granule cells. There is little doubt that both constitute major excitatory systems in the CN. Much evidence suggests that excitatory activity in AN synapses is mediated by an EAA, probably GLU (Godfrey et al., '77, '78; Altschuler et al., 81; Caspary et al., '81; Oliver et al., '83; Potashner, '83; Altschuler et al., '84; Wenthold and Martin, '84; Martin, '85; Wenthold, '85; Altschuler et al., '86; Wickesberg and Oertel, '89; Juiz et al., '90). Recent data also support EAA-mediated neurotransmission at granule cell synapses (Manis, '89; Manis, chapter in this volume; Rubio et al., presentation at this meeting). Obviously, our observation of elevated levels of GLU immunoreactivity in these two types of synaptic endings, by itself is not enough to attribute an specialized excitatory neurotransmitter role to GLU in these synaptic populations. Elevated GLU immunoreactivity might as well be simply an expression of high levels of metabolic activity in synaptic endings like those of the AN, firing at high rates (Caspary, '86). However, in the context of the already mentioned data supporting a neurotransmitter role for GLU in the CN, our findings have at least the value of another piece of evidence which strengthens the possibility of an specialized role for this amino acid in neurotransmission at identified synapses in the CN. Increased levels of GLU immunoreactivity have also been found in putative excitatory endings in the cerebellum (Somogyi, et al. '86, Ottersen, '89), lateral geniculate nucleus (Montero and Wenthold, '89) and substantia gelatinosa of the spinal cord (Maxwell et al., '90). On the other hand, work in progress in our laboratory supplies additional hints to a possible specialized role of GLU in excitatory synapses in the CN. When the number of synaptic vesicles in samples of "R" endings from the AN and "P" endings is statistically correlated with the number of gold particles labeling fixed GLU molecules, significant positive correlations are obtained for "R" endings but not for "P" endings (data not shown). This link between GLU immunoreactivity and synaptic vesicles in "R" endings suggests participation of GLU in neurotransmitter mechanisms in these endings in the CN. Positive correlation between GLU immunoreactivity and synaptic vesicles in putative excitatory endings also has been found by Montero and Wenthold ('89) in endings of the lateral geniculate body. More evidence comes from results obtained by Hackney and co-workers (presentation at this meeting) showing $Ca^{++}$-dependent decrease of GLU immunogold immunolabeling in AN endings from VCN slices after $K^+$-induced depolarization. This may be another indication that at least part of GLU immunolabeling in "R" endings is related to a releasable neurotransmitter pool of GLU. Therefore, although there is no conclusive evidence that elevated levels of GLU immunolabeling in "R" endings are linked to a neurotransmitter function, available data suggest that this is at least a valid hypothesis that deserves further work.

Regardless the role of GLU, it is clear from our findings that the increased level of GLU immunolabeling in populations of synaptic endings constitutes a very useful anatomical marker for subsets of endings which are probably involved in excitatory circuits in the CN. The use of this method alone or in combination with other techniques may provide new interesting insights on the chemical anatomy of putative excitatory systems in the CN. For instance, it will be important to explore the possibility of the presence in the CN of putative excitatory synapses others than the two previously described, which probably originate from spiral ganglion neurons and CN granule cells respectively. In fact, preliminary results suggest the existence of more than two classes of excitatory synapses in the CN. Figure 5 shows an example of a third category of synaptic endings which seem to contain increased levels of GLU immunoreactivity. These endings are found in granule cell domains of the DCN, and they have the all the characteristics of the "mossy-like" endings described by Mugnaini and co-workers in granule cell areas of the CN of rat, cat and mouse ('80) . The origin of this type of ending is not clear (Kane, '77; Mugnaini, '80). However, this finding indicates that at least three classes of putative excitatory endings may be identified in the CN, based on patterns of GLU immunolabeling and generic electron microscopic characteristics. Further work will be necessarily aimed at exploring the possibility of more classes of potentially excitatory endings originating from more than two sources. The question of whether CN neurons receive descending excitatory projections (see chapter by Saldana in this volume) besides the extensive and better known descending inhibition (see chapters in this volume on this issue), is interesting enough to deserve additional work. It will be necessary to combine anterograde tract-tracing and quantitative GLU immunocytochemistry to explore this interesting possibility. At the same time, it will be essential to address with greater detail the distribution of putative excitatory endings on specific neuronal types in the CN. This is one of the keys to understand the organization of the microcircuits responsible for the processing abilities of individual CN neurons. It seems therefore that quantitative immunocytochemical approaches to EAA-mediated neurotransmission in the CN constitute powerful tools that still need to be fully exploited.

ACKNOWLEDGEMENTS

Supported by NIDCD grant DC00383 (R.A. Altschuler) and Spanish Research Council (CICYT) grant PB/0523/CO2/00 (Jose M. Juiz)

# REFERENCES

Adams, J.C. and Mugnaini, E., 1987, Patterns of glutamate decarboxylase immunostaining in the feline cochlear nuclear complex studied with silver enhancement and electron microscopy, *J. Comp. Neurol.*, 262:375-401.

Altschuler, R.A., Neises, G.R., Harmison, G.G., Wenthold, R.J. and Fex, J., 1981, Immunocytochemical localization of aspartate aminotransferase immunoreactivity in the cochlear nucleus of the guinea pig, *Proc. Nat. Acad. Sci. U.S.A.*, 78:6553-6557.

Altschuler, R.A., Wenthold, R.J., Schwartz, A.M., Haser, W.G., Curthoys, N.P., Parakkal, M. and Fex, J., 1984, Immunocytochemical localization of glutaminase-like immunoreactivity in the auditory nerve, *Brain Res.*, 291:173-178.

Caspary, D.M., 1986, Cochlear Nuclei: Functional neuropharmacology of the different cell types, in: "Neurobiology of Hearing: The Cochlea", R.A. Altschuler, R.P. Bobbin, D.W. Hoffman, eds., pp. 303-332, Raven Press, New York.

Conlee, J.W. and Kane, E.S., 1982, Descending projections from the inferior colliculus to the dorsal cochlear nucleus in the cat: an autoradiographic study, *Neuroscience*, 7(1):161-178.

De Mey, J., Moeremans, M., Geuens, G., Nuydens, R. and De Brabander, M., 1981, High resolution light and electron microscopic localization of tubulin with the IGS (immuno gold staining) method, *Cell Biol. Int. Rep.*, 5(9):889-899.

De Mey, R.J., 1983, The preparation of immunoglobulin gold conjugates (IGS reagents) and their use as markers for light and electron microscopic immunocytochemistry, in: "Immunocytochemistry", , A.C. Cuello, ed., pp. 347-372, John Willey and Sons, New York.

Eccles, J.C., 1964, The Physiology of Synapses, Springer-Verlag, Berlin.

Fonnum, F., 1984, Glutamate: A neurotransmitter in mammalian brain, *J. Neurochem.*, 42:1-11.

Gentschev, T. and Sotelo, C., 1973, Degenerative patterns in the ventral cochlear nucleus of the rat after primary deafferentation. An ultrastructural study, *Brain Res.*, 62:37-46.

Godfrey, D.A., Carter, J.A., Berger, S.J., Lowry, O.H. and Matschinsky, F.M., 1977, Quantitative histochemical mapping of candidate transmitter amino acids in cat cochlear nucleus, *J. Histochem. Cytochem.*, 25:417-31.

Godfrey, D.A., Carter, J.A., Lowry, O.H. and Matchinsky, F.M, 1978, Distribution of gamma-aminobutyric acid, glycine, glutamate and aspartate in the cochlear nucleus of the rat, *J. Histochem. Cytochem.*, 26:118-126.

Gray, E.G., 1959, Axo-somatic and axo-dendritic synapses of the cerebral cortex. An electron microscope study, *J. Anat.*, 93:420- 433.

Gray, E.G., 1969, Electron microscopy of excitatory and inhibitory synapses, "Mechanisms of synaptic transmission", Progress in Brain Res., Vol. 37.

Hepler, J.R., Toomim, C.S., Mc Carthy., K.D., Conti, F., Battaglia, G., Rustioni, A. and Petrusz, P., 1988, Characterization of antisera to glutamate and asparate, *J. Histochem. Cytochem.*, 36:13-22.

Hirsch, J.A. and Oertel, D., 1988, Synaptic connections in the dorsal cochlear nucleus of mice "in vitro", *J. Physiol.*, 396:549-562.

Horisberger, M. and Rosset, J., 1977, Colloidal gold, a useful marker for transmission and scanning electron microscopy, *J. Histochem. Cytochem.*, 25:295-305.

Ibata, Y. and Pappas, G., 1976, The fine structure of synapses in relation to the large spherical neurons in the anterior ventral cochlear nucleus, *J. Neurocytol.*, 5:395-406.

Kane, ES, 1977, Descending inputs to the cat dorsal cochlear nucleus: an electron microscopic study, *J. Neurocytol.*, 6(5):583-605.

Lenn, N.J. and Reese, T.S., 1966, The fine structure of nerve endings in the nucleus of the trapezoid body and the ventral cochlear nucleus, *Am. J. Anat.*, 118:375-389.

Liu, C., Grandes, P., Matute, C., Cuenod, M. and Streit, P., 1989, Glutamate-like immunoreactivity revealed in rat olfactory bulb, hippocampus and cerebellum by monoclonal antibody and sensitive staining method, *Histochemistry*, 90:427-445.

Manis, P., 1989, Responses to parallel fiber stimulation in the guinea pig dorsal cochlear nucleus, *J. Neurophysiol.*, 61:149-158.

Martin , M.R., 1985, Evidence for an excitatory amino acid as the transmitter of the auditory nerve in the "in vitro" mouse cochlear nucleus, *Hearing Res.*, 17:215-222.

Maxwell, D.J., Christie, W.M., Short, A.D., Storm-Mathisen, J. and Ottersen, O.P., 1990, Central boutons of glomeruli in the spinal cord of the cat are enriched with L-glutamate-like immunoreactivity, *Neuroscience*, 36:83-104.

Maycox, P.R., Hell, J.W. and Jahn, R., 1990, Amino acid neurotransmitters: spotlight on synaptic vesicles, *Trends in Neurosci.*, 13:83-87.

McGeer, P.L., Eccles, J.C. and McGeer, E.G., 1987, Inhibitory amino acid neurotransmitters, *in*: "Molecular Neurobiology of the Mammalian Brain", P.L. McGeer, J.C. Eccles, E.G. McGeer, eds., pp. 197-234, Plenum Press, New York.

McGeer, P.L., Eccles, J.C. and McGeer, E.G., 1987, Putative excitatory neurons: glutamate and aspartate, *in*: "Molecular Neurobiology of the Mammalian Brain", Mc Geer P.L., Eccles J.C., McGeer E.G., eds., pp. 175-196, Plenum Press, New York.

Montero, V.M. and Wenthold, R.J., 1989, Quantitative immunogold analysis reveals high glutamate levels in retinal and cortical synaptic terminals in the lateral geniculate nucleus of the macaque, *Neuroscience*, 31:639-647.

Moore, J.K. and Moore, R.Y., 1987, Glutamic acid decarboxylase-like immunoreactivity in brainstem auditory nuclei of the rat, *J. Comp. Neurol.*, 260:157-74.

Moore, JK., 1986, Cochlear nuclei: Relationship to the auditory nerve, *in*: "Neurobiology of Hearing: The Cochlea", R.A. Altschuler, R.P. Bobbin, D.G. Hooffman eds., pp. 283-301, Raven Press, New York.

Mugnaini, E., Osen, K.K., Dahl, A.L., Friedrich V.L., Jr. and Korte, G., 1980, Fine structure of granule cells and related interneurons (termed Golgi cells) in the cochlear nuclear complex of cat, rat and mouse, *J. Neurocytol.*, 9:537-570.

Oberdorfer, M.D., Parakkal, M.H., Altschuler, R.A. and Wenthold, R.J., 1988, Ultrastructural localization of GABA-immunoreactive terminals in the anteroventral cochlear nucleus of the guinea pig, *Hearing Res.*, 33:229-238.

Oliver, D.L., Potashner, S.J., Jones, D.R. and Morest, D.K., 1983, Selective labeling of spiral ganglion and granule cells with D-aspartate in the auditory system of cat and guinea pig, *J. Neurosci.*, 3:455-468.

Ottersen, O.P., 1989, Quantitative electron microscopic immunocytochemistry of neuroactive amino acids, *Anat. Embryol.*, 180:1-15.

Ottersen, O.P. and Storm-Mathisen, J., 1987, Localization of amino acid neurotransmitters by immunocytochemistry, *Trends in Neurosci.*, 10:250-255.

Ottersen, O.P., Storm-Mathissen, J., Madsen, S., Skumlien, S. and Sromhaug, J., 1989, Evaluation of the immunocytochemical method for amino acids, *Med. Biol.*, 64: 147-158.

Pappas, G.D. and Waxman, S.G., 1972, Synaptic fine structure-morphological correlates of chemical and electrotonic transmission, *in*: "Structure and Function of Synapses", G.D. Pappas, D.K. Purpura, eds., pp. 1-44, Raven Press, New York.

Potashner, S.J., 1983, Uptake and release of D-aspartate in the guinea pig cochlear nucleus, *J. Neurochem.*, 41:1094-1109.

Potashner, S.J., Morest, D.K., Oliver, D.L. and Jones, D.R., 1985, Identification of glutamatergic and aspartatergic pathways in the auditory pathway, *in*: "Auditory Biochemistry", D.G. Drescher, ed., pp. 141-162, C. Thomas, Springfield.

Schwartz, A.M. and Gulley, R.L., 1978, Non-primary afferents to the principal cells of the rostral anteroventral cochlear nucleus of the guinea pig, *Am. J. Anat.*, 153:489-508.

Schwartz, I.R., 1981, The differential distribution of label following uptake of 3H-labeled amino acids in the dorsal cochlear nucleus of the cat. An autoradiographic study, *Exp. Neurol.*, 73:601-617.

Somogyi, P., Halasy, K., Somogyi, J., Storm-Mathisen, J. and Ottersen, O.P., 1986, Quantification of immunogold labelling reveals enrichment of glutamate in mossy and parallel fibre terminals in cat cerebellum, *Neuroscience*, 19:1045-1550.

Somogyi, P., Hodgson, A.J., Chubb, I.W., Penke, B. and Erdei, A., 1985, Antisera to gamma-aminobutyric acid application to the central nervous system, *J. Histochem. Cytochem.*, 33:240-248.

Somogyi, P. and Hodgson, A.J., 1985, Antiserum to gamma-aminobutyric acid III. Demonstration of GABA in Golgi-impregnated neurons and in conventional electron microscopic sections of cat striate cortex, *J. Histochem. Cytochem.*, 33:249 257.

Tolbert, L.P. and Morest, D.K., 1982, The neuronal architecture of the anteroventral cochlear nucleus of the cat in the region of the cochlear nerve root: electron microscopy, *Neuroscience*, 7:3053-3067.

Uchizono, K., 1965, Characteristics of excitatory and inhibitory synapses in the central nervous system of the cat, *Nature*, 207:642-643.

van den Pol, A.N., 1984, Colloidal gold and biotin-avidin conjugates as ultrastructural markers for neural antigens, *Q. J. Exp. Physiol.*, 69:1-33.

van den Pol, AN, 1989, Neuronal imaging with colloidal gold, *J. Microsc.*, 155 (Pt 1):27-59.

Wanaka, A., Shiotani, Y., Kiyama, H., Matsuyama, T., Kamada, T., Shiosaka, S. and Tohyama, M., 1987, Glutamate-like immunoreactive structures in primary sensory neurons in the rat detected by a specific antiserum against glutamate, *Exp. Brain Res.*, 65:691-694.

Wenthold, R.J., 1985, Glutamate and aspartate as neurotransmitters of the auditory nerve, *in*: "Auditory Biochemistry", D.G. Drescher, ed., pp. 125-140, C. Thomas, Springfield.

Wenthold, R.J., Huie, D., Altschuler, R.A. and Reeks, K.A., 1987, Glycine immunoreactivity localized in the cochlear nucleus and superior olivary complex, *Neuroscience*, 22:897-912.

Wenthold, R.J., Zempel, J.M., Parakkal, M.H., Reeks, K.A. and Altschuler, R.A., 1986, Immunocytochemical localization of GABA in the cochlear nucleus of the guinea pig, *Brain Res.*, 380:7-18.

Wickesberg, R.E. and Oertel, D., 1989, Auditory nerve neurotransmitter acts on a kainate receptor: evidence from intracellular recordings in brain slices from mice, *Brain Res.*, 486:39-46.

# EXCITATORY AMINO ACID RECEPTORS IN THE RAT COCHLEAR NUCLEUS

Robert J. Wenthold, Chyren Hunter and Ronald S. Petralia

Laboratory of Neurochemistry, National Institute on Deafness and Other Communication Disorders, NIH, Bethesda, MD 20892, USA

The neurotransmitters of the cochlear nucleus (CN) have received considerable attention, but probably the most important class of neurotransmitters, the excitatory amino acids (EAAs), is also the least understood not only in the CN, but throughout the central nervous system. They have been difficult to study on essentially all levels, including measurement of the amino acids themselves, their biosynthesis and degradation, release, and postsynaptic receptors. With the exception of neuropeptides, EAAs appear to be the last major class of neurotransmitters to be identified even though it is now commonly believed that most excitatory synapses in the central nervous system use an EAA as a neurotransmitter. Many attempts were made to identify a marker that could be used for selectively localizing EAA releasing neurons. Several useful markers were identified, including glutamate and aspartate themselves, uptake of radioactive EAAs, and enzymes in the metabolism of glutamate and aspartate, but no reliable method, which worked in all cases, was found.

The cloning of members of the EAA receptor family now permits the detailed analysis of the distribution of these receptors in areas of the central nervous system, using either *in situ* hybridization histochemistry or immunocytochemistry with antibodies made to either the whole or part of the receptor molecule. One obvious outcome of the immunocytochemical studies is the identification of synapses which use EAAs as neurotransmitters since the presence of an EAA receptor on the postsynaptic membrane provides compelling evidence that a particular synapse uses an EAA neurotransmitter. Not only does this add to our basic knowledge of the nervous system, but such information could be the basis for the design of a pharmacological treatment of disorders affecting the central nervous system. Multiple other questions await to be addressed. For example, how is the number and position of receptor molecules at a synapse regulated and does this composition of the receptor change under any condition? What is the role of the receptor in the developing synapse or in disease states? Why are there multiple subforms of most receptors, and is the subunit composition of the receptor related to any functional role in the intact animal? There are two primary reasons for studying EAAs and their receptors in the CN. First, EAAs are believed to be the major excitatory neurotransmitters in this nucleus, and a knowledge of their function is a prerequisite to understanding the function of the CN at the molecular level in both the normal and disease states. Second, with its relatively simple structure and the wealth of knowledge concerning its anatomy, physiology, and pharmacology, the CN is an ideal model structure for studying EAAs and addressing many basic questions such as those pointed out above. In this manuscript we focus on EAA receptors in the CN. We briefly review work on these receptors in the CN and summarize the current state of knowledge of the fast-moving EAA receptor field. Finally, we present recent data on the distribution of EAA receptors in the CN using *in situ* hybridization histochemistry and immunocytochemistry.

## Excitatory amino acid receptors

Excitatory amino acid (EAA) receptors can be divided into two general categories, ionotropic and metabotropic receptors. Ionotropic receptors mediate fast neurotransmission and are linked directly to the ion channel while metabotropic receptors function through second messenger systems and are involved in slow neurotransmission or modulation. Electrophysiological studies have identified three classes of ionotropic EAA receptors named after the agonist that preferentially excites the receptor: kainate (KA), quisqualate (QA) or alpha-amino-3-hydroxy-5-methyl-4-isoxazolepropionic acid (AMPA), and N-methyl-D-aspartate (NMDA) receptors (Watkins and Evans, '81; Foster and Fagg, '84; Monaghan et al., '89). Ligand binding studies done either on tissue sections or tissue homogenates with [$^3$H]KA and [$^3$H]AMPA for the KA and AMPA receptors, respectively, and several different ligands for the NMDA receptor have supported the electrophysiological categorization of EAA receptors (Monaghan et al., '89). Each has a distinct tissue distribution and ligand binding properties. However, physiological studies on intact animals, *in vitro* slices, or cultured neurons leave some uncertainty about the distinction between some classes of receptors, especially KA and AMPA receptors. Thus, the KA and AMPA receptor subtypes are often combined and referred to as non-NMDA receptors. The distinctions between the different subtypes are currently being determined at the molecular level through the cloning of genes encoding EAA receptors and their expression *in vitro*.

One of the most significant developments in EAA research has been the cloning of some members of the non-NMDA receptor class (Wada et al., '89; Gregor et al., '89; Hollmann et al., '89) as well as a member of the metabotropic class (Masu et al., '91; Houamed et al., '91). After one member of the AMPA subtype of receptor was cloned (Hollmann et al., '89), several related clones were isolated, and a total of six subtypes have now been cloned (Kainänen et al., '90; Werner et al., '91; Egebjerg et al., '91; Bettler et al., '90; Boulter et al., '90; Nakanishi et al., '90). Four of these (GluR1, 2, 3, and 4 or GluRA, B, C, D) are quite similar based on primary structure, physiological properties and ligand binding properties. They are all relatively large with molecular weights of about 100,000 daltons, they share about 70% amino acid identity, and when expressed in oocytes or cultured cells, respond to KA and AMPA. This reponse can be blocked by 6-cyano-7-nitroquinoxaline-2,3-dione (CNQX), a potent antagonist of non-NMDA receptors. When expressed in cultured cells, GluR1-4 bind [$^3$H]AMPA with high affinity, but not [$^3$H]KA (Kainänen et al., '90). Based on these properties, GluR1-4 may represent all or some of the members of the AMPA subtype of EAA receptor, and since these receptors respond well to KA, they may also represent some of the KA receptors that have been characterized physiologically. Although each subunit alone is functional when expressed in oocytes or cultured cells (Dawson et al., '90), GluR1-4 appear to form hetero-oligomeric complexes *in vivo*, based both on physiological data (Keinänen et al., '90; Boulter et al., '90) and antibody binding data [unpublished observation]. *In situ* hybridization histochemistry shows that the distribution patterns in brain are different, but overlapping, for these subunits (Kainänen et al., '90; Boulter et al., '90), so the stoichiometry of the receptor complex appears to vary within the brain. The subunit composition of the receptor complex may be functionally significant since different combinations of subunits produce receptors with different electrophysiological properties (Boulter et al., '90; Kainänen et al., '90; Hollmann et al., '91; Verdoorn et al.,

'91). The most dramatic difference observed was the ion specificity of the channel (Hollmann et al., '91; Verdoorn et al., '91). Receptors composed of GluR1 and GluR3, when either expressed alone or together, are able to pass calcium ions (Hollmann et al., '91). However, when GluR2 is co-expressed with them, the channel can no longer pass calcium. Therefore, GluR2 containing receptor complexes form calcium-impermeable channels. This is particularly important since it is believed that the neurotoxicity of glutamate and other EAAs is receptor mediated and related to calcium influx with prolonged stimulation (Choi, '87; Garthwaite et al., '86; Olney et al., '86). It has also been shown that each of the GluR1-4 forms exists in two alternatively spliced versions, referred to as flip and flop, with a different sequence in a 38 amino acid segment between the third and fourth transmembrane regions (Sommer et al., '90). The flip and flop molecules have different distributions in brain, different developmental profiles, and different physiological properties.

Two other molecules related to GluR1-4 have been cloned. GluR5 has about a 40% amino acid identity to GluR1-4 and responds only to glutamate when expressed in oocytes (Bettler et al., '90). GluR6 is quite similar to GluR5, but with less than 40% amino acid identity to GluR1-4 (Egebjerg et al., '91). GluR6 is activated by KA, QA and glutamate, but not AMPA. Therefore, GluR6 may correspond to a KA receptor, as measured physiologically in intact tissues, while the pharmacological category to which GluR5 belongs remains unclear.

The high affinity [$^3$H]KA binding site has not been shown to be associated with any of the above receptors. In rat brain this site may be associated with the recently cloned receptor, KA-1 (Werner et al., '91). KA-1 has 53% amino acid identity to GluR1-4, but is not physiologically functional as a ligand-gated ion channel when expressed in cultured cells. However, the expressed receptor binds [$^3$H]KA with high affinity. In lower vertebrates [$^3$H]KA binding is at least partially associated with related, but distinct, molecules, the kainate binding proteins (KBPs). These proteins, cloned from frog (Wada et al., '89) and chick (Gregor et al., '89) brain, are much smaller than GluR (about 50,000 daltons), but are 35-40% identical in amino acid sequence to the C-terminal portions of GluR1-6 or KA-1. The KBPs have not yet been shown to be functional ion channels that respond to KA, glutamate or several other EAAs, but they bind [$^3$H]KA when expressed in cultured cells.

The metabotropic class of EAA receptor is pharmacologically distinct from the ionotropic class with QA, glutamate and ibotenate being potent agonists. A member of this receptor family was cloned and was shown to be structurally similar to other G protein-coupled receptors (Masu et al., '91; Houamed et al., '91). It can be expected that additional members of this family will be identified soon.

## Excitatory amino acids in the cochlear nucleus

EAAs are thought to be the neurotransmitters at most excitatory synapses in the central nervous system. This appears to be true in the CN as well. The major excitatory synapse in the CN is the one formed between auditory nerve fibers and principal cells, and EAAs are the only substances to be proposed as neurotransmitters at this synapse. The data supporting this are extensive (For reviews, see Wenthold and Martin, '84; Caspary et al., '85; Altschuler et al., '86; Aitkin, '89) and will only be summarized here. This evidence includes the presence of glutamate and aspartate, as well as enzymes involved in their biosynthesis, in auditory nerve terminals; demonstration of calcium-dependent release of glutamate and aspartate from auditory nerve terminals; retrograde transport of D-aspartate in the auditory nerve; the degeneration of principal neurons in the CN after application of exogenous EAAs; and the neurophysiological demonstration of EAA receptors in the CN. Glutamate is usually considered the leading candidate for the endogenous ligand of EAA receptors because it is a fairly potent agonist at most EAA receptors and it is the most abundant EAA in the brain.

Immunocytochemical studies have shown glutamate to be present in some presynaptic terminals (Ottersen, '89), including those in the CN, but in most cases it is more abundant in the somata of neurons. There are several other candidates for the endogenous EAA neurotransmitter including aspartate and small peptides, but the identification of the true neurotransmitter will likely not be as forthcoming as the molecular characterization of the EAA receptors. It is not unreasonable to expect that a number of EAA neurotransmitters exist, and the particular one (or combination) used at a synapse may be related to the subtype of the postsynaptic receptor. Therefore, glutamate is the primary candidate for the auditory nerve neurotransmitter, as well as for other excitatory synapses in the CN, but the evidence is not sufficient to rule out the possibility that another EAA or a combination of EAAs is the actual transmitter at these synapses. Since the EAA receptors are able to accommodate several endogenous EAAs, the identification of the postsynaptic receptor will have no immediate impact on determining the nature of the neurotransmitter.

The nature of the EAA receptor present at auditory nerve synapses has not been completely resolved by electrophysiological studies, although the major receptor at these synapses appears to be of the non-NMDA subtype. Some early studies implicated NMDA receptors in auditory nerve neurotransmission in the cat (Martin, '80), but the conclusions of these studies may have been due to the less specific antagonists that were available at that time. It is now known that the NMDA receptor contains a voltage sensitive channel and functions in concert with non-NMDA receptors (Mayer and Westbrook, '87). NMDA receptors are present in the CN based on ligand binding data (Monaghan and Cotman, '85) as well as electrophysiological data (Wickesberg and Oertel, '89). They may play important roles in establishing neuronal connections during development and in synaptic plasticity in the adult brain (Cline and Constantine-Paton, '89). A KA subtype of EAA receptor has been suggested as the receptor mediating auditory nerve neurotransmission in the CN of the mouse (Wickesberg and Oertel, '89) and chick (Nemeth et al., '85; Jackson et al., '85). In the chick studies, QA and AMPA were weak agonists of this receptor, but in the mouse study AMPA was not tested and QA did not give a response. The lack of response to QA is unexpected since QA has been shown to be an agonist for both KA and AMPA subtypes of receptor. In summary, therefore, the neurophysiological studies on the auditory nerve synapse in the CN support a non-NMDA receptor subtype, but are not refined sufficiently to differentiate between a KA or AMPA subtype. The CN has also been studied using binding of labeled ligands of EAA receptors. [$^3$H]KA binding is low in the CN (Monaghan and Cotman, '82), but [$^3$H]glutamate binding is moderately high (Greenamyre et al., '84; Halpain et al., '84). Glutamate is not a selective ligand for any subtype of receptor, so these data probably include both NMDA and AMPA subtypes of receptors. Using selective ligands the NMDA subtype appears to be present throughout the dorsal cochlear nucleus (DCN) at moderately high levels (Monaghan and Cotman, '85).

Information on neurotransmitters and receptor subtypes at synapses in the CN other than those formed with the auditory nerve is limited. Based on D-aspartate uptake (Oliver et al., '83; Potashner, '83) and high levels of glutaminase (Altschuler et al., '84), granule cells are considered candidates for using an EAA as a neurotransmitter. These neurons give rise to the parallel fiber system which synapses on fusiform cells and interneurons including cartwheel, stellate and Golgi cells of the DCN. Physiological studies showed this pathway to be excitatory, and the input can be blocked with kynurenic acid, a general EAA antagonist (Manis, '89). Glutamate is excitatory to cells receiving parallel fiber input (Hirsch and Oertel, '88). Descending inputs to the CN do not appear to use EAAs extensively based on D-aspartate labeling and release (Staatz-Benson and Potashner, '88). Most excitatory projection neurons probably use EAA neurotransmitters, but these pathways have not been studied in detail.

# Localization of GluR subunits in the cochlear nucleus with *in situ* hybridization histochemistry and immunocytochemistry

**Structure of the cochlear nucleus**

Cell types in the CN have been described previously for the rat (Harrison and Warr, '62; Harrison and Irving, '65; Harrison and Irving, '66; Mugnaini et al., '80; Mugnaini et al., '80; Wouterlood et al., '84; Wouterlood and Mugnaini, '84; Mugnaini, '85) (reviewed by Webster, '85). Based mainly on distribution of these defined cell types, the CN can be divided into the acoustic nerve nucleus, ventral cochlear nucleus (VCN) which includes 5 main areas plus granule cell domains, and the DCN which can be subdivided into 3 layers. Thus, major cell types are characteristic of different regions of the cochlear nuclei, with large spherical cells (Harrison type c) concentrated in area III of the VCN; small spherical cells (Harrison type i) in area I just dorsocaudal to area III; globular cells (Harrison type g) in area II; octopus cells (Harrison type k) in area IV; fusiform cells in layer II of the DCN; cartwheel cells found mainly in layer I of the DCN; and granule cells of the DCN and granule cell domains of the VCN. In addition, the CN is divided broadly by the primary bifurcations of the auditory nerve, which sends ascending branches into the anteroventral cochlear nucleus (AVCN) and descending branches into the posteroventral cochlear nucleus (PVCN) and ultimately into the DCN.

*In situ* hybridization histochemistry

The expression of GluR1-4 subunit mRNAs was studied in the auditory brainstem of the rat using synthetic antisense oligonucleotide probes. Sequences, 45 nucleotides in length, were chosen from between the first and second transmembrane regions of the cDNAs, an area of great subtype sequence divergence, to yield selective probes for each subunit (Kainänen et al., '90). Oligonucleotides were synthesized on an Applied Biosystems DNA Synthesizer and purified according to a standard DNA/RNA purification scheme. Purified oligonucleotides were labeled at the 3' terminal end with [$^{35}$S]dATP and used as probes in hybridization buffer containing 50% formamide and 4X sodium chloride/sodium citrate (SSC) for application to tissue sections. Sixteen $\mu$m thick serial sections of fresh frozen rat brainstem mounted onto subbed slides were fixed in 4% paraformaldehyde in phosphate buffered saline (pH 7.4) and rinsed first in phosphate buffered saline (pH 7.4) and then in 2X SSC. As a final step, sections were dehydrated in a series of graded ethanols and allowed to air dry. Hybridization proceeded for a minimum of 17 hours with probe at a concentration of approximately 8 pg/$\mu$l at 42°C. Hybridization controls consisted of sections hybridized with one or more of the following: probe containing a 20 fold excess of unlabeled oligonucleotide, a sense oligonucleotide to GluR2, or an unrelated oligonucleotide sequence. Posthybridization washes were performed in 1X SSC at 56°C. Sections were apposed to X-ray film or dipped in photographic emulsion (Kodak NTB2 or NTB3) to localize specifically bound probe. Sense controls and those containing excess unlabeled probe resulted in diffuse labeling patterns, with a low density of silver grains equally distributed over cellular and noncellular areas. GluR expression in the present study is in general agreement with that described previously (Keinänen et al., '90; Boulter et al., '90; Sommer et al., '90; Pellegrini-Giampietro et al., '91).

*In situ* hybridization of the CN was examined in detail in serial sections cut in the coronal or sagittal plane. Cell types were identified based on Nissl staining patterns seen after thionin staining of developed autoradiograms in conjunction with available morphological data from the rat and cat as described above (Harrison and Warr, '62; Harrison and Irving, '66; Mugnaini, '85; Osen, '69; Cant and Morest, '84; Saldaña et al, '88). Labeling for a particular cell is expressed relative to labeling of other cells in the CN with the same probe.

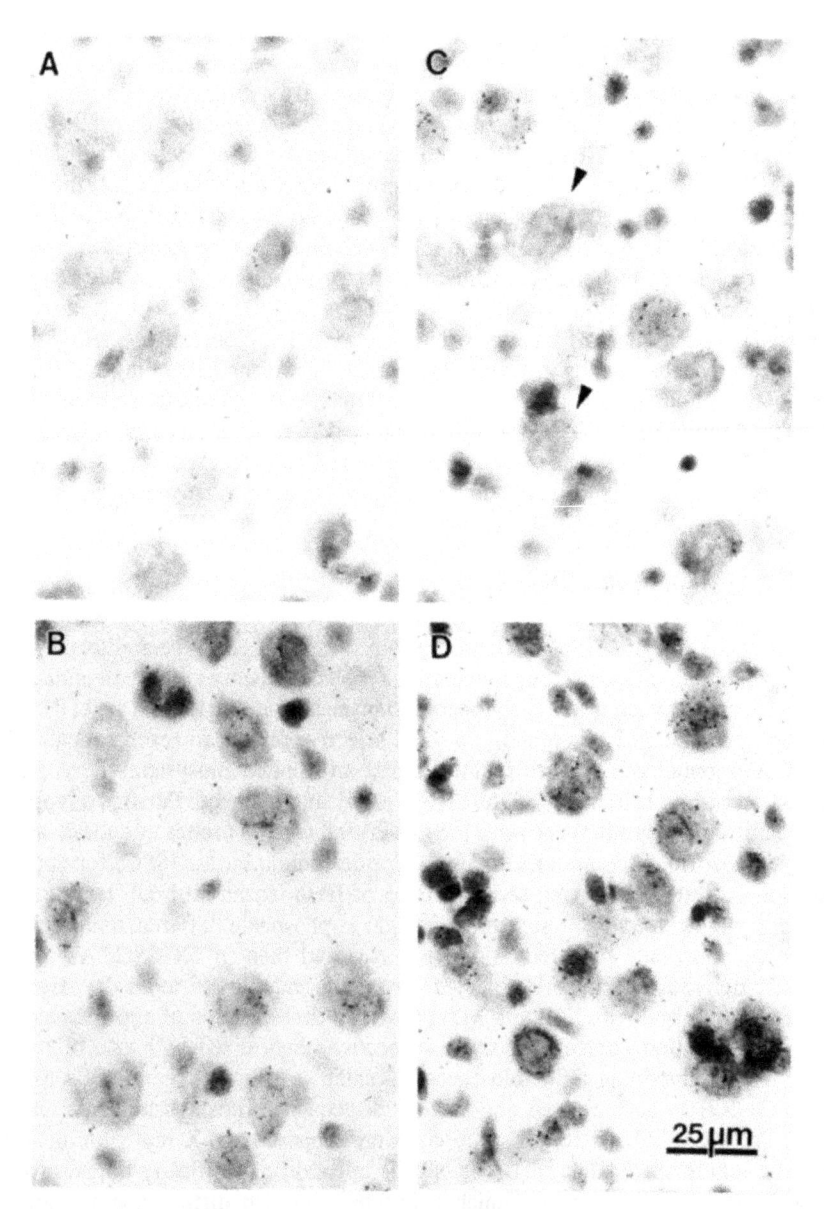

**Figure 1.** Expression of glutamate receptor subtypes GluR1-4 in the VCN. (A) Silver grains are equally distributed over spherical cells and neuropil of the VCN following hybridization with GluR1 probe. (B) Hybridization with GluR2 probe leads to silver grain accumulation over small spherical cells. (C) Moderate hybridization to small spherical cells is seen with GluR3 probe; a subpopulation of spherical cells, however, is unlabeled (arrowhead). (D) Hybridization with GluR4 probe leads to a moderate accumulation of grains over small spherical cells.

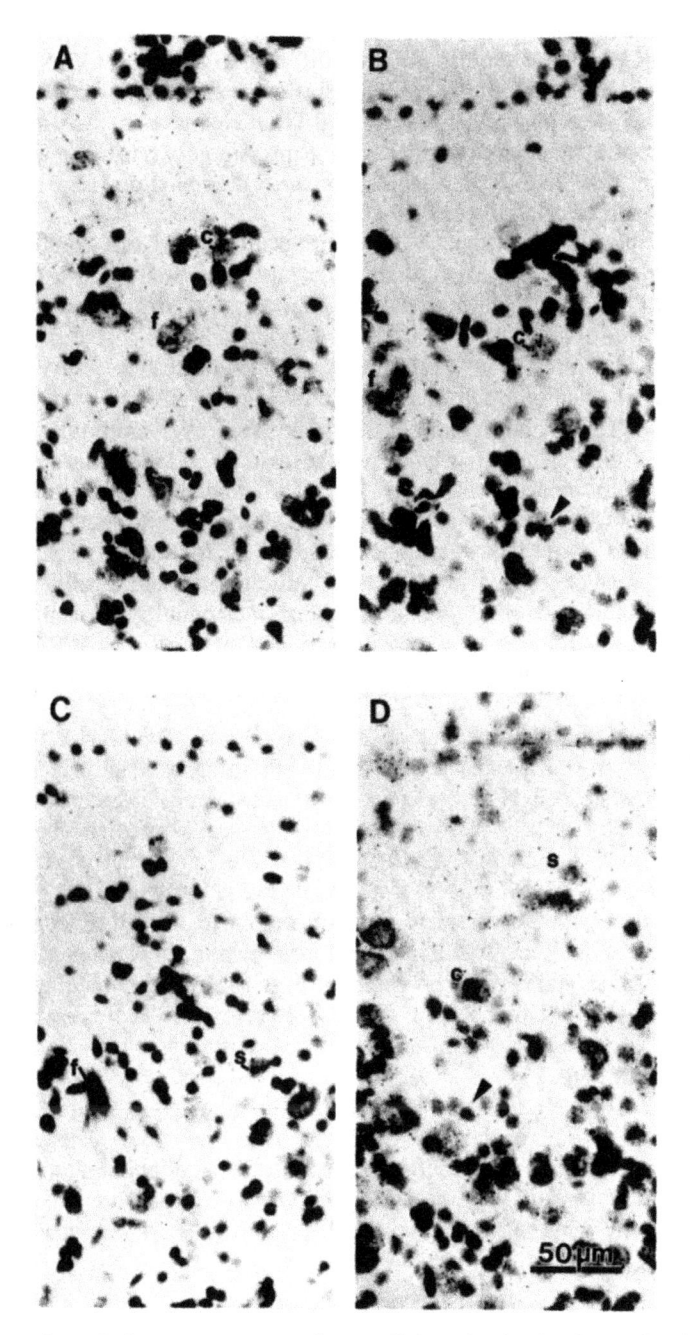

**Figure 2.** Expression of glutamate receptor subtypes GluR1-4 in the DCN. (A) In situ hybridization histochemistry with GluR1 probe leads to accumulation of silver grains over small cells, likely stellate and cartwheel cells in Layer I and superficial Layer II. Fusiform cells are largely unlabeled. (B) Hybridization with GluR2 probe results in silver grains over cartwheel and fusiform cells in Layer I. (C) Hybridization of GluR3 probe is largely absent in cells of the DCN. (D) Moderate accumulations of silver grains are seen over most cell types with GluR4 probe. In B and D, note silver grains over granule cells (arrowheads). Cells are identified as: c, cartwheel cell; s, stellate cell; f, fusiform cell.

Labeling intensities between different probes cannot be directly compared because of their different specific activities.

In the AVCN, expression of GluR2, GluR3 and GluR4 mRNA could be detected in small and large spherical cells with *in situ* hybridization; expression of GluR1 mRNA in these neurons was below detection levels. GluR2 and GluR3 expression was moderate in some spherical cells, while in other spherical cells expression was low or absent; this was especially apparent for GluR3 in the small spherical cell population (Fig. 1). GluR4 mRNA was moderately expressed in all spherical cells.

In the granule cell cap of the AVCN, hybridization with GluR4 probes showed a high density of silver grains. GluR2 transcipts were also observed in the granule cell cap. Hybridization of GluR1 and GluR3 probes, by contrast, was virtually absent in the granule cell cap.

In the PVCN, expression of GluR1 was very low with scattered silver grains observed equally over cells and neuropil. GluR2 expression was generally low to moderate over most cell types in the PVCN; highest grain densities were seen over multipolar and stellate cells present laterally in the PVCN. Lower relative expression of GluR2 was seen in the octopus cell area located ventromedial to the DCN and in the globular cell area present ventrally in the PVCN and AVCN (Harrison and Warr, '62; Harrison and Irving, '66). By contrast, expression of GluR3 transcripts was moderately high in the octopus cell area. Numerous silver grains were seen over globular cell somata; a subset of globular cells, however, showed a slightly reduced number of silver grains. Hybridization to multipolar and stellate cells was comparatively low. GluR4 was expressed in most cell types of the PVCN; no differential density of silver grains could be detected in comparing octopus, multipolar, stellate or globular cells.

The DCN showed a distinctive pattern of expression of GluR1-4. Probes for GluR1 hybridized primarily to mRNA in medium sized cells in layer 1 and superficial regions of layer II, likely cartwheel cells (Fig. 2). Low densities of silver grains were occasionally found over selected fusiform cells; however, most fusiform cells were unlabeled. GluR2 was expressed in the fusiform cells in layer II and in cartwheel and small stellate cells in layers I and II. Low levels of GluR3 mRNA were expressed in cartwheel and stellate cells, and silver grains were observed over a few fusiform cells with the GluR3 probe. GluR4 was abundantly expressed in virtually all identified cells within the DCN. Moderate to dense accumulations of silver grains could be observed over stellate, cartwheel and fusiform cells present in layers I and II. Large cells in the deep layer of the DCN were also labeled with GluR4 probes. Expression of GluR2 and GluR4 transcripts was apparent in granule cells throughout the DCN (Fig. 2).

In summary, in large cells of the VCN, expression of GluR2, GluR3 and GluR4 was moderate to heavy while expression of GluR1 was not seen. Heterogeneity within morphologically defined cell populations may exist since a subpopulation of spherical cells appeared to express less GluR3. In the DCN, GluR2 and GluR4 were abundantly expressed in fusiform neurons, while small cells in the superficial DCN expressed all four GluR subunits.

**Immunocytochemistry**

Several synthetic peptides based on sequences obtained from the cDNA for GluR1-4 were utilized to develop antibodies for immunocytochemical analysis. These antibodies were initially characterized by their immunoprecipitation of solubilized [$^3$H]AMPA binding activity, the selectivity of their recognition of the respective subunits by Western blot analysis of membranes of cultured cells transfected with GluR1,2,3, or 4 cDNAs, and their selectivity on Western blots of brain tissues (Wenthold et al., '90; Wenthold et al., '91; Petralia and Wenthold, '92). Three antibodies which recognize GluR1, GluR2/3, and GluR4 were found to be acceptable for immunocytochemistry. All three antibodies are made to C-terminal pep-

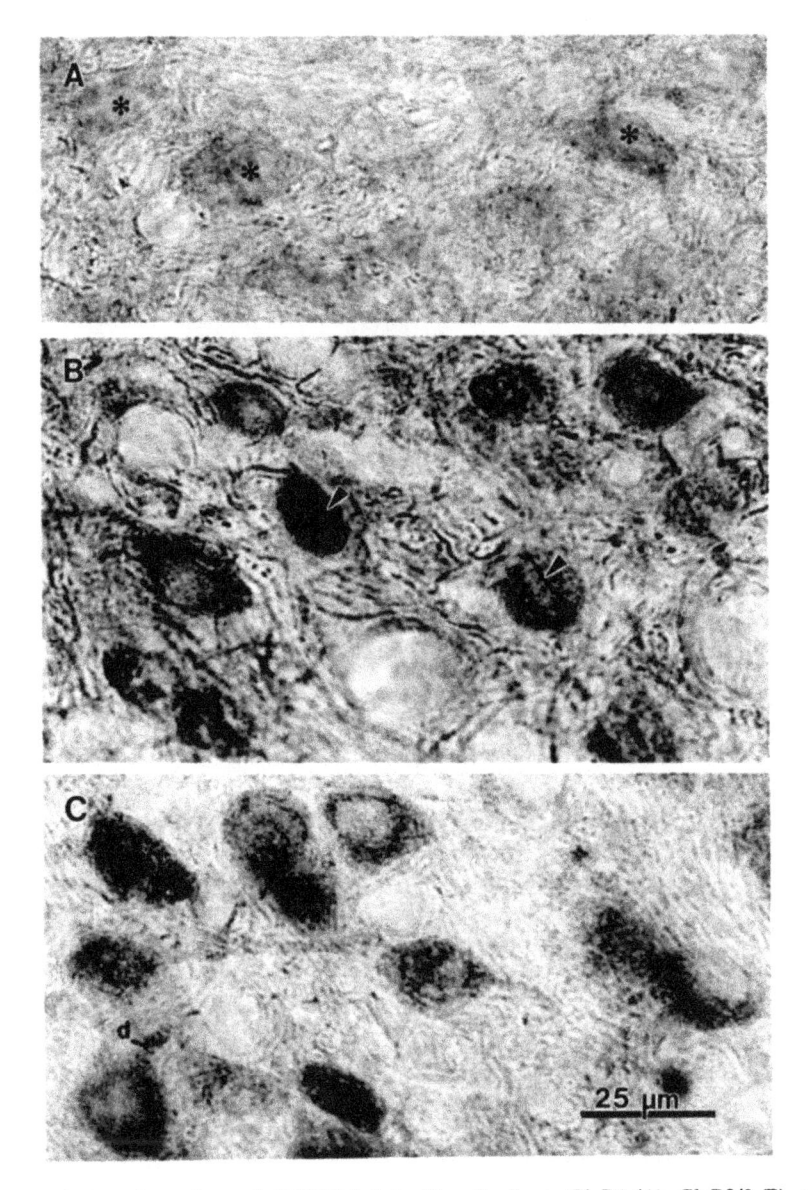

**Figure 3.** Sagittal sections of rostral AVCN labeled with antibodies to GluR1 (A), GluR2/3 (B), and GluR4 (C). Most of the larger cells appear to be spherical bushy cells. Dorsal is at top; rostral is to the left. d, dendrite of probable spherical bushy cell; arrowhead, intranuclear rod; asterisk, cells labeled with antibody to GluR1.

tides of 13 to 14 amino acids, a part of the molecule which is not conserved among the four subunits, except for GluR2 and GluR3, which have identical sequences excepting a single amino acid.

These antibodies were utilized to survey the distribution of GluR1-4 in the rat brain (Wenthold et al., '90; Petralia and Wenthold, '92). Overall distribution of staining in the rat brain with these three antibodies was similar to the binding patterns obtained with *in situ* hybridization histochemistry (Keinänen et al., '90; Bettler et al., '90; Boulter et al., '90; Sommer et al., '90; Pellegrini-Giampietro et al., '91; Gall et al., '90; Monyer et al., '91). In addition, ultrastructural studies of some of the sections of hippocampus and cerebral cortex

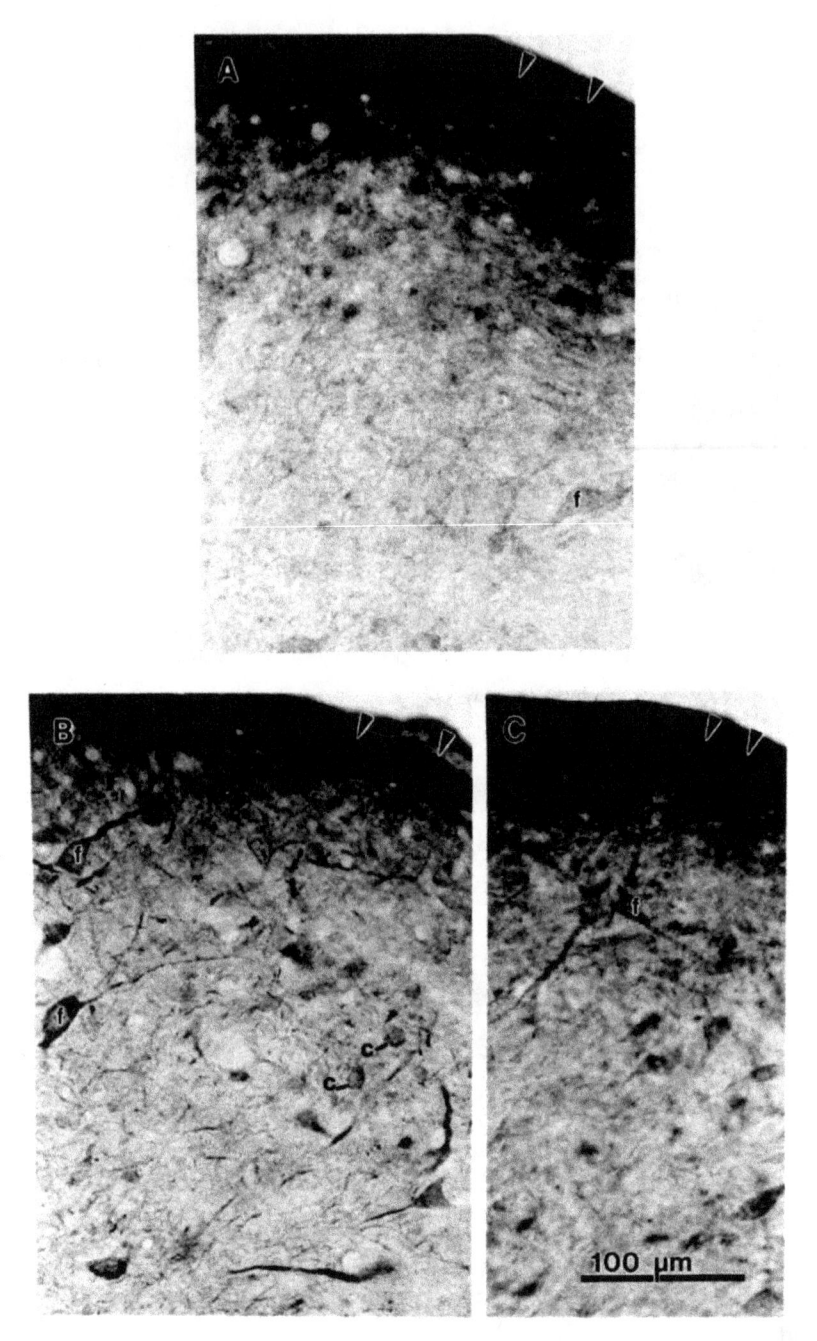

**Figure 4.** Sagittal sections of DCN (in caudal half) labeled with antibodies to GluR1 (A), GluR2/3 (B), and GluR4 (C). Dorsal is at top; rostral is to the left. f, fusiform cell; c, cartwheel-like cell; arrowheads, ependymal surface.

revealed a predominant postsynaptic localization with significant cytoplasmic labeling in dendrites and neuron somata.

The cochlear nuclei were studied in detail utilizing both coronal and sagittal sections prepared for light microscopy as described above. Typically, labeling included staining of neuron cell bodies and major dendrites traced to the soma and/or general neuropil. Staining of the latter included that of both distinct processes and the unresolvable matrix in between. Overall staining of neurons of the VCN was lightest with antibody to GluR1. Probable spherical cells were identified by their round to oval shape with prominent primary dendrite (Harrison and Irving, '65; Saldaña et al., '88). Large spherical cells stained densest with antibody to GluR4 and least with antibody to GluR1 (Fig. 3). Small spherical cells were stained similarly. Distinctive examples of dark puncta within the VCN were sometimes evident on the cell bodies of spherical cells. Few definitive examples of globular cells, which were identified by their eccentric nuclei and dense cytoplasm in the rat (Harrison and Irving, '65) and cat (Osen, '69), were seen. They stained more densely with antibodies to GluR2/3 and GluR4 than with antibody to GluR1. Octopus cells, identified by their characteristic appearance of cell body tipped with long dendrites, and large multipolar neurons stained similarly. In addition, several unidentified types of medium to large neurons throughout the VCN showed a similar overall pattern of staining. Usually staining of smaller cells of the VCN, often of stellate or fusiform structure, was moderately light with all antibodies although slightly less with antibody to GluR1. Granule cells of the VCN were not distinct with any antibodies, probably due largely to their high numerical density and low cytoplasmic content. Neuropil staining was lighter throughout the VCN with antibody to GluR1 than with antibodies to GluR2/3 and GluR4. Many processes including axon-like fibers were stained densely with antibody to GluR2/3.

In contrast, neuropil staining was densest in layers I and II of the DCN with antibody to GluR1 and lightest with antibody to GluR2/3 (Fig. 4). However, the staining pattern of the deeper layers of the DCN was opposite to this, having lightest staining with antibody to GluR1. In general, neuronal staining was light to moderate with all antibodies. The most distinctive neurons were the large fusiform cells of layer II (Fig. 4). Most of these were oriented perpendicular to the ependymal surface although a few were oriented parallel to the surface as described in the cat (Blackstad et al., '84). Typically the long apical and basal dendrites were as well stained as the cell body. Fusiform cell staining varied from light to dense, with lightly staining cells more commonly seen in the caudal DCN. The densest cells were seen with antibodies to GluR2/3 and GluR4. Possible cartwheel cells, identified by their medium-sized, round or oval cell bodies bearing thick, nontapering primary dendrites, were evident in layer I and the boundary of layer II. Staining of these neurons was slightly denser with antibodies to GluR1 and GluR2/3 than with antibody to GluR4. Small stellate neurons of the fusiform type were evident in layer I (Wouterlood et al., '84). They stained moderately with all antibodies, although some denser staining cells were seen with antibody to GluR2/3. Many types of stained, multipolar neurons of layers I and II could not be identified with certainty and probably include stellate neurons and Golgi cells (Mugnaini et al., '80; Wouterlood et al., '84; Wouterlood and Mugnaini, '84). Staining of medium sized multipolar neurons, possibly Golgi cells, of the deeper layers of the DCN was light to moderate with all antibodies. Staining of large multipolar neurons, possibly Giant cells, of the deep layers of the DCN was moderately dense with all antibodies. Occasional granule cells in the DCN were stained lightly with all antibodies. Some of these showed the typical dendritic pattern of this cell type (Mugnaini et al., '80a, '80b; Wouterlood et al., 84).

In addition to the typical pattern of specific staining described above, staining of intranuclear rods (Feldman and Peters, '72; Alvarez-Bolado and Merchán, '88; Dyson et al., '91) was seen with antibody to GluR2/3. These rods were stained best in the large and small spherical cells (Fig. 3), but distinctive examples were evident in octopus, fusiform, cartwheel, granule cells of the DCN, and large multipolar cells of deep layers of the DCN, as well as

**Table I.** GLUR RECEPTOR SUBUNIT *IN SITU* HYBRIDIZATION HISTOCHEMISTRY AND IMMUNOCYTOCHEMISTRY

*IN SITU* HYBRIDIZATION          IMMUNOCYTOCHEMISTRY

|  | GluR1* | GluR2 | GluR3 | GluR4 | GluR1 | GluR2/3 | GluR4 |
|---|---|---|---|---|---|---|---|
| **VENTRAL COCHLEAR NUCLEUS** | | | | | | | |
| Spherical cells | - | + | + | + | +/- | + | + |
| Globular cells | - | + | + | + | + | + | + |
| Octopus cells | - | + | + | + | + | + | + |
| Multipolar/ stellate cells | - | + | + | + | + | + | + |
| Granule cells | - | + | - | + | +/- | +/- | +/- |
| **DORSAL COCHLEAR NUCLEUS** | | | | | | | |
| Fusiform cells | - | + | +/- | + | + | + | + |
| Cartwheel cells/small stellate cells | + | + | + | + | + | + | + |
| Granule cells | - | + | - | + | +/- | +/- | +/- |

+    Selective accumulation of silver grains over cells following autoradiography for *in situ* hybridization or labeling above background with immunohistochemistry.
-    Silver grains present at equally low density over cells and neuropil with *in situ* hybridization or label at background levels with immunocytochemistry.
+/-    Silver grains present over a subpopulation of neurons with *in situ* hybridization or immunolabeling that was not distinct.
*    Hybridization with GluR1 probe resulted in a low accumulation of silver grains over selected large cells of the VCN and fusiform cells of the DCN.

in many kinds of unidentified neurons of the VCN. No rods were identified with antibodies to GluR1 and GluR4. Probably the immunolabeling of intranuclear rods is due to a cross reactivity of the antibody with a protein containing an identical epitope. In addition, the nucleus minus the nucleolus was stained in a few neurons with antibody to GluR4. The latter neurons were considered to be nonspecifically stained.

In summary, neuronal staining was lightest overall and neuropilar staining was lightest in the VCN and deep layers of the DCN with antibody to GluR1. However, neuropilar staining of DCN layers I and II was densest with antibody to GluR1.

## Conclusions

Results of the *in situ* hybridization and immunocytochemistry analyses show that EAA receptors of the GluR family are widely distributed in the CN and suggest that these receptors play a major role in CN synaptic function. Table I summarizes our findings on the distribution of GluR subunit mRNAs and immunoreactivity for selected cell types of the rat CN. In general results from the two studies closely agree. One exception is the labeling of VCN neurons with GluR1 probes. Immunocytochemistry shows light to moderate labeling of large neurons in the VCN while *in situ* hybridization shows very little labeling. An explanation may be that the relatively low level of GluR1 mRNA present in these cells is sufficient to produce the immunolabeling. However, it is possible that additional related members of the GluR family exist and these may be differentially recognized by the antibodies or the *in situ* probes. As discussed above several genes related to GluR have been cloned, and numerous additional related genes, as well as alternatively spliced variants, are expected to be identified.

From these results we can propose three synaptic populations which contain GluRs, and therefore, are likely to use an EAA as a neurotransmitter. These include the auditory nerve/principal cell synapse, the parallel fiber/cartwheel cell synapse, and a synaptic population on granule cells. Neurons receiving auditory nerve input are immunoreactive with GluR antibodies and are heavily labeled with probes for GluR mRNA. Since these neurons form synapses with other inputs, where these receptors may be targeted, analysis of immunoreactivity at the EM level is required to definitively associate GluR with the auditory nerve synapse. Preliminary EM analyses show extensive GluR immunoreactivity postsynaptic to auditory nerve terminals (Petralia and Wenthold, unpublished observation). These results support the extensive data showing that an EAA is the auditory nerve neurotransmitter, as reviewed above. Our results also support previous findings that the parallel fiber pathway uses an excitatory amino acid and suggest that there is an EAA input to granule cells. The origins of inputs to granule cells are not fully known but several sources have been proposed including collaterals of the olivocochlear bundle, neurons of the CN, and type II spiral ganglion cells (Mugnaini et al., '80a, '80b; Benson and Ryugo, '87; Brown et al., '88). In addition to these three synaptic populations, it is likely that GluR is localized at other populations of synapses given the widespread distribution of GluR immunoreactivity and mRNA in the CN.

*In situ* hybridization studies done throughout the central nervous system show a diversity in the expression of GluR subunit mRNAs suggesting that the subunit composition of the functional receptor complex is variable among neuronal populations. Our results on the three GluR-containing synaptic populations in the CN also indicate different subunit compositions. Neurons receiving auditory nerve input express primarily GluR2, GluR3 and GluR4. Those receiving parallel fiber input express all four subunits, and granule cells express GluR2 and GluR4. Since different GluR combinations expressed *in vitro* produce functionally different receptors, GluR on principal neurons, cartwheel cells and granule cells would be expected to have different channel or pharmacological properties. In addition to the different populations of neurons expressing different populations of GluR mRNAs, neurons receiving apparently the same presynaptic input may also express different populations of GluR. For example, our preliminary results show that although most principal neurons in the AVCN express GluR3, some produce relatively little or no GluR3 mRNA. GluR3 production in these neurons does not appear to be strictly divided between morphologically identified cell types. While the functional ramifications remain to be determined, such results suggest that the presynaptic input alone does not control the makeup of the postsynaptic receptor. Alternatively, there may be subtle differences in auditory nerve fibers that have not yet been identified.

ACKNOWLEDGEMENTS

We thank Drs. Keiji Wada and Masayoshi Tachibana for critically reading this manuscript.

REFERENCES

Aitkin, L.M., 1989, The auditory system, in: "Handbook of Chemical Neuroanatomy, Vol. 7: Integrated Systems of the CNS, Part II," A. Björklund, T. Hökfelt and L.W. Swanson, Eds., Elsevier Science Publishers B.V., Amsterdam, pp.165-218.

Altschuler, R.A., Wenthold, R.J., Schwartz, A.M., W.G. Haser, W.G., Curthoys, N.P., Parakkal, M. and Fex, J., 1984, Immunocytochemical localization of glutaminase-like immunoreactivity in the auditory nerve, Brain Res., 291:173.

Altschuler, R.A., Hoffman, D.W. and Wenthold, R.J., 1986, Neurotransmitters of the cochlea and cochlear nucleus: Immunocytochemical evidence, Am. J. Otolaryngol., 7:100.

Alvarez-Bolado, G. and Merchán, J., 1988, Synaptic endfeet in the 'acoustic nerve nucleus' of the rat. An electron microscope study, J. Anat., 159:19.

Benson, T.E. and Ryugo, D.R., 1987, Axons of presumptive type-II spiral ganglion neurons synapse with granule cells of the cat cochlear nucleus, Soc. Neurosci. Abstr. 13:1258.

Bettler, B., Boulter, J., Hermans-Borgmeyer, I., O'Shea-Greenfield, A., Deneris, E.S., Moll, C., Borgmyer, U., Hollmann, M. and Heinemann, S., 1990, Cloning of a novel glutamate receptor subunit, GluR5: Expression in the nervous system during development, Neuron, 5:583.

Blackstad, T.W., Osen, K.K. and Mugnaini, E., 1984, Pyramidal neurones of the dorsal cochlear nucleus: A Golgi and computer reconstruction study in cat, Neuroscience, 13:827.

Boulter, J., Hollmann, M., O'Shea-Greenfield, A., Hartley, M., Deneris, E., Maron, C. and Heinemann, S., 1990, Molecular cloning and functional expression of glutamate receptor genes, Science, 249:1033.

Brown, M.C., Berglund, A.M., Kiang, N.Y.S. and Ryugo, D.K., 1988, Central trajectories of type II spiral ganglion neurons, J. Comp. Neurol. 278:581.

Cant, N.B. and Morest, D.K., 1984, The structural basis for stimulus coding in the cochlear nucleus of the cat, in: "Hearing Science: Recent Advances," C.I. Berlin, ed., College-Hill Press, San Diego, pp.371-421.

Caspary, D.M., Rybak, L.P. and Faingold, C.L., 1985, The effects of inhibitory and excitatory neurotransmission on the response properties of brainstem auditory neurons, in: "Auditory Biochemistry," D.G. Drescher, ed., Charles C. Thomas, Springfield, pp.198-226.

Choi, D.W., 1987, Ionic dependence of glutamate neurotoxicity, J. Neurosci., 7:369.

Cline, H.T. and Constantine-Paton, M., 1989, NMDA receptor antagonists disrupt the retinotectal map, Neuron, 3:413.

Dawson, T.L., Nicholas, R.A. and Dingledine, R., 1990, Homomeric GluR1 excitatory amino acid receptors expressed in Xenopus Oocytes, Mol. Pharmacol., 38:779.

Dyson, S.E., Warton, S.S. and Cockman, B., 1991, Volumetric and histological changes in the cochlear nuclei of visually deprived rats: A possible morphological basis for intermodal sensory compensation, J. Comp. Neurol., 307:39.

Egebjerg, J., Bettler, B., Hermans-Borgmeyer, I. and Heinemann, S., 1991, Cloning of a cDNA for a glutamate receptor subunit activated by kainate but not AMPA, Nature, 351:745.

Feldman, M.L. and Peters, A., 1972, Intranuclear rods and sheets in rat cochlear nucleus, J. Neurocytol., 1:109.

Foster, A. and Fagg, G.E., 1984, Acidic amino acid binding sites in mammalian neuronal membranes: Their characteristics and relationship to synaptic receptors, Brain Res. Rev., 7:103.

Gall, C., Sumikawa, K. and Lynch, G., 1990, Levels of mRNA for a putative kainate receptor are affected by seizures, Proc. Natl. Acad. Sci. U.S.A., 87:7643.

Garthwaite, G., Hajow, F. and Garthwaite, J., 1986, Ionic requirements for neurotoxic effects of excitatory amino acid analogues in rat cerebellar slices, Neuroscience, 18:437.

Greenamyre, T.J., Young, A.B. and Penney, J.B., 1984, Quantitative autoradiographic distribution of L-[$^3$H]glutamate-binding sites in rat central nervous system, J. Neurosci., 4:2133.

Gregor, P., Mano, I., Maoz, I., McKeown, M. and Teichberg, V.I., 1989, Molecular structure of the chick cerebellar kainate binding subunit of a putative glutamate receptor, Nature, 342:689.

Halpain, S., Wieczorek, C.M. and Rainbow, T.C., 1984, Localization of L-glutamate receptors in rat brain by quantitative autoradiography, J. Neurosci., 4:2247.

Harrison, J.M. and Warr, W.B., 1962, A study of the, cochlear nuclei and ascending auditory pathways of the medulla, J. Comp. Neurol., 119:341.

Harrison, J.M. and Irving, R., 1965, The anterior ventral cochlear nucleus, J. Comp. Neurol., 124:15.

Harrison, J.M. and Irving, R., 1966, The organization of the posterior ventral cochlear nucleus in the rat, J. Comp. Neurol., 126:391.

Hirsch, J.A. and Oertel, D., 1988, Synaptic connections in the dorsal cochlear nucleus of mice, in vitro, J. Physiol., 396:549.

Hollmann, M., O'Shea-Greenfield, A., Rogers, S.W. and Heinemann, S., 1989, Cloning by functional expression of a member of the glutamate receptor family, Nature, 342:643.

Hollmann, M., Hartley, M., and Heinemann, S., 1991, Ca$^{2+}$ permeability of KA-AMPA-gated glutamate receptor channels depends on subunit composition, *Science*, 252:851.

Houamed, K.M., Kuijper, J.L., Gilbert, T.L., Haldeman, B.A., O'Hara, P.J., Mulvihill, E.R., Almers, W., and Hagen, F.S., 1991, Cloning, expression, and gene structure of a G protein-coupled glutamate receptor from rat brain, *Science*, 252:1318.

Jackson, H., Nemeth, E.F.,and Parks, T.N., 1985, Non-N-methyl-D-aspartate receptors mediating synaptic transmission in the avian cochlear nucleus: Effects of kynurenic acid, dipicolinic acid and streptomycin, *Neuroscience*, 16:171.

Keinänen, K., Wisden, W., Sommer, B., Werner, P., Herb, A., Verdoorn, T.A., Sakmann, B., and Seeburg, P.H., 1990, A family of AMPA-selective glutamate receptors, *Science*, 249:556.

Manis, P.B., 1989, Responses to parallel fiber stimulation in the guinea pig dorsal cochlear nucleus in vitro, *J. Neurophysiol.*, 61:149.

Martin, M.R., 1980, The effects of iontophoretically-applied antagonists on auditory nerve and amino acid-evoked excitation on the anteroventral cochlear nucleus neurons, *Neuropharmacology*, 19:519.

Masu, M., Tanabe, Y., Tsuchida, K., Shigemoto, R., and Nakanishi, S., 1991, Sequence and expression of a metabotropic glutamate receptor, *Nature*, 349:760.

Mayer, M.L., and Westbrook, G.L., 1987, The physiology of excitatory amino acids in the vertebrate central nervous system, *Prog. Neurobiol.*, 28:197.

Monaghan, D.T., and Cotman, C.W., 1982, The distribution of [$^3$H]kainic acid binding sites in rat CNS as determined by autoradiography, *Brain Res.*, 252:91.

Monaghan, D.R., and Cotman, C.W., 1985, Distribution of N-methyl-D-aspartate-sensitive L-[3H]glutamate-binding sites in rat brain, *J. Neurosci.*, 5:2902.

Monaghan, D.T., Bridge, R.J., and Cotman, C.W., 1989, The excitatory amino acid receptors, *Annu. Rev. Pharmacol. Toxicol.*, 29:365.

Monyer, H., Seeburg, P.H., and Wisden, W., 1991, Glutamate-operated channels: Developmentally early and mature forms arise by alternative splicing, *Neuron*, 6:799.

Mugnaini, E., 1985, GABA neurons in the superficial layers of the rat dorsal cochlear nucleus: Light and electron microscopic immunocytochemistry, *J. Comp. Neurol.* 235:61.

Mugnaini, E., Osen, K.K., Dahl, A.L., Friedrich Jr., V.L., and Korte, G., 1980a, Fine structure of granule cells and related interneurons (termed Golgi cells) in the cochlear nuclear complex of cat, rat, and mouse, *J. Neurocytol.*, 9:537.

Mugnaini, E., Warr, W.B.,and Osen, K.K., 1980b, Distribution and light microscopic features of granule cells in the cochlear nuclei of cat, rat, and mouse, J. *Comp. Neurol.*, 191:581.

Nakanishi, N., Shneider, N.A.,and Axel, R., 1990, A family of glutamate receptor genes: Evidence for the formation of heteromultimeric receptors with distinct channel properties, *Neuron*, 5:569.

Nemeth, E.F., Jackson, H., and Parks, T.N., 1985, Evidence for the involvement of kainate receptors in synaptic transmission in the avian cochlear nucleus, *Neurosci. Lett.*, 59:297.

Oliver, D.L., Potashner, S.J., Jones, D.R., and Morest, D.K., 1983, Selective labeling of spiral ganglion and granule cells with D-aspartate in the auditory system of cat and guinea pig, *J. Neurosci.*, 3:455.

Olney, J.W., Price, M.T., Samson, L., and Labruyere, L., 1986, The role of specific ions in glutamate neurotoxicity, *Neurosci. Lett.*, 65:65.

Osen, K.K., 1969, Cytoarchitecture of the cochlear nuclei in the cat, *J. Comp. Neurol.* 136:453.

Ottersen, O.P., 1989, Quantitative electron microscopic immunocytochemistry of neuroactive amino acids, *Anat. Embryol.*, 180:1.

Pellegrini-Giampietro, D.E., Bennett, M.V.L., and Zukin, R.S., 1991, Differential expression of three glutamate receptor genes in developing rat brain: An in situ hybridization study, *Proc. Natl. Acad. Sci. U.S.A.*, 88:4157.

Petralia, R.S., and Wenthold, R.J., 1992, Light and electron immunocytochemical localization of AMPA-selective glutamate receptors in the rat brain, *J. Comp. Neurol.*, 318:329.

Potashner, S.J., 1983, Uptake and release of D-aspartate in the guinea pig cochlear nucleus, *J. Neurochem.*, 41:1094.

Saldaña, E., Carro, J., Merchan, M., and Collia, F., 1988, Morphometric and cytoarchitectural study of the different neuronal types in the VCN of the rat, *in:* "Auditory Pathway: Structure and Function," J. Syka and R.B. Masterson, eds., Plenum Press, New York, pp.89-93.

Sommer, B., Keinänen, K., Verdoorn, T.A., Wisden, W., Burnashev, N., Herb, A., Köhler, M., Takagi, T., Sakmann, B., and Seeburg, P.H., 1990, Flip and flop: A cell-specific functional switch in glutamate-operated channels of the CNS, *Science*, 249:1580.

Staatz-Benson C., and Potashner, S.J., 1988, Uptake and release of glycine in the guinea pig cochlear nucleus after axotomy of afferent or centrifugal fibers, *J. Neurochem.*, 51:370.

Verdoorn, T.A., Burnashev, N., Monyer, H., Seeburg, P.H., and Sakmann, B., 1991, Structural determinants of ion flow through recombinant glutamate receptor channels, *Science*, 252:1715.

Wada, K., Dechesne, C.J., Shimasaki, S., King, R.G., Kusano, K., Buonanno, A., Hampson, D.R., Banner, C., Wenthold, R.J. and Nakatani, Y., 1989, Sequence and expression of a frog brain complementary DNA encoding a kainate-binding protein, *Nature,* 342:684.

Watkins, J.C. and Evans, R.H., 1981, Excitatory amino acid transmitters, *Annu. Rev. Pharmacol. Toxicol.,* 21:165.

Webster, W.R., 1985, Auditory system, *in*: "The Rat Nervous System," G. Paxinos, ed., Academic Press, New York, pp. 153-184.

Wenthold, R.J. and Martin, M.R., 1984, Neurotransmitters of the auditory nerve and central auditory system, *in*: "Hearing Science: Recent Advances," C. Berlin, ed., College Hill Press, San Diego, pp.341-369.

Wenthold, R.J., Hunter, C., Wada, K. and Dechesne, C.J., 1990, Antibodies to a C-terminal peptide of the rat glutamate receptor subunit, GluR-A, recognize a subpopulation of AMPA binding sites but not kainate sites, *FEBS Lett.* 276:147.

Wenthold, R.J., Yokotani, N., Doi, K. and Wada, K., 1992, Immunochemical characterization of the non-NMDA glutamate receptor using subunit-specific antibodies: Evidence for a hetero-oligomeric structure in rat brain, *J. Biol. Chem.,* 267:501.

Werner, P., Voigt, M., Keinänen, K., Wisden, W. and Seeburg, P.H., 1991, Cloning of a putative high-affinity kainate receptor expressed predominantly in hippocampal CA3 cells, *Nature,* 351:742.

Wickesberg, R.E. and Oertel, D., 1989, Auditory nerve neurotransmitter acts on a kainate receptor: evidence from intracellular recordings in brain slices from mice, *Brain Res.,* 486:39.

Wouterlood, F.G., Mugnaini, E., Osen, K.K. and Dahl, A.L., 1984, Stellate neurons in rat dorsal cochlear nuclear studied with combined Golgi impregnation and electron microscopy: synaptic connections and mutual coupling by gap junctions, *J. Neurocytol.,* 13:639.

Wouterlood, F.G. and Mugnaini, E., 1984, Cartwheel neurons of the dorsal cochlear nucleus: A Golgi-electron microscopic study in rat, *J. Comp. Neurol.,* 227:136.

# GLYCINE AND GABA: TRANSMITTER CANDIDATES OF PROJECTIONS DESCENDING TO THE COCHLEAR NUCLEUS

Steven J. Potashner, Christina G. Benson, E.-Michael Ostapoff,
Nancy Lindberg and D.Kent Morest

Department of Anatomy, University of Connecticut Health Center
Farmington, CT, 06030, USA

Acoustic information, encoded in the cochlea and conveyed to the cochlear nucleus (CN) by cochlear nerve fibers, is processed by cell groups in the CN. Inhibitory neurotransmission appears to play a prominent role at this level of auditory processing (Brugge and Geisler, '78; Voight and Young, '80; Caspary et al., this volume). Information has been emerging recently with regard to the location and transmitters of the inhibitory neurons which synapse in the CN. These neurons may originate in other brain stem nuclei that project to the CN, or could lie within the CN itself (Saint Marie et al., '91, this volume; Oertel and Wickesberg, this volume). These inhibitory projections probably use the amino acid transmitters, glycine and GABA at their synapses in the CN (Whitfield and Comis, '66; Tachibana and Kuriyama, '74; Fex and Wenthold, '76; Fisher and Davies, '76; Godfrey et al, '77, '78; Caspary et al, '79; Wenthold, '79; Martin et al, '82).

## Evidence supporting glycine and GABA as transmitter candidates in the cochlear nucleus

If glycinergic or GABAergic neurons make synapses in the CN, their presynaptic endings should synthesize, store, release, and inactivate glycine or GABA (Werman '66; Orrego, '79). Post-synaptic receptors for these compounds should be present on CN neurons. A variety of studies have demonstrated that high concentrations of glycine and GABA are stored in the CN (Godfrey et al., '77, '78) and many synaptic endings in the ventral CN contain the enzyme which synthesizes GABA (Saint Marie et al., '89). In addition, there is evidence that the CN contains synaptic receptors for glycine and GABA (Martin et al., '82; Caspary et al., '79; Zarbin et al., '81; Frostholm and Rotter, '85; Wenthold et al., '85; Caspary et al, this volume; Oertel and Wickesberg, this volume). We undertook studies to determine if elements in the CN could mediate the synaptic release and the inactivation of synaptically released glycine and GABA.

The inactivation of synaptically released glycine is thought to be mediated by a high affinity uptake mechanism. To determine if the CN contains this transporter, we performed a kinetic analysis of $^3$H-glycine uptake (Fig. 1). The data indicate that the CN contains two uptake mechanisms. One has an apparent $K_m$ of 633 - 718 $\mu$M and is designated *the*

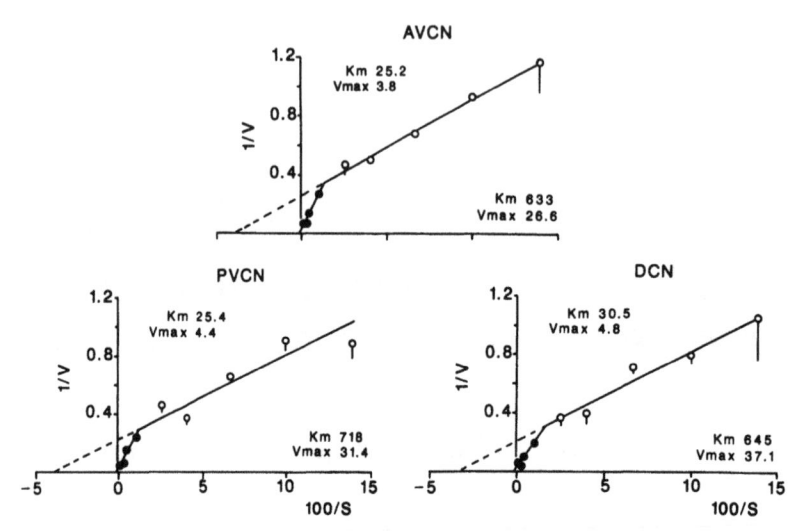

**Figure 1.** Lineweaver-Burk analysis of $^3$H-glycine uptake by subdivisions of the guinea pig cochlear nucleus. Dissected subdivisions were sliced and incubated with $^3$H-glycine. After 5 min, slices were separated from the medium, washed, and homogenized in 85% ethanol. The radioactivity in the homogenate was measured by liquid scintillation spectrometry. The amount of $^3$H-glycine in the medium was between 7 $\mu$M and 1 mM and was expressed as a substrate concentration, or 'S' value. The uptake was expressed as a velocity, or 'V' value, computed as the $^3$H-glycine accumulated in nmols/10 mg cell water/5 min. The points are means ± SEM for 3 - 8 determinations. The equations of the fitted lines were used to compute the $K_m$ and $V_{max}$ for each uptake mechanism. AVCN, anteroventral cochlear nucleus; PVCN, posteroventral cochlear nucleus; DCN, dorsal cochlear nucleus. Reproduced from Staatz-Benson and Potashner ('87).

'*low-affinity*' mechanism. It has a relatively large capacity, with a $V_{max}$ of 26.6 - 37.1 nmols/10 mg cell water/5 min, and is typical of transporters for small neutral amino acids found in many tissues, including the central nervous system (Neame, '68; Blasberg and Lajtha, '65; Lajtha, '67; Blasberg, '68; Logan and Snyder, '72; Davidoff and Adair, '76; Gundlach and Beart, '82). Such transporters probably help to maintain cellular levels of free amino acids useful for protein synthesis and intermediary metabolism. The other uptake mechanism has an apparent $K_m$ of 25.2 - 30.5 $\mu$M and is designated *the 'high-affinity' mechanism*. It has properties like those of transporters found in other parts of the central nervous system where glycine may be a transmitter. Consistent with the observations of others (Neal and Pickles, '69; Johnston and Iversen, '71; Logan and Snyder, '72; Davidoff and Adair, '76; Gundlach and Beart, '82), the mechanism in the CN has an apparent $K_m$ in the 10 - 30 $\mu$M range, is a relatively low capacity system with a $V_{max}$ of 3.8 - 4.8 nmols/10 mg cell water/5 min, is dependent on energy and on the extracellular concentration of Na$^+$, and has a high degree of substrate specificity. Uptake mechanisms with properties such as these are not ubiquitous, but appear to be present in axonal endings and glia of central nervous tissues where glycine may be a transmitter (Neal and Pickles, '69; Hokfelt and Ljungdahl, '71; Johnston and Iversen, '71; Matus and Dennison, '71; Iversen and Bloom, '72; Logan and Snyder, '72; Young and Macdonald, '83).

To determine if the CN contains mechanisms that mediate the synaptic release of glycine and GABA, CN tissues first are allowed to take up 3.4 $\mu$M $^{14}$C-glycine or 1.9 $\mu$M $^{14}$C-GABA, concentrations which limit uptake to that mediated by the appropriate high affinity transporter. Tissues are then superfused with fresh amino acid-free medium, which is collected in fractions. During this superfusion, there is a slow spontaneous loss of $^{14}$C-glycine (Fig. 2A) and $^{14}$C-GABA (Fig. 7) to the superfusion medium. Electrical field stimulation of the tissues for 4 minutes evokes a transient, initially rapid increase in the amount of $^{14}$C-glycine (Fig. 2A) and $^{14}$C-GABA (Fig. 7) released into the medium. This electrically evoked

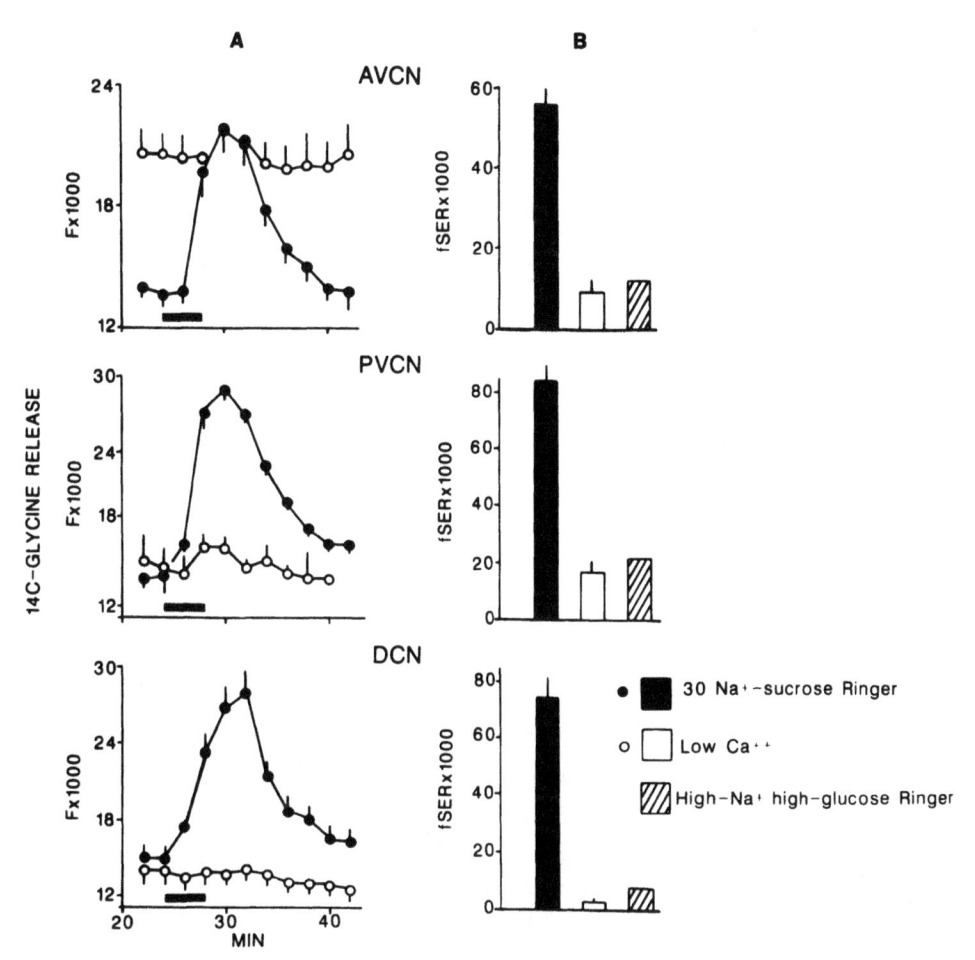

**Figure 2.** Release of $^{14}$C-glycine from subdivisions of the guinea pig cochlear nucleus. Dissected subdivisions took up 3.4 μM $^{14}$C-glycine before being superfused with glycine-free medium, which was collected in fractions. 4 min. of electrical field stimulation (black horizontal bars in A) was applied to evoke release (Potashner '78; Oliver et al., '83). After the superfusion, tissues were homogenized in 85% ethanol. The radioactivity in these extracts, and in each collected superfusate fraction, was measured by liquid scintillation spectrometry. The time course of the spontaneous and electrically evoked release is illustrated in A. Points are means ± SEM of the 'F' value, the fraction of the tissue radioactivity released to the medium per min. The histogram in B illustrates the $^{14}$C-glycine released in response to electrical stimulation, plotted as the mean ± SEM of the 'fSER' value, the fractional stimulus evoked release (see Staatz-Benson and Potashner, '87, for computation of the fSER value). The ordinates for the AVCN, PVCN, and DCN in this and in subsequent release and uptake figures may have different maxima. Control release was measured in a Low Na$^+$ - Low Glucose medium (n = 12 - 14), in a High Na$^+$ - High Glucose medium (n = 2), and in Low Ca$^{++}$ medium where the 1.3 mM CaCl$_2$ normally present was replaced with 0.01 mM CaCl$_2$ and 20 mM MgCl$_2$ (n = 6 - 8). Reproduced from Staatz-Benson and Potashner ('87). Abbreviations as in Fig. 1.

release is expressed quantitatively as the fSER value and plotted as bars in Figs. 2B and 7. Since the presence of Ca$^{++}$ ions in the extracellular space is considered necessary for the synaptic release of transmitters (Rubin, '74), release is compared in 'Control' and in 'Low Ca$^{++}$' media, to determine the fraction of the electrically evoked release originating in synaptic endings. Calcium deprivation reduces the electrically evoked release of $^{14}$C-glycine by 84 - 94% (Fig. 2A & B) and of $^{14}$C-GABA by 74 - 94% (Fig. 7). These findings suggest that the CN contains mechanisms that mediate the synaptic release of glycine and GABA.

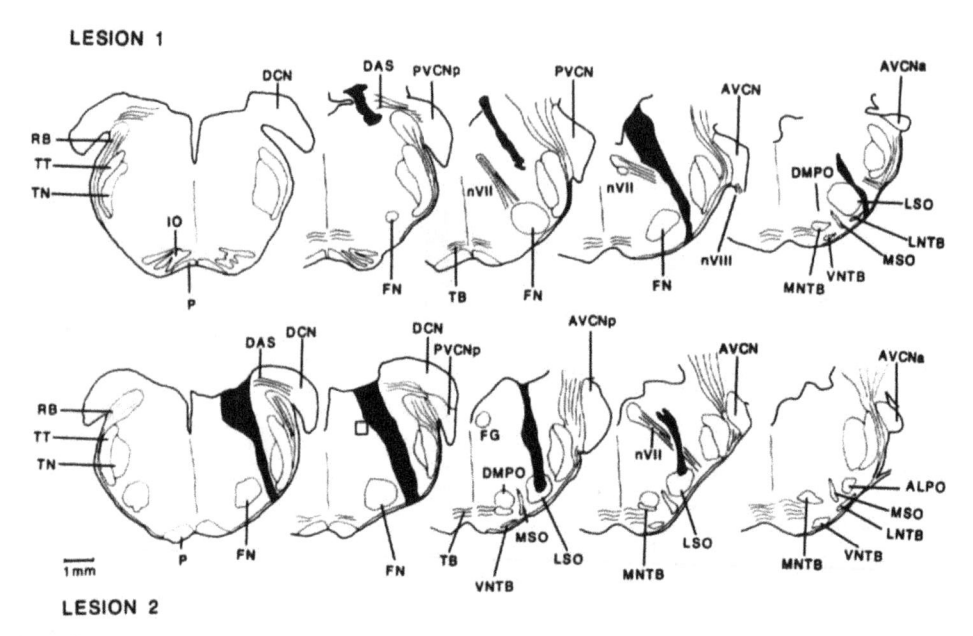

**Figure 3.** Unilateral lesions of the acoustic striae and trapezoid body in the guinea pig. The upper panel contains drawings of a representative series of transverse sections from one case showing the extent of lesion 1, the lower panel illustrates lesion 2. The most caudal section is on the left. Sections were stained with toluidine blue or cresyl violet. ALPO, anterolateral periolivary nucleus; AVCN, anteroventral cochlear nucleus; AVCNa, anterior part of the anteroventral cochlear nucleus; AVCNp, posterior part of the anteroventral cochlear nucleus; DAS, dorsal acoustic stria; DCN, dorsal cochlear nucleus; DMPO, dorsomedial periolivary nucleus; FG, genu of the facial nerve; FN, facial nucleus; IO, inferior olive; LNTB, lateral nucleus of the trapezoid body; LSO, lateral superior olive; MNTB, medial nucleus of the trapezoid body; MSO, medial superior olive; nVII, facial nerve root; nVIII, auditory nerve root; P, pyramidal tract; PVCN, posteroventral cochlear nucleus; PVCNp, posterior part of the posteroventral cochlear nucleus; RB, restiform body; TB, trapezoid body; TN, descending trigeminal nucleus; TT, descending trigeminal tract; VNTB, ventral nucleus of the trapezoid body. Reproduced from Staatz-Benson and Potashner ('88).

These mechanisms are activated by depolarizing electrical stimuli, require the presence of extracellular $Ca^{++}$, and mediate a transient, initially rapid release to the medium of a portion of the glycine or GABA accumulated by high-affinity uptake. The extensive $Ca^{++}$-dependence of these activities suggests that these electrically evoked releases come from synaptic endings (Blaustein et al., '72; Rubin, '74; Sellstrom and Hamberger, '77; Potashner, '78).

If the high affinity uptake mechanism is responsible for inactivating synaptically released transmitter, the uptake should capture the synaptically released transmitter, preventing it from leaving the tissue. The glycine release experiments described above were conducted in 'Low $Na^+$ - Low Glucose' media, conditions which minimize glycine uptake so that synaptically released glycine can leave the tissue and be detected in the superfusion medium. However, if the release is measured in 'High $Na^+$ - High Glucose' medium, conditions which produce maximal rates of glycine uptake (Staatz-Benson and Potashner, '87), the amount of glycine released from the tissue by electrical stimulation is reduced by 77 - 89% (Fig. 2B). Thus, when allowed to function at maximal rates, the high affinity uptake mechanism can capture amounts of glycine practically equivalent to the electrically evoked, $Ca^{++}$-dependent release of glycine.

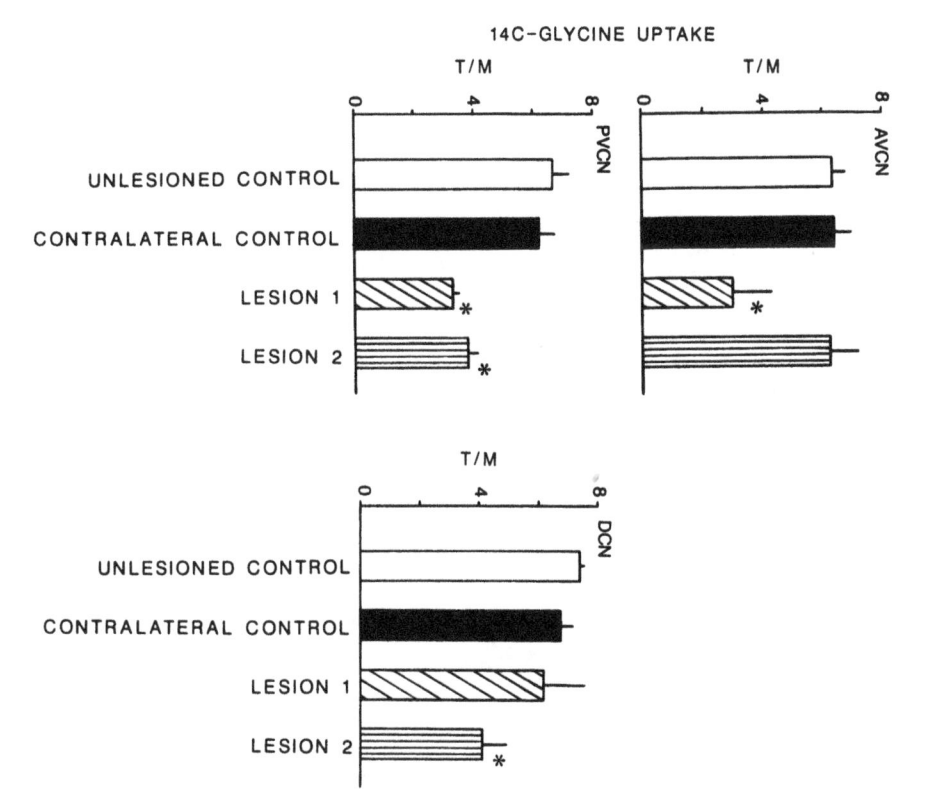

Figure 4. Effect of unilateral acoustic striae lesions on the uptake of $^{14}$C-glycine by subdivisions of the guinea pig cochlear nucleus. Dissected subdivisions were treated as described in Fig. 2. The uptake of $^{14}$C-glycine was estimated as in Potashner and Tran ('84). The uptake of $^{14}$C-glycine was expressed as the tissue/medium ratio (T/M): the ratio of radioactivity in the cellular water of the tissue to that in an equivalent volume of medium. The data plotted are means + SEM from both the left and right CN of 5 - 8 unlesioned animals (unlesioned control), from the right CN of 3 - 4 animals that received lesion 1, and from the right CN of 5 animals that received lesion 2. Values representing the left CN of animals with Lesion 1 or 2 were not significantly different from each other and were therefore pooled (contralateral control). Asterisks denote a significant difference (P < 0.05) from the unlesioned control. Reproduced from Staatz-Benson and Potashner ('88). Abbreviations as in Fig. 1.

# Evidence that glycinergic and GABAergic terminals in the cochlear nucleus are provided by centrifugal projections and by intrinsic neurons

The findings above support the hypothesis that the CN contains glycinergic and GABAergic synaptic endings. Terminals such as these might be contributed by neurons intrinsic to the CN or by extrinsic cells projecting to the CN from other brain stem auditory nuclei. If any of the extrinsic projections use glycine or GABA as a transmitter, the destruction of the projection should depress the release and high affinity uptake of the transmitter in the CN. To test this prediction, we severed the acoustic striae and trapezoid body in the brain stem, thus destroying centrifugal projections to the CN. Using an un-

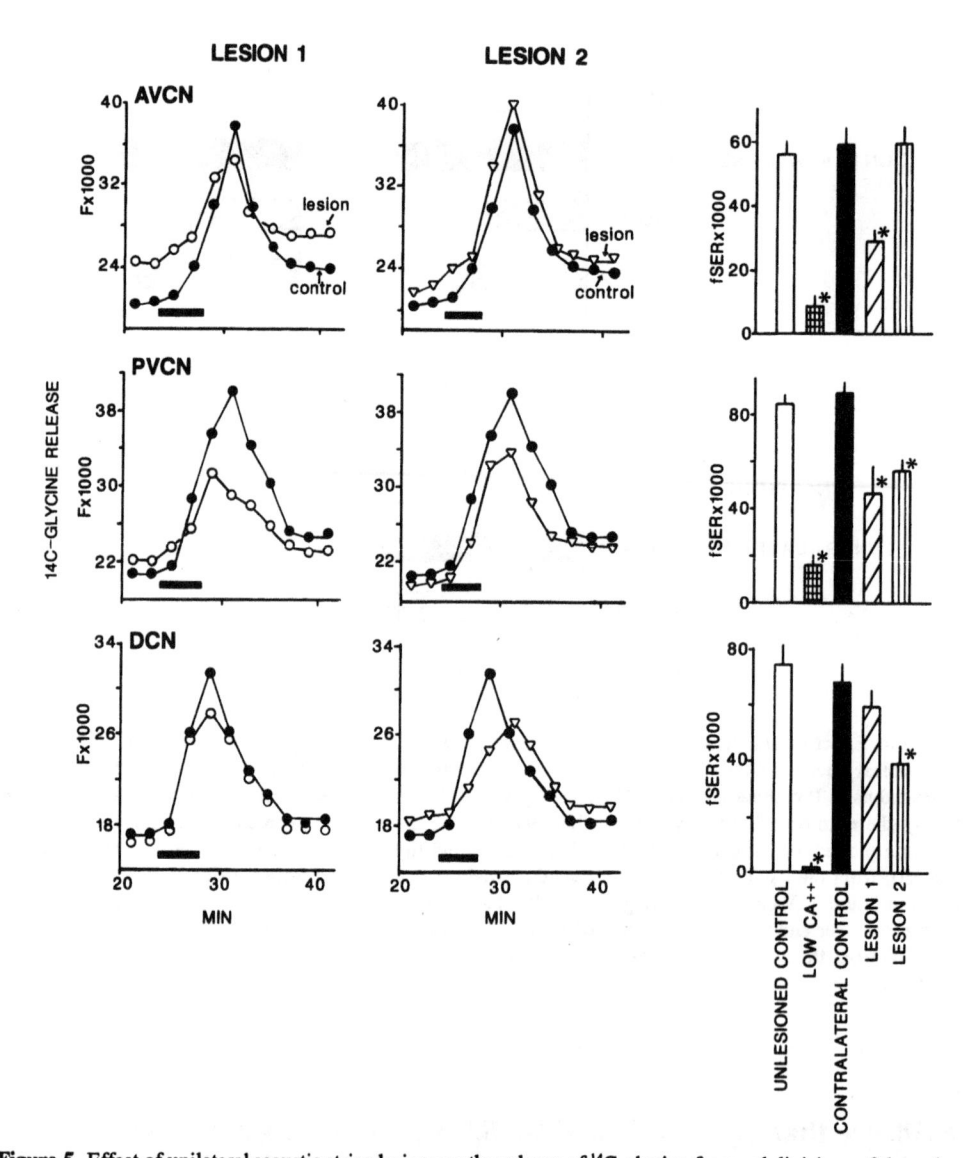

Figure 5. Effect of unilateral acoustic striae lesions on the release of $^{14}$C-glycine from subdivisions of the guinea pig CN. Release was measured as described in Fig. 2. Each subdivision is represented by 3 panels. The left and middle panels show the time course of the spontaneous and the electrically evoked release in tissues from unlesioned controls and from lesioned animals. The points are mean F values. SEM values (not shown) were < 10% of their respective means. Panels on the right illustrate glycine release in response to electrical stimulation, which is represented quantitatively as the fSER value. Plotted values are means + SEM from the left and right subdivisions of unlesioned controls (6 - 10 animals), from the right CN of 3 - 5 animals with lesion 1 and 4 - 7 animals with lesion 2. fSER values from the left CN of animals with lesion 1 or 2 were not significantly different from one another and were therefore pooled (contralateral control). Asterisks denote a significant difference ($P < 0.05$) from the unlesioned control. Reproduced from Staatz-Benson and Potashner ('88). Abbreviations as in Fig. 1.

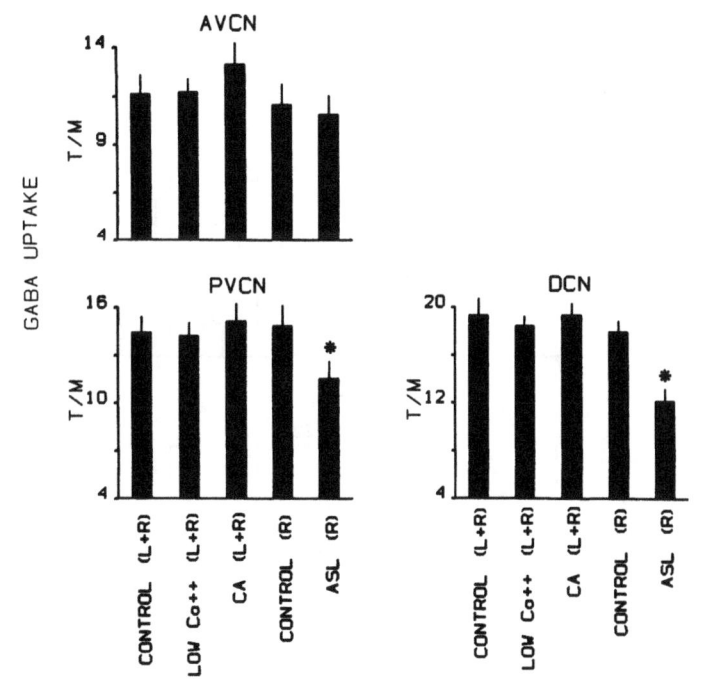

**Figure 6.** Effect of bilateral cochlear ablation compared to unilateral lesions of the dorsal and intermediate acoustic striae on the uptake of $^{14}$C-GABA by subdivisions of the guinea pig CN. Dissected subdivisions were treated as described in Fig. 4, except they took up 1.9 μM $^{14}$C-GABA instead of glycine and all media contained 0.5 mM aminooxyacetate to block the metabolism of GABA. Uptake was measured in subdivisions taken from the left and right CN of 10 [CONTROL (L+R)] and from the right CN of 5 [CONTROL (R)] intact animals. In Low Ca$^{++}$ experiments, uptake was measured in subdivisions from the left and right CN of 5 intact animals. Tissues were taken from the left and right CN of 3 animals which had both cochleas ablated [CA (L+R)] and from the right CN of 4 animals with acoustic striae lesion 2 on the right side of the medulla [ASL (R)]. Asterisks denote a significant difference ($P < 0.05$) from the unlesioned control. Reproduced from Potashner et al. ('85). Abbreviations as in Fig. 1.

lesioned group as controls, we measured the effects of these lesions on the uptake and release of glycine and GABA in subdivisions of the guinea pig CN.

Lesion 1 interrupted the rostral portion of the right trapezoid body at the level of the AVCN and PVCN (Fig. 3). It also severed the intermediate acoustic stria and extended caudally to interrupt some fibers in the rostral portion of the dorsal acoustic stria. Thus, this lesion should have severed many of the centrifugal projections to the AVCN and the PVCN, but many fibers in the dorsal acoustic stria which project to the DCN should have been intact. The lesion significantly depressed glycine uptake (Fig. 4) and release (Fig. 5) in the ipsilateral AVCN and PVCN, but not in the DCN.

Lesion 2 transected the right dorsal acoustic stria, the intermediate acoustic stria, and caudal areas of the trapezoid body, but spared the rostral portions of the trapezoid body connected to the AVCN (Fig. 3). Thus, this lesion should have severed most of the centrifugal projections to the right PVCN and DCN, but left intact many of the fibers to the AVCN. The lesion significantly depressed glycine uptake (Fig. 4) and release (Fig. 5) in the ipsilateral PVCN and DCN, but not in the AVCN. Lesion 2 also significantly depressed GABA uptake (Fig. 6) and release (Fig. 7) in the ipsilateral PVCN and DCN, but not in the AVCN.

The simplest explanation of these findings is that the axonal endings of some centrifugal fibers in the CN mediate the uptake and the release of glycine and GABA, and

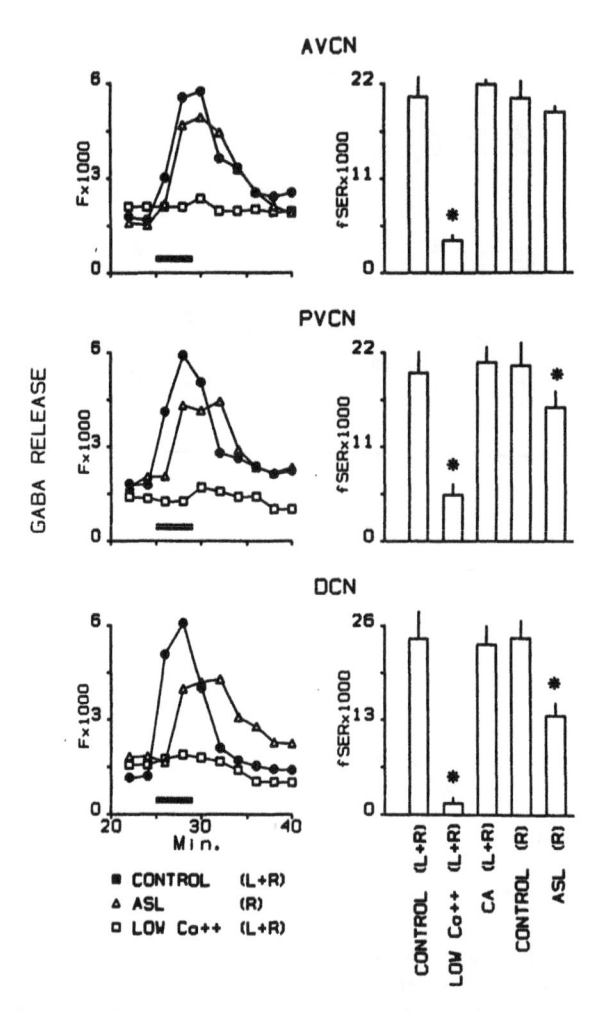

Figure 7. Effect of bilateral cochlear ablation compared to unilateral lesions of the dorsal and intermediate acoustic striae on the release of $^{14}$C-GABA by subdivisions of the guinea pig CN. Release was measured as described in Fig. 2 except that dissected subdivisions took up 1.9 $\mu$M $^{14}$C-GABA instead of glycine and all media contained 0.5 mM aminooxyacetate to block the metabolism of GABA. Each subdivision of the CN is represented by 2 panels. The left panel illustrates the time course of the spontaneous and the electrically evoked release in tissues from unlesioned controls in normal and in low Ca$^{++}$ media, and in tissues from lesioned animals. The data plotted are mean F values. SEM values (not shown) were < 15% of their respective means. Panels on the right illustrate the release in response to electrical stimulation, which is represented quantitatively as the fSER value. Release was measured in subdivisions taken from the left and right CN of 6 [CONTROL (L+R)] and from the right CN of 4 [CONTROL (R)] intact animals. In Low Ca$^{++}$ experiments, release was measured in subdivisions from the left and right CN of 4 intact animals. Release was also measured in tissues taken from the left and right CN of 3 animals which had both cochleas ablated [CA (L+R)] and from the right CN of 4 animals with an acoustic striae lesion 2 on the right side of the medulla [ASL (R)]. Asterisks denote a significant difference (P < 0.05) from the unlesioned control. Reproduced from Potashner et al. ('85). Abbreviations as in Fig. 1.

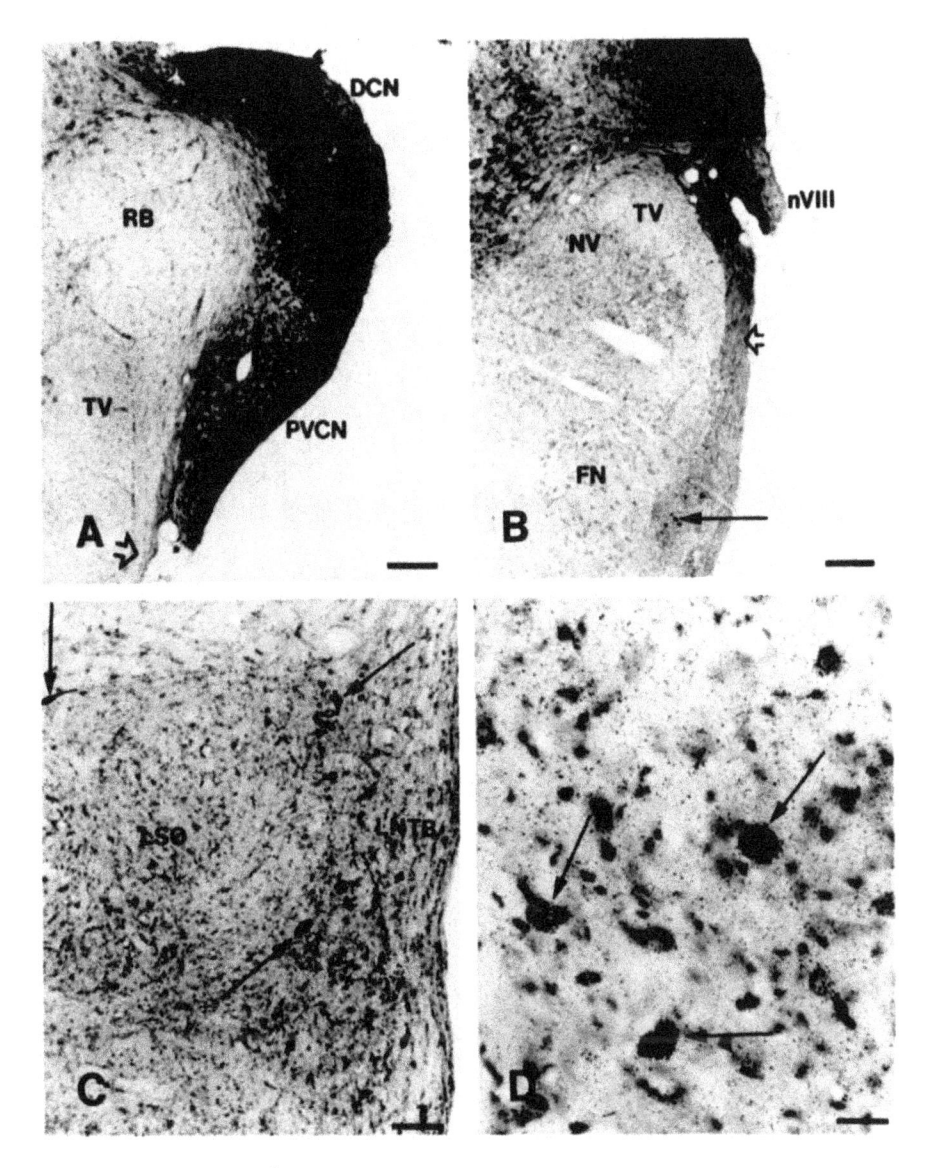

**Figure 8.** Photomicrographs of ³H-glycine autoradiographs (transverse sections). A: The injection site in the right DCN and PVCN of a guinea pig that survived 14 hrs after receiving an injection of 190 μM ³H-glycine. The injection site included all of the CN and parts of the restiform body and vestibular area. Labelled axons leave the cochlear nucleus in the acoustic striae and trapezoid body (clear arrow). Scale: 300 μm. B: Labelled axons in the trapezoid body (clear arrow) and labelled cells in the posterior periolivary nucleus (PPO: solid arrow) in a guinea pig that survived 20 hrs after receiving an injection of 380 μM ³H-glycine into the deep DCN. Scale: 400 μm. C: Labelled neurons in LNTB (arrows) and DPO (vertical arrow). Same case as in B. Scale: 100 μm. D: Labelled cell bodies in the ipsilateral DPO (arrows). Same case as in B. Scale: 22 μm. Reproduced from Benson and Potashner ('90). Abbreviations as in Figs. 3 and 9.

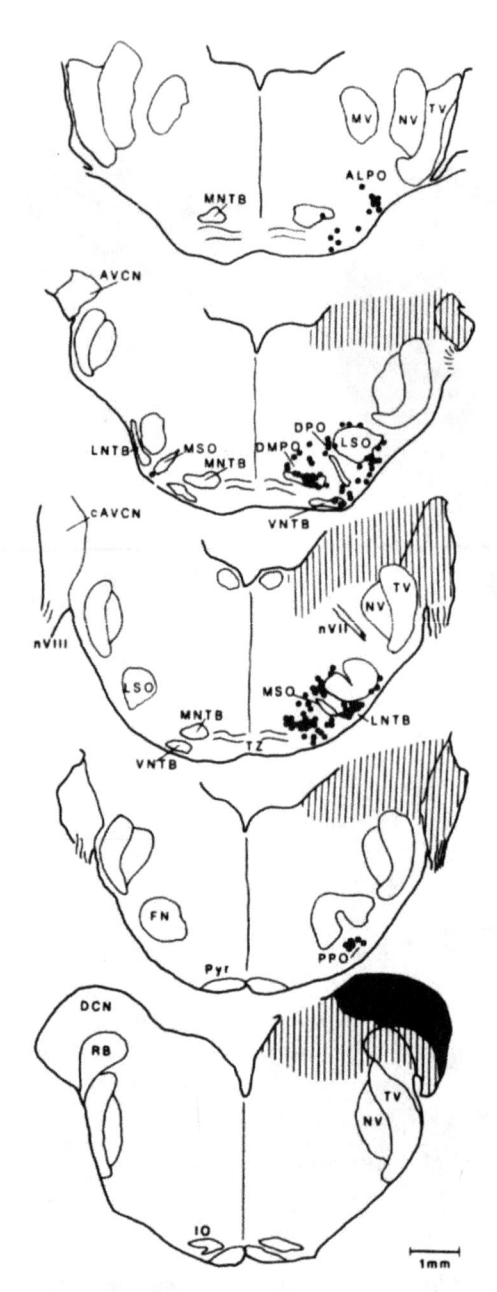

**Figure 9.** Locations of labelled cells in transverse sections through the superior olivary complex 20 hrs after injection of 380 μM ³H-glycine into the right CN. Same case as in Fig. 8B. The most rostral section is at the top. The injection site is indicated by the hatched area and the center of the injection site by solid black. Although the injection was made into the DCN, the injection site extended to include the PVCN, AVCN, and parts of the restiform body and vestibular area. ALPO, anterolateral periolivary nucleus; DNLL, dorsal nucleus of the lateral lemniscus; DMPO, dorsomedial periolivary nucleus; DPO, dorsal periolivary nucleus; FN, facial nucleus; IO, inferior olive; LNTB, lateral nucleus of the trapezoid body; LSO, lateral superior olive; MNTB, medial nucleus of the trapezoid body; MSO, medial superior olive; MV, trigeminal motor nucleus; NV, trigeminal sensory nucleus; nVII, seventh cranial nerve root; nVIII, eighth cranial nerve root; PPO, posterior periolivary nucleus; Pyr, pyramidal tract; RB, restiform body; TV, trigeminal tract; TZ, trapezoid body; VNLL, ventral nucleus of the lateral lemniscus; VNTB, ventral nucleus of the trapezoid body. Reproduced from Benson and Potashner ('90).

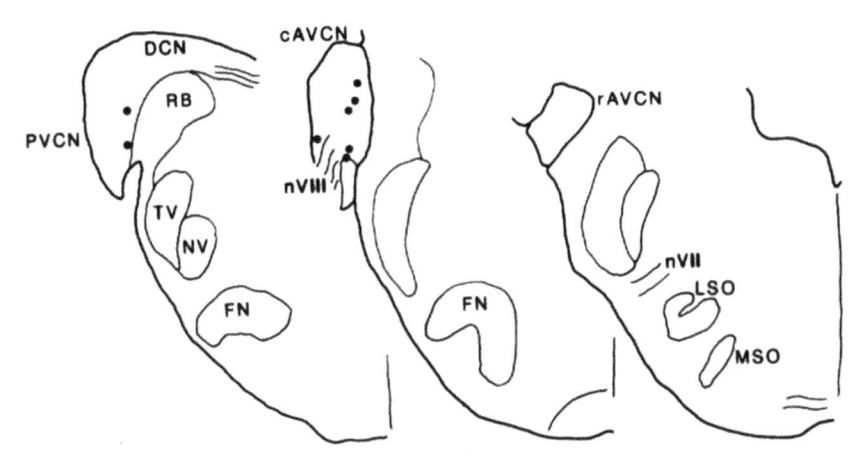

**Figure 10.** Locations of labelled cells in transverse sections through the left CN 20 hrs after injection of 380 µM ³H-glycine into the right CN. Same case as in Fig. 8B. Retrogradely labelled cell bodies were present in the PVCN and caudal AVCN, with only very few appearing in the deep DCN. Reproduced from Benson and Potashner ('90). Abbreviation as in Fig. 9, with additionally: cAVCN and rAVCN, caudal and rostral AVCN.

that these activities are lost when the severed centrifugal fibers degenerate. Thus, a proportion of the glycinergic and GABAergic synaptic endings in the CN may be provided by centrifugal projections. However, the acoustic striae lesions did not completely suppress the uptake and release of these compounds. The residual uptake and release activity remaining in the CN after interruption of the centrifugal projections may be mediated by the synaptic endings of intrinsic glycinergic and GABAergic neurons.

## Locating the origins of the centrifugal glycinergic and GABAergic projections to the cochlear nucleus

The sources of centrifugal projections to the CN include the superior olivary complex, nuclei of the lateral lemniscus, inferior colliculus, and the contralateral CN (Rasmussen, '67; Osen and Roth, '69; van Noort, '69; Adams and Warr, '76; Kane '76, '77a,b; Elverland, '77; Kane and Finn, '77; Kane and Conlee, '79; Cant and Gaston, '82; Spangler et al, '87; Winter et al, '89; Benson and Potashner, '90; Ostapoff et al, '90). We sought to localize a sub-population of these projections which may use glycine or GABA as a transmitter by using a transmitter-specific retrograde radiolabelling technique (Hökfelt and Ljungdahl, '75; Iversen, '78; Streit, '80; Cuenod et al., '82). This technique is based on the assumption that an amino acidergic neuron will take up its transmitter into its axonal endings by a high affinity uptake mechanism and then transport the transmitter retrogradely to the cell body. Neurons that use another transmitter remain unlabelled, as they lack the high affinity uptake mechanism for the first transmitter. For example, when ³H-glycine is used as the tracer, it retrogradely labels only those neurons which, on the basis of independent evidence, are presumed to use glycine as a transmitter (Cuenod et al., '82). Since the guinea pig CN probably contains glycinergic and GABAergic synaptic endings with high affinity uptake mechanisms specific for these amino acids, injections of ³H-glycine or ³H-GABA into the CN should result in the uptake and retrograde axonal transport of these markers, thus labelling neurons which project into the injection site and which may use glycine or GABA as a transmitter.

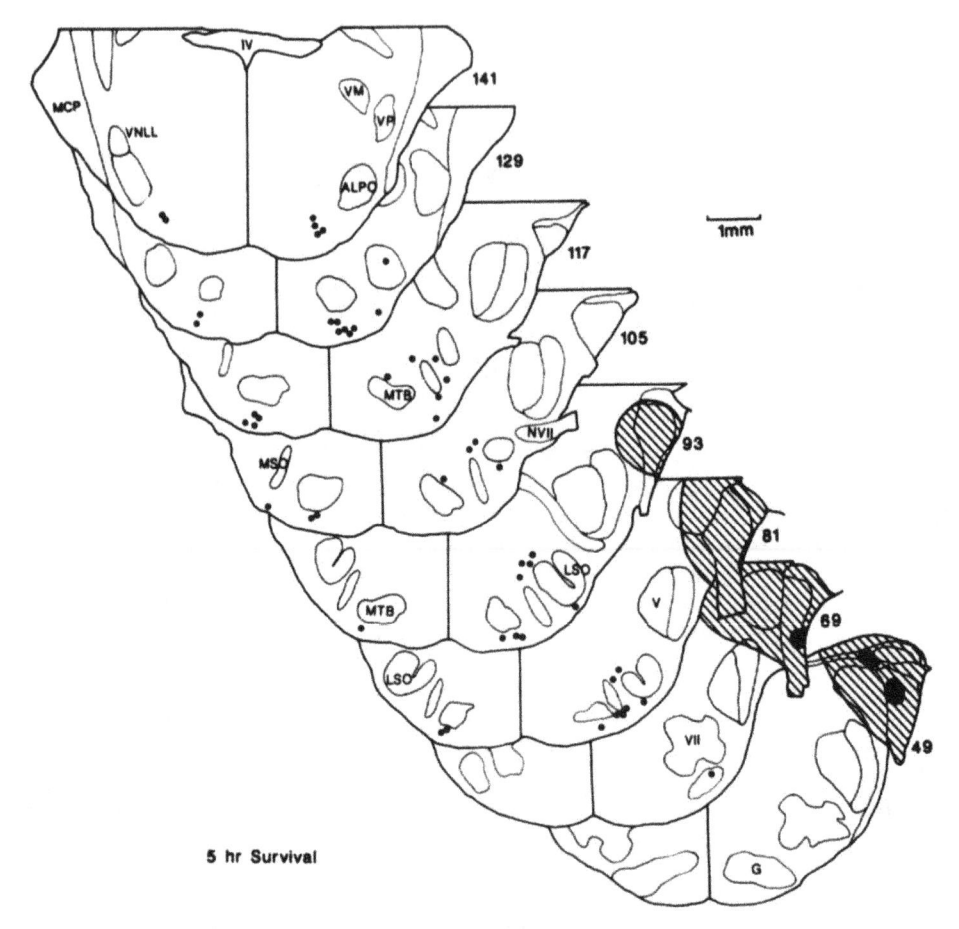

Figure 11. Locations of heavily labelled cells in transverse sections through the superior olivary complex 5 hrs after injection of 120 μM ³H-GABA into the right CN. The most rostral section is at the upper left, section numbers are at the right. The focal points of the injection sites are shown in black, the area over which labelled material spread is shaded. Most labelled cells were observed in the ipsilateral periolivary nuclei, including PPO, LNTB, DPO, VNTB, DMPO, and ALPO. Labelled cells were also observed in the contralateral VNTB. Abbreviations as in Fig. 9, with additionally: G, gigantocellular nucleus of the reticular formation; IV, fourth ventricle; MCP, middle cerebellar peduncle; MTB, medial nucleus of the trapezoid body; V, descending trigeminal nucleus; VII, facial nucleus; VM, motor nucleus of the trigeminal nerve; VP, principal nucleus of the trigeminal nerve. Reproduced from Ostapoff et al. ('90).

After injection of ³H-glycine or ³H-GABA into the right CN of the guinea pig, the injection site filled the entire CN and spread to include parts of the restiform body and the vestibular nuclei (Figs. 8A-B, 9 & 11). Labelled fibers, presumably containing both retrogradely and anterogradely transported radiolabel, radiated from the injected CN in the acoustic striae and trapezoid body (Figs. 8A-B). Ipsilateral to the glycine- or GABA-filled CN, retrogradely labelled cell bodies typically were observed in the periolivary region, including the LNTB, VNTB, DPO, DMPO, PPO, and ALPO (Figs. 8B-D, 9 & 11). After injections of ³H-glycine, but not ³H-GABA, labelled cell bodies were observed in the ipsilateral MNTB (Fig. 9) and VNLL. Contralaterally, significant numbers of retrogradely labelled cell bodies were observed in the VNTB after ³H-GABA injections (Fig. 11). Injections of high concentrations of ³H-glycine labelled a few cells in the contralateral LNTB,

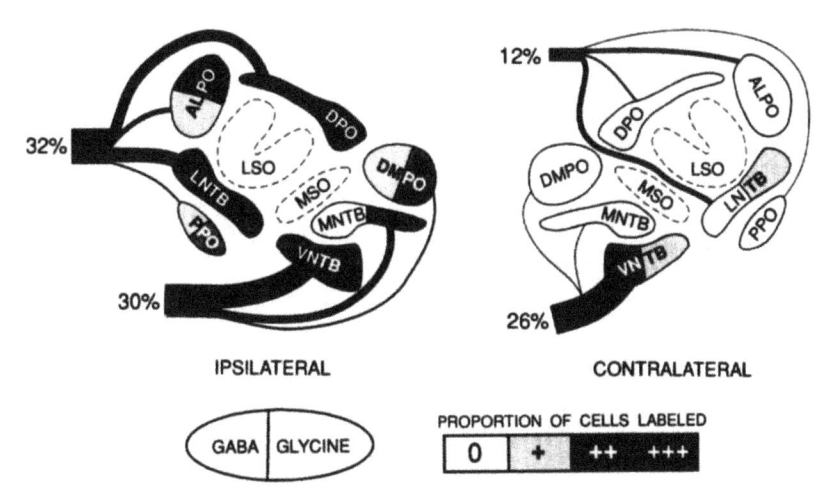

**Figure 12.** Schematic summary of retrograde labelling of the guinea pig superior olivary complex. Nuclei with solid black borders contain cells retrogradely labelled after injections of horseradish peroxidase in the CN. Thus, these nuclei contribute the centrifugal projections to the CN from the superior olivary complex. The thickness of each curved line denotes the relative number of cells in each nucleus that constitutes the projection to the CN. Percentages indicate the relative contributions of the lateral and medial periolivary nuclei. Nuclei with shading contain cells retrogradely labelled after injections of $^3$H-glycine or $^3$H-GABA in the CN. In each nucleus, the shading pertaining to GABA is left-most, that pertaining to glycine is right-most. The proportion of the cells in each nucleus retrogradely labelled by these tracers is provided by the scale at the lower right. For $^3$H-GABA labelling: 0, no labelled cells; +, ≤ 5% of labelled cells; + +, 6 - 25% of labelled cells; + + +, > 25% of labelled cells. For $^3$H-glycine labelling: 0, no labelled cells; +, very few labelled cells; + +, labelled cells; + + +, many labelled cells.

VNTB, and CN (Figs. 9 & 10). $^3$H-Glycine did not label cell bodies in the inferior colliculus. However, because of the lability of the injected $^3$H-GABA, the labelling of structures as distant as the VNLL and the inferior colliculus is difficult to assess.

These findings are summarized in Fig. 12. They indicate that the major sources of centrifugal glycinergic and GABAergic projections to the CN are the neurons in the ipsilateral periolivary nuclei. There is an additional contribution of glycinergic projections from the ipsilateral VNLL and of GABAergic projections from the contralateral VNTB. Other minor sources of glycinergic projections include the contralateral VNTB, LNTB, and CN.

Fig. 12 also illustrates the sources of centrifugal projections to the guinea pig CN originating in the superior olivary complex, as determined with the non-specific retrograde tracer, horseradish peroxidase (Benson and Potashner, '90; Ostapoff et al, '90). The thickness of each curved line denotes the relative number of cells in each nucleus that constitutes the projection to the CN. Four of the periolivary nuclei appear to contribute the major portion of this centrifugal projection; the DPO, LNTB, and VNTB ipsilaterally, and the VNTB contralaterally. Since each of these four nuclei contain neurons retrogradely labelled after injections of $^3$H-glycine or $^3$H-GABA in the CN, it is likely that much of their projection to the CN may use these compounds as transmitters.

In summary, the studies reviewed here illustrate part of an effort to localize glycinergic and GABAergic neurons that synapse in the CN. While some studies focus on projections intrinsic to the CN (e.g., Saint Marie et al., '91), the present findings indicate that the periolivary nuclei of the superior olivary complex are also important sources of centrifugal projections to the CN which may use glycine and GABA as inhibitory transmitters. The periolivary nuclei are thought to be a major source of inhibitory feedback projections to the CN and the cochlea by way of the acoustic striae and the olivocochlear bundle (e.g., Ginzberg and Morest, '83; Wiederhold, '86; Winter et al, '89). These projections may form part of a pathway linking the ascending auditory tracts with a control system utilizing negative feedback circuits.

ACKNOWLEDGEMENTS

This work was supported by grants DC00199 to S.J.P. and DC00127 to D.K.M. from the National Institutes of Deafness and Other Communicative Disorders.

REFERENCES

Adams, J.C. and Warr, W.B., 1976, Origins of axons of the cat's acoustic striae determined by injection of horseradish peroxidase into severed tracts, J. Comp. Neurol., 170:107-122.
Benson, C.G. and Potashner, S.J., 1990, Retrograde transport of [$^3$H]Glycine from the cochlear nucleus to the superior olive in the guinea pig, J. Comp. Neurol., 296:415-426.
Blasberg, R.G., 1968, Specificity of cerebral amino acid transport: A kinetic analysis, in: "Progress in Brain Research, Vol. 29", Lajtha A. and Ford D.H., eds., pp. 245-256, Elsevier, Amsterdam.
Blasberg, R.G. and Lajtha A., 1965, Substrate specificity of steady-state amino acid transport in mouse brain slices, Arch. Biochem. Biophys., 112:361-377.
Blaustein, M.P., Johnson E. M. and Needleman P., 1972, Calcium-dependent norepinephrine release from presynaptic nerve endings in vitro, Proc. Nat. Acad. Sci. USA, 69:2237-2240.
Brugge, J.F. and Geisler, C.D., 1978, Auditory mechanisms of the lower brainstem, Ann. Rev. Neurosci., 1:63-94.
Cant, N.B. and Gaston, K.C., 1982, Pathways connecting the right and left cochlear nuclei, J. Comp. Neurol., 212:313-326.
Caspary, D.M., Havey, D.C. and Faingold, C.L., 1979, Effects of microiontophoretically applied glycine and GABA on neuronal response patterns in the cochlear nuclei, Brain Res., 172:179-185.
Cuenod, M., Bagnoli, P., Beaudet, A., Rustioni, A., Wiklund, L. and Streit, P., 1982, Transmitter-specific retrograde labelling of neurons, in: "Cytochemical Methods in Neuroanatomy", Chan-Palay V. and Palay S.L., eds., A.R. Liss, Inc., New York, pp. 17-44.
Davidoff, R.A. and Adair, R., 1976, GABA and glycine transport in frog CNS: High affinity uptake and potassium-evoked release in vitro, Brain Res., 118:403-415.
Elverland, H.H., 1977, Descending connections between the superior olivary and cochlear nucleus complexes in the cat studied by autoradiographic and horseradish peroxidase methods, Exp. Brain Res., 27:397-412.
Fex, J. and Wenthold, R.J., 1976, Choline acetyltransferase, glutamate decarboxylase, and tyrosine hydroxylase in the cochlea and cochlear nucleus of the guinea pig, Brain Res., 109:575-585.
Fisher, S.K. and Davies, W.E., 1976, GABA and its related enzymes in the lower auditory system of the guinea pig, J. Neurochem., 27:1145-1155.
Frostholm, A. and Rotter, A., 1985, Glycine receptor distribution in mouse CNS:Autoradiographic localization of binding sites, Brain Res. Bull., 15:473-486.
Ginzberg, R.D. and Morest, D.K., 1983, A study of cochlear innervation in the young cat with the Golgi method, Hearing Res., 10:227-246.
Godfrey, D.A., Carter J., Berger S.J., Lowry, O.H. and Matschinsky, F., 1977, Quantitative histochemical mapping of candidate transmitter amino acids in cat cochlear nucleus, J. Histochem. Cytochem., 25:417-431.
Godfrey, D.A., Carter, J., Lowry, O.H. and Matschinsky, F.M., 1978, Distribution of gamma-aminobutyric acid, glycine, glutamate and aspartate in the cochlear nucleus of the rat, J. Histochem. Cytochem., 26:118-126.
Gundlach, A.L. and Beart, P.M., 1982, Neurochemical studies of the mesolimbic dopaminergic pathway: Glycinergic mechanisms and glycinergic-dopaminergic interactions in the rat ventral tegmentum, J. Neurochem., 38:574-581.
Hökfelt, T. and Ljungdahl, A., 1971, Light and electron microscopic autoradiography on spinal cord slices after incubation with labelled glycine, Brain Res., 32:189-194.
Hökfelt, T. and Ljungdahl, A., 1975, Uptake mechanisms as a basis for the histochemical identification and tracing of transmitter-specific neuron populations, in: "The use of axonal transport for studies of neuronal connectivity", Cowan W.M. and Cuenod M., eds., Elsevier, Amsterdam, pp. 249-305.
Iversen, L.L., 1978, Identification of transmitter-specific neurons in the CNS by autoradiography, in: "Handbook of Psychopharmacology, Vol. 9", Iversen L.L., Iversen S.D. and Snyder S.H., eds., Plenum Press, New York, pp. 41-68.

Iversen, L.L. and Bloom, F.E., 1972, Studies of the uptake of $^3$H-GABA and $^3$H-glycine in slices and homogenates of rat brain and spinal cord by electron microscopic autoradiography, Brain Res., 41:131-143.

Johnston, G.A.R. and Iversen, L.L., 1971, Glycine uptake in rat central nervous system slices and homogenates: Evidence for different uptake systems in spinal cord and cerebral cortex, J. Neurochem., 18:1951-1961.

Kane, E.S., 1976, Descending inputs to caudal cochlear nucleus in cats: A horseradish peroxidase (HRP) study, Amer. J. Anat., 146:433-441.

Kane, E.S., 1977a, Descending inputs to the dorsal cochlear nucleus of the cat: An electron microscopic study, J. Neurocytol., 6:587-605.

Kane, E.S., 1977b, Descending inputs to the octopus cell area of the cat cochlear nucleus: An electron microscopic study, J. Comp. Neurol., 173:337-354.

Kane, E.S. and Conlee, J.W., 1979, Descending inputs to the caudal cochlear nucleus of the cat: degeneration and autoradiographic studies, J. Comp Neurol., 187:759-784.

Kane, E.S. and Finn, R.C., 1977, Descending and intrinsic inputs to the cat caudal cochlear nucleus: A horseradish peroxidase study, Neuroscience, 2:897-912.

Lajtha, A., 1967, Transport as control mechanism of cerebral metabolite levels, in: "Progress in Brain Research, Vol. 29", Lajtha A. and Ford D. H., eds., Elsevier, Amsterdam, pp. 201-216.

Logan, W.L. and Snyder, S.H., 1972, High affinity uptake systems for glycine, glutamic and aspartic acids in synaptosomes of rat central nervous tissues, Brain Res., 42:413-431.

Martin, M.R., Dickson, J.W. and Fex, J., 1982, Bicuculline, strychnine, and depressant amino acid responses in the anteroventral cochlear nucleus of the cat, Neuropharmacology, 21:201-207.

Matus, A.I. and Dennison, M.E., 1971, Autoradiographic localization of tritiated glycine at "flat-vesicle" synapses in spinal cord, Brain Res., 32:195.

Neal, M.J. and Pickles, H.G., 1969, Uptake of $^{14}$C-glycine by spinal cord, Nature, 222:679-680

Neame, K.D., 1968, A comparison of the transport systems for amino acids in brain, kidney and tumor, Prog. in Brain Res., 29:185-196.

Oliver, D.L., Potashner, S.J., Jones, D.R. and Morest, D.K., 1983, Selective labelling of spiral ganglion and granule cells with D-aspartate in the auditory system of cat and guinea pig, J. Neurosci., 3:455-472.

Orrego, F., 1979, Criteria for identification of central neurotransmitters and their application to studies with nervous tissue preparations in vitro, Neuroscience, 4:1037-1057.

Osen, K.K. and Roth, K., 1969, Histochemical localization of esterases in the cochlear nuclei of the cat with notes on the origin of acetyl- cholinesterase- positive afferents and the superior olive, Brain Res., 16:165-185.

Ostapoff, E.-M., Morest, D.K. and Potashner, S.J., 1990, Uptake and retrograde transport of [$^3$H]GABA from the cochlear nucleus to the superior olive in the guinea pig, J. Chem. Neuroanat., 3:285-295.

Potashner, S.J., 1978, The effects of tetrodotoxin, calcium and magnesium on the release of amino acids from slices of guinea pig cerebral cortex, J. Neurochem., 31:187-195.

Potashner, S.J., Lindberg, N. and Morest, D.K., 1985, Uptake and release of GABA in the guinea pig cochlear nucleus after axotomy of cochlear and centrifugal fibers, J. Neurochem., 45:1558-1566.

Potashner, S.J. and Tran, P.L., 1984, Decreased uptake and release of D-aspartate in the guinea pig spinal cord after dorsal root section, J. Neurochem., 42:1135-1144.

Rasmussen, G.L., 1967, Efferent connections of the cochlear nucleus, in: "Sensorineural Hearing Processes and Disorders", Graham A. B., ed., Little Brown, Boston, pp. 61-75.

Rubin, R.P., 1974, "Calcium and the secretory process", Plenum Press, New York.

Saint Marie, R.L., Morest, D.K. and Brandon, C.J., 1989, The form and distribution of GABAergic synapses on the principal cell types of the ventral cochlear nucleus of the cat, Hearing Res., 42:97-112.

Saint Marie, R.L., Ostapoff, E.M., Benson, C.G. and Morest, D.K., 1991, Glycine immunoreactive projections from the dorsal to the anteroventral cochlear nucleus, Hearing Res., 51:11-28.

Sellstrom, A. and Hamberger, A., 1977, The uptake and release of putative amino acid transmitters from neurons and glia, Brain Res., 119:189-198.

Spangler, K.M., Cant, N.B., Henkel, C.K., Farley, G.R. and Warr, W.B., 1987, Descending projections from the superior olivary complex to the cochlear nucleus of the cat, J. Comp. Neurol., 259:452-465.

Staatz-Benson, C. and Potashner, S.J., 1987, Uptake and release of glycine in the guinea pig cochlear nucleus, J. Neurochem., 49:128-137.

Staatz-Benson, C. and Potashner, S.J., 1988, Uptake and release of glycine in the guinea pig cochlear nucleus after axotomy of afferent or centrifugal fibers, J. Neurochem., 51:370-379.

Streit, P., 1980, Selective retrograde labelling indicating the transmitter of neuronal pathways, J. Comp. Neurol., 191:429-463.

Tachibana, M. and Kuriyama, K., 1974, Gamma-aminobutyric acid in the lower auditory pathway of the guinea pig, Brain Res., 69:370-374.

van Noort, J., 1969, The anatomical basis for frequency analysis in the cochlear nucleus complex, *Psychiat. Neurolg. Neurochir.*, 72:109-114.

Voight, H.F. and Young, E.D., 1980, Evidence of inhibitory interactions between neurons in dorsal cochlear nucleus, *J. Neurophysiol.*, 44:76-96.

Wenthold, R.J., 1979, Release of endogenous glutamic acid, aspartic acid, and GABA from cochlear nucleus slices, *Brain Res.*, 162:338-343.

Wenthold, R.J., Betz, H., Reeks, K.A., Parakkal, M.H. and Altschuler, R.A., 1985, Localization of glycinergic synapses in the cochlear nucleus and superior olivary complex with monoclonal antibodies specific for the glycine receptor, *Neurosci. Abstr.*, 11:1048.

Werman, R., 1966, Criteria for identification of a central nervous system transmitter, *Comp. Biochem. Physiol.*, 18:745-766.

Whitfield, I.C. and Comis, S.D., 1966, The role of inhibition in information transfer: The interaction of centrifugal and centripetal stimulation on neurones of the cochlear nucleus, *"Final report II AF EOAR"* (U.S. Air Force), 63-115.

Wiederhold, M.L., 1986, Physiology of the olivocochlear system, *in*: "Neurobiology of Hearing: The Cochlea", Altschuler, R.A., Hoffman, D.W. and Bobbin, R.P., eds., Raven Press, New York, pp. 349-370.

Winter, I.M., Robertson, D. and Cole, K.S., 1989, Descending projections from auditory brainstem nuclei to the cochlea and cochlear nucleus of the guinea pig, *J. Comp. Neurol.*, 280:143-147.

Young, A.B. and MacDonald, R.L., 1983, Glycine as a spinal cord neurotransmitter, *in:* "Handbook of the spinal cord", Davidoff R.A., ed,, Marcel Decker, New York, pp. 1-43.

Zarbin, M.A., Wamsley, J.K. and Kuhar, M.J., 1981, Glycine receptor: Light microscopic localization with $^3$H-strychnine, *J. Neurosci.*, 1:532-547.

# INHIBITORY AMINO ACID SYNAPSES AND PATHWAYS IN THE VENTRAL COCHLEAR NUCLEUS

Richard A. Altschuler[1], José M. Juiz[1,2], Susan E. Shore[3],
Sanford C. Bledsoe[1], Robert H. Helfert [1,4], and Robert J. Wenthold[5]

[1]Kresge Hearing Research Institute & Dpt. Anatomy & Cell Biology,
University of Michigan, Ann Arbor, MI 48109, USA,
[2]Dpt. of Histology and Institute of Neuroscience, University of Alicante,
Alicante, Spain,
[3]Dpt. of Otolaryngology, Medical College of Ohio, Toledo, Ohio, USA
[4]Dpt. of Surgery & Pharmacology, Southern Ill. Univ. Sch. Med.,
Springfield, IL, USA and
[5]Laboratory of Neurochemistry, NIDCD, Bethesda, MD., USA

The processing of auditory information requires neurons to fire at a rapid rate. Hair cells can, for example, respond to tones by oscillations of membrane potential at rates as high as many hundred times per second. This response rate firing conveys important information about intensity and frequency and so is maintained throughout the initial processing in the auditory brain stem. The first synapse through which response rate information must be accurately passed is between the inner hair cells and the auditory nerve. It is not surprising that evidence supports an excitatory amino acid transmitter and an ionotropic excitatory amino acid receptor at this synapse (Bledsoe et al., '88; Altschuler et al., '89), since this is one of the few known excitatory transmitter - receptor combinations capable of such speed. The next synapse in the ascending auditory pathway is obligatory, in the cochlear nucleus (CN). Firing rate information transfer is maintained at many auditory nerve - CN synapses, leading to the expectation that these synapses would also have an excitatory transmitter - receptor combination capable of considerable speed. Much evidence suggests that an excitatory amino acid is the auditory nerve transmitter (chapter in this volume by Juiz et al.; Wenthold, '91 and Caspary, '86) and that most CN projection neurons receive this input via an ionotropic excitatory amino acid receptor (see Wenthold et al. and Morest, chapters this volume), currently termed a glutamate receptor (GluR).

In this volume the distribution of excitatory amino acid terminals is discussed in the chapter by Juiz et al., and the distribution of excitatory amino acid receptors is discussed in chapters by Wenthold et al. and Morest. In this chapter we consider how this rapidly conveyed excitatory information may be modulated by inhibitory influences. There are only two known inhibitory transmitter - receptor combinations capable of matching the speed of excitatory information transfer through an excitatory amino acid - GluR synapse. These are the inhibitory amino acids gamma-amino-butyric acid (GABA) and glycine (GLY) acting at ionotropic GABA and GLY receptors, respectively. It is therefore predictable that both

GABA and GLY would be major transmitters in the CN. We have found, in fact, that in the ventral cochlear nucleus (VCN) the great majority of synapses involve a terminal containing either an excitatory amino acid, GABA or GLY. While a major presence of GABA or GLY as auditory brain stem transmitters is not surprising, it is interesting that both are well represented in the CN and that there is a large incidence of co-containment of GABA and GLY within the same CN cells and terminals. Some terminals may also, of course, co-contain an amino acid transmitter and other neurotransmitter candidates, including neuropeptides, for slower modulation.

This chapter will describe the distribution of GABA and GLY terminals and receptors in the VCN, with a particular emphasis on the ultrastructural characterization of these synapses and on the co-localization of GABA and GLY. It will also address the possible origin of GABA, GLY and GABA/GLY terminals as well as consider the role of these synapses in CN processing. Because of the many companion chapters on CN organization and neurotransmitters in this volume, no attempt will be made to completely review the literature and an emphasis will be placed on citing chapters and reviews. The chapter in this volume by Saint Marie et al. contains complementary information on inhibitory amino acid cells and terminals. Chapters by Caspary et al., Evans and Zhao, and Oertel and Wickesberg describe the effects of inhibitory amino acids.

This chapter discusses results in guinea pig using primarily immunocytochemical techniques, with receptor autoradiography also applied to study the distribution of GABA receptors. Antibodies were utilized against GABA and GLY (developed by Wenthold), the GABA receptor (developed by deBlas) and GLY receptor (developed by Betz) and the synthetic enzyme for GABA, GAD (developed by Oertel & Kopin). Immunoperoxidase techniques were used either with pre-embedding staining on free-floating vibratome sections or with post-embedding staining on half-micron plastic (Embed 812) sections, using the Vectastain avidin biotin peroxidase kits and protocols (Vector Laboratories). For electron microscopic evaluations immunoperoxidase techniques were used for pre-embedding staining evaluations and immunogold techniques for post-embedding evaluations. Immunogold staining was often enhanced as described by Juiz et al. ('91). For combined tract-tracing/ immunocytochemistry studies biocytin was iontophoretically injected into DCN and anterogradely labeled terminals visualized with post-embedding staining on sections adjacent to ones immunostained for GABA or GLY. For GABA-receptor autoradiography [$^3$H] labeled GABA was blocked either with the GABA-A receptor agonist isoguvacine or the GABA-B agonist baclofen to differentiate and selectively visualize binding to the two classes of receptors.

## VCN cells and terminals

The projection neurons of the VCN nucleus fall into three major classes (Osen, '69; Cant and Morest, '84; Lorente de No, '81; Moore,, '86; Rhode, '91): bushy cells (spherical bushy and globular bushy) which receive numerous axo-somatic terminals and project primarily to the superior olivary complex; stellate multipolar cells which receive most of their terminals on dendrites and project primarily to the inferior colliculus; and octopus cells which have many axo-dendritic and axo-somatic contacts and project primarily to the ventral lemniscal nuclei (see Cant, '91; Helfert et al., '91 for review). Although the terminals on spherical bushy cells of the rostral anteroventral cochlear nucleus (AVCN) have been most extensively described (Cant and Morest, '84; Schwartz et al., '78, Tolbert and Morest, '82) similar endings are found on all principal cells in the VCN. There are morphologically distinct types of terminals that make axo-somatic contacts which we have attempted to further classify on the basis of the neurotransmitters they contain (Fig. 1) (Altschuler et al., '86). One of these terminal types contains large round vesicles and makes an asymmetric contact (Fig. 1). Most of these degenerate subsequent to lesioning of the auditory nerve and are

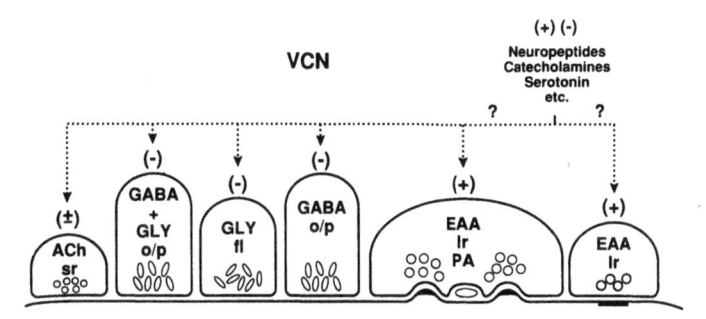

Figure 1. Schematic diagram of different terminal types on VCN somata, their vesicle type (sr = small round, o/p = oval/pleomorphic, fl = flattened, lr = large round) and the neurotransmitter candidate associated with them (ACh = acetylcholine, GABA = gamma aminobutyric acid, GLY = glycine, EAA = an excitatory amino acid). PA = primary afferent auditory nerve terminal. Acetylcholine is also likely to be co-contained in other terminal types.

therefore believed to be the terminals of the auditory nerve. These auditory nerve terminals form large axo-somatic contacts on spherical bushy cells which are also characterized by pre-synaptic invaginations and post-synaptic evaginations. The other two major terminal types make symmetric contacts, one contains oval/pleomorphic vesicles and the other more flattened vesicles (Fig. 1) under our fixation conditions. A fourth class of terminal containing small round vesicles has been described, but these are rarely seen in the guinea pig VCN. Axo-dendritic terminals have similar characteristics.

When we perform immunocytochemical localizations on adjacent sections we find that roughly 95% of VCN terminals immunolabel either with antibodies to GABA, GLY or glutamate. On the basis of our studies we can ascribe an excitatory amino acid transmitter to the population of terminals containing large round vesicles (Altschuler et al., '86; Juiz et al., '90, and this volume). While most of these correspond to auditory nerve terminals, a small population may correspond to the terminals of granule cells or some descending connections (see Juiz et al., this volume). We ascribe GLY as the transmitter for terminals containing flattened vesicles and either GABA only or both GABA and GLY as the transmitter(s) for terminals containing oval pleomorphic vesicles.

## Inhibitory Amino Acid Transmitters and Receptors

GABA and GLY are both small amino acids. GABA is the most common inhibitory transmitter in the CNS, while GLY is particularly common in the auditory brain stem and spinal cord.

The $GABA_A$ receptor is a pentamer now believed to be made made up of two alpha, two beta, and either a delta or gamma subunit depending on subtypes (e.g. Levitan et al., '88; Luddens and Wisden, '91; Pritchett et al., '89; Bormann, '88). The GLY receptor is also a pentamer, made up of alpha and beta membrane spanning subunits, as well as a separate cytoplasmic anchoring region (Malosio et al., '91). All of the subunits of both the $GABA_A$ and GLY receptors appear to have many different subtypes. Depending on which subunits and specific sub-types of subunits are combined, the receptor complex can have different physiological characteristics (e.g. Borman, '88; Levitan et al., '88; Luddens and Wisden, '91; Pritchett et al., '89). Receptors often have different configurations in different brain regions and divisions (e.g. Pritchett et al., '89; Malosio et al., '91). Both the GLY and $GABA_A$ receptors are ionotropic. The binding of GABA or GLY opens up an integral chloride channel causing increased membrane permeability to chloride resulting in a rapid hyperpolarization

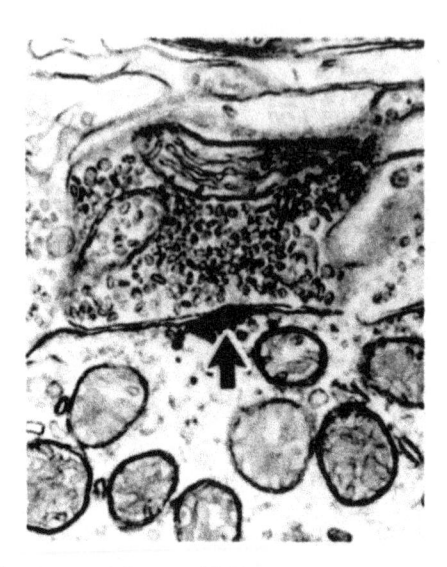

**Figure 2.** Glycine receptor immunoreactivity on a VCN somata (arrow) apposing a terminal with flattened vesicles, using pre-embedding immunoperoxidase methods.

(Sakmann et al., '83). The effect occurs within milliseconds and only lasts for several milliseconds. The transmitter is cleared by diffusion and high affinity uptake (Iversen and Kelley, '75). Since they act by a similar mechanism, the actions of GLY and GABA may be similar, but they are also capable of inducing different conductance states in the chloride channel. Metabotropic receptors such as many neuropeptide receptors and the $GABA_B$ receptor, are not directly channel-linked and operate via second messenger systems. Their action is therefore slower (taking up to several seconds) and the effect is longer lasting. The GABA-B receptor uses a potassium or calcium channel to achieve hyperpolarization. It is often located pre-synaptically where it's activation can inhibit neurotransmitter release, but it can also be post-synaptic.

Evidence supporting GABA and GLY as VCN neurotransmitters includes neurochemical (see Godfey et al., '88, Wenthold and Martin, '84, Potashner et al., this volume for reviews), pharmacological (see Caspary, '86; Martin, '84 and chapters in this volume by Caspary et al.; Evans and Zhao, and Oertel and Wickesberg), uptake studies (Schwartz, '83 and chapter by Potashner et al. in this volume,) and binding studies (Frostholm and Rotter, '86; Glendenning and Baker, '88; Zarbin et al., '81). Immuno-cytochemical studies also provide support for their neurotransmitter role plus information on the localization and distribution of cells and synapses involved in GABA and GLY mediated inhibition (Wenthold, '91 for review).

## Glycine and glycine receptor

We find numerous GLY immunoreactive (IR) terminals in the VCN (Juiz et al., '90, submitted; Wenthold et al., '87). These make axo-somatic and/or axo-dendritic contacts on all VCN projection neurons. GLY receptor IR is also seen on all VCN projection neurons apposing both axo-somatic and axo-dendritic terminals (Altschuler et al., '86; Wenthold et al., '88). GLY IR endings can be oval or dome-shaped and are often grouped in clusters, often with GABA and GABA/GLY IR endings, particularly on spherical bushy cells. Their average diameter ranges from 0.8 - 3.5 $\mu$m, the largest endings being axo-dendritic. GLY IR terminals or terminals apposed by GLY receptor IR form symmetric synapses and usually

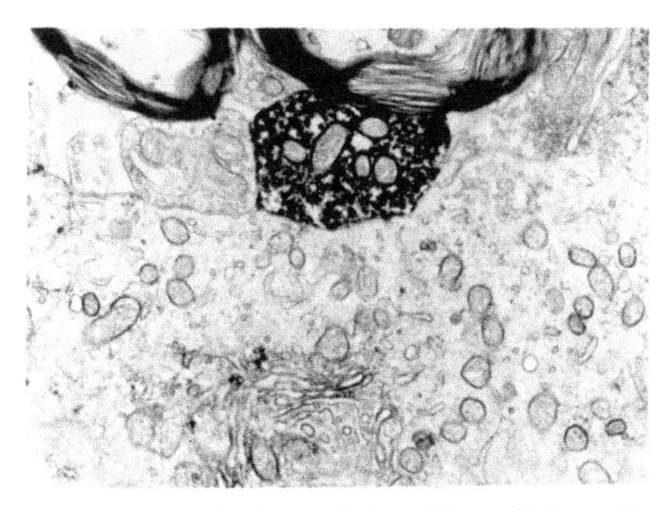

**Figure 3.** GABA immunoreactive labeling in a terminal containing oval/ pleomorphic vesicles making a symmetric synapse on a VCN soma, using pre-embedding immunoperoxidase techniques.

contain flattened vesicles (Figs. 1,2,7). However GLY IR is also found in terminals with oval pleomorphic vesicles and likewise GLY receptor IR apposes some terminals containing oval-pleomorphic vesicle. These probably correspond to terminals which co-contain GABA and GLY (see section below on co-localization).

Synapses with GLY IR terminals and/or post-synaptic GLY receptor IR are abundant on the somata of spherical bushy, globular bushy and octopus cells. Our immunocytochemical results suggest that the majority of inhibitory synapses in the VCN involve GLY either as transmitter or co-transmitter. Distribution patterns of GLY IR terminals on spherical bushy and globular bushy cells are similar. Immunolabeled endings on octopus cells are larger than those on spherical or globular bushy cells. There are few GLY IR terminals and little GLY receptor labeling on multipolar-stellate cell somata; however, the primary dendritic trunks receive many terminals that are GLY IR and apposed by GLY receptor IR.

Spherical bushy, globular bushy or octopus cells are not GLY IR, nor are most multipolar-stellate cells. GLY IR cell bodies, however, are observed in the VCN. Small cells (9- 14 $\mu$m diameter) are found in the medial aspect of the AVCN and the granule cell cap. Some larger (15-24 $\mu$m diameter) GLY IR cells are seen in the central region of the AVCN that could correspond to a sub-class of large stellate multipolar cells (described in this volume). In the posteroventral cochlear nucleus (PVCN), small GLY IR cells are seen in the dorsolateral margins including the granule cell layer. Larger (18-26 $\mu$m) GLY IR cells are located deeper in the PVCN.

## GABA and GABA receptor

We find numerous GABA IR terminals on all major types of projection neurons in the VCN (Juiz et al., '90, submitted; Oberdorfer et al., '88; Wenthold et al., '86, also see chapters by Saint Marie et al., and Adams in this volume for complementary studies). For each cell type the number of axo-somatic GABA IR terminals is less than the number of axo-somatic GLY IR terminals. GABA IR terminals making axo-dendritic contacts are also observed. In general, GABA IR terminals are smaller than GLY IR terminals, with average diameters ranging from 0.8 - 1 $\mu$m. GABA IR terminals contain oval/pleomorphic vesicles (Figs. 1, 3) and form symmetric synapses. Dense-core vesicles are frequently found in GABA

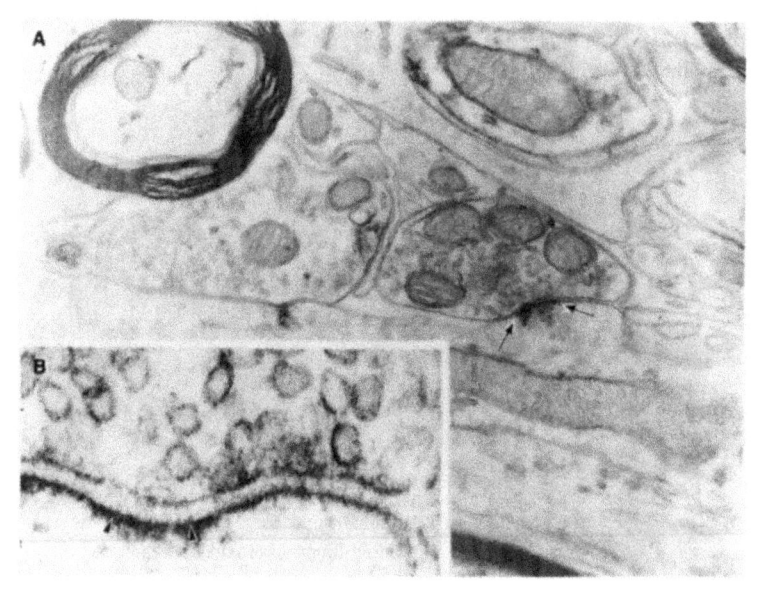

Figure 4. GABA$_A$ receptor immunoreactivity apposing terminals on VCN dendrite (arrows), using pre-embedding immunoperoxidase techniques.

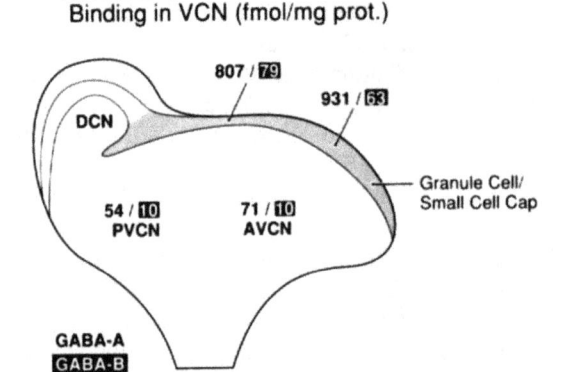

Figure 5. Schematic diagram summarizing GABA$_A$ and GABA$_B$ autoradiographic binding in the VCN.

IR terminals. Despite the large number of GABA IR terminals making axo-somatic contacts on VCN principal cells, when we apply an antibody to the beta subunit of the GABA$_A$/BDZ receptor we do not observe immunolabel apposing any axo-somatic terminals in the VCN. We do see GABA$_A$/BDZ receptor IR apposing some axo-dendritic terminals in VCN (Fig. 4) and both axo-dendritic terminals and axo-somatic terminals in DCN. These results suggest either that GABA IR axo-somatic terminals in VCN are apposed by a GABA$_A$/BDZ receptor non-immunoreactive to the antibody we used, or that they interact with pre- or postsynaptic GABA$_B$ receptors.

Caspary et al., ('84) have shown that baclofen, a GABA$_B$ agonist, reduces tone-evoked activity of CN neurons. Since antibodies to the GABA$_B$ receptor are not yet available we used autoradiographic techniques with [$^3$H] GABA to determine GABA$_A$ versus GABA$_B$ receptor binding in the VCN. Distribution of $GABA_A$ and $GABA_B$ binding sites is similar in the different divisions of the VCN (Juiz et al., '91, in press). The central region of the VCN, containing spherical bushy, globular bushy, octopus and multipolar stellate cells has low levels of $GABA_A$ binding (confirming earlier reports by Frostholm & Rotter, '86 and Glendenning and Baker, '88 with [$^3$H] muscimol) and very low levels of $GABA_B$ binding. In the granule cell/small cell cap of the VCN, however, high densities of binding sites are seen for both $GABA_A$ and $GABA_B$, with the binding of $GABA_A$ higher (Fig 5). In the VCN, in general, the density of $GABA_A$ binding is 5-15 times greater than $GABA_B$. Except in the granule cell/small cell region of the VCN there then appears to be a mismatch between the presence of numerous GABA IR terminals and the absence of $GABA_A$/BDZ receptor IR apposing axo-somatic terminals and the low levels of both $GABA_A$ and $GABA_B$ binding sites.

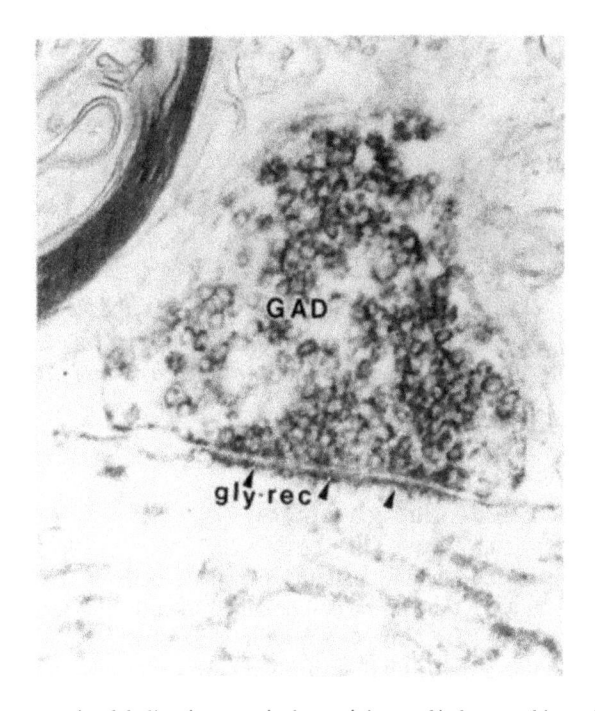

**Figure 6.** GAD immunoreactive labeling in a terminal containing oval/ pleomorphic vesicles apposed by GLY receptor immunoreactivity (arrowheads), using pre-embedding immunoperoxidase techniques.

Pharmacological studies (see chapters by Caspary et al., and Evans and Zhao in this volume) show that GABA has an inhibitory effect in the CN, which is blocked by the $GABA_A$ antagonist bicuculline. Bicuculline antagonizes the effects of GABA at both "low-affinity" and "high-affinity" $GABA_A$ receptors. It may be that the specific subunits and configuration of the $GABA_A$ receptor apposing GABA IR terminals on VCN cell somata is of the low affinity type. This could explain both the poor autoradiographic binding and the lack of binding of the antibody we applied since both techniques favor the "high-affinity" receptor. If there is a low affinity $GABA_A$/BDZ receptor, this raises questions as to the functional consequences of low-affinity versus high-affinity $GABA_A$ receptors in the VCN.

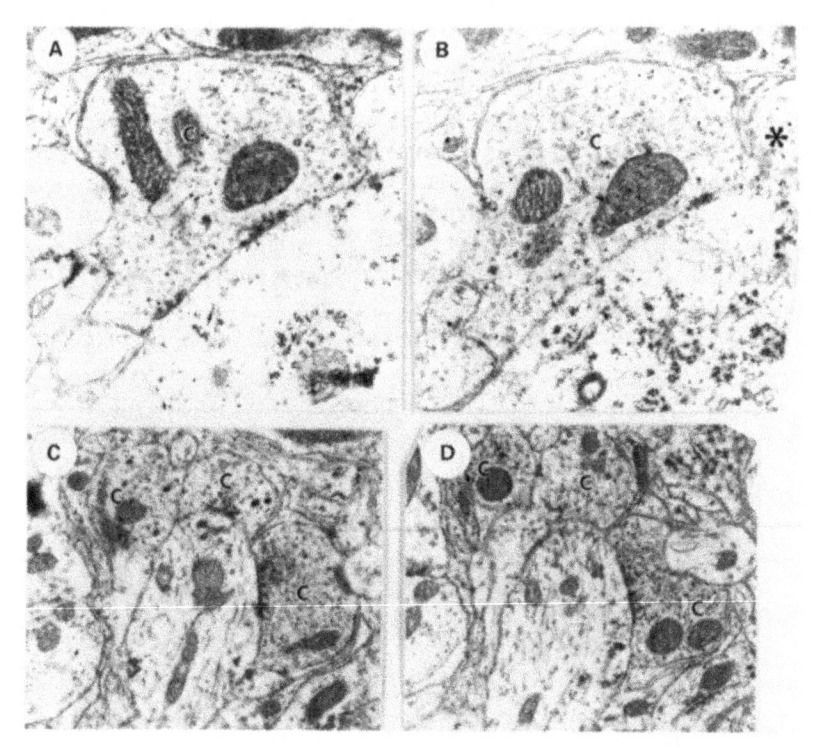

Figure 7. GABA IR (A,C) and GLY IR (B,D) staining on adjacent sections using post-embedding staining immunogold techniques. Asterisk marks a terminal containing only GLY IR, several other terminals (c) co-contain GABA and GLY IR. Note large dense-core vesicles in some of these terminals.

## Co-localization of GABA and glycine

The similarity in the staining patterns of small cells in the DCN was the first indication that GABA and GLY might be co-contained in cells and terminals in the cochlear nucleus (Wenthold et al., '87). This was further suggested by the localization of GLY receptor IR apposing some terminals with oval/pleomorphic vesicles (Oberdorfer et al., '87). GLY receptor IR was then localized apposing terminals immunoreactively labeled for GAD the enzyme of synthesis for GABA (Fig 6). Further studies were able to show the co-containment more directly, (Osen, '90). By immunostaining adjacent sections with antibodies to GABA and GLY, it was possible to directly demonstrate co-localization of GABA and GLY in a large population of CN cells and terminals.

Figure 7 shows adjacent thin sections immunolabeled for GABA and GLY. Using this approach we have been able to identify three categories of inhibitory amino acid cells and terminals in the VCN: "GABA", "GLY" and "GABA/GLY". The GABA/GLY IR terminals are similar to the GLY IR terminals except that they contain oval/pleomorphic vesicles. GABA/GLY terminals often also contain dense core vesicles (Fig. 7). Their distribution varies widely on the different VCN cell types. On the somata of spherical bushy and globular bushy cells "GLY IR" versus "GLY/GABA IR" versus "GABA IR" endings occur at a ratio of approximately 5:2:1. GABA/GLY IR terminals, however, are rarely seen on octopus cells. The primary dendritic trunks of spherical bushy, globular bushy and stellate multipolar cells all receive many GLY, GABA and GABA/GLY IR terminals of similar morphology, size,

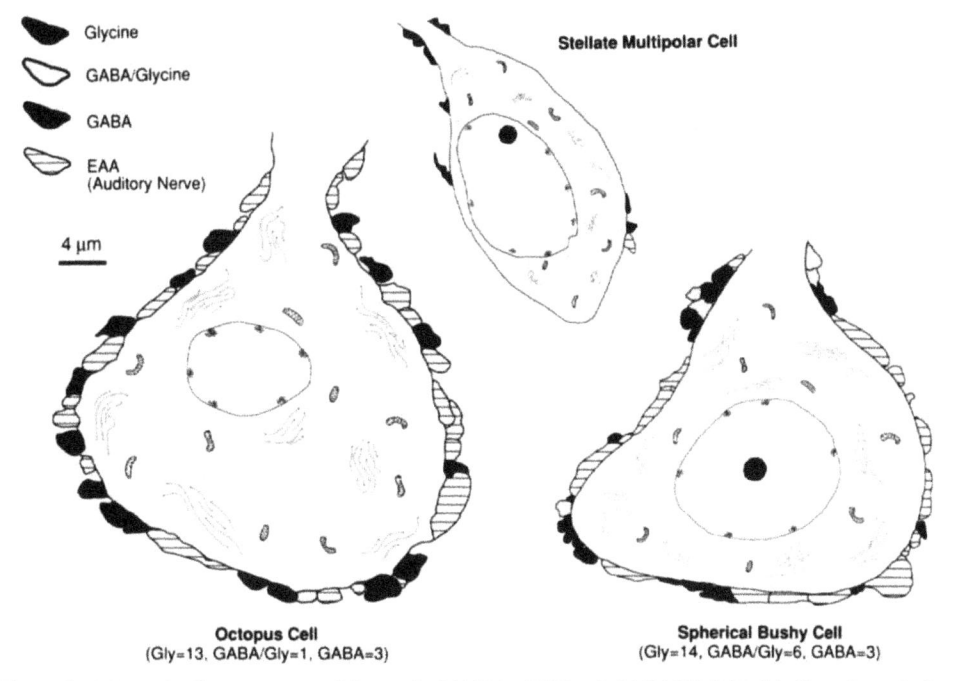

**Figure 8.** Schematic diagram summarizing typical GABA, GLY and GABA/GLY IR labeling of terminals on sections through spherical bushy, multipolar stellate and octopus cells.

distribution and prevalence. The few (often one or two), small (0.6 to 1 μm average diameter), non-cochlear axo-somatic endings observed on stellate multipolar cell profiles are usually GLY or GABA IR.

Small (10-14 μm) cell bodies immunoreactive for both GABA and GLY are concentrated on the dorsolateral margins of the caudal PVCN, including the granule cell layer.

Figure 8 summarizes the pattern of GLY IR, GABA/GLY IR and GABA IR terminals, compared to primary afferent terminals, on VCN principal cells (the patterns on spherical and globular bushy cells are similar). There are roughly twice times as many endings labeled for GLY as GABA on spherical and globular bushy cells of the AVCN, with about 30% of the GLY endings also containing GABA. In PVCN there are 3-4 times more GLY than GABA endings on octopus cells, with GLY/GABA terminals scarce. GLY IR, GABA/GLY IR and GABA IR terminals are often grouped in clusters on the cell somata, particularly on spherical bushy cells. GABA/GLY terminal distribution appears to be the major difference between octopus cells and spherical or globular bushy cells, although source of inputs may also vary. The distribution of terminals on octopus cells may also vary between species, as few GLY IR terminals have been reported on octopus cells in mouse (Oertel and Wickesberg, this volume).

When adjacent sections are examined for GABA, GLY and glutamate, unlabeled terminals are rarely seen, suggesting that the great majority of VCN terminals utilize an excitatory or an inhibitory amino acid as a transmitter. This, however, does not preclude the presence of more neuroactive substances in subpopulations of these terminals (Fig. 1). In fact, large dense core vesicles, typically associated with neuropeptides or catecholamines, are found in many GABA and GABA/GLY IR terminals, particularly in the AVCN. Co-localization of immunoreactivities for GABA and several neuropeptides has been described in periolivary and deep DCN neurons probably projecting to the VCN (Adams and Mugnaini, '85).

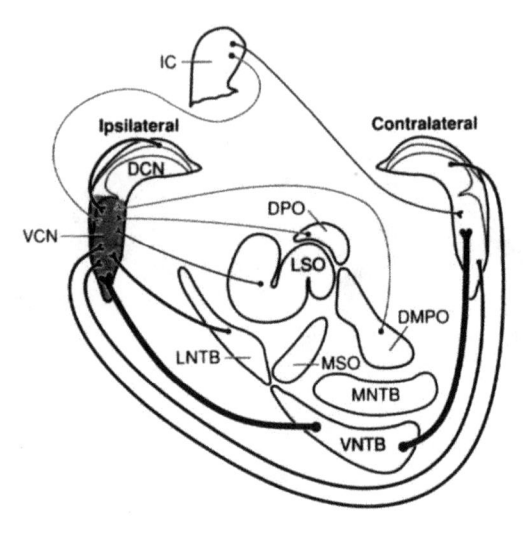

Figure 9. Schematic diagram summarizing major non-cochlear inputs to the VCN.

## Sources of inhibitory amino acid terminals

Figure 9 summarizes potential sources of non-primary VCN terminals based on our studies with retrograde transport of HRP in the guinea pig (Shore et al., '91, in press, unpublished observations). Major non-cochlear inputs to the guinea pig VCN, listed in order of their size, are the 1) ventral nucleus of th trapezoid body (VNTB), 2) lateral nucleus of the trapezoid body (LNTB), 3) dorsal periolivary regions, 4) DCN, 5) contralateral CN, 6) dorsomedial periolivary nucleus (DMPO), 7) inferior colliculus (IC), and 8) lateral superior olivary nucleus (LSO). These projections are consistent with previous reports (see chapters in this volume by Adams; Oertel and Wickesberg; Potashner et al., Saint-Marie et al.; Wickesberg and Oertel).

Neurochemical and immunocytochemical studies in these regions suggest that all these areas are potential sources of GABA, GLY or GABA/GLY inputs, with DCN, VNTB and LNTB likely to have the greatest contributions (see chapters in this volume by Adams; Potashner et al.; Saint Marie et al.; and Wickesberg and Oertel; and Godfrey et al., '88; Helfert et al., '89; Mugnaini and Oertel, '85). Although there are many GABA, GLY and GABA/GLY IR cells in the molecular and fusiform cell layers of the DCN, our HRP studies (Shore et al., '90, in press, unpublished observations) confirm those of Saint Marie et al. ('91 and this volume) which show most cells which project to the VCN are in the deep layer. Saint Marie et al. ('91 and this volume) found that GLY IR cells give the largest projection from DCN to VCN. The chapter in this volume by Potashner et al. discusses other sources of GABA and glycine inputs to VCN besides DCN. We have used anterograde tract tracing combined with immunocytochemistry to examine DCN vs. non-DCN sources of inhibitory amino acid input to individual VCN neurons. Results for a typical globular bushy cell are shown in Figure 10. GLY IR terminals are the major input from DCN, as the results of Saint Marie et al. would suggest, with about half as many GABA/GLY IR terminals as GLY IR, and with few GABA IR terminals. The DCN appears to be the source of roughly half of the inhibitory amino acid input to this globular bushy cell.

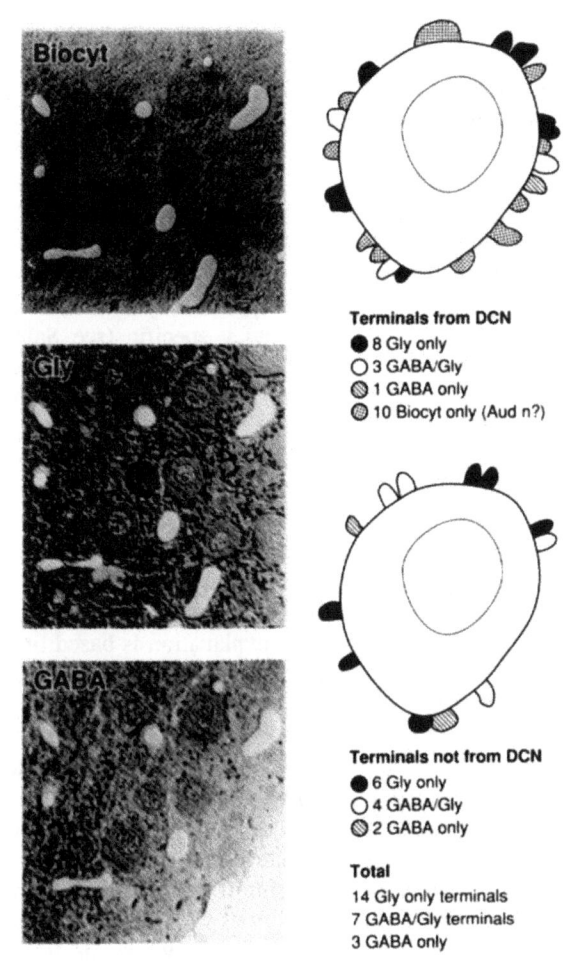

Figure 10. Results from a biocytin injection in DCN showing at left, biocytin (Biocyt), Glycine (GLY) IR and GABA IR labeling on adjacent 0.5 μm plastic sections and at right, schematic diagrams summarizing the GABA, GLY and GABA/GLY input from DCN (co-labeling with biocytin) and not from DCN (without biocytin co-label) on a profile through a typical globular bushy cell. "Biocyt only" terminals might be backlabeled auditory nerve fibers.

## Functional considerations

The immunocytochemical and tract tracing studies reviewed above suggest that most VCN neurons may be influenced by more than one type of inhibitory process. The differential distributions of GABA, GLY and GABA/GLY IR terminals on somata and dendrites of VCN projection neurons as well as the multiple sources of these terminals suggest that inhibitory inputs may play different roles in the processing of acoustic information in the VCN.

Physiological studies have revealed a number of response features of CN neurons that may be subserved by inhibitory influences. These include gaps, pauses or dips in temporal discharge patterns as seen in peristimulus time histograms, changes in discharge patterns as a function of stimulus intensity and/or frequency, delayed first spike latencies, elevated thresholds, non-monotonic rate-intensity functions, and inhibitory surrounds in neuronal response areas. It is not yet possible, however, to assign a specific inhibitory transmitter or input to any of these response features.

GABA and glycine have both been shown to affect response properties of VCN neurons. In contrast to the DCN where effects on non-monotonic rate functions and response areas have been reported, in the VCN the effect of GABA or GLY is generally a supression of spontaneous and stimulus-induced activity with GLY being slightly more potent than GABA (chapters in this volume by Caspary et al.; Evans and Zhao; Oertel and Wickesberg). The effect with either GABA or GLY is greater with on-CF and less with off-CF tones. However, frequencies above CF can also inhibit excitatory amino acid-elicited responses (Caspary, this volume; Martin, '83). The CF response may be mediated by the DCN-VCN pathway, as this pathway appears to be frequency specific (see Saint-Marie et al., and Wickesberg and Oertel, this volume). The off-CF effects may be mediated by the SOC-VCN pathways which are more diffuse (Shore et al., '91) and whose stimulation can lead to inhibition of VCN units (e.g. Comis, '70).

The functional significance of co-containment of GLY and GABA in some endings is not known. Allosteric modulation of GLY receptors by GABA has been shown in Mauthner cells (Werman, '80). Alternatively, both could act independently to selectively filter auditory processing depending on the type of acoustic signal and specific neural networks involved. Two neurotransmitters in the same terminal may act in sequence to achieve a particular temporal pattern of inhibition. Wu & Oertel ('86) see sequential temporal inhibition in CN after stimulation of the auditory nerve. Another explanation is based on the observation that onset and steady-state (adapted) portions of the auditory nerve response behave differently as intensity is increased. One neurotransmitter could possibly have a longer action than the other so that the target neuron may respond only to the onset or to the later portion of the signal. GABA and GLY may also have a role in producing the "adapted" portion of the response. Since fibers in a state of adaptation don't respond as well to presented signals, differences in state of adaptation among individuals in an array of neurons can influence their responses to subsequent components of complex stimuli, such as speech.

To understand VCN information processing it will therefore be important not only to find out more details on the sources and distributions of inhibitory amino acid terminals and the receptor(s) apposing them, but also to determine how specific transmitter and receptor combinations in particular locations influence information transfer. With this information we can take a large step towards understanding how inhibitory influences shape the responses of neurons in the VCN.

ACKNOWLEDGEMENTS

We would like to thank Dr. Michael Oberdorfer, Dr. Roger Albin and Ms. Joann Bonneau for major contributions to studies described in this chapter. We are grateful to Drs. Kopin, deBlas and Betz for providing antibodies, and to NIDCD for support through grant #DC00383.

REFERENCES

Adams, J.C. and Mugnaini, E., 1985, Patterns of immunostaining with antisera to peptides in the auditory brainstem of the cat, *Soc. Neurosci. Abstr.*, 11:32.

Altschuler, R.A., Sheridan, C.E., Horn, J.W. and Wenthold, R.J., 1989, Immunocytochemical localization of glutamate immunoreactivity in the guinea pig cochlea, *Hearing Res.*, 42:167—74.

Altschuler, R.A., Hoffman, D.W. and Wenthold, R.J., 1986, Neurotransmitters of the cochlea and cochlear nucleus: immunocytochemical evidence, *Am. J. Otolaryngol.* 7:100-106.

Altschuler, R.A., Betz, H., Parakkal, M.H., Reeks, K.A. and Wenthold, R.J., 1986, Identification of glycinergic synapses in the cochlear nucleus through immunocytochemical localization of the postsynaptic receptor, *Brain Res.*, 369:316-320.

Bledsoe, S.C., Bobbin, R.P. and Puel, J.-L., 1988, Neurotransmission in the inner ear, *in:* "Physiology of the Ear", A.F. Jahn and J. Santos-Sacchi, eds., Raven Press, New York.

Bormann, J., 1988, Electrophysiology of $GABA_A$ and $GABA_B$ receptor subtypes, *Trends in Neurosci.*, 11:112-116.

Cant, N.B., 1991, Projections to the lateral and medial superior olivary nuclei from the spherical and globular bushy cells of the anteroventral cochlear nucleus, *in:* "Neurobiology of Hearing: The Central Auditory System", R.A. Altschuler, R.P. Bobbin, B.M. Clopton, and D.W. Hoffman, eds., Raven Press, New York.

Cant, N.B. and Morest, D.K., 1984, The structural basis for stimulus coding in the cochlear nucleus, *in:* "Hearing Sciences; Recent Advances", C. Berlin, ed., College Hill Press, San Diego.

Caspary, D.M., 1986, Cochlear nuclei: Functional neuropharmacology of the principal cell types, *in:* "Neurobiology of Hearing: The Cochlea", R.A. Altschuler, D.W. Hoffmann and R.P. Bobbin, eds., Raven Press, New York.

Caspary, D.M., Havey, D.C. and Faingold, C.L., 1979, Effects of microiontophoretically applied glycine and GABA on neuronal response patterns in the cochlear nuclei, *Brain Res.*, 172:179—85.

Caspary, D.M., Rybak, L.P. and Faingold, C.L., 1984, Baclofen reduces tone-evoked activity of cochlear nucleus neurons, *Hearing Res.*, 13:113-122.

Clopton, B.M. and Backoff, P.M., 1991, Spectrotemporal receptive fields of neurons in cochlear nucleus of guinea pig, *Hearing Res.*, 52:329-344.

Comis, S.D., 1970, Centrifugal inhibitory processes affecting neurons in the cat cochlear nucleus, *J. Physiol.*, 210:751-760.

Frostholm, A. and Rotter, A., 1986, Autoradiographic localization of receptors in the cochlear nucleus of the mouse, *Brain Res. Bull.*, 16:189-203.

Glendenning, K.K. and Baker, B.N., 1988, Neuroanatomical distribution of receptors for three potential inhibitory neurotransmitters in the brain stem auditory nuclei of the cat, *J. Comp. Neurol.*, 275:288-308.

Godfrey, D.A., Parli, J.A., Dunn, J.D. and Ross, C.D., 1988, Neurotransmitter microchemistry of the cochlear nucleus and the superior olivary complex, *in:* "Auditory Pathways", J. Syka, and R.L. Masterton, eds., Plenum Press, NY.

Helfert, R.H., Bonneau, J.M., Wenthold, R.J. and Altschuler, R.A., 1989, GABA and glycine immunoreactivity in the guinea pig superior olivary complex, *Brain Res.*, 501:269-286.

Helfert, R.H., Snead, C.R. and Altschuler, R.A., 1991, The ascending auditory pathways, *in:* "Neurobiology of Hearing: The Central Auditory System", R.A. Altschuler, R.P. Bobbin, B.M.Clopton, and D.W. Hoffman, eds., Raven Press, New York.

Iversen, L.L. and Kelly, J.S., 1975, Uptake and metabolism of gamma-aminobutyric acid by neurones and glial cells, *Biochem. Pharmacol.*, 24:933-938.

Juiz, J.M., Helfert, R.H., Wenthold, R.J., De Blas, A.L. and Altschuler, R.A., 1989, Immunocytochemical localization of the $GABA_A$/benzodiazepine receptor in the guinea pig cochlear nucleus: evidence for receptor localization heterogeneity, *Brain Res.*, 504:173-179.

Juiz, J.M., Helfert, R.H., Bonneau, J.M., Wenthold, R.J. and Altschuler, R.A., Synaptic terminals and cell bodies immunoreactive for glycine or co-containing glycine and GABA immunoreactivities in the cochlear nucleus of the guinea pig: An ultrastructural immunogold study, *J. Comp. Neurol.*, submitted.

Juiz, J.M., Helfert, R.H., Bonneau, J.M., Wenthold, R.J. and Altschuler, R.A., 1989, Glycine immunoreactivity in the cochlear nucleus: Postembedding light and electron microscopic immunocytochemistry, *Soc. Neurosci. Abst.*, 15:743.

Juiz, J.M., Helfert, R.H., Bonneau, J.M., Wenthold, R.J. and Altschuler, R.A., 1990, Differential immunostaining of cell bodies and terminals in the cochlear nucleus by glutamate and GABA, *ARO Abst.*, 13:56.

Juiz, J.M., Albin, R.L., Helfert, R.H. and Altschuler, R.A., Distribution of $GABA_A$ and $GABA_B$ binding sites in the cochlear nucleus of the guinea pig, *Brain Res.*, (submitted).

Juiz, J.M., Helfert, R.H. and Altschuler, R.A., A simple procedure for silver intensification of immunocolloidal gold on ultrathin plastic sections, *J. Neurosci. Methods*, submitted.

Levitan, E.S., Schofield, P.R., Burt, D.R., Rhee, L.M., Wisden, W., Kohler, M., Fujita, N., Rodriguez, H.F., Stephenson, A. and Darlison, M.G., 1988, Structural and functional basis for $GABA_A$ receptor heterogeneity, *Nature*, 335:76-79.

Lorente de Nó, R., 1981, "The primary acoustic nuclei", R. Lorente de Nó, ed., Raven Press, New York.

Luddens, H. and Wisden, W., 1991, Function and pharmacology of multiple $GABA_A$ receptor subunits, *TIPS*, 12:49-51.

Malosio, M.L., Marqueze-Pouey, B., Kuhse, J. and Betz, H., 1991, Widespread expression of glycine receptor subunit mRNAs in the adult and developing rat brain, *EMBO* 10:2401-9.

Martin, M.R., 1983, The pharmacology of amino acid receptors and synaptic transmission in the cochlear nucleus, *in*: "Auditory Biochemistry", D.G. Drescher, ed., Charles C. Thomas, Springfield.

Moore, J.K., 1986, Cochlear nuclei: relationship to the auditory nerve, *in*: "Neurobiology of Hearing: The Cochlea", R.A. Altschuler, D.W. Hoffman and R.P. Bobbin, eds., Raven Press, New York.

Mugnaini, E. and Oertel, W.H., 1985, An atlas of the distribution of GABAergic neurons and terminals in the rat CNS as revealed by GAD immunocytochemistry, *in* "Handbook of Chemical Neuroanatomy, vol. 4: GABA and neuropeptides in the CNS", A. Bjorklund and T. Hokfelt, eds., Elsevier, Amsterdam.

Oberdorfer, M.D., Parakkal, M.H., Altschuler, R.A. and Wenthold, R.J., 1987, Co-localization of glycine and GABA in the cochlear nucleus, *Soc. Neurosci. Abstr.*, pp. 544.

Oberdorfer, M.D., Parakkal, M.H., Altschuler, R.A. and Wenthold, R.J., 1988, Ultrastructural localization of GABA-immunoreactive terminals in the anteroventral cochlear nucleus of the guinea pig, *Hearing Res.*, 33:229-238.

Osen, K.K., 1969, The intrinsic organization of the cochlear nuclei, *Acta Otolaryngol.*, 67:352—359.

Osen, K.K., Ottersen, O.P. and Storm-Mathissen, J., 1990, Co-localization of glycine-like and GABA-like immunoreactivities: A Semiquantitative study of individual neurons in the dorsal cochlear nucleus of cat, *in*: "Glycine neurotransmission", O.P.Ottersen and J.Storm-Mathissen, eds., John Willey & Sons, London, pp. 417-45.

Pritchett, D.B., Luddens, H. and Seeburg, P.H., 1989, Type I and type II GABA$_A$. Benzodiazepine receptors produced in transfected cells, *Science*, 245:1389-1392.

Rhode, W.S., 1991, Physiological-morphological properties of the cochlear nucleus, *in:* "Neurobiology of Hearing: The Central Auditory System", R.A. Altschuler, R.P. Bobbin, B.M. Clopton and D.W. Hoffman, eds., Raven Press, New York.

Sakmann, B., Hamill, O.P. and Bormann, J., 1983, Patch-clamp measurements of elementary chloride currents activated by the putative inhibitory transmitters GABA and glycine in mammalian spinal neurons, *J. Neural Transm.*, 18:83-95.

Schwartz, A.M. and Gulley, R.L., 1978, Non-primary afferents to the principal cells of the rostral anteroventral cochlear nucleus of the guinea pig, *Am. J. Anat.*, 153:489-508.

Schwartz, I.R., 1983, Autoradiographic studies of amino acid labeling of neural elements in the auditory brainstem, *in:* "Auditory Biochemistry", D.G. Drescher, ed., Charles C. Thomas, Springfield

Shore S.E., Helfert, R.H., Bledsoe, S.C., Altschuler, R.A. and Godfrey, D.A., 1991, Descending projection to the dorsal and ventral divisions of the cochlear nucleus in guinea pig, *Hearing Res.*, 52:255—268.

Shore, S.E., Godfrey, D.A., Helfert, R.H., Altschuler, R.A. and Bledsoe, S.C. Jr., 1991, Connections between the cochlear nuclei in the guinea pig, *Hearing Res.*, in press.

Spangler, K.M. and Warr, W.B., 1991, The descending auditory system, *in*: "Neurobiology of Hearing: The Central Auditory System", R.A. Altschuler, R.P. Bobbin, B.M. Clopton, and D.W.Hoffman, eds., Raven Press, New York.

Tolbert, L.P. and Morest, D.K., 1982, The neuronal architecture of the anteroventral cochlear nucleus of the cat in the region of the cochlear nerve root: electron microscopy, *Neurosci.*, 7:3053—3067.

Wenthold, R.J., Huie, D., Altschuler, R.A. and Reeks, K.A., 1987, Glycine immunoreactivity localized in the cochlear nucleus and superior olivary complex, *Neurosci.*, 22:897-912.

Wenthold, R.J. and Martin, M.R., 1984, Neurotransmitters of the auditory nerve and central auditory system, *in*: "Hearing Science", C. Berlin, ed., College Hill Press, San Diego.

Wenthold, R.J., Parakkal, M.H., Oberdorfer, M.D. and Altschuler, R.A., 1988, Glycine receptor immunoreactivity in the ventral cochlear nucleus of the guinea pig, *J. Comp. Neurol.*, 276:423—435.

Wenthold, R.J., Zempel, J.M., Parakkal, M.H., Reeks, K.A. and Altschuler, R.A., 1986, Immunocytochemical localization of GABA in the cochlear nucleus of the guinea pig, *Brain Res.*, 380:7-18.

Wenthold, R.J., 1991, Neurotransmitters of brainstem auditory nuclei, *in*: "Neurobiology of Hearing: The Central Auditory System", R.A. Altschuler, R.J. Bobbin, B.M. Clopton, and D.W. Hoffman, eds., Raven Press, New York.

Werman, R., 1980, GABA modulates glycine receptor interaction in Mauthner cells allosterically, *in*: "Transmitters and their receptors", U.Z. Littauer, Y. Dudai, I. Silman, V.I. Teichberg, and Z. Vogel, eds., Wiley and sons.

Wu, S.H. and Oertel, D., 1986, Inhibitory circuitry in the ventral cochlear nucleus is probably mediated by glycine, *J.Neurosci.*, 6:2691-706.

Zarbin, M.A., Warmsley, K.K. and Kuhar, M.J., 1981, Glycine receptor: Light microscopic audoradiographic localization with tritiated strychnine, *J. Neurosci.*, 1:532-547.

# GLYCINERGIC INHIBITION IN THE COCHLEAR NUCLEI: EVIDENCE FOR TUBERCULOVENTRAL NEURONS BEING GLYCINERGIC

Donata Oertel and Robert E. Wickesberg

Department of Neurophysiology, University of Wisconsin, Madison, Wisconsin, USA

The goal of this article is to review evidence that glycine is the neurotransmitter of inhibitory interneurons in the cochlear nuclei. One group of interneurons for which the evidence is particularly strong is the tuberculoventral neurons (Wickesberg and Oertel, '88) in the deep layer of the dorsal cochlear nucleus (DCN). This review will examine how the existing evidence supports the criteria that have been established for the identification of a neurotransmitter (Werman, '66). A review of this evidence leads to the conclusion that tuberculoventral neurons are glycinergic. The evidence is weaker that cartwheel cells of the DCN and D stellate cells of the ventral cochlear nucleus (VCN) are also glycinergic.

All inhibitory postsynaptic potentials (IPSPs) we have recorded in slices seem to be glycinergic because their increases in chloride conductance can be blocked by micromolar strychnine (Wu and Oertel, '86; Hirsch and Oertel, '88). In interpreting the results from slices we make the assumption that the IPSPs we record are not artifactual but we cannot know whether the absence of other types of inhibition reflects the physiology *in situ* or whether it results from the artifactual removal of parts of the circuit. We therefore make no conclusions concerning the absence of other types of IPSPs. Whatever the reason, the relative simplicity of the inhibitory physiology in slices is advantageous in that, with what is known about the anatomy, it becomes possible to identify inhibitory interneurons.

The finding that all detectable IPSPs in the cochlear nuclei seem to be glycinergic on the basis of physiological and pharmacological experiments in slices has come as a surprise since markers of GABAergic inhibition are clearly present in the cochlear nuclei (Mugnaini, '85; Wenthold et al., '86; Adams and Mugnaini, '87; Oberdorfer et al., '88; Osen et al., '90) and since iontophoretic applications of blockers of GABAergic inhibition affect responses to sound *in vivo* (Caspary, this volume; Evans and Zhao, this volume). There are several possible reasons for our not observing GABAergic inhibition in slices. 1) GABAergic pathways may not be activated under conditions *in vitro*. 2) GABA might be washed away by the perfusion saline. Saline flow has to be very rapid to maintain slices in a viable condition. 3) The presence of GABA and GAD may reflect functions other than inhibition.

## Presence of Glycine in the Cochlear Nuclei

The presence of glycine in the cochlear nuclei has been documented both biochemically and immunocytochemically. The biochemical methods provide information about absolute levels but do not allow precise localization of the glycine. Immunocytochemistry reveals where glycine levels are relatively high and where they are relatively low but absolute magnitudes are not known.

Godfrey and his colleagues (Godfrey et al., '77; Godfrey et al., '78) showed analytically that glycine is contained in all regions of the cochlear nuclei. The concentration of glycine as a function of dry weight is highest in the granule cell regions and in the molecular and fusiform cell layers of the DCN. Concentrations in the VCN were about two thirds of those in upper layers of the DCN. Being an amino acid, the presence of glycine is not remarkable. It is noteworthy, however, that the concentration of glycine per unit dry weight in the cochlear nuclei is an order of magnitude higher than in the frontal cortex where it is unlikely that glycine plays a transmitter role (Godfrey et al., '77).

Immunocytochemical assays for glycine became possible with the development of antibodies against glycine (Peyret et al., '87; Wenthold et al., '87; Aoki et al., '88). These antibodies have been used by a number of investigators to examine the distribution of glycine in the cochlear nuclei. Numerous cell bodies and puncta are strongly labeled with these antibodies (Wenthold et al., '87; Aoki et al., '88; Osen et al., '90; Saint Maire et al., '91; Wickesberg et al., '91).

In the DCN the cartwheel cells and tuberculoventral cells are immunopositive. Cartwheel cells in the outer part of the fusiform cell layer are consistently labeled with antibodies against glycine conjugates (Wenthold et al., '87; Osen et al., '90; Saint Marie et al., '91). Smaller neurons in the deep layer have also been found to be labeled (Wenthold et al., '87; Osen et al., '90; Saint Marie et al., '91; Wickesberg et al., '91). Many of these cells are tuberculoventral cells (also called vertical or corn cells by other authors). The experiments by Saint Marie et al., '91 show even more directly that neurons which project from the deep DCN to the VCN are often glycine-positive. Most (96%) of the cells that were labeled with HRP after injections of the ventral cochlear nucleus were also immunopositive for glycine. Labeled puncta are found most frequently in the fusiform cell and deep part of the deep layer in the DCN. Osen et al. '90 also report that fibers of the ventrotubercular tract, the tract that contains the axons of tuberculoventral neurons, are frequently immunopositive.

In the VCN, large cells that are especially common near the nerve root were frequently labeled with the antibody against the glycine-conjugate (Wenthold et al., '87). Some of these glycine-positive neurons have been shown to project to the contralateral cochlear nucleus (Wenthold, '87) and may correspond to D stellate cells (Oertel et al., '90) which in turn probably correspond to the type II stellate cells described in cats by Cant (Cant, '81; Smith and Rhode, '89). Neurons of the small cell cap were also stained. In the VCN many unlabeled neurons were surrounded by labeled puncta. The general staining pattern in the mouse (Fig. 1) resembles that in other mammals. Labeled puncta are considerably less abundant in the octopus cell area (Wickesberg et al., '91) (Fig. 1).

The suggestion was made that antibodies against glycine identify glycinergic neurons by staining their cell bodies (Wenthold, '87; Wenthold et al., '87; Aoki et al., '88). While this conclusion is consistent with results from other types of experiments, it seemed important to test this suggestion. There is no reason, *a priori*, to think that neurons which can accumulate and release glycine at their terminals need necessarily accumulate glycine in their cell bodies. A further difficulty with immunocytochemical localization of amino acids is that it is not clear what the absence of staining with antibodies against glycine means since all cells must contain glycine. It is not known what the difference in the concentration of glycine is between immuno-negative and immuno-positive neurons or whether the staining pattern depends on the details of the immunocytochemical methods. If that difference is small, it

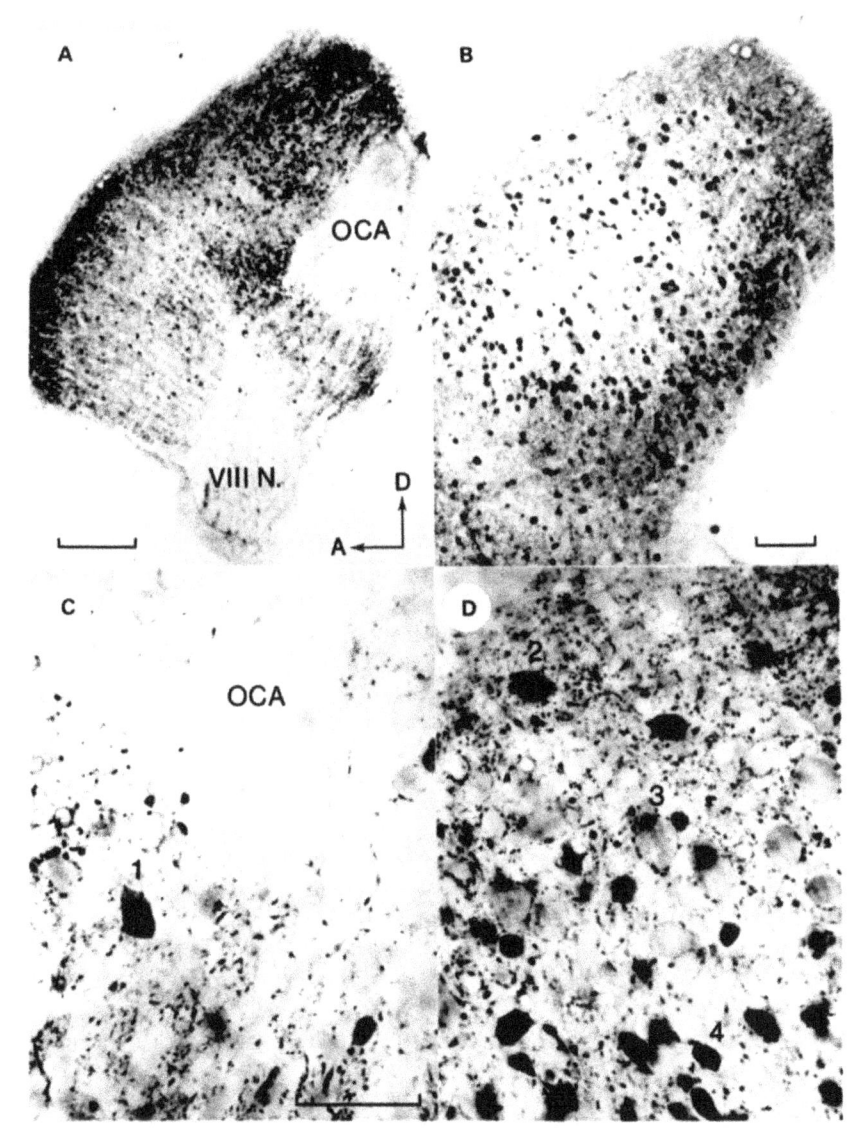

**Figure 1.** Glycine-like immunoreactivity in the cochlear nuclei. Photomicrographs are from two sections of one slice that was maintained "in vitro" and bathed in 50 μM colchicine for 6 hours before it was fixed and processed for immunocytochemistry. The micrographs shown in A and C are from a more medial section than those of B and D. A and C. Stained cell bodies (1, a presumed D stellate cell) are scattered throughout the AVCN and rostral PVCN. These are possibly D stellate cells. Cell bodies were not labeled in the octopus cell area (OCA). Labeled puncta are dense in the AVCN and in the multipolar cell area of the rostral PVCN but they are rare or absent in the octopus cell area. B and D. In this section, where the curved laminae of the DCN appear as concentric rings, one group of labeled cells (2, a presumed cartwheel cell) is in the outer part of the fusiform cell layer (3, a presumed unlabeled fusiform cell) and another (4, a presumed tuberculoventral cell) in the innermost deep layer. Scale bars: A, 250 μm; B, 100μm; C and D, 50μm.

seems possible that staining patterns can differ from experiment to experiment and between investigators. If the staining were inconsistent, the use of this antibody to identify glycinergic neurons would be limited. We therefore embarked on a series of experiments in which we tested whether tuberculoventral neurons, which we expected on the basis of physiological experiments in slices to be glycinergic (Wickesberg and Oertel, '90), stained consistently with the antibody against glycine. The immunocytochemical staining pattern in slices that were cut and then fixed resembled that in perfusion-fixed tissue. Surprisingly, however, the staining pattern of cell bodies changed when the tissue was maintained in vitro. Cell bodies in the deep layer of the DCN, but not cell bodies in the outer fusiform cell layer (probably cartwheel cells), lost their staining after 4 to 6 hours in vitro in the absence of stimulation of the nerve root. The labeling of puncta was not visibly changed. The loss of staining of neuronal cell bodies in the deep DCN was reversible with 15 minutes electrical stimulation of the auditory nerve at the end of a 6 hour incubation (Wickesberg et al., '90). The loss of staining was prevented by bathing the slices in 50 $\mu$M colchicine or 1 $\mu$M strychnine. These experiments show that staining of cell bodies with an antibody to a glycine conjugate depends on the metabolic history of cells. It seems possible, therefore, that immunocytochemical staining is not always consistent in experiments on tissue fixed *in situ* and that immunocytochemical staining does not identify all potentially glycine-positive cell bodies.

The conclusion that antibodies against glycine label glycinergic boutons is supported by the finding that labeled puncta are sparse or absent in the one part of the cochlear nuclear complex, the octopus cell area, where no presumptive glycinergic neurons have been observed to terminate. Tuberculoventral neurons do not project to the octopus cell area (Wickesberg et al., '91). Cartwheel cells are local interneurons whose axons terminate in the fusiform cell layer (Oertel and Wu, '89; Osen et al., '90; Berrebi and Mugnaini, '91). D Stellate cells also have not been seen to terminate in the octopus cell area (Oertel et al., '90).

In conclusion, biochemical and immunocytochemical experiments show that glycine is present in the cochlear nuclei. The results of these experiments indicate that high levels of glycine are found in many, but not all, cell bodies of presumptive glycinergic interneurons.

## Glycine is accumulated in synaptic terminals

The question how glycine is accumulated in neurons and their terminals has been addressed in two ways. Uptake has been measured biochemically in slices of tissue (Johnstone and Iversen, '71; Logan and Snyder, '72; Staatz-Benson and Potashner, '87 and '88; Bergman et al., '89) and it has been visualized with autoradiography at the light and electron microscopic levels.

Upon incubation with radioactively labeled glycine, some synaptic terminals in the cochlear nuclei are strongly labeled. Uptake of glycine into synaptic terminals in both VCN and DCN was demonstrated by Schwartz, '81 and '83. Precise identification of cell types was not possible in these experiments.

Staatz-Benson and Potashner measured uptake of labeled glycine into tissue slices of the cochlear nuclei biochemically. They found evidence for both high and low affinity uptake systems (Staatz-Benson and Potashner, '87). The high-affinity uptake system has an apparent $K_m$ of 25-30 $\mu$M; it depends on extracellular sodium, requires glucose and is similar to uptake systems elsewhere where glycine is thought to act as a neurotransmitter in the spinal cord (Johnstone and Iversen, '71; Logan and Snyder, '72; Pourcho and Goebel, '90). The low-affinity uptake system has an apparent $K_m$ of 630-720 $\mu$M and has a high capacity. Uptake systems like this one are quite general and probably mediate amino acid uptake for protein synthesis and cellular metabolism.

Uptake was reduced by lesions of the acoustic striae, suggesting that some glycinergic inhibition in the cochlear nuclei comes from centrifugal fibers (Staatz-Benson and Potashner, '88). Uptake was not significantly reduced by lesions of the inferior colliculi (Bergman et al., '89). One likely source of extrinsic glycinergic inputs to the cochlear nuclei is from the contralateral cochlear nuclear complex (Mast, '70; Cant and Gaston, '82; Wenthold, '87). These commissural neurons could be D stellate cells (Oertel et al., '90).

## Glycine is released synaptically

Wenthold ('79) measured the release of endogenous glycine from slices that were stimulated with potassium in the presence and absence of extracellular calcium. The calcium-dependent component, about 60% of the total in these experiments, presumably reflects release from synaptic terminals. Staatz-Benson and Potashner ('87) measured the release of radioactively labeled glycine from the anteroventral, posteroventral and dorsal cochlear nuclei in separate tissue slices; they, too, found a significant calcium-dependent component in each of the three parts of the cochlear nuclear complex. The calcium-dependent component was more prominent when the extracellular sodium concentration was lowered, presumably because the high affinity uptake system was slowed.

One difficulty with the interpretation of these experiments is that in measuring release of amino acids in any brain tissue many amino acids, including some that clearly have no neurotransmitter role, are released in a calcium-dependent fashion (Wenthold, '79; Baughman and Gilbert, '80). Also, these measurements of release are necessarily made from bulk tissue so that the sources of release have been localized to each of the three subdivisions of the cochlear nuclei but not to particular cell types.

## Glycine is removed from synapses

Removal of glycine at synapses probably occurs through high-affinity uptake systems in either or both neurons and glial cells. Where glycinergic inhibition is thought to occur, in the retina, in the spinal cord and in the cochlear nuclei, high-affinity uptake systems have been shown to exist (Johnstone and Iversen, '71; Logan and Snyder, '72; Staatz-Benson and Potashner, '87; Pourcho and Goebel, '90). Low-affinity uptake systems are present not only in neurons but in cells generally; they presumably subserve the metabolic needs for amino acids. It is not clear whether uptake by neuronal processes or by glial cell processes plays the major role in clearing glycine from the region of synapses. The autoradiographic finding that some terminals are very obviously labeled with glycine while other processes are not obviously labeled (Schwartz, '81 and '83) suggests that much of the high-affinity uptake is by synaptic terminals.

## Glycine mimics natural response in its action on glycine receptors of target cells

In all ways that have been tested, extrinsically applied glycine mimics IPSPs. Both IPSPs and extrinsically applied glycine produce increases of conductance which drag the voltage to levels slightly hyperpolarized with respect to the resting potentials when recordings are made with potassium acetate-containing electrodes. Both IPSPs and extrinsically applied glycine become depolarizing when chloride is included in the pipette. Both IPSPs and effects of extrinsically applied glycine are blocked by strychnine.

The IPSPs recorded intracellularly in bushy, T stellate, fusiform, tuberculoventral and giant cells in slice preparations are identical in their properties. They reverse at -67 mV on the average, slightly hyperpolarized with respect to the resting potential. Changes in intracellular or extracellular chloride concentration, but not changes in the extracellular potassium concentration, change the reversal potential. These findings indicate that IPSPs are produced by increases in chloride conductance (Wu and Oertel, '86; Hirsch and Oertel, '88).

Bath-applied glycine (as well as GABA) also produces increases in conductance (Wu and Oertel, '86; Hirsch and Oertel, '88). Bath application of glycine reduces the size of the voltage changes produced by current pulses of constant magnitude, reflecting the decreased

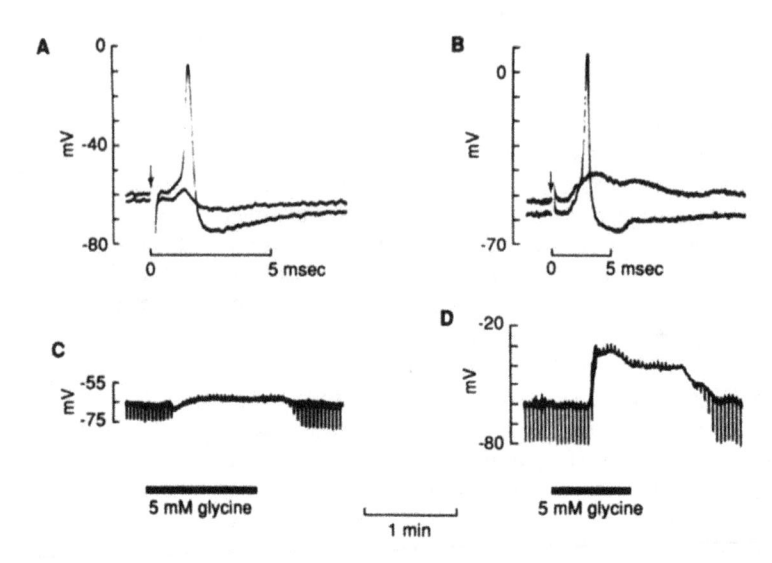

**Figure 2.** Bath-applied glycine mimics IPSPs in that both produce increases in chloride permeability. The upper panels show two superimposed traces of synaptic responses to electrical stimulation with shocks of 0.1 msec duration (arrows) of the auditory nerve root. The lower panels show responses to bath-applied glycine. The recordings on the left were made through electrodes that were filled with 4 M potassium acetate; those on the right were made through electrodes filled with 3 M potassium chloride. Each panel shows recordings from a different cell. A. After a synaptic delay, shocks evoked an excitatory postsynaptic potential that was suprathreshold in one case and that was cut short by an inhibitory synaptic potential in the other in a T stellate cells. B. A similar recording from another T stellate cell with a pipette containing chloride shows that shocks evoke two depolarizing synaptic potentials. C. Chart record from a T stellate cell into which current pulses of constant magnitude were injected to monitor input resistance. When glycine was applied in the bath (bar) the input resistance dropped but the membrane potential changed little. D. In another T stellate cell that was impaled with a chloride containing pipette, bath application of glycine caused the cell to depolarize. (Some traces in this figure have been published previously (Wu and Oertel, '86).

input resistance during the application of glycine (Wu and Oertel, '86; Hirsch and Oertel, '88) (Figs. 2 and 3). When intracellular recordings are made with pipettes that contain high concentrations of chloride, presumably allowing chloride to leak out of the pipette to increase the intracellular concentration of chloride significantly, bath application of glycine depolarized the cells to threshold (Wu and Oertel, '86) (Fig. 2). An intracellular increase in chloride concentration shifts the chloride equilibrium potential positively.

Not all of the tests described above have been carried out on all cell types. Thus it is important to consider for what cell types these findings apply. Intracellular recordings are only possible to make from the larger cells; nothing is known about the smaller cells. Furthermore, at the time some of these experiments were done, the correlation between physiology and anatomical cell type was incomplete. In the ventral cochlear nucleus, the conclusions were tested on bushy and "stellate cells." Subsequent work shows that these tests were done on T stellate cells (Oertel et al., '90). It is not yet known whether inhibition in D stellate cells and in octopus cells resembles that in T stellate cells. In the DCN, subsequent work shows (Oertel and Wu, '89) that the tests reported by Hirsch and Oertel ('88) were mostly on fusiform cells. Less extensive observations show that inhibition in tuberculoventral and giant cells is like that in fusiform cells. We have not observed IPSPs on identified cartwheel cells (Zhang and Oertel, unpublished results).

Tests of the action of glycine on neurons have also been made *in vivo* by iontophoretic injections (Caspary et al., '79; Martin et al., '82; Caspary, '90). Glycine suppresses firing in those neurons tested both in the VCN and in the DCN. Suppression of spontaneous firing is particularly striking (Caspary et al., '79). Tests were made on buildup/pauser units that are probably fusiform cells and primary-like units that are probably bushy cells.

The presence of glycine receptors on cells of the cochlear nuclei has been demonstrated not only physiologically but also biochemically. In an elegant series of studies, Betz and his colleagues have identified, purified, biochemically characterized and cloned the glycine receptor (Betz, '87; Betz and Becker, '88). The glycine receptor complex was initially purified by passing a membrane fraction, solubilized with detergent, of tissue from rat spinal cord, medulla and pons through an affinity column of aminostrychnine agarose and eluting with a buffer that contained glycine (Pfeiffer et al., '82). Three polypeptides of about 48, 58, and 93 kD were isolated with affinity chromatography. These three polypeptides form a pentameric complex (Langosch et al., '88) with the 48 kD component containing the strychnine binding sites (Pfeiffer et al., '82; Graham et al., '83) and the 93 kD protein being a peripheral membrane protein on the cytoplasmic face (Schmitt et al., '87). Expression of the 48 kD component produced functional receptors in Xenopus oocytes, probably as homooligomers (Schmieden et al., '89). These results on the biochemistry of the glycine receptor are significant not only because they provide information about the glycine receptor but also because they support the conclusion that strychnine binds specifically to the glycine receptor.

The ability to purify the glycine receptor made it possible to make antibodies to the receptor complex (Pfeiffer et al., '84). The antibodies have been useful for localizing glycine receptors (Araki et al., '88; Wenthold, et al., '88). Two monoclonal antibodies raised against the 93 kD and the 48 kD components of the receptor complex (Pfeiffer et al., '84) have been used to test for the presence of this protein in the cochlear nuclei (Altschuler et al., '86; Wenthold, '88). These authors report that glycine-receptor-like immunoreactivity was found on all the major cell types except possibly granule cells. Only the neuropil, not the cell bodies of granule cells, contains immunoreactivity (Wenthold et al., '88). Adams ('91) reports that octopus cell bodies and surrounding neuropil have little glycine-receptor-like immunoreactivity but that octopus cells contain glycine receptors in inclusion bodies. Taken together, these experiments show that all major cell types with the possible exceptions of granule and octopus cells have glycine receptors.

A third approach that demonstrates the presence of glycine receptors involves the localization of radioactively labeled strychnine binding by autoradiography. Since strychnine binds the glycine receptor relatively more tightly than it binds other molecules, binding is expected to mark the presence of receptors. Binding to a tissue of tritiated strychnine reveals patterns of differential binding (Frostholm and Rotter, '86; Sanes et al., '87; Glendenning and Baker, '88; Cortes and Palacios, '90). The highest density of labeling was over the deep layer of the dorsal cochlear nucleus; less label was localized over the ventral cochlear nucleus.

## Antagonists act on synaptic responses as they act on exogenously applied agonist

All IPSPs tested in slice preparations of the cochlear nuclei are blocked by strychnine at 0.5 or 1 $\mu$M concentrations whereas they are not blocked by picrotoxin or bicuculline at 100 $\mu$M concentrations (Fig. 3) (Wu and Oertel, '86; Hirsch and Oertel, '88).

Inhibition mediated by increases in chloride permeability is produced not only by glycine receptors but commonly also through $GABA_A$ receptors. Indeed, the bath-application of either glycine or GABA causes an increase in permeability in cells of the ventral (Wu and Oertel, '86) and dorsal cochlear nuclei (Hirsch and Oertel, '88). It is therefore of interest to test whether the action of strychnine is specific for extrinsically applied glycine. Bath-applied strychnine does indeed block the conductance increase produced by bath-applied glycine (Fig. 4).

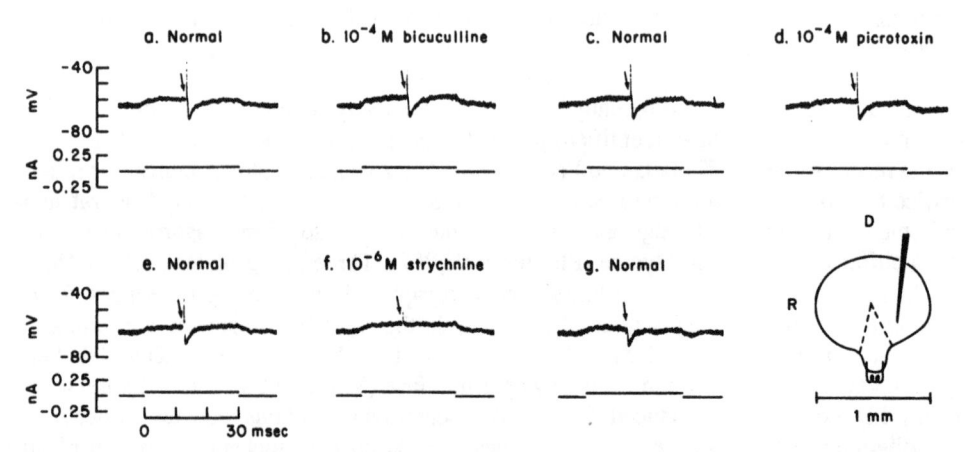

**Figure 3.** IPSPs are blocked by strychnine and not by bicuculline or picrotoxin. All traces were recorded in the order shown from a single cell. The cell was depolarized with current to reveal the IPSP more clearly. The IPSP was unaffected by 100 μM bicuculline and by 100 μM picrotoxin. It was blocked completely and reversibly by 1 μM strychnine. (Figure is reproduced from Wu and Oertel, '86).

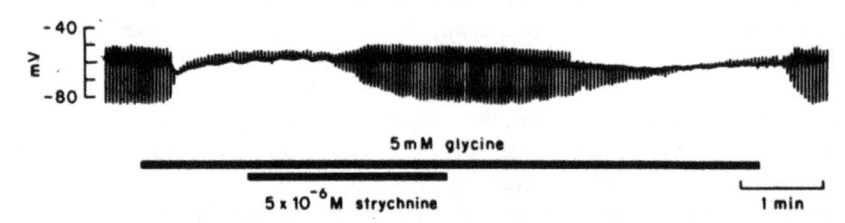

**Figure 4.** The action of glycine is blocked reversibly by strychnine. A chart record from a cell whose input resistance was monitored by voltage changes produced by constant current pulses. Glycine caused the input resistance to drop; strychnine blocked the effect of glycine reversibly. (Figure is reproduced from Wu and Oetel, '86).

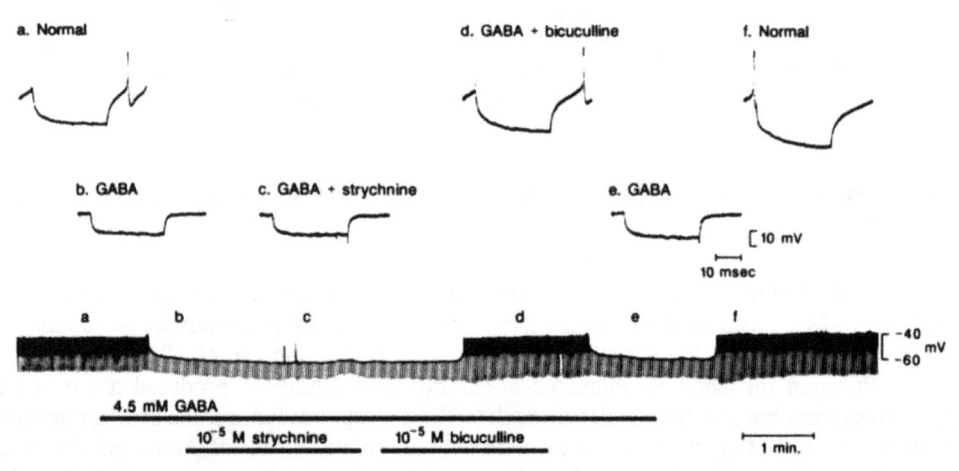

**Figure 5.** GABA produces a change in input resistance that is specifically blocked by bicuculline. In a cell that fired spontaneously (a), the bath application of GABA silenced the firing and reduced the input resistance of the cell (b). 10 μM Strychnine did not affect the response to GABA (c) whereas 10 μM bicuculline blocked the effect of GABA completely (d) and reversibly (e). The effect of GABA was also reversible (f).

Since the action of glycine is similar to that of GABA in that both agonists produce increases in chloride conductance, the specificity of agonists and antagonists was tested (Fig. 5). Bath application of GABA silenced the spontaneous firing and produced the expected decrease in input resistance. The action of GABA was not affected by 10 $\mu$M strychnine but was blocked by bicuculline at the same concentration. The experiments illustrated in Figs. 2, 3, 4 and 5 show physiologically that neurons in the VCN have receptors for both glycine and GABA. The glycine receptors are blocked by strychnine and the GABA receptors are blocked by bicuculline.

The action of these agonists and antagonists is consistent with their actions elsewhere; strychnine specifically blocks glycinergic inhibition and bicuculline blocks GABAergic inhibition. The specificity of action of strychnine was originally described in the spinal cord (Curtis et al., '68) where is has been studied in detail (Homma and Rovainen, '78). In dorsal root ganglia, too, strychnine blocks the action of glycine specifically while bicuculline blocks the action of GABA (Chio and Fischbach, '81). At concentrations at or above 100 $\mu$M the blocking actions of strychnine and picrotoxin are not specific (Choi and Fischbach, '81). In the retina, too, strychnine seems to block the action of glycine (Miller et al., '81; Miller et al., '81; Belgum et al., '84). At high concentrations strychnine can also have curare-like effects (Gilman et al., '80). It makes sense that strychnine affects GABA and nicotinic acetylcholine receptors since glycine, $GABA_A$ and nicotinic receptors belong to a family of molecularly related receptors (Grenningloh et al., '87).

*In situ* it is possible to apply strychnine by iontophoretic application. In such experiments responses to tones of presumptive fusiform cells at the characteristic frequency are accompanied by more firing, presumably because some inhibition had been removed (Caspary et al., '87). In these experiments the concentration of strychnine varies depending on how much is released from the pipette and on the distance from the pipette.

## Identity of Glycinergic Neurons

Which interneurons are glycinergic? To date all inhibition we have recorded electrophysiologically in slices of both the ventral and dorsal cochlear nuclei of mice seems to be glycinergic (Wu and Oertel, '86; Hirsch and Oertel, '88; Zhang and Oertel, '90). All IPSPs we have tested, in recordings from about 1000 neurons of various anatomical types, have had similar and consistent properties. Our conclusions concerning glycinergic inhibition are therefore based on IPSPs that are either spontaneous or that can be driven in slice preparations. Presumably glycinergic inhibition is mediated by interneurons intrinsic to the slice preparations; there are no indications of axoaxonic neuronal circuits in studies of the cochlear nuclei. Interneurons can be identified with a combination of anatomical, immunocytochemical and physiological experiments.

Tuberculoventral cells are very likely to be glycinergic interneurons. Tuberculoventral neurons are the only group of cells that projects from the DCN to the AVCN and rostral PVCN in bats, mice, cats, gerbils and guinea pigs (Feng and Vater, '85; Snyder and Leake, '88; Wickesberg and Oertel, '88; Mueller, '90; Saint Marie et al., '91). The projection pattern of groups of cells is consistent with the anatomy of single tuberculoventral cells (Oertel and Wu, '89). Activating this connection selectively produces inhibition in bushy and T stellate cells that can be blocked with low concentrations of strychnine (Wickesberg and Oertel, '90). The timing of inhibition through the DCN indicates that the connection of tuberculoventral neurons with their targets in the VCN is monosynaptic (Wickesberg and Oertel, '90). Tuberculoventral neurons are labeled immunocytochemically by antibodies to glycine conjugates (Wenthold et al., '87; Saint Marie et al., '91; Wickesberg et al., '91). Immunolabeled puncta are abundant where tuberculoventral neurons are known to end and they are sparse or absent where they do not end (Wickesberg et al., 91).

Cartwheel cells are also possibly glycinergic. Fusiform cells receive bursting, spontaneous, strychnine-sensitive IPSPs (Hirsch and Oertel, '88; Oertel and Wu, '89). The spontaneous IPSPs are intrinsic to the DCN (Hrisch and Oertel, '88). Immunocytochemical experiments show that two cell types are often positive for glycine-like immunoreactivity: tuberculoventral neurons and cartwheel cells. The spontaneous IPSPs probably do not arise from tuberculoventral neurons, glycinergic interneurons that also terminate on fusiform cells (Oertel and Wu, '89), for two reasons. First, other targets of tuberculoventral neurons, the bushy and stellate cells of the VCN, do not have spontaneous IPSPs (Oertel, '83; Wu and Oertel, '84; Wu and Oertel, '86). Second, tuberculoventral neurons are not known to be spontaneously active or to fire in bursts (Oertel and Wu, '89). Cartwheel cells could be the source of these spontaneous, bursty IPSPs. Cartwheel cells terminate on fusiform cells (Berrebi and Mugnaini, '91). They do fire in bursts (Zhang and Oertel, unpublished results; Manis, personal communication). The possibility that some of the smaller cells from which recordings have not been made mediate these spontaneous, bursty IPSPs cannot be completely eliminated.

D Stellate cells are also probably glycinergic. D Stellate cells have axons with collateral endings in the VCN and DCN. Their axons probably project dorsally through the intermediate acoustic stria, just beneath the dorsal acoustic stria (Oertel et al., '90; Smith and Rhode, '89). They may well correspond to type II stellate cells of Cant (Cant, '81) and to the cells that project to the contralateral cochlear nuclear complex (Wenthold, '87; Cant and Gaston, '82). There are glycinergic, inhibitory interneurons in the VCN because disynaptic IPSPs can be evoked in bushy cells from slices that contain no part of the DCN (Wu and Oertel, '84; Wu and Oertel, '86). Large multipolar cells are labeled immunocytochemically with antibodies to glycine-conjugates (Wenthold, '87; Osen et al., '90; Saint Marie et al., '91; Wickesberg et al., '91). Could these neurons correspond to T or D stellate cells? They are unlikely to be T stellate cells because the distribution of T stellate cells, labeled by injections into the inferior colliculus, is different (Osen, '72; Adams, '79; Ryugo et al., '81; Oliver, '87). Furthermore, T stellate cells, probably corresponding to Cant's type I (Cant, '81), are likely to be excitatory (Oliver, '87; Smith and Rhode, '89; Oertel et al., '90). On the other hand, immunocytochemical staining and the projection of D stellate cells to the DCN and further dorsalward is consistent with the projection pattern of the commissural neurons (Cant and Gaston, '82) which have been shown possibly to be glycinergic (Wenthold, '87; Mast, '70). Collaterals of D stellate cells could provide the local inhibition seen in the VCN.

## ACKNOWLEDGEMENTS

Our colleagues in the Department of Neurophysiology have helped with many aspects, both technical and conceptual, of this work. We thank especially Donna Whitlon who helped with immunocytochemical staining shown in Figure 1 and Sam Zhang whose fine work underlies some of the conclusions made here. We also thank Nace Golding and Don Robertson with whom it is always fun to argue. We doubt that we could have done this work without he help of Jo Ann Ekleberry, Joan Meister, Inge Siggelkow, Carol Dizack, Terry Stewart, Paul Luther, Bob Klipstein or Pat Heinritz. This work was supported by grants from the NIH R01 DC00176 and P01 DC00116.

## REFERENCES

Adams, J.C., 1979, Ascending projections to the inferior colliculus, *J. Comp. Neurol.*, 183:519-538.
Adams, J.C. and Mugnaini, E., 1987, Patterns of glutamate decarboxylase immunostaining in the feline cochlear nuclear complex studied with silver enhancement and electron microscopy, *J. Comp. Neurol.*, 262:375-401.

Adams, J.C., (in press), Distribution of some cytochemically distinct cell classes in the ventral cochlear nucleus of cat and human with emphasis on octopus cells and their projections, *Adv. in Speech, Lang. and Hear. Proc.* 3.

Altschuler, R.A., Betz, H., Parakkal, M.H., Reeks, K.A. and Wenthold, R.J., 1986, Identification of glycinergic synapses in the cochlear nucleus through immunocytochemical localization of the postsynaptic receptor, *Brain Res.*, 369:316-320.

Aoki, E., Semba, R., Keino, H., Kato, K. and Kashiwamata, S., 1988, Glycine-like immunoreactivity in the rat auditory pathway, *Brain Res.* 442:63-71.

Araki, T., Yamano, M., Murakami, T., Wanaka, A., Betz, H. and Tohyama, M., 1988, Localization of glycine receptors in the rat central nervous system: an immunocytochemical analysis using monoclonal antibody, *Neurosci.*, 25:613-624.

Baughman, R.W. and Gilbert, C.D., 1980, Aspartate and glutamate as possible neurotransmitters of cells in layer 6 of the visual cortex, *Nature*, 287:848-850.

Belgum, J.H., Dvorak, D.R. and McReynolds, J.S., 1984, Strychnine blocks transient but not sustained inhibition in mudpuppy retinal ganglion cells, *J. Physiol. (Lond.)*, 354:273-286.

Betz, H., 1987, Biology and structure of the mammalian glycine receptor, *Trends Neurosci.*, 10:113-117.

Betz, H. and Becker, C.-M., 1988, The mammalian glycine receptor: biology and structure of a neuronal chloride channel protein, *Neurochem. Int.*, 13:137-146.

Bergman, M., Staatz-Benson, C. and Potashner, S.J., 1989, Amino acid uptake and release in the guinea pig cochlear nucleus after inferior colliculus ablation, *Hear. Res.*, 42:283-291.

Berrebi, A.S. and Mugnaini, E., 1991, Distribution and targets of the cartwheel cell axon in the dorsal cochlear nucleus of the guinea pig, *Anat. Embryol.*, 183:427-454.

Cant, N.B., 1981, The fine structure of two types of stellate cells in the anterior division of the anteroventral cochlear nucleus of the cat, *Neurosci.*, 6:2643-2655.

Cant, N.B. and Gaston, K.C., 1982, Pathways connecting the right and left cochlear nuclei, *J. Comp. Neurol.*, 212:313-326.

Caspary, D.M., Havey, D.C. and Faingold, C.L., 1979, Effects of microiontophoretically applied glycine and GABA on neuronal response patterns in the cochlear nuclei, *Brain Res.*, 172:179-185.

Caspary, D.M., Pazara, K.E., Kossl, M. and Faingold, C.L., 1987, Strychnine alters the fusiform cell output from the dorsal cochlear nucleus, *Brain Res.*, 417:273-282.

Caspary, D.M., 1990, Electrophysiological studies of glycinergic mechanisms in auditory brain stem structures, in: "Glycine Neurotransmission", O.P. Ottersen and J. Storm-Mathisen, eds., John Wiley and Sons, New York, p. 453-483.

Choi, D.W. and Fischbach, G.D., 1981, GABA conductance of chick spinal cord and dorsal root ganglion neurons in cell culture, *J. Neurophysiol.*, 45:605-620.

Cortes, R., and Palacios, J.M., 1990, Autoradiographic mapping of glycine receptors by [$^3$H]strychnine binding, in: "Glycine Neurotransmission", O.P. Ottersen and J. Storm-Mathisen, eds., John Wiley and Sons, New York, p. 239-263.

Curtis, D.R., Hosli, L. and Johnstone, G.A.R., 1968, A pharmacological study of the depression of spinal neurones by glycine and related amino acids, *Exp. Brain Res.*, 6:1-18.

Feng, A.S. and Vater, M., 1985, Functional organization of the cochlear nucleus of rufous horseshoe bats (Rhinolophus rouxi): frequencies and internal connections are arranged in slabs, *J. Comp. Neurol.*, 235:529-553.

Frostholm, A. and Rotter, A., 1986, Autoradiographic localization of receptors in the cochlear nucleus of the mouse, *Brain Res. Bull.*, 16:189-203.

Gilman, A.G., Goodman, L.S., and Gilman, A., 1980, "The Pharmacological Basis of Therapeutics", Macmillan, New York.

Glendenning, K.K. and Baker, B.N., 1988, Neuroanatomical distribution of receptors for three potential inhibitory neurotransmitters in the brainstem auditory nuclei of the cat, *J. Comp. Neurol.*, 275:288-308.

Godfrey, D.A., Carter, J.A., Berger, S.J., Lowry, O.H. and Matchinsky, F.M., 1977, Quantitative histochemical mapping of candidate transmitter amino acids in cat cochlear nucleus, *J. Histochem. Cytochem.*, 25:417-431.

Godfrey, D.A., Carter, J.A., Lowry, O.H. and Matchinsky, F.M., 1978, Distribution of gamma-aminobutyric acid, glycine, glutamate and aspartate in the cochlear nucleus of the rat, *J. Histochem. Cytochem.* 26:118-126.

Graham, D., Pfeiffer, F. and Betz, H., 1983, Photoaffinity-labelling of the glycine receptor of rat spinal cord, *Eur. J. Biochem.*, 131:519-525.

Grenningloh, G., Gundelfinger, E., Schmitt, B. and Betz, H., 1987, Glycine vs GABA receptors, *Nature*, 330:25.

Hirsch, J.A. and Oertel, D., 1988, Synaptic connections in the dorsal cochlear nucleus of mice, in vitro, *J. Physiol. (Lond.)*, 396:549-562.

Homma, S. and Rovainen, C.M., 1978, Conductance increases produced by glycine and gamma-aminobutyric acid in lamprey interneurons, *J. Physiol. (Lond.)*, 279:231-252.

Johnstone, G.A.R. and Iversen, L.L., 1971, Glycine uptake in rat central nervous system slices and homogenates: Evidence for different uptake systems in spinal cord and cerebral cortex, *J. Neurochem.*, 18:1951-1961.

Langosch, D., Thomas, L. and Betz, H., 1988, Conserved quaternary structure of ligand-gated ion channels: The postsynaptic glycine receptor is a pentamer, *Proc.Natl.Acad.Sci.U.S.A.*, 85:7394-7398.

Logan, W.L. and Snyder, S.H., 1972, High affinity uptake systems for glycine, glutamic and aspartic acids in synaptosomes of rat central nervous system tissues, *Brain Res.*, 42:413-431.

Martin, M.R., Dickson, J.W. and Fex, J., 1982, Bicuculline, strychnine and depressant amino acid responses in the anteroventral cochlear nucleus of the cat, *Neuropharmacology*, 21:201-207.

Mast, T.E., 1970, Binaural interaction and contralateral inhibition in dorsal cochlear nucleus of the chinchilla, *J. Neurophysiol.*, 33:108-115.

Miller, R.F., Frumkes, T.E., Slaughter, M. and Dacheux, R.F., 1981, Physiological and pharmacological basis of GABA and glycine action on neurons of the mudpuppy retina. II. Amacrine and ganglion cells, *J. Neurophysiol.*, 45:764-782.

Miller, R.F., Frumkes, T.E., Slaughter, M. and Dacheux, R.F., 1981, Physiological and pharmacological basis of GABA and glycine action on neurons of mudpuppy retina. I. Receptors, horizontal cells, bipolars, and G-cells, *J. Neurophysiol.*, 45:743-763.

Mugnaini, E., 1985, GABA neurons in the superficial layers of the rat dorsal cochlear nucleus: light and electron microscopic immunocytochemistry, *J. Comp. Neurol.*, 235:61-81.

Müller, M., 1990, Quantitative comparison of frequency representation in the auditory brainstem nuclei of the gerbil, Pachyuromys duprasi, *Exp. Brain Res.*, 81:140-149.

Oberdorfer, M.D., Parakkal, M.H., Altschuler, R.A. and Wenthold, R.J., 1988, Ultrastructural localization of GABA-immunoreactive terminals in the anteroventral cochlear nucleus of the guinea pig, *Hearing Res.*, 33:229-238.

Oertel, D., 1983, Synaptic responses and electrical properties of cells in brain slices of the mouse anteroventral cochlear nucleus, *J. Neurosci.*, 3:2043-2053.

Oertel, D. and Wu, S.H., 1989, Morphology and physiology of cells in slice preparations of the dorsal cochlear nucleus of mice, *J. Comp. Neurol.*, 283:228-247.

Oertel, D., Wu, S.H., Garb, M.W. and Dizack, C., 1990, Morphology and physiology of cells in slice preparations of the posteroventral cochlear nucleus of mice, *J. Comp. Neurol.*, 295:136-154.

Oliver, D.L., 1987, Projections to the inferior colliculus from the anteroventral cochlear nucleus in the cat: possible substrates for binaural interaction, *J. Comp. Neurol.*, 264:24-46.

Osen, K.K., 1972, Projection of the cochlear nuclei on the inferior colliculus in the cat, *J. Comp. Neurol.*, 144:355-372.

Osen, K.K., Ottersen, O.P. and Storm-Mathisen, J., 1990, Colocalization of glycine-like and GABA-like immunoreactivities: A semiquantitative study of individual neurons in the dorsal cochlear nucleus of cat, in: "Glycine Neurotransmission", O.P.Ottersen and J.Storm-Mathisen, eds., John Wiley and Sons, New York, p. 417-451.

Peyret, D., Campistron, G., Geffard, M. and Aran, J.M., 1987, Glycine immunoreactivity in the brainstem auditory and vestibular nuclei of the guinea pig, *Acta Otolaryngol.(Stockh).*, 104:71-76.

Pfeiffer, F., Graham, D. and Betz, H., 1982, Purification by affinity chromatography of the glycine receptor of rat spinal cord, *J. Biol. Chem.*, 257:9389-9393.

Pfeiffer, F., Simler, R., Grenningloh, G. and Betz, H., 1984, Monoclonal antibodies and peptide mapping reveal structural similarities between the subunits of the glycine receptor of rat spinal cord, *Proc.Natl.Acad.Sci.USA*, 81:7224-7227.

Pourcho, R.G. and Goebel, D.J., 1990, Autoradiographic and immunocytochemical studies of glycine-containing neurons in the retina, in: "Glycine Neurotransmission", O.P.Ottersen and J.Storm-Mathisen, eds., John Wiley and Sons, New York, p. 356-389.

Ryugo, D.K., Willard, F.H., and Fekete, D.M., 1981, Differential afferent projections to the inferior colliculus from the cochlear nucleus in the albino mouse, *Brain Res.*, 210:342-349.

Saint Marie, R.L., Benson, C.G., Ostapoff, E.-M. and Morest, D.K., 1991, Glycine immunoreactive projections from the dorsal to the anteroventral cochlear nucleus, *Hearing Res.*, 51:11-28.

Sanes, D.H., Geary, W.A., Wooten, G.F. and Rubel, E.W., 1987, Quantitative distribution of the glycine receptor in the auditory brain stem of the gerbil, *J. Neurosci.*, 7:3793-3802.

Schmieden, V., Grenningloh, G., Schofield, P.R. and Betz, H., 1989, Functional expression in Xenopus oocytes of the strychnine binding 48 kd subunit of the glycine receptor, *EMBO J.*, 8:695-700.

Schmitt, B., Knaus, P., Becker, C.-M. and Betz, H., 1987, The Mr 93,000 polypeptide of the postsynaptic glycine receptor complex is a peripheral membrane protein, *Biochem.*, 26:805-811.

Schwartz, I.R., 1981, The differential distribution of label following uptake of $^3$H-labeled amino acids in the dorsal cochlear nucleus of the cat An autoradiographic study, *Exp. Neurol.*, 73:601-617.

Schwartz, I.R., 1983, Differential uptake of $^3$H-amino acids in the cat cochlear nucleus, *Am. J. Otol.*, 4:300-304.

Smith, P.H. and Rhode, W.S., 1989, Structural and functional properties distinguish two types of multipolar cells in the ventral cochlear nucleus, *J. Comp. Neurol.*, 282:595-616.

Snyder, R.L. and Leake, P.A., 1988, Intrinsic connections within and between cochlear nucleus subdivisions in cat, *J. Comp. Neurol.*, 278:209-225.

Staatz-Benson, C. and Potashner, S.J., 1987, Uptake and release of glycine in the guinea pig cochlear nucleus, *J. Neurochem.*, 49:128-137.

Staatz-Benson, C. and Potashner, S.J., 1988, Uptake and release of glycine in the guinea pig cochlear nucleus after axotomy of afferent or centrifugal fibers, *J. Neurochem.*, 51:370-379.

Wenthold, R.J., 1979, Release of endogenous glutamic acid, aspartic acid and GABA from cochlear nucleus slices, *Brain Res.*, 162:338-343.

Wenthold, R.J., 1987, Evidence for a glycinergic pathway connecting the two cochlear nuclei: an immunocytochemical and retrograde transport study, *Brain Res.*, 415:183-187.

Wenthold, R.J., Zempel, J.M., Parakkal, M.H., Reeks, K.A. and Altschuler, R.A., 1986, Immunocytochemical localization of GABA in the cochlear nucleus of the guinea pig, *Brain Res.*, 380:7-18.

Wenthold, R.J., Huie, D., Altschuler, R.A. and Reeks, K.A., 1987, Glycine immunoreactivity localized in the cochlear nucleus and superior olivary complex, *Neuroscience.*, 22:897-912.

Wenthold, R.J., Parakkal, M.H., Oberdorfer, M.D. and Altschuler, R.A., 1988, Glycine receptor immunoreactivity in the ventral cochlear nucleus of the guinea pig, *J. Comp. Neurol.*, 276:423-435.

Wickesberg, R.E. and Oertel, D., 1988, Tonotopic projection from the dorsal to the anteroventral cochlear nucleus of mice, *J. Comp. Neurol.*, 268:389-399.

Wickesberg, R.E. and Oertel, D., 1990, Delayed, frequency-specific inhibition in the cochlear nuclei of mice: A mechanism for monaural echo suppression, *J. Neurosci.*, 10:1762-1768.

Wickesberg, R.E., Whitlon, D., Oertel, D. and Wenthold, R., 1990, Glycine immunoreactivity depends on activity in brain slices of the mouse cochlear nuclear complex, *Ass. Res. Otolaryng. Abstr.*, 13:102-103.

Wickesberg, R.E., D. Whitlon and Oertel, D., (in press), Tuberculoventral neurons project to the multipolar cell area but not to the octopus cell area of the posteroventral cochlear nucleus, *J. Comp. Neurol.*

Wu, S.H. and Oertel, D., 1984, Intracellular injection with horseradish peroxidase of physiologically characterized stellate and bushy cells in slices of mouse anteroventral cochlear nucleus, *J. Neurosci.*, 4:1577-1588.

Wu, S.H. and Oertel, D., 1986, Inhibitory circuitry in the ventral cochlear nucleus is probably mediated by glycine, *J. Neurosci.*, 6:2691-2706.

Zhang, S. and Oertel, D., 1990, Tuberculoventral neurons in the deep DCN are monosynaptically inhibited by shocks to the VCN, *Ass. Res. Otolaryng. Abstr.*, 13:103.

# GABA AND GLYCINE INPUTS CONTROL DISCHARGE RATE WITHIN THE EXCITATORY RESPONSE AREA OF PRIMARY-LIKE AND PHASE-LOCKED AVCN NEURONS

Donald M. Caspary, Peggy S. Palombi, Patricia M. Backoff,
Robert H. Helfert and Paul G. Finlayson

Southern Illinois University School of Medicine, P.O. Box 19230
Springfield, Illinois, 62794-9230, USA

The studies presented in this chapter were designed to identify potential functions for noncochlear inputs onto spherical bushy cells in the anteroventral cochlear nucleus (AVCN). Compounds related to glycine and gamma aminobutyric acid (GABA) were iontophoretically applied in a number of different acoustic paradigms.

## Anatomy and physiology of AVCN

Spherical bushy cells represent one population of principal cells in the AVCN and have been named based on their appearance in Golgi-impregnated and Nissl-stained tissue (Osen, '69; Brawer et al., '74; Morest et al., '90). As one moves rostrolaterally toward the oral pole of the ventral cochlear nucleus (VCN), the density of spherical bushy cells increases. The temporal discharge patterns of these neurons in response to tone burst stimuli at characteristic frequency (CF) have been classified as primarylike (PRI) or phase-locked (PHL) and resemble those of acoustic nerve fibers (Pfeiffer, '66; Caspary, '72; Evans and Nelson, '73; Kiang, '75; Bourk, '76; Rhode et al., '83; Shofner and Young, '85; see Young, '84 for review). PHL neurons display time-locked responses to a particular phase of a stimulus sinusoid and are considered a subset of PRI neurons in this chapter. PRI and PHL neurons both in anesthetized and unanesthetized animals display predominately excitatory responses to tonal stimuli and have been classified as Type I responders (not displaying lateral inhibition) (Evans and Nelson, '73; Goldberg and Brownell, '73; Brownell, '75; Bourk, '76; Young, '84). Spontaneous activity is not inhibited by tones outside the excitatory response area of extracellularly recorded AVCN neurons that exhibit prepotentials (Goldberg and Brownell, '73). In contrast, non-prepotential units display inhibition of excitant amino acid induced activity for tones presented above CF (Martin and Dickson, '83). Based on their similarities in response properties (Pfeiffer, '66; Kiang, '75), AVCN neurons have historically been thought to behave as simple relay cells with outputs reflecting the same or similar information as that transmitted by the dominant acoustic nerve inputs. However, both *in vivo* and *in vitro* studies have suggested that some form of inhibition is present for all AVCN neurons, including bushy cells (Caspary et al., '79; Wu and Oertel, '86.)

# Glycine and GABA as AVCN inhibitory neurotransmitters

Recent information from morphological, immunocytochemical, and pharmacological studies strongly suggests that the inhibitory amino acids glycine and GABA play a role in shaping the output of PRI and PHL neurons, which is sent via the ventral acoustic stria to the superior olivary complex (SOC) (Warr, '66; Caspary '86; Friauf and Ostwald '88; Altschuler et al., this volume; see Helfert et al., '91 for review). Descriptions of non-cochlear inputs onto AVCN bushy cells were originally made from ultrastructural studies, and included significant numbers of Type II bouton endings displaying pleomorphic or flattened synaptic vesicles (putative inhibitory endings) contacting spherical bushy cells (Lenn and Reese, '66). These boutons did not degenerate upon sectioning of the acoustic nerve (Cant and Morest, '78, '79; Ryugo and Fekete, '82), in contrast to the spherical vesicle-containing Type I calyceal endings of acoustic nerve origin. Data from pharmacological, neurochemical and histochemical studies have suggested that many of the Type II endings could release glycine or GABA (Tachibana and Kuriyama, '74; Davies, '75; Fisher and Davies, '76; Godfrey et al., '78; Caspary et al., '79; Wenthold, '79; Canzek and Reubi, '80). Recently, sophisticated neurochemical and immunocytochemical techniques, combined with tract tracing, HRP injection and ultrastructural studies, have been used to detail the significant inhibitory amino acidergic input onto spherical bushy cells of the AVCN. Studies using antibodies against glycine or GABA conjugated to bovine serum albumin or against the GABA synthetic enzyme glutamic acid decarboxylase (GAD) have shown boutons, labeled for one or both amino acids, contacting spherical bushy cells (Mugnaini and Oertel, '85; Altschuler et al., '86; Adams and Mugnaini, '87; Moore and Moore, '87; Saint Marie et al., '89; see Altschuler et al. and Saint Marie et al., this volume). Uptake and release studies from cochlear nucleus (CN) slices further support a role for glycine and GABA in the AVCN (Wenthold, '79; Potashner et al., '85; Staatz-Benson and Potashner, '87, '88; Ostapoff et al., '90; see also Potashner et al., this volume). The presence of specific receptors for glycine and GABA has been demonstrated using specific ligands and receptor antibodies as markers for the receptor subtypes glycine$_1$ (strychnine sensitive), GABA$_A$ (bicuculline sensitive) and GABA$_B$ (bicuculline insensitive). Heavy labeling for GABA receptors has not been observed in AVCN, and the nature of AVCN GABA receptors remains unclear (Altschuler et al., '86; Frostholm and Rotter, '86; Glendenning and Baker, '88; Sanes et al.,'87; Wenthold et al., '88; Juiz et al., '89, '90). *In vivo* and *in vitro* pharmacological studies support the presence of both glycine and GABA receptors in the VCN (Caspary et al., '79; Martin and Dickson, '83; Caspary, '86; Wu and Oertel, '86; see also Oertel and Wickesberg, this volume).

# Sources of glycine and GABA inputs onto AVCN bushy cells

Probable sources of glycinergic inputs onto AVCN bushy cells are the vertical cells of the dorsal cochlear nucleus (DCN) (Lorente de Nó, '81; Feng and Vater, '85; Wickesberg and Oertel, '88; Osen et al., '90). Vertical cells and their projection fibers immunolabel for glycine, with 40% of cells co-labeling for GABA (Osen et al., '90). Evidence from *in vivo* and *in vitro* HRP injection studies suggests that DCN vertical cells receive primary inputs tonotopically matched to their target AVCN neurons (Fig. 1) (Feng and Vater, '85; Wickesberg and Oertel, '88; see Wickesberg and Oertel, this volume). Other major sources of possible glycinergic and GABAergic inputs to bushy cells include cells located in the ipsilateral lateral nucleus of the trapezoid body (LNTB) and bilaterally in the ventral nuclei of the trapezoid body (VNTB). Neurons located in other subnuclei of the SOC may also make contributions to this ventral pathway to the AVCN (Fig. 1) (Elverland, '77; Cant and Morest, '78; Spangler et al., '87; Shore et al., '91). The SOC cell groups of origin immunolabel for glycine and GABA and show retrograde transport of [$^3$H]glycine and [$^3$H]GABA from AVCN

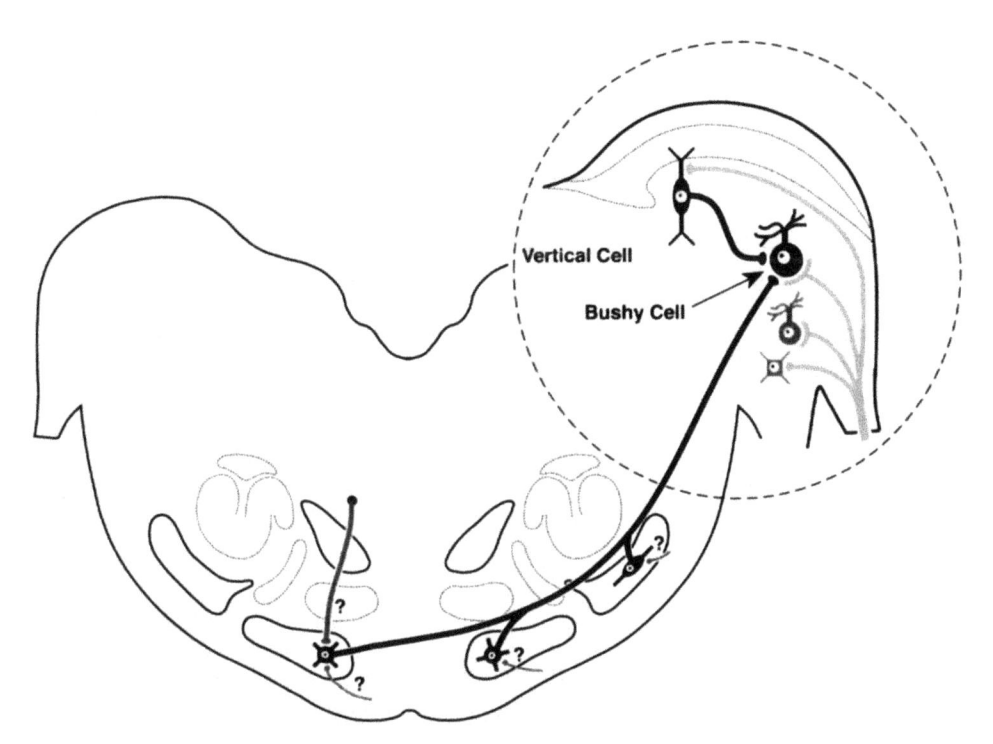

**Figure 1.** Schematic drawing of both cochlear (gray) and non-cochlear (black) inputs to AVCN spherical bushy cells. Recent anatomical studies suggest that inputs from the DCN and from the SOC release inhibitory amino acid neurotransmitters and that these inputs are frequency matched to the primary inputs of their AVCN targets.

injection sites (Moore and Moore, '87; Benson and Potashner, '90; Ostapoff et al., '90; Saint Marie et al., this volume). Axotomy of the ventral input pathway to AVCN decreases both uptake and release of labeled glycine (Staatz-Benson and Potashner, '88; Potashner et al., this volume). Godfrey et al., ('88) found that concentrations of GABA are reduced after sectioning of the trapezoid body. Spangler et al. ('87) suggest that the cells of origin of the descending pathway may also receive inputs of similar frequencies to those of their AVCN targets. Morphological and immunocytochemical data suggest that inhibitory amino acidergic inputs onto AVCN neurons from neurons located in the DCN and the SOC are likely to be activated by frequencies within the excitatory response area of the AVCN neuron(s) onto which they project (Fig. 1).

Despite this impressive support for the presence of inhibitory inputs onto AVCN bushy cells, the functional significance of glycine and GABA inputs onto putative bushy cells remains unknown. However, a number of possible roles have been suggested. Traditionally, one of the most commonly postulated roles for inhibition in the CN has been in the sharpening of frequency tuning through lateral inhibition. Early pharmacological studies provided evidence that inhibitory amino acids might be responsible for lateral inhibition in AVCN neurons without prepotentials, but that for neurons with low levels of spontaneous activity it is difficult to distinguish lateral inhibition from two-tone/noise-tone suppression arising peripherally (Martin, '80; Martin and Dickson, '83). Peripheral suppression and locally-mediated lateral inhibition should be separable using iontophoretic techniques. Martin and Dickson ('83) found that application of the inhibitory amino acid antagonists strychnine and bicuculline could not block lateral inhibition/two-tone suppression or alter near-CF discharge rate in a small number of AVCN prepotential neurons.

## Possible roles for inhibition in AVCN

Potential roles for inhibition in AVCN involve the coding of complex stimuli, such as amplitude-modulated (AM) signals. AVCN neurons display a bandpass gain function in response to AM tones, particularly at higher stimulus intensities, when compared to acoustic nerve fibers (Frisina et al., '85, '90; see also Palmer and Winter, this volume). Inhibition may play a role in determining dynamic range, dynamic range adjustment, echo suppression, temporal adaptation, and post-excitatory suppression in AVCN neurons. Studies comparing the responses of AVCN neurons and acoustic nerve fibers to tones, noise, and tones embedded in background noise have found no major differences between AVCN and acoustic nerve (Møller, '72, '74, '76; Rhode et al., '78; Palmer and Evans, '82; Young et al., '83, '88; Costalupes et al.,'84; Gibson et al., '85; Narins, '87). However, recent studies by May and Sachs ('91), which examined tone-evoked responses recorded from CN neurons in awake behaving animals, have described a resistance to masking by continuous background noise compared to responses obtained with similar paradigms in anesthetized preparations. Echo suppression has been suggested as another possible role of inhibition in the tonotopically matched circuits in AVCN (Wickesberg and Oertel, '90; see Wickesberg and Oertel, this volume).

Preliminary results from our laboratory suggested that glycine and GABA inputs onto AVCN bushy cells do not control lateral inhibition, but instead appear to regulate discharge rate and dynamic range within the excitatory response area of PRI and PHL neurons. Therefore, we focused on determining the role of noncochlear inhibitory inputs onto AVCN neurons by using a number of different acoustic paradigms prior and subsequent to iontophoretic application of agents related to glycine$_1$ and GABA$_A$ receptor function.

## Brief review of methods

Chinchillas (*chinchilla laniger*) were anesthetized with sodium pentobarbital (44 mg/kg) followed 30 minutes later by 44 mg/kg ketamine-HCl and maintained with hourly injections of ketamine-HCl (40 mg/kg). If possible, no additional barbiturate was given to the animal due to its known effects on GABA function (Evans and Nelson, '73; Schulz and MacDonald, '81; Rhode and Kettner, '87). Surgical and recording procedures have been described elsewhere and are only outlined here (Caspary et al., '84, '87). Extracellular recordings were obtained from neurons displaying PRI and PHL temporal responses associated with AVCN spherical bushy cells and localized to the AVCN. Poststimulus-time histograms (PSTH) with a resolution of 10 $\mu$s were collected over a 100 ms period in response to 50 ms (5 ms rise-fall time) digitally-generated tonebursts, presented at 10/s. Experiments utilized a Beyer (DT-48) earphone coupled to specially designed chinchilla ear bars with a probe microphone. Calibration of sound stimulus was performed on line at the level of the tympanum and corrected to dB SPL (re 0.0002 dynes/cm$^2$). A six-barreled piggyback electrode (Havey and Caspary, '80) was used; the recording and balancing barrels were filled with 2M potassium acetate and 4% horseradish peroxidase (HRP) in 0.5 M KCl-Tris buffer, respectively. The remaining four barrels contained: the GABA$_A$ receptor agonist muscimol (10 mM; pH 3.5-4.0), the GABA$_A$ receptor antagonist bicuculline methiodide (10 to 20 mM; pH 3.0), glycine (500 mM; pH 3.0-4.0), and the glycine$_1$ receptor antagonist strychnine-HCl (10 to 20 mM; pH 3.0). Upon isolation of an AVCN neuron, a tuning curve was obtained, and CF as well as threshold determined. An estimate of spontaneous activity was obtained for most neurons. Rate-intensity (RI) functions to CF tones were obtained, and each unit was classified by its PSTH pattern at 25 to 35 dB above threshold.

A number of different acoustic paradigms were utilized to explore selectively the possible involvement of inhibitory mechanisms in the detection of signals in noise. In addition

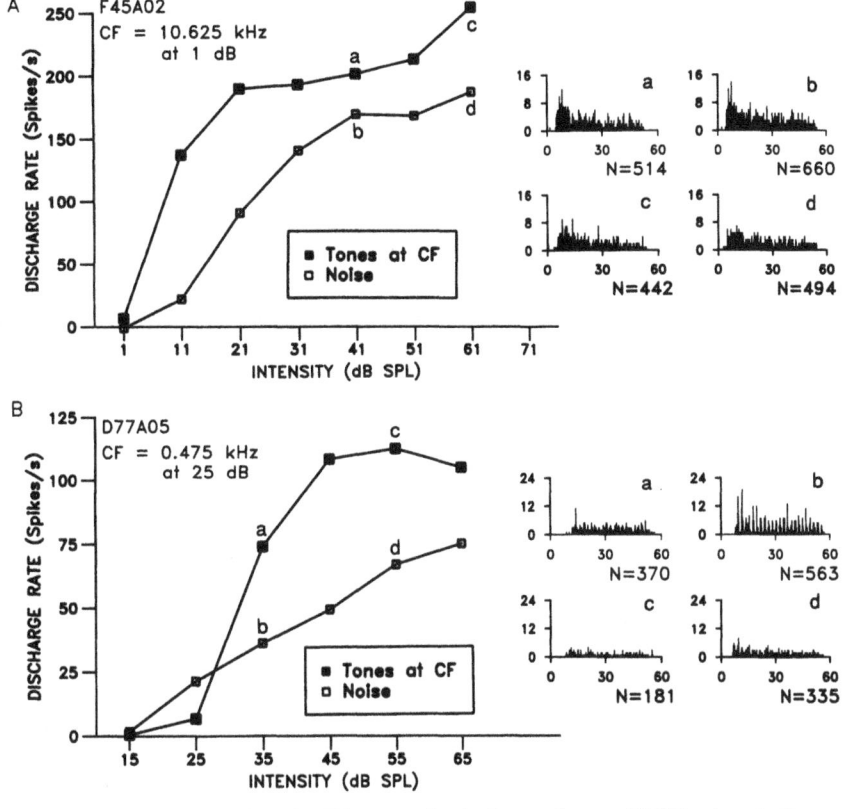

**Figure 2.** Rate-intensity (RI) functions for CF tones and noise bursts for two AVCN units: A: Tone and noise RI functions with similar (parallel) slopes are shown for a primarylike (PRI) unit with a CF of 10.625 kHz; B: Non-parallel tone and noise RI functions are shown for a phase-locked (PHL) unit with a CF of 475 Hz. Insets display PSTHs corresponding to the labeled points (a-d) on the RI curves (in this and subsequent figures, PSTH bin width = 200 μsec; N = total number of spikes).

to CF tone RI functions, RI functions were obtained in response to digitally-synthesized broadband noisebursts. It was theorized that if present, inhibitory circuits might be activated by broadband noise resulting in RI functions with a different slope and/or shape when compared to RI functions generated in response to CF tones. No attempt was made to equate noise/tone energy within the neuron's excitatory response area; therefore, threshold differences between tone-evoked and noise-evoked responses could not be evaluated. Isointensity contours were obtained from all PRI/PHL neurons. Comparative effects of the inhibitory amino acid antagonists were examined under each of these conditions if time permitted. Iontophoretic applications occurred subsequent to obtaining a series of control responses to a particular acoustic paradigm and were followed by a series of controls until responses returned to near predrug levels. Locations of cells were confirmed by examination of electrode tracks and HRP marks in perfused, aldehyde-fixed brains (Caspary et al., '87).

## Functional characteristics of AVCN PHL and PRI neurons

In a population of 102 PRI/PHL AVCN neurons, paired RI functions generated in response to CF tones and broadband noise were classified by visual inspection as displaying either similar (essentially parallel RI functions of similar shape and slope) or different responses to the two acoustic stimuli.

Fifty-six percent (57) of AVCN PRI/PHL neurons displayed RI functions with parallel slopes, while 44% (45) of neurons exhibited different slopes. In Figure 2 an example of a neuron displaying similar CF tone and broadband noise-evoked RI functions is contrasted with a neuron displaying different RI functions. AVCN neurons of similar CF, displaying PRI or PHL temporal response patterns, may exhibit either parallel or very different RI functions in response to CF tones and noise. Slope differences between noise- and tone-evoked RI functions may reflect cochlear suppression observed in recordings from acoustic nerve fibers or may reflect a combination of inhibition occurring centrally and peripheral suppression. These differences suggest that the mechanisms for controlling dynamic range as demonstrated by RI slope and discharge rate may differ from acoustic nerve response for tones and noise or other complex stimuli. Iontophoretic application of inhibitory amino acid antagonists should alter locally-mediated inhibition, while noise/tone suppression originating at the periphery should be unaffected.

Families of isointensity contours from PRI/PHL AVCN neurons were obtained both in quiet and in the presence of continuous broadband noise (Fig. 3). Both PRI and PHL units showed evidence of masking by noise within their excitatory response area and in some cases revealed suppression/inhibition. When tones and a continuous noise masker were presented at approximately equal sound pressure level, maximum discharge rate within the excitatory response area was suppressed (Fig. 3B). For PHL neurons, however, vector strength was relatively unaffected within the excitatory response area in the presence of continuous noise.

## Function of glycine and GABA in AVCN

A population of 73 AVCN neurons overlapping with the above population displaying PRI/PHL temporal response patterns was studied using iontophoretic application of one or more of the previously listed glycine$_I$ and GABA$_A$-related compounds. No significant differences were observed between the percentages of neurons sensitive to application of glycine$_I$-related and those sensitive to GABA$_A$-related compounds (Table I). However, only 58% of AVCN PRI/PHL responders were sensitive to the application of muscimol whereas 82% were sensitive to GABA. Possible explanations for this difference include those related to the iontophoretic technique, differences in transport number, or the possibility that responses to GABA application were mediated in part through GABA$_B$ receptors. Both glycine and GABA reduced discharge rate across frequency in a dose-dependent manner (Fig. 4). In Fig. 4, increasing currents of GABA reduced responses within the excitatory response area, while at the lowest doses of either inhibitory amino acid neurotransmitter spontaneous discharge rate was reduced (not shown).

We initially hypothesized that iontophoretic application of the glycine receptor antagonist strychnine or the GABA$_A$ receptor antagonist bicuculline would alter noise evoked RI functions to a greater extent than CF tone-evoked RI functions. This could be attributed to involvement of centrally-mediated inhibition in shaping noise-evoked responses. This was not observed in these studies. Blockade of either glycine or GABA inputs increased discharge rates both for noise-evoked and tone-evoked RI functions by similar orders of magnitude (Fig. 5). RI changes were greatest at suprathreshold intensities while smaller or minimal effects were observed near threshold (Fig. 5). Inhibitory amino acid antagonists aplied during noise and tone paradigms caused similar changes in slope and magnitude (Fig. 5). Blockade of glycine or GABA inputs did not alter the contour of the RI function, i.e. noise and tone RI curves with different slopes did not become parallel. Noise-evoked RI functions were not altered to a greater degree than CF tone-evoked RI functions. We suggest with these findings that amino acid inhibition may regulate discharge rate at suprathreshold intensities as a way of controlling dynamic range within the excitatory response area of AVCN bushy cells.

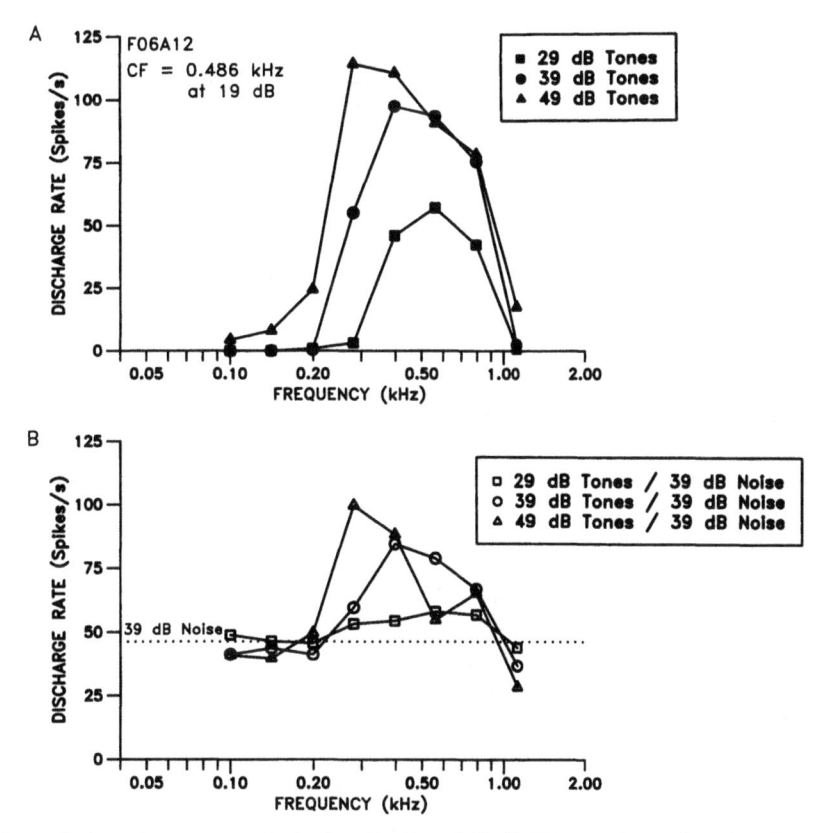

**Figure 3.** A: Isointensity contours obtained at 29, 39 and 49 dB SPL are shown for a low-frequency PHL neuron (CF = 486 Hz). B: Contours obtained to 50 ms tone bursts presented in a background of 39 dB SPL continuous noise. Discharge rate to the continuous noise alone is indicated by the dotted line.

Table I. PRI and PHL Neurons

| Compound | Neurons Tested | Positive Effects | Percentage |
|---|---|---|---|
| GABA | 34 | 28 | 82.4% |
| Muscimol | 31 | 18 | 58.1% |
| Bicuculline | 51 | 40 | 78.5% |
| Glycine | 38 | 29 | 76.3% |
| Strychnine | 42 | 30 | 71.4% |

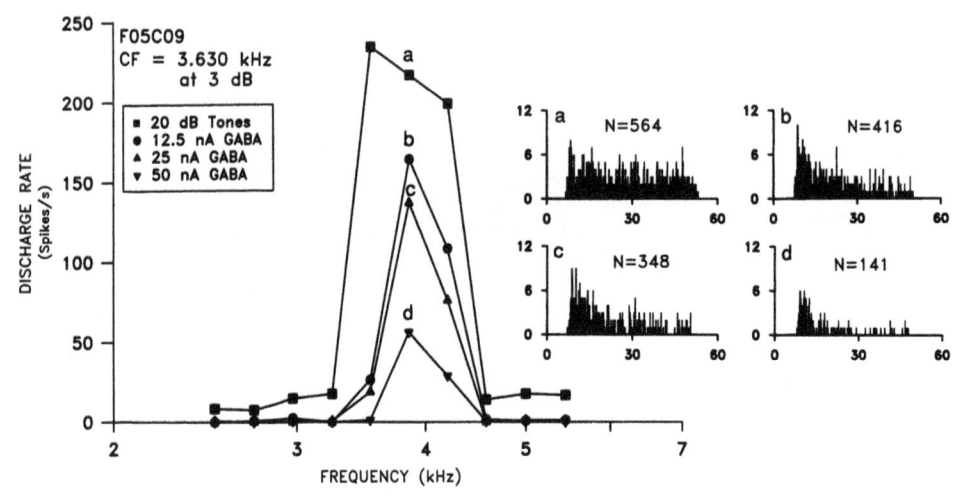

**Figure 4.** Iontophoretic application of increasing doses (currents) of GABA reduces discharge rate across frequency. The control 20 dB isointensity contour from an AVCN PRI neuron (CF=3.630 kHz) is progressively reduced in a dose-dependent manner. Insets display PSTHs corresponding to the labeled (a-d) points on the curves and show decreasing spike counts with increasing doses of GABA.

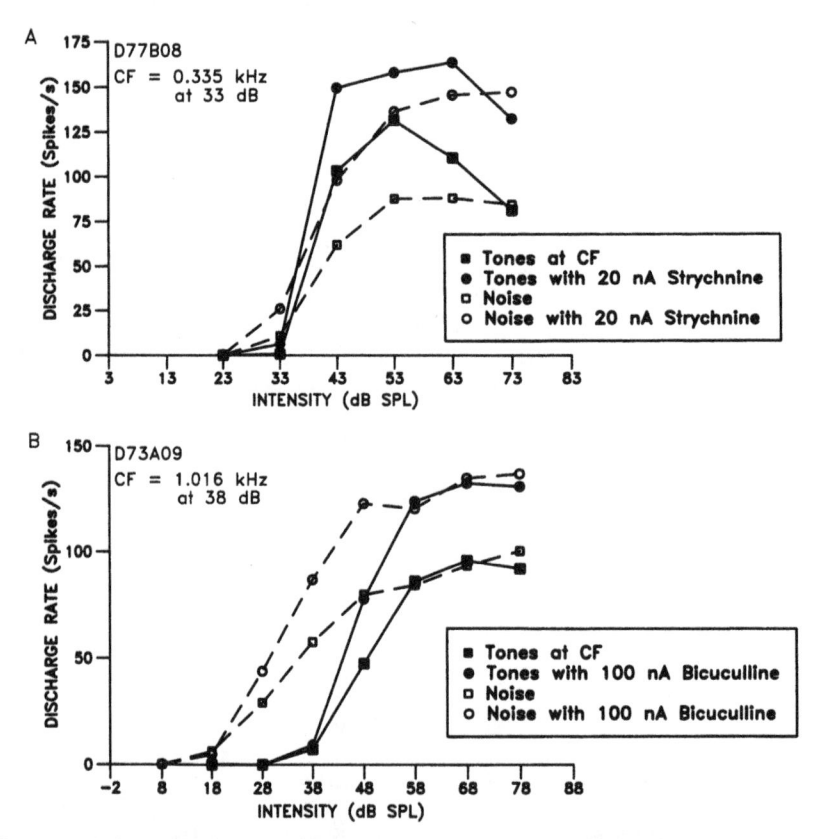

**Figure 5.** A: Strychnine application onto a PHL neuron similarly alters the RI function evoked by CF tone and noise bursts. Strychnine effects on both functions are greatest suprathreshold with no change in the contour. B: Bicuculline application has a similar effect on a second PHL neuron. The RI functions for CF tones and noise bursts show similar suprathreshold increases in discharge rate. In both examples, response thresholds are unchanged and drug effects increase with stimulus intensity.

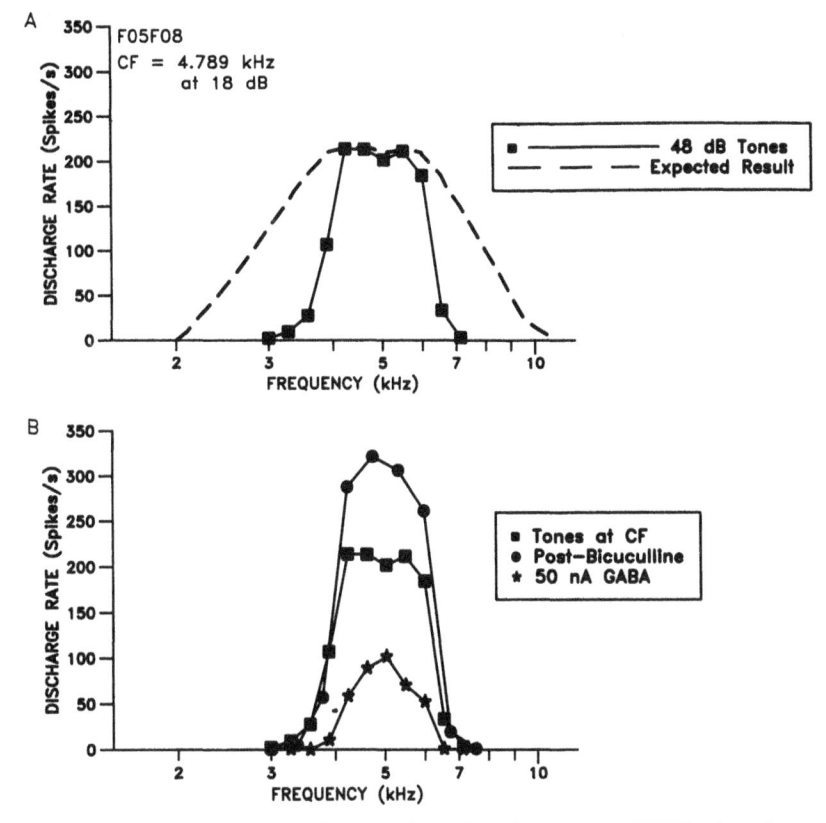

**Figure 6.** A: The theoretical response to blockade of inhibitory inputs onto AVCN bushy cells assuming the presence of inhibitory sidebands mediated by inhibitory amino acid (IAA) inputs. Application of the IAA antagonist would be expected to produce a widening of the excitatory response area as shown by the dashed line. B: The actual data are shown with bicuculline application increasing discharge rate only within the excitatoryeffects were observed near threshold (Fig. 5). Inhibitory amino acid antagonists applied response across frequency.

This same population of AVCN PRI/PHL neurons was examined across frequency. It was hypothesized that blocking glycine or GABA inputs would result in a broadening of the isointensity function indicating a loss in sharpness of tuning as illustrated theoretically in Figure 6A. However, as can be seen from the actual data from this PRI neuron modeled in Figure 6A, blockade of GABA inputs by antagonism of $GABA_A$ receptors resulted in an increased discharge rate only within the excitatory response area (Fig. 6B). Only rarely did glycine or GABA blockade alter discharge rate outside the excitatory frequency response area. Application of either strychnine or bicuculline onto PRI or PHL AVCN neurons increased discharge rate throughout the excitatory response area while having little apparent effect on the sharpness of tuning (Figs. 6, 7). Some PRI or PHL AVCN neurons were sensitive to either glycine or GABA blockade while other PRI or PHL AVCN neurons were sensitive to both the inhibitory amino acid antagonists (Fig. 7). The PRI neuron in Figure 7A displayed greater sensitivity to glycine blockade than to $GABA_A$ blockade. Other neurons, including the one shown in Figure 7B, were nearly equally sensitive to blockade of glycine and $GABA_A$ inputs (54 dB isointensity contour). Blocking inhibitory amino acid inputs onto AVCN neurons increased discharge rate throughout much of the excitatory response area (Fig. 7B). $GABA_A$ blockade of the inputs to this PHL neuron at three different intensities greatly increased the discharge rate throughout the excitatory response area (Fig. 7B). When applied

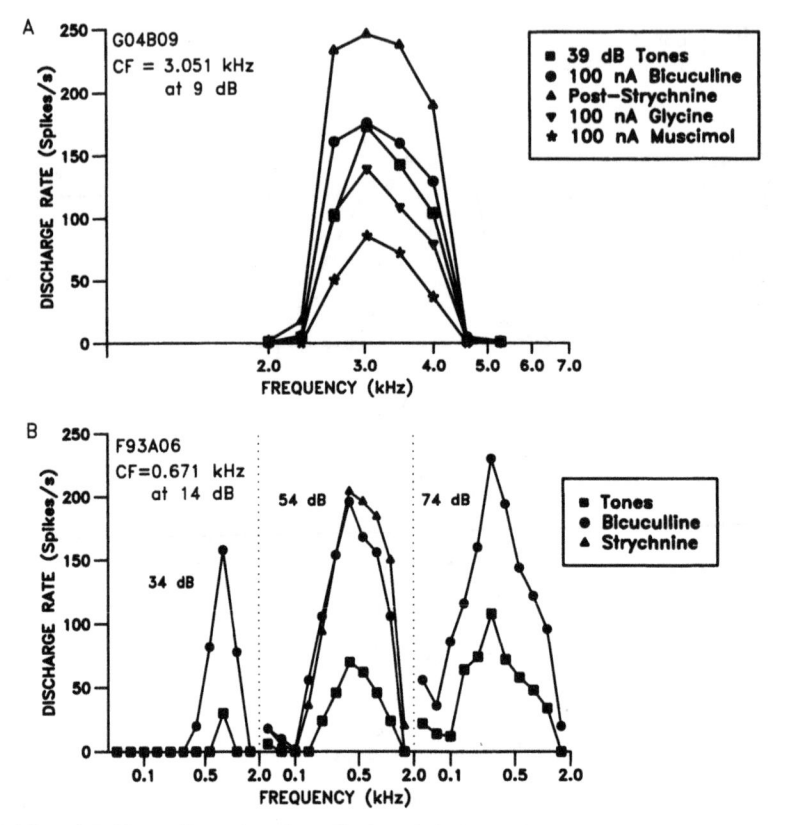

**Figure 7.** Differential effects of iontophoretic application of glycine- and GABA-related compounds on frequency response area are shown for two AVCN neurons. A: Strychnine application increases discharge rate to a greater degree than bicuculline for this AVCN PRI neuron. B: The glycine-receptor antagonist strychnine and the GABA$_A$-receptor antagonist bicuculline have similar effects on discharge rate within the excitatory response area of an AVCN PHL neuron (54 dB isointensity contour). As seen in this series of response contours at 34, 54 and 74 dB SPL (threshold at CF was 14 dB SPL), the increases in discharge rate produced by inhibitory amino acid blockade remain restricted to the unit's excitatory response area.

onto sensitive neurons, the inhibitory amino acid agonists suppressed discharge rate within the excitatory response area as is seen for GABA (Figs. 4 and 6B) and for glycine and muscimol (Fig. 7A).

In this study we expected that blockade of glycine and/or GABA inputs onto AVCN neurons would alter the shape and/or slope of noise-evoked RI functions to a greater extent than RI functions obtained using CF tones. Instead, the parallel shifts observed suggested that the inhibitory effects blocked circuits which control discharge rate within the excitatory response area of the neurons examined.

This was further supported by examining the responses of these cells across frequency. In all cases in which neurons were sensitive to strychnine and/or bicuculline, the largest effects were observed within the neuron's excitatory response area. The finding of a smaller percentage of putative bushy cells sensitive to muscimol relative to GABA supports the presence of and a role for GABA$_B$ as well as GABA$_A$ receptors in AVCN. GABA$_B$ receptors may be present presynaptically on acoustic nerve fibers, acting to suppress the release of the excitatory acoustic nerve neurotransmitter (Caspary et al., '84; Bowery, '89). An alternative explanation is that GABA$_B$ receptors are present postsynaptically and function to suppress responses via a second messenger system (Bowery, '89; Dutar and Nicoll, '88).

The lack of a distinct sensitivity difference between glycine and GABA during specific acoustic paradigms was unexpected. A number of studies have suggested that the lateral ventrotubercular tract from the vertical cells is likely to be glycinergic and may also co-contain GABA (Osen et al., '90; Juiz et al., '90; Altschuler et al., and Oertel and Wickesberg, this volume). The nature of the neurotransmitter from the ventral pathway is also likely to utilize an inhibitory amino acid, but whether it is glycine and/or GABA is not presently known. A number of investigators suggest that markers both for neurotransmitter glycine and GABA are reduced in AVCN upon sectioning of the ventral pathway from SOC (Godfrey et al., '88; Potashner et al., '85; Staatz-Benson and Potashner, '87, '88).

## Conclusions

Glycinergic and/or GABAergic inputs onto AVCN bushy cells displaying primary-like or phase-locked temporal response patterns modulate or control discharge rate within the excitatory response area. These inputs may determine dynamic range within the neuron's excitatory response area. These findings are consistent with anatomical observations of inhibitory projections from neurons which receive acoustic nerve inputs from similar cochlear locations. Inputs onto inhibitory projection neurons, which provide inputs onto AVCN bushy cells, could depress the responses of the projection neurons under certain acoustic conditions. This could allow for increased bushy cell discharge rate, possibly increasing signal detection in the presence of noise. Additional studies using complex stimuli such as AM signals may reveal differences in the CN circuitry and allow a distinction to be made between the glycinergic and GABAergic inputs onto AVCN bushy cells.

#### ACKNOWLEDGEMENTS

Ms. Barbara Armour participated in many of the experiments. Ms. Barbara Armour and Dr. Flint Boettcher provided critical comments and assistance on the manuscript. Supported by NIH DC 00151-10.

#### REFERENCES

Adams, J.C. and Mugnaini, E., 1987, Patterns of glutamate decarboxylase immunostaining in the feline cochlear nuclear complex studied with silver enhancement and electron microscopy, *J. Comp. Neurol.*, 262:375-401.

Altschuler, R.A., Betz, H., Parakkal, M.H., Reeks, K.A., and Wenthold, R.J., 1986, Identification of glycinergic synapses in the cochlear nucleus through immunocytochemical localization of the postsynaptic receptor, *Brain Res.*, 369:316-320.

Benson, C.G. and Potashner, S.J., 1990, Retrograde transport of [$^3$H]glycine from the cochlear nucleus to the superior olive in the guinea pig, *J. Comp. Neurol.*, 296:415-426.

Bourk, T.R., 1976, Electrical responses of neural units in the anteroventral cochlear nucleus of the cat, *Doctoral Thesis*, Dept. of Electrical Engineering and Computer Science, Massachusetts Institute of Technology.

Bowery, N., 1989, GABA$_B$ receptors and their significance in mammalian pharmacology, *TIPS*, 10(10):401-407.

Brawer, J.R., Morest, D.K. and Kane, E.C., 1974, The neuronal architecture of the cochlear nucleus of the cat, *J. Comp. Neurol.*, 155(3):251-283.

Brownell, W.E., 1975, Organization of the cat trapezoid body and the discharge characteristics of its fibers, *Brain Res.*, 94:413-433.

Cant, N.B. and Morest, D.K., 1978, Axons from non-cochlear sources in the anteroventral cochlear nucleus of the cat. A study with the rapid golgi method, *Neurosci.*, 3:1003-1029.

Cant, N.B. and Morest, D.K., 1979, Organization of the neurons in the anterior division of the anteroventral cochlear nucleus of the cat. Light microscopic observations, *Neurosci.*, 4:1909-1923.

Canzek, V. and Reubi, J.C., 1980, The effect of cochlear nerve lesion on the release of glutamate, aspartate, and GABA from cat cochlear nucleus, in vitro, *Exp. Brain Res.*, 38:437-441.

Caspary, D.M., 1972, Classification of sub-populations of neurons in the cochlear nuclei of the kangaroo rat, *Exp. Neurol.*, 37:131-151.

Caspary, D.M., 1986, Cochlear nuclei: functional neuropharmacology of the principal cell types, in: "Neurobiology of Hearing: The Cochlea," R. Altschuler, R. Bobbin, and D. Hoffman, eds., Raven Press, New York, pp. 303-332.

Caspary, D.M., Havey, D.C. and Faingold, C.L., 1979, Effects of microiontophoretically applied glycine and GABA on neuronal response patterns in the cochlear nuclei, Brain Res., 172:179-185.

Caspary, D.M., Pazara, K.E., Kössl, M. and Faingold, C.L., 1987, Strychnine alters the fusiform cell output from the dorsal cochlear nucleus, Brain Res., 417:273-282.

Caspary, D.M., Rybak, L. and Faingold, C.L., 1984, Baclofen reduces tone-evoked and spontaneous activity of cochlear nucleus neurons, Hearing Res., 13:113-122.

Costalupes, J.A., Young, E.D. and Gibson, D.J., 1984, Effects of continuous noise backgrounds on rate response of auditory nerve fibers in cat, J. Neurophysiol., 51:1326-1344.

Davies, W.E., 1975, The distribution of GABA transaminase-containing neurones in the cat cochlear nucleus, Brain Res., 83:27-33.

Dutar, P. and Nicoll, R.A., 1988, A physiological role for $GABA_B$ receptors in the central nervous system, Nature, 332:156-158.

Elverland, H.H., 1977, Descending connections between the superior olivary and cochlear nuclear complexes in the cat studied by autoradiographic and horseradish peroxidase methods, Exp. Brain Res., 27:397-412.

Evans, E.F. and Nelson, P.G., 1973, On the functional relationship between the dorsal and ventral divisions of the cochlear nucleus of the cat, Exp. Brain Res., 17:428-442.

Feng, A.S. and Vater, M., 1985, Functional organization of the cochlear nucleus of rufous horseshoe bats (rhinolophus rouxi): Frequencies and internal connections are arranged in slabs, J. Comp. Neurol., 235:529-553.

Fisher, S.K. and Davies, W.E., 1976, GABA and its related enzymes in the lower auditory system of the guinea pig, J. Neurochem., 27:1145-1155.

Friauf, E. and Ostwald, J., 1988, Divergent projections of physiologically characterized rat ventral cochlear nucleus neurons as shown by intra-axonal injection of horseradish peroxidase, Exp. Brain Res., 73:263-284.

Frisina, R.D., Smith, R.L. and Chamberlain, S.C., 1990, Encoding of amplitude modulation in the gerbil cochlear nucleus: II. Possible neural mechanisms, Hearing Res., 44:123-142.

Frisina, R.D., Smith, R.L. and Chamberlain, S.C., 1985, Differential encoding of rapid change in sound amplitude by second-order auditory neurons, Exp. Brain Res., 60:417-422.

Frostholm, A. and Rotter, A., 1986, Autoradiographic localization of receptors in the cochlear nucleus of the mouse, Brain Res. Bull., 16:189-203.

Gibson, D.J., Young, E.D. and Costalupes, J.A., 1985, Similarity of dynamic range adjustment in auditory nerve and cochlear nuclei, J. Neurophysiol., 53:940-958.

Glendenning, K.K. and Baker, B.N., 1988, Neuroanatomical distribution of receptors for three potential inhibitory neurotransmitters in the brainstem auditory nuclei of the cat, J. Comp. Neurol., 275:288-308.

Godfrey, D., Parli, J., Dunn, J. and Ross, C., 1988, Microchemistry of the cochlear nucleus and superior olivary complex, In: "Auditory Pathway Structure and Function," J. Syka and R.B. Masterton, eds., Plenum Press, New York, pp. 107-121.

Godfrey, D.A., Carter, J.A., Lowry, O.H., and Matschinsky, F.M., 1978, Distribution of gamma-aminobutyric acid, glycine, glutamate and aspartate in the cochlear nucleus of the rat, J. Histochem. and Cytochem., 26:118-126.

Goldberg, J.M. and Brownell, W.E., 1973, Discharge characteristics of neurons in anteroventral and dorsal cochlear nuclei of cat, Brain Res., 64:35-54.

Havey, D.C. and Caspary, D.M., 1980. A Simple technique for constructing "piggy- back" multibarrel microelectrodes, Electroencephalograph. Clin. Neurophysiol., 48:249-251.

Helfert, R.H., Snead, C.R. and Altschuler, R.A., 1991, The ascending auditory pathways, in: "Neurobiology of Hearing: The Central Auditory System," R.A. Altschuler, R.P. Bobbin, B.M. Clopton, and D.W. Hoffman eds., Raven Press, New York, pp. 1-26.

Juiz, J., Helfert, R., Wenthold, R., De Blas, A., and Altschuler, R., 1989, Immunocytochemical localization of the $GABA_A$/benzodiazepine receptor in the guinea pig cochlear nucleus: evidence for receptor localization heterogeneity, Brain Res., 504:173-179.

Juiz, J.M., Helfert, R.H., Bonneau, J.M., Wenthold, R.J., and Altschuler, R. A., 1990, Synaptic terminals and cell bodies immunoreactive for glycine or co-containing glycine and GABA immunoreactivities in the cochlear nucleus of the guinea pig: an ultrastructural immunogold study, (submitted).

Kiang, N.Y.-S., 1975, Stimulus representation in the discharge patterns of auditory neurons, in: "The Nervous System. Human Communication and Its Disorders", D.B. Tower, ed., Raven Press, New York, pp. 81-96.

Lenn, N.J. and Reese, T.S., 1966, Fine structure of nerve endings in the nucleus of the trapezoid body and the ventral cochlear nucleus, *Am. J. Anat.*, 118:375-390.

Lorente de No, R., 1981, "The Primary Acoustic Nuclei", Raven Press, New York.

Martin, M. and Dickson, J.W., 1983, Lateral inhibition in the anteroventral cochlear nucleus of the cat: A microiontophoretic study, *Hearing Res.*, 9:35-41.

Martin, M.R., 1980, The effects of iontophoretically applied antagonists on auditory nerve and amino acid evoked excitation of anteroventral cochlear nucleus neurons, *Neuropharmacol.*, 19:519-528.

May, B.J. and Sachs, M.B., 1991, Unit classification and dynamic range properties of neurons in the ventral cochlear nucleus of awake and behaving cats, *ARO Abstr.*, 14:142.

Møller, A.R., 1972, Coding of amplitude and frequency modulated sounds in the cochlear nucleus of the rat, *Acta Physiol. Scand. (Stockh)*, 86:223-238.

Møller, A.R., 1974, Dynamic properties of cochlear nucleus units in response to excitatory and inhibitory tones, *in:* "Facts and Models in Hearing", E. Zwicker and E. Terhardt, eds., Springer-Verlag pp. 227-240.

Møller, A.R., 1976, Dynamic properties of excitation and two-tone inhibition in the cochlear nucleus studied using amplitude tones, *Exp. Brain Res.*, 25:307-321.

Moore, J.K. and Moore, R.Y., 1987, Glutamic acid decarboxylase-like activity in brainstem auditory nuclei, *J. Comp. Neurol.*, 260:157-174.

Morest, D.K., Hutson, K.A. and Kwok, S., 1990, Cytoarchitectonic atlas of the cochlear nucleus of the chinchilla, *Chinchilla laniger*, *J. Comp. Neurol.*, 300:230-248.

Mugnaini, E. and Oertel, W.H., 1985, An atlas of the distribution of GABAergic neurons and terminals in the rat CNS as revealed by GAD immunohistochemistry, *in:* "Handbook of Chemical Neuroanatomy: GABA and Neuropeptides in the CNS", A.Bjorklund and T.Hokfelt, eds., Elsevier Science Publishers, Amsterdam/New York, pp. 436-608.

Narins, P.M., 1987 Coding of signals in noise by amphibian auditory nerve fibers, *Hearing Res.*, 26:145-154.

Osen, K.K., 1969, Cytoarchitecture of the cochlear nuclei in the cat, *J. Comp. Neurol.*, 136:453-484.

Osen, K.K., Ottersen, O.P., and Storm-Mathisen, J., 1990, Colocalization of glycine-like and GABA-like immunoreactivities. A semiquantitative study of individual neurons in the dorsal cochlear nucleus of cat, *in:* "Glycine Neurotransmission", O.P. Ottersen and J. Storm-Mathisen, eds., John Wiley and Sons, New York, pp. 417-452.

Ostapoff, E.-M., Morest, D.K., and Potashner, S.J., 1990, Uptake and retrograde transport of [$^3$H]GABA from the cochlear nucleus to the superior olive in the guinea pig, *J. Chem. Neuroanat.*, 3:285-295.

Palmer, A.R. and Evans, E.F., 1982, Intensity coding in the auditory periphery of the cat: Responses of cochlear nerve and cochlear nucleus neurons to signals in the presence of bandstop masking noise, *Hearing Res.*, 7:305-323.

Pfeiffer, R.R., 1966, Classification of response patterns of spike discharges for units in the cochlear nucleus: tone-burst stimulation, *Exp. Brain Res.*, 1:220-235.

Potashner, S.J., Lindberg, N. and Morest, D.K., 1985, Uptake and release of gamma-aminobutyric acid in the guinea pig cochlear nucleus after axotomy of cochlear and centrifugal fibers, *J. Neurochem.*, 45:1558-1566.

Rhode, W.S., Geisler, C.D. and Kennedy, D.T., 1978, Auditory nerve fiber responses to wide-band noise and tone combinations, *J. Neurophysiol.*, 41:692-704.

Rhode, W.S. and Kettner, R.E., 1987, Physiological study of neurons in the dorsal and posteroventral cochlear nucleus of the unanesthetized cat, *J. Neurophysiol.*, 57(2):414-442.

Rhode, W.S., Oertel, D. and Smith, P.H., 1983, Physiological response properties of cells labeled intracellularly with horseradish peroxidase in cat ventral cochlear nucleus, *J. Comp. Neurol.*, 213:448-463.

Ryugo, D.K. and Fekete, D.M., 1982, Morphology of primary axosomatic endings in the anteroventral cochlear nucleus of the cat: A study of the endbulbs of Held, *J. Comp. Neurol.*, 210:239-257.

Saint Marie, R.L., Morest, D.K., and Brandon, C.J., 1989, The form and distribution of GABAergic synapses on the principal cell types of the ventral cochlear nucleus of the cat, *Hearing Res.*, 42:97-112.

Sanes, D.H., Wooten, W.A., Geary, G.F. and Rubel, E.W., 1987, Quantitative distribution of the glycine receptor in the auditory brain stem of the gerbil, *J. Neurosci.*, 7(11):3793-3802.

Schulz, D.W., and MacDonald, R.L., 1981, Barbiturate enhancement of GABA-mediated inhibition and activation of chloride ion conductance: Correlation with anticonvulsant and anesthetic actions, *Brain Res.*, 209:177-188.

Shofner, W.P. and Young, E.D., 1985, Excitatory/inhibitory response types in the cochlear nucleus: Relationships to discharge patterns and responses to electrical stimulation of the auditory nerve, *J. Neurophysiol.*, 54:917-939.

Shore, S.E., Helfert, R.H., Beldsoe, Jr., S.C., Altschuler, R.A. and Godfrey, D.A., 1991, Descending projections to the dorsal and ventral divisions of the cochlear nucleus in guinea pig, *Hearing Res.*, 52:255-268.

Spangler, K.M., Cant, N.B., Henkel, C.K., Farley, G.R. and Warr, W.B., 1987, Descending projections from the superior olivary complex to the cochlear nucleus of the cat, *J. Comp. Neurol.*, 259:452-465.

Staatz-Benson, C. and Potashner, S.J., 1987, Uptake and release of glycine in the guinea pig cochlear nucleus, *J. Neurochem.*, 49:128-137.

Staatz-Benson, C. and Potashner, S.J., 1988, Uptake and release of glycine in the guinea pig cochlear nucleus after axotomy of afferent or centrifugal fibers, *J. Neurochem.*, 51:370-379.

Tachibana, M. and Kuriyama, K., 1974, Gamma-aminobutyric acid in the lower auditory pathway of the guinea pig, *Brain Res.*, 69:370-374.

Warr, W.B., 1966, Fiber degeneration following lesions in the anterior ventral cochlear nucleus of the cat, *Exp. Neurol.*, 14:453-474.

Wenthold, R.J., 1979, Release of endogenous glutamic acid, aspartic acid, and GABA from cochlear nucleus slices, *Brain Res.*, 162:338-343.

Wenthold, R.J., Parakkal, M.H., Oberdorfer, M.D. and Altschuler, R.A., 1988, Glycine receptor immunoreactivity in the ventral cochlear nucleus of the guinea pig, *J. Comp. Neurol.*, 276:423-435.

Wickesberg, R.E. and Oertel, D., 1988, Tonotopic projection from the dorsal to the anteroventral cochlear nucleus of mice, *J. Comp. Neurol.*, 268:389-399.

Wickesberg, R.E. and Oertel, D., 1990, Delayed, frequency-specific inhibition in the cochlear nuclei of mice: A mechanism for monaural echo suppression, *J. Neurosci.*, 10(5):1762-1768.

Wu, S.H. and Oertel, D., 1986, Inhibitory circuitry in the ventral cochlear nucleus is probably mediated by glycine, *J. Neurosci.*, 6:2691-2706.

Young, E.D., 1984, Response characteristics of neurons of the cochlear nuclei, *in*: "Hearing Science", C.Berlin, ed., College-Hill Press, San Diego, CA, pp. 423-459.

Young, E.D., Costalupes, J.A., and Gibson, D.J., 1983, Representation of acoustic stimuli in the presence of background sounds: adaptation in the auditory nerve and cochlear nucleus, *in:* "Hearing-Physiological Basis and Psychophysics", R.Klinke and R.Hartman, eds., Springer-Verlag, New York, pp. 119-127.

Young, E.D., Shofner, W.P., White, J.A., Robert, J.-M., and Voigt, H.F., 1988, Response properties of cochlear nucleus neurons in relationship to physiological mechanisms, *in*: "Auditory Function: Neurobiological Bases of Hearing", G.M. Edelman, W.E. Gall, and W.M. Cowan, eds., John Wiley and Sons, New York, pp. 277-312.

# NEUROPHARMACOLOGICAL AND NEUROPHYSIOLOGICAL DISSECTION OF INHIBITION IN THE MAMMALIAN DORSAL COCHLEAR NUCLEUS

Edward F. Evans and Wei Zhao

Department of Communication and Neuroscience, Keele University
Keele, Staffs. ST5 5BG, U.K.

The cochlear nucleus (CN) is the first stage in the auditory pathway at which inhibition (mediated synaptically in response to single stimuli), is encountered, as was first shown by Galambos ('44) - in recordings probably from outlying cells of the dorsal cochlear nucleus (Galambos and Davis, '48).

It is now well recognised that the units of the *dorsal* division of the cochlear nucleus (DCN) exhibit most of the inhibitory effects encountered in the cochlear nucleus (Evans and Nelson '73a). This inhibition is easily modified or eliminated by depressant anaesthetics (Evans and Nelson '73a), a feature not always considered in studies of inhibition in the cochlear nucleus.

The distribution of inhibition in the receptive field has been used as a basis for classification of cochlear nucleus cells into five functional types (Evans and Nelson '73a) and this classification has been further elaborated by Young and colleagues (e.g., Young, '85).

The question arises, what is the nature of the inhibition encountered in the cochlear nucleus and for what aspects of auditory function can it be held responsible?

In the present work, an automated receptive field mapping paradigm (Evans '74, '79b) has been exploited, together with neuropharmacological techniques adapted from the pioneering methods of Caspary (e.g., Caspary et al. '79) to attempt to dissect out different varieties of inhibition in the dorsal cochlear nucleus of the chloralose anaesthetised guinea pig. The guinea pig has been chosen for convenience and low cost, and because extensive anatomical and immunohistochemical data are becoming available (e.g., Hackney et al., '90).

Chloralose has been used as the maintenance anaesthesia following surgery under neuroleptanaesthesia (Evans, '79a,b), because its actions, at least in the cat (Evans and Nelson, '73a) had least effect on inhibition in the cochlear nucleus, and it provides a highly stable anaesthetic regime in the guinea-pig. It is preferred to a decerebrate preparation, which would be difficult if not impossible in the guinea-pig. Even so, caution may be required because of our finding of lower and higher than expected proportions of type IV receptive fields in the granule cell and deep layers respectively of the dorsal cochlear nucleus in this preparation, compared with the cat (Evans and Nelson, '73a). This may be related to differences between chloralose anaesthesia and the unanaesthetised state or may represent species differences between the guinea pig and cat (Hackney, et al., '90).

For neuropharmacological dissection of putative inhibitory transmitters, we have studied iontophoretic application of the *specific antagonists* of the transmitters besides the agonists themselves and the interactions between agonists and antagonists. This paper, however, will concentrate on the results obtained with the antagonists *alone* in order to elucidate the effects of *naturally occuring inhibition*. This avoids drawing conclusions potentially based on non-specific effects of the iontophoretic application of the agonists themselves. To this end, we have adapted and extended the techniques of Caspary (e.g., Caspary et al., '79) as described in more detail in Zhao and Evans ('90, '91) and Evans and Zhao ('92).

Briefly, extra-cellular recordings were made by 7-barrel micropipettes in guinea-pigs anaesthetised with neuroleptanaesthesia for surgery (Evans, '79a) and maintained by alpha-chloralose (40 mg/kg) given every two hours or so. The central barrel of the 7-barrelled micro-electrode is NaCl filled for recording; two barrels are used respectively for current balancing (NaCl) and for localisation of the electrode position (Fast Green). The remaining four barrels contained four agonists or antagonists drawn from: GABA, bicuculline methiodide (antagonist to $GABA_A$ receptors), Baclofen (agonist of $GABA_B$ receptors); glycine, strychnine (antagonist to glycine receptors); acetylcholine, atropine (antagonist to muscarinic cholinergic receptors), curarine (antagonist to nicotinic cholinergic receptors). The responses of cochlear nucleus cells recorded extra-cellularly were collected to pure tone stimuli of generally 80 ms duration presented 4 times a second. The tones were presented in a pseudo-random sequence of frequencies and intensities so that a range of 60 dB and typically 3 octaves was covered evenly in about 4 minutes of data collection (Evans, '74, '79b). The resolution in the intensity domain was typically 4 dB; in the frequency dimension: 1/64 of 3 octaves. At the same time, the number of spikes evoked by the cell during the tone burst was counted and plotted (as in Fig. 1a) as the length of a vertical bar, the centre of which indicates the respective tone frequency and intensity. The receptive field map can also be represented as a contour map (Fig. 1b) constructed after spline smoothing, and sections can be cut through the contours: vertically to produce rate-level functions (as in Fig. 2b), and horizontally (as in Fig. 2c, where the stimulus intensity is marked as A in Fig. 2a) to produce *iso-level functions*. Separately, peristimulus time histograms were collected at characteristic (best) frequency (on-CF) and at off-CF frequencies using typically 80 ms tones (see Fig. 2d). All tones were shaped with 5 ms linear rise and fall times. In all cases, for example as shown in Figure 2a, data were collected before (left hand panel) during (middle panel) and following (right hand panel) administration of the antagonist to ensure that the drug effects were fully reversible. One advantage of the automated receptive field mapping paradigm is that by subtracting the control receptive field map (e.g., Fig. 1a left panel) from the receptive field map obtained by blocking the inhibition with the specific antagonist (Fig. 1a middle panel), a receptive field map of the inhibition is obtained (Fig. 1a right hand panel).

With these methods, at least five varieties of inhibition have been dissected and investigated: a) 'lateral/sideband' inhibition; b) 'background' inhibition; c) 'off' inhibition; d) presumed pre-synaptic inhibition; e) contralateral inhibition. These will be considered in turn.

## 'Lateral/side-band' type inhibition

This is the classical tone-evoked inhibition of the dorsal cochlear nucleus, exemplified in the receptive field map of a type IV cell in Figure 1a, left hand panel, and in the corresponding three dimensional contour map (b).

We have now shown this inhibition to be glycinergic (Zhao and Evans, '90; Evans and Zhao, '92): it can be completely blocked by strychnine, e.g. Figure 1 centre panel, where the receptive field is converted into an excitatory response area, similar to that of a cochlear nerve

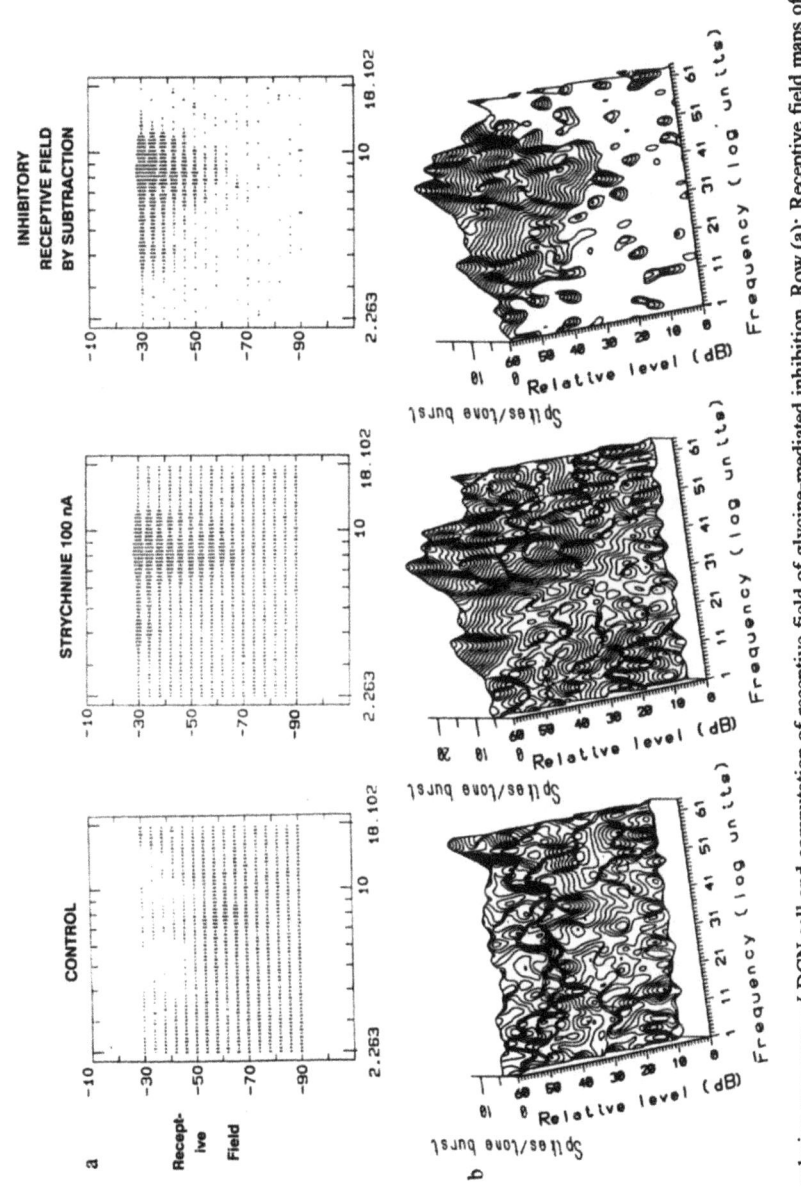

Figure 1. Effect of strychnine on a type IV DCN cell and computation of receptive field of glycine-mediated inhibition. Row (a): Receptive field maps of a Type IV cell, before and during iontophoretic application of strychnine. Row (b) shows three-dimensional representations of the receptive field maps in (a). Left-hand column: control receptive field, showing extensive inhibition. Middle column shows effect of 100 nA strychnine abolishing the inhibition and revealing the excitatory input to the cell. Right-hand column shows receptive field of inhibition computed by substracting control receptive field (left column) from that obtained during block of glycinergic receptors with strychnine (middle column). Note that in the right-hand column alone, inhibition is indicated by positive values; maximum inhibition is at the CF, but extends widely beyond the excitatory region revealed in the middle column.

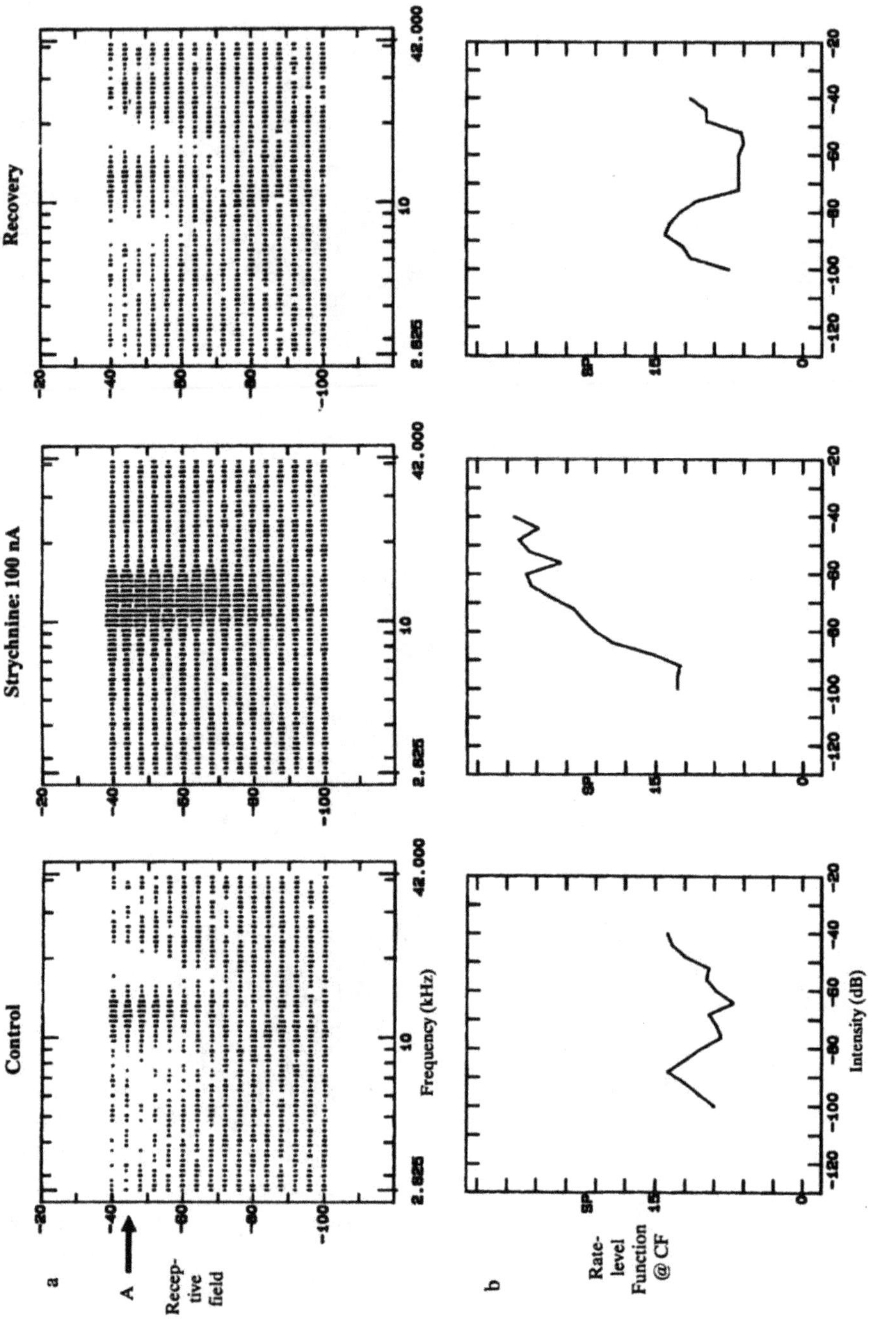

a
A
Recep-
tive
field
Control
Strychnine: 100 nA
Recovery
Frequency (kHz)
2.825
10
42.000
-20
-40
-60
-80
-100
b
Rate-
level
Function
@ CF
SP
15
0
Intensity (dB)
-120
-100
-80
-60
-40
-20

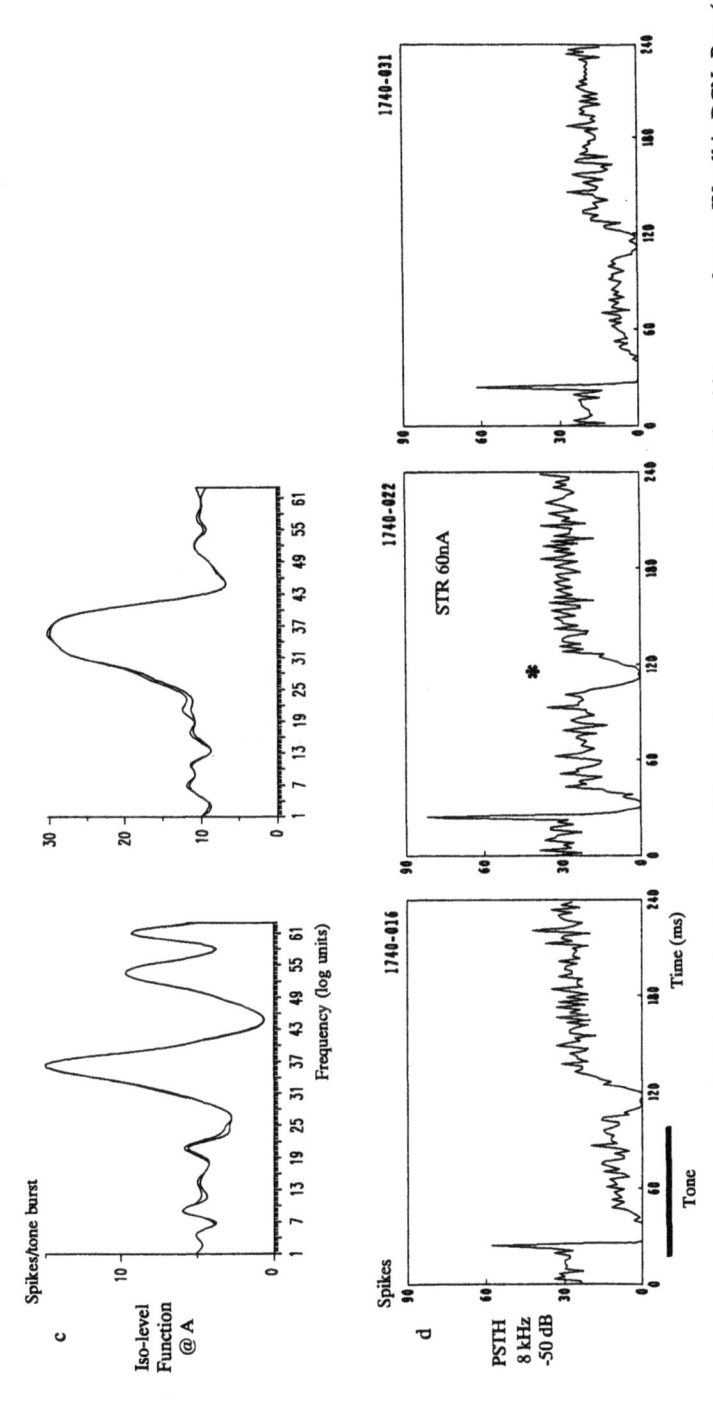

**Figure 2.** Effect of strychnine on (a) receptive field, (b) rate-level function, (c) iso-level function, and (d) peristimulus time histograms of a type IV cell in DCN. Row (a) receptive fields, before, during and following application of strychnine. Row (b) rate-level functions obtained at CF (12 kHz), from the receptive fields of row (a). Note conversion by strychnine of non-monotonic rate-level function (control and recovery) into a steep monotonic function (middle panel). Row (c) shows iso-level functions taken at -44 dB, indicated by the arrow A in (a). Note extensive inhibitory side-bands in control case (left column), surrounding narrowed excitatory response region, and removed by the strychnine blockade (middle column) with widening of the excitatory area and reduction in response contrast. Row (d) shows peristimulus time histograms of the cell at 8 kHz i.e. in the inhibitory side-band below CF. Note strychnine blockade of the sustained inhibition only, leaving unaffected the transient inhibition at the onset of the stimulus, and off-inhibition (indicated by asterisk).

fibre in shape, but wider in width (see below). Subtracting the control receptive field from that in which the glycinergic input is blocked gives, in the right hand panel of Figure 1, the glycinergic *inhibitory receptive field*, also shown as a contour plot below. This shows the important result that while the inhibition may *appear* to be predominantly 'lateral' or 'sideband', in reality it extends *throughout* the receptive field and is *strongest* at the excitatory CF. The previous iontophoretic experiments of Caspary et al., ('87) with strychnine blockade have revealed the existence of glycinergic inhibition at CF; we have now shown that glycinergic inhibition extends beyond the CF region to account for the inhibitory sidebands also.

Evidently, the width of the excitatory receptive area after strychnine blockade is wider than a cochlear nerve fibre of corresponding CF. In experiments reported elsewhere (Zhao and Evans, '92), we have measured the width of the excitatory response area 10dB above threshold and compared this width with that of cochlear nerve fibres from the guinea-pig with corresponding CF (7.5-15 kHz) from the data of Evans et al. ('91). The excitatory response areas are about 2.5 times wider than those of corresponding cochlear nerve fibres. Thus the convergent input from at least three cochlear nerve fibres on average must be involved in determining the excitatory input to a type IV cell in the dorsal cochlear nucleus of the guinea-pig.

Similar comparisons have been made between the bandwidths of the inhibitory receptive fields (again measured 10dB above threshold) and the excitatory receptive fields of cochlear nerve fibres. The inhibitory receptive fields have been determined by the subtraction method outlined above, or following clear delineation of the inhibitory area by application of bicuculline (which enhances the inhibitory receptive field as in Fig. 3b), or by blocking the excitatory input within the inhibitory receptive field by Baclofen as in Figure 5, central panel. Comparing the bandwidth of the inhibitory area 10dB above threshold with the excitatory bandwidth of cochlear nerve fibres of similar CF, the receptive field was some 4 times larger. This implies that at least 4 cochlear nerve fibres on average converge upon the interneurones responsible for the inhibitory receptive field. It needs to be emphasized that these type IV cells were found predominantly in the central region of the DCN and probably represent Giant cells.

Figure 2 shows another type IV cell with its receptive field in row (a) before, during and following blockade of the inhibitory receptive field with strychnine. In row (d) are shown peristimulus time histograms in response to tone bursts at 8 kHz (i.e., in the lower inhibitory sideband). These show that the glycinergic inhibition, blocked by strychnine, is sustained. The transient onset inhibition (and off-inhibition: see later) is not so readily blocked.

Figure 2 also shows the effect of blocking the glycinergic inhibition on information processing in the DCN.

The *rate-level functions* (row b) show that the strychnine blockade converts the non-monotonic rate-level function characteristic of type IV cells into a monotonic rate-level function (middle panel) with subsequent recovery (right hand panel). It also shows an increase in the slope of the rate-level function. Similar effects have previously been reported by Caspary et al. ('87). For units with relatively monotonic rate-level functions (e.g. type I and type II cells in the DCN), the effects of strychnine blockade are comparable: the slope of the rate-level function is increased and also the *discharge range* (i.e., the range of discharge rates between threshold and saturation; Zhao and Evans, '91).

Row (c) shows the effect of strychnine blockade on the *iso-level rate function* i.e., a horizontal section through the receptive field map, 4 dB below the maximum level (-44 dB, shown by the arrow at A in row a). The left hand iso-level function demonstrates the strong inhibitory side-bands on either side of a narrow excitatory response area. During strychnine blockade of the inhibition (right-hand iso-level function, corresponding to the centre panel of Fig. 2a), the lower inhibitory side-band is completely blocked while the high frequency inhibitory side-band is almost completely blocked. The loss of this inhibition reveals the true width of the excitatory response, being substantially wider than in the control case, as has been noted above. It also reveals how the *spectral contrast* (i.e., the ratio of discharge rate between the excitatory centre of the receptive field and the surrounding frequencies) is dependent upon the inhibitory side-bands. 'Lateral' or 'side-band' inhibition therefore serves to *enhance spectral contrast*.

From these results, we conclude that the glycinergic system responsible for the 'lateral/side-band' inhibition in dorsal cochlear nucleus units is frequency specific and is responsible for the non-monotonic rate-level function. It is also necessary for the enhancement of spectral contrasts, and narrows the response area. It is also presumably responsible for the phenomenon described by Palmer and Evans ('82) in which maskers, surrounding a narrow-band signal and falling in the inhibitory side-bands, can 'bias' the working point of a DCN cell, thus extending enormously its dynamic range.

This tone-related sustained inhibition is presumably mediated by the specific inhibitory pathway ending on the basal dendrites of Giant cells, probably derived from the activity of the glycinergic vertical cells (e.g. Osen et al., '90), with additional input from the ventro-tubercular tract arising from the VCN (Evans and Nelson '73b; Young et al. '88).

## Background/Tonic inhibition

We have shown the existence of a tonic 'background' inhibition on most of the cell types in the DCN, which inhibition is $GABA_A$-ergic and is therefore blocked by bicuculline (Zhao and Evans, '90; Evans and Zhao, '92). An example of this, for a type IV cell, is shown in Figure 3. The upper row shows the receptive field maps characteristic of a type IV cell (a) before, (b) during, and (c) following, bicuculline blockade. The middle panel shows how the bicuculline blockade has increased the background spontaneous activity, without affecting, and even *enhancing* the inhibitory receptive field.

We have some evidence that this inhibitory blockade by bicuculline does not produce a non-specific increase *regardless* of stimulus frequency and intensity. Using our subtraction technique (Evans and Zhao, '92) to compute the effect of the transmitter, its overall inhibititory effect is found to be interrupted at frequencies and intensities corresponding to the excitatory response area, in at least type I cells. Surprisingly, part of this region shows actual *excitation*. This could occur by a disinhibitory process involving $GABA_A$ action on a glycinergic inhibitory pathway, by interaction between $GABA_A$ and excitatory amino acid systems or, less likely, by a non-specific effect of iontophoretic current on the cell.

The second row in Figure 3 demonstrates some of the effects of bicuculline blockade of the background inhibition on the rate-level functions. The effects are opposite to those from strychnine. The non-monotonicity of the rate level function is enhanced. In studies of other types of DCN cells (Zhao and Evans, '91; Evans and Zhao, '92), the slopes of the rate-level functions are *decreased* by application of bicuculline. In several cases, the reduction in slope is accompanied by a reduction in the driven firing rate, thus representing a decrease in the *gain* of the cell response.

In the lower row in Figure 3, the peristimulus time histograms of the type IV cell show the enhancement of the spontaneous activity (middle panel, h). This reveals a sustained inhibition during the tone, compared with the mild excitatory response (panels g,i) in the control and recovery conditions respectively.

The $GABA_A$-ergic system, therefore, appears to be responsible for a 'non-specific' inhibition of background (spontaneous) activity *except* in the region of the receptive field occupied by the excitation (for type I cells at least) and/or surround inhibition (for type II, III and type IV cells). It thus can serve to enhance spectral contrast by reducing background spontaneous activity, and by increasing the gain of the cell's response. It is important to emphasize that the apparent adjustment of the 'background' tonic inhibitory influence by the

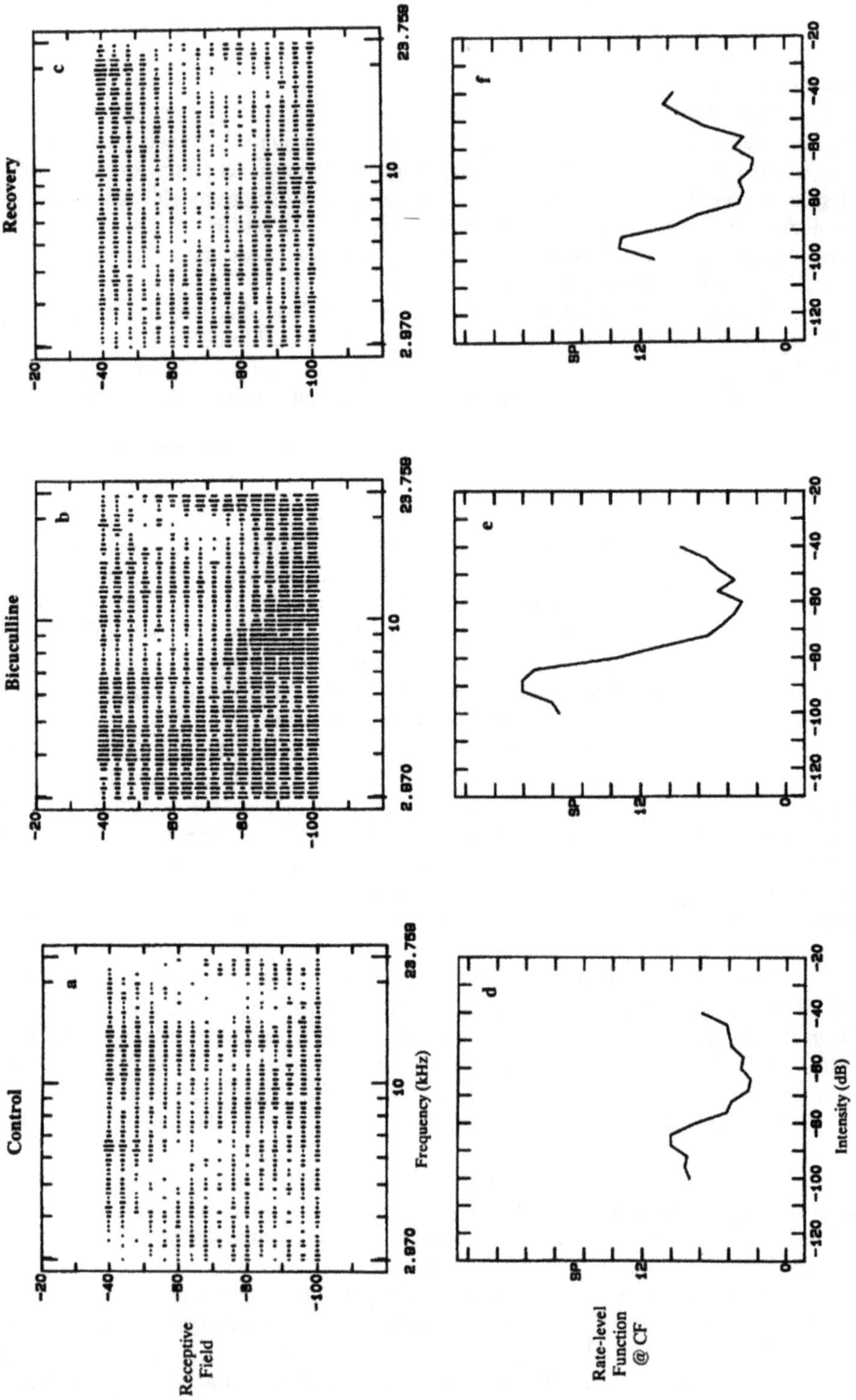
Control
Bicuculline
Recovery
a
b
c
d
e
f
Receptive Field
Rate-level Function @ CF
Frequency (kHz)
Intensity (dB)
2.970
10
23.758
-20
-40
-60
-80
-100
-120
SP
12
0

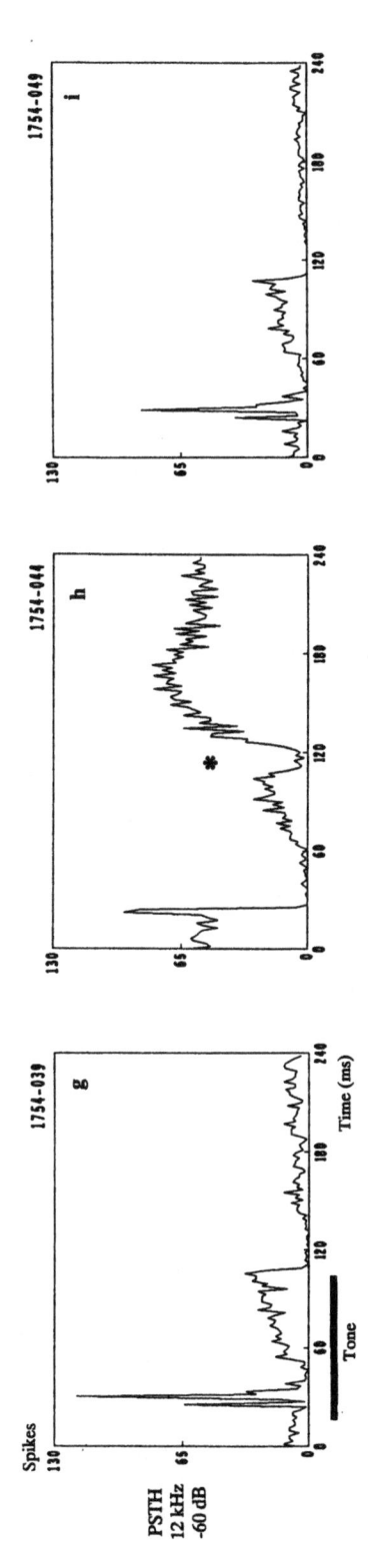

**Figure 3.** Effect of iontophoretically applied bicuculline on receptive field of a type IV DCN cell. (a) - (c): Receptive field maps showing increase in spontaneous activity under bicuculline blockade, revealing and enhancing wide field of inhibition (b). (d) - (f): Rate-level function showing enhancement of the non-monotonic function under bicuculline blockade (e). (g) - (i): Peristimulus time histogram at CF (12 kHz). Note increase in spontaneous activity under bicuculline blockade (h), revealing and enhancing transient and sustained inhibition and converting excitation at CF to net inhibition, leaving intact off-inhibition (indicated by asterisk).

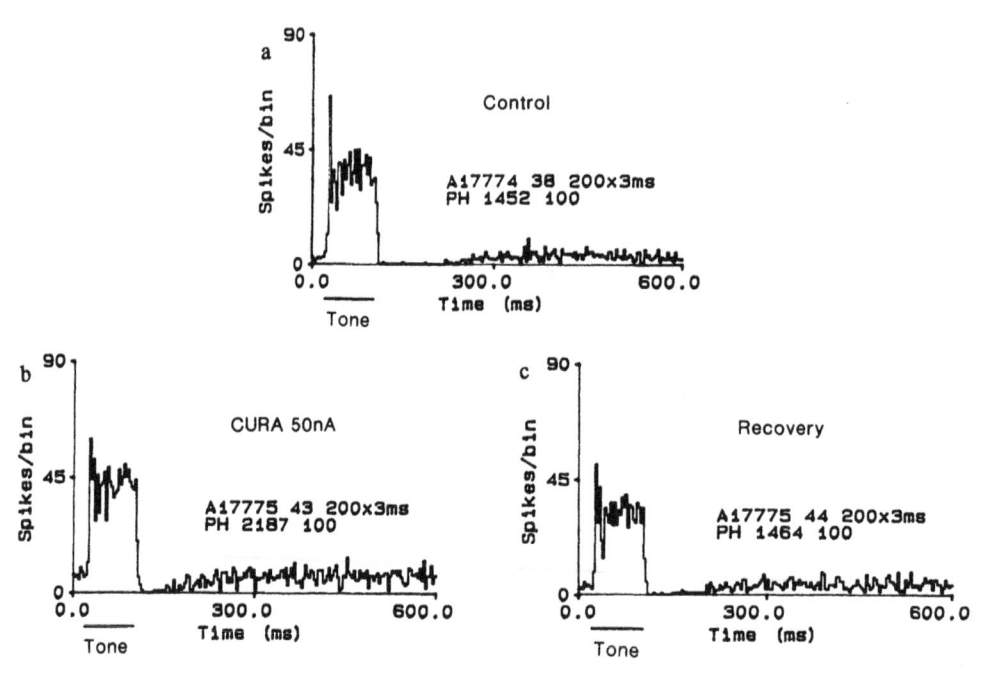

**Figure 4.** Effect of iontophoretically applied curarine on the peristimulus time histograms of a type I DCN cell. Note curarine (b) shortened the duration of the off-inhibition and additionally increased the spontaneous activity.

GABA$_A$-ergic system does not affect the glycinergic stimulus-evoked inhibitory system; rather it serves to enhance or even reveal it.

It seems likely that the GABA$_A$-ergic effects are mediated by the less stimulus-specific input (from the descending pathways) *via* the granule cell-parallel fibre system on to the apical dendrites of the pyramidal/fusiform and giant cells by way of the Stellate and/or Cartwheel cells (e.g., Osen et al., '90). However, the more specific effects described above in the region of the excitatory response and lateral inhibitory response regions can only easily be explained by assuming that there is a GABA$_A$-ergic inhibitory input onto stimulus-specific glycinergic cells that in turn inhibit the cells recorded from.

## Off-inhibition

This is the inhibition following the termination of a stimulus, examples of which are shown in Figure 2 (d) and Figure 3 (h), indicated by the asterisk in both figures. Neither strychnine nor bicuculline had any effect on this off-inhibition, except that the latter served to enhance it by increasing the background activity. Nor is the off-inhibition sensitive to blockers of muscarinic cholinergic receptors (atropine): (see Zhao and Evans, '91; Evans and Zhao, '92). However, we have some preliminary evidence of some effects of nicotinic receptor blockers (curarine) as shown in Figure 4. The top and bottom rows show the peristimulus time histograms of a type I cell in the DCN before and after recovery from iontophoretic application of curarine (middle panel). During the application, there is a marked reduction, but not complete blockade, of the off-inhibition. Unfortunately, the curarine also increases somewhat the spontaneous activity. Nevertheless, the degree of off-inhibition seems much less than might be accounted for by the increase in spontaneous activity alone, but this does need to be confirmed.

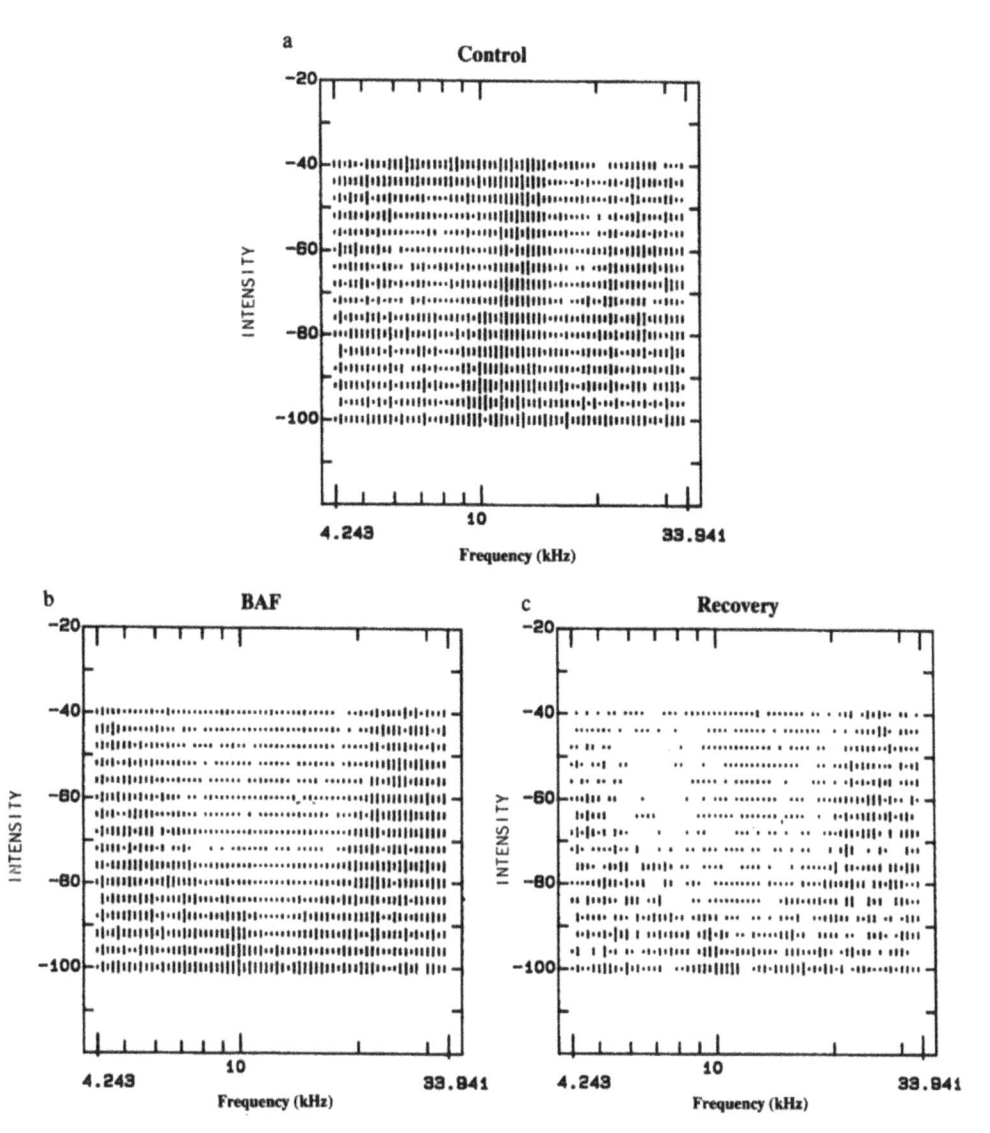

**Figure 5.** Effect of GABA$_B$ agonist, Baclofen, on a type IV cell in DCN. Control and recovery (a,c) receptive field maps show mixture of extensive inhibition and central excitatory patches in the type IV receptive field. Panel b shows the disappearance of the excitatory input under Baclofen (30 nA). Note that Baclofen blocked the excitatory response in the receptive field with little effect on the spontaneous activity.

The question arises of course whether the off-inhibition *is* a synaptic phenomenon rather than simply a membrane-based phenomenon. That the latter is unlikely to be the case, has been argued by Evans and Nelson ('73a), who presented evidence that the bandwidth of the off-inhibition was more closely related to that of the *on-inhibition* than to that of the on-excitation. That the off-inhibition can occur *without preceding excitation* is also shown in Figures 2 (d: middle panel) and 3 (h), again inconsistent with a wholly membrane-determined origin of the off-inhibition.

Since off-inhibition enhances the contrast between the difference in discharge rate corresponding to stimulus 'on' to stimulus 'off', off-inhibition is presumably responsible for enhancement of temporal contrast (Evans, '85).

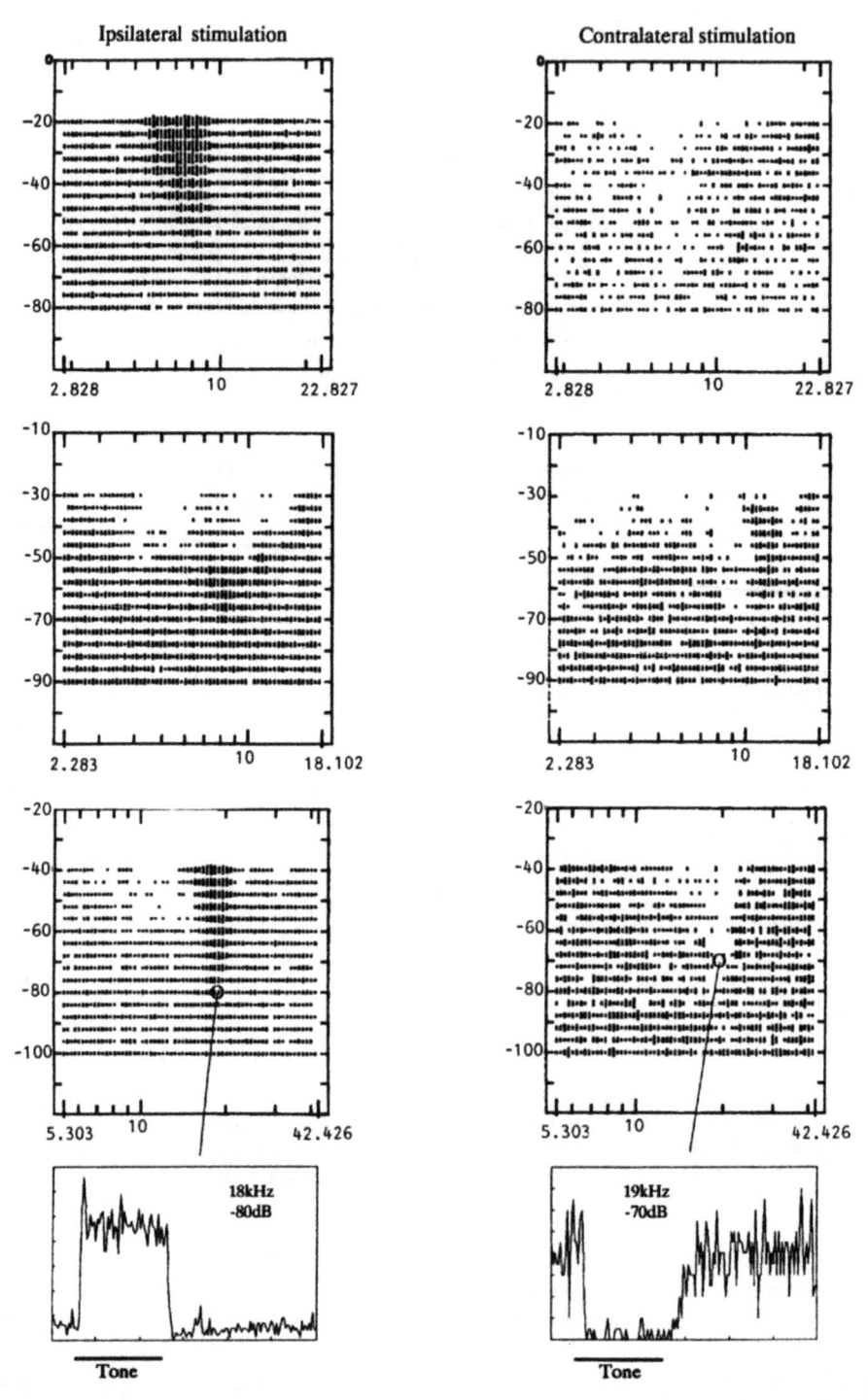

**Figure 6.** Contralateral inhibition. The upper three pairs of receptive field maps represent the receptive fields obtained from ipsilateral tone stimulation (left column) and contralateral tone stimulation (right column) for three cells: top row: type I; middle row: type IV; and lower row: type III cells in the DCN. Note the contralateral stimulation primarily evokes an inhibitory receptive field at or just above the ipsilateral excitatory CF. Lowest row: peristimulus time histograms indicating the ipsilateral excitation and contralateral inhibition obtained with similar tone stimulus parameters.

264

## Presumed pre-synaptic inhibition

Figure 5 shows an example of the island of excitation in the CF region of a type IV DCN cell almost entirely blocked by the application of Baclofen (middle panel), an agonist of $GABA_B$, thought to be a pre-synaptic inhibitory transmitter.

Similar, although not identical, effects have been reported by Caspary et al., ('84). In particular, in at least one of our cases (Fig. 5), Baclofen could block excitatory input virtually completely, *without* effect on the spontaneous activity. The effects are presumably related to the action of presynaptic terminals on the primary afferent dendrites on the Type IV cells, hence regulating the degree of their afferent input.

## Contralateral inhibition.

Figure 6 shows in the upper two thirds, receptive field maps from type I, type IV and type III cells (top to bottom) respectively under stimulation from the ipsilateral ear (left column) compared with contralateral stimulation (right column). In each case, stimulation of the contralateral ear evokes an inhibitory receptive field at or just above (in frequency) the ipsilateral *excitatory* receptive field, but, most importantly, having a similar threshold to the ipsilateral responses (thus eliminating the possibility of involvement of contralateral acoustic (bone-conducted) cross-talk.

The latency of this contralateral inhibition is short, being about 5 ms greater than ipsilateral excitation, in other words it is in the region of 10 ms from stimulus onset. This is much shorter than the latencies given in the earliest reports of contralateral inhibition (e.g. 30-60 ms; reviewed by Klinke et al., '70), but is similar to that found by Mast ('70: 8 - 25 ms). It would therefore be more consistent with input from the commissural cells in the contralateral VCN (e.g., Cant and Gaston, '82; Osen et al., '90). We are investigating whether this inhibition is glycinergic, as suggested on anatomical grounds by Wenthold et al. ('87) and Osen et al. ('90).

The combined use of our automated receptive field mapping technique and iontophoretic application of a variety of pharmacological putative inhibitory antagonists has allowed us to dissect out at least five types of inhibition in the DCN: a glycinergic 'lateral/side-band' inhibition, probably mediated by vertical cells and the ventro-tubercular tract; a $GABA_A$-ergic tonic 'background' inhibition, presumably mediated by intrinsic and extrinsic (descending) pathways; $GABA_B$ pre-synaptic inhibition, capable of blocking at least some of the excitatory input to most if not all types of DCN cells receiving primary afferent inputs; off-inhibition, mediated possibly in part by nicotinic receptors, but where the identity of the transmitter is at present unknown; contralateral inhibition, probably mediated by the commissural cells. To these we can presumably add *onset (transient) inhibition,* as yet not characterized pharmacologically.

Adjustment of these separable systems confers the potential of exquisitely fine control of information processing, particularly spectral and temporal contrast enhancement, in the cochlear nucleus.

ACKNOWLEDGEMENTS

Supported in part by the Medical Research Council and by the Wellcome Trust.

# REFERENCES

Cant, N. B. and Gaston, K. C., 1982, Pathways connecting the right and left cochlear nuclei. *J. Comp. Neurol.*, 212: 313-326.

Caspary, D. M., Havey, D. C. and Faingold, C. L., 1979, Effects of microiontophoretically applied glycine and GABA on neuronal response patterns in the cochlear nuclei, *Brain Res.*, 172: 179-185.

Caspary, D. M., Rybak, L. P. and Faingold, C. L., 1984, Baclofen reduces tone-evoked activity of cochlear nucleus neurons, *Hear. Res.*, 13: 113-122.

Caspary, D. M., Pazara, K. E., Kossl, M. and Faingold, C. L., 1987, Strychnine alters the fusiform cell output from the dorsal cochlear nucleus, *Brain Res.*, 417: 273-282.

Evans, E. F., 1974, Auditory frequency selectivity and the cochlear nerve, in: "Facts and Models in Hearing", E. Zwicker and E. Terhardt, eds., Springer-Verlag, Heidelberg, 118-129.

Evans, E. F., 1979a, Neuroleptanaesthesia for the guinea pig: an ideal anaesthetic procedure for long-term physiological studies of the cochlea, *Arch. Otolaryngol.*, 105: 185-186.

Evans, E. F., 1979b, Single unit studies of the mammalian auditory nerve, in: "Auditory Investigations: The Scientific and Technological Basis", H.A. Beagley, ed., Oxford University Press, Oxford, 324-367.

Evans, E. F., 1985, Aspects of the neuronal coding of time in the mammalian peripheral auditory system relevant to temporal resolution, in: "Time Resolution in Auditory Systems: 11th Danavox Symposium", A. Michelsen, ed., Springer Verlag, Berlin Heidelberg New York, 74-95.

Evans, E. F. and Nelson, P. G., 1973a, The responses of single neurones in the cochlear nucleus of the cat as a function of their location and the anaesthetic state, *Exp. Brain Res.*, 17: 402-427.

Evans, E. F. and Nelson, P. G., 1973b, On the functional relationship between the dorsal and ventral divisions of the cochlear nucleus of the cat, *Exp. Brain Res.*, 17: 428-442.

Evans, E. F., Pratt, S. R., Spenner, H. and Cooper, N. P., 1991, Comparisons of physiological and behavioural properties: auditory frequency selectivity, in: "9th International Symposium on Hearing: Auditory Physiology and Perception", Y. Cazals, ed., Pergamon Press Oxford, 1-6.

Evans, E. F. and Zhao, W., 1992, Inhibition in the dorsal cochlear nucleus: pharmacological dissection, varieties, nature and possible functions, in: "Cochlear Nucleus: Structure and Function in Relation to Modelling", Vol. 3 of Advances in Speech Hearing and Language Processing, W.A. Ainsworth, ed., (In Press).

Galambos, R., 1944, Inhibition of activity in single auditory nerve fibers by acoustic stimulation. *J. Neurophysiol.*, 7: 287-303.

Galambos, R. and Davis, H., 1948, Action potentials from single auditory nerve fibers?, *Science*, 108:513-513.

Hackney, C. M., Osen, K. K. and Kolston, J., 1990, Anatomy of the cochlear nuclear complex of guinea pig, *Anat. Embryol.*, 182:123-149.

Klinke, R., Boerger, G. and Gruber, J., 1970, The influence of the frequency relation in dichotic stimulation upon the cochlear nucleus activity, in: "Frequency Analysis and Periodicity Detection in Hearing", Plomp R and Smoorenburg GF, eds., Sijthoff Leiden, 162-167.

Mast, T. E., 1970, Binaural interaction and contralateral inhibition in dorsal cochlear nucleus of the chinchilla, *J. Neurophysiol.*, 33:108-115.

Osen, K. K., Ottersen, O. P. and Storm-Mathisen, J., 1990, Colocalization of glycine-like and GABA-like immunoreactivities: a semiquantitative study of individual neurons in the dorsal cochlear nucleus of cat, in: "Glycine Transmission", P. Ottersen and J. Storm-Mathisen, eds. 417-451.

Palmer, A. R. and Evans, E. F., 1982, Intensity coding in the auditory periphery of the cat: Responses of cochlear nerve and cochlear nucleus neurons to signals in the presence of Bandstop masking noise, *Hearing. Res.*, 7:305-323.

Wenthold, R. J., 1987, Evidence for a glycinergic pathway connecting the two cochlear nuclei: an immunocytochemical and retrograde transport study, *Brain. Res.*, 415:183-187.

Young, E. C., 1985, Response Characteristics of Neurons of the Cochlear Nuclei, in: "Hearing Science", Berlin, C. I., Ed., San Diego, College-Hill Press, 423-460.

Young, E. D., Shofner, W. P., White, J. A., Robert, J. M. and Voigt, H. F., 1988, Response Properties of Cochlear Nucleus Neurones in Relationship to Physiological Mechanisms, *Annual Symposium of The Neurosciences Institute*, Wiley and Sons, 277-312.

Zhao, W. and Evans, E. F., 1990, Pharmacological microiontophoretic investigation of receptive-field and temporal properties of units in the cochlear nucleus, *British Journal of Audiology*, 24: 193-193.

Zhao, W. and Evans, E. F., 1991, Dorsal cochlear nucleus units: neuropharmacological effects on dynamic range and off-inhibition, *British Journal of Audiology*, 25:53-54.

Zhao, W. and Evans, E. F., 1992, Bandwidths of excitatory and inhibitory receptive fields in the dorsal cochlear nucleus, *British Journal of Audiology*, 26-179-180.

# COMPARISON OF QUANTITATIVE AND IMMUNOHISTO-CHEMISTRY FOR CHOLINE ACETYLTRANSFERASE IN THE RAT COCHLEAR NUCLEUS

Donald A. Godfrey

Department of Otolaryngology - Head & Neck Surgery and Departments of Physiology & Biophysics and Anatomy, Medical College of Ohio, Toledo, Ohio 43699-0008, USA

## Early work on acetylcholine in the cochlear nucleus

Assessment of the role of a putative neurotransmitter in a region of the brain requires a multifaceted approach. The transmitter, as well as enzymes related to its metabolism, should be present in significant amounts in the neurons using it, and it must be released when these neurons are active. Neurons acted upon by the transmitter must have receptors for it. There should be a mechanism for stopping the action of the transmitter, usually involving uptake or enzymatic degradation to remove it from the synaptic cleft. Application of the transmitter to neurons should give the same effects as seen when it is released through physiological stimulation. Pharmacological agonists and antagonists of the transmitter should affect the postsynaptic neurons in ways predictable from their effects at synapses where the transmitter has become well established (Werman, '66).

When we began our studies of neurotransmitters in the cochlear nucleus about 20 years ago, acetylcholine seemed a good candidate to start with. It was well established as the neurotransmitter of motoneurons, preganglionic autonomic neurons, and postganglionic parasympathetic neurons. The information obtained during the study of these cholinergic neuronal systems (Werman, '66; Taylor and Brown, '89) would facilitate our studies of acetylcholine in the cochlear nucleus. The metabolism of acetylcholine is relatively simple, involving only two enzymes: choline acetyltransferase (ChAT) for synthesis and acetylcholinesterase (AChE) for degradation. Further, acetylcholine apparently has little function in neurons other than neurotransmission, and no known function in glial cells. So far as is currently known, only cholinergic neurons have significant amounts of ChAT, and therefore the ability to synthesize acetylcholine. Thus, compelling evidence for a neuron being cholinergic can be provided by demonstrating a high concentration of ChAT.

---

ABBREVIATIONS: AChE - acetylcholinesterase; AVCN - anteroventral cochlear nucleus; ChAT - choline acetyltransferase; DCN - dorsal cochlear nucleus; PVCN - posteroventral cochlear nucleus; VCN - ventral cochlear nucleus.

At the time we began our studies, there was evidence from AChE slide histochemistry (Rasmussen, '67; Osen and Roth, '69; McDonald and Rasmussen, '71), from pharmacological studies (Comis and Whitfield, '68; Comis, '70), and from indirect measurement of release (Comis and Davies, '69), that acetylcholine functioned to some extent as a neurotransmitter in the cochlear nucleus. A portion of the cholinergic innervation was thought to derive from branches of the olivocochlear bundle, for which available evidence suggested acetylcholine as transmitter (Rasmussen, '67; Brown and Howlett, '68; Osen and Roth, '69; Klinke and Galley, '74).

## Quantitative histochemical results

Our studies utilized the biochemical and histochemical methods available at the time: quantitative histochemistry (Lowry and Passonneau, '72), involving microdissection and microassay for ChAT and AChE activities, along with histochemical staining for AChE activity. The microdissection method, with incorporation of a mapping approach (Godfrey and Matschinsky, '76), provided reliable quantitation at a spatial resolution as small as 20 $\mu$m. The slide histochemistry provided higher spatial resolution, but without reliable quantitation. These methods enabled a detailed examination of the prominence and distribution of probable cholinergic elements in the cochlear nucleus and related fiber tracts (Godfrey et al., '77, '81, '84, '87c, '88). Their combination with a surgical lesioning approach provided insights about sources and routes of cholinergic fibers innervating the cochlear nucleus (Godfrey et al., '83, '87a,b, '90).

The detailed distributions of ChAT and AChE activities in the cochlear nucleus differ between the two major mammalian families studied, cats and rats. However, in both, ChAT activities in cochlear nucleus subregions are low-to-moderate in magnitude and decline dramatically after section of centrifugal pathways to the nucleus. These centrifugal ChAT-containing pathways enter via two major routes: from the olivocochlear bundle via collaterals and from the trapezoid body. The olivocochlear collateral route is more prominent in cats, whereas the trapezoid body route is much more prominent in rats. The most likely origins for the centrifugal ChAT-containing pathways to the cochlear nucleus are in or near the superior olive.

These studies were hampered by lack of a high resolution marker for ChAT. The resolution available through the histochemical reaction for AChE has been valuable in certain cases, such as marking the olivocochlear bundle (Godfrey et al., '84). However, AChE activity is often unimpressive in cholinergic axons, such as those of the facial motor root (Godfrey et al., '84). Also, AChE might not always be prominent in cholinergic somata (Levey et al., '83; Adams, '89) and exists in noncholinergic neurons, especially those receiving cholinergic synapses. ChAT, on the other hand, is found in high concentrations throughout cholinergic neurons and is virtually absent from noncholinergic neurons. For example, the ratio of ChAT enzyme activities between the facial motor and auditory nerve roots in the rat is over 300, whereas the ratio of AChE activities is 13.

## Immunohistochemistry for ChAT

Availability of antibodies to purified ChAT has enabled high resolution immunohistochemical localization of this enzyme. We have used a polyclonal antibody against ChAT in the rat cochlear nucleus. Immunoreaction obtained with this antibody in a previous study of rat brain (Tago et al., '89) compared qualitatively well with our microchemical results for several brain regions. In our procedure, brains from rats perfused with cold (4°C), buffered (pH 7.4) 4% paraformaldehyde were sectioned at 30 $\mu$m in a cryostat (Hacker-Bright) at

−20°C. Every third section was put onto frozen 100 mM phosphate buffer, pH 7.4. The sections were allowed to melt into the buffer and rinse for 2 hr, then were transferred to a 1:1000 dilution of ChAT antiserum (Chemicon), containing 0.4% Triton X-100, and incubated for about 16 hr at 4°C. After the incubation, the sections were rinsed with phosphate-buffered saline, incubated with biotinylated secondary antiserum (Vectastain) for 1 hr, rinsed, incubated with Avidin-Biotin-HRP complex (Vectastain) for 30 min, rinsed, then incubated for approximately 3 min with 0.5 g/l diaminobenzidine (DAB) in 50 mM phosphate buffer, pH 7.4, containing 0.03% hydrogen peroxide. As controls, some sections were incubated with buffer containing normal rabbit serum, but no primary antibody. Also, some sections were incubated with a 1:10,000 dilution of ChAT antiserum.

Although immunohistochemistry is a powerful approach for studying localization of neurotransmitter systems, there are potential artifacts that may complicate interpretation of the results. False positives may result from cross-reactivity of the antibody with peptide sequences on molecules other than that against which the antibody was made (Polak and Van Noorden, '84; Tago et al., '89). False negatives may result from failure of the antibody to reach antigenic sites. Another difficulty with immunohistochemistry is the subjectivity involved in its interpretation. At first glance, the stained sections seem to convey a straightforward message. However, our experience so far suggests that criteria must be carefully chosen to correctly discern the relation between what is seen and the underlying chemistry. Besides variation in results depending on fixation procedures, we have noted some gradients in staining intensity within sections which seem unlikely to be related to concentrations of ChAT. There are general differences in staining across preparations, and even among different sections in the same preparation. Nevertheless, we obtain little staining when the primary antibody is omitted and no obvious improvement in specificity with reduced antibody concentration. Within sections, not all structures appear as either darkly labeled or unlabeled; some have intermediate labeling, thus complicating the decision about which to consider immunopositive. We have used some strategies to aid our interpretations of the ChAT immunohistochemistry. Firstly, we have compared the results to those obtained previously with the more objective microdissection and microassay method. Since the two methods take such different approaches, agreement between them provides powerful confirmation that conclusions are correct. Also, we have used computer imaging of stain optical densities to confirm our visual criteria for immunopositivity.

Comparisons between quantitative and immunohistochemistry would best be made by applying both to the same series of sections. However, the two methods require different processing of tissue. Quantitative measures of enzyme activity are seriously compromised by fixation of tissue, while ChAT immunohistochemistry requires perfusion fixation. Our attempts at ChAT immunohistochemistry on fresh frozen sections, fixed after sectioning, have not given satisfactory results. Thus, our studies so far have compared immunohistochemistry with quantitative histochemistry in different rats.

## ChAT immunoreactivity in the rat cochlear nucleus: overall observations

Based on the microdissection and assay results, we expected a moderate amount of ChAT immunoreactivity in the cochlear nucleus, much less than in the facial or trigeminal motor nuclei, but more than in cerebellum and rather similar to the spinal trigeminal nucleus (Godfrey and Matschinsky, '81; Godfrey et al., '83). Within the cochlear nucleus, ChAT immunoreactivity should be especially prominent in fibers and terminals, namely those of the centrifugal ChAT-containing pathways (Godfrey et al., '87a,b). These should be distributed throughout the nucleus, except for very few in the interstitial nucleus (auditory, or cochlear, nerve root). We expected few labeled somata, most of which should be in granular regions since these have the highest residual ChAT activity after transection of centrifugal fibers (Godfrey et al., '83). Of the non-primary fiber bundles with connections to the cochlear

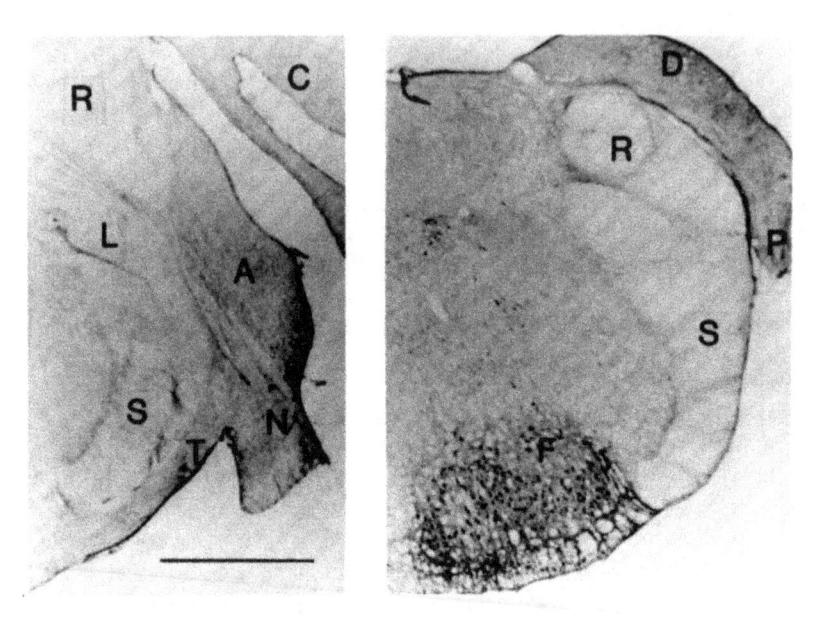

Figure 1. Photomicrographs of two transverse sections of rat brain stem immunoreacted for ChAT. Scale bar at bottom represents 1 mm and applies to both photomicrographs. Dorsal is up, lateral to right. Abbreviations: A, AVCN; C, cerebellum (flocculus), D, DCN; F, facial nucleus; L, centrifugal labyrinthine bundle, containing olivocochlear and centrifugal vestibular fibers (thin, darkly stained tract located just below L); N, auditory nerve root; P, PVCN; R, restiform body; S, spinal trigeminal tract; T, trapezoid body.

nucleus, the centrifugal labyrinthine bundle, containing the olivocochlear fibers as well as those to the vestibular labyrinth, should have prominent immunoreactivity, comparable to that of the facial motor root (Godfrey et al., '84). The trapezoid body and acoustic striae (a term used here to refer collectively to the dorsal and intermediate acoustic striae) should have modest immunoreactivity, much less than the facial motor root, but more than the spinal trigeminal tract or the auditory nerve root (Godfrey and Matschinsky, '81).

In general, the findings matched our expectations (Godfrey and Heaney, '91). Immunoreactivity for ChAT was less prominent in the cochlear nucleus than in the facial nucleus (Fig. 1). The immunoreactivity of the trapezoid body was much less than those of the centrifugal labyrinthine bundle and facial motor root, but more than that of the spinal trigeminal tract. We looked unsuccessfully for bundles of labeled fibers running within the trapezoid body, especially its more superficial part near the cochlear nucleus (Godfrey et al., '88); there were only hints of individual thin immunopositive fibers. Bundles of fibers in the trapezoid body which stain darkly for AChE activity have also not been observed (Osen et al., '84). Immunopositive fibers were found corresponding in location to all three AChE-positive collateral routes - strial, subpeduncular, and ventral - from the olivocochlear bundle (White and Warr, '83; Osen et al., '84; Godfrey et al., '87a). Many fibers entering the cochlear nucleus from the subpeduncular route were seen to turn dorsalward into the medial part of the nucleus. Based on our previous results (Godfrey et al., '87a,b), some of the fibers in the acoustic striae may not be strial-route olivocochlear collaterals, but may instead be such dorsally directed fibers.

At a more detailed level, the optical density of immunoreactivity correlated poorly with our previous quantitative measurements of ChAT enzyme activity (Fig. 2). The optical densities varied over a narrow range among regions of the cochlear nucleus, and differences in optical density were consistently found only where there were large differences in enzyme activity, as between interstitial nucleus and DCN fusiform soma layer (20-fold range). When the facial nucleus, with enzyme activity 10 times that of the DCN fusiform soma layer,

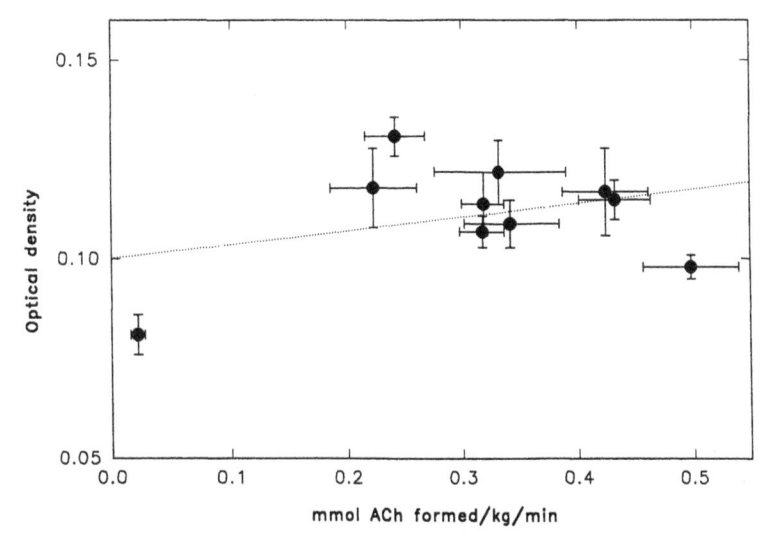

**Figure 2.** Optical density of ChAT immunoreactivity plotted against measured ChAT enzyme activity (mmol acetylcholine formed/kg/min) for regions of rat cochlear nucleus. The optical densities were measured using a computer imaging system for many outlined areas within each region (The thin artifactual superficial layer of staining (Fig. 1) was excluded from the outlined areas). Means and standard errors of the data for 4 rats are plotted against means and standard errors of enzyme activities for the same regions published previously (Godfrey et al., '83). The dotted line is a linear regression fit to the data; the correlation coefficient is +0.34. Regions represented are, from left to right, interstitial nucleus (auditory nerve root), granular region lateral to PVCN, DCN molecular layer, DCN deep layer, caudal PVCN, granular region dorsolateral to AVCN, caudoventral AVCN, rostral AVCN, DCN fusiform cell layer, and caudodorsal AVCN.

was included in the plot of Figure 2, the correlation coefficient increased from 0.34 to 0.70. Thus, as found previously for amygdala (Hellendall et al., '86), although attractive digitized images can be generated by computer, the optical density of the ChAT immunoreactivity represents only grossly the preponderance of ChAT.

## ChAT-immunoreactive terminals in the rat cochlear nucleus

The most prominent form of ChAT immunoreactivity in the cochlear nucleus was as small oval spots, or puncta, 0.5 to 2 μm in diameter, having the appearance of synaptic terminals (Fig. 3). Probably few of these were fibers cut in cross-section, since there was little evidence of length. Some could represent fiber varicosities, which were seen along some fibers oriented relatively parallel to the plane of section. Since the quantitative histochemical results had indicated that most ChAT activity in the cochlear nucleus represents fibers and terminals of centrifugal pathways, these ChAT-immunoreactive puncta seemed likely to be accurate representations of ChAT localization in the nucleus. Counts of areal densities of immunoreactive puncta in the various cochlear nucleus regions showed a reasonably good correlation with measured ChAT activities (Fig. 4).

In the VCN, the puncta occurred in clusters, corresponding mainly to the locations of lightly staining somata (Fig. 3). In the DCN, the puncta were more diffusely arranged, with some clustering that was not as clearly related to individual neurons. There was variation in puncta areal density across the layers of the DCN, with lowest densities in the superficial part of the molecular layer and deep part of the deep layer and highest densities in the fusiform soma layer (Fig. 5). This distribution correlates fairly well with the distribution of ChAT activity measured across these layers, except for the deep part of the deep layer.

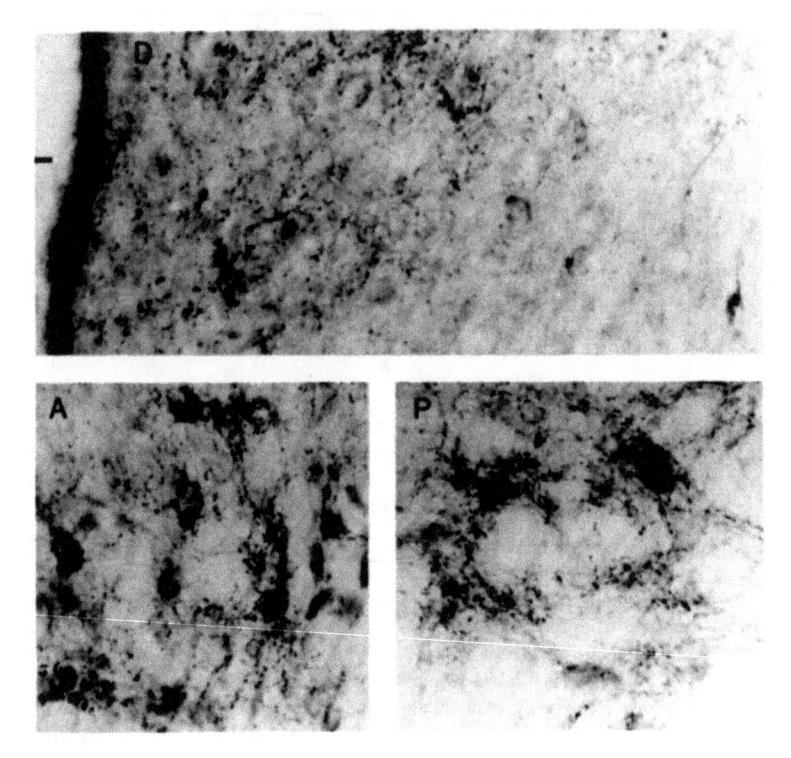

Figure 3. Photomicrographs of transverse sections of rat brain stem immunoreacted for ChAT, showing distributions of puncta (presumed terminals) in DCN (D) at top, oriented perpendicular to surface (at left), and AVCN (A) and PVCN (P) at bottom, oriented so that dorsal is up. Scale bar at upper left represents 10 μm and applies to all photomicrographs.

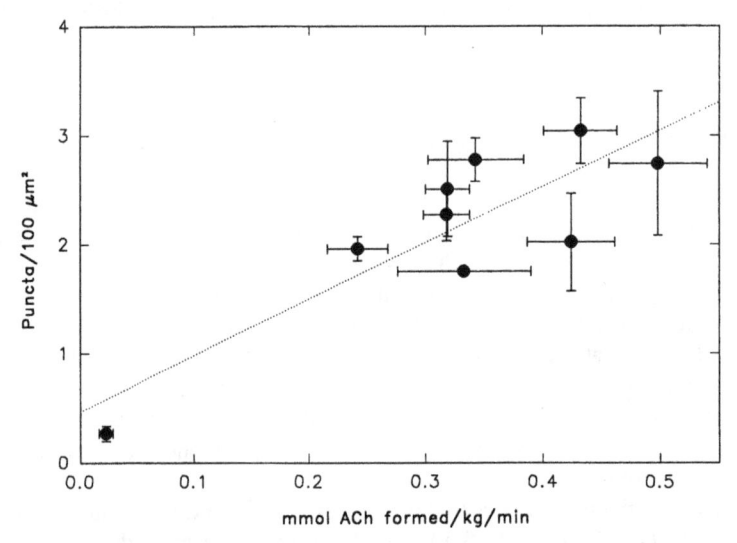

Figure 4. Areal density of ChAT-immunoreactive puncta in rat cochlear nucleus regions, plotted against ChAT enzyme activity (mmol acetylcholine formed/kg/min). The enzyme activities and regions are the same as in Fig. 2, except for omission of granular region lateral to PVCN. The ChAT-immunoreactive puncta were counted in a central portion of each region in 3 rats, at 400X magnification using a Leitz Orthoplan II microscope. The means and standard errors across the rats are presented. The dotted line is a linear regression fit to the data; the correlation coefficient is +0.86.

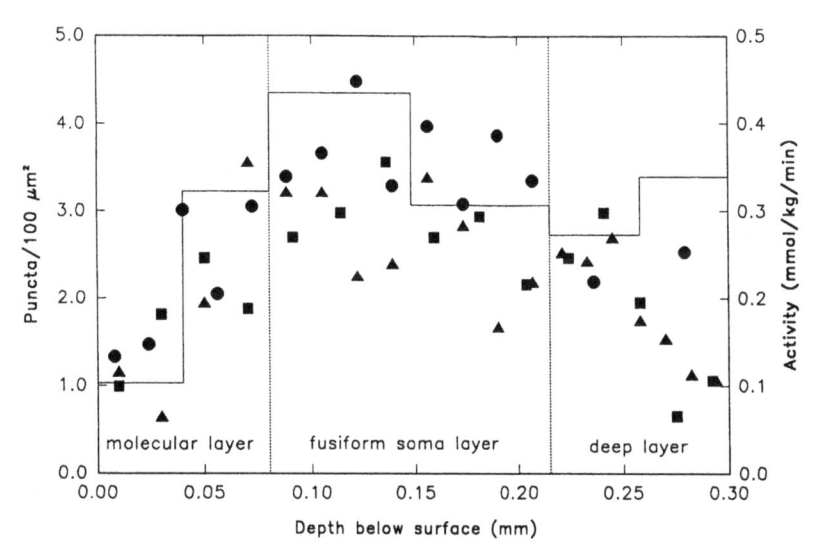

**Figure 5.** Areal density of ChAT-immunoreactive puncta plotted against depth below DCN surface in 3 rats. The vertical dotted lines represent the average depths of the boundaries between molecular, fusiform soma, and deep layers. The puncta densities for the 3 rats, measured by counting immunolabeled puncta in a band, orthogonal to the layers, approximately halfway between the ventrolateral and dorsomedial ends of the DCN, are shown by different symbol shapes. To fit the data for the 3 rats onto the same graph, actual depth measurements were adjusted so that data occur at the proper relative depths within each layer. The depth measurements were not corrected for tissue shrinkage. The solid horizontal lines represent mean measured ChAT activities (scale at right) for superficial and deep portions of each layer in 3 rats, at a comparable medial-lateral position in the DCN (Godfrey et al., '87c).

Unfortunately, the ChAT-immunoreactive puncta do not show well in all preparations. Where they do, however, they may represent fairly accurately the locations of ChAT-containing, and therefore likely cholinergic, terminals.

## ChAT-immunoreactive somata in the rat cochlear nucleus

The identification of ChAT-immunoreactive somata in the rat cochlear nucleus was complicated by intermediate staining of some somata, between those reacting as darkly as known cholinergic neurons and those obviously unreactive. Gradations in immunoreactivity of neuronal somata for the transmitter amino acids glycine and $\gamma$-aminobutyrate (GABA) have previously been documented by optical density measurements in thin sections (Osen et al., '90). While our thicker sections preclude such accurate quantitation of somata immunoreactivities, optical density measurements for several neurons in each category served as an objective check of our visual criterion for choosing immunopositive neurons (Table I). The immunoreactivity of the cholinergic facial motoneurons was used as our standard for immunopositivity. Three main groups of immunopositive neurons were identified on this basis in the cochlear nucleus: neurons with small somata in granular regions (Mugnaini et al., '80a,b), neurons with small elongated somata in the region of the acoustic striae, and cochlear root neurons (Harrison et al., '62; Harrison and Irving, '65; Merchan et al., '88; Osen et al., '91) (Fig. 6).

The immunopositive neurons in granular regions were expected from our previous quantitative histochemical measurements, as mentioned above. About 70% of these were located in medial granular regions, which are less easily defined than the dorsolateral regions specifically sampled in the quantitative histochemical studies. A reexamination of our quantitative results for lesioned rats showed that medial regions often had more intrinsic ChAT acti-

**Table I.** Optical Density of ChAT Immunoreactivity for Neuron Somata
Mean ± SEM (Number of cells)

| Neuron group | Cytoplasm density | Nucleus density |
| --- | --- | --- |
| Facial motoneurons | 0.66 ± 0.05 (9) | 0.09 ± 0.01 (4) |
| VNTB neurons | 0.77 ± 0.04 (19) | 0.15 ± 0.02(19) |
| LSO neurons | 0.57 ± 0.02 (20) | 0.14 ± 0.02(19) |
| Acoustic striae neurons | 0.56 ± 0.03 (28) | 0.17 ± 0.01(26) |
| Cochlear root neurons | 0.61 ± 0.01 (5) | 0.27 ± 0.02 (4) |
| Granular regions neurons | 0.68 ± 0.03 (19) | 0.18 ± 0.03(15) |
| Non-reactive CN neurons | 0.23 ± 0.01 (24) | 0.08 ± 0.01(22) |

All measurements are for the same rat. Using a computer imaging system, several point measurements were made for somatic cytoplasm and for the nucleus of each neuron, avoiding any immunoreactive puncta, and were averaged to obtain the values for that neuron. Average cytoplasm density for neurons considered non-immunoreactive is significantly less than for any of those considered ChAT-immunopositive ($p<0.001$). Abbreviations: VNTB, ventral nucleus of trapezoid body; LSO, lateral superior olivary nucleus; CN, cochlear nucleus.

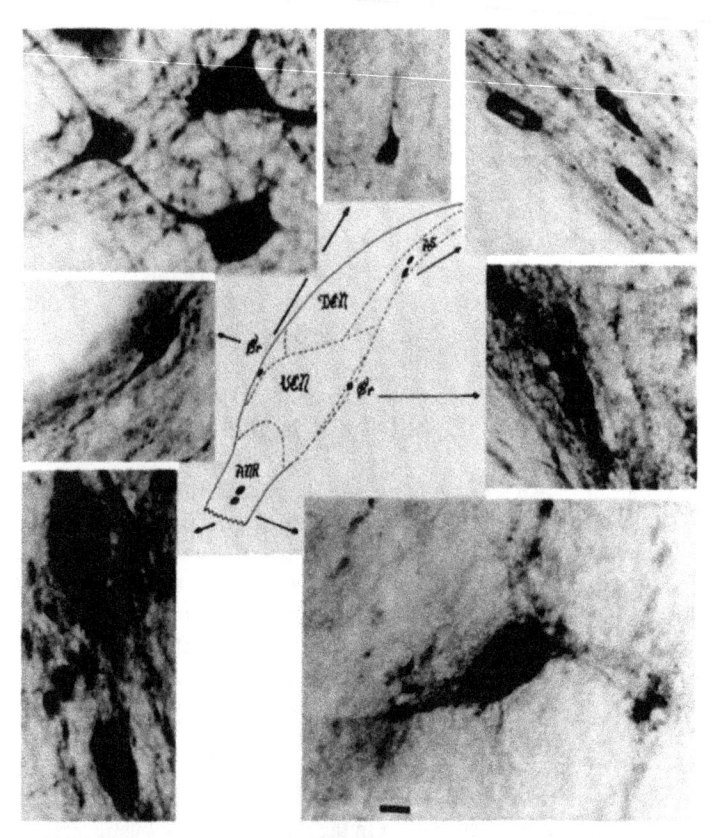

**Figure 6.** Photomicrographs of transverse sections of rat brain stem immunoreacted for ChAT. Scale bar at bottom of lowest right photomicrograph represents 10 μm and applies to all photomicrographs. Shown are facial motoneurons (upper left) and 3 categories of immunopositive neurons in the cochlear nucleus, whose general locations are schematically illustrated in the center diagram. Three neurons of granular regions (Gr) include one, at left, located in the granular region dorsolateral to AVCN, one, at top, located more dorsally in the granular region dorsolateral to AVCN, and one, at right, located in the granular region medial to PVCN, near the entrance of the subpeduncular olivocochlear branch. Three immunopositive neurons of the acoustic striae region (AS) are shown at upper right (the restiform body occupies the lower left corner of this view). At bottom are shown 3 neurons of the auditory nerve root (ANR) (cochlear root neurons).

**Table II.** Average Number of ChAT-Immunopositive Neuron Somata per Rat Cochlear Nucleus: Mean ± SEM (Number of rats)

| Location | Number |
| --- | --- |
| Acoustic striae region | 44 ± 12 (5) |
| Auditory nerve root | 55 ± 11 (4) |
| Granular regions | 85 ± 23 (5) |
| Dorsal cochlear nucleus | 12 ± 5 (5) |
| Ventral cochlear nucleus | 6 ± 3 (5) |
| Total | 201 ± 28 (5) |

Counts in 10-43 sections of either cochlear nucleus in 5 rats were multiplied by appropriate factors to obtain estimates of total immunopositive neuron somata per cochlear nucleus.

vity (i.e., remaining after transection of centrifugal innervation of the nucleus) than did central portions of the AVCN and PVCN (Godfrey et al., '83, '87b). The immunopositive neurons in the granular regions resemble in size the Golgi cells described by Mugnaini et al. ('80a); many resemble in shape the spindle-shaped cells reported by McDonald and Rasmussen ('71) to stain for AChE activity. The other two groups of ChAT-immunopositive somata were not predicted from our quantitative studies, but are not inconsistent with them. The neurons in the acoustic striae region could account for the considerable range of ChAT activities often measured for samples of this tract in individual rats. Samples with particularly high ChAT activities (Godfrey et al., '87a,b) might correspond to those containing ChAT-immunopositive somata. Some of these neurons are located just superficial to the acoustic striae in the deepest part of the deep DCN and might account for the disparity here between average measured ChAT activity and areal densities of ChAT-immunoreactive puncta (Fig. 5). The cochlear root neurons, seen as darkly immunoreactive somata in the auditory nerve root (Fig. 1), occur peripherally enough in the root that they would not have been included in our previous microdissections.

We have estimated the numbers of ChAT-immunopositive somata in regions of the cochlear nucleus in 5 rats (Table II). The numbers are small. The number of immunopositive somata in the auditory nerve root corresponds well with the recently reported number of cochlear root neurons (Merchan et al., '88). Occasionally, ChAT-immunopositive somata were observed in the main bodies of the DCN and VCN.

The occurrence of ChAT-immunoreactive puncta on ChAT-immunopositive somata was examined in 4 rats. Puncta, generally few in number, were found on about a third of approximately 50 somata in each group - acoustic striae, cochlear root, and granular regions. These observations suggest that the proposed cholinergic neurons of the cochlear nucleus may receive much less cholinergic innervation than the non-cholinergic neurons.

Vetter and Mugnaini (this volume), using a different antibody against ChAT, have reported results similar to ours as regards ChAT-immunoreactive puncta and as regards immunopositive somata in granular and acoustic striae regions. However, the cochlear root neurons are immunonegative in their preparations. This discrepancy, which must result from a technical difference between the studies, such as the different antibodies or fixation procedures, serves again to emphasize the need for independent evaluation of the results of the ChAT immunohistochemistry. Toward this end, we have begun direct quantitative assay of small tissue samples containing cochlear root neurons (Yao and Godfrey, '92). If these neurons actually are cholinergic, we would expect samples containing them to have consistently high ChAT activities, comparable to those of the facial nucleus. The results so far do not show such high ChAT activities and so are not consistent with the cochlear root neurons being cholinergic. Some samples have moderate ChAT activities, which could be consistent with some cholinergic terminals on cochlear root neurons (Vetter and Mugnaini, this volume). Had the results supported the possibility that the cochlear root neurons were cholinergic, this might have represented a rare case of a pool of ascending (Lopez et al., this volume) cholinergic neurons in a sensory system. Likely cholinergic neurons identified so far in mammalian sensory pathways are either interneurons or descending neurons (Godfrey et al., '80).

## Generalizations from the comparisons of quantitative and immunohistochemistry

In general, the immunohistochemical results for ChAT agree well with the quantitative histochemical results in suggesting that most cholinergic elements of the rat cochlear nucleus are centrifugal fibers and their terminals. The intrinsic ChAT activity of the nucleus identified by quantitative histochemistry appears from the immunohistochemistry to be associated with neurons having their somata located near its external boundaries. Some issues remain unresolved. ChAT-containing centrifugal fibers in the trapezoid body, suggested from the quantitative assays, were not well visualized with the immunohistochemistry. The light immunoreactivity of most somata of the cochlear nucleus probably represents an artifact, but might suggest a low concentration of ChAT within them. Based on our recent quantitative assays, the immunoreactivity of cochlear root neurons seen in our preparations also appears to represent a false positive, but, again, we can not rule out a low concentration of ChAT within them. Such issues will require further work for their resolution. A strength of combining the quantitative and immunohistochemical methods is to focus attention onto the conclusions which they jointly support.

ACKNOWLEDGEMENTS

I am grateful to Dr. Carol Bennett-Clarke for instruction in the immunohistochemical technique for choline acetyltransferase, to Dr. Bennett-Clarke and Dr. Hardress J. Waller for critical comments on the manuscript, to the Medical College of Ohio Data Services Department for instruction in and use of their computer imaging system, and to Michelle Heaney and Tim Godfrey for technical assistance. Supported by NIH (NIDCD) grant DC00172 and by intramural funds from Medical College of Ohio.

REFERENCES

Adams, J.C., 1989, Non-olivocochlear cholinergic periolivary cells, Soc. Neurosci. Abstr., 15:1114.
Brown, J.C. and Howlett, B., 1968, The facial outflow and the superior salivatory nucleus: an histochemical study in the rat, J. Comp. Neurol., 134:175-192.
Comis, S.D. and Davies, W.E., 1969, Acetylcholine as a transmitter in the cat auditory system, J. Neurochem., 16:423-429.
Comis, S.D. and Whitfield, I.C., 1968, Influence of centrifugal pathways on unit activity in the cochlear nucleus, J. Neurophysiol., 31:62-68.
Comis, S.D., 1970, Centrifugal inhibitory processes affecting neurons in the cat cochlear nucleus, J. Physiol., 210:751-760.
Godfrey, D.A. and Heaney, M.L., 1991, Immunoreactivity for choline acetyltransferase in the rat cochlear nucleus and superior olive, Abstr. Assoc. Res. Otolaryngol.: 10.
Godfrey, D.A. and Matschinsky, F.M., 1976, Approach to three-dimensional mapping of quantitative histochemical measurements applied to studies of the cochlear nucleus, J. Histochem. Cytochem., 24:697-712.
Godfrey, D.A. and Matschinsky, F.M., 1981, Quantitative distribution of choline acetyltransferase and acetylcholinesterase activities in the rat cochlear nucleus, J. Histochem. Cytochem., 29:720-730.
Godfrey, D.A., Williams, A.D. and Matschinsky, F.M., 1977, Quantitative histochemical mapping of enzymes of the cholinergic system in cat cochlear nucleus, J. Histochem. Cytochem., 25:397-416.

Godfrey, D.A., Ross, C.D., Herrmann, A.D. and Matschinsky, F.M., 1980, Distribution and derivation of cholinergic elements in the rat olfactory bulb, *Neuroscience*, 5:273-292.

Godfrey, D.A., Park, J.L., Rabe, J.R., Dunn, J.D. and Ross, C.D., 1983, Effects of large brain stem lesions on the cholinergic system in the rat cochlear nucleus, *Hearing Res.*, 11:133-156.

Godfrey, D.A., Park, J.L. and Ross, C.D., 1984, Choline acetyltransferase and acetylcholinesterase in centrifugal labyrinthine bundles of rats, *Hearing Res.*, 14:93-106.

Godfrey, D.A., Park-Hellendall, J.L., Dunn, J.D. and Ross, C.D., 1987a, Effect of olivocochlear bundle transection on choline acetyltransferase activity in the rat cochlear nucleus, *Hearing Res.*, 28:237-251.

Godfrey, D.A., Park-Hellendall, J.L., Dunn, J.D. and Ross, C.D., 1987b, Effects of trapezoid body and superior olive lesions on choline acetyltransferase activity in the rat cochlear nucleus, *Hearing Res.*, 28:253-270.

Godfrey, D.A., Carlson, L. and Ross, C.D., 1987c, Quantitative inter-strain comparison of the distribution of choline acetyltransferase activity in the rat cochlear nucleus, *Hearing Res.*, 31:203-210.

Godfrey, D.A., Carlson, L., Parli, J.A. and Ross, C.D., 1988, Distribution of choline acetyltransferase activity in the trapezoid body of the rat, *Soc. Neurosci. Abstr.*, 14:490.

Godfrey, D.A., Beranek, K.L., Carlson, L., Parli, J.A., Dunn, J.D. and Ross, C.D., 1990, Contribution of centrifugal innervation to choline acetyltransferase activity in the cat cochlear nucleus, *Hearing Res.*, 49:259-280.

Harrison, J.M., Warr, W.B. and Irving, R.E., 1962, Second order neurons in the acoustic nerve, *Science*, 138:893-895.

Harrison, J.M. and Irving, R., 1966, The organization of the posterior ventral cochlear nucleus in the rat, *J. Comp. Neurol.*, 126:391-402.

Hellendall, R.P., Godfrey, D.A., Ross, C.D., Armstrong, D.M. and Price, J.L., 1986, The distribution of choline acetyltransferase in the rat amygdaloid complex and adjacent cortical areas, as determined by quantitative micro-assay and immunohistochemistry, *J. Comp. Neurol.*, 249:486-498.

Klinke, R. and Galley, N., 1974, Efferent innervation of vestibular and auditory receptors, *Physiol. Rev.*, 54:316-357.

Levey, A.I., Wainer, B.H., Mufson, E.J. and Mesulam, M.-M., Colocalization of acetylcholinesterase and choline acetyltransferase in the rat cerebrum, *Neuroscience*, 9:9-22.

Lowry, O.H. and Passonneau, J.V., 1972, "A Flexible System of Enzymatic Analysis," Academic Press, New York.

McDonald, D.M. and Rasmussen, G.L., 1971, Ultrastructural characteristics of synaptic endings in the cochlear nucleus having acetylcholinesterase activity, *Brain Res.*, 28:1-18.

Merchan, M.A., Collia, F., Lopez, D.E. and Saldaña, E., 1988, Morphology of cochlear root neurons in the rat, *J. Neurocytol.*, 17:711-725.

Mugnaini, E., Osen, K.K., Dahl, A.-L., Friedrich, V.L. Jr. and Korte, G., 1980a, Fine structure of granule cells and related interneurons (termed Golgi cells) in the cochlear nuclear complex of cat, rat, and mouse, *J. Neurocytol.*, 9:537-570.

Mugnaini, E., Warr, W.B. and Osen, K.K., 1980b, Distribution and light microscopic features of granule cells in the cochlear nuclei of cat, rat, and mouse, *J. Comp. Neurol.*, 191:581-606.

Osen, K.K., Mugnaini, E., Dahl, A.-L. and Christiansen, A.H., 1984, Histochemical localization of acetylcholinesterase in the cochlear and superior olivary nuclei. A reappraisal with emphasis on the cochlear granule cell system, *Arch. Ital. Biol.*, 122:169-212.

Osen, K.K., Ottersen, O.P. and Storm-Mathisen, J., 1990, Colocalization of glycine-like and GABA-like immunoreactivities: A semiquantitative study of individual neurons in the dorsal cochlear nucleus of cat, *in*: "Glycine Neurotransmission," O.P. Ottersen and J. Storm-Mathisen, eds., John Wiley & Sons, London, pp. 417-451.

Osen, K.K., Lopez, D.E., Slyngstad, T.A., Ottersen, O.P. and Storm-Mathisen, J., 1991, GABA-like and glycine-like immunoreactivities of the cochlear root nucleus in rat, *J. Neurocytol.*, 20:17-25.

Osen, K.K. and Roth, K., 1969, Histochemical localization of cholinesterases in the cochlear nuclei of the cat, with notes on the origin of acetylcholinesterase-positive afferents and the superior olive, *Brain Res.*, 16:165-185.

Polak J.M. and Van Noorden, S., 1984, "An Introduction to Immunocytochemistry: Current Techniques and Problems," Oxford University Press, Oxford.

Rasmussen, G.L., 1967, Efferent connections of the cochlear nucleus, *in*: "Sensorineural Hearing Processes and Disorders," A.B. Graham, ed., Little, Brown, & Co., Boston, pp. 61-75.

Tago, H., McGeer, P.L., McGeer, E.G., Akiyama, H. and Hersh, L.B., 1989, Distribution of choline acetyltransferase immunopositive structures in the rat brainstem, *Brain Res.*, 495:271-297.

Taylor, P. and Brown, J.H., 1989, Acetylcholine, *in*: "Basic Neurochemistry," G. Siegel, B. Agranoff, R.W. Albers, and P. Molinoff, eds., Raven Press, New York, pp. 203-231.

Werman, R., 1966, Criteria for identification of a central nervous system transmitter, *Comp. Biochem. Physiol.*, 18:745-766.

White, J.S. and Warr, W.B., 1983, The dual origins of olivocochlear neurons in the albino rat, *J. Comp. Neurol.*, 219:203-214.

Yao, W. and Godfrey, D.A., 1992, Quantitative evidence that cochlear root neurons are not cholinergic, *Soc. Neurosci. Abstr.* 18: (in press).

# CHOLINE ACETYLTRANSFERASE IN THE RAT COCHLEAR NUCLEI: IMMUNOLOCALIZATION WITH A MONOCLONAL ANTIBODY

Douglas E. Vetter[1], Costantino Cozzari[2],
Boyd K. Hartman[3] and Enrico Mugnaini[1]

[1]Laboratory of Neuromorphology, Graduate Degree Program in
Biobehavioral Sciences, University of Connecticut, Storrs,
CT 06269-4154, USA
[2]Istituto di Biologia Cellulare, C.N.R., Viale Marx 43, 00137, Roma, Italy
[3]Department of Psychiatry, University of Minnesota, 672 Diehl Hall, 505 Essex Street S.E., Minneapolis, MN, 55455, USA

The cochlear nuclear complex (CoN) contains several populations of neurons that are highly diverse not only with respect to morphological features but also physiological responses to sound. With few exceptions, morphologically and functionally distinct cell types are differentially distributed in the anteroventral (AVCoN), posteroventral (PVCoN) and dorsal (DCoN) divisions of the CoN. With the advent of intracellular injection of tracers coupled with microelectrode recording, physiological response patterns can be more precisely correlated with individual cell classes within these subdivisions, increasing the need for extensive information on neurocytological features. In order to more fully describe the transformations of sound stimuli performed in the auditory centers, it is helpful to correlate the ultrastructural and chemical phenotypes of the neurons, especially those related to synapses, and the molecules involved in neurotransmission. Immunocytochemical methods have proven invaluable in this respect, because they afford the localization of special molecules not only at the regional level, as obtained with biochemical micromethods, but also at the cellular and subcellular levels.

A number of putative neurotransmitters and/or their synthetic enzymes have been explored in detail within the acoustic brain stem. The putative excitatory amino acids glutamate and aspartate, as well as the inhibitory amino acids GABA and glycine are localized in the CoN (reviewed by Ottersen and Storm-Mathisen, '84; Schwartz, '84; Caspary, '86;

---

List of abbreviations: ACh - acetylcholine; AChE - acetylcholinesterase; AVCoN - anteroventral cochlear nucleus; ChAT - choline acetyltransferase; CoN - cochlear nuleus; DAS - dorsal acoustic stria; DCoN - dorsal cochlear nucleus; GC - granule cell layer; IAS - intermediate acoustic stria; lam - lamina of granule cells; MS - medial sheet; MVPO - medioventral periolivary region; OCA - octopus cell area; PV-type - pleomorphic vesicle containing type bouton; RV-type - round vesicle containing type bouton; SPC - subpeduncular corner; VCoN - ventral cochlear nucleus.

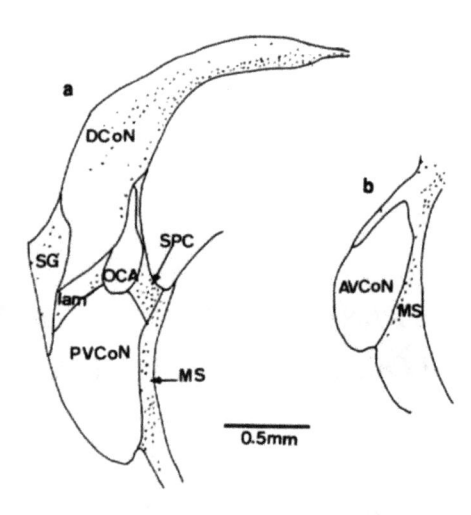

**Figure 1.** Camera lucida drawings of two sections of the rat cochlear nuclei showing the DCoN-PVCoN (a) and the AVCoN (b). All ChAT-positive cells (dots), present throughout a complete unilateral series, were approximately projected onto these two sections to emphasize their mode of distribution.

Adams and Mugnaini, '87; Moore, '88; Osen et al., '90; and others in this volume). There has been long standing biochemical evidence showing that acetylcholine (ACh) and its synthetic enzyme choline acetyltransferase (ChAT) are also present in the CoN, along with both muscarinic (Wamsley et al., '84; Whipple and Drescher, '84; Frostholm and Rotter, '86) and nicotinic (Morley et al., '77) ACh receptors. Deafferentation studies (Comis and Davies, '69; McDonald and Rasmussen, '71; Osen et al., '84; Godfrey et al., '87, '88, '90) indicate that the cholinergic innervation derives in part from descending fiber systems and in part from an intrinsic component yet to be identified with certainty, and that species differences exist in the relative contributions of these components. Acetyl-cholinesterase (AChE) histochemistry reveals few densely stained neurons in the CoN, with the exception of the granular neurons, which appear densely stained in the cat but not in other species (Osen and Roth, '69; Osen et al., '84). This enzyme, however, represents an uncertain correlate of cholinergic transmission. It is of interest, therefore, to localize all neural elements that have the capacity to synthetize ACh and map any regional variations in their distribution. In this study we have utilized a monoclonal antibody against ChAT and standard peroxidase anti-peroxidase immunocytochemistry, coupled with heavy metal intensification of the DAB reaction product for light microscopy (Adams, '81) and normal DAB for preembedding electron microscopy. The antibody specificity is demonstrated elsewhere (Cozzari et al., '90; Cozzari and Hartman, '92). We report here the distribution of ChAT-like immunoreactive neural structures in the rat CoN, and for brevity these are referred to as ChAT-positive. A quantitative light microscopic study with a polyclonal antiserum, in essential agreement with our data, is presented by Godfrey et al. (this volume). We uncovered a substantial population of ChAT-positive cell bodies situated within the CoN itself, and numerous ChAT-positive axons and terminals whose intrinsic or extrinsic nature has not yet been precisely determined. Therefore, cell bodies and axons are described separately. Immunoelectron microscopy, although time consuming and labor intensive, offers the great advantage of identifying not only the cholinergic neurons, but also cholinoceptive targets. This recently begun baseline study (Vetter et al., '91b) only partially exploits the latter potential.

Table I. Number of ChAT-positive cells counted in each region of the rat CoN

| region of CoN | left | right | mean |
|---|---|---|---|
| SPC | 28 (12%) | 28 (12%) | 28 (12%) |
| sup.g.c.l. + lamina MS | 19 (7%) | 28 (12%) | 24 (10%) |
| (AVoCN+PVCoN) | 44 (18%) | 57 (25%) | 51 (22%) |
| AVCoN+PVCoN magnocellular | 3 (1%) | 1 (0.4%) | 2 (0.8%) |
| total VCoN | 94 (39%) | 114 (50%) | 105 (45%) |
| DAS | 94 (39%) | 71 (31%) | 83 (35%) |
| DCoN | 52 (22%) | 43 (19%) | 48 (21%) |
| total DCoN | 146 (61%) | 114 (50%) | 131 (55%) |
| *GRAND TOTAL* | *240* | *228* | *234* |

The number of ChAT-positive cells situated in the CoN of the rat was estimated by counting all immunopositive cells with a visible nucleus. Cells in both the left and right cochlear nucleus were counted.

## ChAT-positive cell bodies

Immunopositive somata, in many cases including their proximal dendrites, were present in the CoN, albeit with different frequency of occurrence throughout the three main subdivisions (Fig. 1). Cells were counted on each side of the brain stem of one animal after collecting serial coronal sections through the entire CoN. Only cell profiles containing the nucleus were counted, and no correction was made for split nuclei. ChAT-positive cells numbered 240 and 228 on the left and the right side respectively (Table I). The DCoN contained more than one half of the immunopositive cells. Within the DCoN, the majority of ChAT-positive cells were located in the deep region, in the dorsal acoustic stria (DAS), in the strial corner adjacent to the DAS at the dorsomedial pole of the nucleus, and at the most ventromedial border of the nucleus adjacent to the intermediate acoustic stria (IAS). ChAT-positive cells were not encountered in the molecular layer. In the VCoN the majority of ChAT-positive cells were situated in the subpeduncular corner, the medial sheet, and the cap area. Occasional immunostained cells were present in the lamina of granule cells that separates the DCoN and the PVCoN, and in the superficial granular layer. Only rare immunostained cells were present in the magnocellular portions of the AVCoN and the PVCoN. Thus, most of the ChAT-positive cells are situated within territories of the CoN rich in granule cells and small cells (Osen, 69; Mugnaini et al., '80a and '80b; Lorente de Nó, '81). For evaluation of the cell shape at the light microscopic level, we used wet mounted sections and semithin sections of osmicated, resin embedded slices, to avoid distortion brought about by the drying process. Within the striae and the medial sheet, the cells tended to be spindle-shaped, with the main axis oriented parallel to the coursing fibers, while in all other locations they were rounded or ovoidal. The mean cell body diameter measured 13.5 μm

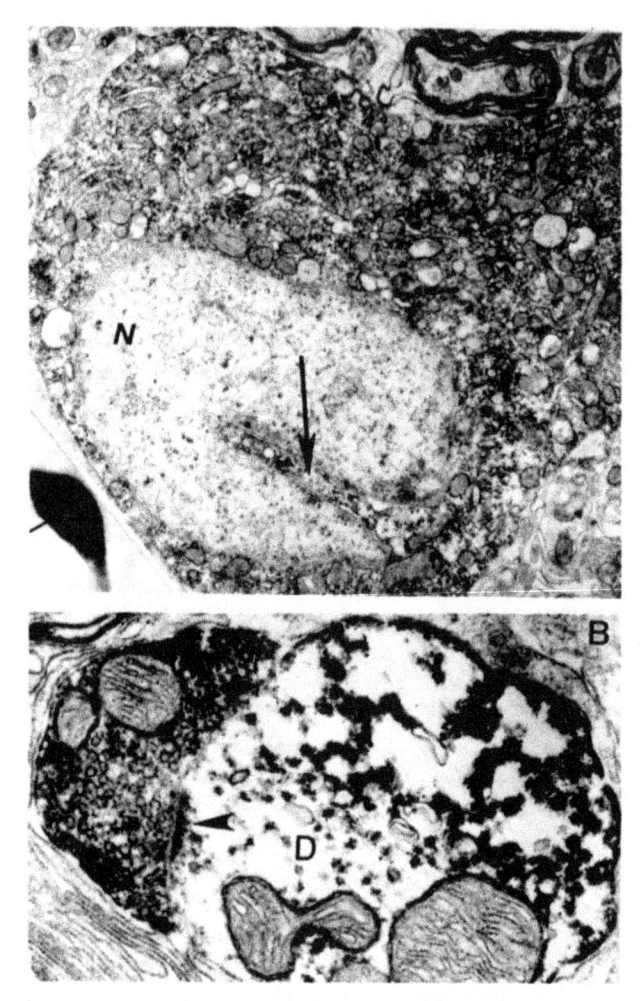

**Figure 2.** A: A ChAT-positive DCoN neuron contains numerous mitochondria. The nucleus (N) is deeply indented (arrow). Original magnification, X12,800. Inset: A ChAT-positive DCoN neuron, from a resin embedded section, has a deep nuclear indentation (arrow) visible even in the light microscopic. Magnification, X549. B: A ChAT-positive bouton containing round synaptic vesicles forms an asymmetric synapse (arrowhead) with a ChAT-positive dendrite (D) in the medial sheet of the AVCoN. Magnification, X41,811.

($\pm$ 2.5 $\mu$m). The cells therefore fall into the small cell categories described by Osen ('69) and Lorente de Nó ('81). In general, the nucleus of the immunoreactive cells was eccentrically placed in the rounded cells and centrally placed in the spindle-shaped cells. Immunoreactive boutons were seen in close proximity to some of the ChAT-positive somata and their principal dendrites. The main stem dendrites of the spindle-shaped cells were predominantly oriented along the fiber bundles, while those of the rounded cells extended in all directions, giving the neurons a stellate configuration.

At the electron microscopic level, the ChAT-positive cells situated in the various subdivisions of the CoN appeared to share many cytological features. The perimeter of the cell nucleus usually presented at least one deep infolding, which gave the nucleus a lobated appearance (Fig. 2A). These infoldings, which were penetrated by polyribosome rich cytoplasm, were visible also in light microscopic sections of osmicated material (Fig. 2A, inset). The perikarya contained numerous mitochondria and small, unremarkable Nissl bodies. There were few axosomatic boutons and the cell surface was mainly covered by astrocytic

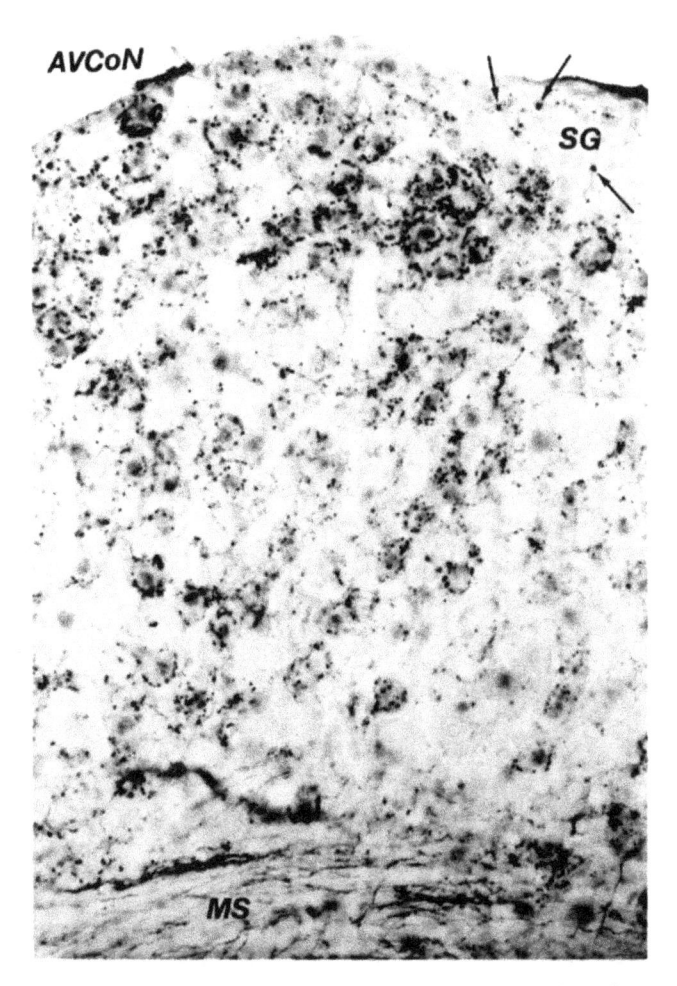

**Figure 3.** ChAT-positive boutons surround the cell bodies and main stem dendrites of presumed bushy cells in the AVCoN. Arrows point to large endings in the superficial granular layer (SG). MS, medial sheet with ChAT-positive fibers. Magnification, X244.

processes. Some of the axosomatic and axodendritic (Fig. 2B) boutons were ChAT-positive. The fine structural analysis suggests that the ChAT-positive cells are members of a homogeneous population of neurons. It must be pointed out, however, that their varying location may imply activation by different inputs. Because it has been reported that some ChAT activity remains in the surgically isolated rat CoN (Godfrey et al., '87), we assume that these neurons give rise to an intrinsic cholinergic innervation.

## ChAT-positive fibers

The immunopositive fibers were broadly classified as thin ($< 1$ $\mu$m in diameter) or thick ($> 1$ $\mu$m). The thin fibers were nearly ubiquitous (Figs. 3-5) and appeared particularly numerous in the medial sheet. A few thin fibers, presumably representing olivocochlear fibers with an aberrant course, were also present in the auditory nerve distal to the glial border. The majority of the thin fibers entered the CoN from the rostral portion of the trapezoid body, while the remainder entered from the striae. Their different routes suggest that they

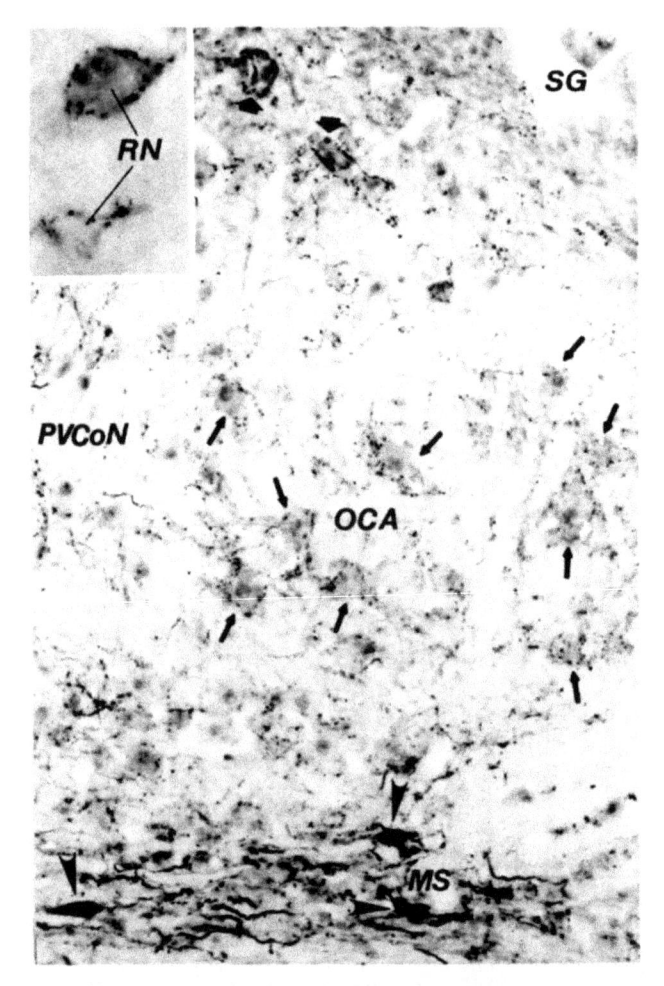

Figure 4. Small ChAT-positive boutons surround cell bodies and dendrites of octopus cells (small arrows) in the octopus cell area (OCA) of the PVCoN. Large arrows point to pericellular arrays of bigger boutons in the cap area. SG, superficial granular layer; MS, medial sheet with thick and thin fibers and three positive cells (arrowheads). Magnification, X208. Inset: ChAT-positive boutons surround cell bodies and dendrites of immunonegative root neurons (RN). Magnification, X416.

may derive from different groups of neurons in the superior olivary complex. These observations are in accord with biochemical evidence for the presence of a major trapezoid route for descending cholinergic fibers to the rat CoN (Godfrey et al., '87). The thick fibers, the number of which was less than that of thin fibers, also occurred in several locations. They were seen most frequently in the strial corner, the medial sheet, the cap area, and the subpeduncular corner; they occurred less frequently in the more ventrolateral portions of the DAS, as well as in the ventral tip of the VCoN, where they seemed to cross from medial to lateral toward the superficial granule cell layer. An occasional thick fiber was also encounterd in the molecular layer of DCoN. Some of the thick fibers appeared highly branched and gave rise to large en passant and terminal boutons (Fig. 5B). The thick fibers were often seen to branch off the olivocochlear bundle, but they were rare or absent in the trapezoid body. Presumably, the thick and thin fibers that enter the CoN from the striae originate from medial and lateral olivocochlear neurons (Osen et al, '84; Benson and Brown, '90; Ryan et al., '90; Brown et al, '91), while the thin fibers that arrive via the trapezoid body may

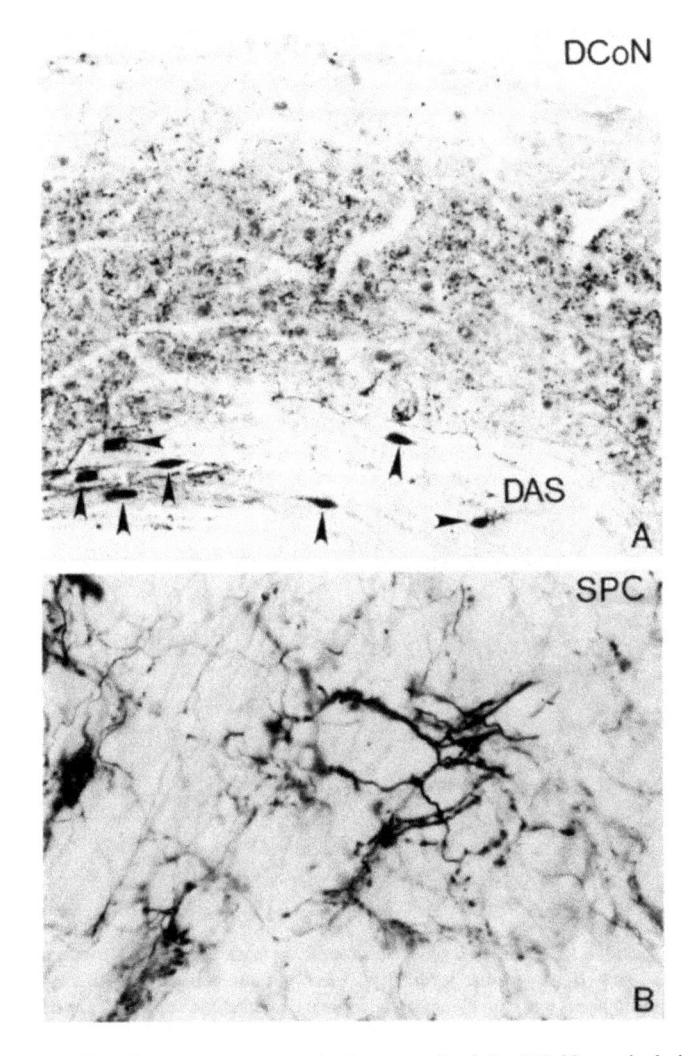

**Figure 5.** A: ChAT-positive boutons are present in the neuropil of the DCoN, particularly in the portion approximately corresponding to layers 2-3. Several ChAT-positive neurons (arrowheads) are located in the dorsal acoustic stria (DAS). Magnification, X155. B: Thick and thin ChAT-positive fibers, along with nests of boutons, are situated in the subpeduncular corner (SPC). Magnification, X416.

originate from small ChAT-positive cells located in the medial part (MVPO) of the ventral periolivary region (Faye-Lund, '86; Vetter et al., '91a). By immunoelectron microscopy, thick fibers usually appeared myelinated, while thin fibers were either myelinated or unmyelinated.

## ChAT-positive terminals

All subdivisions of the CoN contained ChAT-positive terminals (Figs. 3-5). However, the frequency, mode of distribution and size of these boutons showed consistent topographic variations. The magnocellular portions of the AVCoN and PVCoN and the cap area were densely innervated, while the superficial granular cell layer, the lamina and the DCoN were moderately to densely innervated. In the magnocellular portions of the AVCoN and PVCoN

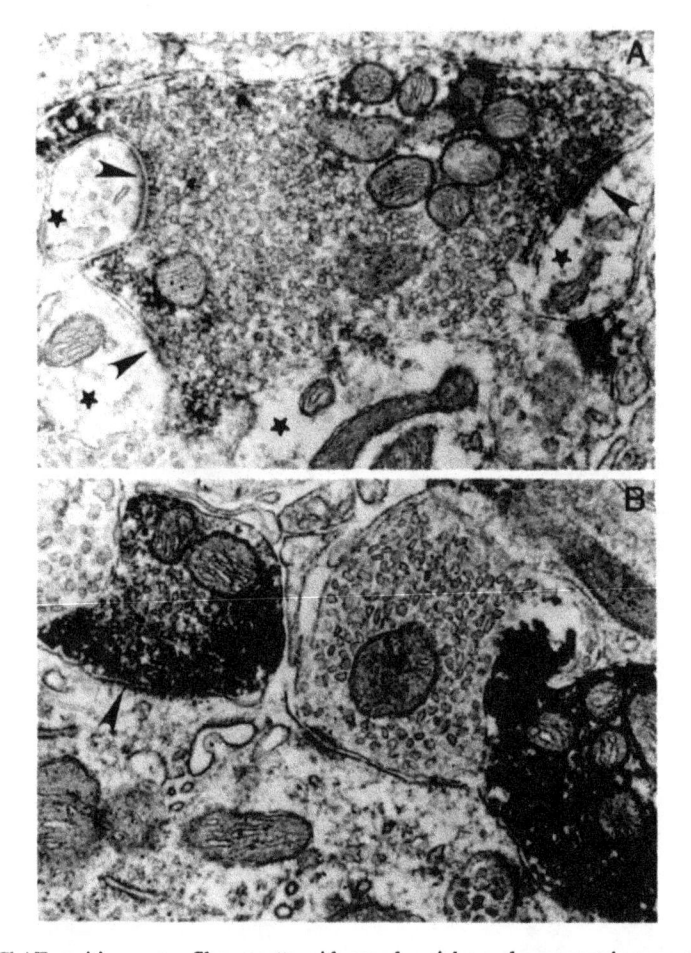

Figure 6. A: ChAT-positive mossy fiber rosette with round vesicles and asymmetric synaptic specializations (arrowheads) from the superficial granular layer. Stars mark granule cell dendrites. Magnification, X32,472.
B. Two ChAT-positive boutons, one with pleomorphic vesicles, contact the soma of a spherical cell in the rostral AVCoN. One of the boutons forms a symmetric synapse (arrowhead). Magnification, X35,970.

and in the cap area, immunostained boutons showed a distinct pericellular distribution, with strings of boutons also aligned with emerging dendrites (Figs. 3 and 4). Most of the large cell bodies in these regions were surrounded by a ring of immunostained terminals, but some neurons, presumably corresponding to giant and multipolar cells, were nearly devoid of axosomatic boutons. A neuropil distribution of immunostained boutons predominated in the granule cell domain and in the DCoN (Figs. 3-5). The octopus cells were innervated by a homogeneous population of axosomatic and axodendritic ChAT-positive boutons (Fig. 4). These were the smallest immunoreactive boutons within the entire CoN. Boutons in the other areas varied considerably in size. The largest boutons, some of which measured up to 4 $\mu$m in diameter, were observed in the granule cell domain (Fig. 3, arrows). Some varisized ChAT-positive terminals were present within the striae and the fiber rich medial sheet, situated in close proximity to the dendrites of immunostained cells. ChAT-positive terminals were also apposed in pericellular arrays to the ChAT-negative root neurons (Ross and Burkel, '71; Merchán et al. '88; Osen et al. '91) and extended over their dendrites, which run parallel and perpendicular to the cochlear nerve fibers (Fig. 4, inset). In the DCoN, the majority of immunostained puncta occurred below the molecular layer, in the upper portion of the deep region that includes layers 2 and 3 (Fig. 5A).

By immunoelectron microscopy, we identified some ChAT-positive mossy rosettes in the centers of glomerular synaptic fields within granule cell domains (Fig. 6A). The rosettes contained round synaptic vesicles and formed asymmetric synaptic junctions (RV-type bouton). In the magnocellular regions of the VCoN and in the DCoN, immunoreactive boutons varied not only in shape and size, but also in their internal structure. Some of the ChAT-positive boutons displayed round vesicles and asymmetric synapses (RV-type bouton). These features, generally associated with excitatory functions, place these terminals in the same general class as the mossy rosettes and are in accord with previous microiontophoretic studies in the cat and the chinchilla (Whitfield, '68; Whitfield and Comis, '68; Caspary et al., '83), advocating an excitatory role for ACh in the VCoN. Other ChAT-positive boutons, however, including many of the axosomatic boutons on spherical cells of the VCoN, showed pleomorphic vesicles and symmetric synapses (PV-type bouton) (Fig. 6B), which are features usually associated with inhibitory functions. This would imply that an inhibitory role for part of the cholinergic innervation is not restricted to the DCoN (Caspary et al., '83). Studies are in progress to ascertain whether there is a differential distribution of ChAT-positive, RV-type and PV-type boutons in the various subdivisions of the CoN in relation to different neuron classes. It should be noted, however, that the shape of synaptic vesicles in aldehyde-fixed neural tissues is easily altered by procedural parameters and the present results require confirmation and further careful analysis. Recent morphological and biochemical studies on the cerebral cortex (Beaulieu and Somogyi, '91), spinal cord (Todd, '91), and retina (O'Malley and Masland, '89) have suggested the coexistence of GABA and acetylcholine in the same neuron. This possibility should also be explored for the ChAT-positive neurons of the CoN, their axon terminals, as well as the terminals of descending ChAT-positive fibers.

In general, the ChAT-positive terminals seem to conform to the different principles of organization of the various subdivisions of the CoN. In the VCoN, they form pericellular arrays, filling the gaps left by the other axosomatic terminals, primarily those of the auditory nerve fibers. The immunostained terminals appear relatively large around presumed spherical and globular bushy cells, known to receive large primary endings, but appear small around the octopus cells, which receive smaller primary endings. In the laminated DCoN they show a layered distribution. In the granule cell domain the immunostained terminals are large and constitute the central elements of some of the glomeruli, which are the main synaptic arrays in this region.

## Comparison with acetylcholinesterase (AChE) staining

Previous studies by Osen et al. ('84) have shown that histochemical staining for AChE reveals four main classes of positive elements in the rat CoN: a few neuronal cell bodies (large multipolar neurons in the VCoN and smaller neurons in the granule cell domain and its proximity), a fiber plexus formed by thick and thin axons, a diffuse neuropil staining, and dense patches situated in the DCoN and the granule cell domain. Except for the large multipolar neurons and the patches, these components may coincide, at least in part, with the ChAT-positive structures described above. The large AChE-positive multipolar neurons are definitely ChAT-negative, although they seem to be cholinoceptive. The smaller AChE-positive neurons and the fiber plexus may be part of the densely ChAT-immunoreactive somata and fibers. The diffuse neuropil staining could be related to the ChAT-positive boutons and their postsynaptic targets. However the neuropil staining appears in ascending order of density in regions II, IV, III, V, and I of the VCoN (terminology of Harrison and Irving, '65, '66), the granule cell domain, the cap area, and the molecular layer of the DCoN, whereas the ChAT-positive boutons are most conspicuous in regions III, I, and V and the cap area of the VCoN. The presence of RV-type and PV-type AChE-positive boutons in the CoN (McDonald and Rasmussen '71) is in accord with the immunocytochemical observations reported here. The patches, which remain after deafferentation of the CoN, presumably represent AChE that is situated on intrinsic elements, whose identity and function remain to be clarified. It should be stressed that the numbers of ChAT immunoreactive neurons in the CoN and thin fibers in the striae are higher than those of their presumed AChE-positive counterparts, and that AChE staining did not clearly reveal the numerous ChAT-positive thin fibers in the trapezoid body. Therefore, although AChE in the rat CoN may be largely involved in synaptic transmission, ChAT-like immunoreactivity remains the most reliable index of cholinergic function.

Our immunocytochemical results obtained with a monoclonal antibody directed against ChAT, the synthetic enzyme for the production of ACh, indicate that there are three main classes of putative cholinergic structures within the rat CoN. 1) Approximately 200 small and intensely immunoreactive neurons are situated primarily in the deep DCoN, the striae, the subpeduncular corner, the medial sheet and, to a lesser extent, the cap area. These cells represent a novel neuron class with a characteristic chemical phenotype. The ChAT-positive neurons could be either projection neurons with local axon collaterals or local circuit neurons intrinsic to the DCoN, the VCoN, or both. It remains to be ascertained whether they give rise to RV-type or PV-type boutons. 2) ChAT-positive axons, both myelinated and unmyelinated, course in the striae and the trapezoid body and occur throughout the CoN. Because the immunostained axons in the fiber pathways are much more numerous than the immunostained intrinsic neurons, we assume that they mainly represent afferent fibers. The extrinsic fibers may originate from three different groups of cells in the superior olive. 3) Finally, varisized ChAT-positive terminals are located throughout the CoN. The size and shape of the immunopositive terminals seem to be dictated largely by the postsynaptic elements and/or by competition with other fibers. The ChAT-positive boutons also seem to be heterogeneous with respect to their synaptic fine structure, some being provided with round vesicles and asymmetric synapses and others with pleomorphic vesicles and symmetric synapses (RV-type and PV-type, respectively). These aspects of morphology, presumably determined by gene expression in the respective parent neurons, fall into functionally opposite, excitatory and inhibitory, classes. The presence of different types of ACh receptor (Wamsley et al., '84; Morley, '77) and the demonstration that microiontophoresis of ACh has either excitatory or inhibitory effects upon CoN neurons (Caspary et al., '83) also lend support to a notion of multiple postsynaptic effects brought about by the release of ACh. The immunocytochemical demonstration of a rich complement of ChAT-like immunoreactive terminals suggests that Ach plays a much larger role in the CoN than previously recognized. The vast majority, if not all, of the neurons in the CoN appear to be cholinoceptive, including the intrinsic, ChAT-positive neurons themselves. One may suggest that the cholinergic innervation, because of its widespread distribution, represents a counterpart of the noradrenergic innervation (Kromer and Moore, '76, '80; Levitt and Moore'79, '80; Moore '88). This descending input is involved in modulating neuronal responses to sounds. The cholinergic fiber system, however, has at least one additional role, that of activating the granule cells through mossy fibers (Dunn et al., '92), and requires a more sophisticated explanation, for which there is still insufficient knowledge. Renewed emphasis on microiontophoretic studies of CoN neurons, together with further investigations utilizing neuroanatomical and molecular probes could therefore provide valuable results.

ACKNOWLEDGMENTS

This work was supported by US-PHS grant DC 01805-01 and funds from the C.N.R. (Progr. Bil., Italy-USA).

# REFERENCES

Adams, J.C., 1981, Heavy metal intensification of DAB-based HRP reaction product., *J. Histochem. Cytochem.*, 29:775.

Adams, J.C., and Mugnaini, E., 1987 Patterns of glutamate decarboxylase immunostaining in the feline cochlear nuclear complex studied with silver enhancement and electron microscopy., *J. Comp. Neurol.*, 262:375-401.

Beaulieu, C. and Somogyi, P., 1991, Enrichment of cholinergic terminals on GABAergic neurons and coexistence of immunoreactive GABA and choline acetyltransferase in the same synaptic terminals in the striate cortex of the cat., *J. Comp. Neurol.*, 304:666-680.

Benson, T.E. and Brown, M.C., 1990, Synapses formed by olivocochlear axon branches in the mouse cochlear nucleus., *J. Comp. Neurol.*, 295:52-70.

Brown, M.C., Pierce, S., and Berglund, A.M., 1991, Cochlear-nucleus branches of thick (medial) olivocochlear fibers in the mouse: A cochleotopic projection., *J. Comp. Neurol.*, 303:300-315.

Caspary, D., 1986, Cochlear nuclei: Functional neuropharmacology of the principle cell types., *in*: "Neurobiology of Hearing: The Cochlea", R.A. Altschuler, D.W. Hoffman, and R.P. Bobbin, eds., Raven Press, New York.

Caspary, D.M., Havey, D.C. and Faingold, C.L., 1983, Effects of acetylcholine on cochlear nucleus neurons., *Exp. Neurol.*, 82:491-498.

Comis, S.D. and Davies, W.E., 1969, Acetylcholine as a transmitter in the cat auditory system., *J. Neurochem.*, 16:423-429.

Cozzari, C., Howard, J., and Hartman, B., 1990, Analysis of epitopes on choline acetyltransferase (ChAT) using monoclonal antibodies (Mabs)., *Soc Neurosci. Abst.*, 16:200.

Cozzari, C. and Hartman, B.K., 1992, High yield purification of choline acetyltransferase and production of high affinity monoclonal antibodies., *Eur. J. Biochem.*, in press.

Dunn, M.E., Vetter, D.E., Berrebi, A.S., Krider, H.M., and Mugnaini, E., 1992, The mossy fiber-granule cell-cartwheel cell system in the mammalian cochlear nuclear complex, *in*: "Advances in Speech Hearing and Language Processing, Volume III, Cochlear Nucleus: Structure and Function in relation to Modelling", W.A. Ainsworth, E.F. Evans, and C.M. Hackney, eds., JAI Press Ltd, London.

Faye-Lund, H., 1986, Projection from the inferior colliculus to the superior olivary complex in the albino rat., *Anat. and Embryol.*, 175:35-52.

Frostholm, A. and Rotter, A., 1986, Autoradiographic localization of receptors in the cochlear nucleus of the mouse, *Brain Res. Bull.*, 16:189-203.

Godfrey, D.A., Park-Hellendall, J.L., Dunn, J.D., and Ross, C.D., 1987, Effect of olivocochlear bundle transection on choline acetyltransferase activity in the rat cochlear nucleus, *Hearing Res.*, 28:237-251.

Godfrey, D.A., Parli, J.A., Dunn, J.D., and Ross, C.D., 1988, Neurotransmitter microchemistry of the cochlear nucleus and superior olivary complex, *in*: "Auditory Pathways", J. Syka and R.B. Masterton, eds., Plenum Publishing, New York.

Godfrey, D.A., Beranek, K.L., Carlson, L., Parli, J.A., Dunn, J.D., and Ross, C.D., 1990, Contribution of centrifugal innervation to choline acetyl-transferase activity in the cat cochlear nucleus, *Hearing Res.*, 49:259-280.

Harrison, J.M., and Irving, R., 1965, The anterior ventral cochlear nucleus, *J. Comp. Neurol.*, 124:15-42.

Harrison, J.M., and Irving, R., 1966, The organization of the posterior ventral cochlear nucleus in the rat, *J. Comp. Neurol.*, 126:391-402.

Kromer, L.F., and Moore, R.Y., 1976, Cochlear nucleus innervation by central norepinephrine neurons in the rat, *Brain Res.*, 118:531-537.

Kromer, L.F., and Moore, R.Y., 1980, Norepinephrine innervation of the cochlear nuclei by locus coeruleus neurons in the rat, *Anat. Embryol.*, 158:227-244.

Levitt, P. and Moore, R. V., 1979, Origin and organization of brain stem catecholamine innervation in the rat, *J. Comp. Neurol.*, 186:505-528.

Levitt, P. and Moore, R. V., 1980, Organization of the brain stem noradrenalin hyperinnervation following neonatal 6-hydroxy-dopamine treatment in rat, *Anat. Embryol.*, 158:133-150.

Lorente de Nó, R., 1981, The Primary Acoustic Nuclei, Raven Press, New York, N.Y.

McDonald, D.M. and Rasmussen, G.L., 1971, Ultrastructural characteristics of synaptic endings in the cochlear nucleus having acetylcholinesterase activity, *Brain Res.*, 28:1-18.

Merchán, M.A., Collia, F., López, D.E., and Saldaña, E., 1988, Morphology of cochlear root neurons in the rat, *J. Neurocytol.*, 17:711-725.

Moore, J.K., 1988, Cholinergic, GABA-ergic, and noradrenergic input to cochlear granule cells in the guinea pig and monkey, *in*: "Auditory Pathways", J. Syka and R.B. Masterton, eds., Plenum Publishing, New York.

Morley, B.J., Lorden, J.F., Brown, G.B., Kemp, G.E., and Bradley, R.J., 1977, Regional distribution of nicotinic acetylcholine receptor in rat brain, *Brain Res.*, 134:161-166.

Mugnaini, E., Osen, K.K., Dahl, A-L., Friedrich, V.L.jr., and Korte, G., 1980a, Fine structure of granule cells and related interneurons (termed Golgi cells) in the cochlear nuclear complex of cat, rat and mouse, *J. Neurocytol.*, 9:537-570.

Mugnaini, E., Warr, B.W., and Osen, K.K. 1980b, Distribution and light microscopic features of granule cells in the cochlear nuclei of cat, rat, and mouse, *J. Comp. Neurol.*, 191:581-606.

O'Malley, D., and Masland, R.H., 1989, Co-release of acetylcholine and gamma-aminobutyric acid by a retinal neuron, *Proc. Natl. Acad. Sci. USA*, 86:3414-3418.

Osen, K.K., 1969, Cytoarchitecture of the cochlear nuclei in the cat, *J. Comp. Neurol.*, 136:453-484.

Osen, K.K., López, D.E., Slyngstad, T.A., Ottersen, O.P., and Storm-Mathisen, J., 1991, GABA-like and glycine-like immunoreactivities of the cochlear root nucleus in rat, *J. Neurocytol.*, 20:17-25.

Osen, K.K., Mugnaini, E., Dahl, A-L., and Christiansen, A.H., 1984, Histochemical localization of acetylcholinesterase in the cochlear nuclear and superior olivary nuclei. A reappraisal with emphasis on the cochlear granule cell system, *Arch. Ital. Biol.*, 122:169-212.

Osen, K.K., Ottersen, O.P., and Storm-Mathisen, J., 1990, Colocalization of glycine-like and GABA-like immunoreactivities. A semiquantitative study of individual neurons in the dorsal cochlear nucleus of cat, *in*: "Glycine neurotransmission", O.P. Ottersen and J.Storm-Mathisen, eds., J. Wiley and Sons, New York, N.Y.

Osen, K.K., and Roth, K., 1969, Histochemical localization of cholinesterases in the cochlear nuclei of the cat, with notes on the origin of acetyl-cholinesterase-positive afferents and the superior olive, *Brain Res.*, 16:165-185.

Ottersen, O P., and Storm-Mathisen, J., 1984, Neurons containing and accumulating transmitter aminoacids, *in*: "Handbook of Chemical Neuroanatomy", A. Björklund, T. Hökfelt, and M.J. Kuhar, eds., vol 3, part II, Elsevier, Amsterdam.

Ross, M.D., and Burkel, W., 1971, Electron microscopic observations of the nucleus, glial dome, and meninges of the rat acoustic nerve, *Am. J. Anat.*, 130:73-92

Ryan, A.F., Keithley, E.M., Wang, Z-X., Schwartz, I.R., 1990, Collaterals from lateral and medial olivocochlear efferent neurons innervate different regions of the cochlear nucleus and adjacent brainstem, *J. Comp. Neurol.*, 300:572-582.

Schwartz, I.R., 1984, Autoradiographic studies of amino acid labeling of neural elements in the auditory system, *in*: "Auditory Biochemistry", D.G. Drescher, ed., Charles Thomas, Springfield.

Todd, A.J., 1991, Immunohistochemical evidence that acetylcholine and glycine exist in different populations of GABAergic neurons in lamina III of the rat spinal dorsal horn, *Neuroscience*, 44:741-746.

Vetter, D.E., Adams, J.C., and Mugnaini, E., 1991a, Chemically distinct rat olivocochlear neurons, *Synapse*, 7:21-43.

Vetter, D.E., Cozzari, C., Hartman, B.K., and Mungnaini, E., 1991b, Comparison between AChE histochemical staining and ChAT immunocytochemistry in the rat cochlear nuclei, *Soc. Neurosci. Abstr.*, 17:302.

Wamsley, J.K., Zarbin, M.A., and Kuhar, M.J., 1984, Distribution of muscarinic cholinergic high and low affinity agonist binding sites: A light microscopic autoradiographic study, *Brain Res. Bull.*, 12:233-243.

Whipple, M.R. and Drescher, D.G., 1984, Muscarinic receptors in the cochlear nucleus and auditory nerve of the guinea pig, *J. Neurochem.*, 43:192-198.

Whitfield, I.C. 1968, Centrifugal control mechanisms of the auditory pathway, *in*: "Hearing Mechanisms in Vertebrates", A.V.S. De Reuck and J. Knight, eds., Churchill, London.

Whitfield, I.C., and Comis, S.O., 1968, A reciprocal gating mechanism in the auditory pathway, *in*: "Cybernetic Problems in Bionics", H.L. Oestricher and D.R. Moore, eds., Gordon and Breach, New York.

# VI. PROJECTIONS AND RESPONSE PROPERTIES OF COCHLEAR NUCLEUS NEURONS

# THE COCHLEAR ROOT NEURONS IN THE RAT, MOUSE AND GERBIL

Dolores E. López, Miguel A. Merchán, Victoria M. Bajo and Enrique Saldaña

Departamento de Biología Celular y Patología, Facultad de Medicina
Universidad de Salamanca, Salamanca, Spain

## Generalities

The auditory portion of the eighth cranial nerve (the cochlear nerve) contains a neuronal population which has been well documented only in a few species of small rodents belonging to the Muridae family.[1]

The neurons of the cochlear nerve were first described in detail by Lorente de Nó ('26) in the mouse (Fig. 1). Almost forty years went by before Harrison and Warr ('62) rediscovered large cells scattered along the cochlear nerve. These large neurons, called "b-cells" by Harrison and Irving ('66), form what Harrison and Warr ('62) designated the "acoustic nerve nucleus" of the rat. Later studies using rats (Ross and Burkel, '71; Merchán et al., '88; Osen et al., '91; El Barbary, '91; Merchán et al., '92), mongolian gerbils, (Chamberlain, '77; McGinn and Faddis, '87; Brown et al., '88; Ostapoff and Morest, '89; Statler et al., '90) and mice (Webster and Trune, '82; Willard and Ryugo, '83; Brown et al., '88) have provided additional information about the morphological features, topography and connections of these neurons. A detailed account of their morphology was published by Merchán et al. ('88) using cell and myelin staining, Golgi impregnation, retrograde transport of peroxidase (HRP) and conventional electron microscopy. According to these authors, the neurons of the cochlear nerve are situated in both the intranuclear and extranuclear portions of the cochlear nerve root. In their study, these neurons are referred to as "cochlear root neurons" (RN). This term will be used throughout this chapter.

In the rat, mouse and gerbil, the RN are large (they are probably the largest neurons of the entire cochlear nuclear complex -CNC-), with the longest diameter of about $35\mu$m. The RN have an oval somata with distinct, darkly stained Nissl bodies, and an eccentric, typically infolded nucleus. Topographically, RN are usually arranged in rows parallel to the fibers of the cochlear nerve. Most RN are situated in the anterior and lateral half of the nerve root,

---

[1]Tarlov ('37) described ectopic ganglion cells in the central or glial segment of several human cranial nerves, including the acoustic nerve. According to him, these cells are "of the central nervous system type". However, his description of the ectopic ganglion cells of the acoustic nerve is very general and does not allow for detailed comparisons between the cells reported by him and those described in rodents.

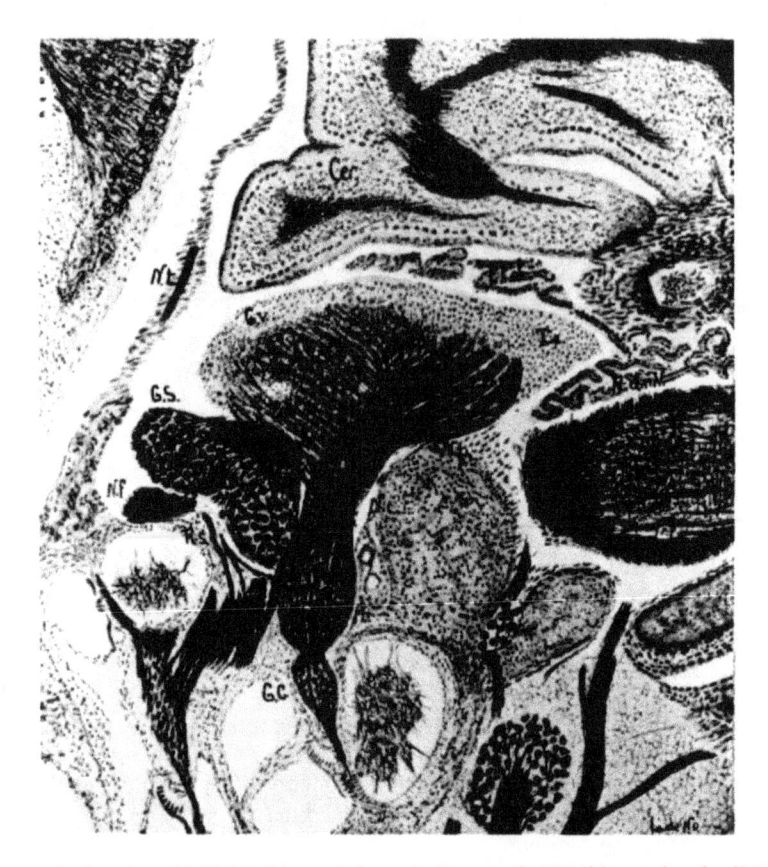

**Figure 1.** Parasagittal section of a 15 day old mouse drawn by Lorente de Nó with amazing detail. Note a group of cochlear root neurons in the distal portion of the cochlear nerve. This figure is a reproduction of figure 14 of the article "Etudes sur l'Anatomie et la Phisiologie du labyrinthe de l'oreille et du VIIIe nerf. Deuxiéme partie: Quelques données au sujet de l'anatomie des organes sensoriels du labyrinthe", published in "Travaux du Laboratorie de Recherches biologiques de L'Université de Madrid", 24:53-153, 1926, when Lorente de Nó was working at the Cajal Institute in Madrid.

This article contains a clear description of these neurons: "..il existe dans le tronc du cochléaire, peu avant son entrée dans le bulbe, des cellules nerveuses qui correspondent, samble-t-il, au ganglion ventral (portion postérieure). Or, ce que l'on ignore, c'est que cela ait lieu sur tout le trajet du cochléaire, et principalement près du "porus acusticus internus". Ce sont là dés cellules semblables à celles de la portion caudale du ganglion ventral (noyau de la racine descendante du cochléaire), monopolaires ou bipolaires, à corps piriforme ou fusiforme, et dont les prolongements sont orientés dans des plans longitudinaux (parallèles au plan médian de la tête). Dans une coupe transversale, elles apparaissen à contours arrondis, alors que dans les coupes longitudinales (Fig. 14) où elles sont coupées en long, elles apparaisent porvues de prolongements. Leur lieu de répartition est, comme il a été dit plus haut, tout le tronc du cochléaire, mais de préférence aux extremités de celui-ci, près du bulbe et près du "porus acusticus internus". Il existe en ce dernier point un amas de 30 ou 40 cellules; pour leur faire de la place, les petits faisceaux de nerf se trouvent légèrement séparés. On aperçoit entre ces cellules des fibres ramifiées dont l'origine nous est encore inconnue jusqu'à présent; nous ne croyons pas cependant qu'il soit aventuré de supposer qu'il s'agisse là de brances de bifurcation du nerf".

where the fibers from the basal coil of the cochlea are located. As seen by the Golgi method, they possess few dendritic trunks, which are thick and oriented mainly parallel or perpendicular to the cochlear nerve fibers (Fig. 2). The RN receive sparse synaptic inputs on the cell body (Merchán et al., '88). Their axon is thick and projects outside the CNC via the trapezoid body (tz) (Harrison and Warr, '62; Merchán et al., '88).

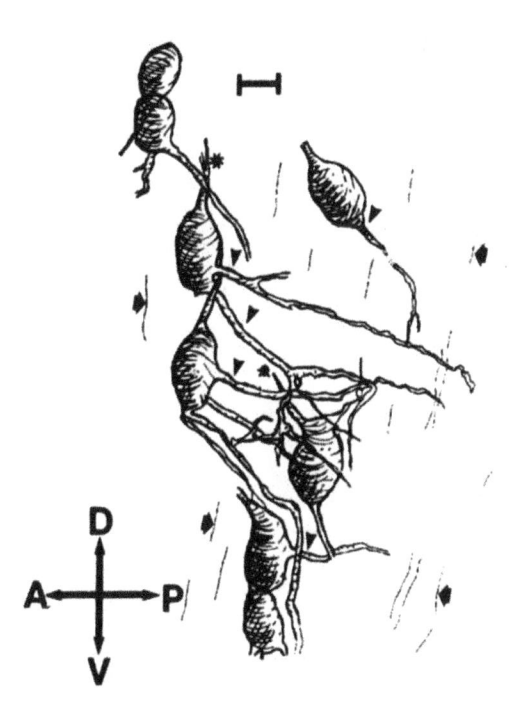

Figure 2. Camera lucida drawings of rat RN impregnated by the Golgi method. These neurons are located in the extranuclear portion of the cochlear root. Sagittal section. Note the thick dendritic trunks perpendicular (arrowheads) or parallel (asterisks) to the fibers of the cochlear nerve (solid arrows). The dendrites perpendicular to the fibers are oriented mostly in the anteroposterior direction. Calibration bar, 25 μm.

The cochlear nerve root does not contain a homogeneous population of cells. Most cells are RN, and share the above mentioned features. However, a few notably smaller cells can be found among the RN (Ross and Burkel, '71; Osen et al., '91).

This chapter briefly summarizes our current knowledge of the RN. New data about their ontogeny and the trajectories of their axons are also included. Most of the data described in this chapter are from albino rats (Rattus norvergicus). Nonetheless, where available, data from mice (Mus musculus) and gerbils (Meriones unguiculatus) have also been included.

## Ontogeny

We have studied the development of RN in the rat. Sections from animals at different stages of development, ranging from prenatal day 15 (E15) to postnatal day 15 (P15), were processed by the Nissl or reduced silver methods.

The RN are first identified on P1; at this time, they are located near the glial border, close to the neurons of the Scarpa ganglion (Fig. 3A). At this stage, all neurons in the CNC are relatively small, densely packed and undifferentiated as compared to those in the adult animal. Although the RN are more differentiated than other neuronal types of the CNC, they show signs of cellular indifferentiation, such as a large nucleus and two nucleoli.

As the cochlear nerve grows, neurons easily recognizable as RN undergo a process of reorganization in the cochlear nerve root, and they reach their typical arrangement in rows on P7 (Fig. 3B). During development, the RN somata seem to be consistently located in close relationship with the fibers innervating the basal coil of the cochlea (López et al., '89).

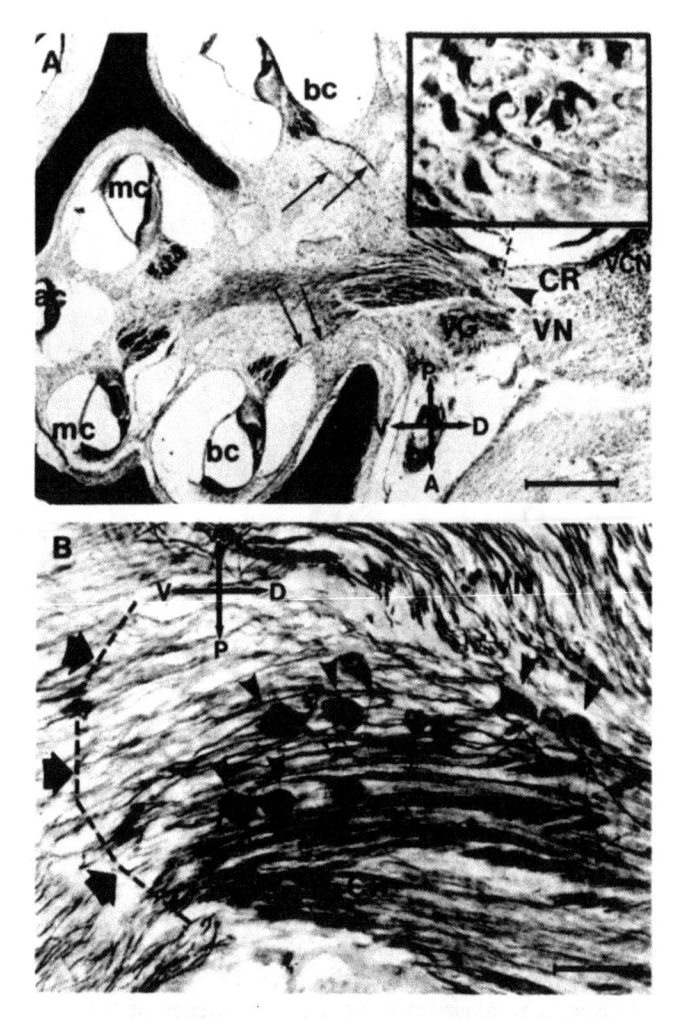

**Figure 3.** A. Parasagittal section of a 1 day old (P1) rat stained by the Nissl method (Cresyl violet). Note RN (arrowhead) in the distal portion of the cochlear root (CR), near the glial border (asterisk). The thin arrows point to the fibers of the basal coil of the cochlea (bc). Calibration bar, 500 μm. The inset shows a detail of the RN. Other abbreviations: ac, apical coil; mc, medial coil; VG, vestibular or Scarpa ganglion; VN, vestibular nerve; VCN, ventral cochlear nucleus. B. Parasagittal section of a P7 rat processed with the reduced silver method. The RN (arrowheads) are arranged in rows that follow the trajectory of the fibers of the cochlear nerve. These neurons are located in the anterior portion of the cochlear root (CR). The solid arrows indicate the glial border. Calibration bar, 60 μm.

We were not able to observe neurons of either the spiral ganglion, or the Scarpa ganglion appended to the central portion of the cochlear nerve at any time during development. While these results suggest that the RN could have a central origin, some authors have proposed other possibilities (Ross and Bourkel, '71). In our opinion, the RN stem from neuroblasts of the CNC and, like other neuron types of the CNC, migrate from the ventral portion of the embryonic CNC. Thus, our results support the notion put forth by Harrison and Warr ('62), who postulated a non-ganglionary origin for these neurones.

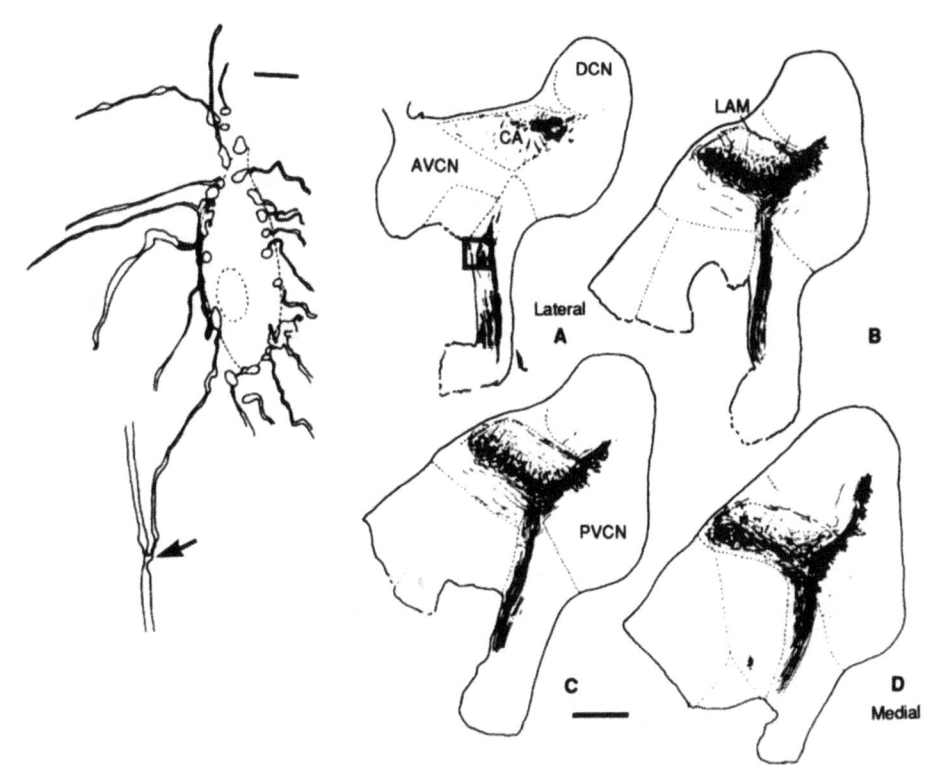

Figure 4. Camera lucida drawings of parasagittal sections illustrating the distribution of cochlear fibers labeled after HRP injections in the basal regions of the spiral ganglion. A is lateral, D is medial. Calibration bar 500 μm. The inset shows a RN from section A intermingled among labeled fibers. Collaterals of labeled cochlear fibers end in small knobs in close apposition to the RN somata. The arrow indicates the point of emergence of one of these collaterals. Calibration bar, 5 μm. Abbreviations: AVCN, anteroventral cochlear nucleus; CA, peripheral cap of small cells; DCN, dorsal cochlear nucleus; LAM, lamina; PVCN, posteroventral cochlear nucleus.

## Inputs to RN

The RN receive inputs from cochlear (primary) as well as non cochlear axons. Only about 40% of the cell body surface is covered with synaptic terminals (Merchán et al., '88).

The primary inputs are formed by small synaptic boutons arising from short axon collaterals that emerge at right angles from auditory nerve fibers prior to their main bifurcation, (Harrison and Warr, '62; Osen et al., '91). These boutons, whose ultrastructural features resemble those of the terminals formed by cochlear fibers in other regions of the CNC, contain large round vesicles and form asymmetric axosomatic and axodendritic synaptic contacts (type 1 terminals of Merchán et al., '88). These boutons are the most numerous on the RN cell body (75%). In general, each cochlear fiber gives off only one collateral, but each RN is contacted by many type 1 boutons (Merchán et al., '88).

The fact that the RN are located mostly in the anterior and lateral half of the cochlear nerve, where the fibers from the basal coil of the cochlea are found (Sando '65; Arnesen and Osen '78), suggests that they could receive selective inputs from specific cochlear regions. To test this hypothesis, we injected HRP into different regions of the rat spiral ganglion to label cochlear fibers (see also Osen et al., '91). As expected, after small injections in basal regions, labeled fibers are seen close to the RN on their way to the cochlear nuclei. (Fig. 4).

Many of the labeled fibers give off a single collateral that usually ends in a single knob, in close apposition with the somata or proximal dendrites of RN (Fig. 4, inset). Unlike injections in basal regions, injections in the middle or apical turns of the spiral ganglion label fibers that run at some distance from the RN somata. In this case, only a few fibers give off collaterals in the cochlear nerve. Although these collaterals are not usually seen in the vicinity of RN somata, the possibility that they might establish a sparse contact on the RN dendrites, some of which extend over long distances within the cochlear nerve, cannot be ruled out.

These results indicate that, at least in the rat, the RN are innervated by primary afferents mostly, if not exclusively, originating in basal regions of the cochlea. These primary afferents are known to convey information of high frequency sounds. This suggests that the RN could be involved in the processing of sounds in the upper part of the frequency spectrum of the rat. This could also be the case in the mouse and the gerbil, in which collaterals from labeled type I axons ending on or near RN have been seen after HRP injections into basal regions of the spiral ganglion (Brown et al.,'88).

In addition to inputs from cochlear nerve fibers, the RN receive synaptic contacts from other boutons whose ultrastructural features suggest a non-cochlear origin. These non-cochlear boutons, which are few in number, seem to be morphologically heterogeneous, and contain either round or pleomorphic vesicles (Merchán et al., 88). Light microscopic immunocytochemistry in the rat has shown that, unlike the primary boutons, some of the non-primary boutons contain GABA. Others, less numerous, contain glycine or colocalize GABA and glycine. The GABA and/or glycine immunopositive boutons mainly synapse on dendritic profiles. Therefore, the cell body appears to be for the most part devoid of presumptive inhibitory contacts (Osen et al., '91).

Our preliminary data obtained with postembedding immunoelectron microscopy show that the gerbil could be somewhat different from the rat in that the cell body of the RN may be contacted by slightly higher numbers of GABA-containing boutons with pleomorphic vesicles.

The origin of non-primary inputs to the RN is unknown. Potential sources include higher auditory centers known to project to the cochlear nuclei (i.e. the superior olivary complex, the nuclei of the lateral lemniscus and the inferior colliculus -IC-) as well as neurons within the CNC (Osen et al., '91). After injections of an anterograde neuroanatomical tracer, *Phaseolus vulgaris* leucoagglutinin (PHA-L), into the IC (Saldaña, unpublished observations) or the nuclei of the lateral lemniscus (Bajo '91) many labeled fibers were seen in different regions of the CNC, but not in the extranuclear portion of the cochlear nerve root on either side, which suggests that these auditory centers do not innervate the RN. Similarly, preliminary experiments with PHA-L and WGA-HRP injections in the dorsal or ventral cochlear nucleus have failed to reveal anterogradely labeled fibers in the cochlear nerve root, thus suggesting that non primary inputs to the RN originate to a large extent from outside the CNC. Therefore, it seems reasonable to assume that at least some of the non-primary inputs must come from the superior olivary complex, which, on the other hand, is known to contain many GABAergic and glycinergic cells that project to the CNC (Adams and Wenthold '87). A further possibility that needs to be explored is that non-primary inputs to RN originate from other, possibly non-auditory, structures yet to be identified.

## Axon target

The central projections of the RN have not been defined with certainty. It is known that they project outside the CNC via the tz (Harrison and Warr, '62; Merchán et al., '88) but their final target remains to be identified. We have used two different approaches to address this question. In the first, HRP was injected at different levels of the rat auditory pathway in an attempt to label the RN retrogradely. The second approach consisted of stain-

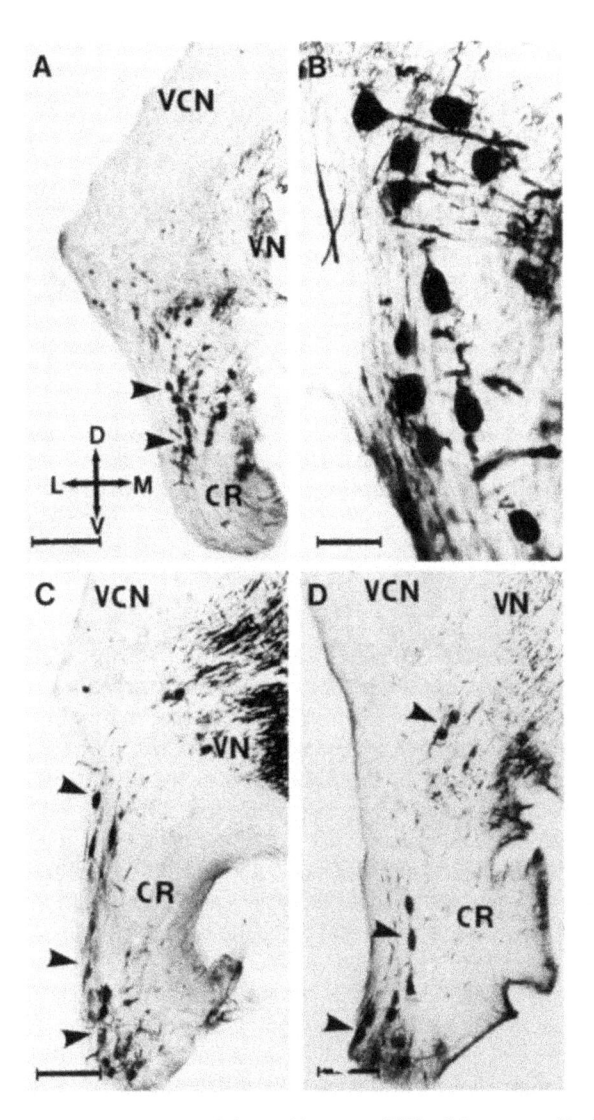

**Figure 5.** Micrographs of coronal sections of the cochlear root (CR) of the mouse (A, B), gerbil (C) and rat (D) immunostained for CaBP. Note that in all three species, the RN (arrowheads) are strongly immunopositive, as are the fibers of the vestibular nerve (VN). B. The CaBP-antibody used by us stains the RN in a Golgi-like fashion, thus allowing the visualization of cell bodies and dendrites. Calibration bar in A, 250 μm, also applies to C and D. Calibration bar in B, 65 μm.

ing with a monoclonal antibody against the calcium-binding protein, calbindin D-28K (CaBP), which strongly stains the RN and their axons (Merchán et al., '92). Preliminary studies from our laboratory have shown that the RN are retrogradely labeled after HRP injections in the contralateral nuclei of the lateral lemniscus, but not after injections in the IC. These results indicate that the targets of the RN could include the region of the contralateral lateral lemniscus and suggest that their axons do not extend beyond the level of the lateral lemniscus itself (Merchán et al., '92). In these experiments the number of labeled RN in the contralateral CR was higher the more rostral and ventral the injection was placed.

With antibodies against CaBP the RN, including their cell body, dendrites and axon, are strongly immunostained in the three species studied (Fig 5). In all three species, the

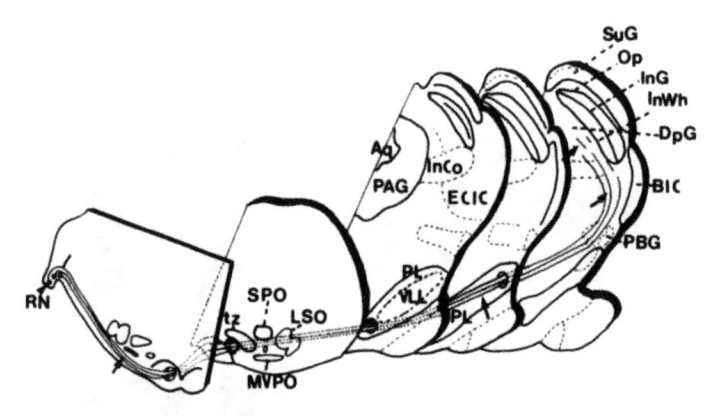

**Figure 6.** Schematic diagram showing the trajectory or RN axons (arrows) in the rat. See text for details. Abbreviations: Aq, aqueduct; BIC, brachium of the IC; DpG, deep gray layer of SC; ECIC, external cortex of IC; InCo, intercollicular nucleus; InG, intermediate gray layer of SC; InWh, intermediate white layer of SC; LSO, lateral superior olive; MVPO, medioventral periolivary nucleus; Op, optic nerve layer of SC; PAG, periaqueductal gray; PBG, parabigeminal nucleus; PL, paralemniscal regions; SPO, superior paraolivary nucleus; SuG, superficial gray layer of SC; VLL, ventral nucleus of the lateral lemniscus; tz, trapezoid body.

immunostained axons of the RN are extremely thick (up to 7μm), and this makes it possible to follow their trajectories in serial sections.[2] These thick axons leave the CNC via the tz. In the mouse and the gerbil, several neuron types in the ventral cochlear nucleus, whose analysis goes beyond the scope of this chapter, stain for CaBP. Although many of these cells project through the tz, the axons of the RN can be distinguished because of their extreme thickness. Unlike the mouse and the gerbil, in the rat only one other neuron type of the ventral cochlear nucleus, the octopus cell, is immunostained with the antibody against CaBP. Since the lightly immunostained axon of the octopus cells is about 1.5-2.5 μm thick and courses through the intermediate acoustic stria (see also Friauf and Ostwald, '88), one can feel confident that the very thick axons that leave the ventral cochlear nucleus through the tz belong to the RN.

In what follows, the trajectories of the RN axons will be described only for the rat although the data also apply to the mouse and the gerbil. These trajectories are shown schematically in figure 6. Once in the tz, the thick axons run medially and cross the midline, reaching the contralateral nucleus of the tz, where they are found immediately ventral to the nucleus. Although the neurons of the nucleus of the tz are CaBP-positive, their axons are considerably thinner than the axons of the RN, and the latter can be readily distinguished. Upon reaching the level of the contralateral nucleus of the tz, the thick axons curve rostrally, and run rostrolaterally towards the lateral lemniscus (LL). In the LL, most thick axons ascend ventrodorsally in the periphery of the tract. In general, the axons are not seen to course among the cells of the nuclei of the LL, but rather in the paralemniscal regions (PL) located rostrally to them. At the level of the parabigeminal nucleus, the axons become progressively more rostral; then, they turn slightly medially and continue their ascending trajectory through the intercollicular tegmentum until they enter the deep layers of the superior colliculus (SC). In this region, they become thinner and intermingle with axons from other sources, so that they can no longer be confidently identified. Very few, if any, thick axons seem to enter the

---

[2]The remarkable thickness of the RN axons has already been pointed out with different techniques, such as the reduced silver method (2.75-5.5 μm) (Harrison and Irving, '66), and standard electron microscopy (4-5 μm) (Ross and Burkel, '71).

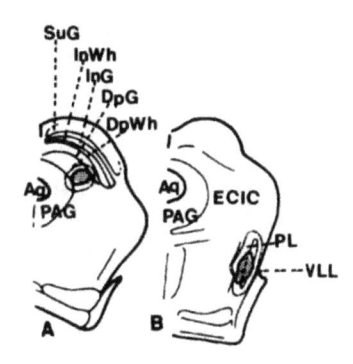

Figure 7. Schematic diagram showing injection sites of HRP that resulted in retrograde labeling of contralateral RN in the rat. The injection site in A affected the deep layers of the superior colliculus and the intercollicular region. The injection site in B included the VLL and the PL region. Abbreviations: Aq, aqueduct; DpG, deep gray layer of SC; DpWh, deep white layer of SC; ECIC, external cortex of IC; InG, intermediate gray layer of SC; InWh, intermediate white layer of SC; PAG, periaqueductal gray; PL, paralemniscal region; SuG, superficial gray layer of SC; VLL, ventral nucleus of the lateral lemniscus.

IC; most RN axons circumvent the IC ventrally and rostrally on their way to the SC. It is possible that the labeling of contralateral RN reported by some authors after injections of retrograde tracers into the IC (Cannon and Giesler, '78; Willard and Ryugo, '83; Merchán et al., '88) could have been due to the spreading of tracer into regions surrounding it.

To ascertain whether the RN innervate the SC, new experiments consisting of injections of HRP in tectal regions of the rat were undertaken. RN were retrogradely labeled in the contralateral side following injections that involved the deep layers of the SC (Fig. 7), thus confirming the conclusion suggested by the CaBP staining to the effect that the RN axon reaches the SC. Although in some of our cases the tracer diffused into pretectal areas, in no case was the IC affected by the spread of tracer.

The fact that in a previous study the RN were retrogradely labeled after HRP injections into the VLL (Merchán et al., '92) could have two possible explanations. First, it is possible that the thick axons of the RN send collaterals that innervate the lemniscal nuclei (and/or PL regions) which were not revealed by the CaBP immunocytochemistry. It is also possible that in such experiments the HRP diffused outside the VLL into the rostral paralemniscal regions and was taken up by *en passant fibers* (see also Kandler and Herbert, '91). A combination of both possibilities could have also occurred. New experiments are needed to accurately delimit the target(s) of the RN axons.

## Role

Very little is known about the functional role of the RN. However, some of their morphological and connectional features can be used as a basis to discuss of their possible functions.

Since the RN receive inputs from collaterals of the auditory nerve fibers that emerge prior to their bifurcation, it is safe to assume that they are stimulated before any other neuronal types of the CNC. Therefore, this neuronal type is the first one in the central nervous system to receive auditory information.

While each cochlear fiber innervating the RN gives off only one collateral, each RN receives axosomatic contacts from many collaterals. Since the boutons formed by these collaterals are usually small, the postsynaptic effect of individual auditory nerve inputs is unlikely to cause the cell to fire. Rather, input from many fibers seems to be required. This summatory effect of inputs would be better achieved by loud high frequency sounds, since the RN receive innervation from a group of fibers within a restricted range of frequencies.

In the rat (Merchán et al. '92), mouse and gerbil, these neurons exhibit very strong immunoreactivity for parvalbumin (unpublished observations) and calbindin D-28K. This indicates that the RN have a complex calcium regulation system. The ability to rapidly reduce the concentration of free intracellular calcium is of vital importance for neurons bearing excitatory amino acid receptors and calbindin helps to protect neurons against excitotoxicity (Christakos et al., '89). Some authors have proposed that calcium-binding proteins are functionally related to the discharge properties of the cells: by binding calcium ions they may affect the timing of the action potential and neurotransmitter release (Celio '89). Accordingly, this complex calcium regulation system might confer the RN with the ability to respond rapidly to auditory nerve inputs. Moreover, the RN must have a fast conduction velocity, as inferred from the thickness of their axon. These data combine to suggest that the RN provide a rapid mechanism for informing higher centers of the functional state of the nerve.

Our results indicate that the RN axons reach the SC. In mammals, the SC is the neural structure responsible for the elaboration of orientation movements of the eyes, head, pinnae and whiskers towards a novel sensory stimulus. This tectal structure has been specifically related to the spatial localization of external stimuli (Aitkin, '86). Neurons in the SC have been shown to respond to auditory, visual or somatosensory stimuli, and a polysensorial integration is established at this level. Responses to auditory stimuli are located in the intermediate and deep layers of the SC (Middlebrooks, '88). It has been reported that the deep layers of the SC are involved in potentiating the acoustic startle reflex (Tischler and Davis, '83). Sensorimotor integration carried out by the SC is also necessary to guide pinna movements (Henkel, '81); the SC innervates a paralemniscal zone in the lateral midbrain tegmentum that is a premotor region that supplies the facial motor nucleus (Henkel and Edwards, '78). The direct projection from the RN to the SC could be a fast way of informing this nucleus for coordination of the directional orientation of the head when a sudden, loud noise, perhaps with energy at high frequencies, is heard. Moreover, if the RN were to send axon collaterals to the VLL and/or PL, this would be consistent with such a role, as these structures seem to be part of the anatomical circuit of the startle reflex (Davis et al., '82).

The previous data suggest that the RN could be associated with auditory stimuli that would generate startle reflexes in rodents. Given the short latency that characterizes this reflex (Davis et al., '82), the fast conduction velocity of the RN axon is not without interest in this context.

All the species in which RN have been described so far belong to the Muridae family of rodents. This raises the question of whether these neurons represent an etological specialization in relation with biologically significant sounds. These biological sounds may include call sounds between animals of the same species or sounds produced by other species. Ecological and phylogenetic studies are needed before conclusions can be drawn.

ACKNOWLEDGMENTS

We would like to express our gratitude to Dr. Marco Celio, University of Fribourg, for providing the antibodies against calbindin D-28K and parvalbumin. This research was supported by the Spanish DGICYT (PB 88-0372 and PB 90-0523-CO2-01).

REFERENCES

Adams, J.C., and Wenthold, R.J., 1987, Immunostaining of GABA-ergic and glycinergic inputs to the anteroventral cochlear nucleus, *Soc. Neurosci. Abstr.*, 13:1259.

Aitkin, L., 1986, "The auditory midbrain: Structure and function in the central auditory pathway," Humana Press., Clifton, New Jersey, p. 1-246.

Arnesen, A.R., and Osen, K.K., 1978, The cochlear nerve in the cat: Topography, cochleotopy, and fiber spectrum, *J. Comp. Neurol.*, 178: 661-678.

Bajo, V.M., 1991, "Morfología y proyecciones de los núcleos del lemnisco lateral de la rata," Doctoral Dissertation, University of Salamanca, Salamanca.

Brown, M.C., Berglund, A.M., Kiang, N.Y.S., and Ryugo, D.K.,1988, Central trajectories of type II spiral ganglion neurons, *J. Comp. Neurol.*, 278:581-590.

Cannon, J.T., and Giesler, G.J., 1978, Projections of the cochlear nucleus and superior olivary complex to the inferior colliculus of the rat, *Soc. Neurosci. Abstr.*, 4:4.

Celio, M.C., 1989, Calcium binding proteins in the brain, *Arch. Ital. Anat. Embriol.*, 94:227-236.

Chamberlain, S.C., 1977, Neuroanatomical aspects of the gerbil inner ear: Light microscope observations, *J. Comp. Neurol.*, 171:193-204.

Christakos, S., Gabrielides, C., and Rhoten, W.B., 1989, Vitamin D-dependent calcium binding proteins. Chemistry, distribution, functional considerations, and molecular biology, *Endocrine Rev.*, 10:3-26.

Davis, M., Gendelman, D.S., Tischler, M.D. and Gendelman, P.M., 1982, A primary acoustic startle circuit: Lesion and stimulation studies, *J. Neurosci.*, 2:791-805.

EL Barbary, A., 1991, Auditory nerve of the normal and jaundiced rat. I. Spontaneous discharge rate and cochlear nerve histology, *Hearing Res.*, 54:75-90.

Harrison, J.M., and Irving, R., 1965, The anterior ventral cochlear nucleus, *J. Comp. Neurol.*, 124:15-42.

Harrison, J.M., and Warr W.B., 1962, A study of the cochlear nuclei and ascending auditory pathways of the medulla, *J. Comp. Neurol.*, 119:341-379.

Henkel, C.K., 1981, Afferent sources of a lateral midbrain tegmental zone associated with the pinnae in the cat as mapped by retrograde transport of horseradish peroxidase, *J. Comp. Neurol.*, 203:213-226.

Henkel, C.K., and Edwards, S.B., 1978, The superior colliculus control of pinna movements in the cat: Possible anatomical connections, *J. Comp. Neurol.*, 182:763-776.

Kandler, K., and Herbert, H., 1991, Auditory projections from the cochlear nucleus to pontine and mesencephalic reticular nuclei in the rat, *Brain Res.*, 562:230-242.

López, D.E., Angulo, M.A., Saldaña, E., and Pérez de la Cruz, M..A., 1989, Distribution pattern of root neurons in the extracellular portion of the rat cochlear root. A silver reduced and HRP transganglionic study, *Eur. J. Neurosci.*, suppl. 2:262.

Lorente de Nó, R., 1926, Etudes sur l'anatomie et la physiologie du labyrinthe de l'oreille et du VIII nerf. Deuxième partie: Quelques données au sujet de l'anatomie des organes sensoriels du labyrinthe., *Travaux du Lab. de Recherches biologiques de l'Université de Madrid*, 24:53-153.

McGinn, M.D., and Faddis, B.T., 1987, Auditory experience affects degeneration of the ventral cochlear nucleus in mongolian gerbils, *Hearing Res.*, 31:235-244.

Merchán, M.A., López, D.E., Alonso, J.R. and Bajo, V.M., 1992, Cochlear root neurones of the rat: New findings, *in*: "Advance in speech, hearing and language processing, Vol. III, Cochlear nucleus: Structure and function in relation to modeling," W.A. Ainsworth, E.F. Evans, and C.M. Hackney, eds., JAI Press Ltd., London, (in press).

Merchán, M.A., Collía, F., López, D.E., and Saldaña, E.,1988, Morphology of cochlear root neurons in the rat, *J. Neurocytol.*, 17:711-725.

Middlebrooks, J.C., 1988, Auditory mechanisms underlying a neural code for space in the cat's superior colliculus, *in*: "Auditory function. Neurobiological bases of hearing," G.M. Edelman, W.E. Gall and W.M. Cowan, eds., John Wiley & Sons, New York, p. 431-455.

Osen, K.K., López, D.E., Slyngstad, T.A.,.Ottersen, O.P., and Storm-Mathisen, J., 1991, GABA-like and glycine-like immunoreactivities of the cochlear root nucleus in rat, *J. Neurocytol.*, 20:17-25.

Ostapoff, E.M., and Morest, D.K., 1989, A degenerative disorder of the central auditory system of the gerbil, *Hearing Res.*, 37:141-162.

Ross, M.D., and Burkel, W., 1971, Electron microscopic observations of the nucleus, glial dome, and meninges of the rat acoustic nerve, *Am. J. Anat.*, 130:73-92.

Sando, I., 1965, The anatomical interrelationships of the cochlear nerve fibers, *Acta Otolaryngol.*, 59:417-436.

Statler, K.D., Chamberlain, S.C., Slepecky, N.B., and Smith, R.L., 1990, Development of mature microcystic lesions in the cochlear nuclei of the Mongolian gerbil, Meriones unguiculatus, *Hearing Res.*, 50:275-288.

Tischler, M.D., and Davis, M., 1983, A visual pathway mediates fear-conditioned enhancement of acoustic startle, *Brain Res.*, 276:55-71.

Webster, D.B., and Trune, D.R., 1982, Cochlear nucleus complex of mice, *Am. J. Anat.*, 163:103-130

Willard, F.H., and Ryugo, D.K., 1983, Anatomy of the central auditory system, *in*: "The auditory psychobiology of the mouse," J.F. Willot, ed., Charles C. Thomas, Springfield, Illinois, p. 201-304.

# PROJECTIONS OF COCHLEAR NUCLEUS TO SUPERIOR OLIVARY COMPLEX IN AN ECHOLOCATING BAT: RELATION TO FUNCTION

John H. Casseday[1], John M. Zook[2] and Nobuyuki Kuwabara[2]

[1]Department of Neurobiology, Duke University Medical Center, Durham, NC 27710, USA
[2]Department of Zoological and Biomedical Sciences & OUCOM, Ohio University, Athens, OH 45701, USA

Studies of comparative neuroanatomy provide information about structural variations that in turn can be clues to function. It is in this spirit that we use the mustached bat *Pteronotus parnellii* as our example in this paper on the inputs from the cochlear nucleus to the superior olivary complex. We will show that the projections from the cochlear nucleus to the lateral superior olive (LSO) are virtually identical to those of other mammals, but the projections to the medial superior olive (MSO) are different in that the ipsilateral input is extremely sparse. We shall propose that this difference is in the proportion of the ipsilateral to contralateral inputs, not that there is a unique input to the MSO of the mustached bat. We will focus on the functional consequences of the unusual pattern of inputs to MSO. We will argue that one of the consequences of the bat's adaptations for echolocation is the evolution of the MSO into a structure in which a difference in projections from one cochlear nucleus produces great functional differences.

The pattern of inputs from the cochlear nucleus to the superior olive was one of the first clues that the superior olives have a fundamental role in encoding the cues for localizing sound in space (Jeffress; '48; Stotler, '53). Early behavioral studies indicated that the crossed input from the cochlear nucleus to the superior olives was necessary for animals to lateralize sound (Masterton et al., '67) or to localize sound in space (Moore et al., '74; Casseday and Neff, '75). The generally accepted view is that differences in the binaural level of sound are first encoded in the LSO (Galambos et al., '59; Goldberg and Brown, '68; Caird and Klinke, '83), whereas differences in binaural timing are first encoded in MSO (Goldberg and Brown, '69). The anatomical and physiological bases for binaural hearing have been reviewed elsewhere (Casseday and Covey, '87; Kuwada and Yin, '87) and need only brief review here. The LSO receives direct excitatory input from the anteroventral cochlear nucleus (AVCN) of the ipsilateral side (Tsuchitani and Boudreau, '66; Guinan et al., '72; Tsuchitani, '77; Caird and Klinke, '83; Zook and DiCaprio, '88; Covey et al., '91), but the input from the contralateral AVCN is relayed from MNTB via neurons that provide inhibitory input to LSO (Moore and Caspary, '83; Spangler et al., '85; Zook and DiCaprio,'88; Adams and Mugnaini '90; Guinan and Li, '90; Sanes, '90; Smith et al., '91; Kuwabara and Zook, '91). The MSO receives direct excitatory input from AVCN of both sides (Hall, '65; Clark, '69; Goldberg

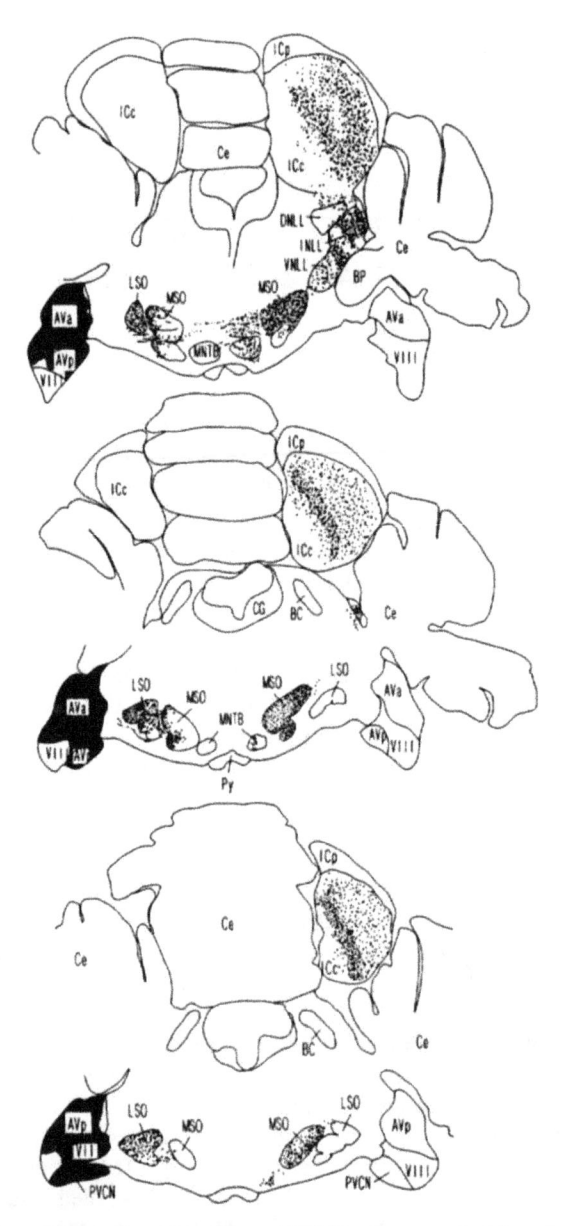

Figure 1. Drawing to show anterograde transport from AVCN in the mustached bat, *Pteronotus parnellii*. An injection of WGA-HRP fills the anterior and posterior parts of AVCN (AVa and AVp) and part of PVCN. The dots representing terminal-like labeling indicate that LSO receives input only from the ipsilateral cochlear nucleus. The input to the MSO is bilateral; however the contralateral MSO is much more densely labeled than the ipsilateral MSO. The MNTB is labeled only contralateral to the injection. Terminal-like labeling is also shown in the intermediate and ventral nuclei of the lateral lemniscus (INLL and VNLL) and in the central nucleus of the inferior colliculus (ICc). Little or no labeling is found in the dorsal nucleus of the lateral lemniscus (DNLL). The bottom section is posterior, and the sections are 40 $\mu$m thick and spaced about 320 $\mu$m apart. Other abbreviations: BP, brachium pontis; BC, brachium conjunctivum; CG, central grey; Ce, cerebellum; ICp, pericentral area of the inferior colliculus; Py; pyramids; VIII, 8th nerve.

and Brown, '69; Guinan et al., '72; Perkins, '73; Lindsey, '74; Inbody and Feng, '81; Kiss and Majorossy, '83; Yin and Chan, '90; Brunso-Bechtold et al., '90). Animals that have very small heads and localize high frequency sounds may be considered as a special case, because they have little or no usable range of binaural time differences. One might argue that this is the primitive condition, because the first mammals were certainly very small creatures. This limitation of high frequency listeners was consistent with the idea that bats have little or no MSO (Harrison and Irving, '66; Irving and Harrison, '67). However another point of view is that bats could have special requirements for localizing sound since they must often do so in pursuit of flying prey in three-dimensional space. That is, the demands for localizing sound include great accuracy at vertical localization as well as horizontal localization, and for catching prey they must measure distance with great accuracy.

From the point of view that MSO operates as a binaural time comparator, it seems puzzling that the mustached bat has a large differently organized structure in the position where MSO is usually seen in other mammals. On the basis of several kinds of comparisons, we eventually concluded that this structure is MSO. First the cytoarchitecture was more like MSO than any periolivary structure (Zook and Casseday, '82a). Second, it projected only to the ipsilateral inferior colliculus (Zook and Casseday, '82b) as in other mammals (e.g., Adams, '79; Brunso-Bechtold et al., '90), and these ascending projections followed the same tonotopic pattern as seen in other mammals, e.g., ventral MSO projects to low frequency IC, and dorsal MSO projects to high frequency IC (Zook and Casseday, '82b; '85; Ross et al, '88; Ross and Pollak, '89). Finally, the pattern of input from the cochlear nucleus is in many respects like the input from cochlear nucleus in non-echolocating mammals. For example the input is mainly from spherical bushy cells of the AVCN (Warr, '66; Harrison and Feldman, '70; Zook and Casseday, '85; Cant and Casseday, '86; Casseday et al., '88), as we shall demonstrate here. However, the ascending projections from cochlear nucleus reveal some differences between *Pteronotus* and other mammals. The purpose of this report is to describe these differences and their physiological consequences.

First it is useful to point out that, in echolocating bats, the basic structure of the cochlear nucleus is the same as is seen in other mammals. The AVCN can be divided into two parts, an anterior part that consists mainly of spherical cells and a posterior part that has a mixture of cell types including a large proportion of multipolar cells (Zook and Casseday, '82a). It is mainly the spherical cells that project to LSO and MSO (Zook and Casseday, '85; Casseday et al., '88). One difference between bats and other mammals is that the spherical cells are among the smallest cells in AVCN; nevertheless, their other features are the same as spherical cells in other species. The posteroventral cochlear nucleus (PVCN) also shares cytoarchitectural features with other mammals, including a distinct area that contains octopus cells. Some of these project to LSO and MSO in bats (Vater and Feng, '90; Casseday et al., '88). The dorsal cochlear nucleus is small and has poorly developed lamination; these are features that are shared by man and other primates (Moore, '80).

The general picture of the projections from AVCN can be seen from anterograde transport of wheat-germ agglutinin conjugated to horseradish peroxidase (WGA-HRP). Figure 1 shows the results of an injection of WGA-HRP that includes most of AVCN and extends into a small part of PVCN. The anterograde transport to the LSO is, as would be expected from observations on other mammals, to the ipsilateral LSO only. However, the transport to the MSO is unusual in two respects.

First, most of the transport is to the contralateral MSO; very little goes to the ipsilateral MSO. As a consequence of this unusual projection, the pattern of transport within MSO is not predominantly within the half of MSO ipsilateral to the injection, as in other mammals (e.g., Stotler, '53; Warr, '82; Moore; Strominger and Strominger, '71), but rather it fills the entire MSO on the contralateral side. The projections to the ipsilateral MSO are mainly to the lateral edge, a pattern resembling that in other mammals. In short the projections to LSO are essentially like those in other animals, but the projections to MSO are predominantly to the contralateral side. The projections to MNTB are exclusively contra-

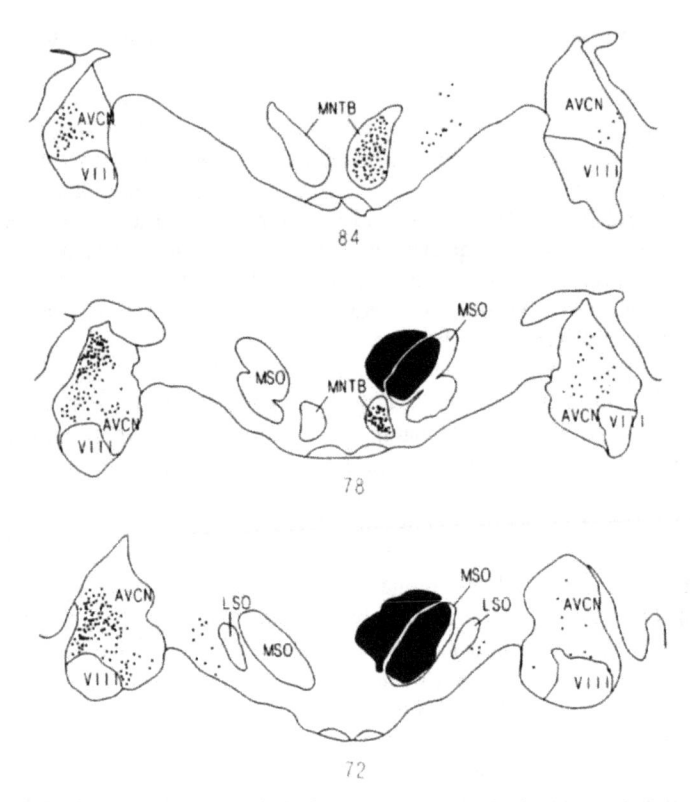

**Figure 2.** Drawing of a frontal section through the brainstem to show the locations of labeled cells after an injection of WGA-HRP in MSO. There are two major sources of projections to MSO, the contralateral AVCN and the ipsilateral MNTB. There are a few labeled cells in the ipsilateral cochlear nucleus. The solid black area shows the injection site in MSO; dots indicate the locations of labeled cells. Because the density of labeled cells in the contralateral AVCN and ipsilateral MNTB was extremely high, the number of dots in these structures represents only a fraction of the true number of labeled cells (from Covey et al., '91).

lateral, as in other mammals (Friauf and Ostwald, '88; Kuwabara et al., '91a; Smith et al., '91).

Figure 2 shows this asymmetry of projections from AVCN as seen by retrograde transport from MSO; there are many more labeled cells in contralateral AVCN than in the ipsilateral AVCN. However, the details of innervation of cells in MSO seem to be like those of other mammals (Clark, '69; Schwartz, '80; Kiss and Majorossy, '83; Schwartz, '84; Zook and Leake, '89). The fibers form varicosities on the cell body and along one of the two main dendrites--usually the one nearest the origin of the fiber (Fig. 3). Figure 2 makes the further point that a large number of cells are labeled in the MNTB.

The results just presented raise the question of whether this pattern of projections has physiological consequences. We recorded the responses to stimuli presented binaurally via earphones from single units in the superior olive of awake mustached bats (Covey et al., '91). We obtained best frequencies (BF) from 60 single units in LSO and 94 units in MSO and binaural response characteristics from 55 single units in LSO and 72 single units in MSO. The basic results can be summarized very briefly. The frequency organization is basically like that in the cat (Guinan et al., '72; Tsuchitani, '77). In LSO, low frequencies are located laterally and high medially (Fig. 4A) ; in MSO low frequencies are located dorsally and high

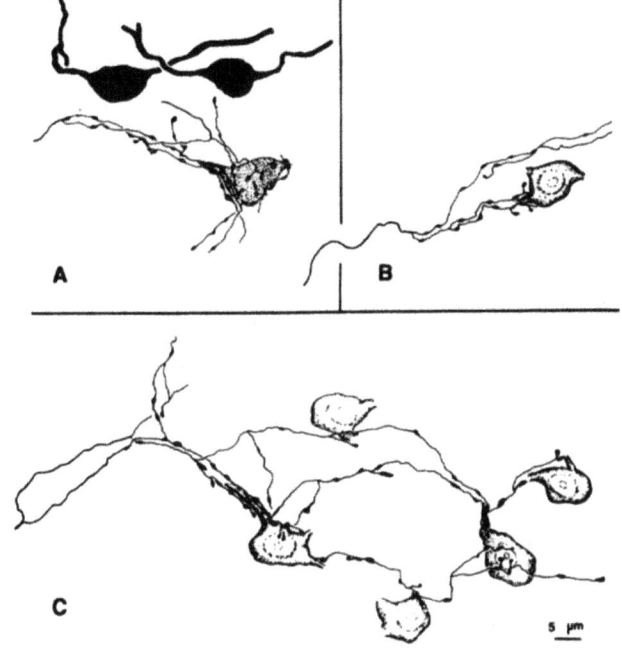

**Figure 3.** Drawings to show cell types and pattern of terminals in the superior olive of the mustached bat. A. In Golgi preparations, the cells in MSO have the typical bipolar appearance found in other mammals, as seen by the cells illustrated in solid black. Impregnated fibers show typical pattern of varicosities along dendrites and around a counterstained cell body in MSO. B. An axon filled intracellularly with HRP bifurcates to form varicosities along the dendrites of two adjacent cells in MSO. C. An extracellular injection of HRP in AVCN fills several fibers in MSO. The pattern of labeled varicosities, in relation to counterstained cells in MSO, is essentially the same as shown with the other two techniques (from Zook and Leake, '89).

ventrally (Fig. 4B). The only difference in the frequency organization is that certain frequencies have an expanded representation. These are the frequencies in the constant frequency (CF) component of the echolocation signal. Specifically, the second harmonic ($CF_2$), about 61 kHz, has a very large representation, and the third harmonic ($CF_3$), about 92 kHz, has a somewhat expanded representation (Fig. 5). Other frequencies in the hearing range of the bat, including those in the frequency modulated part of the echolocation signal, have a much smaller representation. Thus the tonotopy of LSO and MSO follows the general mammalian plan; however, within this plan, there is an expanded representation of certain biologically significant sounds, the dominant harmonics in the CF part of the echolocation signal. The results so far reinforce the view that LSO and MSO are homologous to the same-named structures in other mammals.

The tests with binaural stimuli also established that LSO is similar to other mammals. Ninety-one percent of the LSO cells were excited by sounds at the ipsilateral ear, and the excitatory response could be inhibited by simultaneously presenting sound at the contralateral ear (EI) (Fig. 6A). The remainder of LSO cells were excited by the ipsilateral ear without any effect of sound at the contralateral ear (EO). Examples of binaural cells encountered in several electrode penetrations through LSO of one animal are shown in Figure 4C, and a composite of three cases is illustrated in Figure 5B.

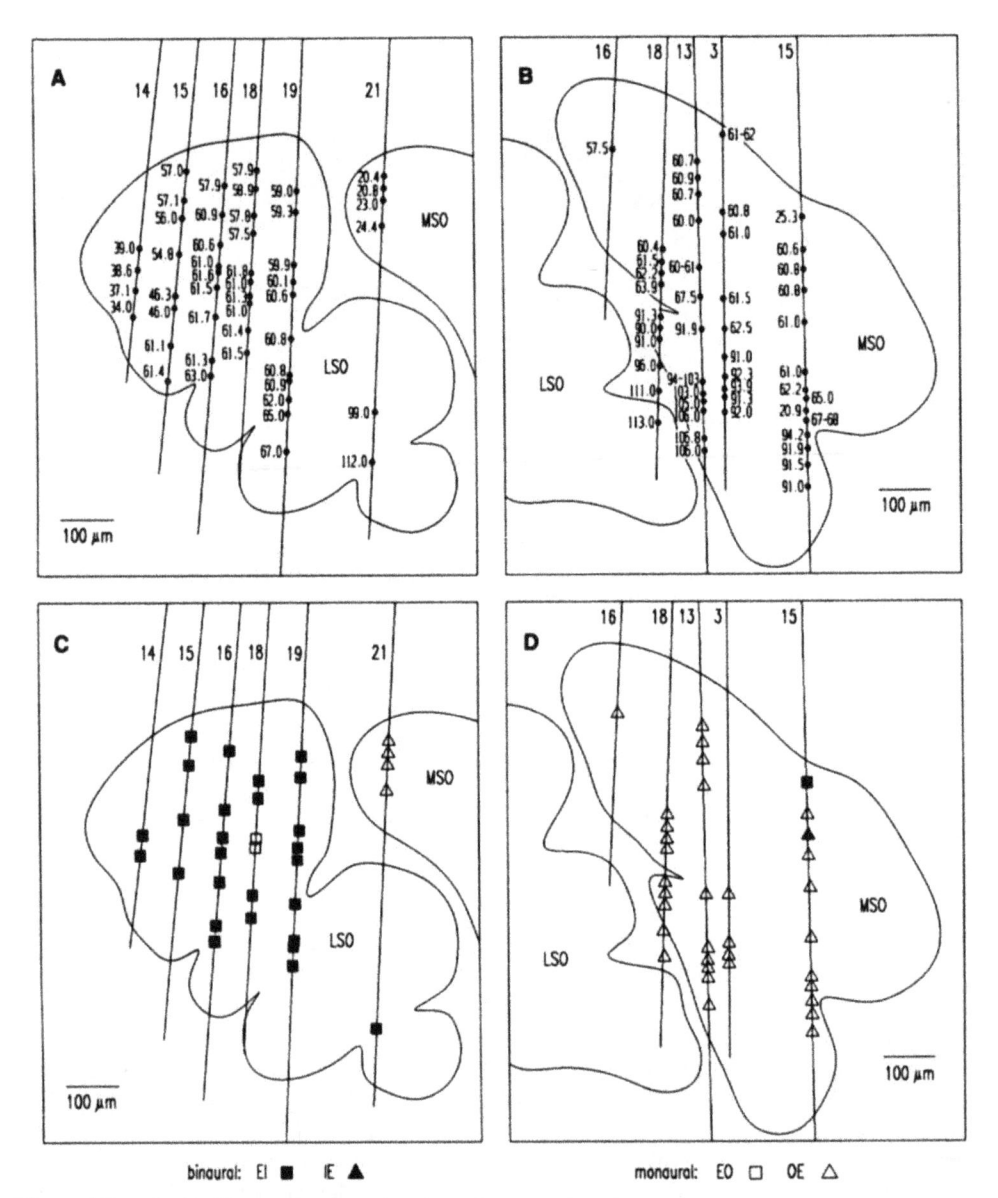

**Figure 4.** Drawings to illustrate representative electrode penetrations through the superior olive in two mustached bats. A. Best frequencies recorded along six different penetrations, mainly through LSO but including the dorsomedial tip of MSO. Best frequencies in both LSO and MSO increase dorsolaterally to ventromedially. B. Best frequencies recorded along five different penetrations, mainly through MSO. Note the shift in BFs from the 60 kHz range to the 90 kHz range at the border between the dorsal and ventral limbs of MSO. C. Responses to binaural stimuli for the same recording sites shown in A. Almost all neurons in LSO are binaural. D. Responses to binaural stimuli for the same recording sites shown in B. Almost all neurons in MSO are monaural (from Covey et al., '91).

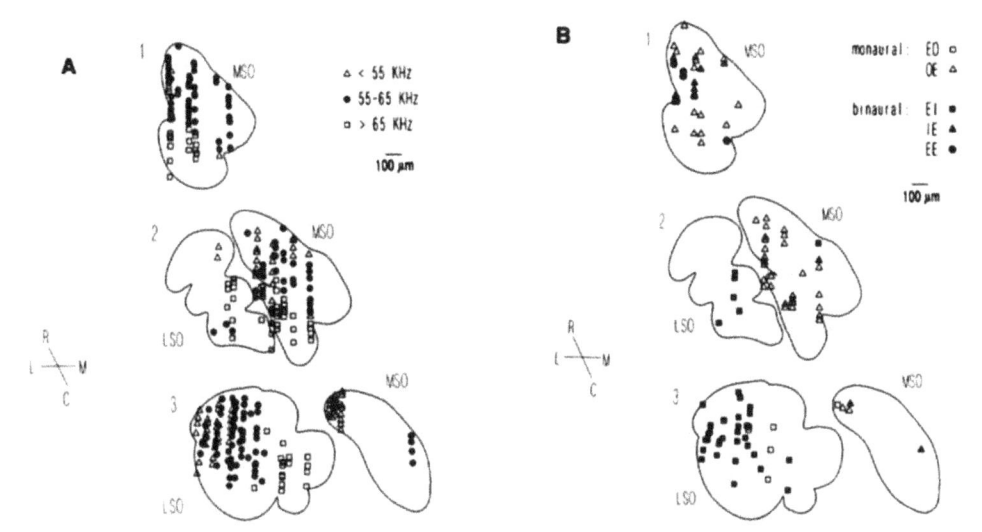

**Figure 5.** A. Composite drawings to illustrate the tonotopic organization of LSO and MSO. Recording sites from five cases have been pooled and plotted on standardized sections. Data from several sections have been collapsed onto rostral (1) middle (2) and caudal (3) thirds of the superior olives, so that the recording sites shown in any one section may be at slightly different rostro-caudal levels. The frequency range has been divided into three bands: a narrow middle band (55-65 kHz) that includes the $CF_2$, a lower band that includes all frequencies below 55 kHz, and an upper band that includes all frequencies above 65 kHz. The 10 kHz band around the $CF_2$ is greatly expanded in both LSO and MSO. In LSO it fills approximately the middle one-third of the nucleus, and in rostral MSO it fills the entire dorsal limb. B. Composite drawings to illustrate the distribution of binaural responsiveness in LSO and MSO. Recording sites from five cases have been plotted on standardized sections; the locations of recording sites have been collapsed in the rostrocaudal dimension as described for Fig. 5A. There is a clear division into binaural responses in LSO and monaural responses in MSO (from Covey et al., '91).

In contrast, the cells in MSO did not behave like most cells in MSO of other mammals. First, the aural response properties were different; 72% of MSO cells responded to monaural stimuli, only to the contralateral ear. Of the remainder, 21% were excited by sound at the contralateral ear, and that response was inhibited by sound at the ipsilateral ear (IE) (Fig. 6B). Only 6% were excited by sound at both ears (EE), and one cell responded to the ipsilateral ear only. This result is different from that seen in the MSO of nonecholocating mammals, where most neurons are EE, but this finding is not unique, as a small proportion of IE units are also found in MSO of other mammals (Goldberg and Brown, '69; Inbody and Feng, '81; Yin and Chan, '90).

A second difference is seen in the temporal patterns of responses. Whereas LSO cells typically (94%) respond tonically (Fig. 7A), the cells in MSO are mostly (71%) phasic (Fig. 7B). Only 28% of MSO cells responded tonically, some at the onset (Fig. 7C) and some at the offset (Fig. 7D) of sound. For many of the phasic neurons, the response changed from onset to offset with small changes of frequency. This change in pattern was seen most commonly in neurons that had a BF near the $CF_2$ or $CF_3$. For example the neuron in Fig. 8 responded phasically at frequencies above the BF of 60.7 kHz (Fig. 8A). As the frequency was decreased to 60.15 kHz, the onset response decreased (Fig. 8B); at 60.05 kHz a small *phasic-on* response was accompanied by a larger *phasic-off* response (Fig. 8C), and at 59.9 kHz only a *phasic-off* response remained (Fig. 8D). Later we shall discuss the possible functional consequences of this pattern of response.

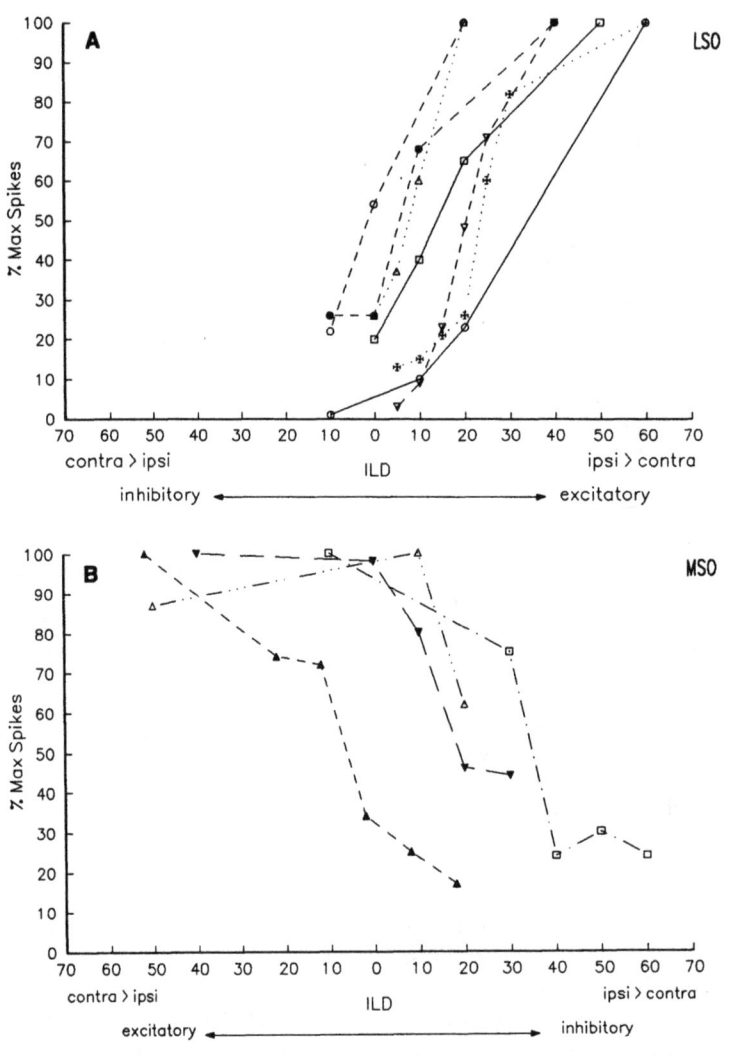

**Figure 6.** Discharge rate is plotted as a function of interaural level difference (ILD). A. The responses of neurons in LSO are almost completely suppressed when sound level at both ears is equal; the slopes of the curves are relatively steep. The firing rates of different individual neurons decrease to 1/2 max at different ILDs. B. The discharges of binaural MSO neurons are never completely suppressed even when the sound level is much higher at the inhibitory ear than at the excitatory ear, and discharge decreases to 1/2 max when the level at the ipsilateral ear exceeds that at the contralateral (from Covey et al., '91).

We can now return to the details of the asymmetrical input to MSO to ask whether there are specializations in the projections from the cochlear nucleus that correspond to the unusual response properties in MSO. Figure 9A shows the results of intraaxonal injection of a fluorescent dye into two axons. The fiber that arises from a spherical bushy cell (Fig. 9A right), projects only to the contralateral MSO. The fiber that arises from a stellate cell (Fig. 9A left), projects bilaterally. Figure 10 indicates that these are examples of a rule, namely that the bushy cells project to the contralateral side, whereas the stellate cells contribute the ipsilateral projection. Taking this information together with that in Figures 1 and 2, we can conclude that a small proportion of the projections to MSO arise from stellate cells in AVCN

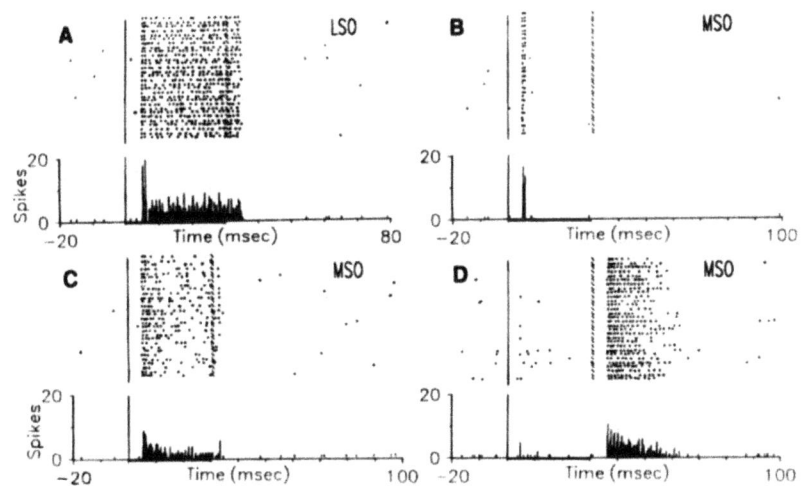

**Figure 7.** Dot raster diagrams and post-stimulus time histograms (PSTHs) to illustrate typical discharge patterns in LSO and MSO. Each PSTH represents the number of spikes that occurred in each 0.2 msec time bin, totalled over 40 stimulus presentations. A. Most neurons in LSO had a relatively high level of tonic discharge throughout the duration of the sound. B. The majority of neurons in MSO responded phasically with one or two spikes at the onset of the sound. C. Some neurons in MSO responded in a primary-like pattern. D. A few neurons in MSO responded to the offset of the sound, but little or not at all to the onset. In this and the following figures showing dot rasters and post-stimulus time histograms, the following conventions are used: The space between the y-axis and the vertical line shows 20 msec of spontaneous activity. The vertical line indicates the stimulus onset, and the hatched marks in the dot raster represent stimulus offset. The thick horizontal line below the x-axis indicates the stimulus duration. The stimuli were 30 msec. tone bursts at BF and 10 dB above threshold, presented at a rate of 2/sec. (from Covey et al., '91).

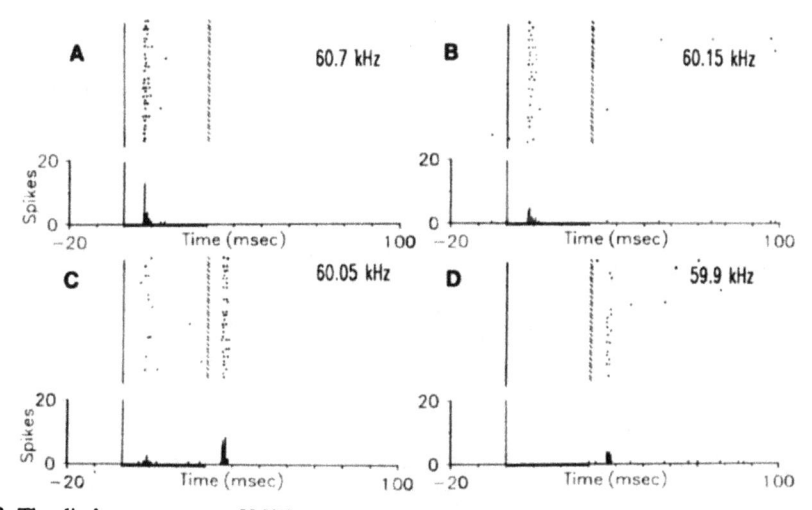

**Figure 8.** The discharge patterns of MSO neurons can change as frequency of a pure tone stimulus is changed. The series of dot raster diagrams and PSTHs shown here was collected from a neuron in MSO with a BF of 60.7 kHz. All stimuli are at 10 dB above threshold. A. At BF, the neuron discharges phasically to the onset of the sound. B. Just below BF the onset response is slightly diminished. C. Further below BF, an offset discharge appears in addition to the onset response. D. As frequency is further lowered the onset response disappears and only the offset discharge remains. See caption for Fig. 7 for explanation of graphs (from Covey et al., '91).

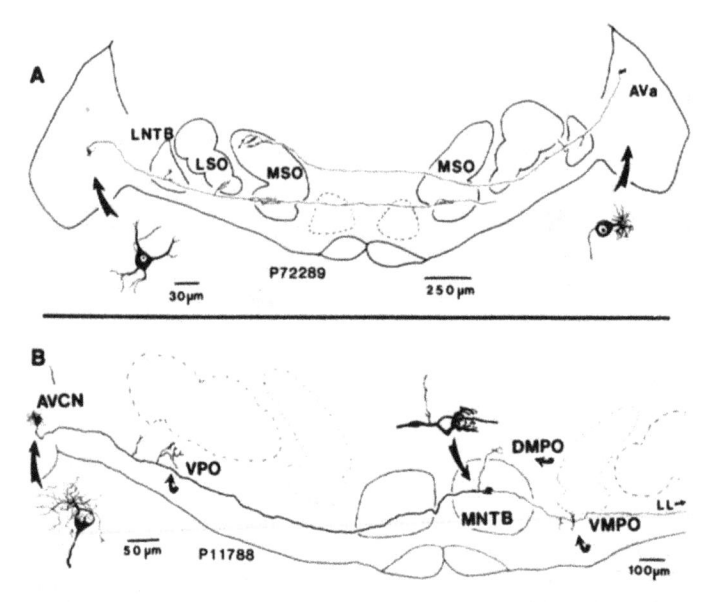

**Figure 9.** Drawings to show projections from axons labeled intracellularly in the mustached bat. A. Axons that project to MSO arise from two sources and are of two types. At the left, the axon of a labeled stellate cell projects to the ipsilateral lateral nucleus of the trapezoid body (LNTB), to LSO, as well as to MSO of both sides. At the right, the axon of a spherical bushy cell in the anterior part of AVCN (AVa) projects to the ipsilateral LNTB and to the contralateral MSO. This axon does not project to the ipsilateral MSO. B. The projection to MNTB arises from a globular bushy cell in the contralateral AVCN. Note that the ending in MNTB is a large calyx of Held. Axon collateral branches terminate in the ventral periolivary nucleus (VPO) ipsilaterally and in the ventromedial and dorsomedial periolivary nuclei (VMPO and DMPO) contralaterally. A collateral axon continues to the contralateral lateral lemniscus (LL). Drawings are from unpublished observations (Zook and Kuwabara). Nomenclature for brainstem auditory nuclei is based on cytoarchitectural similarities with other species (Zook and Casseday, '82a).

of both sides, but most of the projections arise from spherical bushy cells in the contralateral AVCN. In short, the mustached bat lacks all or most of the ipsilateral input to MSO from bushy cells. This observation is the anatomical corollary of the monaural response properties of most MSO cells in *Pteronotus*.

We now suggest that the projections to MSO via the MNTB provide a corollary for the phasic response properties. Figure 9B illustrates that in *Pteronotus*, as in other mammals (Warr, '72; Friauf and Ostwald, '88; Kuwabara et al., '91a), the calyces of Held in MNTB arise from globular bushy cells in AVCN. The cells that receive the calyces have long been known to project to the LSO and are the presumptive inhibitory inputs to EI cells in LSO (Spangler et al., '85; Zook and DiCaprio, '88; Adams and Mugnaini, '90; Bledsoe et al., '90 Kuwabara and Zook, '91). There is now neuropharmacological and immunocytochemical evidence that this input is inhibitory (Moore and Caspary, '83; Adams and Mugniani, '90; Bledsoe et al., '90). An important recent finding is that these cells also project to MSO (Kuwabara and Zook '91; Adams and Mugnaini, '90). This projection follows the tonotopic arrangement in MSO (Fig. 11) just as it does in LSO (Kuwabara and Zook, '91). Figure 11 shows that these cells project to MSO in *Pteronotus*. Thus the MNTB supplies a major source of input to MSO from the contralateral AVCN. Grothe ('90) has recently confirmed by neuropharmacological methods that this input is inhibitory.

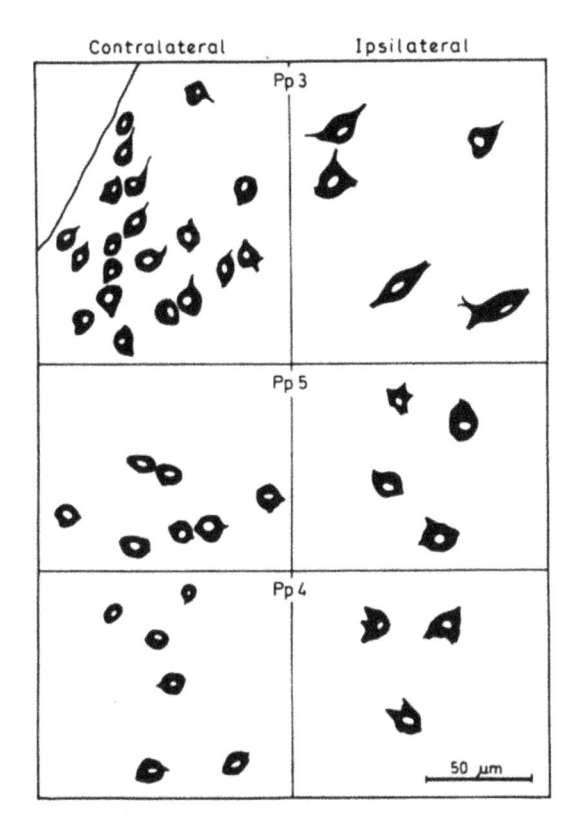

**Figure 10.** Labeled cells in AVCN after injections in MSO of three mustached bats. The labeled cells ipsilateral to the injection are larger and multipolar, consistent with the idea, suggested in Fig. 9A, that the projections to ipsilateral MSO arise from stellate cells. The labeled cells contralateral to the injection are small and round, consistent with the idea that these are spherical bushy cells like that shown in Fig. 9A.

Still other, newly recognized inputs to MSO may also account for the proportion of MSO units having IE responses. Figure 12 shows cells in the lateral nucleus of the trapezoid body (LNTB) that have been intracellularly labeled. The axons of these cells project to LSO as well as to MSO. These LNTB cells are good candidates for transforming excitatory input from the cochlear nucleus to inhibitory input to LSO and MSO for two reasons. First, the LNTB cells are contacted by fibers from the cochlear nucleus (Warr, '72; Tolbert et al., '82; Roullier and Ryugo, '84; Spirou et al., '90; Smith et al., '91; Kuwabara et al., 91). Thus the LNTB is another relay of input to these structures from the cochlear nucleus. Second, cells in LNTB stain for glycine (Peyret et al., '86; '87; Wenthold et al. '87; Helfert et al, '89; Bledsloe et al., '90). Thus the LNTB connections may help explain the IE responses in MSO, although by the same reasoning, their role in LSO is unclear. In any case, the projections from LNTB to MSO and LSO have also been found in the mouse and gerbil and are apparently part of the general mammalian pattern of relays from the cochlear nucleus to the superior olives.

We can now summarize the inputs to MSO in a way that helps explain the response properties of its cells. The direct input from spherical bushy cells in AVCN˙ to the contralateral MSO is almost certainly excitatory. The indirect input arises from globular bushy cells that project to the contralateral MNTB. The cells in MNTB provide inhibitory

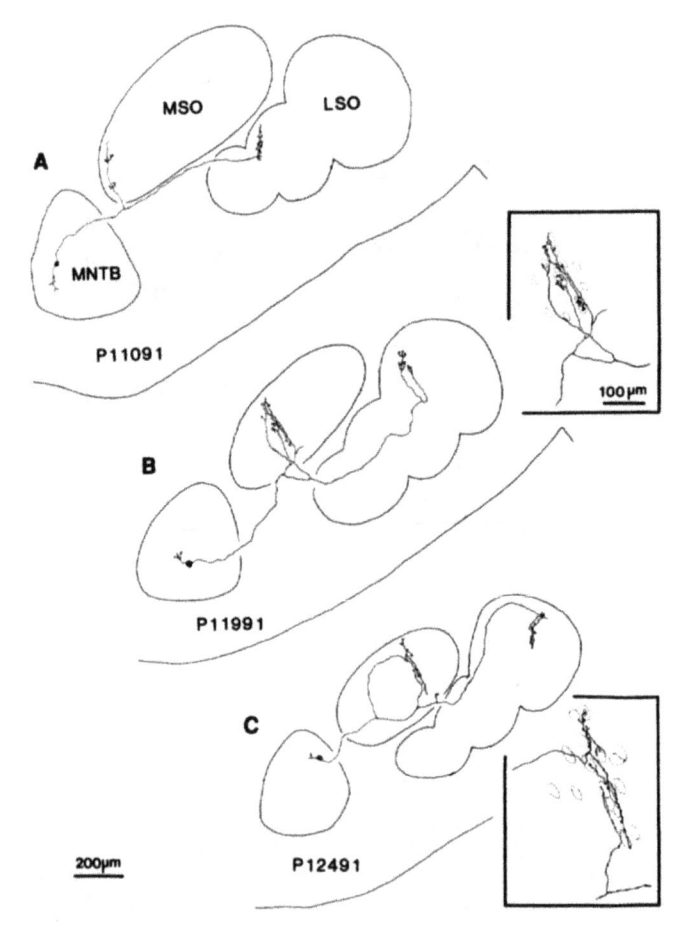

**Figure 11.** Projections from MNTB provide the contralateral input to LSO and MSO. These projections follow the tonotopic pattern in both MSO and LSO. A. A cell in medial MNTB projects to high frequency regions of MSO and LSO. B. A cell in the middle of MNTB projects to middle regions of MSO and LSO. C. A cell in lateral MNTB projects to low frequency regions of MSO and LSO. Cells were filled with Lucifer yellow (from Kuwabara and Zook, '92).

input to MSO cells. Thus cells in MSO receive, from the contralateral AVCN, direct excitatory input and indirect inhibitory input. It is not a great speculative step to imagine that the excitatory input gets there before the inhibitory input and that it is this sequence of events that produces the phasic response properties in the MSO cells. A similar interaction could account for the responses of IE units in the MSO. Ipsilateral inhibitory input, relayed by LNTB cells, could reach MSO cells before or concurrent with the direct excitatory input from the contralateral cochlear nucleus.

The results showing emphasis on monaural inputs to the MSO are significant for ideas on the evolution of the superior olive and projections from the cochlear nucleus. If we accept the idea that MSO in *Pteronotus* is homologous with MSO in other mammals, then the MSO in *Pteronotus* has diverged from the general mammalian plan during evolution. The divergence is expressed as a diminished input from AVCN to the ipsilateral MSO. It should be noted that the monaural emphasis continues up to the level of the midbrain; electrophysiolo-

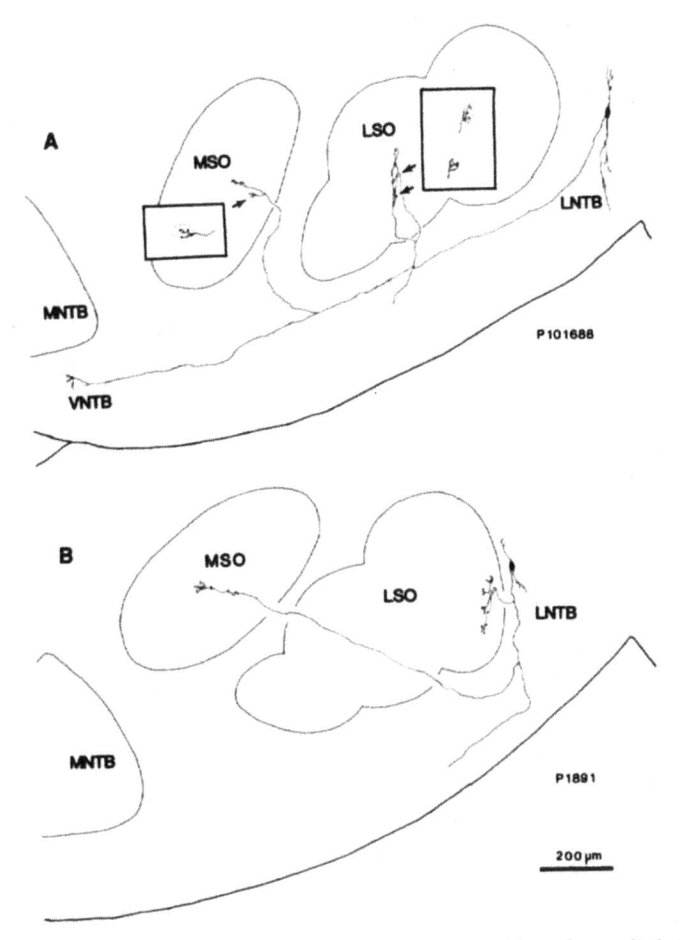

**Figure 12.** Illustration to show the pattern of axon projections to MSO and LSO from cells in LNTB. The cells have been filled intracellularly with lucifer yellow. A. An elongate cell in LNTB sends axon branches to cells in LSO and MSO. The two inserts illustrate the axosomatic patterns of terminal boutons in relation to counterstained cell somata (dashed outlines). A third axon branch formed a terminal spray within VNTB. B. Another example of an elongate cell in LNTB that sends axon branches to LSO and MSO. In this case a third collateral branch courses medially, but its destination could not be traced as it went outside the cut surface of the tissue slice. Axons from LNTB cells were not found to project to MNTB (from Kuwabara and Zook, '92).

gical experiments combined with retrograde transport of HRP show that the MSO is the main source of the input to the monaural region of the enlarged 60 kHz contour of the inferior colliculus (Ross and Pollak, '89).

Of what significance is this altered input for the bat? We can only speculate about the answer, but two possibilities can be tested. The first involves localization in the vertical plane. Bats have to be as adept at localizing sound in the vertical plane as they are in the horizontal plane. A possible mechanism for this would be encoding of elevation-dependent changes in frequency and intensity. That is, the sound level spectral contents of echoes are altered by the structure of the external ear as the echo source changes in elevation (Simmons and Lawrence, '82; Fuzessery and Pollak, '85). These spectral and temporal differences are potential cues that could be encoded in the central auditory system. In the inferior colliculus of the mustached bat, all binaurally excited neurons have the lowest thresholds to sounds at the midline in the horizontal dimension, but in the vertical dimension, neurons tuned to different harmonics of the echolocation call are differentially sensitive to different signal elevations. For example, neurons tuned to 60 kHz are most sensitive to about 0° elevation, whereas neurons tuned to 90 kHz are most sensitive to elevations about -40° (Fuzessery and Pollak, '85). This spatial sensitivity is a direct consequence of the directional properties of the external ear (Fuzessery and Pollak, '85). Exactly how the MSO or LSO might contribute to this processing is now an open question. There is no information on spatial selectivity for monaural units in the inferior colliculus. For units in MSO and LSO, we do know that units with BFs of 60 kHz have both the lowest thresholds as well as the widest range of thresholds, whereas units with BFs of 90 kHz have higher thresholds. As there is an increase in intensity in the frequencies around 90 kHz at lower elevations, the units in LSO and MSO with BFs around 90 kHz would be more likely to be activated at lower elevations of the echo source. However, the range of thresholds of units with BFs around 60 kHz overlap with the thresholds of 90 kHz units, and the lowest thresholds for units with BFs around 60 kHz are about 15 dB below the lowest thresholds of units with BFs around 90 kHz. Thus the 60 kHz units in MSO and LSO would be excited over a wide range of elevations. The sample of units was not large enough to address the question of whether or not there is a topography of thresholds in LSO or MSO. The answer to this question would be important to an analysis of mechanisms for encoding sound in space (Reed and Blum, '90) and should be addressed in future experiments.

It now seems unlikely that MSO neurons encode the fine frequency structure of the pulse or echo. Grothe ('90) recently found that MSO neurons in the mustached bat do not respond to frequency modulations of a pure tone.

The second possibility concerns distance estimation or target ranging. This is suggested by two observations, 1) that for some neurons in MSO, the response pattern changes from on to off as the frequency of the stimulus decreases and 2) that this on-off pattern was seen mainly in neurons tuned to the intense second harmonic of the constant frequency component of the echolocation call, around 60 kHz. To see how the on-off pattern might be related to echolocation we need to briefly mention how the mustached bat changes its echolocation pulse. During flight, the bat compensates for Doppler shifts in echoes by lowering the frequency of the constant frequency component of its echolocation call. Because of Doppler effects, the returning echoes are always slightly higher in frequency than the emitted pulse. The Doppler-shift compensating mechanism of the bat keeps the echoes within the frequency range of the narrow filter neurons that have best frequencies around 60 kHz. This means that the on-off MSO neurons could respond at the offset of the pulse and at the onset of the echo; presumably if the two overlapped, the response would be facilitated. The reason such facilitation might have functional importance is that the bat systematically decreases the duration of the constant frequency component as it approaches its prey. Thus the facilitated response could be the mechanism whereby the bat adjusts the duration of its echolocation signal in direct relation to the distance from its target. These considerations make it worthwhile to design future experiments to determine whether there is a facilitated response when the end of a vocalized pulse overlaps with the beginning of a Doppler-shifted echo. If so, the MSO could provide a mechanism whereby the bat adjusts the duration of its cholocation pulse in relation to target distance--that is, the bat utilizes a vocal yardstick.

ACKNOWLEDGEMENTS

The authors' work is supported by NIH grants DC-287, DC-503 and DC-38. We are indebted to Drs. Ellen Covey and Gerhard Neuweiler for their useful comments on an earlier draft of the manuscript.

# REFERENCES

Adams, J.C., 1979, Ascending projections to the inferior colliculus, *J. Comp. Neurol.* 183:519-538.

Adams, J.C., and Mugnaini, E., 1990, Immunocytochemical evidence for inhibitory and disinhibitory circuits in the superior olive, *Hearing Res.*, 49: 281-298.

Bledsoe, S.C.J., Snead, C.R., Helfert, R.H., Prasad, V., Wenthold, R.J., and Atlshuler, R.A., 1990, immunocytochemical and lesion studies support the hypothesis that the projection from the medial nucleus of the trapezoid body to the lateral superior olive is glycinergic, *Brain Res.*, 517:189-194.

Brunso-Bechtold, J.K., Henkel, C.K., and Linville, C., 1990, Synaptic organization in the adult ferret medial superior olive., *J. Comp. Neurol.*, 294:389-398.

Caird, D., and Klinke, R., 1983, Processing of binaural stimuli by cat superior olivary complex neurons, *Exp. Brain Res.*, 52:385-399.

Cant, N.B., and Casseday, J.H., 1986, Projections from the anteroventral cochlear nucleus to the lateral and medial superior olivary nuclei, *J. Comp. Neurol.*, 247:457-476.

Casseday J.H., and Covey E., 1987, Central auditory pathways in directional hearing, *in*: "Directional Hearing," W. Yost, and G. Gourevitch, ed., Springer, New York.

Casseday, J.H., Covey, E., and Vater, M., 1988, Connections of the superior olivary complex in the Rufous Horseshoe Bat, *Rhinolophus rouxi*, *J. Comp. Neurol.*, 278:313-329.

Casseday, J.H., and Neff, W.D., 1975, Auditory localization: Role of auditory pathways in the brain stem of the cat, *J. Neurophysiol.*, 38:842-858.

Clark, G. M., 1969, The ultrastructure of nerve endings in the medial superior olive of the cat, *Brain Res.*, 14:293-305.

Covey, E., Vater, M., and Casseday, J.H., 1991, Binaural properties of single units in the superior olivary complex of the mustached bat, *J. Neurophysiol.*, 66:1080-1094.

Friauf, E., and Ostwald, J., 1988, Divergent projections of physiologically characterized rat ventral cochlear nucleus neurons as shown by intra-axonal injection of horseradish peroxidase, Exp. Brain. Res., 73:263-284.

Fuzessery, Z.M., and Pollak, G.D., 1985, Determinants of sound location selectivity in bat inferior colliculus: A combined dichotic and free-field stimulation study, *J. Neurophysiol.*, 54:757-781.

Galambos, R., Schwartzkopff, J., and Rupert, A., 1959, Microelectrode study of superior olivary nuclei, *Am. J. Physiol.*, 197:527-536.

Goldberg, J.M., and Brown, P.B., 1968, Functional organization of the dog superior olivary complex: An anatomical and electrophysiological study, *J. Neurophysiol.*, 31:639-656.

Goldberg, J.M., and Brown, P.B., 1969, Response of binaural neurons of dog superior olivary complex to dichotic tonal stimuli: Some physiological mechanisms of sound localization., *J. Neurophysiol.*, 32:613-636.

Grothe, B., 1990, "Versuch einer Definition des medialen Kernes des oberen Olivenkomplexes bei der Neuwelt fledermaus *Pteronotus p. parnellii*," Ph.D. dissertation, Ludwig-Maximilians-Universität, Munich.

Guinan, J.J. Jr., Norris, B.E., and Guinan, S.S., 1972, Single auditory units in the superior olivary complex. II: Locations of unit categories and tonotopic organization, *Int. J. Neurosci.*, 4:147-166.

Guinan, J.J., and Li, R.Y.-S., 1990, Signal processing in brainstem auditory neurons which receive giant endings (calyces of Held) in the medial nucleus of the trapezoid body of the cat, *Hearing Res.*, 49:321-334.

Hall, J.L., 1965, Binaural interaction in the accessory superior olivary nucleus of the cat, *J. Acoust. Soc. Am.*, 37:814-823.

Harrison J.M., and Feldman M.M., 1970, Anatomical aspects of the cochlear nucleus and superior olivary complex, *in*: "Contributions to Sensory Physiology, Vol. 4," W. D. Neff, ed., Academic, New York.

Harrison, J.M., and Irving, R., 1966, Visual and nonvisual auditory systems in mammals, *Science*, 154:738-743.

Helfert, R.H., Bonnueau, J.M., Wenthold, R.J., and Altschuler, R. A., 1989, GABA and glycine immunoreactivity in the guinea pig superior olivary complex, *Brain Res.*, 501:269-286.

Inbody, S.B., and Feng, A.S., 1981, Binaural response characteristics of single neurons in the medial superior olivary nucleus of the albino rat, *Brain Res.*, 210:361-366.

Irving, R., and Harrison, J.M., 1967, Superior olivary complex and audition: A comparative study, *J. Comp. Neurol.*, 130:77-86.

Jeffress, L.A., 1948, A place theory of sound localization, *J. Comp. Physiol. Psychol.*, 41:35-39.

Kiss, A., and Majorossy, K., 1983, Neuron morphology and synaptic architecture in the medial superior olivary nucleus, *Exp. Brain Res.*, 52:315-327.

Kuwabara, N., DiCaprio, R.A., and Zook, J.M., 1991, Afferents to the medial nucleus of the trapezoid body and their collateral projections, *J. Comp. Neurol.*, 314:684-706.

Kuwabara, N., and Zook, J.M., 1992, Projections to the medial superior olive from the medial and lateral nuclei of the trapezoid body in rodents and bats, *J. Comp. Neurol.*, in press.

Kuwada, S., and Yin, T.C.T., 1987, Physiological studies of directional hearing, in: "Directional Hearing," W. A. Yost, and G. Gourevitch, ed., Springer-Verlag, New York.

Lindsey, B.G., 1974, Fine structure and distribution of axon terminals from the cochlear nucleus on neurons in the medial superior olivary nucleus of the cat, *J. Comp. Neurol.*, 16:81-104.

Masterton, B., Jane, J.A., and Diamond, I.T., 1967, Role of brain stem auditory structures in sound localization. I. Trapezoid body, superior olive, and lateral lemniscus, *J. Neurophysiol.*, 30:341-359.

Moore, C.N., Casseday, J.H., and Neff, W.D., 1974, Sound localization: The role of the commissural pathways of the auditory system of the cat, *Brain Res.*, 82:13-26.

Moore, J.K., 1980. The primate cochlear nuclei: loss of lamination as a phylogenetic process, *J. Comp. Neurol.*, 193:609-629.

Moore, M.M., and Caspary, D.M., 1983, Strychnine blocks binaural inhibition in lateral superior olivary neurons, *J. Neurosci.*, 3:237-242.

Perkins, R.E., 1973, An electron microscopic study of synaptic organization in the medial superior olive of normal and experimental chinchillas, *J. Comp. Neurol.*, 148:387-416.

Peyret, D., Campistron, G., Geffard, M., and Aran, J.-M.,1987, Glycine immunoreactivity in the brainstem auditory and vestibular nuclei of the guinea pig, *Acta Otolaryngol.*, 104:71-76.

Peyret, D., Geffard, M., and Aran, J.-M., 1986, GABA immunoreactivity in the primary nuclei of the auditory central nervous system, *Hearing Res.*, 23:115-121.

Reed, M.C., and Blum, J.J., 1990, A model for the computation and encoding of azimuthal information by the lateral superior olive, *J. Acoust. Soc. Am.*, 88:1442-1453.

Ross, L.S., and Pollak, G.D., 1989, Differential ascending projections to aural regions in the 60 kHz contour of the mustache bat's inferior colliculus, *J. Neurosci.*, 9:2819-2834.

Ross L.S., Pollak G.D., and Zook J.M., 1988, Origin of ascending projections to an isofrequency region ofthe mustache bat's inferior colliculus, *J. Comp. Neurol.*, 270:488-505.

Rouiller, E.M., and Ryugo, D.K., 1984, Intracellular marking of physiologically characterized cells in the ventral cochlear nucleus of the cat, *J. Comp. Neurol.* 225:167-186.

Sanes, D.H., 1990, An in vitro analysis of sound localization mechanisms in the gerbil lateral superior olive, *J. Neurosci.*, 10:3494-3506.

Schwartz, I.R., 1980, The differential distribution of synaptic terminal on marginal and central cells in the cat medial superior olivary nucleus, *Am. J. Anat.*, 159:25-31.

Schwartz I.R., 1984, Axonal organization in the cat medial superior olivary nucleus, in: "Contributions to Sensory Physiology, Vol. 8," W. D. Neff, ed., , New York.

Simmons, J.A., and Lawrence, B.D., 1982, Echolocation in bats: the external ear and perception of the vertical positions of targets, *Science*, 218:481-483.

Smith, P.H., Joris, P.X., Carney, L.H., and Yin, T.C.T., 1991, Projections of physiologically characterized globular bushy cell axons from the cochlear nucleus of the cat, *J. Comp. Neurol.*, 304:387-407.

Spangler, K.M., Warr, W.B., and Henkel, C.K., 1985, The projections of principal cells of the medial nucleus of the trapezoid body in the cat, *J. Comp. Neurol.*, 238:249-262.

Spirou, G.A., Brownell, W.E., and Zidanic, M., 1990, Recordings from cat trapezoid body and HRP labeling of globular bushy cell axons, *J. Neurophysiol.*, 63:1169-1190.

Stotler, W.A., 1953, An experimental study of the cells and connections of the superior olivary complex of the cat, *J. Comp. Neurol.*, 98:401-432.

Strominger, N.L., and Strominger, A.I., 1971, Ascending brain stem projections of the anteroventral cochlear nucleus in the rhesus monkey, *J. Comp. Neurol.*, 143:217-242.

Tolbert, L.P., Morest, D.K., and Yurgelun-Todd, D.A., 1982, The neuronal architecture of the anteroventral cochlear nucleus of the cat in the region of the cochlear nerve root: Horseradish peroxidase labelling of identified cell types, *Neurosci.*, 7:3031-3052.

Tsuchitani, C., 1977, Functional organization of lateral cell groups of cat superior olivary complex, *J. Neurophysiol.*, 40:296-318.

Tsuchitani, C., and Boudreau, J.C., 1966, Single unit analysis of cat superior olive S segment with tonal stimuli, *J. Neurophysiol.*, 29:684-697.

Vater, M., and Feng, A.S., 1990, Functional organization of ascending and descending connections of the cochlear nucleus of horseshoe bats, *J. Comp. Neurol.*, 292:373-395.

Warr, W.B., 1966, Fiber degeneration following lesions in the anterior ventral cochlear nucleus of the cat, *Exp. Neurol.*, 14:453-474.

Warr, W.B., 1972, Fiber degeneration following lesions in the multipolar and globular cell areas in the ventral cochlear nucleus of the cat, *Brain Res.*, 40:247-270.

Warr W.B., 1982, Parallel ascending pathways from the cochlear nucleus: Neuroanatomical evidence of functional specialization, in: "Contributions to Sensory Physiology, Vol. 7," W. D. Neff, ed., Academic, New York.

Wenthold, R.J., Huie, D., Altschuler, R.A., and Reeks, K.A., 1987, Glycine immunoreactivity localized in the cochlear nucleus and superior olivary complex, Neurosci., 22:897-912.

Yin, T.C.T., and Chan, J.C.K., 1990, Interaural time sensitivity in medial superior olive of cat, J. Neurophysiol., 64:465-488.

Zook, J.M., and Casseday, J.H., 1982a, Cytoarchitecture of auditory system in lower brainstem of the mustache bat, *Pteronotus parnellii*, J. Comp. Neurol., 207:1-13.

Zook J.M., and Casseday J.H., 1982b, Origin of ascending projections to inferior colliculus in the mustache bat, *Pteronotus parnellii*, J. Comp. Neurol., 207:14-28.

Zook, J.M., and Casseday, J.H., 1985, Projections from the cochlear nuclei in the mustache bat, *Pteronotus parnellii*, J. Comp. Neurol., 237:307-324.

Zook, J.M., and DiCaprio, R.A., 1988, Intracellular labeling of afferents to the lateral superior olive in the bat, *Eptesicus fuscus*, Hearing Res., 34:141-148.

Zook, J.M., and Leake, P.A., 1989, Connections and frequency representation in the auditory brainstem of the mustache bat, *Pteronotus parnelli*, J. Comp. Neurol., 290:243-261.

# THE MONAURAL NUCLEI OF THE LATERAL LEMNISCUS: PARALLEL PATHWAYS FROM COCHLEAR NUCLEUS TO MIDBRAIN

Ellen Covey

Department of Neurobiology, Duke University
Medical Center, Durham, NC 27710, USA

The mammalian cochlear nucleus is made up of several different divisions, each of which receives input from the fibers of the auditory nerve, and each of which performs some different transformation of this input. These different divisions of the cochlear nucleus in turn give rise to different parallel pathways. Some of these pathways ascend directly to the midbrain, some provide interconnections among the divisions of the cochlear nucleus, and others innervate cell groups in the brainstem where the signals originating in the cochlear nucleus undergo further transformation. These transformations may be accomplished through various mechanisms, for example, different patterns of axonal termination on cell bodies or dendrites, different biophysical properties of the target cells, or convergence of inputs from more than one pathway. The present review focuses on the transformations that occur in one system of brainstem auditory nuclei, the intermediate (INLL) and ventral (VNLL) nuclei of the lateral lemniscus.

It is clear that these cell groups must play an important role in hearing. They are a major target of the crossed ascending pathways from the cochlear nucleus, and a major source of input to the central nucleus of the inferior colliculus (ICc). The projections to the ICc from INLL and VNLL alone are at least equal in magnitude to the projections from the superior olivary nuclei (Adams, '79).

Because the INLL and VNLL occupy a position midway between the cochlear nucleus and the ICc, they, like the nuclei of the superior olivary complex, represent a stage of specialized analysis that is accomplished before the outputs of the different brainstem pathways converge at the ICc.

This review summarizes anatomical and physiological evidence that bears on the question of what this analytical task might be and how the different inputs to INLL and VNLL could contribute to its performance. This evidence includes the cytoarchitecture of the lateral lemniscal nuclei, the response properties of neurons in INLL and VNLL, the patterns of projections to INLL and VNLL from the cochlear nucleus, and the response properties of the neurons that give rise to these ascending projections. This information leads to two hypotheses, first, that the function of this system of monaural pathways is to transmit to the ICc basic information about the temporal structure of sounds, and second, that within this system there are two different streams of processing, one for precise signalling of the onset time of a sound and one for transmitting information about ongoing properties of a sound such as intensity or duration.

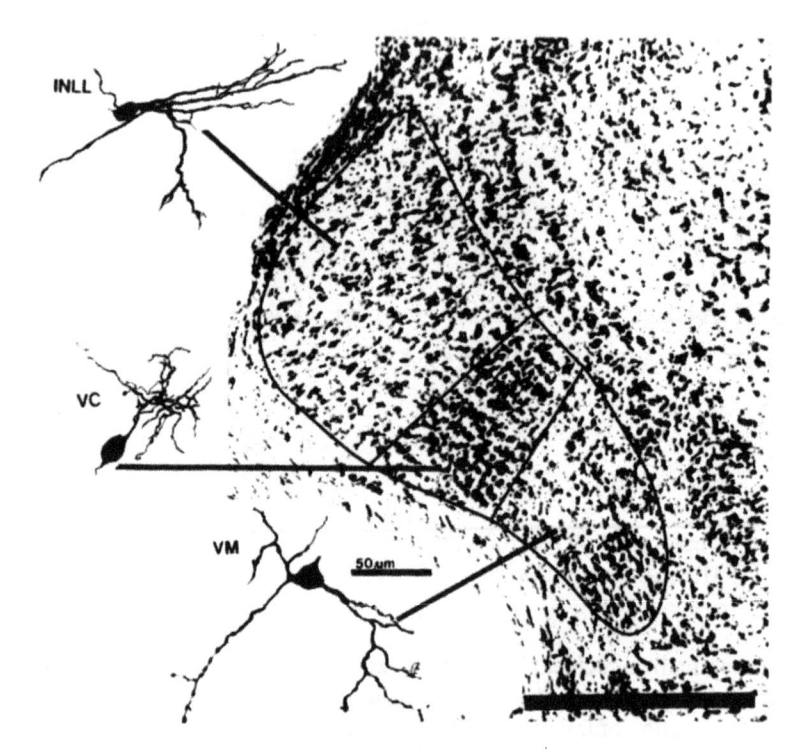

**Figure 1.** In the big brown bat, *Eptesicus fuscus*, three clearly distinct cell groups lie ventral to the dorsal nucleus of the lateral lemniscus (DNLL). These are the intermediate nucleus (INLL), the columnar division of the ventral nucleus (VC in the figure) and the multipolar cell division of the ventral nucleus (VM in the figure). They are seen here in a Nissl-stained frontal section through the left side of the brainstem. Each division contains a characteristic cell type, as seen in the drawings of Golgi impregnated cells. Calibration bar for photomicrograph = 500 μm.

## Cytoarchitecture

In all mammals there are groups of cell bodies located between the ascending fibers of the lateral lemniscus. With the exception of the dorsal nucleus (DNLL), which clearly is part of the binaural system, all of these cell groups receive projections from the contralateral cochlear nucleus (e.g., Cajal, '09; Warr, '66; '69; '82; Glendenning et al., '81; Zook and Casseday, '85; Covey and Casseday, '86) and project heavily to the ipsilateral ICc (e.g. Roth et al., '78; Adams, '79; Brunso-Bechtold et al., '81; Schweizer '81; Zook and Casseday, '82b; '87; Whitley and Henkel, '84; Covey and Casseday, '86; Saint-Marie and Baker, '90). These monaural cell groups reach their highest degree of development and differentiation in mammals that use audition as the main modality for orientation and identification of objects in their environment. For example, INLL and VNLL are especially prominent in echolocating bats and dolphins (Poljak, '26; Zook and Casseday, '82a; Covey and Casseday, '86; Zook et al., '88). Even in small animals such as cats and gerbils, the size of the monaural lateral lemniscal nuclei is comparable to or larger than that of the superior olivary complex, suggesting that they play an important role in audition in all mammals.

Figure 1 shows a frontal section through the lateral lemniscal nuclei in the echolocating bat *Eptesicus fuscus*. The bat is used as an example because the segregation of cell types can be seen very clearly. There are at least four separate cell groups embedded between the fibers of the lateral lemniscus: the dorsal nucleus (DNLL), the INLL, and two parts of the VNLL, the columnar division (VNLLc) and the multipolar cell division (VNLLm).

Each nucleus contains morphologically distinct cell types. For example, in *Eptesicus*, many cells in INLL are elongate, with their dendrites oriented orthogonal to the ascending fibers of the lateral lemniscus. Cells in the VNLLc are tightly packed in columns separated by fiber bundles; they are small and round in shape, with one thick dendrite that branches extensively at some distance from the cell body. They are very similar in appearance to spherical bushy cells in the anteroventral cochlear nucleus (AVCN). The VNLLm, as its name implies, contains mainly multipolar cells.

Although the cell types in the cat are not as well segregated, they appear to be essentially the same as those in the bat. Adams ('79), in his description of the region ventral to DNLL mentions multipolar cells, "horizontal cells" similar to the elongate cells in INLL of the bat, and small oval cells similar to those in VNLLc of the bat.

## Response properties of neurons in INLL and VNLL

To date, there have only been four published studies which provide any data on the response properties of neurons in the INLL and VNLL. Two of these are in the cat (Aitkin et al., '70; Guinan et al., '72a;b) and two are in echolocating bats (Metzner and Radtke-Schuller, '87; Covey and Casseday, '91). Although there are a few inconsistencies between the results reported in these studies, the data, on the whole, present a reasonably coherent picture of the response properties of neurons in the mammalian INLL and VNLL, and one that is consistent with the known anatomical connections.

**Monaurality.** Most available evidence indicates that the responses of neurons in the INLL and VNLL are monaural, driven by sound at the contralateral ear. Aitkin et al. ('70) first distinguished between the DNLL, where responses are binaural, and the region ventral to DNLL, where responses are nearly all monaural. This is consistent with the anatomical connections in the cat, where nearly all input from the cochlear nucleus to the INLL and VNLL is from the contralateral side (e.g., Warr, '66; '69; '82; Glendenning et al., '81; Covey and Casseday, '86).

Guinan et al. ('72a;b) described the responses of units that were encountered during experiments to record from the cat superior olivary complex. Approximately half of the units that they identified as being in the VNLL were binaural; however, it seems likely that these binaural units were either in the small medial region of VNLL that receives binaural projections in the cat (Glendenning et al., '81), or in the lateral tegmental (paralemniscal) area where the major input is from the superior olive (Glendenning et al., '81) and responses are binaural (Covey and Casseday, '91).

In the big brown bat, all units in VNLLc and VNLLm, and nearly all units in INLL are monaural, excited by sound at the contralateral ear. The few binaural units found in INLL are at marginal locations, mainly near the border with DNLL (Covey and Casseday, '91). These results are almost identical to those of Aitkin et al. ('70) in the cat, match the connectional data (Zook and Casseday, '85, Covey and Casseday, '86), and support the idea that the INLL and VNLL are almost exclusively monaural nuclei.

**Multiple tonotopic representations.** Aitkin et al. ('70) showed that DNLL of the cat contains a complete tonotopic representation, and that this tonotopy is clearly separate from that in the more ventral parts of the lateral lemniscal nuclei. However, neither this study nor that of Guinan et al. ('72a;b) addressed the question of whether there might be more than one tonotopically organized area ventral to DNLL. Multiple tonotopic representations have been seen in this region in both species of echolocating bat that have been studied. In the horseshoe

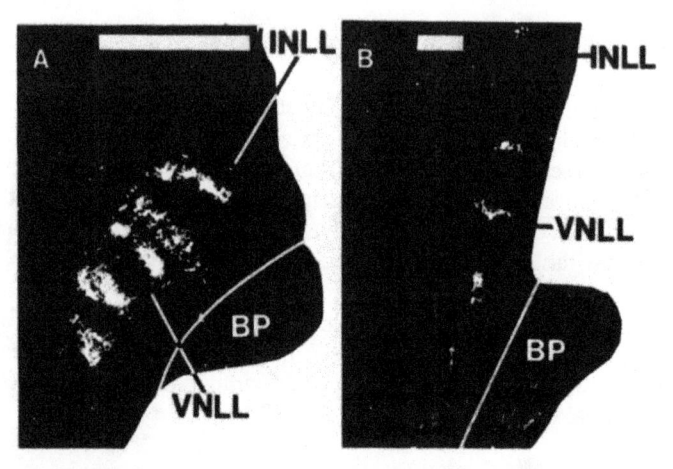

**Figure 2.** Dark-field photomicrographs to illustrate the banded patterns of anterograde transport that result from injections of WGA-HRP in the ventral cochlear nucleus of the big brown bat (A) and cat (B). Both are frontal sections through the right VNLL. BP = brachium pontis. Calibration bars for both sections are 500 μm.

bat, the tonotopic map in INLL is separate from that in VNLL (Metzner and Radtke-Schuller, '87). In the big brown bat there are at least three separate tonotopic representations in the region ventral to DNLL, corresponding to the three cytoarchitectonic divisions, INLL, VNLLc and VNLLm (Covey and Casseday, '91).

Injections of WGA-HRP in the cochlear nucleus of both cat and bat result in multiple patches of anterograde label in the region ventral to DNLL (Figure 2). In addition, injection of HRP in axons of cells in the cochlear nucleus of the rat shows that each axon provides several collaterals that terminate in the region below DNLL (Friauf and Ostwald, '88). Taken together, this evidence strongly suggests that the region below DNLL contains multiple tonotopic representations in all mammalian species, and that these originate, at least partly, from collateralization of single axons.

**Broad frequency tuning.** Frequency tuning in the INLL and VNLL appears to largely reflect the frequency tuning of neurons at lower levels. There is certainly no evidence for increased frequency selectivity at this level; in fact, there is some evidence that frequency tuning becomes broader, especially in VNLLc. Figure 3 compares the broad tuning curves of neurons in VNLLc with the V-shaped tuning curves of neurons in INLL, and with the "filter" and "closed" type tuning curves seen in the inferior colliculus of *Eptesicus*, where sharpening of frequency specificity is known to occur (Casseday and Covey, '91).

The results in the horseshoe bat and the cat are similar to those in *Eptesicus* in that they show no evidence for sharpening of tuning at the level of the INLL or VNLL. The tuning properties described in the INLL and VNLL of the horseshoe bat by Metzner and Radtke-Schuller ('87) indicate that frequency tuning in the nuclei of the lateral lemniscus is very similar to that seen at the level of the cochlear nucleus (Neuweiler and Vater, '77). Guinan et al. ('72b) reported that most units recorded in the cat VNLL had V-shaped tuning curves but that several had "wide" tuning curves, possibly similar to those seen in VNLLc of the big brown bat.

**Discharge patterns.** Aitkin et al ('70) first classified the responses of units in the lateral lemniscal nuclei of the cat as "sustained" or "onset", but did not give relative proportions. Guinan et al. ('72b) reported that over half of the units in VNLL of the cat were "choppers"; the remainder were mostly "phasic on", but also included a few "tonic" or

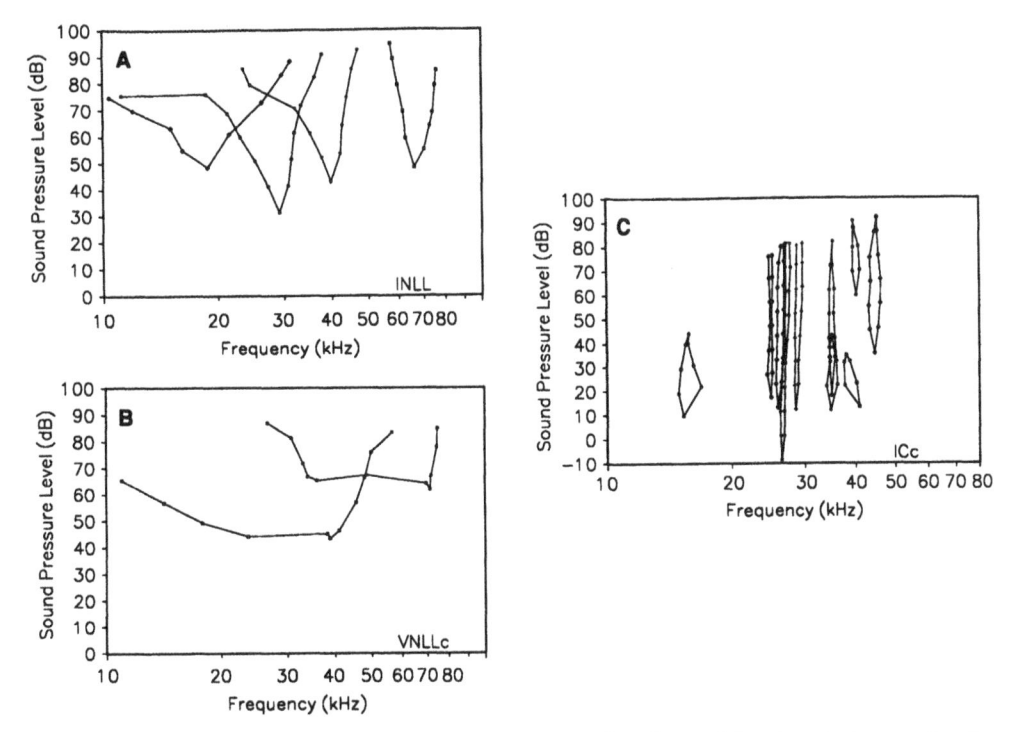

**Figure 3.** Neurons in INLL and VNLL are, on the average, more broadly tuned than those in the central nucleus of the inferior colliculus (ICc). (A) In INLL virtually all neurons have V-shaped tuning curves similar to the ones illustrated. The tuning curves of neurons in VNLLm are virtually identical to those in INLL. (B) In VNLLc, neurons are responsive to an extremely broad frequency range, beginning at only a few dB above BF threshold. (C) Narrow frequency selectivity of the type shown by many neurons in the ICc is never seen in the nuclei of the lateral lemniscus.

"phasic-tonic" responses. Figure 4 illustrates the most common discharge patterns of neurons in the INLL, VNLLc and VNLLm of the big brown bat and Figure 5 shows their distribution in each division (Covey and Casseday, '91). As in the cat, discharge patterns can be classified as sustained or onset. Based on regularity of firing, the sustained responses can further be classified as chopper (regular ISI), or tonic (irregular ISI). The tonic responses differ from primary-like in the absence of an initial transient and in a relatively constant firing rate throughout the tone burst. The phasic responses typically consist of a single spike, the latency of which is correlated with the onset of the stimulus. They can be further classified on the basis of the variability in the latency of this spike as "phasic variable latency" or "phasic constant latency" responders. Virtually all neurons in VNLLc are phasic constant latency responders, that is, the standard deviation in the latency of the first spike is less than 0.1 msec. Phasic constant latency responses are discussed in more detail below. In INLL and VNLLm, responses are about equally divided between chopper, tonic and phasic variable latency.

As will be discussed below, it seems unlikely that any of these response patterns is strictly the result of input from a single cell type in the cochlear nucleus, but rather that they result from convergence of multiple direct and indirect inputs.

**Latencies.** Only one study reports latencies for neurons in the monaural lateral lemniscal nuclei (Covey and Casseday, '91). In the big brown bat, latencies in INLL range between 2.7-11.1 msec (mean 5.1); in VNLLc between 1.7-5.6 msec (mean 3.3); in VNLLm

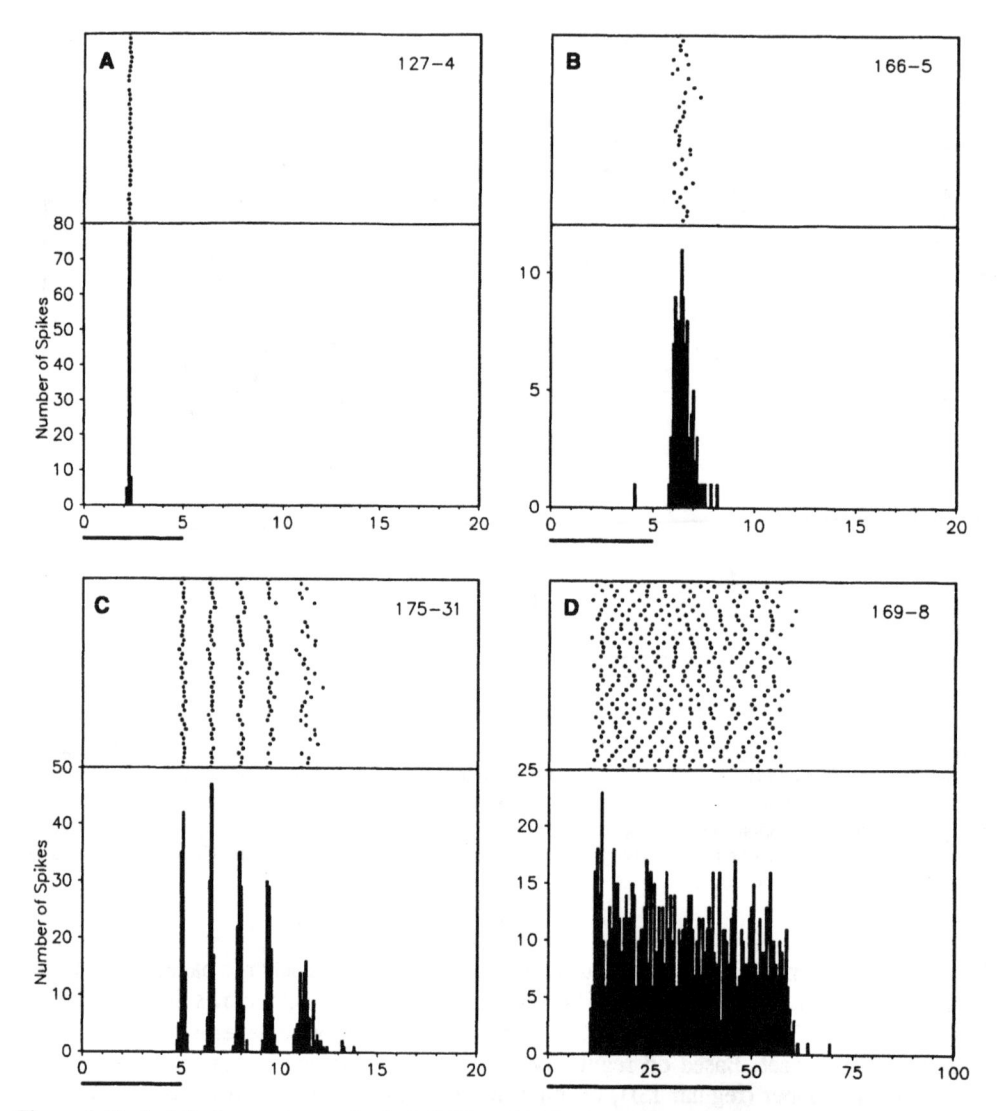

**Figure 4.** Typical discharge patterns of units in INLL and VNLL in response to pure tones at BF, 20 dB above threshold. Above each poststimulus time histogram (PSTH) is a dot raster showing 40 of the 100 responses that make up the PSTH in the panel below. (A) Phasic, constant latency neuron in VNLLc; (B) phasic variable-latency neuron in INLL; (C) chopper in VNLLm; (D) tonic response of a neuron in INLL. The bar beneath the PSTH indicates the stimulus duration (adapted from Covey and Casseday, 1991).

between 3.3-6.3 msec (mean 4.4). The relationship between latency and stimulus frequency or amplitude is discussed below.

The short latencies of neurons in VNLLc are consistent with the fact that their cell bodies are contacted by large calyx-like endings from axons originating in the cochlear nucleus (Figure 6; Covey and Casseday, '86; Zook and Casseday, '85).

**Properties that show suitability for encoding temporal events.** There are certain properties shared by all neurons in the INLL, VNLLc and VNLLm of the big brown bat that make them ideally suited to transmit information about the timing of auditory events (Covey and Casseday, '91). First, neurons in these nuclei have little or no spontaneous activity. Thus,

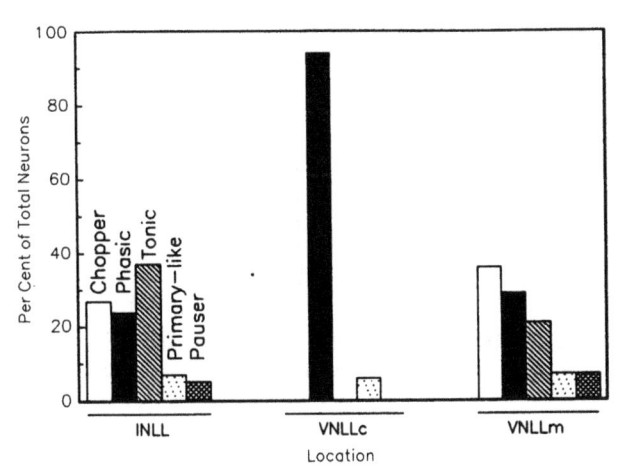

**Figure 5.** Distributions of response types in INLL, VNLLc and VNLLm. Discharge patterns in INLL and VNLLm are approximately evenly divided between chopper, tonic, and phasic variable latency responses. Virtually all units in VNLLc are phasic constant latency responders (from Covey and Casseday, 1991).

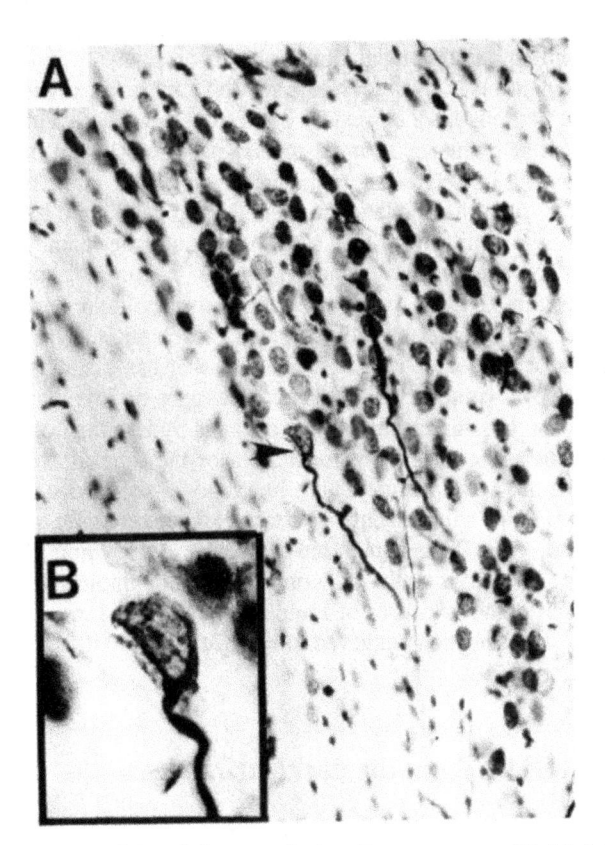

**Figure 6.** Calyceal ending in VNLLc of the mustache bat, *Pteronotus parnellii*, labeled with HRP from an injection in the ventral cochlear nucleus. (Photomicrograph by J. Zook).

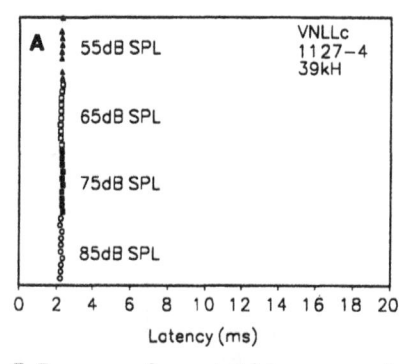

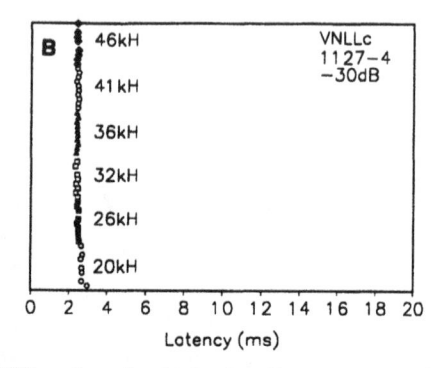

**Figure 7.** Responses of a constant latency neuron in VNLLc when stimulus level or frequency are varied. (A) Dot raster diagram to show latency of responses when frequency is held constant but sound pressure level is varied. (B) Dot raster diagram to show latency of responses when sound pressure level is held constant but frequency is varied.

when one or more spikes occur, it is an unambiguous signal that an auditory stimulus is present.

Second, nearly all neurons in these structures have very short integration times; that is, they respond robustly to stimuli a millisecond or less in duration. This means that they should be capable of responding to transient auditory events such as a frequency modulated (FM) sound sweeping rapidly through their range of frequency sensitivity or that they should be able to follow rapid amplitude modulations.

Third, none of the neurons in INLL, VNLLc or VNLLm are narrowly tuned to frequency; in fact many VNLLc neurons have extremely broad tuning curves. This suggests that there may be integration across frequency in order to increase precision in the temporal domain.

Finally, all neurons in VNLLc and some neurons in INLL and VNLLm exhibit latency constancy. This means, first, that the latency of the first spike elicited by the onset of a sound has an extremely low variability from trial to trial; second, the latency of the first spike is virtually independent of stimulus level and frequency. Figure 7 shows an example of a constant latency neuron in VNLLc. At a few dB above threshold the neuron consistently fires one spike in response to the onset of each stimulus presentation. The latency of this spike under constant conditions (10 dB above threshold, at BF) has a standard deviation of 0.03 msec. When stimulus level is increased, the latency changes by less than 1 msec over 30 dB. Similarly, when frequency is varied, the latency changes by less than one msec over 30 kHz. Such a neuron provides an extremely precise marker of the time of onset of a sound, either for the onset of a tone at any frequency within its range of sensitivity, or for the time when a frequency modulated (FM) stimulus enters its range of sensitivity (Suga, '70; Pollak et al., '77; Bodenhamer et al., '79; Bodenhamer & Pollak, '81; Covey and Casseday, '91).

Phasic neurons in VNLLc are not the only ones that exhibit latency constancy. These properties are also seen in the responses of some neurons in INLL and VNLLm, particularly choppers. The types of information conveyed to the ICc by these different classes of neurons are considered in detail below.

## Patterns of projections from the cochlear nucleus

Anatomical evidence from lesions, anterograde transport and retrograde transport indicates that the INLL and VNLL are major targets of the ventral cochlear nucleus, and that this projection is almost entirely contralateral (e.g., Warr, '66; '69; '82; Glendenning et al., '81; Zook and Casseday, '85). Further, there is evidence that the projection to each target

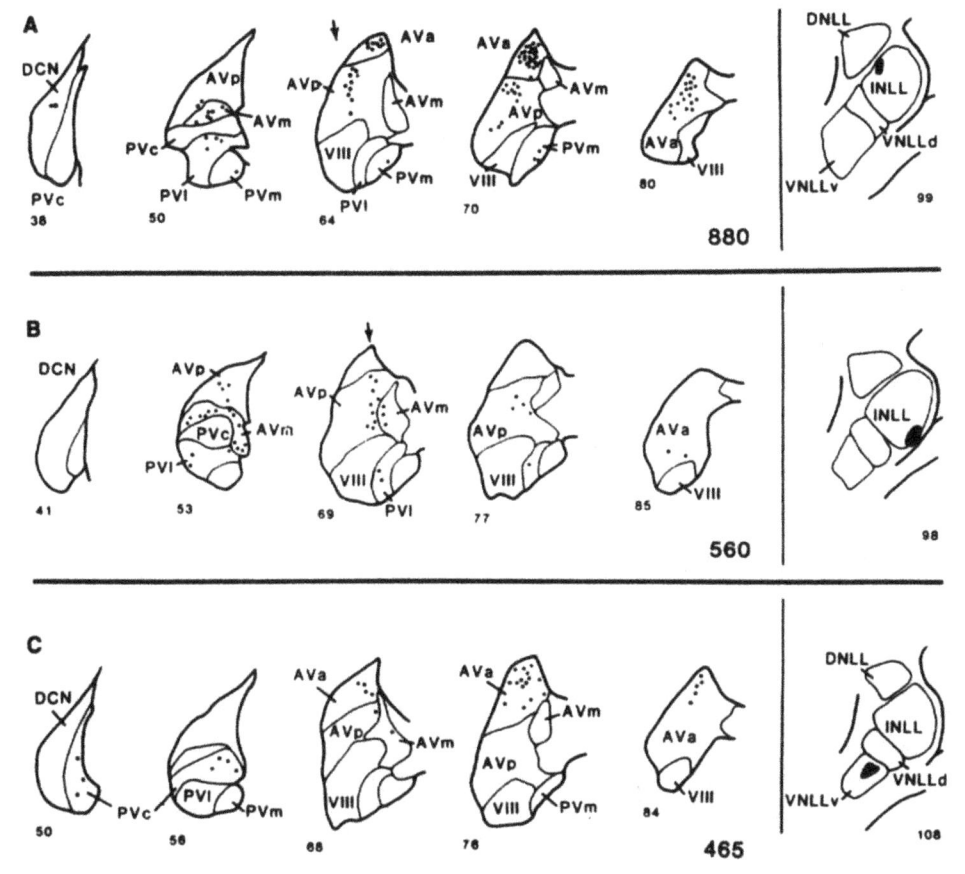

Figure 8. Patterns of input from the ventral cochlear nucleus to INLL and VNLL in the mustache bat, *Pteronotus parnellii* as seen by retrograde transport from small HRP injections. In all three cases the labeled cells are in the contralateral cochlear nucleus. (A) and (B) show injections in INLL; there are many labeled cells throughout AVCN, and a few in PVCN. (C) shows an injection in the columnar division of VNLL (VNLLv). There are labeled cells in both AVCN and PVCN. DCN: dorsal cochlear nucleus; PVc: central part of PVCN; PVl: lateral part of PVCN; PVm: medial part of PVCN; AVp: posterior part of AVCN; AVm: medial part of AVCN; AVa: anterior part of AVCN; VIII: auditory nerve; VNLLd: multipolar cell division of VNLL; VNLLv: columnar division of VNLL. From Zook and Casseday (1985).

structure probably originates in more than one division of the ventral cochlear nucleus, with convergence from AVCN and PVCN, possibly from several different cell types with different response properties (Zook and Casseday, '85; Covey and Casseday, '86; Friauf and Ostwald, '88). In addition, INLL and VNLL receive indirect projections from the ventral cochlear nucleus via the MNTB and periolivary nuclei (Glendenning et al., '81; Zook and Casseday, '85).

Injections of retrograde tracers in INLL and VNLL result in extensive labeling of cells in the ventral cochlear nucleus (Glendenning et al., '81; Zook and Casseday '85). After HRP injections in INLL of the cat, a little over half of the labeled cells are found in the contralateral cochlear nucleus, with the remainder mainly in the ipsilateral MNTB. In the cochlear nucleus about 3 times as many labeled cells are in AVCN as in PVCN. After HRP injections in VNLL of the cat, 90% of all labeled cells are in the contralateral cochlear nucleus, with the remaining 10% in the ipsilateral MNTB and ipsilateral periolivary nuclei. In the cochlear nucleus, both AVCN and PVCN contain labeled cells, with about twice as many in AVCN as in PVCN (Glendenning et al., '81).

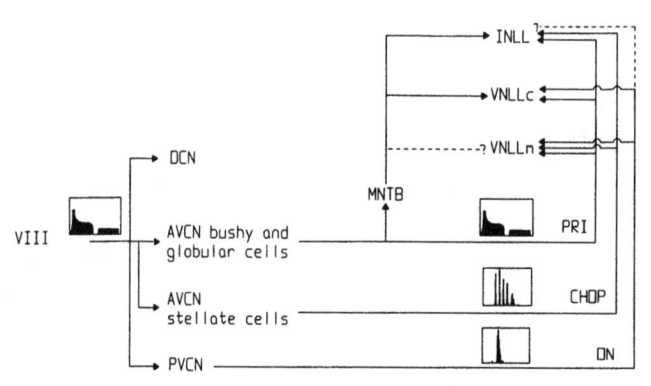

**Figure 9.** Schematic diagram to summarize the projections to INLL, VNLLc and VNLLm from different cell types in the ventral cochlear nucleus.

The results in the bat are very similar (Fig. 8); HRP injections in INLL result in many labeled cells in AVCN and MNTB and a few in PVCN. Injections in VNLLc and VNLLm result in many labeled cells in AVCN and PVCN and some in MNTB (Zook and Casseday '85; Covey and Casseday, unpublished).

These results suggest that AVCN projects heavily throughout the entire region of INLL and VNLL, but that the MNTB and PVCN projections are distributed in a graded fashion with the most dense MNTB projection dorsally, in INLL, and the most dense PVCN projection ventrally, in VNLLm. This arrangement is shown diagrammatically in Fig. 9.

There are two sources of information about what cell types project from the cochlear nucleus to the nuclei of the lateral lemniscus. First, there is the retrograde labeling of neurons after HRP injections in the lateral lemniscal nuclei. A second method is HRP labeling of physiologically characterized axons originating from neurons in the cochlear nucleus and the tracing of their patterns of termination. Although there have been a number of such studies in the cat (e.g., Spirou et al., '90; Smith et al., '91) it has generally been reported that the label in the axons fades before they reach the VNLL.

In the rat, however, Friauf and Ostwald ('88) were consistently able to see labeled terminals in the VNLL and INLL after injection of HRP in axons originating in AVCN or PVCN. They showed that the VNLL received collaterals from every one of the 26 axons that they injected, and that the projections were all contralateral, with no evidence of any ipsilateral projections. They concluded from this that the projections from the ventral cochlear nucleus to the contralateral VNLL represent a major efferent pathway of CN neurons.

AVCN is the major source of input to both INLL and VNLL; projections originate from spherical and globular bushy cells, and possibly also from stellate cells. The discharge patterns of globular and spherical bushy cells in AVCN are typically primary-like or primary-like with notch (e.g., Rouiller and Ryugo, '84; Smith and Rhode, '87; Spirou et al., '90; Smith et al., '91). These cells typically have low spontaneous activity (Friauf and Ostwald, '88; Smith et al., '91), low standard deviations in the time of occurrence of the first spike (Spirou et al., '90), and narrow V-shaped tuning curves (Friauf and Ostwald, '88). For globular cells with low BF, phase-locking is even better than in the auditory nerve (Smith et al., '91). In the rat and cat, the axons of globular cells ascend lateral to the nuclei of the VNLL where each axon gives rise to several horizontal collaterals with perisomatic non-calyceal endings (Friauf and Ostwald, '88; Smith et al., '91).

Stellate cells in AVCN are choppers (Rhode et al., '83; Rouiller and Ryugo, '84; Wu and Oertel, '84; Smith and Rhode, '90; Spirou et al., '90). The standard deviations of the first spikes are as small as those of globular cells (Smith and Rhode, '90; Spirou et al., '90). Although there are no published data regarding their patterns of axonal arborization in the lateral lemniscal nuclei, the fact that they are labeled after HRP injections in INLL and VNLL suggests that they also provide input to these nuclei.

PVCN projects heavily to VNLL and sparsely to INLL. These projections appear to originate from both stellate and octopus cells. Multipolar or stellate cells in PVCN of the cat are reported to be choppers (Smith and Rhode, '89). In the rat, they respond with a phasic "on" discharge pattern, have low spontaneous activity and symmetrical V-shaped tuning curves. Their axons ascend lateral to or within the VNLL and give rise to long, vertically extending collaterals. (Friauf and Ostwald, '88). The same axons terminate in calyx-like endings in a region that Friauf and Ostwald ('88) term "rostral periolivary nucleus", but which could possibly correspond to ventral VNLL in other species. In the remainder of VNLL the axons of PVCN multipolar cells terminate in boutons distributed along a vertical plane. This vertical pattern of termination suggests that the projection from the multipolar cells of PVCN is non-tonotopic.

Octopus cells in PVCN are reported to be phasic "on" responders in the cat (Godfrey et al., '75; Rhode et al., '83; Rouiller and Ryugo, '84). In the rat, however, single axons tentatively identified as coming from octopus cells in PVCN respond with a primarylike discharge pattern, have a high spontaneous firing rate, and very wide tuning curves. These axons ascend lateral to the VNLL and provide 4-5 horizontal collaterals that branch extensively and terminate in bead-like boutons (Friauf and Ostwald, '88).

## Selective enhancement of temporal information in INLL and VNLL

Both connectional and physiological evidence suggest that INLL and VNLL receive convergent parallel inputs from different cell populations in the cochlear nucleus, and that these inputs are then transformed in a way that not only preserves the temporal pattern of sounds as encoded in the responses of auditory nerve fibers, but selectively enhances certain features of this pattern.

Like auditory nerve fibers and certain neurons in the ventral cochlear nucleus, neurons throughout INLL and VNLL respond robustly to transient stimuli of short duration. This means that they can follow very rapid frequency or amplitude modulations. It has been shown that low BF globular cells in the cochlear nucleus are more precise phase lockers than auditory nerve fibers (Smith et al., '91); it therefore seems likely that further enhancement of phase-locking or locking to rapid amplitude modulations could occur at the level of INLL and VNLL. Most neurons in the INLL and VNLL have little or no spontaneous activity. Although the specific mechanisms responsible for this improvement in signal-to-noise ratio remain to be discovered, it is clearly an important step in producing an unambiguous timing signal for the occurrence of a specific sound.

**Two streams of temporal processing.** In general, the neurons in INLL and VNLL can be divided into two broad populations (Fig. 10). The first population of neurons signal very precisely the time of onset of a sound; the second population of neurons respond continuously throughout the time a sound is present and thus can transmit information about its intensity and duration. At least in the bat, these two streams of processing can clearly be distinguished from one another on the basis of their neuronal morphology, the type of terminals supplied by axons originating in the ventral cochlear nucleus, breadth of frequency tuning, and possibly the subsets of cells that provide their input. In addition, they differ in their patterns of termination in the ICc.

The population of onset encoders include virtually all of the neurons in VNLLc. These are small, round cells that resemble bushy cells in AVCN. All of the inputs from axons originating in the cochlear nucleus appear to terminate on the cell body, some as large, calyx-like endings. In addition, there appear to be glycine-positive terminals on the cell body, possibly originating in MNTB. These cells have shorter response latencies than others in

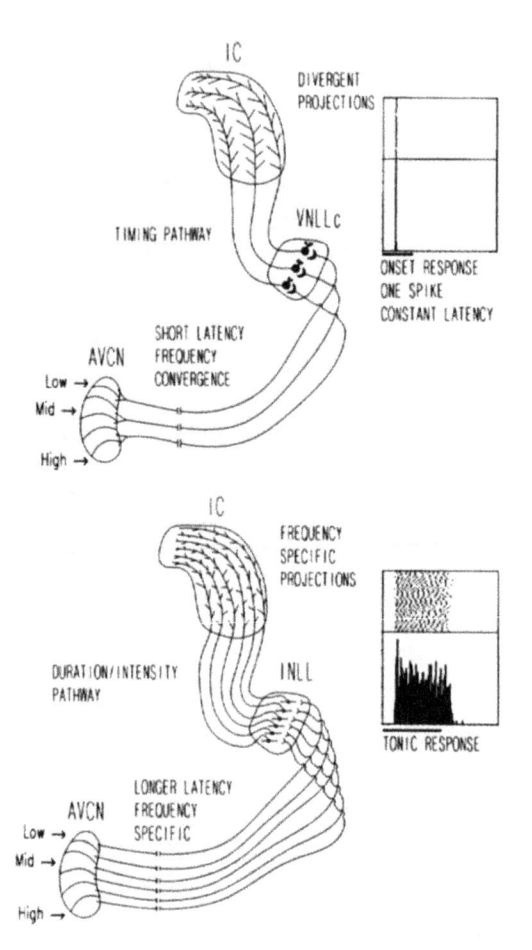

**Figure 10.** Schematic diagram to illustrate the hypothesis that there are two different streams of processing within the monaural pathways, with each encoding different information about the temporal structure of sounds.

INLL or VNLL. They are very broadly tuned to frequency, and have an extremely low variability in the timing of the single onset spike. Furthermore, the timing of this spike is nearly independent of changes in frequency or sound pressure level. These neurons provide a very precise all-or-none signal of the time of onset of a sound anywhere within their range of sensitivity. They provide heavy and widespread projections to the ICc.

The population of intensity/duration encoders include the choppers and tonic units in INLL and VNLLm. These units probably vary somewhat in their morphology, but all have multiple large dendrites that are contacted by punctate terminals of axons originating in the ventral cochlear nucleus. They respond with latencies that are one to several msec longer than those of VNLLc neurons. Many of the tonic units have a relatively wide dynamic range over which firing rate changes in response to sound pressure level. Both tonic and chopper units respond with little adaptation throughout the entire time a stimulus is present. Their responses would therefore provide a signal either of stimulus duration for a tone, or for an FM signal, the dwell time within their range of frequency sensitivity.

Although the monaural nuclei of the lateral lemniscus have until now been largely neglected by auditory researchers, there is finally enough information available so that it is possible to formulate hypotheses such as those presented here regarding their function and organization. It is to be hoped that future studies will test these and other hypotheses, and lead to a complete understanding of the functional role of these important structures in hearing.

## ACKNOWLEDGEMENTS

I thank B. Rosemond, C. Vance and S. Keros for their technical help in preparing the manuscript, and Dr. G. Neuweiler for his critical reading and helpful comments. Research supported by NIH grants DC-607 and DC-287.

## REFERENCES

Adams, J.C., 1979, Ascending projections to the inferior colliculus, *J. Comp. Neurol.*, 183:519-538.
Aitkin, L.M., Anderson, D.J., and Brugge, J.F., 1970, Tonotopic organization and discharge characteristics of single neurons in nuclei of the lateral lemniscus of the cat, *J. Neurophysiol.*, 33:421-440.
Bodenhamer, R.D., and Pollak, G.D., 1981, Time and frequency domain processing in the inferior colliculus of echolocating bats, *Hearing Res.*, 5:317-335.
Bodenhamer, R.D., Pollak, G.D., and Marsh, D.S., 1979, Coding of fine frequency information by echoranging neurons in the inferior colliculus of the Mexican free-tailed bat, *Brain Res.*, 171:530-535.
Brunso-Bechtold, J.K., Thompson, G.C., and Masterton, R.B., 1981, HRP study of the organization of auditory afferents ascending to central nucleus of inferior colliculus in cat, *J. Comp. Neurol.*, 197:705-722.
Cajal, S.R. y., 1909, Histologie du Systeme Nerveux de l'Homme et des Vertebres. Tome I., Instituto Ramon y Cajal, 1952, Madrid, pp. 774-848.
Casseday, J.H., Covey E., 1991, Frequency tuning properties of neurons in the inferior colliculus of an FM bat, *J. Comp. Neurol.*, 319:34-50.
Covey, E., and Casseday, J.H., 1986, Connectional basis for frequency representation in the nuclei of the lateral lemniscus of the bat, *Eptesicus fuscus*, *J. Neurosci.*, 6:2926-2940.
Covey, E., and Casseday, J.H., 1991, The monaural nuclei of the lateral lemniscus in an echolocating bat: Parallel pathways for analyzing temporal features of sound, *J. Neurosci.*, 11:3456-3470.
Friauf, E., and Ostwald, J., 1988, Divergent projections of physiologically characterized rat ventral cochlear nucleus neurons as shown by intra-axonal injection of horseradish peroxidase, *Exp. Brain. Res.*, 73:263-284.
Glendenning, K.K., Brunso-Bechtold, J.K., Thompson, G.C., and Masterton, R.B., 1981, Ascending auditory afferents to the nuclei of the lateral lemniscus, *J. Comp. Neurol.*, 197:673-704.
Godfrey, D.A., Kiang, N.Y.S., and Norris, B.A., 1975, Single unit activity in the posteroventral cochlear nucleus of the cat, *J. Comp. Neurol.*, 162:247-268.
Guinan, J.J., Guinan, S.S., and Norris, B.E., 1972a, Single auditory units in the superior olivary complex. I: Responses to sounds and classifications based on physiological properties, *Int. J. Neurosci.*, 4:101-120.
Guinan, J.J., Norris, B.E., and Guinan, S.S., 1972b, Single auditory units in the superior olivary complex. II: Locations of unit categories and tonotopic organization, *Int. J. Neurosci.*, 4:147-166.
Metzner, W., and Radtke-Schuller, S., 1987, The nuclei of the lateral lemniscus in the rufous horseshoe bat, *Rhinolophus rouxi*, *J. Comp. Physiol.*, 160:395-411.
Neuweiler, G., and Vater, M., 1977, Response patterns to pure tones of cochlear nucleus units in the bat, *Rhinolophus ferrumequinum*, *J. Comp. Physiol.*, 115:119-133.
Poljak, S., 1926, Untersuchungen am Oktavussystem der Saugetiere und an den mit diesem koordinierten motorischen Apparaten des Hirnstammes, *J. Psychol. Neurol.*, 32:170-231.
Pollak, G., Marsh, D., Bodenhamer, R., and Souther, A., 1977, Echo-detecting characteristics of neurons in inferior colliculus of unanesthetized bats, *Science*, 196:675-678.
Rhode, W.S., Oertel, D., and Smith, P.H., 1983, Physiological response properties of cells labeled intracellularly with horseradish peroxidase in cat ventral cochlear nucleus, *J. Comp. Neurol.*, 213:448-463.
Roth, G.L., Aitkin, L.M., Anderson, R.A., and Merzenich, M.M., 1978, Some features of the spatial organization of the central nucleus of the inferior colliculus of the cat, *J. Comp. Neurol.*, 182:661-680.
Rouiller, E.M., and Ryugo, D.K., 1984, Intracellular marking of physiologically characterized cells in the ventral cochlear nucleus of the cat, *J. Comp. Neurol.*, 225:167-186.
Saint Marie, R.L., and Baker, R.A., 1990, Neurotransmitter- specific uptake and retrograde transport of [3H]⁻ glycine from the inferior colliculus by ipsilateral projections of the superior olivary complex and nuclei of the lateral lemniscus, *Brain Res.*, 524:244-253.
Schweizer, H., 1981, The connections of the inferior colliculus and the organization of the brainstem auditory system in the greater horseshoe bat (*Rhinolophus ferrumequinum*), *J. Comp. Neurol.*, 201:25-49.

Smith, P.H., Joris, P.X., Carney, L.H., and Yin, T.C.T., 1991, Projections of physiologically characterized globular bushy cell axons from the cochlear nucleus of the cat, *J. Comp. Neurol.*, 304:387-407.

Smith, P.H., and Rhode, W.S., 1987, Characterization of HRP-labeled globular bushy cells in the cat anteroventral cochlear nucleus, *J. Comp. Neurol.*, 266:360-375.

Smith, P.H., and Rhode, W.S., 1989, Structural and functional properties distinguish two types of multipolar cells in the ventral cochlear nucleus, *J. Comp. Neurol.*, 282:595-616.

Spirou, G.A., Brownell, W.E., and Zidanic, M., 1990, Recordings from cat trapezoid body and HRP labeling of globular bushy axons, *J. Neurophysiol.*, 63:1169-1190.

Suga, N., 1970, Echo-ranging neurons in the inferior colliculus of bats, *Science*, 170:449-452.

Warr, W.B., 1966, Fiber degeneration following lesions in the anterior ventral cochlear nucleus of the cat, *Exp. Neurol.*, 14: 453-474.

Warr, W.B., 1969, Fiber degeneration following lesions in the posteroventral cochlear nucleus of the cat, *Exp. Neurol.*, 23: 140-155.

Warr W.B., 1982, Parallel ascending pathways from the cochlear nucleus: Neuroanatomical evidence of functional specialization, *in*: "Contributions to Sensory Physiology", Vol. 7, D. W. Neff, ed., Academic, New York.

Whitley, J., and Henkel, C., 1984, Topographical organization of the inferior collicular projection and other connections of the ventral nucleus of the lateral lemniscus in the cat, *J. Comp. Neurol.*, 229:257-270.

Wu, S.H., and Oertel, D., 1984, Intracellular injection with horseradish peroxidase of physiologically characterized stellate and bushy cells in slices of mouse anteroventral cochlear nucleus, *J. Neurosci.*, 4:1577-1588.

Zook, J.M., and Casseday, J.H., 1982a, Cytoarchitecture of auditory system in lower brainstem of the mustache bat, *Pteronotus parnellii*, *J. Comp. Neurol.*, 207:1-13.

Zook, J.M., and Casseday, J.H., 1982b, Origin of ascending projections to inferior colliculus in the mustache bat, *Pteronotus parnellii*, *J. Comp. Neurol.*, 207:14-28.

Zook, J.M., and Casseday, J.H., 1985, Projections from the cochlear nuclei in the mustache bat, *Pteronotus parnellii*, *J. Comp. Neurol.*, 237:307-324.

Zook, J.M., and Casseday, J.H., 1987, Convergence of ascending pathways at the inferior colliculus in the mustache bat, *Pteronotus parnellii*, *J. Comp. Neurol.*, 261:347-361.

Zook, J.M., Jacobs, M.S., Glezer, I., and Morgane, P.J., 1988, Some comparative aspects of auditory brainstem cytoarchitecture in echolocating mammals: Speculations on the morphological basis of time-domain processing, *in*: "Animal Sonar: Processes and Performance," P. E. Nachtigall, and P. W. B. Moore, ed., Plenum, New York.

# ASCENDING PROJECTIONS FROM THE COCHLEAR NUCLEUS TO THE INFERIOR COLLICULUS AND THEIR INTERACTIONS WITH PROJECTIONS FROM THE SUPERIOR OLIVARY COMPLEX

Douglas L. Oliver and Gretchen E. Beckius

Department of Anatomy, University of Connecticut Health Center
Farmington, CT 06030, USA

The cochlear nucleus represents the beginning of several pathways with distinct functional properties within the central auditory system. By far, the majority of ascending pathways eventually terminate in the inferior colliculus (IC). Projections to the IC originate from fusiform cells in the dorsal cochlear nucleus (DCN) and from stellate cells in the ventral cochlear nucleus (for recent reviews see Oliver and Shneiderman, '91; Oliver and Huerta, '92; Morest, et al., this volume). This direct projection may carry monaural information to the midbrain. In contrast, the pathways that synapse in the superior olivary complex (SOC) may transmit binaural information to the IC. The projections from the bushy cells of the ventral cochlear nucleus to the SOC represent the beginning of this indirect pathway (Casseday et al., this volume; Morest et al., this volume; Smith, this volume). In turn, the principal nuclei of the SOC, the medial (MSO) and lateral superior olive (LSO), each project to the IC (Oliver and Shneiderman, '91; Oliver and Huerta, '92). Both the pathways from the cochlear nucleus and the SOC terminate in the IC in a similar tonotopically organized fashion. Yet, we still know very little about how the efferent projections of the cochlear nucleus interact with other projections at the level of the auditory midbrain.

In this chapter, we will compare the morphology of the projections from the cochlear nucleus to those from the superior olive. Three types of experiments will be presented. In the first, we will demonstrate the axonal morphology of the cochlear nucleus efferents with a new anterograde tracer, dextran. This will allow direct light microscopic comparisons of the axonal endings from the cochlear nucleus as they terminate in the SOC, nuclei of the lateral lemniscus, and IC. In the second experiment, we will evaluate the synaptic morphology of axonal endings in the IC. The axonal endings from the DCN and the anteroventral cochlear nucleus (AVCN) have been identified by anterograde axonal transport of tritiated amino acids and electron microscopic (EM) autoradiography (Oliver, '84, '85, '87). Now, we can compare the synaptic morphology of the axons from the cochlear nucleus to axonal endings from the LSO also labeled with EM autoradiographic methods. The third experiment will address the issue of where the cochlear nucleus efferents terminate in the IC relative to the inputs from the LSO and MSO. Two different anterograde axonal markers are injected into two different sites in the same animal. This allows the axonal arbors of the two inputs to be mapped relative to each other.

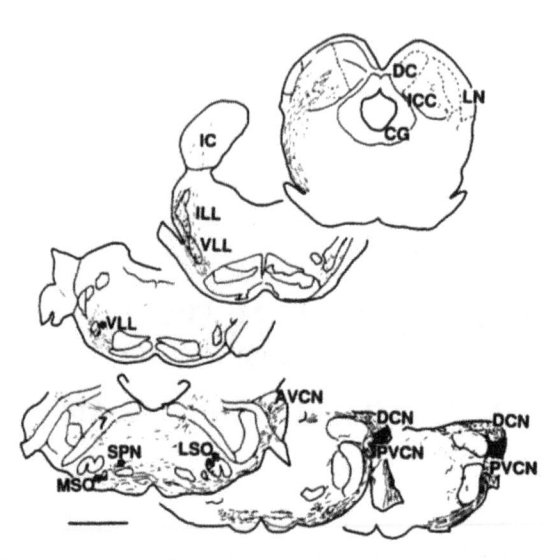

Figure 1. Injection of dextran into the cochlear nucleus and resulting transport in the rat. Transverse sections are serially arranged with caudalmost at lower right. Abbreviations: AVCN, anteroventral cochlear nucleus; CG, central gray; DCN, dorsal cochlear nucleus; DC, dorsal cortex; IC, inferior colliculus; ICC, central nucleus of the inferior colliculus; ILL, intermediate nucleus of the lateral lemniscus; LN, lateral nucleus of the inferior colliculus; LSO, lateral superior olive; MSO, medial superior olive; PVCN, posteroventral cochlear nucleus; SPN, superior paraolivary nucleus; VLL, ventral nucleus of the lateral lemniscus; 7, facial nerve root. Scale = 2 mm.

## Experiment one: Anterograde transport of dextrans from the cochlear nucleus

In this experiment, we will illustrate the types of axonal endings made by the cells of the cochlear nucleus as they terminate in the SOC, the nuclei of the lateral lemniscus, and the IC. Since different cell types in the cochlear nucleus may project to different targets, this methodology may allow us to relate the morphology of the terminal portion of the axon to the parent cell types. Solutions of 10% dextran (Molecular Probes) are dissolved in saline and injected into the cochlear nucleus through a beveled glass micropipette with a 1-20 $\mu$m diameter tip. The best results have been obtained with a 10% solution of dextran conjugated with biotin (D-1956), although this is usually mixed with 10% fluorescent dextran (D-1817 or D-1820) to make a total solution of 20% dextran. Since the transport rate of dextran is moderately slow, around 2-2.4 mm per day, the animal is allowed to survive 6 days before it is perfused with 2.5% paraformaldehyde buffered in phosphate (0.1 M, Ph 7.4). To reveal the biotin-labeled dextran, an avidin-biotin reaction (Vector Labs) is run overnight in the presence of Triton X-100 and followed by a histochemical reaction for horseradish peroxidase (HRP) (Adams, '81). This histochemical reaction is similar to that developed to reveal Lucifer Yellow coupled to biotin in intracellular injections (Hill and Oliver, '91).

Figure 1 shows an injection of dextran in the cochlear nucleus of the rat. Although the injection is centered caudally, it involves the DCN, the rostral posteroventral cochlear nucleus (PVCN), and probably the caudal AVCN. Labeled axons emerge from the DCN into the dorsal acoustic stria, then they travel across the reticular formation and turn into the medial part of the contralateral lateral lemniscus without giving off terminals in the SOC.

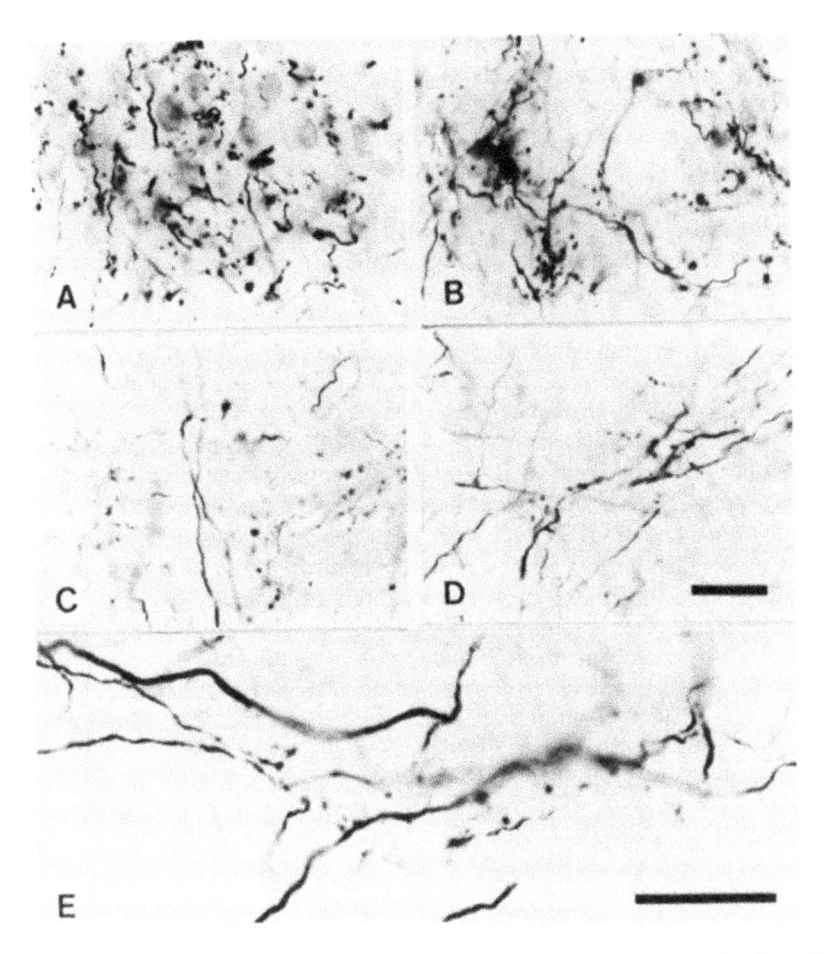

**Figure 2.** Photomicrographs of dextran labeling in the lateral superior olive (A), the ventral nucleus of the lateral lemniscus (B), the intermediate nucleus of the lateral lemniscus (C), and the central nucleus of the inferior colliculus (D, E). Case 91-14 from Fig. 1. Scale bars = 100 μm.

Fibers from ventral CN emerge into the lateral part of the trapezoid body. These branch to innervate the LSO and MSO ipsilateral to the injection. Some fibers continue across the trapezoid body and terminate in the contralateral MSO and superior periolivary nucleus (SPN). Based upon the location of the terminals in the medial limb of the LSO, it is likely that the injection was centered primarily in the high frequency parts of the ventral cochlear nucleus.

The most intriguing part of this case is the morphology of the axonal endings in the various targets of the cochlear nucleus. For example, the axonal branches in the ipsilateral LSO are prominent and form frequent medium and large swellings and terminal boutons (Fig. 2A). Presumably, these endings are from spherical bushy cells in the AVCN and PVCN. These endings form a dense neuropil near the cell bodies stained with neutral red. The few fibers that terminate in the ipsilateral MSO are similar. In contrast, the fibers in the contralateral SOC are found mostly in the SPN and are small and sparse by comparison.

These are thinner and have small numbers of small endings in this case. Although thick fibers are seen to enter the contralateral MNTB, no labeled calyx-type endings are seen in the sections processed for immunocytochemistry.

The fibers from the trapezoid body ascend in the lateral part of the lateral lemniscus (Fig. 1). Fibers emerge from the lateral part of the lateral lemniscus to terminate in the ventral nucleus of the lateral lemniscus (VLL). Once again, the fibers are characterized by large terminal boutons and swellings (Fig. 2B). Each fiber branches many times and gives off many endings. These large fibers and their endings may correspond to the endbulb-like endings seen in other species (Adams, '78; see also Covey, this volume). In addition, fine fibers are seen. The fibers of several sizes could represent inputs from different cell types in the cochlear nucleus. We can only speculate as to the cell types that give rise to these fibers. They could include octopus cells, stellate cells, or bushy cells (Adams, '78; Cant, '90; Covey, this volume; Smith, this volume).

As one ascends to the more dorsal parts of the lateral lemniscus, thin fibers predominate. In the intermediate (ILL, Fig. 1) and dorsal nuclei of the lateral lemniscus, thinner axons emerge at right angles to the main lemniscal axons and give rise to smaller and more delicate axonal endings (Fig. 2C). Once again, we do not know the cell types of origin. However, it is likely that many of the cell types that project to the IC also project to the ILL.

At the level of the IC, the fibers enter the central nucleus (ICC, Fig. 1) where they form several bands. Two prominent bands are seen that run from ventrolateral to dorsomedial. In this case, the fibers do not appear to continue as far dorsal as the dorsal cortex of the IC (DC). Fibers in the central nucleus include both thick and thin axons, and both give off side branches with clusters of terminals (Fig. 2D, 2E). These axons run for some length along the bands. It seems likely that a single axon may innervate more than one neuron in a single fibro-dendritic lamina. Thus, the innervation from the cochlear nucleus may be distributed to groups of neurons within a similar frequency range. The sources of these axons include the fusiform and giant cells in the DCN and the stellate cells of the AVCN and PVCN (Oliver, '84; Oliver '87). Although many large axons emerge from the DCN, it is not certain precisely which cells give rise to the thick or thin axons in the IC.

Additional fibers are found laterally in a region that has been called the external cortex of the IC (Faye-Lund and Osen, '85). Some terminals are seen deep to the largest laterally placed cells. Once again, the precise source of these axons from the cochlear nucleus is not certain. Although the banding pattern in this region is not as obvious as it is more medially, the appearance of the projection to this region is not altogether different from the projections of the cochlear nucleus to the lateral part of the central nucleus seen in the cat (Oliver, '87).

In summary, this experiment shows that the projections of the cochlear nucleus in the rat are similar to those reported for the cat. Moreover, the morphology of axonal endings from the cochlear nucleus may be different in each of its target nuclei in the brainstem. The transport of dextrans yields a high resolution picture of the axonal endings and may be useful in identifying the innervation patterns attributed to different cell types in the cochlear nucleus. Future experiments are required to refine these observations.

The transport of dextrans may offer some technical advantages over other anterograde labeling techniques. It gives complete filling of labeled axons in paraformaldehyde fixed tissue. This is compatible with fluorescent markers used in double labeling studies and with some immunocytochemistry procedures. It is also compatible with intracellular injections of Lucifer Yellow in both living and fixed brain slices (see Zook, this volume). Unlike intracellular filling methods (see Smith, this volume), a population of axons is labeled, and this may make regular patterns of termination more evident. The anterograde transport of dextran also allows the entire trajectory of a projection to be labeled, a feat not often accomplished in intracellular studies.

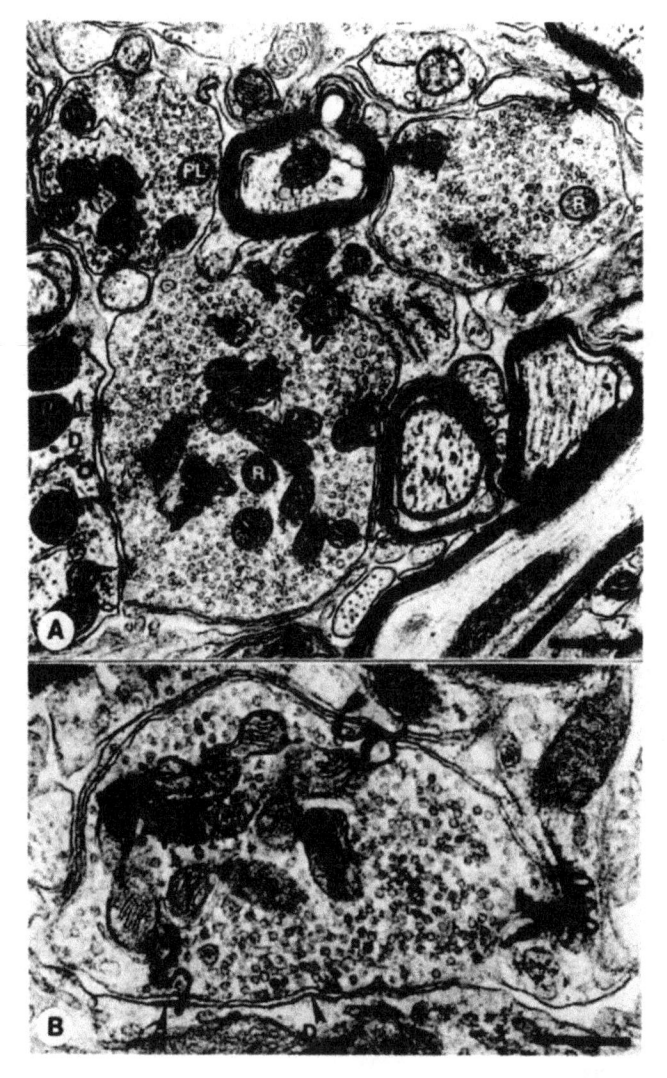

Figure 3. Electron microscopic autoradiographs of axonal endings in the inferior colliculus labeled after injections of $^3$H-amino acids into the cochlear nucleus of the cat. A, from an injection into the dorsal cochlear nucleus. B, from an injection into the anteroventral cochlear nucleus. Scale bars = 0.5 μm.

## Experiment two: Projections from the cochlear nucleus and superior olivary complex to the IC compared at the electron microscopic level

In previous studies of the projections of the cochlear nucleus to the IC of the cat (Oliver, '84, '85, '87), it has been possible to identify populations of axonal endings that arise from the DCN and AVCN at the EM level. The present experiment will allow us to evaluate the relative contribution of the superior olive to the synaptic organization of the IC. In both the previous and current experiments, injections of tritiated leucine and proline are made into a location in the lower auditory brainstem. After a survival of 1-2 days, the location of the labeled axons is revealed by autoradiography at both the light and electron microscopic levels.

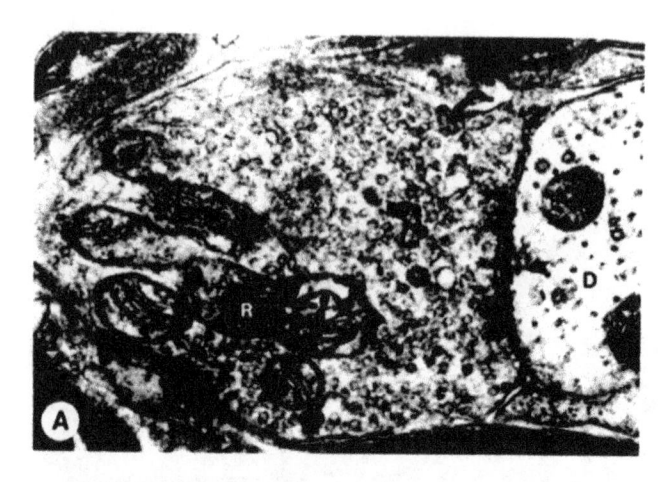

**Figure 4.** Electron microscopic autoradiography of an axonal ending in the cat inferior colliculus labeled after injection of $^3$H-amino acids into the "contralateral" lateral superior olivary nucleus. Labeled endings contain round synaptic vesicles (R). Synaptic contacts (arrowheads) are found on dendrites (D). Endings with pleomorphic vesicles are not labeled. Scale bars = 0.5 μm.

The results of our previous EM autoradiography studies are summarized in Figure 3. The projections from the CN to the IC exhibit the morphology associated with excitatory synaptic transmission. For example, axons from DCN (Fig. 3A) contain small, round synaptic vesicles and make asymmetrical synaptic contacts (*arrowheads*). These are found primarily in the contralateral IC. In contrast, axons from AVCN are found in both the contralateral and ipsilateral IC, albeit only in the lower frequency ranges. Only the contralateral projection from AVCN is found to the higher frequency parts of the IC. The morphology of the axonal endings from the AVCN is similar to that of endings from the DCN (Fig. 3B). If the axons from the DCN and AVCN converge into the same part of the IC, they could represent from 25-33% of the presumed excitatory endings. Most of the endings terminate on dendrites, but some large cells receive multiple synaptic inputs on their somata from the cochlear nucleus.

The current experiments are the first to examine the synaptic inputs to the IC from the superior olivary complex. Although these results are preliminary, the projections to the IC from the LSO and MSO exhibit some features that may distinguish them from the projections of the cochlear nucleus.

The projection from the LSO is bilateral and consists of separate populations of neurons with different targets (Glendenning and Masterton, '83; Shneiderman and Henkel, '87; Saint Marie et al., '89). Because of this bilateral projection, it is important to compare the LSO projections to both sides.

Endings from the contralateral LSO contain round vesicles and make asymmetric synapses, the morphology often associated with excitatory amino acid neurotransmitters (Fig. 4). Although a full analysis of this population is still in progress, it is clear that most of the endings from the contralateral LSO fit this description. While similar to endings seen in previous studies, some of the synaptic contacts display presynaptic densities not observed in endings from the cochlear nucleus.

In contrast, the ipsilateral LSO sends at least two types of endings to the IC. Many of the endings from ipsilateral LSO contain flat vesicles and make symmetrical synapses (Fig. 5A), the morphology often associated with inhibitory neurotransmitters such as GABA and glycine. However, endings with round synaptic vesicles that make asymmetrical synapses (Fig. 5B) also are found and may prove to be equally common. These data suggest that at least two populations of cells are involved in the ipsilateral projection to the IC from the

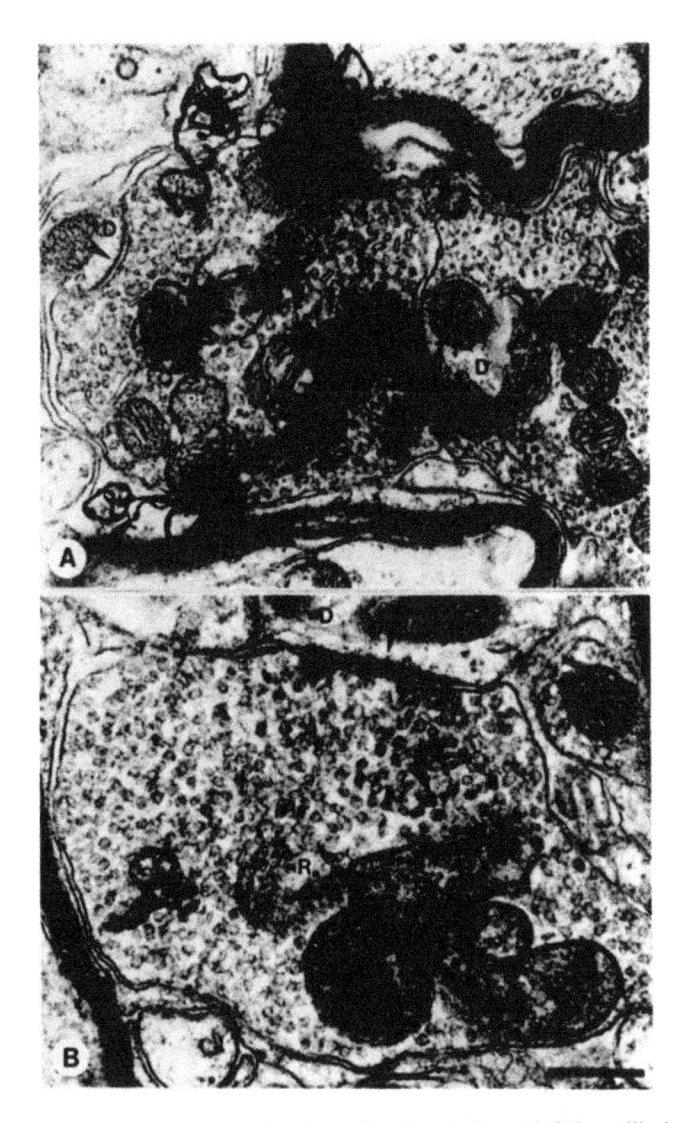

**Figure 5.** Electron microscopic autoradiography of axonal endings in the cat inferior colliculus labeled after an injection of $^3$H-amino acids into the "ipsilateral" lateral superior olivary nucleus. A, endings with pleomorphic vesicles (PL) are sometimes well labeled. B, endings with round vesicles (R) also are labeled. D, dendrite. Scale bars = 0.5 μm.

LSO. This is consistent with evidence that suggests some LSO cells with ipsilateral projections may use glycine as a neurotransmitter (Saint Marie et al, '89; Saint Marie and Baker, '90). A second population of cells may project ipsilaterally and use another possibly excitatory transmitter. The MSO, unlike the LSO, projects primarily to the ipsilateral IC (e.g. Henkel and Spangler, '83). Its synaptic inputs to the IC are simpler than those from the LSO. Synaptic endings from MSO typically contain round vesicles and have asymmetrical synapses. In summary, these experiments suggest that the synaptic inputs to the IC from the superior olive are complex relative to those from the cochlear nucleus. While most inputs from the cochlear nucleus have a morphology linked to excitatory transmitters, projections from the olives have several different morphologies and may represent inputs from different cell types.

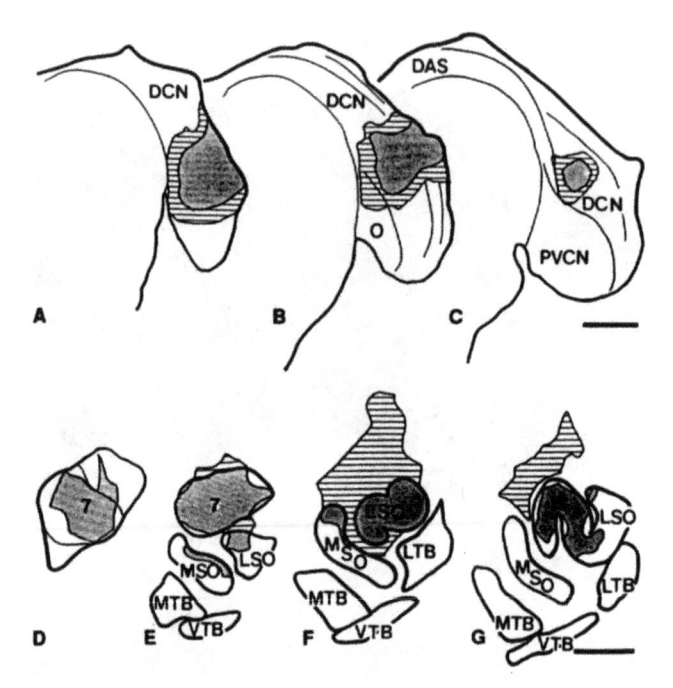

**Figure 6.** Injections of wheat germ agglutinin-HRP in the dorsal cochlear nucleus (A-C) and ³H-leucine into the superior olivary complex (D-G) of the cat. Abbreviations: DCN, dorsal cochlear nucleus; DAS, dorsal acoustic stria; LSO, lateral superior olive; LTB, lateral nucleus of the trapezoid body; MTB, medial nucleus of the trapezoid body; MSO, medial superior olive; O, octopus cell area; PVCN, posteroventral cochlear nucleus; VTB, ventral nucleus of the trapezoid body; 7, facial nucleus. Scale bars = 1.0 mm.

## Experiment three: Modes of termination of cochlear nucleus and superior olivary projections to the IC

To better understand the relationship of CN and SOC projections to the IC, we have used two anterograde tracers to label both inputs in the same cat. Wheat germ agglutinin-HRP (WGA-HRP, L-7017, Sigma) is injected into one site and ³H-leucine is injected into the other. Both compounds are dissolved in saline and injected through a glass micropipette with a 1-2 µm diameter tip. The saline allows us to record single and multiple unit activity to pure tones, FM tones, and sinusoidally amplitude modulated tones. Physiological controls are important in these cases because we wish to compare the inputs to the same frequency regions of the IC. The location of the WGA-HRP is revealed by HRP histochemistry (see Shneiderman et al., '88), and the ³H-leucine is demonstrated by light microscopic autoradiography (see above). To analyze this material, adjacent sections processed for the two histological procedures are superimposed and aligned using blood vessels to determine the relationship of the two labels. As in the first two experiments, the results are preliminary, but they suggest some fundamental features of the relationship of monaural and binaural inputs to the IC.

In the first case, a WGA-HRP injection was made in the 17 kHz range of the DCN (Figs. 6A, 6B, 6C). This injection only slightly invaded the posterior PVCN which does not project to the IC in the cat (Oliver, '87). The second injection of ³H-leucine was placed in the LSO on the same side. Units at the second injection site responded to 17-18 kHz tones from the ipsilateral ear but were unresponsive or suppressed when the tone was applied to the contralateral ear. Although the injection is centered in the LSO there was efflux of ³H-leucine

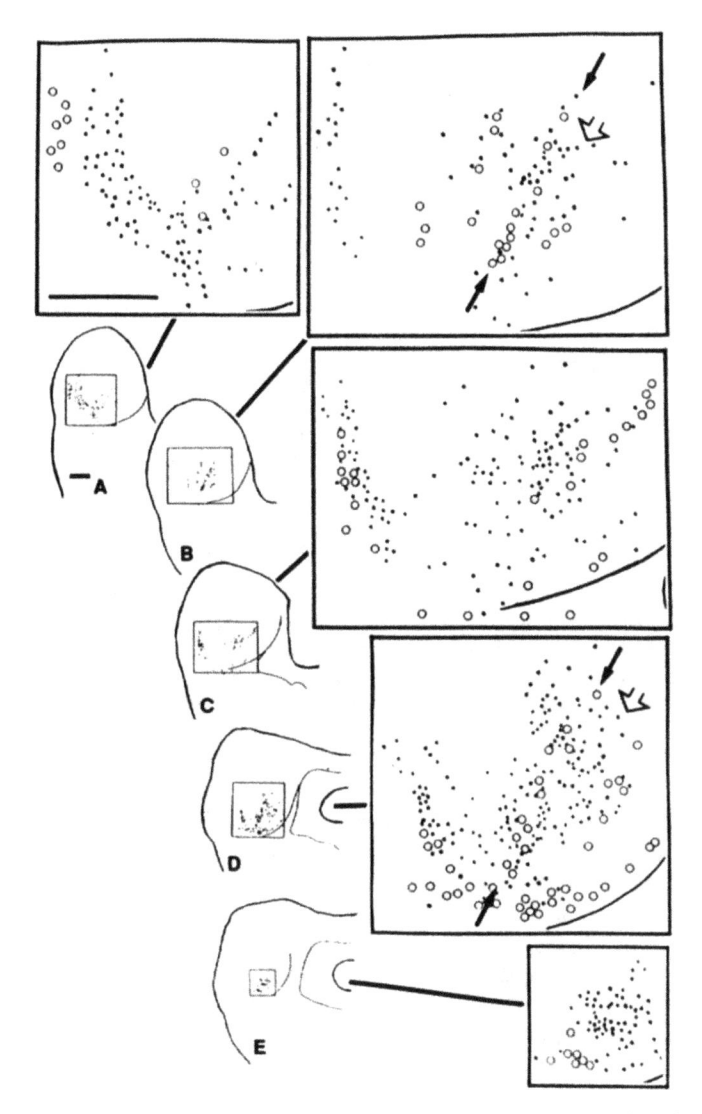

**Figure 7.** Transport of labels from "contralateral cochlear nucleus" and "contralateral lateral superior olive" to the inferior colliculus. In each drawing, the labeling in two adjacent transverse serial sections is superimposed. Open circles, WGA-HRP label from dorsal cochlear nucleus; dots, $^3$H-label from lateral superior olive. Black arrow, area of overlap; hollow arrow, adjacent band of non-overlapping label. Scale bars = 1 mm.

into the facial nucleus and some spread into the adjacent, low-frequency part of the MSO. This spread of injected tracer does not affect the interpretation of this case since the adjacent nuclei do not project to the contralateral IC.

Figure 7 shows how the projection from the DCN is related to the projection from the LSO in the *contralateral* IC. An enlarged view of the label is displayed for each section (Fig. 7A-E) where the WGA-HRP label is symbolized by open circles and the $^3$H-leucine label is shown as black dots. Both inputs terminate in bands in the same frequency zones in the IC. Two patterns of termination are evident. In the first pattern, the two inputs do not overlap. For example in panel A, in the caudal part of the lateral central nucleus, there is little overlap. Likewise in panel E, rostrally. In the second pattern, there are zones where both inputs overlap. The medial part of the central nucleus is interesting. Here, the two banded

343

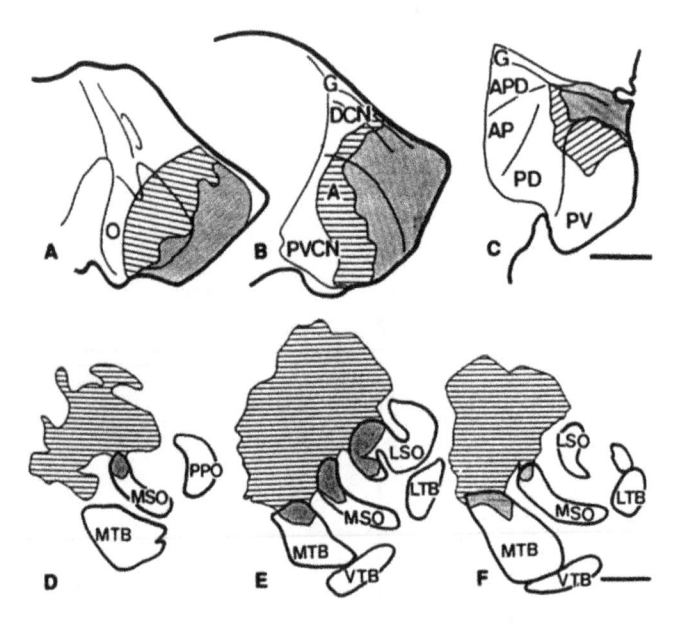

**Figure 8.** Injection of wheat germ agglutinin-HRP into the dorsal cochlear and posteroventral cochlear nucleus (A-C) and injection of $^3$H-leucine into the superior olivary complex (D-F) of the cat. Abbreviations as in Figure 6 with additionally: A, anterior PVCN; AP, posterior part of AVCN; APD, posterodorsal part of AVCN; G, granule cell layer; PD, dorsal part of posterior AVCN; PPO, posterior periolivary nucleus; PV, ventral part of posterior AVCN. Scale bars = 1.0 mm.

inputs appear to terminate along the same fibrodendritic layer (Figs. 7B, 7D, *black arrows*). There are definite zones where the two inputs are superimposed. Even so, within this same band, there may be small zones where one input predominates over the other. Beside this layer, there is a band of label from the contralateral DCN (Figs. 7B, 7D, *open arrows*) that does not appear to overlap with the inputs from the contralateral LSO.

A second case (Fig. 8), shows two injections that involve low frequency (1 kHz) parts of the DCN/PVCN (Fig. 8A-C) and MSO (Fig. 8D-F). The injection sites are somewhat larger than in the previous case. But the present results will focus only on the projections from the MSO and the cochlear nucleus to the *ipsilateral* IC.

Figure 9 shows that both patterns of termination are found in the low-frequency parts of the ipsilateral IC. Both inputs converge in the dorsolateral central nucleus and dorsal cortex. There is one prominent band at the dorsolateral edge of the central nucleus. Labeling is also seen in one or more bands ventromedially. In this case, the lack of overlap is as striking as its presence. The overlap of the two inputs is most evident in the caudal central nucleus (Figs. 9A-B, *black arrow*) where the two labels are mixed in the lateral band. The lack of overlap is most evident in the dorsal cortex (Figs. 9A-C, *black triangle* marks the border) where the input from the cochlear nucleus predominates. Rostrally, to the mid-levels of IC, the input from MSO predominates (Fig. 9C, *hollow arrow*).

In summary, these experiments suggest that the inputs from the cochlear nucleus to the IC can remain segregated from inputs ascending from other brainstem nuclei. Projections from the contralateral cochlear nucleus to the IC can remain separate from the projections of the contralateral LSO. Since these represent monaural and binaural functions, respectively, this could represent a substrate for separation of monaural and binaural processing at the level of the midbrain. Our results also show that the projection of the MSO may remain separate from the projection of the ipsilateral cochlear nucleus. Since the MSO may be particularly

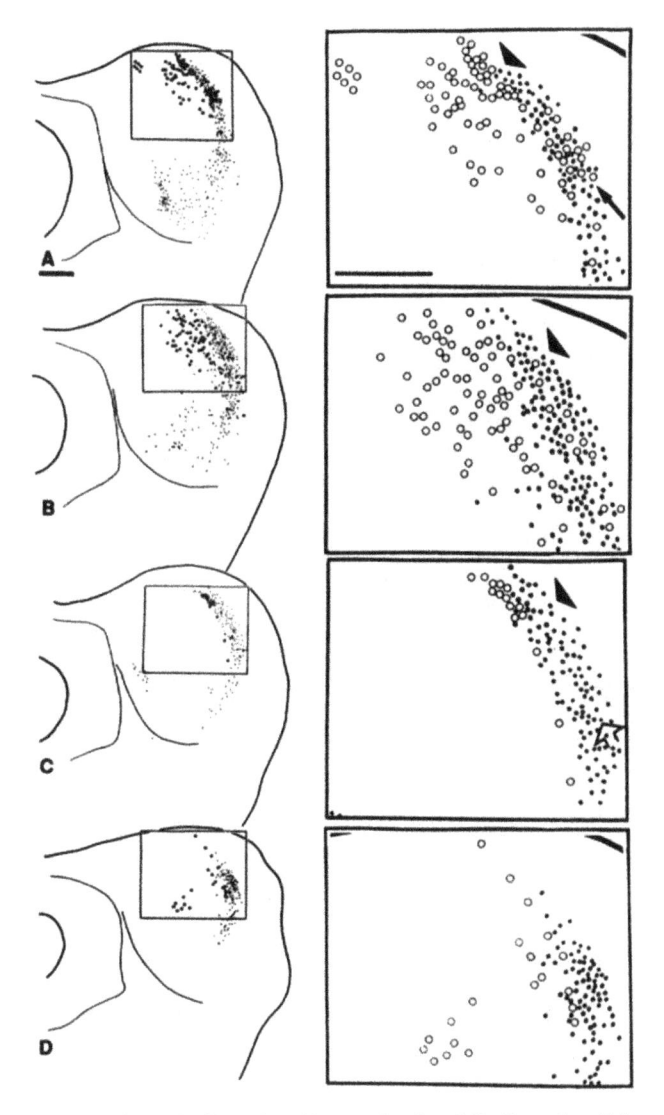

**Figure 9.** Transport of labels from "ipsilateral cochlear nucleus" and "ipsilateral medial superior olive" to the inferior colliculus. In each drawing, the labeling in two adjacent transverse serial sections is superimposed. Open circles, WGA-HRP label from cochlear nucleus; dots, $^3$H-label from superior olive; black arrow, area of overlap; hollow arrow, adjacent band of non-overlapping label; black triangle, border between central nucleus and dorsal cortex. Scale bars = 1 mm.

concerned with interaural time sensitivity, it suggests that time sensitive cells in the IC do not generally receive an input from the ipsilateral cochlear nucleus.

These experiments also show that inputs from the cochlear nucleus may converge with inputs from the SOC to the IC. Some neurons with binaural responses by virtue of their inputs from the SOC may receive additional inputs from the cochlear nucleus. It is likely that this convergence could alter the binaural properties of neurons in the IC.

# Summary

The morphology of the projections from the cochlear nucleus is compared to those from the superior olive. Three types of experiments are presented. First, the axonal morphology of the efferents from cochlear nucleus in the rat is demonstrated by dextran transport and histochemistry. These projections are similar to those seen in the cat. The morphology of axonal endings from the cochlear nucleus that terminate in the SOC, nuclei of the lateral lemniscus, and IC may be different. This methodology may help relate specific cell types in the cochlear nucleus to their mode of termination at their target nuclei. Second, the synaptic morphology of axonal endings in the IC was evaluated by electron microscopic autoradiographic methods. The axonal endings from the DCN and the AVCN are compared to the endings from the LSO and MSO. The synaptic inputs from the SOC are more complex than those from the cochlear nucleus and provide evidence for both excitatory and inhibitory inputs from the SOC. Third, the convergence of inputs from the cochlear nucleus and the SOC is addressed. Two different anterograde axonal markers are injected into two different sites in the same animal. Evidence is presented that these two inputs can remain segregated or can converge at the level of the IC. Although these results are preliminary, these experiments certainly argue for continued study of the efferents from cochlear nucleus and their interaction with the superior olive at the level of the IC.

ACKNOWLEDGEMENTS

Supported by grant R01-DC00189 from the National Institutes of Health, USA. Thanks to D. Bishop for technical support.

REFERENCES

Adams, J.C., 1978, Morphology and physiology in the ventral nucleus of the lateral lemniscus, *Soc. Neurosci. Abstr.*, 4:1.
Adams, J.C., 1981, Heavy metal intensification of DAB-based HRP reaction product., *J. Histochem. Cytochem.*, 29:775.
Cant, N.B., 1990, Projection patterns of the different neuronal types in the ventral cochlear nucleus of the gerbil, *In*: "Cochlear nucleus: structure and function in relation to modelling", Conf. Proceedings, Univ. Keele.
Faye-Lund, H., Osen, K.K., 1985, Anatomy of the inferior colliculus in rat, *Anat. Embryol.*, 171:1-20.
Glendenning, K.K., Masterton, R.B., 1983, Acoustic chiasm: Efferent projections of the lateral superior olive, *J. Neurosci.*, 3:1521-1537.
Henkel, C.K., Spangler, K.M., 1983, Organization of the efferent projections of the medial superior olivary nucleus in the cat as revealed by HRP and autoradiographic tracing methods, *J. Comp. Neurol.*, 221:416-428.
Hill, S.J. and Oliver, D.L., 1991, Intracellular injections of biotinylated lucifer yellow in combination with retrograde tracing in lightly fixed sections of the inferior colliculus (IC), *Soc. Neurosci. Abstr.*, 17:1510.
Oliver, D.L., 1984, Dorsal cochlear nucleus projections to the inferior colliculus in the cat: A light and electron microscopic study, *J. Comp. Neurol.*, 224:155-172.
Oliver, D.L., 1985, Quantitative analyses of axonal endings in the central nucleus of the inferior colliculus and distribution of $^3$H-labeling after injections in the dorsal cochlear nucleus, *J. Comp. Neurol.*, 237:343-359.
Oliver, D.L., 1987, Projections to the inferior colliculus from the anteroventral cochlear nucleus in the cat: Possible substrates for binaural interaction, *J. Comp. Neurol.*, 264:24-46.
Oliver, D.L. and Shneiderman A., 1991, The anatomy of the inferior colliculus. A cellular basis for integration of monaural and binaural information, *in*: "Neurobiology of Hearing, Vol II: The Central Auditory System," Altschuler et al., eds, Raven Press, N.Y.
Oliver, D.L. and Huerta, M., 1992, The anatomy of the inferior and superior colliculi, *in*: "Springer Series in Auditory Research, Vol. 1, The Anatomy of Mammalian Auditory Pathways," R.R. Fay and A.N. Popper, eds, Springer-Verlag, N.Y.

Saint Marie, R.L., Baker, R.A., 1990, Neurotransmitter-specific uptake and retrograde transport of [$^3$H]glycine from the inferior colliculus by ipsilateral projections of the superior olivary complex and nuclei of the lateral lemniscus, *Brain Res.*, 524:244-253.

Saint Marie, R.L., Ostapoff, E-M., Morest, D.K., Wenthold, R.J., 1989, A glycine-immunoreactive projection of the cat lateral superior olive: possible role in midbrain ear dominance, *J. Comp. Neurol.*, 279:382-396.

Shneiderman, A. and Henkel, C.K., 1987, Banding of lateral superior olivary nucleus afferents in the inferior colliculus: A possible substrate for sensory integration, *J. Comp. Neurol.*, 266:519-534.

Shneiderman, A., Oliver, D.L., Henkel, C.K., 1988, The connections of the dorsal nucleus of the lateral lemniscus. An inhibitory parallel pathway in the ascending auditory system?, *J. Comp. Neurol.*, 276: 188-208.

# RESPONSES OF COCHLEAR NUCLEUS CELLS AND PROJECTIONS OF THEIR AXONS

Philip H. Smith, Philip X. Joris, Matthew I. Banks and Tom C.T. Yin

Department of Neurophysiology, University of Wisconsin, Madison, Wisconsin, USA

The advent of intracellular recording using horseradish peroxidase-filled glass electrodes offered a new and exciting approach to the *in vivo* cochlear nucleus (CN) preparation. Sharp glass microelectrodes, filled with a standard salt solution containing positively charged HRP molecules, made it feasible to record and characterize supra- and subthreshold intracellular responses of individual neurons to auditory stimuli, inject and subsequently recover the same HRP-labeled cell and ask the following basic questions; 1) Do morphologically defined cell types in the cochlear nucleus respond in a certain way to simple auditory stimuli at both the sub- and suprathreshold level? 2) Are the synaptic inputs of different cell types different, in terms of location, type and concentration, and are they arranged in ways that might help to explain the cell type's unique responsiveness? 3) What is the projection pattern of individual axons originating from a given cell type, what are the shapes of vesicles within the terminals of these axons and can we make educated guesses about their influence on other cell populations based on this information?

Although fruitful, the method has proven to be difficult for many reasons including the surgically difficult location of the CN, the respiratory and cardiac-induced pulsations that are prevalent in the brainstem, susceptibility of the nucleus to edema and electrode "plugging" during penetrations deep into the nucleus. As a consequence, the number of cells recorded from and recovered after injection is small. Never-the-less, results from this lab, using this method in the cat CN (Rhode et al., '83; Rhode et al., '83; Smith and Rhode, '85; Smith and Rhode, '87; Smith and Rhode, '89; Smith et al., '91) have indicated that there is a good correlation between certain unique physiological features and a characteristic set of anatomical traits displayed by a given principal cell type.

One of the other shortcomings of this method, when employed on cells or their axons within the cochlear nucleus, is a failure to fill the axon sufficiently such that its course and termination sites outside the CN complex can be identified. For this reason, our recent approach has been to expose a major fiber tract emanating from the CN, penetrate an axon in that tract, record intra-axonally, characterize the response to simple stimulus paradigms and inject with tracer. This method eliminates the possibility of studying synaptic events seen in cell body or proximal axonal recordings but is advantageous in that the spike output of the cell can be monitored more stably, alteration of response behaviors due to injury is less likely, and more of the axon can be filled. To further characterize the influence of some of these CN axons on third order cells we have begun to look at the synaptic responses of cells in other auditory brainstem nuclei, using the *in vitro* brain slice method in rats, when an output pathway of the CN is electrically stimulated (Banks and Smith, '90). In addition, we have begun using the biotin/lysine conjugate known as neurobiotin (Vector Labs; Horikawa and Armstrong, '88) as the intracellular marker in both the *in vivo* and *in vitro* experiments. Its advantages over HRP include improved electrode recording characteristics, ease of injection from unbeveled electrodes, and improved filling of the injected neuron. Our initial attempts at strial recordings have focused on the ventral acoustic stria (VAS) or trapezoid body (TB) and it is only recently that we have initiated work on the dorsal and intermediate acoustic striae (DAS and IAS). As a consequence, the data presented here will deal primarily with cells whose axons use the VAS.

## Bushy Cells

As the name implies, bushy cells are distinguished by the bushy nature of their dendritic tree (Brawer et al., '74) and are subdivided into spherical (SBC) and globular bushy cell (GBC) categories. Their axons are the primary component of the VAS, the disruption of which affects sound localization, (Casseday and Neff, '75; Jenkins and Masterton, '82; Masterton, et al., '81; Moore et al., '74; Neff, '62) the startle response (Davis et al., '82) and pinna reflexive behavior (Cassella and Davis, '86).

Globular bushy cells show unique physiological features. Intracellular records from GBC somata or the proximal axon (Smith and Rhode, '87) reveal what appears to be very large and very fast excitatory synaptic events that do not necessarily elicit a spike. Subsequent work by others, in mouse brain slice (Wu and Oertel, '84) and the dissociated cell preparation, (Manis and Marx, '92) has shown that the speed of repolarization after a synaptic event is due to a potassium conductance active around the resting potential. In the cat, short tones at the cell's characteristic frequency (STCF, with CF being the frequency which elicits an increase in spike output at the lowest stimulus intensity) generate characteristic peristimulus time histogram (PSTH) patterns (Fig. 1). PSTHs of globular bushy cells with CFs above 3 kHz resemble the auditory nerve (AN) PSTH but differ at higher stimulus intensities in that the PSTH patterns show a precisely timed onset peak which may be followed by a brief depression of activity and then a resumption of sustained activity. This PST pattern is called primary-like-with-notch ($PL_N$) if the notch is prominent or $O_L$ (onset with low sustained activity) if the notch is not as obvious and the sustained rate not as high. Our more recent experiments (Smith et al., '91), recording from high CF globular bushy axons in the TB, have confirmed these results. In addition, a number of axons of GBCs with CFs < 3 kHz were labeled. Those with CFs between 1 and 3 kHz also show the $PL_N$ response with spikes in the sustained portion of the response phase-locked (spikes occurring at a particular phase angle of a sinusoidal stimulus) albeit not as well as eighth nerve (AN) fibers with the same CF (Johnson, '80). Curiously, these same 1-3 kHz GBCs will often phase-lock very accurately with vector strengths (a measure of phase-locking, Goldberg and Brown, '69) >0.9 when driven by a low frequency tone (less than 1 kHz). GBCs with CFs below 1 kHz also phase-lock better than AN fibers with similar CFs, both in terms of their vector strengths (which were above 0.9) and the degree to which they entrain (respond to every cycle) to the stimulus. The above observations imply that the large, fast (yet sometimes subthreshold) synaptic events generated by AN inputs can, under appropriate stimulus conditions, generate globular bushy cell outputs with enhanced onset components and/or enhanced phase-locking. In addition, the spontaneous spike rates (SR) of our labeled GBCs with CFs below 2-3 kHz were usually lower than AN fibers (less than 2 spks/sec.). When taken with the observation of Liberman (presentation at this meeting), that GBC-AN input is dominated by high SR units, this feature strongly suggests that AN fiber-evoked synaptic events can be subthreshold in the GBC population.

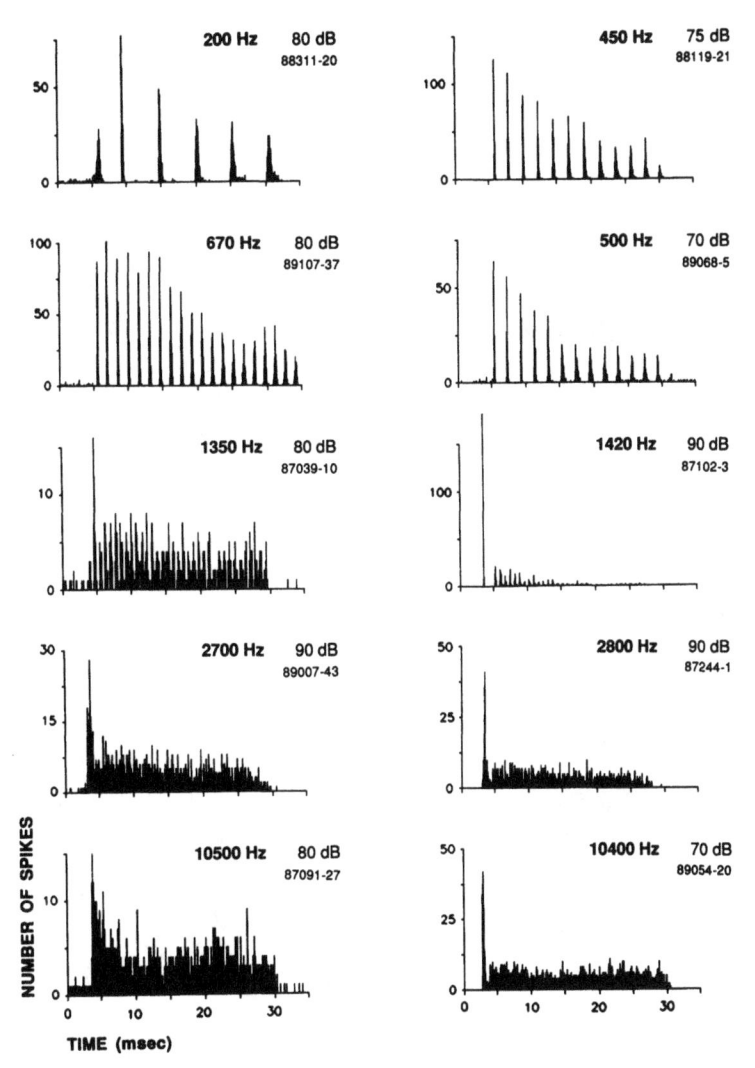

**Figure 1.** Representative short tone responses (200 - 25 msec tones every 105 msec) from axons of positively identified spherical (left column: Spherical bushy cells. PL. dorsal component) and globular (right column: Globular bushy cells. PL$_N$. Ventral component) bushy cells, with similar CFs. Dorsal and ventral component refers to the location of the axons in the trapezoid body as they crossed the midline. CF, stimulus intensity level and unit number are indicated for each response.

GBCs also show characteristic morphological features. At the electron microscopic (EM) level, most of the surface of the GBC body and proximal dendrites is covered with terminals. Almost half of the terminals are presumably from AN fibers based on their large round vesicle contents. The remainder contain flat or pleomorphic vesicles and rarely small round vesicles. In contrast, the distal dendritic tree is sparsely innervated with only about 1 in 10 of the terminals of AN origin. GBC axons do not give off collaterals in the CN before entering the ventral stria but do display numerous myelinated collaterals in both the ipsilateral and contralateral olivary complex (Fig. 2; in all subsequent descriptions the use of ipsilateral and contralateral, referring to axonal locations, will be relative to the location of the parent cell body). Ipsilaterally, branches are almost always sent to innervate the posterior periolivary nucleus (PPO) or its rostral extension, the lateral nucleus of the trapezoid body (LNTB).

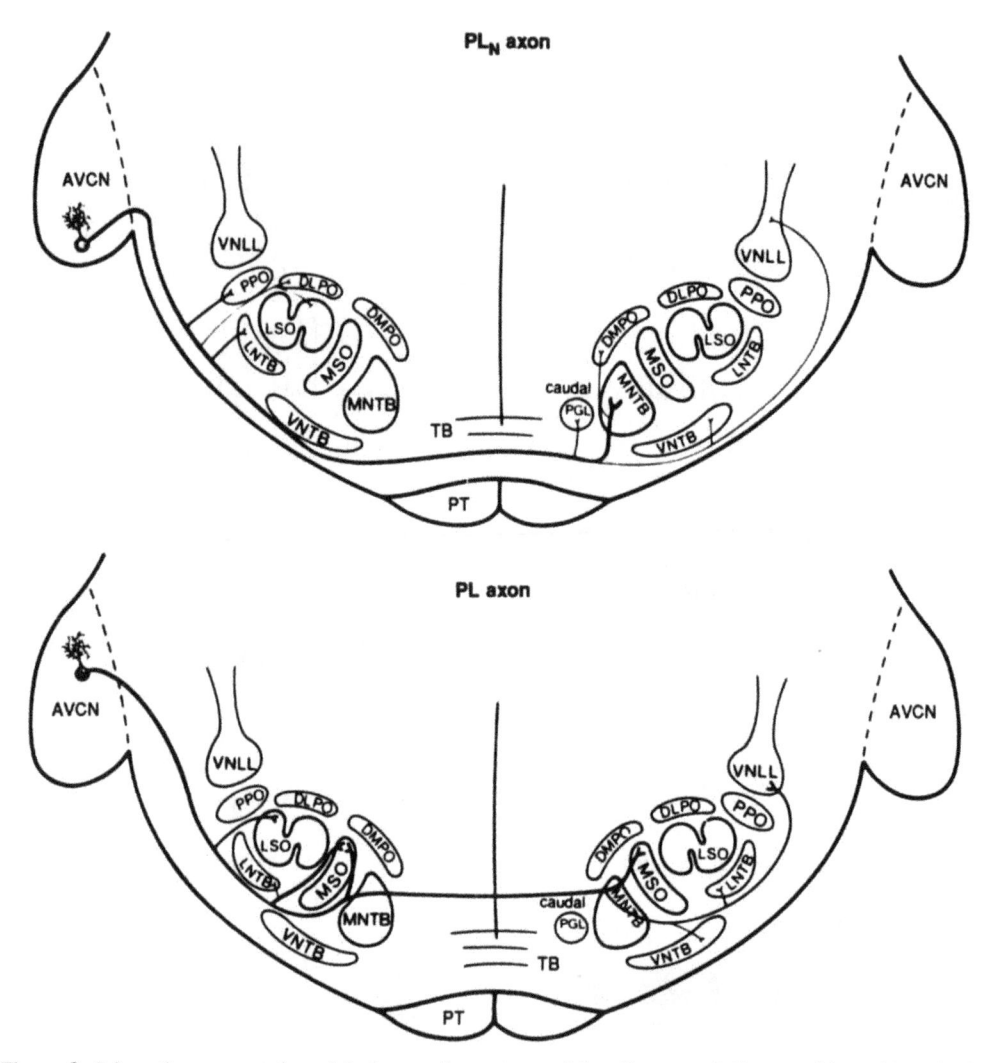

**Figure 2.** Schematic representation of the innervation patterns of the olivary, periolivary and lateral lemniscal nuclei by globular (upper, PL$_N$ axon) and spherical (lower, PL axon) bushy cell axons. For simplicity, all nuclei are collapsed onto a single rostrocaudal level in the coronal plane. Thickness of the collaterals intended to qualitatively represent the "strength" of the innervation.

**Abbreviations:** AVCN - anteroventral cochlear nucleus; DLPO and DMPO - dorsolateral and dorsomedial periolivary nuclei; LNTB, MNTB and VNTB - lateral, medial and ventral nuclei of the trapezoid body; LSO and MSO - lateral and medial superior olive; PGL - nucleus paragigantocellularis lateralis located at the caudal extreme of the olivary complex; PPO - posterior periolivary nucleus; PT - pyramidal tract; TB - trapezoid body; VNLL - ventral nucleus of the lateral lemniscus.

Others (Adams, '83; Ostapoff et al., '85; Spangler et al., '87; Winter et al., '89) have shown that many cells in this region provide GABAergic feedback to the CN. Ipsilateral GBC collaterals can also end in a region designated ventrolateral periolivary nucleus (VLPO) in the form of large terminals known as diminuitive endbulbs (Adams, '83) on cells that presumably project back to CN. Occasionally an ipsilateral collateral extended into the LSO and DLPO. This appears to differ from the rat (Friauf and Ostwald, '88) where GBCs provide a major input to the LSO. The GBC axons cross the midline in the ventral half of the

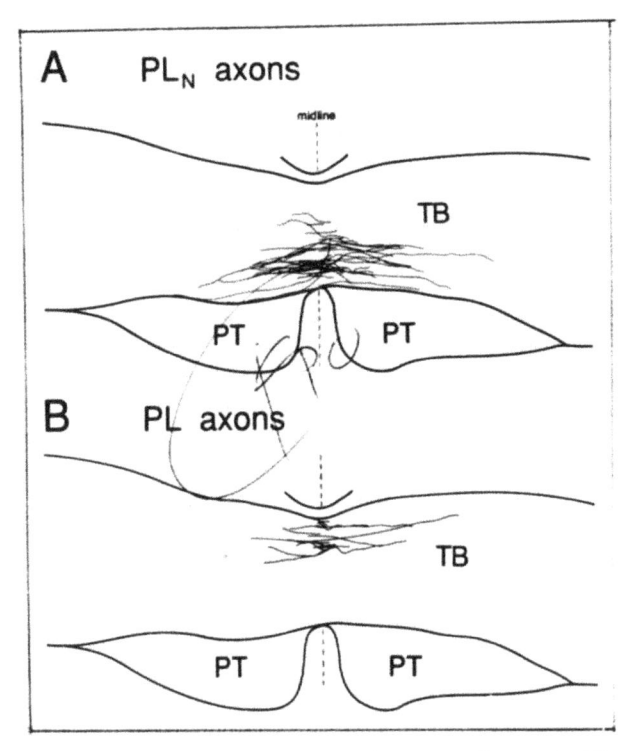

**Figure 3.** Summary diagram of the locations of labeled globular bushy cell axons (A, $PL_N$ axons) and spherical bushy cell axons (B, PL axons) in a coronal section of the trapezoid body at the midline of the brainstem. In each case the distance between the dorsal surfaces of the pyramidal tract (PT) and the TB was measured and normalized and the injected axon placed accordingly as it crossed the midline.

trapezoid body (Fig. 3) and always have one or two calyces of Held terminating on a principal cell(s) in the contralateral medial nucleus of the trapezoid body (MNTB). At the EM level these terminals contain large round vesicles and cover over 25% of the surface of the MNTB cell body. The great majority of the remaining terminals on the soma contain non-round vesicles indicating that one MNTB cell receives only one calyx. Our recordings from MNTB cells in the rat brain slice suggest that the calycean ending is glutamatergic. Shock stimulation of the TB at the midline generates a huge fast suprathreshold synaptic event that can be blocked by the non-NMDA glutamate antagonist CNQX. Contralaterally the GBC axon also can have, but need not have, collaterals coming off the main axon that innervate the dorsomedial periolivary nucleus (DMPO), the ventral nucleus of the lateral lemniscus (VNLL), and a region at the caudal extreme of the olivary complex, ventral and medial to the facial motor nucleus.

Good intracellular records from spherical bushy cell somata *in vivo* are not available but our trapezoid body experiments have revealed that the spike output of these cells differs in some ways from GBCs (Fig. 1). For SBCs with high CFs (> 3 kHz) peristimulus time histogram (PSTH) closely resemble those of AN fibers and are thus called "primary-like" (Peiffer, '66) (PL). Cells with 1 to 3 kHz CFs also are primary-like but, like the GBC population, show reduced synchrony (compared to AN fibers) during the sustained portion of the response. Like the GBC population, some SBCs with low best frequencies (< 1 kHz) show enhanced phase-locking capabilities compared to AN fibers. Unlike our injected GBC population, intraaxonal recordings from SBCs fibers in the VAS tend to show higher spontaneous spike rates which are typically greater than 20 spikes/sec. The larger size of AN endbulb terminals on spherical compared to globular bushy cells probably makes them more likely to drive these cells during spontaneous AN spike activity.

Quantification of the number and distribution of synaptic inputs on the somata and dendritic tree of individually labeled spherical bushy cells has not been done. Our TB experiments have shown that spherical bushy cell axons, like GBC axons, have no collaterals within the CN, but branch considerably within the olivary complex (Fig. 2). This innervation pattern as well as the location of these axons as they cross the midline (Fig. 3) differs from that of GBCs. Ipsilaterally the axon sends collaterals primarily to innervate both the ipsilateral lateral and medial superior olive (LSO and MSO). We have yet to record from and inject SBC axons with CFs > 10 kHz from this set of experiments where our electrodes penetrate the TB medial to the MSO. It is possible that the set of SBC axons innervating the high frequency limb of the LSO does not cross the midline. Those with CFs < 10 kHz innervate the approriate regions of the LSO and MSO given the frequency maps that have been reported for these nuclei (Guinan et al., '72). The axons travel across the midline rostrally in the dorsal half of the trapezoid body (Fig. 3). Upon reaching the contralateral side, the axon heads rostrally to innervate the VNLL but not before sending a number of collaterals caudally, just medial to the MSO where they innervate this structure in a ladder-like fashion. This is pertinent in light of the Jeffress model of sound localization (Jeffress, '48). One requirement for the cells in this model, that present day scientists presume to reside in the MSO, is that at least one of the afferent inputs to these binaural cells have a delay line configuration. The input of our labeled SBC axons to the contralateral MSO fulfills this requirement.

## Multipolar cells

Type I multipolars, initially classified in the AVCN based on the sparse synaptic coverage of their cell bodies (Cant, '81), do not seem to respect the boundaries of the subregions of the VCN. Information is available on members of this class in both the AVCN and PVCN. The axons of those in the AVCN project to the inferior colliculus (Adams, '79; Cant, '82) (IC) by way of the TB (Tolbert et al., '82), and send collaterals to the DCN (Adams, '83). Our early intracellular recordings from a small number of AVCN multipolars whose axons left the CN via the TB (Rhode et al., '83), showed that short tones produce a chopper response (a regular firing pattern whose regularity is not related to the stimulus frequency). Subsequent recordings from multipolar cells in slices of the mouse AVCN (Wu and Oertel, '84) showed that they were intrinsically capable of generating a "chopping" response to passive current injection. Recordings from a small population of multipolars in the cat posterior PVCN/nerve root area, whose axons sent collaterals heading toward or into the DCN, then entered the TB (Smith and Rhode, '89), indicated that these cells also displayed tone-induced chopper PSTHs. EM analysis of members of this population shows a sparse "Type 1" innervation of the soma and distal dendritic tree with only about 1/5 of the surface covered with terminals. Close to half of this coverage is from AN fibers. In contrast, close to 1/2 of the proximal 100 $\mu$m of the dendritic tree is covered with terminals, 2/5 of them from AN fibers. Recently devised compartmental models of stellate cells (Banks and Sachs, '91; Sachs, this volume) indicate that the dendritic location and number of converging inputs onto this cell type are important factors in determining the response regularity. The labeled terminals of the myelinated axon collaterals of these injected multipolar cells, within the PVCN, contain small (compared to vesicles in AN terminals) round vesicles.

In our recent recordings from the trapezoid body we have attempted to trace the connections of this axonal population in the brainstem. Results thus far indicate that the multipolar cells/choppers have a very different course and innervation pattern from that of the bushy cell axons. Recording from and injection of their axons has proven to be difficult, presumably due to their small diameter. It is hoped that the recent change to the unbeveled

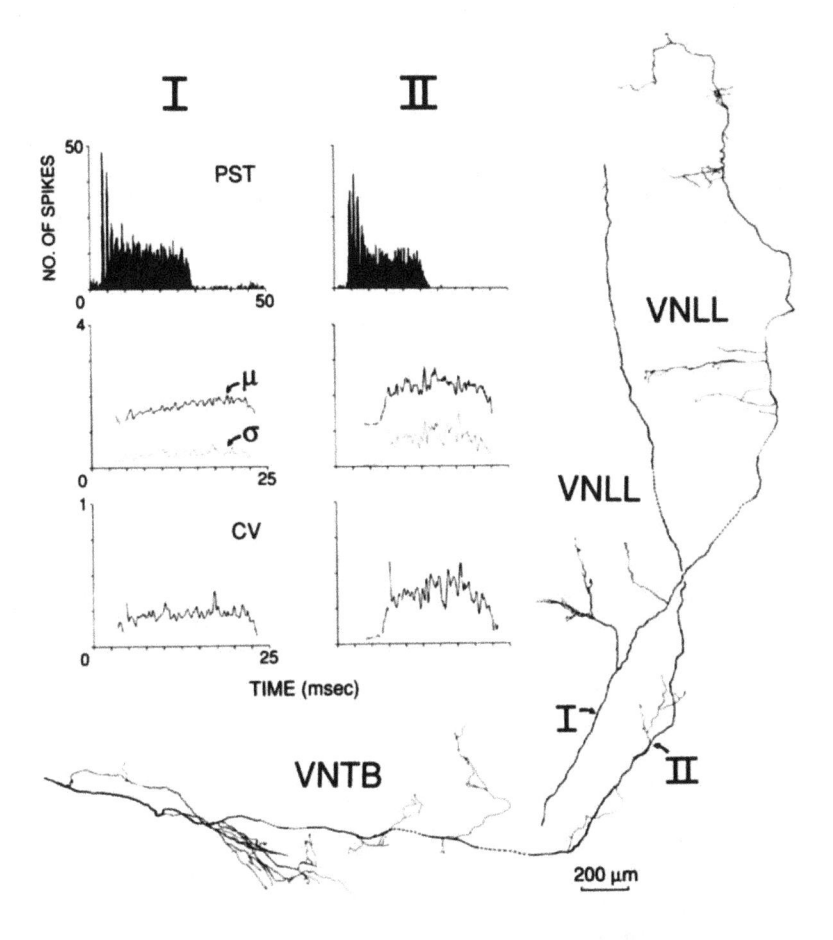

**Figure 4.** Contralateral axonal projection pattern and short-tone physiology of two chopper axons from two separate experiments. Column I. Upper; "Sustained" chopper PST to 200 - 25 msec, 55 dB SPL tones at CF (20 kHz). Middle; mean ($\mu$) and standard deviation ($\sigma$) of the interspike intervals for spikes occurring during the short tone response. Lower; coefficient of variation (CV), a measure of spike regularity, equal to the ratio of standard deviation to mean interspike interval (see Young et al., '88 for details). Column II. Same measures for a "transient" chopper response to 40 dB SPL tones at CF (28 kHz). Camera lucida drawings of the contralateral innervation pattern of these two axons. Both crossed the midline and innervated the VNTB (for simplicity, this part of the innervation pattern of axon I has been omitted). They then headed dorsally and rostrally into the lateral lemniscus. One sent numerous side-branches into the ventral nucleus of the lateral lemniscus (VNLL) while the other did not. Both continued dorsally in the lemniscus, past the VNLL, but faded before reaching the inferior colliculus.

electrodes and neurobiotin tracer will help to alleviate the problem. Thus far most of our injections have been at a site on the axon contralateral to the CN containing the cell body so our data on the location of the soma and the innervation pattern of the axon in the ipsilateral brainstem is sparse. We have been able to follow only one labeled axon retrogradely back to its parent soma, a multipolar cell in the rostral PVCN. Ipsilaterally, an axon collateral of this cell branched from the main axon, caudal to the LSO, innervated the PPO and continued rostrally and dorsally into the DMPO. Axons of other choppers showed no collaterals ipsilaterally but were lightly filled and could not be followed for the entire extent of their ipslateral course from CN to the midline. All our axons with "chopper" responses cross the midline. At the midline, the thin myelinated axons travel either at the ventral extreme of the TB or about midway up the TB between the more ventrally situated globular and dorsally situated spherical bushy axons. Contralaterally these axons frequently give rise to numerous collateral branches which terminate ventral and lateral to the MNTB but medial to the MSO, a periolivary region generally designated the ventral nucleus of the trapezoid body (VNTB) in the cat (Fig. 4). Our recordings from VNTB cells in the rat brain slice consistently show an epsp in response to shock stimulation of the TB at the midline, perhaps resulting from the activation of this axonal population. Electron microscopy of VNTB terminals from two chopper axons in the cat revealed the anticipated small round vesicle profiles. The main axon continues rostrally and dorsally into the lateral aspect of the lateral lemniscus (LL) where some axons send collaterals medially to innervate the VNLL while others do not (Fig. 4). Whether this distinction is a feature of any particular type of chopper (e.g. sustained versus transient, Blackburn and Sachs, '89; Young et al., '88) remains to be determined. The axons always faded at the level of the VNLL or just dorsal to it in the LL, so the projection pattern beyond this point is unknown. The results of others, described above, would indicate that these axons must continue into the IC, but the influence of their terminal arbors here is unknown.

Many questions remain concerning the multipolar cells whose axons use the TB as their pathway to the auditory CNS. Included among these are: 1) Are there differences in the projection patterns and or influences of axons of multipolars situated in the PVCN compared to those in the AVCN? 2) Are there differences in those multipolars whose chopper patterns can be subdivided based on the regularity of their sustained firing patterns? 3) What are the specific termination patterns of these cells within their primary target, the inferior colliculus?

## Intermediate stria axons

Very recently we have directed our attention toward the two other major pathways into and out of the CN, namely the intermediate and dorsal acoustic striae (IAS and DAS). The functional role and specific connections of cells in the DCN and VCN, using the DAS and IAS respectively, are unknown. One ablation-behavioral study using a combined DAS/IAS lesion (Masterton and Moreland-Granger, '88) indicated a lesion-induced loss of tone and noise "detection reliability". Recent speculation, based on responses of DCN cells indicates that the DAS axons could be carrying pertinent cues used in monaural localization tasks. Because of space limitations and the very preliminary nature of our data we will not describe data from DAS axons and only briefly describe some of our results on cells whose axons use the IAS.

At least three types of cells in the VCN are known to send their axons out of the CN via the IAS, the octopus cells of the octopus cell area (Adams and Warr, '76; Osen, '69) (OCA), the large multipolar cells in the VCN which innervate the contralateral CN (Cant and Gaston, '82; Wenthold, '87) and the large multipolar cells in the nerve root area that have been designated onset-choppers (Smith and Rhode, '89) ($O_C$s) based on their STCF response.

Good intracellular data have yet to be recorded from octopus cell bodies either *in vivo* or *in vitro*. Early extracellular responses from the OCA were classified as $O_L$ or $O_I$ (Godfrey et al., '75) ($O_I$; response virtually only at the stimulus onset). Short tone responses from the only octopus cell recorded from and successfully injected within the CN (Rhode et al., '83) were classified as $O_L$, given the historical perspective, but could as easily have been called $PL_N$. Thus far, in the strial experiments, we have injected two axons running in the IAS contralateral to the excitatory ear. Both had what could be classified as $O_L$ responses (Fig. 5) but close examination of the PST reveals a pause or notch in the activity immediately after the onset peak. One of these cells had a response area that extended below 1 kHz. When presented with tones in this region (Fig. 5) its response was highly phase-locked with a

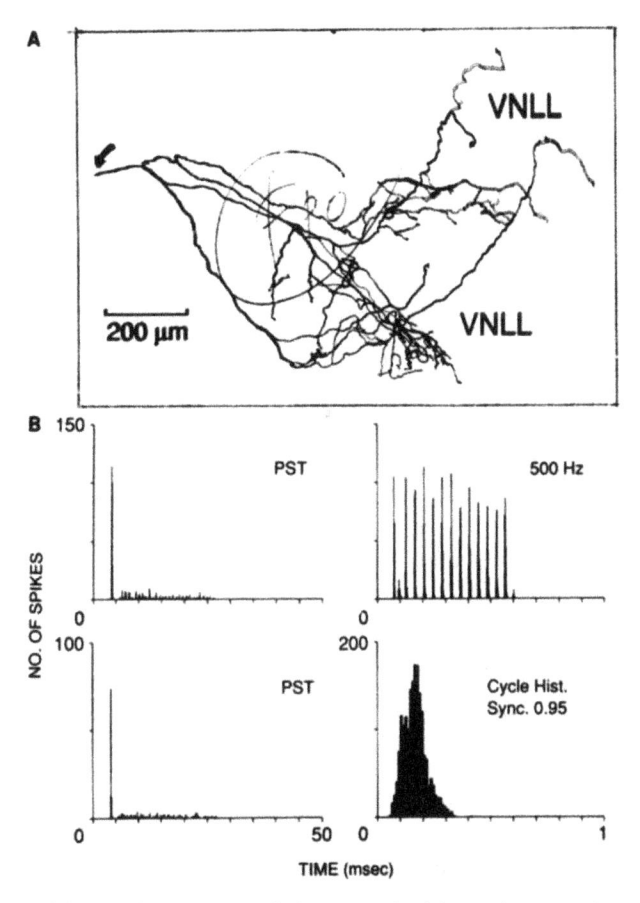

**Figure 5.** Contralateral innervation pattern and short-tone physiology of presumed octopus cell axons. A. Camera lucida drawing of an IAS axon contralateral innervation pattern of the VNLL. The axon came across the midline in the IAS and gave off a single collateral branch to the dorsomedial periolivary nucleus (not shown) then continued laterally and dorsally (curved arrow indicates medial-most point of drawing) to primarily innervate the VNLL. B. Upper panels; PST generated by 200 - 25 msec tones at CF (1360 Hz, left) and at 500 Hz (right) for the axon shown in A. Lower left panel; PST generated by 100 - 25 msec tones at CF (2500 Hz) for another IAS axon with a similar contralateral innervation pattern. Lower right panel; cycle histogram of the spikes during the sustained portion of the PST above indicating the precise phase-locking to the 500 Hz tone. Sync. = synchrony coefficient.

synch. coefficient well over the upper limit of AN fibers with similar CFs. Thus, many of the features of these responses, namely the well timed onset, the low level of sustained activity which may be preceeded by a notch, and the degree of phase-locking certainly resemble features of the globular bushy cell population. Anatomically, the cell body and proximal dendritic tree of the octopus cell have been shown to be covered with synaptic terminals many of which are of AN origin (Kane, '73). No systematic study has been done on the synaptic inputs to the distal dendritic tree. Lesions confined primarily to the OCA (Warr, '82) show a primary axonal projection to the contralateral VNLL. Fast blue injections into contralateral CN (Schofield and Cant, '88) labeled terminals in the VNLL, many of which were large "calyx -like" endings that contained large round vesicles. These could potentially provide a fast disynaptic pathway to the IC, as virtually every cell in the VNLL projects there. Our injected IAS axons show a similar heavy innervation of the VNLL (Fig. 5). Both axons crossed the midline just dorsal to the TB and headed laterally and rostrally over the dorsal edge of the MSO. Each gave off a single collateral branch to the DMPO then continued rostrolaterally where it sent a profusion of branches into the VNLL as well as what appeared to be the rostral-most extent of the LNTB (Guinan Jr., et al., '72). Similar to Schofield and Cant's description, some of the VNLL collaterals gave rise to very large terminal specializations. We were unable to follow either axon back to its cell body.

We have not recorded from or labeled any axons of members of the large multipolar cell class that display onset chopper responses. Intracellular recordings within the CN from subsequently labeled $O_C$ cells (Smith and Rhode, '89) were from large multipolar cells located in the nerve root region. Our limited number of labeled cells makes it plausible that this cell type might be located in other areas of the CN as well. Anatomical and response features of these cells differed considerably from the type I multipolar - sustained chopper population located in this region (see above). Physiologically, the $O_C$s respond over a wider range of frequencies and have wider dynamic ranges to CF tones in both spike output and level of depolarization. Curiously, they do not show regular sustained firing at levels of depolarization that would surely elicit such a response from the sustained chopper. Equally curious is the observation that those multipolar cells in the mouse slice, proposed to be the rodent homologue of this cell type (Oertel et al., '90), show regular firing patterns to sustained depolarizing current. At the EM level the cell body and proximal dendrites of these large multipolars in the cat are heavily innervated with over 4/5 of the surface covered, a feature used by Cant ('81) to classify "type II" multipolars in the AVCN. Vesicle shape indicated that over 1/3 of these terminals originated from AN fibers. The axons of the $O_C$ cells send myelinated collateral branches into the PVCN as well as the DCN before entering the IAS. Unlike the round synaptic vesicles contained within the collateral terminals of the sustained chopper population, the $O_C$ terminals have non-round synaptic vesicles implying an inhibitory function.

We have not recorded from or injected tracer into any commissural fibers that innervate the contralateral CN and whose cell bodies are located in all three major regions of the nucleus. Nothing is known about the physiology or the ultrastructural anatomy of the commissural cells. However, certain features common to these cells (Cant and Gaston, '82; Osen et al., '90; Wenthold, '87) and $O_C$s have led to the speculation (Osen, personal communication) that the $O_C$ population may, in fact, be a part of the commissural population. These features are: 1) The labeled $O_C$ cells are large multipolar cells in the nerve root region as are some of the commissural cell population. 2) The axons of both cells utilize the IAS. 3) The terminals of $O_C$s contain non-round, presumably inhibitory vesicles, and the somata, axons and axon terminals of the commissural cells are glycinergic in nature.

We shall continue to use these approaches to further elaborate the basic structure and function of this nucleus that serves as a gateway to the auditory central nervous system.

REFERENCES

Adams, J.C., 1979, Ascending projections to the inferior colliculus, *J. Comp. Neurol.*, 183:519-538.

Adams, J.C, 1983, Cytology of periolivary cells and the organization of their projections in the cat, *J. Comp. Neurol.*, 215: 275-289.

Adams, J.C., 1983, Multipolar cells in the ventral cochlear nucleus project to both the dorsal cochlear nucleus and the inferior colliculus, *Neurosci. Lett.*, 37:205-208.

Adams, J.C. and Warr, W.B., 1976, Origins of axons in the cat's acoustic striae determined by injection of horseradish peroxidase into severed tracts, *J. Comp. Neurol.*, 170:107-122.

Banks, M.I. and Sachs, M.B., 1991, Regularity analysis and a compartmental model of chopper units in the anteroventral cochlear nucleus, *J. Neurophysiol.*, 65:606-629.

Banks, M.I. and Smith, P.H., 1990, Intracellular recordings from cells in the rat superior olivary complex (SOC) labeled with biocytin, *Soc. Neurosci. Abstr.*, 16:722.

Brawer, J.R., Morest, D.K. and Kane, E., 1974, The neuronal architecture of the cochlear nucleus of the cat, *J. Comp. Neurol.*, 155:251-300.

Blackburn, C.C. and Sachs, M.B., 1989, Classification of unit types in the anteroventral cochlear nucleus: PST histograms and regularity analysis. *J. Neurophysiol.*, 62:1303-1329.

Cant, N.B., 1981, The fine structure of two types of stellate cells in the anterior division of the anteroventral cochlear nucleus, *Neuroscience*, 6:2643-2655.

Cant, N.B., 1982, Identification of cell types in the anteroventral cochlear nucleus that project to the inferior colliculus, *Neurosci. Lett.*, 32:241-246.

Cant, N.B., and Gaston, K.G., 1982, Pathways connecting the right and left cochlear nuclei, *J. Comp. Neurol.*, 212:313-326.

Casseday, J.H. and Neff, W.D., 1975, Auditory localization: role of auditory pathways in the brainstem of cat, *J Neurophysiol.*, 38:842-858.

Cassella, J.V. and Davis, M., 1986, Habituation, prepulse inhibition, fear conditioning, and drug modulation of the acoustically elicited pinna reflex in rats, *Behav. Neurosci.*, 100: 39-44.

Davis, M.,, Gendelman, D.S., Tischler, M.D. and Gendelman, P.M., 1982, A primary acoustic startle circuit: lesion and stimulation studies, *J. Neurosci.*, 2:791-805.

Friauf, E. and Ostwald, J., 1988, Divergent projections of physiologically characterized rat ventral cochlear nucleus neurons as shown by intra-axonal injection of horseradish peroxidase, *Exp. Br. Res.*, 73:263-285.

Godfrey, D.A., Kiang, N.Y.S., and Norris, B.E., 1975, Single unit activity in the posteroventral cochlear nucleus of the cat, *J. Comp. Neurol.*, 162:247-268.

Goldberg, J.M. and Brown, P.B.,1969, Responses of binaural neurons of dog superior olivary complex to dichotic tonal stimuli: Some physiological mechanisms of sound localization, *J. Neurophysiol.*, 32:613-636.

Guinan Jr., J.J., Norris, B.E., and Guinan, S.S., 1972, Single auditory units in the superior olivary complex. II. Location of unit categories and tonotopic organization, *Int. J. Neurosci.*, 4:147-166.

Horikawa, K. and Armstrong, W.E., 1988, A versatile means of intracellular labeling: injection of biocytin and its detection with avidin conjugates, *J. Neurosci. Methods*, 25:1-11.

Jeffress, L.A., 1948, A place theory of sound localization, *J. Comp. Physiol., Psychol.* 41:35-39.

Jenkins, W.M. and Masterton, R.B., 1982, Sound localization: effects of unilateral lesions in central auditory system, *J. Neurophysiol.*, 47:987-1016.

Johnson, D.H., 1980, The relationship between spike rate and synchrony in responses of auditory-nerve fibers to single tones, *J. Acoust. Soc. Am.*, 68:1115-1122.

Kane, E.C., 1973, Octopus cells in the cochlear nucleus of the cat: Heterotypic synapses on homeotypic neurons, *Int. J. Neurosci.*, 5:251-279.

Manis, P.B. and Marx, S.O., 1991, Outward currents in isolated ventral cochlear nucleus neurons, *J. Neurosci.*, (in press).

Masterton, R.B., Glendenning, K.K. and Nudo, R.J., 1981, Anatomical-behavioral analyses of hindbrain sound localization mechanisms, *in*: "Neural Mechanisms of Hearing", J.Syka and L. Aikin, eds. Plenum, New York, pp 263-275.

Masterton, R.B. and Moreland-Granger, E., 1988, Role of the acoustic striae in hearing: contribution of the dorsal and intermediate striae to detection of noises and tones, *J. Neurophysiol.*, 60:1841-1860.

Moore, C.N., Casseday, J.H. and Neff, W.D., 1974, Sound localization: the role of the commissural pathways of the auditory system of the cat, *Brain Res.*, 82:13-26.

Neff, W.D., 1962, Neural structures concerned in localization of sounds in space, *Psychol. Beitr.*, 6:492-500.

Oertel, D., Wu, S.H., Garb, M.W., and Dizack, C., 1990, Morphology and physiology of cells in slice preparations of the posteroventral cochlear nucleus of mice, *J. Comp. Neurol.*, 295:136-154.

Osen, K.K., 1969, Cytoarchitecture of the cochlear nuclei of the cat, *J. Comp. Neurol.*, 136:453-585.

Osen, K.K.,, Ottersen, O.P., and Storm-Mathisen, J., 1990, Colocalization of glycine-like and GABA-like immunoreactivities: A semiquantitative study of individual neurons in the dorsal cochlear nucleus of the cat, *in*: "Glycine Neurotransmission", O.P.Ottersen and J. Storm-Mathisen, eds., John Wiley and Sons Ltd., N.Y., pp 417-451.

Ostapoff, E.M., Morest, D.K. and Potashner, S.J., 1985, Retrograde transport of $^3$H-GABA from the cochlear nucleus to the superior olive in guinea pig, *Soc. Neurosci. Abstr.*, 11:1051.

Pfeiffer, R.R., 1966, Classification of response patterns of spike discharges for units in the cochlear nucleus: Tone burst stimulation, *Exp. Brain. Res.*, 1:220-235.

Rhode, W.S., Smith, P.H., and Oertel, D., 1983, Physiological response properties of cells labeled intracellularly with horseradish peroxidase in cat dorsal cochlear nucleus, *J. Comp. Neurol.*, 213:426-447.

Rhode, W.S., Oertel, D., and Smith, P.H., 1983, Physiological response properties of cells labeled intracellularly with horseradish peroxidase in the cat ventral cochlear nucleus, *J. Comp. Neurol.*, 213:448-463.

Schofield, B.R. and Cant, N.B., 1988, Structure of synaptic calyces and their post-synaptic targets in the ventral nucleus of the lateral lemniscus in the guinea pig, *Soc. Neurosci. Abstr.*, 14:487.

Smith, P.H., and Rhode, W.S., 1985, Electron microscopic features of physiologically characterized, HRP-labeled fusiform cells in the cat dorsal cochlear nucleus, *J. Comp. Neurol.*, 237:127-143.

Smith, P.H., and Rhode, W.S., 1987, Characterization of HRP-labeled globular bushy cells in the cat cochlear nucleus, *J. Comp. Neurol.*, 266:360-376.

Smith, P.H., and Rhode, W.S., 1989, Structural and functional properties distinguish two types of multipolar cells in the ventral cochlear nucleus, *J. Comp. Neurol.*, 282:595-616.

Smith, P.H.,, Joris, P.X,, Carney, L.H. and Yin, T.C.T., 1991, Projections of physiologically characterized globular bushy cell axons from the cochlear nucleus of the cat, *J. Comp. Neurol.*, 304:387-407.

Spangler, K.M., Cant, N.B.,, Henkel, C.K., Farley, G.R. and Warr, W.B., 1987, Descending projections from the superior olivary complex to the cochlear nucleus of the cat, *J. Comp. Neurol.*, 259: 452-465.

Tolbert, L.P., Morest, D.K., and Yurgelun-Todd, D.A., 1982, The neuronal architecture of the anteroventral cochlear nucleus of the cat in the region of the cochlear nerve root: Horseradish peroxidase labelling of identified cell types, *Neuroscience*, 7:3031-3052.

Warr, W.B., 1982, Parallel ascending pathways from the cochlear nucleus: Neuroanatomical evidence of functional specialization, *in:* "Contributions To Sensory Physiology. Vol. 7", Academic Press., pp. 1-38.

Wenthold, R.J., 1987, Evidence for a glycinergic pathway connecting the two cochlear nuclei: An immunocytochemical and retrograde transport study, *Brain Res.*, 415:183-187.

Winter, I.M., Robertson, D., and Cole, K.S., 1989, Descending projections from auditory brainstem nuclei to the cochlea and cochlear nucleus of guinea pig, *J. Comp. Neurol.*, 280:143-157.

Wu, S.H., and Oertel, D., 1984, Intracellular injection with horseradish peroxidase of physiologically characterized stellate and bushy cells in slices of mouse anteroventral cochlear nucleus, *J. Neurosci.*, 4:1577-1588.

Young, E.D., Robert, J.-M., and Shofner, W.P. 1988, Regularity and latency of units in ventral cochlear nucleus: Implications for unit classification and generation of response properties, *J. Neurophysiol.*, 60:1-29.

# PHYSIOLOGY OF THE DORSAL COCHLEAR NUCLEUS MOLECULAR LAYER

Paul B. Manis, John C. Scott and George A. Spirou*

Departments of Otolaryngology-Head and Neck Surgery, Neuroscience and Biomedical Engineering, and The Center for Hearing Sciences, The Johns Hopkins University School of Medicine, Baltimore, MD 21205, USA

*Department of Otolaryngology, West Virginia University School of Medicine, Morgantown, WV 26506, USA

The dorsal cochlear nucleus (DCN) is a complex but elegantly organized division of the cochlear nucleus. Lorente de Nó ('81) recognized that the cortex of the nucleus (layers 1 and 2) bore a remarkable resemblance to the cerebellum, in both the structural characteristics of the cells and their organization. The chief organization of the nucleus is based on a concentric laminar arrangement of cells, their dendrites, and fiber systems, somewhat like a single cerebellar folium. A local cartesian framework can be imposed on the nucleus, where the principal axes are depth (perpendicular to the local surface), the strial axis (parallel to the long axis of the nucleus and to the unmyelinated parallel fibers in the molecular layer), and the transstrial axis (parallel to the isofrequency planes formed by the auditory nerve fibers). Subsequent anatomical studies have exposed further similarities between the outer layers of the DCN and the cerebellum. Although the two structures serve different functions, the common ontogenetic origins and parallel expression of proteins in corresponding cell types (Mugnaini and Morgan, '87; Mugnaini et al., '87; Berrebi and Mugnaini, this volume; Berrebi et al., '90) suggests that the two structures share some mechanisms of information processing.

We are studying how activity in molecular layer cells and fibers influences the neuronal processing that takes place in the principal final common pathway of the DCN, the pyramidal cells. Our approach involves studying specific physiological mechanisms related to cellular information processing, including membrane conductances and synaptic function. We will first present a description of the properties of the principal cells, or pyramidal cells, followed by a description of the properties of the most common inhibitory interneuron found in this region, the cartwheel cells. Both of these cell types receive a major portion of their excitatory input from the axons of granule cells, or parallel fibers (Osen and Mugnaini, '80). In the last section, we examine the role that protein kinase C may play in modulating synaptic transmission at parallel fiber synapses in the molecular layer. All experiments were carried out in an *in vitro* brain slice preparation of the guinea pig DCN (Manis, '89a, '90b), with intracellular recording, field potential recording and pharmacological manipulations.

# Membrane conductances of DCN cells

Mammalian central neurons differ in the types of ion channels in their membranes (see Llinás, '88). The presence of different sets of channels in specific neuronal types results in distinct and varied repertoires of intrinsic discharge patterns, even in the absence of interacting cell networks. Within a given local region or nucleus different morphological cell types have been shown to exhibit differing physiological properties (for example, Wu and Oertel, '84; Mason and Larkman, '90). Cell responses to excitatory or inhibitory conductances then depend upon several factors, including the cell's membrane potential at the initiation of the synaptic conductance, the recent (tens of milliseconds to seconds) history of the membrane potential, and the biochemical state of the ion channels participating during the ensuing voltage changes. We have begun to characterize the membrane properties of two cell types contacted by parallel fibers, the pyramidal cells and the cartwheel cells.

**Pyramidal cells.** Intracellular recordings from morphologically identified pyramidal cells in the brain slice show them to be "simple-spiking" cells that discharge action potentials overshooting 0 mV, and frequently exhibit distinctive two-component afterhyperpolarizations (Oertel and Wu, '89; Manis, '90b). For depolarizing current injections from rest, the cells fire regularly and the interspike interval shows little time dependent change over the first 300 msec.

Hirsch and Oertel ('88a) showed that simple spiking DCN cells often exhibited two important properties. The first was that these cells responded nonlinearly to very small depolarizing current injections, with a slowly increasing depolarization that could outlast the current pulse and generate an action potential. This nonlinearity was shown to be sensitive to tetrodotoxin, and was therefore postulated to be due to a non-inactivating sodium conductance similar to that characterized in cortical neurons (Stafstrom et al., '85) and Purkinje cells (Llinás and Sugimori, '80). Such a conductance produces inward rectification in the region just positive to the cell resting potential, and makes the cell behave as a leaky integrator in this voltage range. As a result, the cell can respond (although with long latency) to a barrage of subthreshold excitatory synaptic currents because the persistent conductance enhances the temporal summation of depolarizing inputs. The second property was that pyramidal cells exhibit a long duration afterhyperpolarization after a period of intense discharge produced by depolarizing currents. Such slow hyperpolarizations are present in other central neurons, where they are generated by one or more potassium conductances and are frequently under the control of neuromodulatory agents.

Although the normal mode of discharge of DCN pyramidal cells at rest is a regular train of action potentials, the first 100 msec of the discharge to a current step is sensitive to the membrane potential prior to the current pulse (Manis, '90b). This is illustrated for one cell in Figure 1. For small depolarizing steps from rest (i.e., near -56 mV for this example; Fig. 1A), the cell fires regularly. The same step delivered to the cell after it has been hyperpolarized to -65 mV results in a long latency to the first spike (Fig. 1B) although the steady state discharge rate will be nearly the same as from rest. During the depolarizing current step, the cell membrane potential very slowly depolarizes at a mean rate near 0.1 mV/ms. When the cell reaches the threshold for action potential generation, it begins firing regularly at nearly the same rate as when it is depolarized from rest. Larger depolarizing pulses from rest cause the cells to discharge at a higher rate (Fig. 1C). Hyperpolarization of the cell to -70 mV prior to the step then results in a response consisting of an initial action potential, followed by a long first interspike interval, and finally a resumption of regular firing (Fig. 1D). During the first interspike interval the cell is significantly depolarized, and also often shows a slow depolarizing shift in membrane potential. Interestingly, the afterhyperpolarizations of the first few action potentials reach a more negative potential than when the cell is driven from rest (also see Manis, '90b, figure 8B). The voltage range over which long latency and long first interspike intervals can be evoked is -60 to -85 mV,

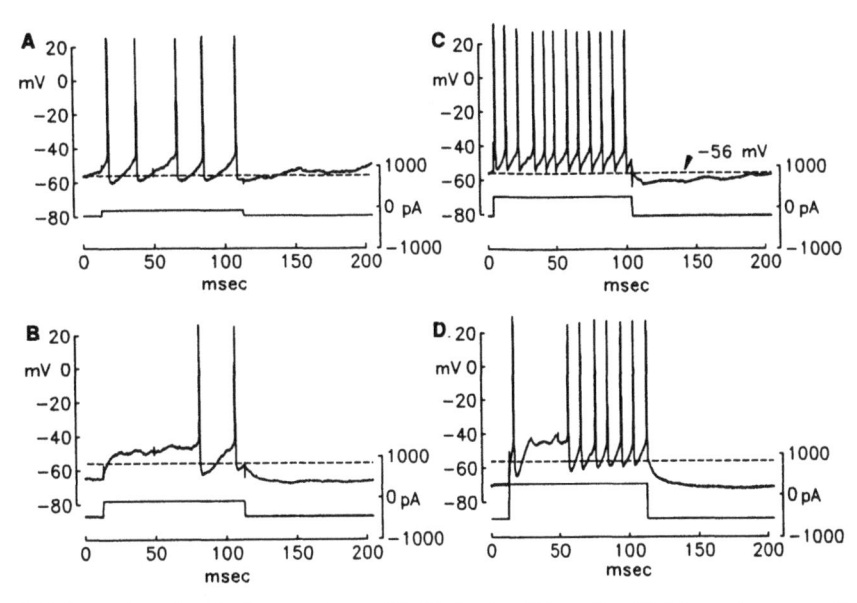

**Figure 1.** Voltage-dependent discharge patterns of DCN simple spiking cells. All data are from one cell. A: Cell is held with -200 pA of current to stop spontaneous activity (holding potential -56 mV as indicated in panel C). Cell responds to 120 pA step (to -80 pA) with a regular discharge of action potentials. B: Cell hyperpolarized to -65 mV, using -445 pA; step to -80 pA now results in discharge with long latency and slow subthreshold voltage rise. C: Injection of larger current (278 pA from -200 pA holding) results in higher discharge rate. D: When cell is first hyperpolarized to -70 mV with -566 pA, then stepped to +278 pA, discharge shows a long first interspike interval followed by regular discharge. Note substantial depolarization during the first interspike interval. In all panels, the dashed line indicates -56 mV (see panel C), corresponding to the membrane potential at which spontaneous activity was stopped by hyperpolarizing current through the recording electrode. Vertical mark on traces at 50 msec is an electronic artifact. Cell: 13 Jun 89 I.

whereas the mean resting potential of non-spontaneously active cells in the slices is -56 mV (Manis, '90b).

These observations indicate that the intrinsic membrane conductances of DCN neurons could contribute to their response patterns to tones *in vivo*. The regular firing, long latency and long first interspike interval patterns resemble the "chopper", "buildup" and "pauser" responses to tonal stimuli (Pfeiffer, '66), respectively, associated with the pyramidal cells (Godfrey et al., '75; Rhode et al., '83). Previously these response patterns had been hypothesized to be generated by specifically timed interactions between excitatory and inhibitory synaptic inputs to the cells, with the cessation of discharge specifically associated with an inhibitory input (e.g., Kane, '74). Our observations suggest that the requirement for such timed inhibitory conductances can be relaxed: hyperpolarization occurring some time prior to an excitatory input, such as from auditory nerve fibers, can also alter the response pattern. Since the response patterns shown here resulted from intracellular current injection, it follows that *any* change in membrane conductance which hyperpolarizes a cell could cause the cell to undergo a transition from one discharge pattern to another. Specific mechanisms for such hyperpolarization could include synaptic inhibition (not necessarily associated with auditory stimulation), slow afterhyperpolarizations (e.g., Hirsch and Oertel, '88a; Manis, '90b), or activation of potassium or chloride conductances by calcium or other second messenger systems.

One explanation (Manis, '90b) for the voltage-dependent discharge patterns of DCN pyramidal cells requires the involvement of a transient potassium conductance similar to the "A" current (Connor and Stevens, '71). Most, but not all transient potassium currents are

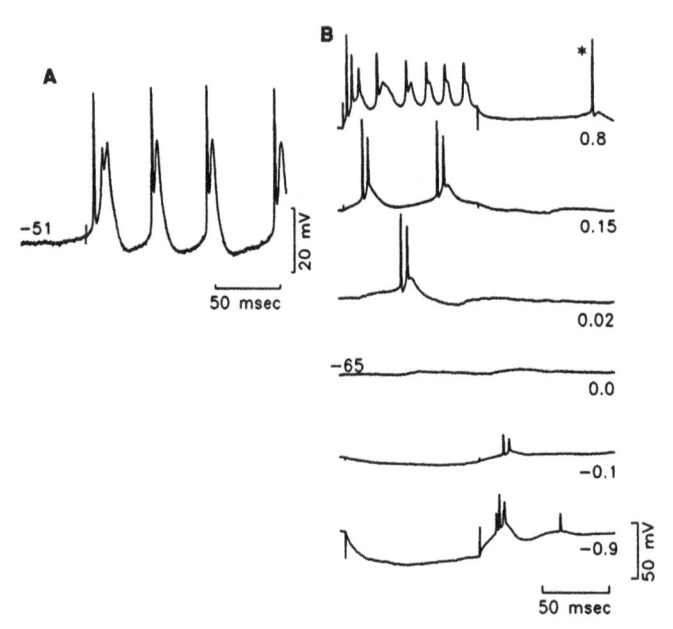

**Figure 2.** Complex spiking cell in DCN. A: Spontaneous activity in complex spiking cell immediately after impalement; resting potential -51 mV. Activity consists of bursts with an approximate interval of 50 msec. Vertical mark before first burst is an electronic artifact. B: Same cell as in A, responses to current pulses (current level indicated beneath traces at right, in nA) after hyperpolarizing the cell to stop spontaneous activity (holding potential -65 mV). Responses to hyperpolarizing pulses (lower 3 traces) are averages of 4 presentations; cell discharged bursts after the pulse was turned off. Action potential amplitude on anodal break is reduced due to lack of precise synchrony. Small depolarizing pulses give rise to one or two clear bursts; large pulses give rise to a series of oscillations with a maximal rate near 60 Hz, and with superimposed fast action potentials. Occasional single action potentials are seen, as in the top trace, marked with asterisk. Cell: 13 Jun 89 E.

sensitive to the potassium channel blocker 4-aminopyridine (4-AP), and DCN cells appear to posses a conductance that is sensitive to 4-AP (Hirsch and Oertel, '88a). A transient potassium conductance would be de-inactivated (i.e., more of the channels become available for activation) by membrane hyperpolarization, increasing the number of channels available for activation during a subsequent depolarization. When depolarizing currents are injected, the transient potassium conductance will be activated and create an outward current that tries to hyperpolarize the cell. Membrane depolarization would proceed slowly only as the transient potassium conductance inactivates. In this way, a transient K current could contribute to the generation of long latency discharges or to a long first interspike interval.

**Cartwheel cells.** The cartwheel cells, after the granule cells, are the most numerous interneuronal type in the DCN (Wouterlood and Mugnaini, '84; Berrebi and Mugnaini, '91a). Cartwheel cells are biochemically and to some extent morphologically similar to cerebellar Purkinje cells (Wouterlood and Mugnaini, '84; Berrebi et al., '90), although they are not identical. One of the hallmarks of Purkinje cell physiology is the "complex spike" burst that can be recorded extracellularly in response to stimulation of climbing fiber afferents (Eccles et al., '67) or evoked with intracellular depolarization (Llinás and Sugimori, '80). It was noted in earlier experiments (Hirsch and Oertel, '88a; Manis, '90b) that some cells in the DCN exhibited complex action potential profiles. We therefore hypothesized that this particular physiology was associated with the cartwheel cells. A combined intracellular recording and staining study was undertaken to test this hypothesis. Seven complex spiking cells with unambiguous physiological correlations were recovered; every complex spiking cell was a cartwheel neuron (Spirou et al., '91). In contrast, all morphologically identified pyramidal cells discharged simple spikes (see also Oertel and Wu, '89; Manis, '90b). Morphologically, the cartwheel cells were very similar to those described by Hackney et al. ('90) and Berrebi and Mugnaini ('91a).

The complex spike bursts consist of 2-3 fast action potentials superimposed on a slow depolarization. Recordings from one complex spiking cell (morphologically unidentified) are shown in Figure 2. This cell exhibited spontaneous complex spikes arising from a resting potential of -51 mV, with an interspike interval of about 50 msec. The response of the cell to current pulses is shown in Figure 2B. The cell exhibited anodal break bursts (bottom two traces). For the hyperpolarizing current steps the data shown are averages of 4 presentations. The anodal break bursts were not well synchronized and not all presentations resulted in a burst, resulting in small averaged spike amplitudes. However, spikes on each individual response were full amplitude. With small depolarizing current pulses the cell fired distinct bursts, whereas larger depolarizing currents resulted in a series of slow potential oscillations superimposed with fast action potentials.

The slow depolarization underlying the burst response is probably produced by a calcium conductance (Hirsch and Oertel, '88a; Manis, unpublished observations). There may be more than one type of voltage-sensitive calcium conductance in these cells, because the bursting phenomenon can be evoked under two different conditions. First, bursts are evoked at the anodal break of hyperpolarizing pulses, which would be consistent with activation of T-type calcium channels (Fox et al., '87; also Wang et al., '91). Second, bursts are evoked during large depolarizing pulses where the membrane potential does not return to the resting level between bursts. This probably precludes the involvement of somatically located T-type calcium channels, since these channels require hyperpolarization to remove inactivation. T-type channels could become involved only if they are located at an electrotonicically remote site in the cell that undergoes greater hyperpolarization between the slow depolarizations than the somatic recording site.

It is important to recognize that the physiology of the cartwheel cells is not the same as that of Purkinje cells. Although the cartwheel cells show simple bursts for small current injections, and trains of bursts for larger injections, only 3 cells out of our sample of 30 complex-spiking cells have exhibited a "plateau potential", which is characteristic of Purkinje cells (Llinás and Sugimori, '80). Also, Purkinje cells do not produce complex spike discharges on anodal break. The cartwheel cells also have very rare and limited "simple-spiking" modes of behavior (see asterisk in Fig. 2B); when isolated action potentials do occur they are followed by afterdepolarizations. In contrast, the simple spike firing mode is common in Purkinje cells. Interestingly, the overall features of the bursting response of the cartwheel cells closely resemble those found in some cortical pyramidal cells (McCormick et al., '85; Mason and Larkman, '90).

The rapid discharge of 2-4 action potentials may make cartwheel cells more effective in producing transmitter release at their terminals. Since one of the primary targets of cartwheel cells appears to be the pyramidal cells (Berrebi and Mugnaini, '91b), they should have a strong influence on the pyramidal cells. In intracellular recordings, parallel fiber stimulation produces an EPSP-IPSP sequence in most cells (Hirsch and Oertel, '88b; Manis, unpublished observations); the IPSP is almost completely blocked by 2-4 $\mu$M strychnine, suggesting that it is mediated by a glycinergic synapse. As the cartwheel cells are immunoreactive for glycine (Wenthold et al., '87), they are a major candidate for this inhibitory input. However they also stain for gamma-aminobutyric acid and glutamic acid decarboxylase (Mugnaini, '85; Peyret et al., '86; Wenthold et al., '86), although so far inhibitory potentials attributable to GABAergic transmission have been elusive in slice experiments.

To summarize, we have shown that two different cell types in the outer layers of the DCN have different physiological properties. The pyramidal cells exhibit regular firing when depolarized from rest, but the pattern of firing can be modified by prior hyperpolarization in

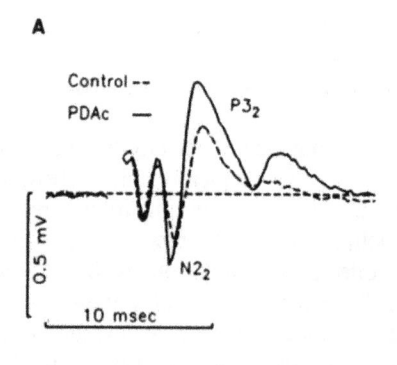

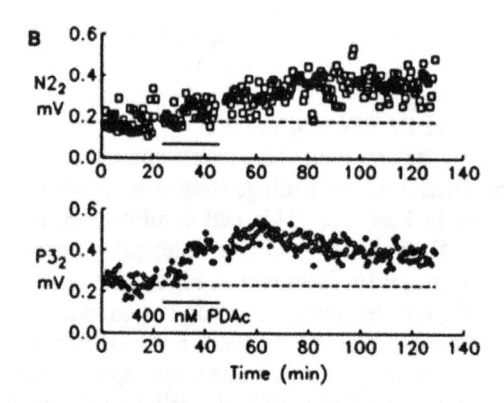

Figure 3. Effects of 400 nM phorbol 12,13 diacetate (PDAc) on parallel fiber evoked field potentials recorded in layer 2 of DCN in strial slice. A: Control response (dashed line) and response at end of 20 minute wash with PDAc (solid line). The phorbol ester did not change the afferent fiber volley, but increased the amplitudes of the population spike ($N2_2$) and synaptic current ($P3_2$). B: Upper plot shows the amplitude of the $N2_2$, recorded once every 30 seconds, as a function of time during the experiment. The PDAc was washed across the slice only during the time indicated by the bar between 22 and 42 minutes. The dashed line indicates the mean amplitude of the response during the period prior to PDAc application. The lower plot shows the amplitude of the $P3_2$ determined simultaneously in the same experiment. Both events undergo an amplitude increase, but have different time courses. Experiment: 9 May 91 G.

the voltage range that can be attained by inhibitory synaptic potentials or slow after hyperpolarizations. In contrast, the cartwheel cells show complex action potential discharge, a portion of which may be mediated by calcium channels. The cartwheel cells are one potential source of inhibitory input that can hyperpolarize pyramidal cells and thereby influence their response to other excitatory stimuli.

## Parallel fibers and protein kinase C in the DCN

The molecular layer of the DCN contains many components of second messenger systems, in contrast to the deep layers of the DCN and other regions of the cochlear nucleus. This suggests that these second messenger system components are important to specific functions carried out by cells in the molecular layer. Many components of second messenger systems identified in the DCN are also present in the cerebellum. Adenylate cyclase concentrations (as measured by forskolin binding; Worley et al., '86b) are significantly elevated in the DCN molecular layer; by analogy with the cerebellum this is likely to be in the highest concentration in parallel fibers and in lower concentrations in postsynaptic cells. A cyclic-GMP dependent protein kinase has been identified by immunocytochemistry in displaced Purkinje cells in the molecular layer of rat DCN (De Camilli et al., '84). Inositol trisphosphate ($IP_3$) receptor-like immunoreactivity is found in high concentration in both cerebellar Purkinje cells and in DCN cartwheel cells (Mignery et al., '89; Ryugo et al., '92). Protein kinase C (PKC) has been localized to the molecular layer of the DCN by both phorbol ester binding autoradiography (Worley et al., '86a) and by immunocytochemistry (Saito et al., '88). The immunocytochemistry suggests that PKC may be localized both in the parallel fibers and in postsynaptic cells, although the identity of these cells is unclear.

In order to determine whether the anatomically localized PKC has a significant functional role, we examined the effects of phorbol esters, potent and somewhat selective stimulators of PKC, on synaptic transmission at parallel fiber synapses in the DCN. Phorbol ester stimulation of PKC resulted in enhanced synaptic currents at parallel fiber synapses. Figure 3A shows the effects of 400 nM phorbol 12,13 diacetate (PDAc), on field potentials

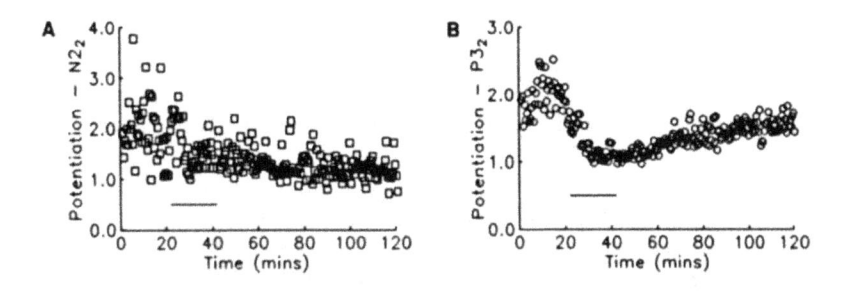

**Figure 4.** Effects of PDAc on paired pulse potentiation. A: Potentiation of the population spike ($N2_2$), determined by comparing responses to pairs of equal amplitude stimuli delivered 40 msec apart, plotted as a function of time in the experiment (as in Fig. 3B). After challenge with PDAc, the potentiation decreases. B: Potentiation of the synaptic currents ($P3_2$). Although the potentiation is initially depressed by the PDAc, potentiation returns during the wash phase. Experiment: 9 May 91 G.

produced in the layer of pyramidal cells by parallel fiber stimulation. The field potentials consist of 3 major components (Manis, '89a). First is the $N1_2$ (not labeled in the figure), which is produced by the synchronous volley of action potentials in the parallel fibers. Second is the $N2_2$, produced by synchronous firing of cells in layer 2 (e.g., population spike). Finally, the large positive wave, $P3_2$, corresponds to the passive component of the excitatory currents produced in the superficial molecular layer by the parallel fiber synapses. PDAc caused an increase in the amplitudes of both the population spike ($N2_2$) and the postsynaptic currents ($P3_2$), and a slight decrease in the latency of the population spike. In Figure 3B, the time course of the effect of PDAc is shown. The dashed line indicates the mean response amplitude over the first 20 minutes of recording, prior to application of PDAc (which is indicated by bar at 22-42 mins.). PDAc increased the $P3_2$, and this effect shows partial reversal during the wash (80-130 min.). The increase in $N2_2$ amplitude occurs more slowly and is stable at the 80-130 min. time points where the $P3_2$ declines. Similar effects on all response components were observed with 200 nM and 1 $\mu$M PDAc. The inactive phorbol ester, 4-$\alpha$-phorbol, had no effects on any part of the response at 1 $\mu$M.

The phorbol application also produced a dramatic decrease in paired-pulse potentiation. This is shown in Figure 4, for the same experiment as shown in Figure 3. The potentiation of the population spike, measured at 40 msec interstimulus interval, was initially large and variable, but decreased to a steady level just above 1.0 after challenge with PDAc (Fig. 4A). The potentiation of the synaptic currents was also initially large, and decreased rapidly upon challenge with PDAc; however the potentiation of the synaptic currents returned as the phorbol was washed out of the slice (Fig. 4B). The changes in the amplitudes and potentiation of the synaptic currents are consistent with a presynaptic site of action of the phorbol ester (Harris and Cotman, '83).

These results demonstrate that direct stimulation of protein kinase C has physiological effects in the DCN and indicate that the actions of the PDAc may differ in the presynaptic terminal (increase in response amplitude with decrease in paired pulse potentiation) and postsynaptic cell (dissociation between effects on synaptic current, $P3_2$, and the population spike, $N2_2$, in Figs. 3B and 4). However, we cannot conclude from these results that PDAc has induced long term potentiation, since PDAc is a lipophilic compound and is difficult to wash out of the slice.

At present, it is not known which membrane receptors on DCN cells are coupled to any of the second messenger systems, and, other than experiments like those shown in Figures 3 and 4, the effects of activating those second messenger systems are equally unknown. We can look to the cerebellum to provide clues about possible pathways. In adult rat cerebellum, quisqualate can stimulate a metabotropic receptor that drives phosphatidylinositol turnover (Blackstone et al., '89), and this effect is blocked by non-N-methyl-D-aspartate (NMDA) receptor antagonists. Recently, it has been shown that activation of both the metabotropic quisqualate receptor and an ionotropic receptor ($\alpha$-amino-3-hydroxy-5-methyl-4-isoxazole-propionic acid, AMPA) is necessary to produce persistent decreases in the sensitivity to glutamate (but not to aspartate) in cultured Purkinje cells (Linden et al., '91); phorbol ester stimulation of PKC also produces a decrease in glutamate sensitivity in cerebellar slices (Crepel and Krupa, '88). Since the other arm of $IP_3$ production is the generation of diacylglycerol, an endogenous activator of PKC, it can be postulated that metabotropic receptor activation also activates PKC. In hippocampus, activation of one type of metabotropic quisqualate receptor (2-amino-3-phosphonopropionic acid (AP3) insensitive; 1-aminocyclopentane-1,3-dicarboxylic acid (ACPD) activated) results in a decrease in the strength of a slow afterhyperpolarization usually attributed to calcium-dependent potassium channels (Stratton et al., '91); such an action would be consistent with enhancement of transmitter release at the parallel fibers if calcium-dependent potassium channels are present in or near the terminals. However, direct evaluation of the effects of PKC or metabotropic receptor stimulation on ionic channels has not been done in DCN.

The presence of functional components of second messenger systems as well as NMDA receptors (Manis, '89a; '90b) associated with the circuitry of the DCN molecular layer leads to the hypothesis that these systems are involved in the generation of long term changes in synaptic or voltage-dependent conductances, as they are in other brain regions. Thus, NMDA receptor activation is necessary (but not sufficient) for the activation of some types of hippocampal and cortical long term potentiation (Collinridge et al., '83; Kimura et al., '89; Komatsu et al., '88; Nicoll et al., '88). Increased intracellular calcium is a critical step in this process (Malenka et al., '88), as is the activation of second messenger systems (Malinow et al., '88, '89). On the other hand, NMDA receptor activation is not a necessary step for other kinds of long term synaptic changes to occur (Abraham et al., '91; Fields et al., '91; Ito, '89; Komatsu et al., '91). For example, cerebellar parallel fiber synapses on Purkinje cells undergo depression when stimulated simultaneously with climbing fibers (Crepel and Krupa, '88; Ito et al., '82; Kano and Kato, '87; Sakuri, '87), and this depression appears to depend on climbing-fiber induced calcium entry into Purkinje cells (Crepel and Jaillard, '91; Crepel and Audinat, '91; Hirano, '90; Sakuri, '90). Calcium release from intracellular stores may also be important, as metabotropic receptor activation (in conjunction with non-NMDA ionotropic receptor activation) appears necessary for the generation of long term changes in receptor sensitivity in cerebellar Purkinje cells in culture (Linden et al., '91). Although many of the biochemical mechanisms implicated in cerebellar plasticity appear to be present in the DCN, the DCN lacks climbing fiber innervation, which is thought to be a critical component in the cerebellar system.

We have shown that pyramidal cells in the DCN exhibit voltage-dependent discharge patterns, that the cartwheel cells exhibit complex action potential bursts that are probably partially calcium-dependent, and that synaptic transmission at parallel fiber synapses is modulated by action of a protein kinase. These results indicate only some of the complexity of information processing mechanisms in the DCN molecular layer. Other second messenger pathways are also likely to modulate parallel fiber synaptic transmission and the ionic conductances involved in the discharge characteristics of pyramidal and cartwheel cells. It will be interesting in the future to see how specific types of receptors and second messenger systems are coupled in the DCN, and to determine their contribution to information processing by cells in the molecular layer.

ACKNOWLEDGEMENTS

We are indebted to Dr. David K. Ryugo, Debora Wright and Sussan Paydar for their major contributions to the morphological identification of the complex spiking cells. This work was supported by NIH-NIDCD grants R01-DC00425 and an RCDA K04-DC00048 to PBM.

# REFERENCES

Abraham, W.C. and Wickens, J.R., 1991, Heterosynaptic long-term depression is facilitated by blockade of inhibition in area CA1 of the hippocampus, *Brain Res.*, 546:336-340.

Berrebi, A.S. and Mugnaini, E., 1991a, Distribution and targets of the cartwheel cell axon in the dorsal cochlear nucleus of the guinea pig, *Anat. Embryol.*, 183:427-454.

Berrebi, A.S., Morgan, J.I., and Mugnaini, E., 1990, The Purkinje cell class may extend beyond the cerebellum, *J. Neurocytol.*, 19:643-654.

Blackstone, C.D., Supattapone, S., and Snyder, S.H., 1989, Inositolphospholipid-linked glutamate receptors mediate cerebellar parallel-fiber-Purkinje-cell synaptic transmission, *Proc. Natl. Acad. Sci.*, 86:4316-4320.

Collinridge, G.L., Kehl, S.J., and McLennan, H., 1983, The antagonism of amino acid-induced excitation of rat hippocampal CA1 neurons in vitro, *J. Physiol.*, 344:33-46.

Connor, J.A. and Stevens, C.F., 1971, Voltage clamp studies of a transient outward membrane current in gastropod neural somata, *J. Physiol.*, 312:21-30.

Crepel, F. and Krupa, M., 1988, Activation of protein kinase C induces a long term depression of glutamate sensitivity of cerebellar Purkinje cells. An in vitro study, *Brain Res.*, 458:397-401.

Crepel, F. and Jaillard, D., 1991, Pairing of pre- and postsynaptic activities in cerebellar Purkinje cells induces long-term changes in synaptic efficacy in vitro, *J. Physiol.*, 432:123-141.

De Camilli, P., Miller, P.E., Levitt, P., Walter, U., and Greengard, P., 1984, Anatomy of cerebellar Purkinje cells in the rat determined by a specific immunohistochemical marker, *Neuroscience*, 11:761-817.

Eccles, J.C., Ito, M., and Szentagothai, J., 1967, "The Cerebellum as a Neuronal Machine," Springer-Verlag, New York.

Fields, R.D., Yu, C., and Nelson, P.G., 1991, Calcium, network activity, and the role of NMDA channels in synaptic plasticity in vitro, *J. Neurosci.*, 11:134-146.

Fox, A.P., Nowycky, M.C., and Tsien, R.W., 1987, Kinetic and pharmacological properties distinguish three types of calcium currents in chick sensory neurones, *J. Physiol.*, 394:149-172.

Godfrey, D.A., Kiang, N.Y.S., Norris, B.E., 1975, Single unit activity in the dorsal cochlear nucleus of the cat, *J. Comp. Neurol.*, 162:269-284.

Hackney, C.M., Osen, K.K., Kolston, J., 1990, Anatomy of the cochlear nuclear complex of guinea pig, *Anat. Embryol.*, 182:123-149.

Harris, E.W. and Cotman, C.W., 1983, Effects of acidic amino acid antagonists on paired-pulse potentiation at the lateral perforant path, *Exp. Brain Res.*, 52:455-460.

Hirano, T., 1990, Depression and potentiation of the synaptic transmission between a granule cell and a Purkinje cell in rate cerebellar culture, *Neurosci. Lett.*, 119:141-144.

Hirsch, J.A. and Oertel, D., 1988a, Intrinsic properties of neurones in the dorsal cochlear nucleus of mice in vitro, *J. Physiol.*, 396:535-548.

Hirsch, J.A. and Oertel, D., 1988b, Synaptic connections in the dorsal cochlear nucleus of mice, in vitro, *J. Physiol.*, 396:549-562.

Ito, M., 1989, Long-term depression, *Ann. Rev. Neurosci.*, 12:85-102.

Ito, M., Sakurai, M., and Tongroach, P., 1982, Climbing fibre induced depression of both mossy fibre responsiveness and glutamate sensitivity of cerebellar of cerebellar purkinje cells, *J. Physiol.*, 324:113-134.

Kane, E.S., 1974, Synaptic organization in the dorsal cochlear nucleus of the cat: A light and electron microscopic study, *J. Comp. Neurol.*, 155:301-330.

Kano, M. and Kato, M., 1987, Quisqualate receptors are specifically involved in cerebellar synaptic plasticity, *Nature*, 325:276-279.

Kimura, F., Nishigori, A., Shirokawa, T., and Tsumoto, T., 1989, Long term potentiation and N-methyl-D-aspartate receptors in the visual cortex of young rats, *J. Physiol.*, 414:125-144.

Komatsu, Y., Fujii, K., Sakaguchi, H., and Toyama, K., 1988, Long-term potentiation of synaptic transmission in kitten visual cortex, *J. Neurophysiol.*, 59:124-141.

Komatsu, Y., Nakajima, S., and Toyama, K., 1991, Induction of long-term potentiation without participation of N-methyl-D-aspartate receptors in kitten visual cortex, *J. Neurophysiol.*, 65:20-32.

Linden, D.J., Dickinson, M.H., Smeyne, M., and Connor, J.A., 1991, A long-term depression of AMPA currents in cultured cerebellar Purkinje neurons, *Neuron*, 7:81-89.

Llinás, R. and Sugimori, M., 1980, Electrophysiological properties of in vitro Purkinje cell somata in mammalian cerebellar slices, *J. Physiol.*, 305:171-195.

Llinás, R., 1988, The intrinsic electrophysiological properties of mammalian neurons: Insights into central nervous system function, *Science*, 242: 1654-1664.

Lorente de Nó, R., 1981, "The Primary Acoustic Nuclei", Raven Press, New York.

Malenka, R.C., Kauer, J.A., Zucker, R.S., and Nicoll, R.A., 1988, Postsynaptic calcium is sufficient for potentiation of hippocampal synaptic transmission, *Science*, 242:81-84.

Malinow, R., Madison, D.V., and Tsien, R.W., 1988, Persistent protein kinase activity underlying long-term potentiation, *Nature*, 335:820-824.

Malinow, R., Schulman, H., and Tsien, R.W., 1989, Inhibition of postsynaptic PKC or CaMKII blocks induction but not expression of LTP, *Science*, 245:862-866.

Manis, P.B., 1989a, Responses to parallel fiber stimulation in the guinea pig dorsal cochlear nucleus in vitro, *J. Neurophysiol.*, 61:149-161.

Manis, P.B., 1989b, Evidence for functional NMDA-type receptors in the guinea pig dorsal cochlear nucleus, *Assoc. for Research in Otolaryngology Abstr.*, 12:60.

Manis, P.B., 1990a, Pharmacology of synaptic transmission at parallel fiber synapses in the dorsal cochlear nucleus, *Assoc. for Research in Otolaryngology Abstr.* 13:95.

Manis, P.B., 1990b, Membrane properties and discharge characteristics of guinea pig dorsal cochlear nucleus neurons studied in vitro, *J. Neurosci.*, 10:2338-2351.

Mason, A. and Larkman, A., 1990, Correlations between morphology and electrophysiology of pyramidal neurons in slices of rat visual cortex. II. Electrophysiology, *J. Neurosci.*, 10:1415-1428.

McCormick, D.A., Connors, B.W., Lighthall, J.W., and Prince, D.A., 1985, Comparative electrophysiology of pyramidal and sparsely spiny stellate neurons of the neocortex, *J. Neurophysiol.*, 54:782-806.

Mignery, G.A., Sudhof, T.C., Takei, K., and DeCamilli, P., 1989, Putative receptor for inositol 1,4,5-trisphosphate similar to ryanodine receptor, *Nature*, 342:192-195.

Mugnaini, E., 1985, GABA neurons in the superficial layers of the rat dorsal cochlear nucleus: Light and electron microscopic immunocytochemistry, *J. Comp. Neurol.* 235:61-81.

Mugnaini, E. and Morgan, J.I., 1987, The neuropeptide cerebellin is a marker for two similar neuronal circuits in rat brain, *Proc. Natl. Acad. Sci.*, 84:8692-8696.

Mugnaini, E., Berrebi, A.S., Dahl, A.-L., and Morgan, J.I., 1987, The polypeptide PEP-19 is a marker for Purkinje neurons in cerebellar cortex and cartwheel neurons in the dorsal cochlear nucleus, *Arch. Ital. Biol.*, 126:41-67.

Nicoll, R.A., Kauer, J.A., and Malenka, R.C., 1988, The current excitement in long-term potentiation, *Neuron*, 1:97-103.

Oertel, D., Wu, S-H., 1989, Morphology and physiology of cells in slice preparations of the dorsal cochlear nucleus of mice, *J. Comp. Neurol.*, 283:228-247.

Osen, K.K. and Mugnaini, E., 1981, Neuronal circuits in the dorsal cochlear nucleus, in: "Neuronal Mechanisms in Hearing," J. Syka and L. Aitkin, eds., Plenum, New York.

Peyert, D., Geffard, M., and Aran, J.-M., 1986, GABA immunoreactivity in the primary nuclei of the auditory central nervous system, *Hearing Res.*, 23:115-121.

Pfeiffer, R.R., 1966, Classification of response patterns of spike discharges for units in the cochlear nucleus: Tone burst stimulation, *Exp. Brain Res.* 1:220-235.

Rhode, W.S., Smith, P.H., and Oertel, D., 1983, Physiological response properties of cells labeled intracellularly with horseradish peroxidase in cat dorsal cochlear nucleus, *J. Comp. Neurol.*. 213:426-447.

Ryugo, D.K., Sharp, A.H., Wright, D.D., and Snyder, S.H., 1992, Immunocytochemical localization of the inositol 1,4,5-trisphosphate receptor in cartwheel cells of the mammalian dorsal cochlear nucleus, *Assoc. for Ressearch in Otolaryngology Abstr.*, 15:76.

Saito, N., Kikkawa, U., Nishizuka, Y., and Tanaka, C., 1988, Distribution of protein kinase C-like immunoreactive neurons in rat brain, *J. Neurosci.*, 8:369-382.

Sakurai, M., 1987, Synaptic modification of parallel fibre-Purkinje cell transmission in in vitro guinea pig cerebellar slices, *J. Physiol.*, 394:463-480.

Sakurai, M., 1990, Calcium is an intracellular mediator of the climbing fiber in induction of cerebellar long-term depression, *Proc. Natl. Acad. Sci.*, 87:3383-3385.

Spirou, G.A., Wright, D.D., Ryugo, D.K., and Manis, P.B., 1991, Physiology and morphology of cells from slice preparations of the guinea pig dorsal cochlear nucleus, *Assoc. for Research in Otolaryngology Abstr.*, 14:142.

Stafstrom, C.E., Schwindt, P.C., Chubb, M.C. and Crill, W.E., 1985, Properties of persistent sodium conductance and calcium conductance of layer V neurons from cat sensorimotor cortex in vitro, *J. Neurophysiol.*, 53:153-170.

Stratton, K.R., Worley, P.F., and Baraban, J.M., 1990, Pharmacological characterization of phosphoinositide-linked glutamate receptor excitation of hippocampal neurons, *Eur. J. Pharmacol.*, 186:357-361.

Wang, X-J., Rinzel, J., and Rogawski, M.A., 1991, A model of the T-type calcium current and the low-threshold spike in thalamic neurons, *J. Neurophysiol.*, 66:839-850.

Wenthold, R.J., Zempel, J.M., Parakkal, K.A., Reeks, K.A., and Altschuler, R.A., 1986, Immunocytochemical localization of GABA in the cochlear nucleus of the guinea pig, *Brain Res.*, 380:7-18.

Wenthold, R.J., Huie, D., Altschuler, R.A., and Reeks, K.A., 1987, Glycine immunoreactivity localized in the cochlear nucleus and superior olivary complex, *Neuroscience*, 22:897-912.

Worley, P.F., Baraban, J.M., and Snyder, S.H., 1986a, Heterogenous localization of protein kinase C in rat brain: Autoradiographic analysis of phorbol ester receptor binding, *J. Neurosci.*, 6:199-207.

Worley, P.F., Baraban, J.M., DeSouza, E.B., and Snyder, S.H., 1986b, Mapping second messenger systems in the brain: Differential localizations of adenylate cyclase and protein kinase C, *Proc. Natl. Acad. Sci.*, 83:4053-4057.

Wouterlood, F.G. and Mugnaini, E., 1984, Cartwheel neurons of the dorsal cochlear nucleus: A Golgi-electron microscopic study in rat, *J. Comp. Neurol.*, 227:136-157.

Wu, S.H. and Oertel, D., 1984, Intracellular injection with horseradish peroxidase of physiologically characterized stellate and bushy cells in slices of mouse anteroventral cochlear nucleus, *J. Neurosci.*, 4:1577-1588.

# CODING OF THE FUNDAMENTAL FREQUENCY OF VOICED SPEECH SOUNDS AND HARMONIC COMPLEXES IN THE COCHLEAR NERVE AND VENTRAL COCHLEAR NUCLEUS.

Alan R. Palmer and Ian M. Winter

MRC Institute of Hearing Research, University of Nottingham
University Park, Nottingham NG7 2RD, U.K.

Voiced vowel sounds are periodic and consist of harmonics of a fundamental frequency (F0). This fundamental frequency is determined by the rate at which the vocal folds are opening and closing and varies over about an octave for a speaker (from about 80 to 150 Hz in males, 160 to 300 Hz in females and 200 to 400 Hz in children). Harmonic series (of which voiced speech sounds are a special case) are characterized by a low-pitch (the voice pitch in speech sounds), which has been extensively studied psychophysically (see Evans, '78; Greenberg, '80). The voice pitch provides a powerful cue for the identification of individual speakers and for the grouping together of the elements of a single sound source allowing them to be segregated from other interfering sounds, including other voices (Assmann and Summerfield, '90; Scheffers, '83). Variations in voice pitch produce intonation which is used to convey several types of suprasegmental or segmental contexts in speech (Rosen et al., '81).

The representation of the fundamental frequency of complex sounds has been extensively studied at the level of the cochlear nerve (Smoorenburg and Linschoten, '77; Evans, '78; Young and Sachs, '79; Delgutte and Kiang, '84; Miller and Sachs, '84; Palmer et al., '86; Horst et al., '90; Palmer, '90; Delgutte and Cariani, '91). These studies have shown that harmonic series (including speech sounds) produce modulation of the discharge at the F0 of fibres at frequencies where their response area is wide enough to allow more than one component to interact, provided a single intense component is not dominating the output. An alternative cue for F0 is evident in the timing of the discharges of the whole population of nerve fibres. The phase-locking of all fibres in response to voiced speech sounds is locked to harmonics of the F0 (Miller and Sachs, '84; Palmer, '90). For inharmonic series it has been shown that the phase-locking is capable of representing the fine time structure of the stimulus waveform and therefore conveys information consistent with the perceived pitch of these signals (Evans, '78; Delgutte and Cariani, '91).

At the level of the cochlear nucleus, Kim and his colleagues (Kim et al., '86; Kim and Leonard, '88) have demonstrated that many units show a strong synchronization of their discharges to the F0 of complex sounds. In the case of 'onset' units (i.e. units which only respond to a constant level tone burst with one or a few spikes at the stimulus onset), the locking to the voice pitch of speech sounds was particularly marked, leading this group to describe the responses as a "pitch-period following response".

The number of onset units in these studies was somewhat limited and their best frequencies (BFs) fell between the maxima in the spectra (at the formant frequencies) of the vowel stimuli. It was, therefore, unclear what aspect of the responses of the cochlear nerve fibres led to the strong locking of the onset units to the F0. Were they, for example, selective for the pitch-period as a result of sensitivity to modulation of the cochlear fibre input at F0 (as occurs at frequencies remote from the formants (Young and Sachs, '79; Miller and Sachs, '84; Palmer et al., '86; Palmer, '90) or was the sensitivity occurring "de novo" as a result of processing by the onset units?.

We have addressed this question by measuring the responses of both cochlear nerve fibres and onset units in the ventral cochlear nucleus to a range of stimuli which are periodic (and thus evoke a strong low-pitch), but which collectively produce a range of different patterns of modulation across the array of cochlear nerve fibres. The recordings were made from anaesthetized guinea pigs and details of the methods used may be found in previous publications (Palmer et al., '86; Palmer, '90); only those concerned with the specific stimuli will be given here. All stimuli were presented at approximately 80 dB SPL.

We routinely measure the discharge characteristics of each cochlear nucleus neurone and classify these responses according to the usual schemes (e.g. Rhode and Smith, '86). The majority of our sample are of the onset-chopper (ON-C) type with a few onset-with-a-low-level-of-sustained-discharge (ON-L) and one onset-inhibition (ON-I). For the present purposes we have included all these types and simply termed them onset units. In Figures 1, 2, 3 and 5 it will be seen that there is a paucity of onset units with BFs below 1 kHz. We do not know whether this is simply a result of uneven sampling, because these BFs occupy a relatively inaccessible part of the ventral cochlear nucleus (lateral and ventral), or whether there are indeed a paucity of onset units with low BFs in the guinea pig.

## Responses to harmonic series which include the fundamental.

Our major interest in these studies was the timing of the neural discharges and how these related to the period of the F0. For this reason we have computed, for each cochlear nerve fibre or cochlear nucleus onset unit, an autocorrelation function (which reveals periodicities in the spike train) from the times of occurrence of the neural discharges. To do this, we first constructed period histograms and computed the autocorrelation function by cross multiplying the first and second halves with progressive time offsets. This procedure produces a smooth function (Palmer, '91) very similar (except for some minor differences at long delays due to adaptation) to the function obtained using intervals between each spike and successive spikes (which is the more usual form of autocorrelation). Autocorrelation functions, plotted on the y-axis at the BF for each fibre and each onset unit, to the first 50 harmonics of 100Hz all in cosine phase and with equal amplitude, are shown in Figure 1. In this and subsequent figures we indicate cochlear nerve fibres with spontaneous rate below 0.5/s with thicker lines. Above each of the 'autocorrelograms' is a summary autocorrelation function obtained by summing the functions of all fibres or onset units. Given the uneven sampling this summary is likely to be distorted, but nevertheless it serves to emphasize the overall patterns observed. We have used harmonic series with both 100 Hz and 300 Hz F0s, but only the 100Hz results will be shown here as the pattern generated by the 300 Hz series was similar. In Figure 1a a strong modulation of the discharge of cochlear nerve fibres of all BFs is evidenced by a peak at the time delay (10 ms) corresponding to the period of the F0. Fibres with BFs of 1-3 kHz are able to resolve individual harmonics so their autocorrelation functions show evidence of periodicities close to 1/BF, but still contain a prominent peak at 10 ms. The summary autocorrelation function shows a well defined peak at 10 ms.

A plot of the responses of onset units to the cosine harmonic series is shown in Figure 1B. It can be seen from Figure 1B that all of the onset units show strong locking of their discharges to the F0 with few discharges occuring at other times. For the onset units with

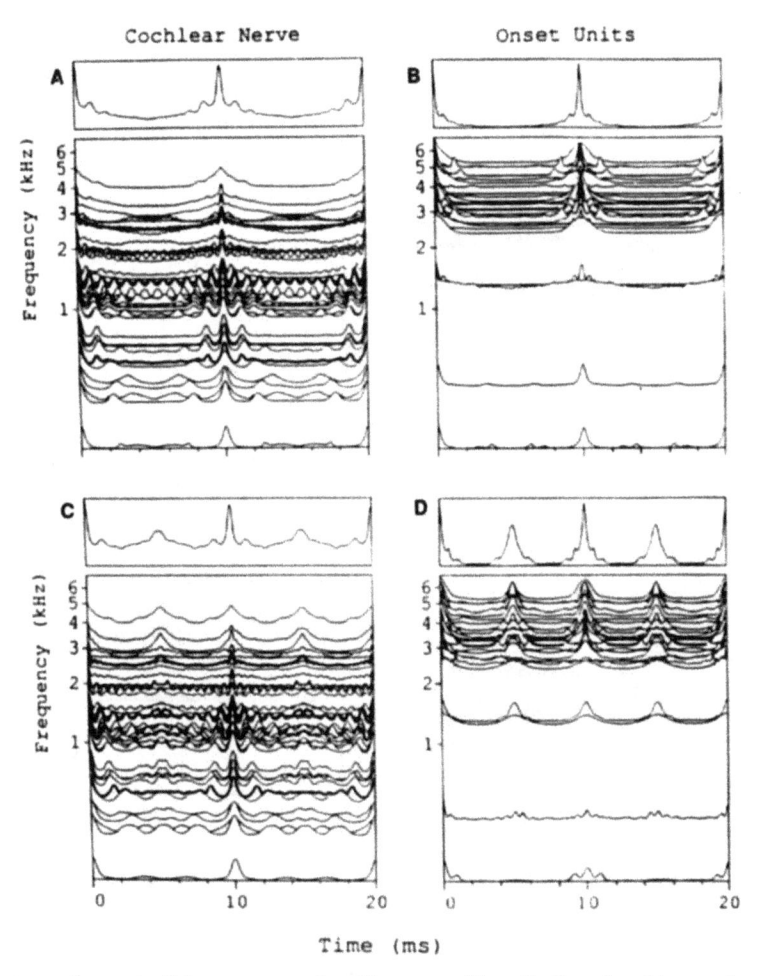

**Figure 1.** Autocorrelograms of the responses of cochlear nerve fibres (A,C) and cochlear nucleus onset units (C,D) to stimuli consisting of the first 50 harmonics of 100 Hz. All harmonics were in cosine phase for Figs. 1A and 1B. The phase of the harmonics for Figs. 1C and 1D were alternating cosine and sine. Each line in the major part of the figure represents the autocorrelation function of the spike discharges of a single fibre or onset unit, normalized to the biggest value, and plotted on the abscissa at the BF. The thicker lines show fibres with spontaneous rates less than 0.5/s. The function above each panel is the summary autocorrelation function obtained by summing all of the individual functions.

BFs between 1-3 kHz the phase locking to the individual harmonics seen in the nerve fibres is effectively suppressed. The summary autocorrelogram shows a well defined peak at 10 ms with virtually no discharges occurring at other delays.

In Figures 1C and 1D we show similar plots for cochlear nerve fibres and onset units in response to a harmonic series with exactly the same spectrum, but with alternating sine and cosine phase harmonics.

More than one pitch (F0, 2F0 etc.) may be identified by listeners for both the cosine and alternating phase stimuli (Lundeen and Small, '84). However, the most salient of the pitches does change from F0 for the cosine phase stimulus to 2F0 for the alternating phase stimulus. The modulation pattern which these stimuli evoke at the output of a bank of bandpass filters is also markedly different (Lundeen and Small, '84) and this is reflected in the pattern of responses in the cochlear nerve fibres as shown in Figure 1C.

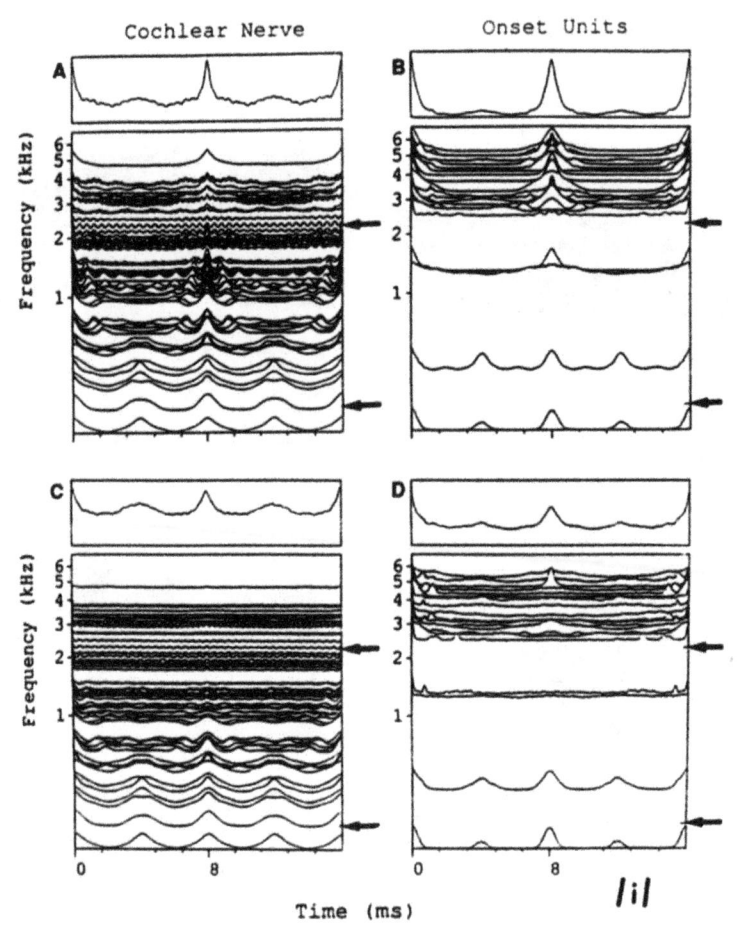

**Figure 2.** Autocorrelograms as in Fig.1 for the responses to the vowel /i/ either presented in quiet (Figs. 2A, 2B) or in noise at a signal-to-noise ratio of 10 dB (Figs. 2C, 2D). The arrows next to the abscissa indicate the formant frequencies.

At the lowest BF there is modulation at the F0, but in higher BF fibres the major component of the modulation is now at 2F0 and the summary autocorrelation function shows peaks at 5 ms intervals. The doubling of modulation frequency in the cochlear nerve response has been previously reported for harmonic series with alternating sine/cosine phase components, which did not include the fundamental component and which generally had low F0s (Horst et al., '90). The responses of all but the lowest BF onset unit are securely locked to 2F0 (i.e. peaks at 5 ms in Fig. 1D and in the summary autocorrelation function) not F0. This result suggests that onset units may respond to the pitch period of speech sounds mainly as a result of the modulation of their cochlear fibre input at the F0 rather than by intrinsic processing.

## Responses to synthetic voiced vowel sounds in quiet and in noise

To investigate the proposition that the voice-pitch following of onset units is a response to modulation in the cochlear nerve fibre input we have used noise to reduce the F0 modulation. The virtual abolition of F0 modulation by noise in the majority of nerve fibres

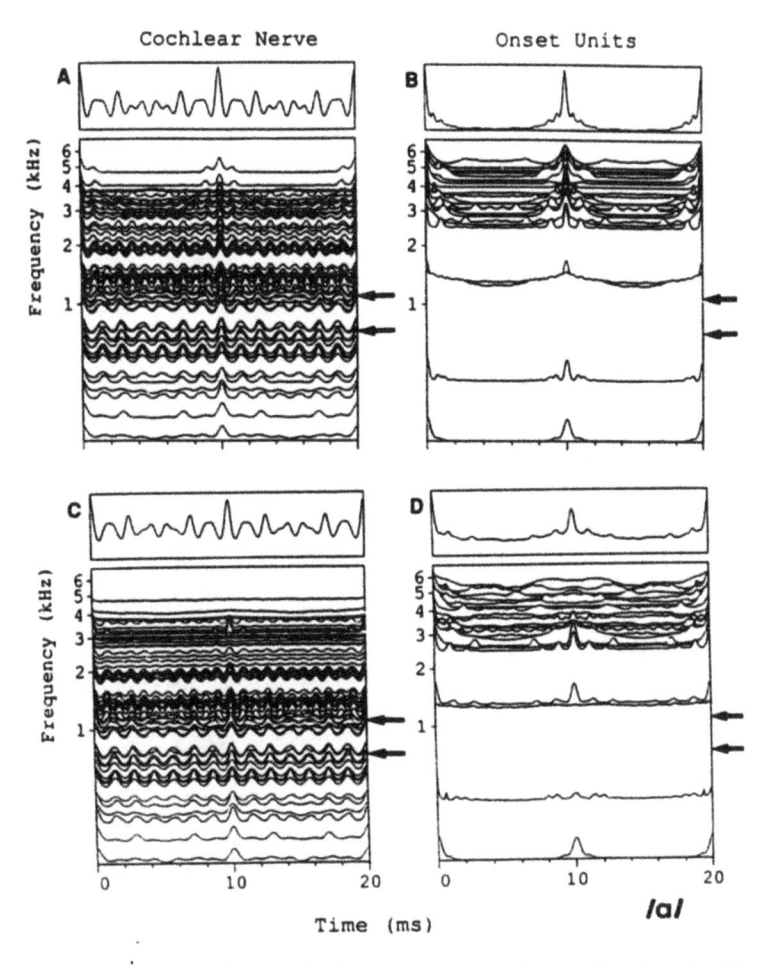

**Figure 3.** Autocorrelograms as in Figs. 1 and 2 for the responses to the vowel /a/ in quiet (Figs. 3A, 3B) or in noise at a signal-to-noise ratio of 10 dB (Figs. 3C, 3D).

with BFs remote from the formants of vowel sounds has been well documented in cats, (Miller and Sachs, '84). In Figures 2 and 3 we show a replication of this result in the guinea pig for the vowels /i/ and /a/ with F0s of 125 and 100Hz respectively. In Figures 2A and 3A are shown the autocorrelation functions to the vowel presented in quiet and in 2C and 3C are the responses to the vowels presented in a white noise at a signal-to-noise ratio of 10 dB. For both vowels the modulation at F0 in the cochlear nerve fibres with BFs remote from the formants (between 1 and 2 kHz and above 3 kHz in Fig. 2 and above 2 kHz in Fig. 3) is reduced or abolished by the noise. The very notable exception which is visible at 3.19 kHz in Figure 3C is a low-spontaneous rate fibre, a point to which we shall return later. The summary autocorrelation functions show a good response to F0 in Figure 3A and a weaker response in Figure 3C. Most of the F0 peak in Figure 2C is derived from the region of the first formant and the peaks at 4 ms are also indicative of a first formant response (250 Hz). The equivalent pictures for the onset units are shown in Figures 2B, 2D and 3B and 3D. The response of the onset units to the vowels in quiet shows strong locking to the period of the F0, which is a replication of the data presented by Kim and his collegues (Kim et al., '86; Kim and Leonard, '88).

As demonstrated for the harmonic series (see Fig. 1 in the 1-2 kHz region), many onset units appear to be responding to the modulation and suppressing the carrier. In the

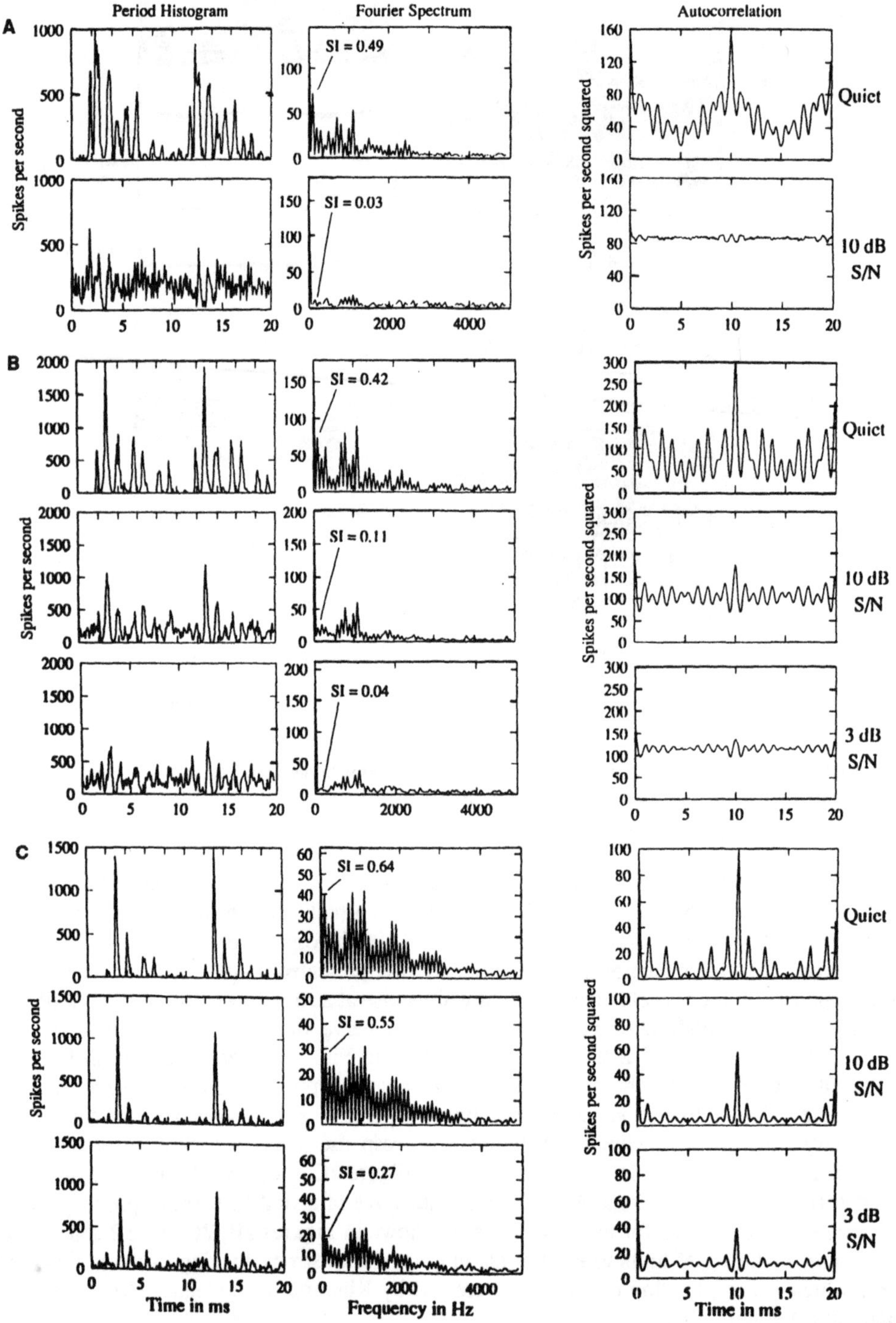

Period Histogram
Fourier Spectrum
Autocorrelation
A
B
C
Spikes per second
Spikes per second squared
SI = 0.49
SI = 0.03
SI = 0.42
SI = 0.11
SI = 0.04
SI = 0.64
SI = 0.55
SI = 0.27
Quiet
10 dB S/N
3 dB S/N
Time in ms
Frequency in Hz

presence of the noise the F0 locking of many of the onset units is reduced, but there are significant numbers (particularly to the /a/) which still show locking to the F0 despite very little remaining modulation in cochlear nerve fibres with the same BF (see Fig. 3 C and D in the 3-5 kHz region). We have confirmed this by measuring the synchronization index to the F0 in quiet and in noise and these data for the vowel /a/ have been previously published (Palmer and Winter, '92).

The degree to which the noise reduces the F0 modulation of the cochlear nerve fibre response depends upon its spontaneous rate. This is illustrated in Figure 4 for fibres with high medium and low rates of spontaneous discharge (i.e. above 18/s, 0.5-18/s and below 0.5/s see (Liberman, '78)).

All three fibres have similar BFs near 3 kHz and show strong discharge modulation to the vowel /a/ in quiet. This is evident in the period histograms (quantified as the synchronization indices shown) and in the autocorrelation functions. The F0 modulation in the high spontaneous rate fibre (Fig. 4A) is not statistically significant ($p > 0.05$, see Buunen and Rhode, '78) in the noise at 10 dB S/N ratio. The medium spontaneous rate fibre (Fig. 4B) retains significant F0 modulation at 10 dB S/N ratio, but the modulation becomes non statistically significant ($p > 0.05$) in noise at 3 dB S/N. The low-spontaneous rate fibre (Fig. 4C, which is the 3.19 kHz fibre mentioned previously) discharge is strongly modulated at the F0 ($p < 0.001$) even in the presence of noise at 3 dB S/N ratio. It is notable that the modulation at F0 in the low-spontaneous rate unit in 3 dB S/N ratio exceeds that for the other two fibres in 10 dB S/N ratio.

## Responses to three component (missing fundamental) complexes

Sinusoidally amplitude-modulated tones represent a special case of a partial harmonic series which evokes a low-pitch since they consist of a carrier frequency with two side bands separated from it by a frequency corresponding to the modulation rate. We have used two such complexes which have also been used extensively before in both psychophysical and physiological experiments (de Boer, '56; Evans, '78; Greenberg, '80; Delgutte and Cariani, '91). The stimuli were synthesized as three separate components of frequency 1000, 1200 and 1400 Hz for the first and 900, 1100, 1300 Hz for the second.

The sideband amplitudes were 6 dB below the central component and in the same phase thereby producing 100% sinusoidal amplitude modulation of the central carrier component. The pitch evoked by the first complex is 200 Hz while that of the second is ambiguous between frequencies slightly higher and slightly lower than the 200 Hz difference frequency of the complex (de Boer, '56). If the cochlear nucleus onset units are driven by modulation at their input they should respond to both of these stimuli in the same way, as both are modulated at 200 Hz. For the output of the onset units to be consistent with the perceived pitch they would need to respond differently to these stimuli. Figure 5 shows the autocorrelation function of cochlear nerve fibres and onset units to the two stimuli. The data of Figures 5A and 5B are a replication of extensive data with these types of stimuli already

Figure 4. Responses of three cochlear nerve fibres with high (Fig. 4A) medium (Fig. 4B) and low (Fig. 4C) spontaneous discharge rates to the vowel /a/ in quiet and in noise at 10 and 3 dB signal-to-noise ratios. The first column is the period histogram locked to the F0 and its Fourier transform is shown in the second column. The synchronization index to the F0 is shown (the value of the 100 Hz component divided by the mean discharge rate given by the value at zero frequency). The third column shows the autocorrelation function computed from the same data. The instantaneous discharge values on the abscissa of the autocorrelation function have been reduced by a factor of $10^4$ to give convenient labels. Figure 4A shows data from a fibre with BF of 3.11 kHz and with spontaneous rate of 83.7 sp/s. Figure 4B is from a fibre with BF 3.51 kHz and spontaneous rate of 6 sp/s and Fig. 4C shows a fibre with BF 3.19 kHz and spontaneous rate of 0 sp/s.

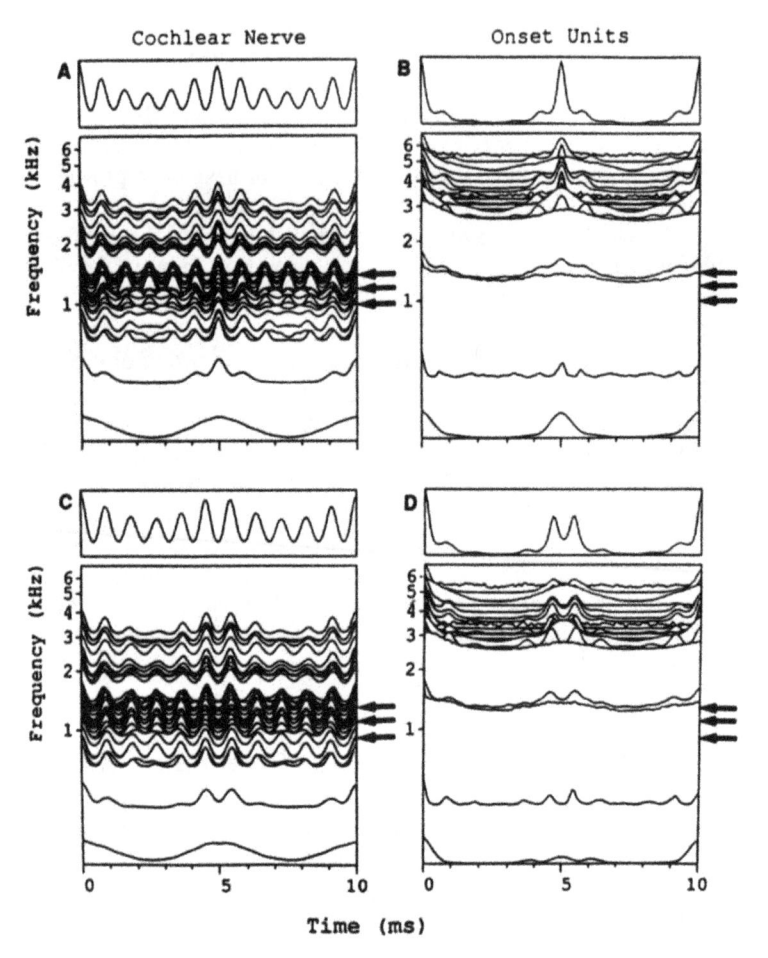

**Figure 5.** Autocorrelograms (as in Figs. 1-3) of the responses to three component complexes. Figs. 5A and 5B show the responses to the three components 1000, 1200 and 1400 Hz and Figs. 5C and 5D show the responses to the three components 900, 1100 and 1300 Hz. The position of the three components are shown by arrows.

published (Evans, '78; Horst et al., '90; Delgutte and Cariani, '91). Cochlear nerve fibres with a wide range of BFs respond to the fine-time structure of the stimuli. The most prolific periodicities in their discharges correspond to the pitches which are heard as can be seen particularly clearly in the summary autocorrelation functions (5 ms or 200 Hz in Fig. 5A; 4.69 and 5.31 ms 213 and 188 Hz in Fig. 5B) and the smaller peaks indicate phase locking to the individual components.

The majority of the onset units did not simply respond at the modulation rate of these stimuli, but responded at the most prolific periodicities in the nerve fibre discharges. For the harmonic series this resulted in a clear peak at 5 ms in individual units and in the summary autocorrelation function (which of course would have been the same whether responding to the fine structure or the modulation envelope). For the inharmonic series two different types of response are visible. The response shown by the majority of units (and hence clearly represented in the summary function) is a response to the most prolific periodicities in the input fibres and results in a double peak either side of 5 ms corresponding to the periods of the ambiguous perceived pitches. The response of the remaining few units appears to be simply a result of the envelope modulation and produces a single peak at 5 ms. We have

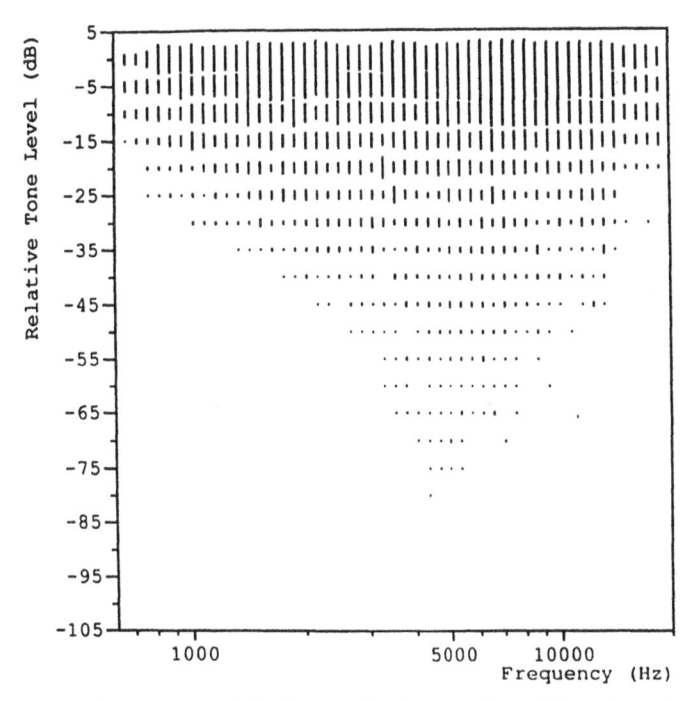

**Figure 6.** Response area for an onset unit in the ventral cochlear nucleus of the guinea pig. The zero decibel reference for the abscissa is approximately 105 dB SPL. The response evoked from the unit by a single presentation of a 50 ms tone pip at each of a series of frequencies and sound levels is shown by a line at the appropriate position; the length of the line indicates the number of spikes (scale 3 spikes/dB). The tones were presented in pseudorandom order.

presented these stimuli to a few onset units with BFs up to 12 kHz and these all responded as if to the modulation envelope and not to the fine structure.

## Discussion

The cochlear nerve data which we have presented here are extensions and confirmations of earlier studies. Here, we have used them as a comparison, in the same species, for the responses of the onset units in the ventral cochlear nucleus. Onset units show more precise phase-locking to the F0 of speech sounds than do other unit types (see, for example, Figs. 2 and 4 in Kim and Leonard, '88). In the course of these experiments we have also recorded from chopper units which are located in the same area of the lateral-ventral part of the ventral cochlear nucleus. The response of the chopper units to the stimuli we have used was somewhat variable, but many choppers gave responses which were similar to those of the onset units we have described here. A major deficiency in the sample of onset units which we present here is the paucity of onset units with BFs below 1 kHz (as in the earlier study of Kim et al.). We do, however, show some low-frequency onset units which respond in the same way as those with higher BFs. While we shall endeavour to fill these gaps, it seems likely that the responses which we have described for onset units with BFs above 1 kHz will also apply to those with lower BFs.

The responses of the onset units to the cosine harmonic series and to the alternating phase harmonic series, which both included all harmonics of F0 down to and including the F0, were strikingly different (being locked to F0 and 2F0 respectively). This result suggests that a large part of the ability of onset cells to follow F0 of harmonic complexes (including speech), resides in their sensitivity to the modulation of the discharge of the cochlear nerve fibres at their input. This view was reinforced for many of the onset units by the fact that the strong locking to the F0 of speech sounds was considerably attenuated by the presence of noise, which reduces or eliminates the F0 modulation in the majority of cochlear nerve fibres (Delgutte and Kiang, '84; Miller and Sachs, '84). Also consistent with a simple response to modulation were the minority of our onset unit sample which responded to the inharmonic three-tone complex with a single peak at the period of the modulation envelope.

It is notable, however, that in the presence of the noise there were onset units which still phase-locked to the F0 of the speech sounds. This was more marked for the /a/ than for the /i/. In view of the data shown in Figure 4, that the noise did not abolish the modulation at F0 in the responses of low-spontaneous rate fibres to the vowel /a/ it seems possible that these onset units, still locked to the F0 of /a/ in noise, may be receiving their input selectively from the low-spontaneous rate fibres. We cannot on the basis of our own data rule out this possibility and such a selective response to different groups of cochlear nerve fibres has been proposed for the chopper unit population (Blackburn and Sachs, '90). However, some recent data (Wang and Sachs, '91) has indicated that in response to single formant speech-like sounds onset units maintain locking to the F0 over dynamic ranges which exceed those of even the low-spontaneously active cochlear nerve fibres.

In response to the inharmonic three-tone complex the majority of onset units responded to periodicities in the fine time structure of the cochlear nerve fibre responses, not simply to the modulation envelope. This observation is consistent with the view that onset units are acting as coincidence detectors receiving inputs from cochlear nerve fibres with a range of BFs. Indeed, many of our observations can be accommodated within this hypothesis. In response to stimuli which generate modulation across a range of cochlear nerve fibre BFs the output of the onset unit will be locked to the most frequent interval, which is the F0 for speech and cosine harmonic series, but 2F0 for the alternating phase stimulus. High BF onset units (above 5 kHz) will receive only the locking to the modulation envelope and not the fine time structure and therefore will only respond to the inharmonic three-tone complex at the modulation frequency. Onset units generally have response areas which are more extensive than those of cochlear nerve fibres of similar BF (Godfrey el al., '75), as can be seen in the typical example shown in Figure 6. The onset unit shown in this figure has a BF of 5 kHz and shallow high- and low- frequency slopes to its response area. At the level we have presented our stimuli (which corresponds to approximately -20 dB on the abscissa of Fig.6), the response area of this onset unit extends up to above 20 kHz and downwards to about 700 Hz. The relatively wide response areas of such onset units will extend to the region of the formants of vowels. Thus in background noise they will be able to respond to the coincidences generated by the large proportion of cochlear nerve fibres responding to the first and second formants (the common interval amongst these fibres will be F0). The fewer onset units still locked to the F0 of /i/ in noise probably is due to the fact that the /i/ has such a low first formant (270 Hz) and a high second formant (2300 Hz). The responses to the first formant fall outside the majority of the onset unit response areas, while the number of coincidences due to the second formant will be limited by the poorer phase-locking to these frequencies in the guinea pig (Palmer and Russell, '91). What is not obvious is whether the onset units are indeed using the common intervals (and hence coincidences) across nerve fibre inputs which are completely unmodulated at the F0 or whether they are exquisitely sensitive to the small degree of modulation at F0 which exists in fibres at or near to the formant frequencies. At the very least we can suggest that it is not a neccessary requirement that onset units in the ventral cochlear nucleus receive input at their BF which is strongly modulated at F0 in order for their discharges to be locked to F0. For the range of stimuli which we have used here, the temporal discharge patterns of the onset units in the ventral cochlear nucleus are consistent with the low-pitch perceived in all cases (de Boer, '56; Lundeen and Small, '84).

ACKNOWLEDGEMENTS

This work was supported by the MRC. We would like to thank Ms. Padma Moorjani for technical assistance and Dr. A.Q. Summerfield for helpful comments.

REFERENCES

Assmann, P. and Summerfield, A.Q., 1990, Modeling the perception of concurrent vowels: vowels with different fundamental frequencies, *J. Acoust. Soc. Amer.*, 88: 680-697.
Blackburn, C.C. and Sachs, M.B., 1990, The representation of the steady-state vowel /Σ/ in the discharge patterns of cat anteroventral cochlear nucleus neurons, *J. Neurophysiol.*, 63: 1191-1212.
Boer, E. de, 1956, Pitch of Inharmonic Signals, *Nature*, 178: 535-536.
Buunen, T.J.F and Rhode, W.S., 1978, Responses of fibers in the cat's auditory nerve to the cubic difference tone, *J. Acoust. Soc. Amer.*, 64: 772-781.
Delgutte, B. and Kiang, N.Y.S., 1984, Speech coding in the auditory nerve: I. Vowel-like Sounds, *J. Acoust. Soc. Am.*, 75: 866-878.
Delgutte, B. and Kiang, N.Y.S. 1984, Speech coding in the auditory nerve: V. Vowels in Background Noise, *J. Acoust. Soc. Am.*, 75: 908-918.
Delgutte, B. and Cariani, P., 1991, Coding of fundamental frequency in the auditory nerve: a challenge to rate-place models, in: "The Psychophysics of Speech Perception", M.E.H. Schouten, ed., in press.
Evans, E.F., 1978, Place and time coding in the peripheral auditory system: Some physiological pros and cons, *Audiol.*, 17: 369-420.
Godfrey, D.A., Kiang, N.Y.S. and Norris, B.E., 1975, Single unit activity in the posteroventral cochlear nucleus of the cat, *J. Comp. Neurol.*, 162: 247-268.
Greenberg, S., 1980, Temporal neural coding of pitch and vowel quality. UCLA Working papers in Phonetics, 52.
Horst, J. W., Javel, E. and Farley, G.R., 1990, Coding of spectral fine structure in the auditory nerve .2. Level-dependent nonlinear responses. *J. Acoust. Soc. Amer.*, 88: 2656-2681.
Kim, D. O., Rhode, W.S. and Greenberg, S.R., 1986, Responses of cochlear nucleus neurons to speech signals: neural encoding of pitch, intensity and other parameters, in: "Auditory frequency selectivity", B.C.J. Moore and R.D. Patterson, eds., Plenum Press, New York, pp. 281-288.
Kim, D.O. and Leonard, G., 1988, Pitch-period following response of cat cochlear nucleus neurones to speech sounds, in: "Basic issues in hearing", H. Duifhuis, J.W. Horst and H.P. Wit, eds., Academic Press, London, pp. 252-260.
Liberman, M.C., 1978, Auditory nerve response from cats raised in a low noise chamber, *J. Acoust. Soc. Amer.*, 63: 442-455.
Lundeen, C. and Small, A.M., 1984, The influence of temporal cues on the strength of periodicity pitches, *J. Acoust. Soc. Am.*, 75: 1578-1587.
Miller, M. I. and Sachs, M.B., 1984, Representation of voice pitch in discharge patterns of auditory nerve fibers, *Hearing Res.*, 14: 257-279.
Palmer, A.R., 1990, The representation of the spectra and fundamental frequencies of steady-state single- and double-vowel sounds in the temporal discharge patterns of guinea pig cochlear nerve fibers, *J. Acoust. Soc. Am.*, 88: 1412-1426.
Palmer, A.R., 1991, Segregation of the responses to paired vowels in the auditory nerve of the guinea pig using autocorrelation, in: "The psychophysics of speech perception", M.E.H. Schouten, ed.,in press.
Palmer, A.R. and Russell, I.J., 1986, phase-locking in the cochlear nerve of the guinea pig and its relation to the receptor potential of inner hair cells, *Hearing Res.*, 24: 1-15.
Palmer, A.R. and Winter, I.M., 1992, Cochlear nerve and cochlear nucleus responses to the fundamental frequency of voiced speech sounds and harmonic complex tones, in: "Auditory Physiology and Perception", Y. Cazals, L. Demany and K. Horner, eds., Advances in the biosciences, 83:231-239.
Palmer, A.R., Winter, I.M. and Darwin, C.J., 1986, The representation of steady-state vowel sounds in the temporal discharge patterns of guinea pig cochlear nerve and primarylike cochlear nucleus neurons, *J. Acoust. Soc. Am.*, 79: 100-113.
Rhode, W.S. and Smith, P.H., 1986, Encoding timing and intensity in the ventral cochlear nucleus of the cat, *J. Neurophysiol.*, 56: 261-286.
Rosen, S.M., Fourcin, A.J. and Moore, B.C.J., 1981, Voice pitch as an aid to lipreading, *Nature*, 291: 150-152.
Scheffers, M.T.M., 1983, Sifting vowels: Auditory pitch analysis and sound segregation. *Doctoral Thesis*, University of Groningen.

Smoorenburg, G. F., and Linschoten, D.H.,1977, A neuro-physiological study on auditory frequency analysis of complex tones, *in*: "Psychophysics and physiology of hearing", E.F. Evans and J.P. Wilson, eds., (Academic Press, London), pp.175-183.

Wang, X. and Sachs, M.B., 1991, Amplitude modulation of avcn units. Responses to single formant stimuli. Proceedingss of the 14th Midwinter Research Meeting of the Association for Research in Otolaryngology, p.137.

Young, E.D. and Sachs, M.B., 1979, Representation of steady-state vowels in the temporal aspects of the discharge patterns of populations of auditory nerve fibers, *J. Acoust. Soc. Am.*, 66: 1381-1403.

# VII. COMPUTER MODELLING OF THE COCHLEAR NUCLEUS

# COMPUTER MODELLING OF THE COCHLEAR NUCLEUS

Ray Meddis and Michael J. Hewitt

Speech and Hearing Laboratory, University of Technology
Loughborough, UK

Computer modelling represents a promising but not yet fully established methodology for studying the complex systems found in the auditory nervous system. In this article we shall be reviewing some of the methods used by modellers and outlining some ideas concerning good practice which, if implemented, should lead to its greater acceptance. Our aim is to dispel some of the skepticism surrounding this activity and make researchers aware of the potential benefits of computer modelling for anatomists and physiologists working in and around the cochlear nucleus.

Modelling the cochlear nucleus using computers is still a relatively new enterprise. As a consequence, there is a very small literature and much of that is in manuscript form. To simplify and encourage access, we have prepared a bibliography of extant papers, irrespective of whether they have been published or not (see appendix A). However, this volume contains two detailed accounts of important modelling exercises for the interested reader (see chapters by Sachs and Young).

## The issue of complexity

It is in the nature of experimental sciences, that experiments can only be carried out within very limited domains, even though the experimentalists see the value of their work in terms of their contribution to the 'big picture'. When systems are simple, we can carry the common agreed 'big picture' in our heads or represent the situation in terms of a few simple equations or a flow diagram. However, when systems become complex and multifaceted we need more powerful representational techniques. When systems contain many independent components, are nonlinear or involve feedback processes, it is simply beyond the power of the human mind to predict the response of that system through intuition or simple calculation. The cochlear nucleus has all of these properties and the modeller has the job of capturing and representing the complex interactions which occur within its narrow physical compass.

The cochlear nucleus is complex; Figure 1 illustrates our concept of the wiring diagram of the system. It has been based on a large number of individual anatomical studies reported in the literature. It should not be taken as definitive but suggestive! This particular diagram has been seen and improved by many researchers in the area, most of whom have made suggestions for changes and all of whom have emphasised the tentative nature of the enterprise. Whatever the merits of this particular diagram, the nature of the modeller's

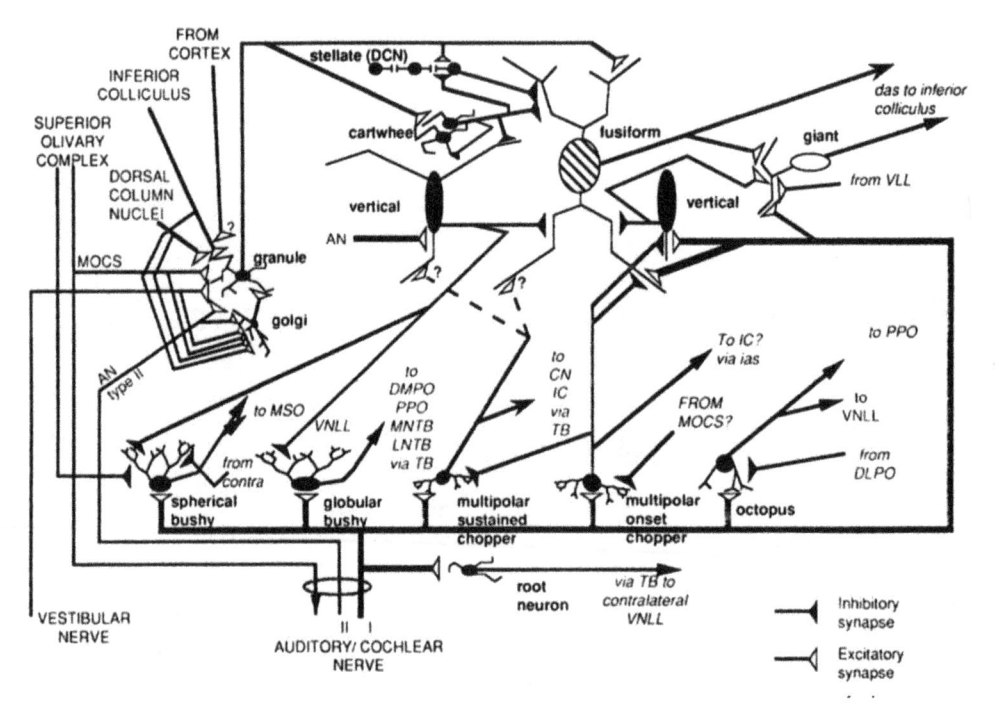

**Figure 1.** Tentative outline wiring diagram of cochlear nucleus.

problem is clear. How can we characterise the physiological functioning and psychological relevance of such a complex system? We believe that only computer modelling is likely to allow us to achieve this goal.

## *In vitro* modelling

To provide a context for this discussion, we would like to describe briefly some work of our own. This is a model of a multipolar (or stellate), sustained chopper in the VCN. As psychologists, we are interested in the functioning of this cell because it appears to be involved in the neural coding of the perceived pitch of complex signals. While these cells are largely unresponsive to signal level (dynamic range is 20-30 dB), they do faithfully reproduce, and even sometimes exaggerate, the amplitude modulation (AM) characteristics of an acoustic signal. This is particularly true if its AM frequency is close to the intrinsic frequency of the cell (Kim et al., '90; Frisina et al., '90). A detailed description of the model is given elsewhere (Hewitt et al., '92).

The cell itself is modelled using a mathematical model of a nerve cell due to McGregor ('87), known as a 'point neuron'. This is one of the simplest models of nerve cell action. It works by predicting the change in a cell's membrane potential in response to depolarising and hyperpolarising current. When the membrane potential exceeds a certain value, an action potential is generated. Figure 2 shows the response of the model neuron to a depolarising and a hyperpolarising direct current applied intracellularly. The depolarising current causes the cell to fire repetitively; a characteristic known as "chopping". The performance of this simple system has much in common with the *in vitro* recordings made by Oertel ('83) in multipolar cells; their results are shown along side the model results.

Figure 3 shows the effect of delivering short current pulses in rapid succession. The inertia of the model is such that the effect of one pulse has not completely died down before

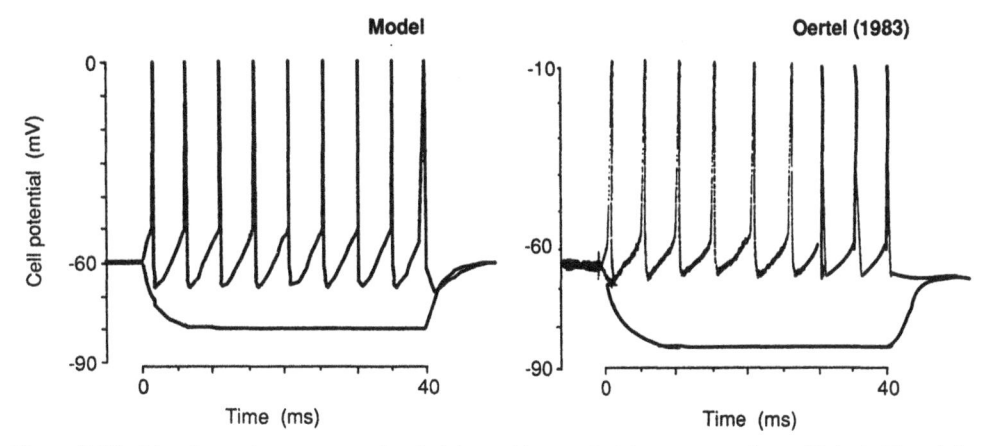

**Figure 2.** Model and neural responses to depolarising and hyperpolarising current pulses of 0.6 nA. Neural data redrawn from Oertel ('83).

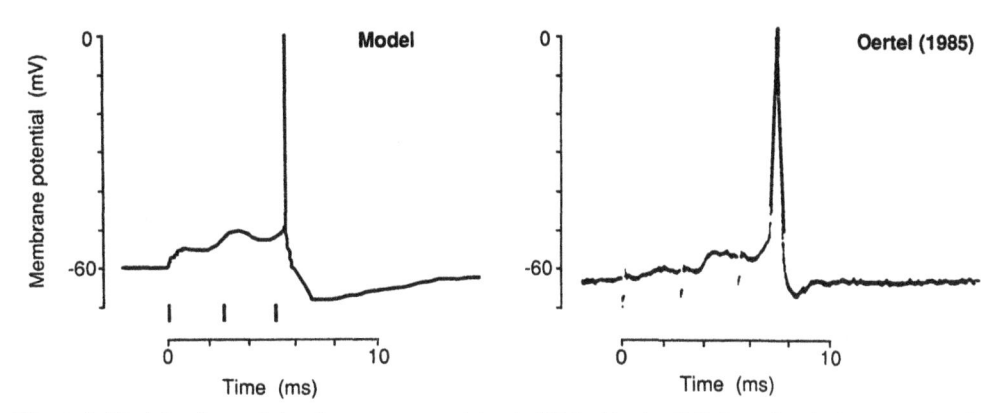

**Figure 3.** Model and neural data in response to subthreshold 'shock' stimuli (0.1 ms duration) presented to the auditory nerve at a rate of 360/s. Neural data redrawn from Oertel ('85).

another one arrives. This leads to the *temporal summation* of inputs so that individual pulses which would not have caused an action potential when delivered alone, can produce action potentials in association with other pulses which are delivered soon before or after. Again these results are directly comparable with Oertel's *in vitro* data for shocks applied to the auditory nerve root (Oertel, '85).

Whenever *in vitro* data are available for a given cell type, the modeller is well advised to simulate this first. This is because the cell is operating with a minimum of extrinsic input and therefore no assumptions need to be made about connectivity between the cells. Similarly, intracellular current injection can be simulated without any assumptions concerning synaptic transmission. As a consequence, the basic parameters of the cell's functioning can be established with a minimum of complication.

## *In vivo* modelling

To simulate the results of *in vivo* auditory stimulation of multipolar nerve cells we need a more sophisticated simulation of the input to the cell. To do this we use a peripheral

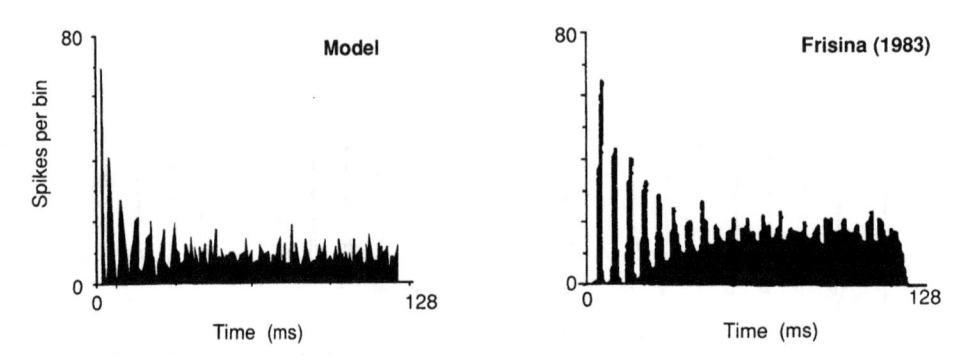

**Figure 4.** Model and neural PSTHs from responses to 128 ms best-frequency tones presented at 32 dB above threshold. Neural data redrawn from Frisina ('83).

model which delivers a stream of action potentials closely resembling the activity normally recorded from the auditory nerve (AN). The most important component of such a model is the inner hair cell which transduces the mechanical motion of the basilar membrane into the electrical activity of the auditory nerve fibres. A number of useful digital models are available and have recently been evaluated (Hewitt and Meddis, '91). We used a model which replicates a large number of properties and does not involve a heavy computational load (Meddis, '88).

Real multipolar neurons receive many AN inputs although the exact number is unknown. For this exercise we simulated the activity of 60 simultaneously active AN fibres each firing independently and delivering an input to the multipolar cell model. These model nerve fibres were configured to have the same characteristic frequency (5 kHz) but act as independent fibres in all other respects. Note that the simulation of these AN fibres contains a stochastic element and this introduces an element of unpredictability into the results.

Our model multipolar neuron assumes that the inputs are delivered to the dendrites, i.e. at some distance from the soma. This has the effect, in real neurons, of low pass filtering the input, i.e. the excitatory post-synaptic potentials (EPSPs) generated at the nerve terminals are smeared in time before they arrive at the soma. The effects of dendritic filtering are complex (Rall, '89) but, in this implementation, we were able to approximate reality more simply without loss of accuracy by applying a low pass digital filter to the temporal pattern of current changes introduced by the EPSPs from the model fibres. The significance of this filtering effect is discussed below.

The low-pass filtered EPSPs were then passed to the McGregor ('87) soma model and the response of the model to a sustained 5 kHz tone is a chopping response similar to that shown in Figure 2 for a direct current depolarisation. The action potentials are regularly spaced and sustained for the duration of the stimulus. A PSTH based on a large number of such tests of the model shows the characteristic chopping pattern which is widely reported in the physiology literature (Fig. 4). The chopping pattern in the PSTH fades slightly during stimulus presentation, but this is caused by the noisiness of the input from the peripheral model rather than any reduction in the stability of the inter-spike intervals (ISIs). This modelling result has been reported previously by other authors (Arle and Kim, '91; Banks and Sachs, '91).

## Simplification and parameter variation

The model is very much simpler than the real thing but this is not necessarily a defect. Its simplicity allows us the opportunity of demonstrating that the complex behaviour of the cell can be explained in terms of a subset of the total set of properties of the cell. In this case, we can say that the chopping behaviour of a cell may be a function of its cell membrane characteristics and its potassium channel conductance characteristics in particular. No additional mechanism such as inhibitory feedback loops are necessary to generate this kind of response. A more elaborate model with detailed simulation of dendritic effects, a full range of other excitatory and inhibitory inputs and a complete representation of all the voltage controlled ion-channels would be very interesting but it would not help us to hypothesise as to which factors are controlling which aspects of the response.

In so far as the researcher is interested in knowing which aspects of the physical world determine which aspects of its behaviour, modelling has an obvious role in defining the possibilities. Young's paper in this volume uses a highly simplified model of bushy cells in the AVCN. This is similar in many respects to the point neuron described above but characterised by a *nonlinear* voltage-controlled potassium channel. The introduction of nonlinearity into the potassium channel behaviour changes the response of the model radically. For example, Young's model can be configured to produce primary-like responses with notch ($PL_N$) to pure tone input. By systematically varying (i) the number of modelled AN inputs and (ii) the conductance parameters, Young and his colleagues were able to show that a $PL_N$ responses occurred with only certain combinations of these parameters. Other combinations produced different response patterns such as primary-like, onset and onset-L response patterns. The establishment of a possible link between these parameters and the response patterns requires confirmation by experimentation. However, it is a link that would have been uncovered only slowly using experimentation alone.

A further example can be given in the context of our model of multipolar cell activity. The chopping pattern shown in Figure 4 resulted from a model with 60 inputs applied to the model dendrites. Only further anatomical research could advise as to the correct number of fibres to use but modelling allows us to devise hypotheses about the consequences of having different numbers of inputs. Not surprisingly, if the number of inputs is reduced, we find that the firing pattern of the cell becomes more irregular. However, if we reduce the number of fibres, relocate them on the soma (by omitting the low pass filter element from the model) and raise the threshold, we still get a chopping pattern but it is more short lived. In other words the model begins to behave like a transient chopper (CHOP-T).

In this way the model has yielded a new hypothesis concerning the origin of CHOP-T behaviour in cochlear nucleus neurons. However, there are other models which deliver the same performance. For example, Arle and Kim ('91) and Banks and Sachs (91) have shown that transient chopping behaviour can also be induced by adding a noisy inhibitory input to the model cell. Now, the modelling process has produced two alternative hypotheses. Only experiment can decide which of these two alternatives is correct. If a transient chopper becomes a sustained chopper when inhibition is blocked in an animal preparation, then the inhibitory model becomes the best candidate. If it remains a transient chopper, then the best available model is the one which identifies somatic input as the crucial morphological feature of CHOP-T multipolar cells. So far, this experiment has not been attempted.

## Exhaustive testing

Before resorting to experiment a model needs to be exhaustively studied against all known characteristics of the system under investigation. In the case of sustained choppers, we know that they have a number of defining properties in addition to those described above. Firstly, the ISI remains constant over time (as measured by the coefficient of variation). Secondly, they have limited dynamic ranges in response to signal level; there is little increase in firing rate in response to signal level increases beyond 30 dB above threshold. Thirdly, the sustained chopper has a characteristic bandpass amplitude modulation transfer function (AMTF) which becomes low-pass at very low signal levels. Fourthly, it shows a charac-

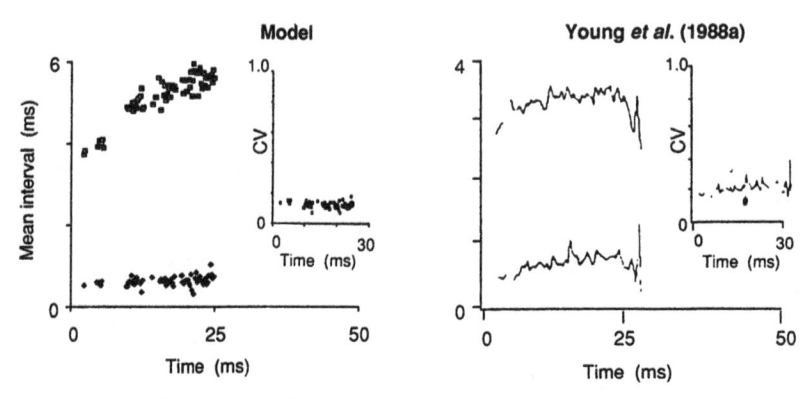

Figure 5. Regularity analysis on model and neural PSTHs. Each panel shows the mean (upper line) and standard deviation (lower line) of interspike intervals (ISI) as a function of time during a best frequency tone burst. The inset shows the coefficient of variance (CV = $\sigma$ ISI/$\mu$ ISI). Chop-S neurons are characterised by a low (< 0.35) CV. Neural data from Young et al., ('88).

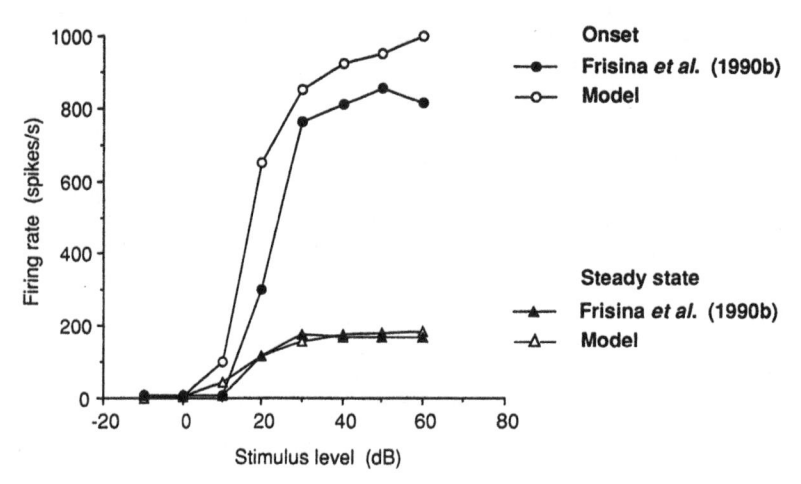

Figure 6. Model and neural rate-level functions. The onset function represents the maximal firing rate during the first (or highest) millisecond of response at stimulus onset. The steady-state function represents the average firing rate over a 20-ms period, 25 ms after stimulus onset. Neural data redrawn from Frisina et al. ('90).

teristic intrinsic oscillation frequency in response to continuous high frequency tones presented at moderate to high intensities. Figures 5, 6, 7, and 8 show the successful performance of the computer model when tested for all of these properties and compared with empirical data.

Clearly, the more properties which can be simulated by a model, the more useful the model will be. Ideally, the model should simulate all known properties. At this point the model needs no further modification until experimenters generate new results which cannot be simulated by the model. The modeller can set the pace, however, by exploring the response of the model to situations which have not yet been studied experimentally. For example, the model under discussion has minor peaks in the AMTF at 2 and 3 times the best modulation frequency. These have not been reported yet in the literature; indeed the resolution of the AMTFs reported in the experimental literature is not great enough to show such subsidiary peaks. The model makes a clear prediction that such peaks would be found if experimenters looked for them.

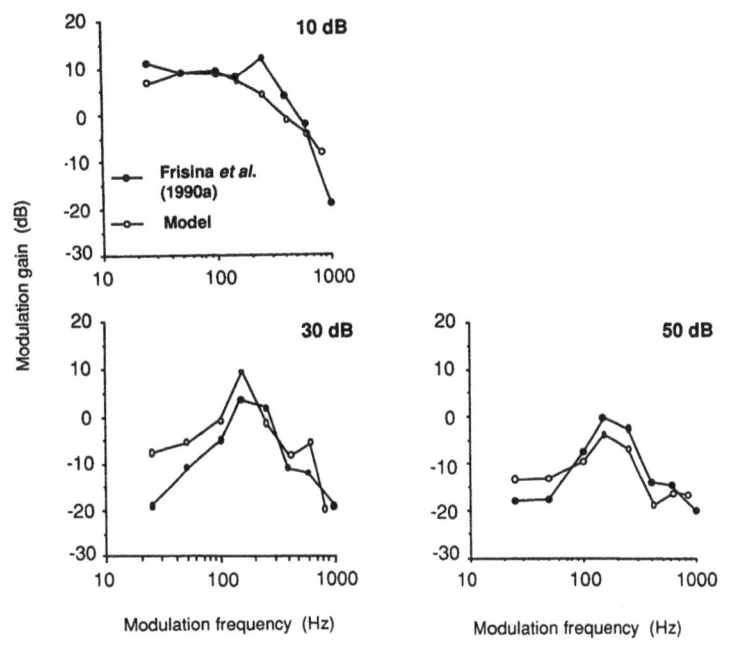

**Figure 7.** Model and neural modulation transfer functions for three input levels. Model parameters: 200-ms duration AM signals; 35% AM; 40 repetitions. Neural data redrawn from Frisina et al. ('91).

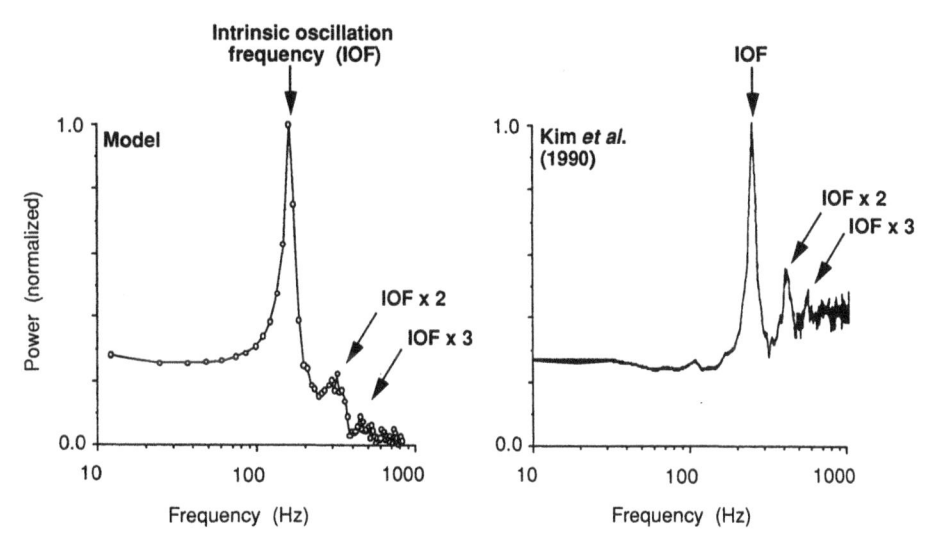

**Figure 8.** Model and neural intrinsic oscillations. Each panel shows the normalised power spectrum of autocorrelated spike trains in response to best-frequency tones. Arrows indicate response components at the harmonics of the cell intrinsic oscillation frequency. Neural data adapted from Kim et al. ('90).

## Complex systems

So far, the discussion has concentrated on the simulation of single cell responses. A fuller understanding of cochlear nucleus functioning must await studies which look at the complex interactions among systems of cells. This has already been attempted in the dorsal cochlear nucleus by Pont and Damper ('91) where the many aspects of the activity of type

IV neurons in the DCN have been successfully simulated using simplified artificial neurons. The anatomy of the DCN and the complex response patterns of cells in this area make it clear that single cell modelling studies will be of limited value here. The complex interactions of granule, stellate, cartwheel, vertical, giant and fusiform cells offer an exciting challenge for modellers of the future. However, each of these cells has a different morphology and physiology so that a comprehensive modelling exercise will need to begin with single cell models of each cell type.

It is also increasingly clear that we need to develop strong and productive hypotheses concerning the functions of the many inhibitory cells which operate in the cochlear nucleus and the inhibitory inputs which originate in other nuclei. Obviously, progress can be made using purely experimental methods. However, it is likely that the modeller also has a role to play here by constructing models which are faithful to both the anatomy (in terms of connectivity and morphology) and the cells' known physiology. These can then act as a basis for exploring the potential behaviour of the system. It is likely that analysis will reveal a number of distinct types of functions for inhibitory systems.

The list of possible functions for inhibitory systems seems endless. Researchers are already familiar with the computational potential of fast lateral inhibition; this has the effect of emphasising peaks of activity across a sheet of nervous tissue and of highlighting areas of local variation across the sheet. Neighbouring inhibitory neurons with the same input but slightly delayed activity relative to a projection neuron will allow the neuron to respond to the onset of stimulation but not to a steady state signal. The same inhibitory neuron, if it has a slightly higher threshold, may simply offer the principal neuron protection from input overload; low level inputs do not trigger the inhibitory action but high level inputs do. Descending inhibition from a nucleus which receives excitatory input from the cell in question could be acting as a simple negative feedback loop, but if the descending inhibition reflects activity in other cell groups then the effects are less predictable. Slow acting tonic inhibition offers another world of possibilities and inhibition of inhibitory cells produces yet another situation which will prove difficult to predict without computer modelling assistance. Similarly, inhibition applied to the soma has qualitatively different effects from inhibition applied to remote locations on dendrites.

Modellers have yet to make a strong contribution in this area but the possibilities are obvious both when modelling known circuits and also when modelling types of circuits so that an anatomist can make more sense of the functional significance of the patterns that are reconstructed from the light and electron microscope.

We hope to have shown that modelling techniques have a potentially important future role to play in making sense of the functioning of the cochlear nucleus. As the techniques become more specialised and more powerful, it is likely that modelling will emerge as a specialism in its own right alongside anatomy, physiology and psychology just as theoretical physics has come to operate alongside the hard sciences. Its function will be to synthesise and make sense of the enormous amount of data generated within its sister disciplines and to produce theoretical systems which can drive experimentation along an optimally productive course.

REFERENCES

Arle, J.E., and Kim, D.O., 1991, Neural modeling of intrinsic and spike-discharge properties of cochlear nucleus neurons, *Biol. Cyber.*, 64:273-283.

Banks, M.I., and Sachs, M.B., 1991, Regularity analysis in a compartmental model of chopper units in the anteroventral cochlear nucleus, *J. Neurophysiol.*, 65:606-629.

Frisina, R.D., 1983, "Enhancement of responses to amplitude modulation in the gerbil cochlear nucleus: single-unit recordings using an improved surgical approach," Dissertation and special report ISR-S-23, Institute for Sensory Research, Syracuse, New York.

Frisina, R.D., Smith, R.L., and, Chamberlin, S.C., 1990, Encoding of amplitude modulation in the gerbil cochlear nucleus: I. A hierarchy of enhancement, *Hearing Res.*, 44:99-122.

Frisina, R.D., Smith, R.L., and, Chamberlin, S.C., 1990, Encoding of amplitude modulation in the gerbil cochlear nucleus: II. Possible neural mechanisms, *Hearing Res.* 44:123-142.

Hewitt, M. J., and Meddis, R., 1991, An evaluation of eight computer models of mammalian inner haircell function, *J. Acoust. Soc. Am.*, 90:904-917.

Hewitt, M.J., Meddis, R., and Shackleton, T.M., 1992, A computer model of a cochlear nucleus stellate cell: Responses to pure-tone and amplitude-modulated stimuli, *J. Acoust. Soc. Am.*, 91:2096-2109.

Kim, D.O., Sirianni, J.G., and Chang, S.O., 1990, Responses of DCN-PVCN neurons and auditory nerve fibers in unanaesthetized decerebrate cats to AM and pure tones: Analysis with autocorrelation/power-spectrum, *Hearing Res.*, 45:95-113.

MacGregor, R.J., 1987, "Neural and Brain Modeling", Academic Press, San Diego.

Meddis, R., 1988, Simulation of auditory-neural transduction: Further studies, *J. Acoust. Soc. Am.*, 83:1056-1063.

Oertel, D., 1983, Synaptic responses and electrical properties of cells in brain slices of the mouse anteroventral cochlear nucleus, *J. Neurosci.*, 3:2043-2053.

Oertel, D., 1985, Use of brain slices in the study of the auditory system: spatial and temporal summation of synaptic inputs in cells in the anteroventral cochlear nucleus of the mouse, *J. Acoust. Soc. Am.*, 78:328-333.

Pont, M.J. and Damper, R.I., 1991, A computational model of afferent neural activity from the cochlea to the dorsal acoustic stria, *J. Acoust. Soc. Am.*, 89:1213-1228.

Rall, W., 1989, Cable theory for dendritic neurons, *in*: "Methods in neuronal modeling: From synapses to networks," C.Koch and I.Segev eds., MIT press Cambridge, Mass. pp. 9-62.

Young, E.D., Robert, J.-M., and Shofner, W.P., 1988, Regularity and latency of units in the ventral cochlear nucleus: Implications for unit classification and generation of response properties, *J. Neurophysiol.*, 60:1-29.

APPENDIX A

Computer models of cochlear nucleus function - a bibliography

Arle, J.E. and Kim, D.O., 1991, A modeling study of single neurons and neural circuits of the ventral and dorsal cochlear nucleus, *in*: "Analysis and Modeling of Neural Systems", (proceedings of symposium held in Berkeley, CA, in July 1990), F.Eeckman, ed., Kluwer, Boston.

Arle, J.E. and Kim, D.O., 1991, Neural modeling of intrinsic and spike-discharge properties of cochlear nucleus neurons, *Biol. Cyber.*, 64:273-283.

Arle, J.E. and Kim, D.O., 1991, Simulations of cochlear nucleus neural circuitry: excitatory-inhibitory response-area types I-IV, *J. Acoust. Soc. Am.*, 90:3106-3121.

Banks, M.I. and Sachs, M.B., 1991, Regularity analysis in a compartmental model of chopper units in the anteroventral cochlear nucleus, *J. Neurophysiol.*, 65:606-629.

Berthommier, F., 1989, A model of the relation between tonotopy and synchronisation in the auditory system, *CR.Acad.Sc.*, Serie III, 309:695-701.

Berthommier, F., Schwartz, J.L. and Escudier, P., 1989, "Auditory processing in a post-cochlear neural network: vowel spectrum processing based on spike synchrony", Eurospeech, Paris.

Berthommier, F., 1990, "Reseaux de neurones et traitement des signaux dans le systeme auditif, technical report", Faculte de Medecine de Grenoble, France.

Blackwood, N., Meyer, G. and Ainsworth, 1990, A model of the processing of voiced plosives in the auditory nerve and cochlear nucleus, *Proceedings of the Institute of Acoustics*, 12:423-430.

Damper, R.I., Pont, M.J. and Elenius, K, 1991, Representation of Initial Stop Consonants in a Computational Model of the Dorsal Cochlear Nucleus, *in*: "Advances in Speech, Hearing and Language Processing", Vol. 3, W.A.Ainsworth, ed., JAI Press Ltd, London, in press.

Diaz, J.M., 1989, "Chopper firing patterns in the mammalian anterior ventral cochlear nucleus: A computer model", Masters Thesis, Boston University, Boston, Mass.

Davis, K.A., and Voigt, H.F., 1991, Neural modelling of the DCN: PST and cross-correlation analysis using shortduration tone-burst stimuli, *Abstracts of the XIVth Midwinter Meeting, Association for Research in Otolaryngology*.

Ghoshal, S., Kim, D.O. and Northrop, R.B., 1991, Modeling amplitude-modulated (AM) tone encoding behaviour of cochlear nucleus neurons, Proc. The 17th Annual IEEE Northeast Bioengineering Conference, Hartford, CT. pp. 5-6.

Ghoshal, S., Kim, D.O. and Northrop, R.B., 1992, Amplitude modulated tone encoding behavior of cochlear nucleus neurons: Modeling study, *Hear. Res.*, 58:153-165.

Hewitt, M.J., Meddis, R. and Shackleton, T.M., 1992, A computer model of a cochlear-nucleus stellate cell: responses to amplitude-modulated and pure-tone stimuli, *J. Acoust. Am.*, 91:2096-2109.

McMullen, T. and Voigt, H., 1984, Neuronal circuitry of dorsal cochlear nucleus: a computer model, *Soc. Neuroscience*, 10:842, (Abstracts).

Mashari, S.J. and Pont, M.J., 1990, A hybrid neural network model with applications in the study of language acquisition, *Proceedings of the Institute of Acoustics*, 12:315-321.

Meddis, R., Hewitt, M. and Shackleton, T.M., (in press), An anatomical/ physiological approach to auditory selective attention, *in*: "Auditory physiology and perception" Cazals, L.Demany, and K.Horner, eds., (proceedings of the 9th international symposium on hearing, Carcans, France).

Meddis, R., 1991, A physiological model of auditory selective attention, *in*: "Advances in Speech, Hearing and Language Processing" Vol 3., W.A.Ainsworth, ed., JAI Press, in press.

Meyer, G.F., Blackwood,N. and Ainsworth, W.A., 1990, A computational model of the auditory nerve and cochlear nucleus, *in*: "Modelling and Simulation", B.Schmidt, ed., (fourth European Multiconference, Nuremberg), American Society of Computer Simulation SCS.

Meyer, G.F. and Ainsworth, W.A., 1991, Modelling response patterns in the cochlear nucleus using simple units, *in:* "Advances in Speech, Hearing and Language Processing", Vol 3., W.A.Ainsworth, ed., JAI Press Ltd, London, in press.

Meyer, G.F., Morris, A., Ainsworth, W.A. and Schwartz, J.L., 1991, Processing of Plosives by models of cells in the CN, *Proceedings of the Institute of Acoustics*, 13: 485-492.

Neti, C., 1990, "Neural network models of sound localisation based on directional filtering of the pinna", Ph.D. thesis, Baltimore, Maryland.

Pont, M.J., 1988, A neural model of the mechanisms underlying infant perception of voice-onset time, *Neural Networks*, Supplement 1:270, (Abstracts).

Pont, M.J. and Damper, R.I., 1988, A neural model of infant speech perception, *Proceedings SPEECH U88*, Edinburgh, p.515-522.

Pont, M.J. and Damper, R.I., 1989, A possible neural basis for the categorical perception of the English voiced / voiceless contrast, *Proceedings Eurospeech T89*, Paris, pp.239-242.

Pont, M.J. and Damper, R.I., 1989, The representation of synthetic stop consonants in a computational model of the dorsal cochlear nucleus, *J. Acoust. Soc. Am.*, Suppl. 1:S45-S46 (Abstract).

Pont, M.J., 1990, "The role of the dorsal cochlear nucleus in the perception of voicing contrasts in initial English stop consonants: a computational modelling study", PhD Thesis, Department of Electronics and Computer Science, University of Southampton. See also (published abstracts): (Pont, M.J. (1990) Journal of the Acoustical Society of America 87, p.1817; Pont, M.J. (1990) Speech Communication 9:p.95).

Pont, M.J. and Mashari, S.J., 1990, Modelling the acquisition of voicing contrasts in English and Thai, *Proceedings of the Institute of Acoustics*, 12(10):323-329.

Pont, M.J. and Damper, R.I., 1991, A computational model of afferent neural activity from the cochlea to the acoustic stria, *J. Acoust. Soc. Am.*, 89:1213-1228.

Pont, M.J. and Damper, R.I., 1991, Exploring the role of the dorsal cochlear nucleus in the perception of voice-onset time, *in*: "Advances in Speech, Hearing and Language Processing", Vol 2, W.A.Ainsworth, ed., JAI Press Ltd, London, in press.

Sanders, D.J. and Green, C.G.R., 1991, Properties of modelled and networked cochlear nucleus neurons, *in*: "Advances in Speech, Hearing and Language Processing", Vol 3., W.A.Ainsworth, ed., JAI Press Ltd, London, in press.

Voigt, H.F. and Davis, K.A., 1991, Computer simulations of neural correlations in dorsal cochlear nucleus, *in*: "Advances in Speech, Hearing and Language Processing" Vol. 3., W.A.Ainsworth, ed., JAI Press Ltd, London, in press.

Wu, Z.L., Schwartz, J.L. and Escudier, P., 1991, Physiologically-plausible modules and articulatory-based acoustic events, *in*: "Advances in Speech, Hearing and Language Processing" Vol. 3., W.A.Ainsworth, ed., JAI Press Ltd, London, in press.

# REGULARITY OF DISCHARGE CONSTRAINS MODELS OF VENTRAL COCHLEAR NUCLEUS BUSHY CELLS

Eric D. Young, Jason S. Rothman, and Paul B. Manis

Departments of Biomedical Engineering and Otolaryngology/Head and Neck Surgery and Center for Hearing Sciences, The Johns Hopkins University, Baltimore MD, 21205, USA

## Properties of bushy cells

The bushy cells of the ventral cochlear nucleus (VCN) are characterized by two unusual features. The first is the large synaptic contact, the endbulb of Held, made by auditory nerve (AN) fibers on bushy cell somata (Brawer and Morest, '75; Lorente de Nó, '81; Ryugo and Fekete, '82). In the rostral anteroventral cochlear nucleus (AVCN), spherical bushy cells receive a small number of large endbulbs, as few as one to four per cell (Lorente de Nó, '81; Ryugo and Sento, '91). In the posterior AVCN and the anterior part of the posteroventral cochlear nucleus (PVCN), endbulbs are smaller (Brawer and Morest, '75; Lorente de Nó, '81; Rouiller et al., '86). It is clear that the globular bushy cells in this region receive endbulbs from a larger number of AN fibers than the spherical cells of rostral AVCN, although current estimates of the number of endbulbs per cell are indirect (Spirou et al., '90; Liberman, '91). Based on estimates of the total number of endbulbs in the globular bushy cell area (Ryugo and Rouiller, '88) and the number of globular bushy cells (Brownell, '75; Osen, '70), the number of endbulbs per cell has been estimated as about 17 (Spirou et al., '90). In addition to endbulbs, both spherical and globular bushy cells receive smaller bouton terminals from AN fibers (Cant and Morest, '79; Lenn and Reese, '66; Lorente de Nó, '81; Ryugo and Sento, '91; Liberman, '91); these terminals increase the numbers of AN fibers per cell, but the contribution of these small terminals to postsynaptic processing is unclear, given their relatively small size compared to endbulbs.

Each endbulb contains a number of individual synaptic terminal zones (Cant and Morest, '79; Lenn and Reese, '66; Tolbert and Morest, '82). This fact suggests that an endbulb produces a large postsynaptic conductance change, so that there is a secure synaptic coupling between AN fiber and bushy cell. Several lines of physiological evidence support this conclusion. First, AN fiber input produce large EPSPs in bushy cells, which frequently lead directly to postsynaptic action potentials (Oertel, '83; Smith and Rhode, '87). Second, bushy cells give response properties which are similar to those of AN fibers in many ways (Rhode et al., '83; Rouiller and Ryugo, '84; Smith and Rhode, '87). These response properties have been classified as primarylike, primarylike-with-notch (pri-N), and onset-L (on-L), based on PST histogram shape (Pfeiffer, '66a; Bourk, '76; Rhode and Smith, '86; Young et al., '88a; Blackburn and Sachs, '89); these response types are collectively called bushy cell response types (BCRTs) below. Third, the action potentials of many BCRTs, recorded with extracellular metal electrodes, show prepotentials, a waveform which precedes the action potential of the cell by 0.5-1 ms (Bourk, '76; Pfeiffer, '66b). Prepotentials are thought to be action potentials invading the presynaptic end bulb and analyses of prepotential occurrence suggest that a prepotential is almost always followed by a postsynaptic action potential, as is consistent with a secure synapse (Pfeiffer, '66b; Goldberg and Brownell, '73; Bourk, '76; Guinan and Li, '90).

The second unusual feature of bushy cells is their membrane conductance (Manis and Marx, '91; Oertel, '83; Oertel, '85). Bushy cells show a strong membrane rectification, so that when the cell is depolarized from rest, its membrane conductance increases dramatically compared to its value at rest or with hyperpolarization (Oertel, '83). A voltage-clamp study of bushy cell somata *in vitro* has identified a low threshold potassium channel which is partially activated at rest, and which activates fully over a 40 mV range above rest (Manis and Marx, '91); this channel appears to be sufficient to explain the membrane rectification. It has been suggested that the effect of the membrane rectification is to decrease the membrane time constant when a bushy cell is depolarized, thus increasing the rate at which EPSPs decay and preventing temporal summation of successive EPSPs (Oertel, '85).

The combination of these two features, the large size of AN inputs to bushy cells and the rectifying properties of bushy cell membranes, are thought to account for the principal physiological property of BCRTs, which is their ability to preserve temporal features of AN discharge, such as phase locking, at frequencies in the kHz range (Bourk, '76; Rhode and Smith, '86; Blackburn and Sachs, '89). The large size of the AN input serves to bypass the low-pass filtering effect of postsynaptic membrane capacitance (Young et al., '88b) and the membrane rectification serves to decrease temporal summation; both effects increase the cell's ability to follow temporal patterns in the AN input at high frequencies.

In this paper, we discuss the regularity properties of BCRT discharge patterns; regularity is another way in which AN fibers and bushy cells have similar response characteristics. We then consider models of input/output processing in bushy cells to show the extent to which the large size of AN inputs and the membrane rectification are important in producing BCRT characteristics. In particular, we show that analysis of regularity of discharge provides a powerful constraint on the range of models which can successfully account for BCRT characteristics.

## Regularity in bushy cells

Figure 1 shows a comparison of the regularity of AN fibers and BCRTs. This figure shows the standard deviation of interspike intervals (ordinate) plotted versus the mean interspike interval (abscissa). The points show results from primarylike and related response types in the CN. The computation is identical to that described by Young, Robert and Shofner ('88a), except that the results in Figure 1 are computed from responses to 50 ms tone bursts (as opposed to 25 ms bursts), which eliminates a computational artifact and reduces the spread of the data (Young et al., '88a).

The response types which are included in Figure 1 are listed in the legend. Primarylike, pri-N, and on-L types are the usual, well-defined BCRT subgroups (Pfeiffer, '66a; Bourk, '76; Smith and Rhode, '87). The unusual and pri?? response types include units whose PST histograms are not strictly primarylike or pri-N, but are nevertheless similar to primarylike PSTs, and are clearly not chopper or onset PSTs. The onset category includes all onset units except on-L and onset-C. Onset units have a reliable, well-timed first spike in response to BF tone bursts, followed by little maintained discharge. On-L units are a subclass of onset units whose properties resemble those of pri-N units, except that the maintained discharge rate is low (Smith and Rhode, '87). Onset-C units are a group defined by Rhode

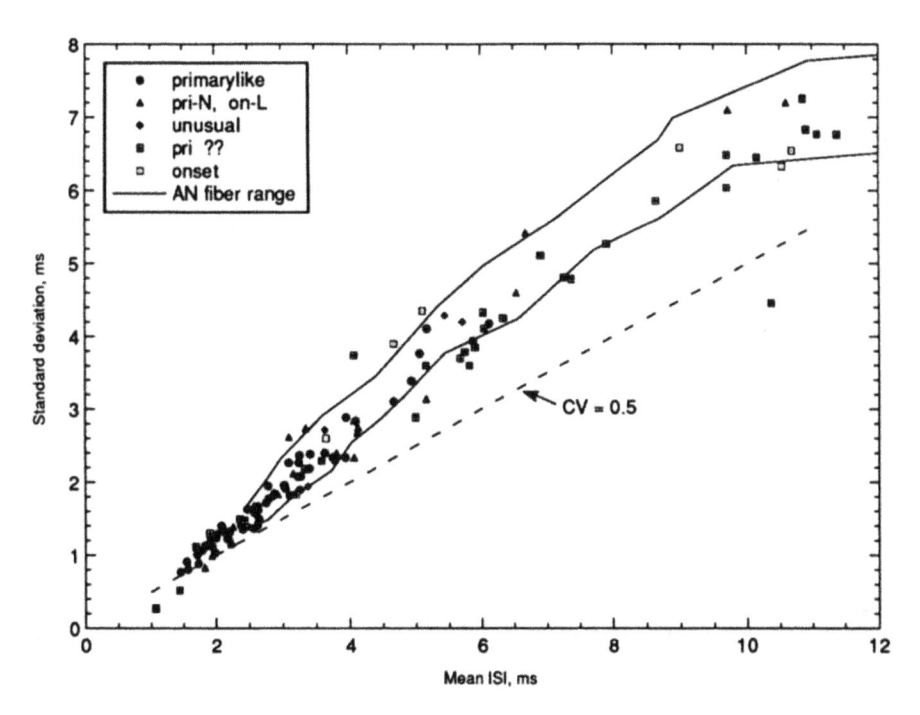

**Figure 1.** Regularity analysis of bushy cell response types compared to auditory nerve fibers. Symbols show standard deviation versus mean interspike interval for five groups of CN primarylike and related response types, defined in legend. Solid lines show range of auditory nerve regularity. Note that lines end at shortest auditory nerve interspike interval, about 2.5 ms. Dashed line shows where standard deviation is half of mean (CV=0.5). Mean and standard deviation were computed for intervals whose first spike occurred between 12 and 20 ms after the onset of a 50 ms best-frequency tone burst. CN data and some AN data from an unpublished study of Young and Sachs; other auditory nerve data from Li, '91.

and Smith ('86) which are recorded from multipolar cells rather than from bushy cells (Smith and Rhode, '89); onset-C units are not included in Figure 1, but are slightly more regular than BCRTs. There is good evidence that primarylike, pri-N, and on-L responses are recorded from bushy cells (Rhode et al., '83; Rouiller and Ryugo, '84; Smith and Rhode, '87; Friauf and Ostwald, '88). Although there is no evidence that unusual and pri?? units are recorded from bushy cells, they are included in Figure 1 because they sometimes have prepotentials in their action potentials and their regularity and latency and, to some extent, their PST histograms are identical to those of the response types that have been shown to originate in bushy cells.

The data in Figure 1 show that BCRTs are irregular, in that their standard deviations exceed half their mean interspike intervals; that is, virtually all the data in Figure 1 are above the CV=0.5 line, which approximately separates regular units (choppers) below the CV=0.5 line and irregular units (BCRTs) above the line (Young et al., '88a; Blackburn and Sachs, '89; Spirou et al., '90). The most important aspect of Figure 1 is that the BCRT data have the same regularity as AN fibers. The solid lines show the range in which data from 119 AN fibers, analyzed in the same way as the CN units, scatter (Young and Sachs, unpublished; Li, '91). There is no systematic displacement of BCRT regularity from AN regularity.

The similarity of regularity of AN fibers and BCRT units shown in Figure 1 is an important result, because it provides a powerful constraint on the properties of successful models of bushy cell input/output processing. In the next sections, we show that if the output spike trains from a bushy cell model are to be as irregular as the input spike trains, then each EPSP evoked by an AN input must be large enough to exceed threshold, i.e. to cause a spike

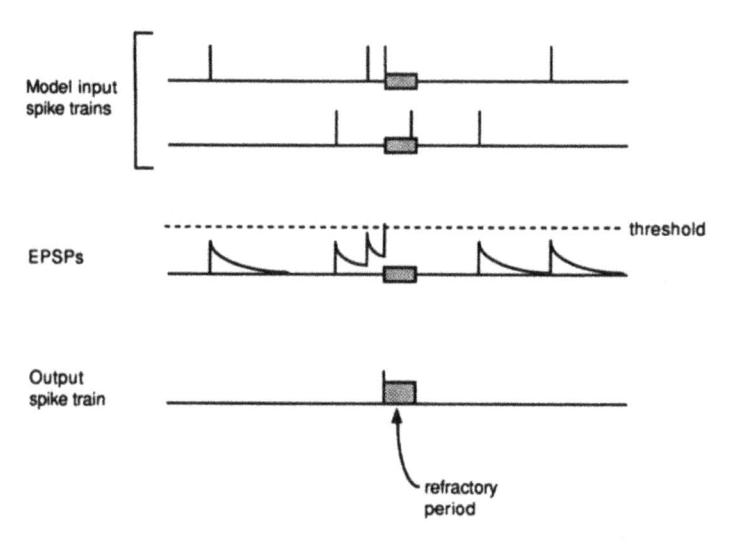

**Figure 2.** Description of shot-noise threshold model. Top two lines show two input spike trains; each vertical line is a spike. Third line shows EPSPs produced by input spikes. The EPSP following a spike at time $t_1$ is given by $w \cdot exp[(t-t_1)/\tau]$, for $t > t_1$. Bottom line shows output spike train; output spikes produced when summated EPSPs exceed threshold (threshold equals 1.0). Shaded boxes show refractory period following output spikes, during which input spikes are ignored. EPSP value starts at 0 following a refractory period.

in the postsynaptic cell. That is, summation of EPSPs or coincidence of EPSPs cannot be required to produce an output spike. This point is made first with a simplified neural model which does not attempt to accurately reproduce the electrical properties of bushy cells and then with a more sophisticated model which is based on the membrane properties of bushy cells (Manis and Marx, '91).

## Regularity of a shot-noise threshold model

Figure 2 shows a simple EPSP summation model which has been used previously to study the regularity properties of auditory neurons (Molnar and Pfeiffer, '68; Tuckwell and Richter, '78). A number of input spike trains, representing AN fibers, are applied to the model (top two lines). Whenever a spike occurs on one of the inputs, an EPSP is produced (third line). The EPSP rises instantaneously to an amplitude $w$ and decays exponentially with time constant $\tau$. EPSPs from successive inputs sum linearly until threshold is exceeded, at which time an output spike is produced (bottom line). Following an output spike, the model is refractory for a deadtime $t_D$, during which input spikes are ignored. At the end of the refractory period, the model is restarted with 0 EPSP value. For the simulations reported here, $t_D = 0.7$ ms and $\tau = 5$ ms. The EPSP amplitude $w$ is given as a fraction of threshold; that is, the threshold value is 1.0 and a single EPSP is capable of exceeding threshold by itself if $w > 1$. The value of $\tau$ used here is similar to values reported for bushy cells (Manis and Marx, '91; Wu and Oertel, '87); the value of $t_D$ is based on examination of the hazard functions of BCRTs (unpublished observations). The results to be reported are not sensitive to the value of t over the range investigated (2 to 10 ms).

The input spike trains are modelled after the refractory properties of AN fibers, as described by Li ('91). Input (AN) spikes are produced by a deadtime-modified Poisson process (DTMPP; Teich et al., '78; Young and Barta, '86) in which a Poisson process is modified by deleting spikes that occur within a deadtime $t_A$ of previous spikes. Following each spike, the DTMPP is refractory for a time period $t_A = t_{arp} + t_{rrp}$, where $t_{arp}$ is the

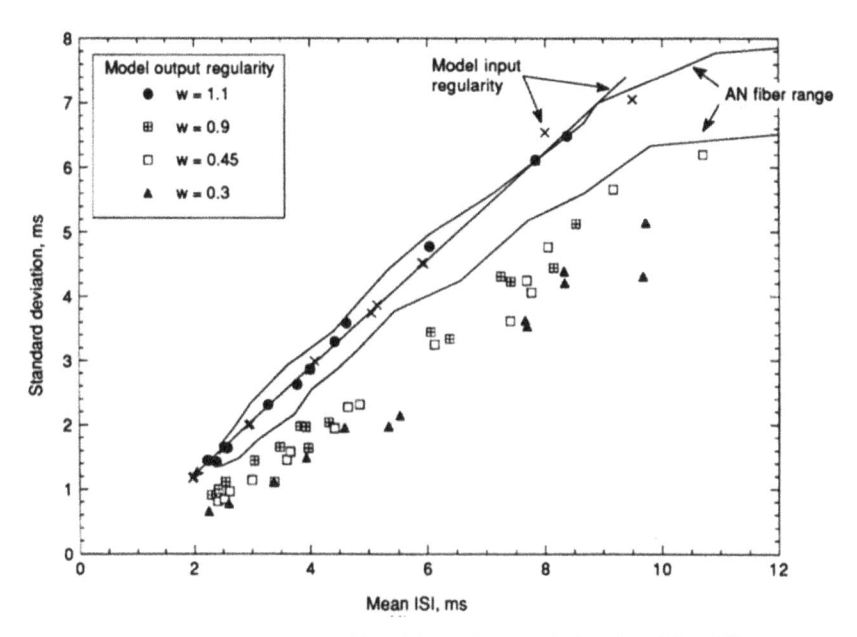

**Figure 3.** Regularity of the shot-noise threshold model. Xs show regularity of model AN fiber spike train inputs. Other symbols show regularity of outputs of the model for four values of EPSP amplitude *w* (see legend). Model parameters: threshold amplitude = 1.0, $\tau$ = 5 ms; $t_D$ = 0.7 ms. Range of regularity of actual AN fibers shown as in Fig. 1. Straight line drawn by eye through Xs.

absolute refractory period (fixed at 0.7 ms; Li, '91) and $t_{rp}$ is a random variable with an exponential distribution (Young and Barta, '86; Li, '91). This simple model is capable of reproducing the refractory properties of AN fibers in the absence of phase locking. Note that the DTMPP does not attempt to account for the physiological properties of refractoriness in AN fibers; rather, it is designed to produce spike trains with statistical properties that are similar to those of AN fibers.

Figure 3 shows the regularity of the inputs and outputs of the shot-noise threshold model for four different values of EPSP amplitude *w*. The Xs show the regularity of the DTMPP inputs; a straight line is drawn through these inputs to show their trend. A variety of input spike trains were produced, with average rates varying from 67.5 to 500 spikes/s. The range of actual AN fiber regularity is also shown, as in Figure 1. The DTMPP spike trains are slightly more irregular than real AN fibers for interspike intervals longer than about 7 ms (i.e. the Xs are near the upper end or above the AN fiber range). This irregularity is caused by inaccurate parameter adjustment for the $t_{rp}$ distribution at low rates and does not affect the conclusions that are drawn below.

From 1 to 6 input spike trains were applied to the shot-noise threshold model at each of four different EPSP amplitudes (see legend in Fig. 3). The rates of the input spike trains were adjusted in each case to produce a range of output rates from around 100 spikes/s to around 500 spikes/s. The regularities of these output spike trains are shown by the symbols defined in the legend of Figure 3. When the EPSP amplitude exceeds threshold, the output spike train is essentially identical to the superposition of the input spike trains, except for the effects of the refractoriness of the shot-noise threshold model. It is not surprising, therefore, that the regularity of the output of the model is essentially equal to the regularity of the input for EPSP amplitude 1.1 (●). When the EPSP amplitude is smaller than threshold, however, the output spike trains are substantially more regular than the input spike trains. Increased regularity is seen even if the EPSP amplitude is 0.9 times threshold (⊞), i.e. even if only a minimum amount of EPSP summation is required. As the EPSP amplitude is made smaller, there is some increase in regularity; however, the main change occurs as the EPSP amplitude drops below threshold.

The model shown in Figure 2 is a considerable simplification of what goes on in a bushy cell. However, the model demonstrates that the effect on regularity of EPSP amplitude is determined primarily by whether EPSPs must sum in order to reach threshold. The model shows that the effects on regularity of requiring a coincidence of input spikes in order to produce an output spike are a statistical property of the coincidence itself, and not some hidden property of the more complex bushy cell model that will be considered in the next section.

It is clear that the difference in Figure 3 between the regularity of model input and model output with EPSP amplitude less than one is substantially larger than the difference in regularity between AN fibers and BCRT units in Figure 1. This result suggests that the EPSPs produced by AN fibers in bushy cells must be larger than threshold, at least for the AN fibers whose spike trains are dominant in driving the cell. In the next section, we describe a more physiological model of bushy cells which is based on current knowledge of their membrane properties. The results of regularity analysis of this more complicated and more realistic model support the conclusions drawn from Figure 3.

## Model of bushy cell membrane properties

The model described in this section is based on the voltage clamp studies of isolated bushy cell somata by Manis and Marx ('91). Three channels were found in bushy cell somata in that study: a fast sodium conductance responsible for the action potential; a high-threshold potassium conductance, with the properties of a delayed rectifier (Hille, '84); and a low-threshold potassium conductance with voltage-sensitive gating properties similar to those of the M channel described in bullfrog sympathetic ganglion by Adams, Brown and Constanti ('82). The low-threshold potassium channel is the one mentioned in the Introduction above which is thought to account for the membrane rectification in bushy cells. The high threshold potassium and fast sodium channels activate at higher membrane potentials than the M channel and contribute little to the membrane conductance near resting potential.

Figure 4A shows an electrical equivalent circuit of the model. The model consists of the three voltage-activated channels described in the previous paragraph ($G_{Na}$, $G_K$, and $G_M$) along with a leakage channel $G_L$ and a synaptic channel $G_E$. The conductance of $G_L$ is small compared to the others and is used to set the resting membrane potential. The voltage-activated channels are modelled with Hodgkin-Huxley type equations (Hodgkin and Huxley, '52). The conductance $G_i$ for current flow of ion i is given by

$$G_i = G_i^0 \, a_i(V,t) \, b_i(V,t) \qquad \text{for } i = \text{Na, K, or M} \tag{1}$$

and the current flow through this conductance is

$$I_i = G_i^0 \, a_i(V,t) \, b_i(V,t) \, [V - E_i] \tag{2}$$

If there is no external applied current ($I_{ext} = 0$), then the membrane potential V obeys the usual differential equation,

$$C_S \frac{dV}{dt} = G_L(E_L - V) + G_{Na}(E_{Na} - V) + G_K(E_K - V) +$$
$$G_M(E_M - V) + G_E(E_E - V) \tag{3}$$

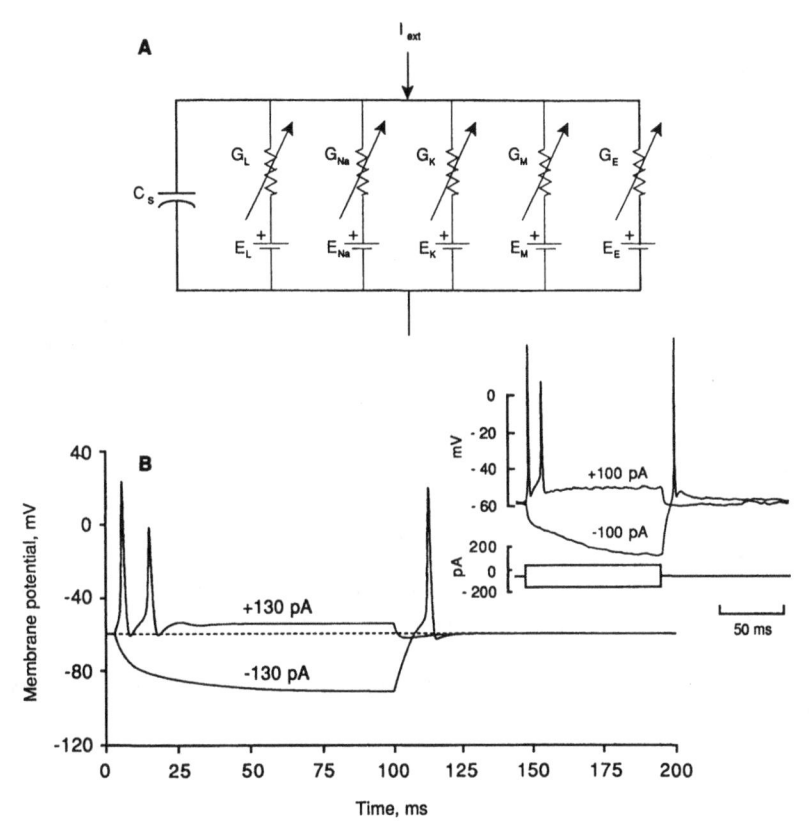

**Figure 4.** Properties of the bushy cell model. The model consists of a single compartment representing the soma of the cell only. A. Equivalent circuit of the model showing membrane capacitance $C_S$, leakage conductance $G_L$, fast sodium conductance $G_{Na}$, high threshold potassium conductance $G_K$, low threshold potassium conductance $G_M$, and a synaptic conductance $G_E$. $G_L$ is a constant; $G_{Na}$, $G_K$, and $G_M$ are voltage dependent and are modelled with Hodgkin-Huxley style differential equations; $G_E$ is zero except when an input action potential arrives, after which $G_E$ increases transiently according to Eqn. 5. There may be from 1 to 20 parallel independent $G_E$s in the model. $E_{Na}$ = 55 mV, $E_K$ = -77 mV, $E_E$ = -10 mV, and $E_L$ is adjusted to make the resting potential -60 mV; typically $E_L$ is near 0 mV. B. Responses of the model to depolarizing and hyperpolarizing current clamps of $\pm 130$ pA. Inset shows current clamp records from a bushy cell for $\pm 100$ pA currents (reprinted with permission from Manis and Marx, '91).

In Eqns. 1 and 2, $a_i(V,t)$ is the Hodgkin-Huxley activation variable for $G_i$ and $b_i(V,t)$ is the inactivation variable. Power functions of the activation and inactivation variables frequently appear in equations like Eqns. 1 and 2; however, for the bushy cell model, only the activation variable for Na, $a_{Na}$, is taken to a power and the power in this case is 2. For $G_K$ and $G_M$, $b_i(V,t) = 1$; i.e. inactivation applies only to $G_{Na}$. $G_i^0$ is the maximum conductance that $G_i$ can assume (325 nS for $G_{Na}$, 40 nS for $G_K$, and 20 nS for $G_M$; $G_L$ is fixed at 1.7 nS). The activation variables obey first-order differential equations of the form

where $a_{i\infty}(V)$ is the voltage-dependent steady-state value of $a_i$ and $\tau_{ia}(V)$ is the voltage-dependent time constant. The inactivation variables $b_i$ obey similar equations. There is a different $a_{i\infty}$ (or $b_{i\infty}$) and $\tau_{ia}$ (or $\tau_{ib}$) function for each activation and inactivation variable.

These functions determine the properties of the channels and of the model. Details of the pertinent equations and parameters of the activation and inactivation functions are lengthy and are given elsewhere (Rothman, '91).

The steady-state ($a_{i\infty}$ or $b_{i\infty}$) and time constant ($\tau_{ia}$ or $\tau_{ib}$) functions are fit to data from the literature. The model for $G_M$ is based on the low-threshold potassium channel in isolated guinea pig bushy cell somata (Manis and Marx, '91). The model for $G_K$ is based on the delayed rectifier of bullfrog sympathetic ganglion (Adams et al., '82), which resembles the delayed rectifier in guinea pig bushy cells (Manis and Marx, '91). The model for $G_{Na}$ is based on the fast sodium channel of the node of Ranvier in toad myelinated nerve fibers (Frankenhaeuser and Huxley, '64). $G_K$ and $G_{Na}$ are not based on data from CN bushy cells because Manis and Marx did not obtain a full characterization of these channels.

The data upon which the model is based were not taken at mammalian body temperature. In order to correct for the temperature difference, the time constants of the equations were decreased by a temperature factor computed by assuming a $Q_{10}$ of 3 (i.e. the time constants decrease by a factor of 3 for every 10°C increase in temperature). The time constant and steady-state functions were further modified to make the absolute refractory period of the model about 0.7 ms and to give the model a sharp threshold (Rothman, '91).

A comparison of the electrophysiological properties of the final model and of a bushy cell soma is shown in Figure 4B. These plots show the membrane potential response to hyperpolarizing and depolarizing current pulses. In response to the depolarizing pulse, both model and cell give two spikes at stimulus onset followed by a small maintained depolarization. In response to the hyperpolarizing pulse, the model and cell give larger hyperpolarizing responses that increase slowly in amplitude, reflecting the deactivation of $G_M$ and the consequent increase in membrane resistance. At the termination of the hyperpolarizing pulse, there is an anode-break spike in both model and cell. The general similarity of model and cell behavior supports the idea that the model accurately reproduces bushy cell membrane characteristics.

In the next section, the effects of applying simulated AN inputs to the bushy cell model are discussed. AN inputs are applied by activating the excitatory synaptic conductance $G_E$. If an AN spike arrives at time $t_1$, then $G_E$ undergoes a transient increase of the form

$$G_E(t) = G_E^0 \frac{t - t_1}{t_p} \exp\left[-\frac{t - t_1 - t_p}{t_p}\right] \quad \text{for } t > t_1 \tag{5}$$

This function peaks when $t = t_1 + t_p$ at a value of $G_E^0$; for the simulations reported here, $t_p = 0.1$ ms, so that the increase in $G_E$ lasts about 0.5 ms.

There are two parameters of the synaptic input: the first parameter is the amplitude of the conductance increase $G_E^0$; the second parameter is the number of independent synaptic inputs to the cell. Between 1 and 20 parallel $G_E$ circuits are included in the model; each parallel $G_E$ circuit is activated by a separate AN fiber, and the spike trains of the input fibers are independent. In the next section we demonstrate that both parameters, the strength of the synaptic input $G_E^0$ and the number of inputs, affect the response characteristics of the model.

The input AN spike trains were simulated using a model discussed by Johnson and Swami (Johnson and Swami, '83) in which the instantaneous discharge rate $\lambda(t;t_{-1})$ is given by the product of a stimulus dependent factor $s(t)$ and a refractory factor $r(t-t_{-1})$, where

$$\lambda(t;t_{-1}) = s(t) \cdot r(t-t_{-1}) \tag{6}$$

$s(t)$ is adjusted to produce a primarylike PST histogram patterned after those of AN fibers (Kiang et al., '65; Rhode and Smith, '85), using the model described by Westerman and Smith (Westerman and Smith, '84). Although results on phase-locking are not presented in this paper, $s(t)$ can also be used to produce an accurate model of phase-locking. $r(t;t_{-1})$ is the refractory function which describes the recovery of discharge after the previous spike, which occurred at time $t_{-1}$. $r(t-t_{-1})$ is estimated from the hazard functions of AN fibers (Li, '91), using a model similar to that described in the section on the shot-noise threshold model. The simulations were done using a method of Johnson and colleagues (Johnson et al., '86) for simulating a non-homogeneous Poisson process from its instantaneous rate function.

All simulations were done with the system simulation package ACSL (Advanced Continuous Simulation Language, Mitchell and Gauthier Associates, Boston, MA). Differential equations were integrated numerically using a fifth order Runge-Kutta-Fehlberg method with variable step size.

## Response characteristics of the bushy cell model

Figure 5 shows PST histograms of the input (Fig. 5A) and the output (Fig. 5B-F) of the bushy cell model for various combinations of synaptic strength and number of inputs. The simulated AN inputs are patterned after responses to 50 ms BF tone bursts for high BF fibers, where there is no phase locking. All AN inputs have PST histograms like the one in Figure 5A; their steady state discharge rates are approximately 167 spikes/s and their spontaneous rates are 10 spikes/s. Figure 5B shows the response of the bushy cell model to one input with a peak conductance of 18 nS. The PST histogram in this case is primarylike. It is easy to understand what is going on in this case, because an 18 nS conductance is large enough that each input spike is capable of producing an output spike by itself. That is, the output spike train should be identical to the input spike train, except for slight differences in the refractory behavior of input and output. The model produces primarylike responses for one or two inputs and conductances large enough that input spikes exceed threshold ($G_E^0$ > 6 nS). However, with two inputs, the PST histograms are intermediate in character and could be called either primarylike or pri-N.

Figure 5C shows the response of the bushy cell model to 5 inputs, each with a conductance of 10 nS. The response in this case is pri-N with a clear notch following the sharp onset peak. There is a spike in the output at the latency of the onset peak on essentially every trial, and the notch corresponds to the refractory period that follows those onset spikes. The sharp onset peak results from the convergence of several inputs on the cell; because there are several inputs, each of which has a high instantaneous discharge rate at stimulus onset, there is a high probability, approaching 1, of producing an output spike during the onset peak of the PST histogram (Young et al., '88a). Pri-N responses are seen for cases with between 3 and about 7 inputs and conductances above threshold; they are also seen at lower (subthreshold) conductances with a larger number of inputs.

When inputs with weaker synaptic conductances are applied (3 and 2 nS, Figs. 5D and E), the response takes on an onset character. The PST histogram in Figure 5D is classed as on-L and the one in Figure 5E as onset-I. Onset responses occur with synaptic conductances that are below threshold (i.e. < 6 nS) and with a varying number of synaptic inputs (from 1 to 20). The mechanism of action here may be similar to that of pri-N units in that the input is strong enough at stimulus onset to produce a spike, but the inputs are too weak, because of lower input discharge rate and the weakness of the synapses, to produce a sustained response.

Finally, Figure 5F shows a response which occurs for large numbers of inputs ($\geq 8$) and large (suprathreshold) synaptic strength. This is called a dip response and, unlike the four response types described above, dip responses are rarely if ever seen in the CN.

The PST histograms in Figure 5 show that the bushy cell model is able to produce most of the range of BCRTs observed in single unit studies. The mechanisms discussed above by which these PST response types are produced from AN primary responses are not novel;

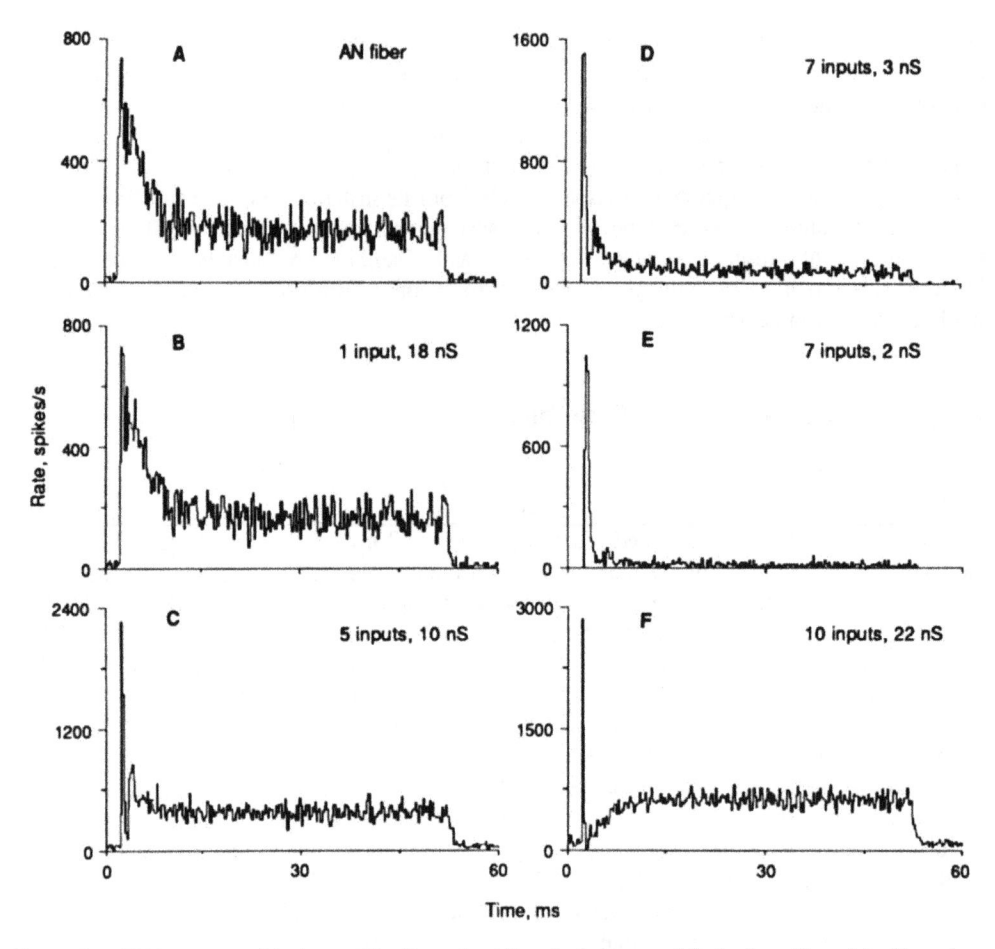

**Figure 5.** PST histograms of the input AN spike trains (A) and of outputs of the bushy cell model with various combinations of number of independent inputs and conductance per input (B-F). Input configurations are given in the legends. Responses to 50 ms tone bursts are shown. PST histograms computed with 0.2 ms bins.

indeed, the potential importance for CN response patterns of convergence, refractoriness, and the strength of synaptic inputs have been discussed in many single unit surveys of CN response types (Pfeiffer, '66a; Bourk, '76; Rhode and Smith, '86; Smith and Rhode, '87; Young et al., '88a). The bushy cell model results in Figure 5 confirm, in a quantitative fashion, that the mechanisms that have been widely hypothesized to account for BCRT behavior do in fact work. In the next section, the regularity behavior of the bushy cell model with various input configurations is considered; we show that regularity is a stronger constraint on the model than PST histograms are.

## Regularity of the bushy cell model

The regularity of the bushy cell model for various numbers of inputs and various synaptic conductances is shown in Figure 6. Each symbol shows the regularity of the model for one input conductance, as defined in the legend. At each conductance, there are several data points, corresponding to different numbers of synaptic inputs to the model. As mentioned above, all inputs had the same steady-state firing rate, and the average regularity of the inputs

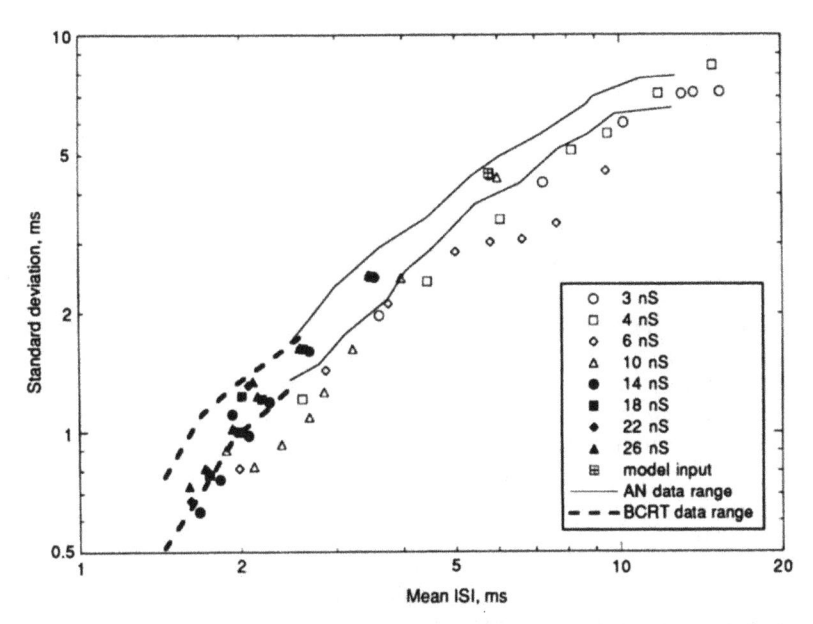

**Figure 6.** Regularity of the bushy cell model compared to AN fibers and BCRT units. Symbols show regularity of bushy cell model for various synaptic conductances (see legend). At each conductance, results with several different numbers of inputs are shown. Regularity of model input shown by ⊞. Range of AN fiber regularity shown by solid lines and range of BCRT regularity at short ISIs shown by heavy dashed lines. Note that axes are logarithmic to show range of data more clearly.

is shown by the ⊞. The unfilled symbols show results for low synaptic conductance ($\leq 10$ nS). Over most of the range of mean interspike intervals, the low synaptic conductances give results that are substantially more regular than the AN fiber data (solid lines). Recall from Figure 1 that BCRT units mostly fall within the range of AN fiber regularity. Because BCRT units fire at higher discharge rates than AN fibers, the BCRT data extend to shorter interspike intervals than the AN data, and the heavy dashed lines in Figure 6 show the range of BCRT unit regularity at short interspike intervals. The unfilled symbols fall below the BCRT (and AN) data except for the longest interspike intervals.

The filled symbols in Figure 6 show the regularity of the model for large input conductances ($\geq 14$ nS). With large conductances, the model's spike trains have regularities that are within the AN/BCRT range for most cases. The principal exceptions are the results for 14 nS conductance at short interspike intervals (●s for mean interval $\leq 2.1$ ms) which lie at or just below the BCRT data at the same interval lengths.

The data in Figure 6 lead to the same conclusion as was drawn from the results of the simpler model in Figure 3. In order to produce a bushy cell model that is as irregular as its auditory nerve inputs, it is necessary that the strength of the individual synaptic inputs be large. In the case of the shot-noise threshold model, the EPSP size has to be greater than threshold. In the case of the bushy cell model, the EPSP size required to make the model sufficiently irregular has to be somewhat larger than the threshold size (6 nS). However, the threshold in the bushy cell model is more complex than in the shot-noise threshold model. 6 nS is the minimum conductance that is sufficient to produce a spike in the model's resting state. When the model is receiving inputs at a substantial rate, its threshold is higher and varies with time, because of the effects of the refractory mechanisms (increased potassium conductance, increased sodium inactivation). Nevertheless, the behaviors of the two models are quite similar and can be simply stated as follows: in order to produce a sufficiently irregular output, the inputs to the model must be individually capable of producing an output spike for each input spike.

## Discussion and conclusions

The conclusion that inputs to CN bushy cells should produce large EPSPs which are capable of leading, by themselves, to output spikes is generally consistent with current concepts of bushy cell physiology. However, the quantification of bushy cell input/output characteristics provided by the model raises some new questions that cannot be answered at present. The first of these is the means by which onset responses are generated. In the model, onset responses are seen only with low synaptic conductances ($\leq 6$ nS), and as a result give spike trains that are more regular than those seen in BCRTs (Fig. 6). Whether the regularity of model onset units is a problem is not clear, because onset units have sustained discharge rates that are too low to allow their regularity to be determined in most cases. However, when the regularity of onset units can be determined, they are usually as irregular as primarylike units (Fig. 1; Young, et al., '88a). Furthermore, it is clear that only a portion of the onset units, including those with onset-L PSTs, are recorded from bushy cells (Rouiller and Ryugo, '84; Smith and Rhode, '87); other onset responses are recorded from octopus cells or multipolar cells (Rhode, Oertel et al., '83; Friauf and Ostwald, '88; Smith and Rhode, '89). Nevertheless the model, in its present form, does not allow the generation of irregular units with onset response properties.

Second, the model produces a non-physiological response type, the dip response pattern, over a substantial range of input parameters (more than about 8 inputs with suprathreshold conductances). The dip response appears to be generated by a mechanism similar to depolarization block; that is, during the onset phase of the input spike trains, the input rate is high and depolarizes the cell sufficiently to block action potential production by some combination of sodium inactivation and potassium activation. Such a mechanism has been suggested to account for onset responses in the octopus cells of PVCN (Ritz and Brownell, '82). The cell recovers from the block as the input rate subsides in the steady state.

A third problem is the means by which low spontaneous rate primarylike and pri-N units are produced. The model predicts that, for input synaptic conductances that are large enough to produce irregular output discharge patterns, the spontaneous rate of the output should be something like the sum of the spontaneous rates of the input spike trains. Thus, in order to produce a low spontaneous rate bushy cell, all the inputs would have to have low spontaneous rates. If auditory nerve fibers with different spontaneous discharge rates assort randomly onto CN bushy cells, the number of low spontaneous rate bushy cells should be very small (Molnar and Pfeiffer, '68). That is, the probability of getting a cell with only low spontaneous rate inputs is small, given that low spontaneous rate fibers (spontaneous rate $\leq 1$ spike/s) account for only 10-20% of the AN population (Liberman, '78). In fact, there are clearly more low spontaneous rate primarylike and pri-N units than such a random assortment model would predict (Bourk, '76; Blackburn and Sachs, '89; Spirou et al., '90).

In the case of the spherical bushy cells of rostral AVCN, which receive inputs from a small number of AN fibers ($\leq 4$; Lorente de Nó, '81; Ryugo and Sento, '91; Liberman, presnetation at this meeting), it has been shown that AN fibers distribute preferentially onto bushy cells, in the sense that low and medium spontaneous rate AN fibers (spontaneous rate $\leq 18$ spikes/s) end on the same cells as other low and medium spontaneous rate fibers, whereas high spontaneous rate fibers terminate with other high spontaneous rate fibers (Ryugo and Sento, '91). With this sorting of input fibers by spontaneous rate, the model is capable of producing low or medium spontaneous rate primarylike units.

For globular bushy cells, the situation is less clear. It is appropriate to compare these cells with pri-N responses in the model, because most evidence suggests that globular bushy cells give pri-N responses (Bourk, '76; Smith and Rhode, '87). The minimum spontaneous rate of model configurations that give irregular, pri-N outputs is 27 spikes/s, with inputs whose spontaneous rate is 10 spikes/s. Because 10 spikes/s is a low value for the average spontaneous rate of AN fibers (Liberman, '78; Rhode and Smith, '85, '86), the minimum spontaneous rate of pri-N responses produced by the model, with random assortment of spontaneous rates at its input, will be at least 27 spikes/s. By contrast, substantial numbers of pri-N units in the CN have spontaneous rates less than 18 spikes/s (77% of pri-N units with BFs below 2 kHz and 41% of pri-N units with BFs above 2 kHz; Blackburn and Sachs, '89).

Selective assortment of inputs with different spontaneous rates would help reduce the spontaneous rate of model pri-N units. No one has yet studied directly the distribution of different spontaneous rate groups on a globular bushy cell. However, Spirou, Brownell and Zidanic (Spirou et al., '90) and Liberman (presentation at this meeting) have estimated that the average number of AN fibers that contact a globular bushy cell is between 17 (Spirou et al. for modified endbulbs as defined by Rouiller et al., '86) and $\approx 50$ (Liberman for all types of terminals). Because of the large number of fibers synapsing on a globular bushy cell, it is not clear that even selective distribution of different AN spontaneous rate groups could produce low or medium spontaneous rate globular bushy cells. For example, if there were 50 inputs to a globular bushy cell, all of which have spontaneous rates of 0.5 spikes/s, then the output spontaneous rate of the cell, with the large synaptic strength required by our model, would be something like 25 spikes/s. Thus it is not clear how low or medium spontaneous rate pri-N responses can be produced.

There are additional problems with pri-N responses, if they are produced by globular bushy cells with 17-50 inputs. With this large number of inputs, the model predicts dip responses, not pri-N responses; moreover, the dip responses with large numbers of inputs look less and less like anything that is commonly seen in the CN. Thus in our current state of knowledge, it is not possible to give a satisfactory account of the generation of pri-N responses from globular bushy cells. Two explanations that might suffice suggest themselves. The first is that the inhibitory inputs which are known to exist on bushy cells (Adams and Mugnaini, '87; Altschuler et al., '86; Saint Marie et al., '89; Wenthold et al., '87) might serve to regulate the synaptic input received by globular bushy cells. Inhibitory inputs might respond to the average output from globular bushy cells in a feedback fashion to keep their average rates or spontaneous rates below a certain level; such inhibition might also convert dip responses into pri-N responses by counteracting the large depolarization near stimulus onset. Inhibitory inputs might also be capable of producing onset responses from bushy cells with large conductance synaptic inputs, by counteracting the steady-state depolarization produced by the inputs. It is generally consistent with these ideas that the inhibitory input to VCN cells is on-BF (Caspary, this volume). Whether inhibitory control is a workable hypothesis can be determined by studying the responses of the bushy cell model with inhibitory as well as excitatory synaptic channels. Whether the hypothesis holds for the real CN requires identification and experimental study of the response properties of the sources of inhibitory input to the globular bushy cells.

A second possibility for explaining the responses of globular bushy cells, given the large number of AN fibers that contact them, is that not all the inputs to a cell have large synaptic conductances. That is, it may be that bushy cells receive a small number (less than 8) of large synaptic inputs, each of which is capable of producing an output spike by itself. The remainder of the inputs to the cell (between $\approx 9$ with Spirou's estimate and $\approx 42$ with Liberman's estimate) would be smaller and not capable of firing the cell by themselves. Such an arrangement is suggested by the anatomical results of Liberman (presentation at this meeting), in that many of the AN inputs to globular bushy cells are small terminals. Heterogeneity of input synaptic strength appears capable of solving the problems of spontaneous discharge rate and dip responses because the total synaptic input to the cell is reduced by making many of the cell's inputs subthreshold. However, it is not yet clear whether such an arrangement will give sufficiently irregular response patterns. Clearly the 8 or fewer dominant inputs will tend to produce an irregular output, but it is not clear what contribution to regularity will be made by the additional smaller inputs. This question can be answered by further modelling. Of course, regardless of the regularity of such a model, it is unclear why such a synaptic arrangement would be useful in the CN.

Our conclusions appear to be inconsistent with those of Carney (Carney, '90), who interpreted results on the threshold sensitivity of CN units as demonstrating that pri-N (but not primarylike) units require coincidence of spikes from different AN inputs to reach threshold. Her argument is based on data showing that the thresholds of low-BF pri-N units, but not primarylike units, are sensitive to the phase spectrum of a broadband stimulus. The sensitivity to phase arises, in her argument, from the necessity, in order to reach threshold, for coincidence of inputs of slightly different BFs and from the fact that the degree of coincidence of AN fibers with different BFs can be manipulated by manipulating the phase spectrum of the signal. Our results may not be inconsistent with hers in that we show results primarily from high BF units ($> 2$ kHz) and she showed results primarily from low BF units ($< 3$ kHz). As the BF decreases, regularity becomes less meaningful as the strength of phase locking increases and the degree to which phase locking influences regularity increases; at the same time, the behavior of phase locking becomes increasingly important as a constraint on bushy cell models (Rothman, '91). The phase locking of primarylike units with very low BFs ($<1$ kHz) is very strong, stronger than AN fiber phase locking (Blackburn and Sachs, '89; Joris et al., '90). When the bushy cell model is applied to the prediction of phase locking in BCRTs, the strong phase locking behavior of low BF units is obtained only in models with small input synaptic conductances, whereas the phase locking behavior of high BF units, which is slightly weaker than AN fibers (Bourk, '76; Rhode and Smith, '86; Blackburn and Sachs, '89), is obtained in models with large input synaptic conductances. Thus, there may be a difference in behavior of low and high BF BCRTs.

ACKNOWLEDGEMENTS

The work reported here was supported by NIH grants DC00115 (EDY) and DC00048 (PBM) and by a grant from the Deafness Research Foundation (PBM). Phyllis Taylor assisted in preparing the graphics.

Note added in proof: The conductances for the bushy cell model in Figures 5 and 6 were inadvertently reported as their values at 22°C, without a temperature correction factor of 3. The actual synaptic input conductances are 3-times larger than those reported.

REFERENCES

Adams, J.C., and Mugnaini, E., 1987, Patterns of glutamate decarboxylase immunostaining in the feline cochlear nuclear complex studied with silver enhancement and electron microscopy, *J. Comp. Neurol.*, 262:375-401.

Adams, P.R., Brown, D.A., and Constanti, A., 1982, M-currents and other potassium currents in bullfrog sympathetic neurons, *J. Physiol. (Lond.)*, 330:537-572.

Altschuler, R.A., Betz, H., Parakkal, M.H., Reeks, K.A., and Wenthold, R.J., 1986, Identification of glycinergic synapses in the cochlear nucleus through immunocytochemical localization of the postsynaptic receptor, *Brain Res.*, 369:316-320.

Blackburn, C.C., and Sachs, M.B., 1989, Classification of unit types in the anteroventral cochlear nucleus: PST histograms and regularity analysis, *J. Neurophysiol.*, 62:1303-1329.

Bourk, T.R., 1976, Electrical Responses of Neural Units in the Anteroventral Cochlear Nucleus of the Cat, Ph.D. Dissertation, Massachusetts Institute of Technology.

Brawer, J.R., and Morest, D.K., 1975, Relations between auditory nerve endings and cell types in the cat's anteroventral cochlear nucleus seen with the Golgi method and Nomarski optics, *J. Comp. Neurol.*, 160:491-506.

Brownell, W.E., 1975, Organization of the cat trapezoid body and the discharge characteristics of its fibers, *Brain Res.*, 94:413-433.

Cant, N.B., and Morest, D.K., 1979, The bushy cells in the anteroventral cochlear nucleus of the cat. A study with the electron microscope, *Neuroscience*, 4:1925-1945.

Carney, L.H., 1990, Sensitivities of cells in anteroventral cochlear nucleus of cat to spatiotemporal discharge patterns across primary afferents, *J. Neurophys.*, 64:437-456.

Frankenhaeuser, B., and Huxley, A.F., 1964, The action potential in the myelinated nerve fiber of Xenopus Laevis as computed on the basis of voltage clamp data, *J. Physiol. (Lond.)*, 171:302-315.

Friauf, E., and Ostwald, J., 1988, Divergent projections of physiologically characterized rat ventral cochlear nucleus neurons as shown by intra-axonal injection of horseradish peroxidase, *Exp. Brain Res.*, 73:263-284.

Goldberg, J.M., and Brownell, W.E., 1973, Discharge characteristics of neurons in anteroventral and dorsal cochlear nuclei of cat, *Brain Res.*, 64:35-54.

Guinan Jr., J.J., and Li, R.Y.-S., 1990, Signal processing in brainstem auditory neurons which receive giant endings (calyces of Held) in the medial nucleus of the trapezoid body of the cat, *Hearing Res.*, 49:321-334.

Hille, B., 1984, Ionic channels of excitable membranes, Sinauer, Sunderland MA.

Hodgkin, A.L., and Huxley, A.F., 1952, A quantitative description of membrane current and its applicaton to conduction and excitation in nerve, *J. Physiol. (Lond.)*, 117:500-544.

Johnson, D.H., and Swami, A., 1983, The transmission of signals by auditory-nerve fiber discharge patterns, *J. Acoust. Soc. Am.*, 74:493-501.

Johnson, D.H., Tsuchitani, C., Linebarger, D.A., and Johnson, M.J., 1986, Application of a point process model to responses of cat lateral superior olive units to ipsilateral tones, *Hearing Res.*, 21:135-159.

Joris, P.X., Smith, P.H., and Yin, T.C.T., 1990, Projections of spherical bushy cells to MSO in the cat: Evidence for delay lines, *Soc. Neurosci. Abst.*, 16:723.

Kiang, N.Y.S., Watenabe, T., Thomas, E.C., and Clark, L.F., 1965, Discharge Patterns of Single Fibers in the Cat's Auditory Nerve, MIT Press, Cambridge, MA.

Lenn, N.J., and Reese, T.S., 1966, The fine structure of nerve endings in the nucleus of the trapezoid body and the ventral cochlear nucleus, *Am. J. Anat.*, 118:375-390.

Li, J., 1991, Estimation of the Recovery of Discharge Probability in Cat Auditory Nerve Spike Trains and Computer Simulations, Ph.D. dissertation, The Johns Hopkins Univ.

Liberman, M.C., 1978, Auditory-nerve response from cats raised in a low-noise chamber, *J. Acoust. Soc. Am.*, 63:442-455.

Lorente de Nó, R., 1981, The Primary Acoustic Nuclei, Raven Press, New York.

Manis, P.B., and Marx, S.O., 1991, Outward currents in isolated ventral cochlear nucleus neurons, *J. Neurosci.*, 11:2865-2880.

Molnar, C.E., and Pfeiffer, R.R., 1968, Interpretation of spontaneous spike discharge patterns of cochlear nucleus neurons, *Proc. I.E.E.E.*, 56:993-1004.

Oertel, D., 1983, Synaptic responses and electrical properties of cells in brain slices of the mouse anteroventral cochlear nucleus, *J. Neuroscience*, 3:2043-2053.

Oertel, D., 1985, Use of brain slices in the study of the auditory system: Spatial and temporal summation of synaptic inputs in cells in the anteroventral cochlear nucleus of the mouse, *J. Acoust. Soc. Am.*, 78:328-333.

Osen, K.K., 1970, Afferent and efferent connections of three well-defined cell types of the cat cochlear nucleus, in: "Excitatory Synaptic Mechanisms", P.Anderson and J.K.S. Jansen, ed., Universitetsforlaget, Oslo.

Pfeiffer, R.R., 1966a, Classification of response patterns of spike discharges for units in the cochlear nucleus: tone burst stimulation, *Exp. Brain Res.*, 1:220-235.

Pfeiffer, R.R., 1966b, Anteroventral cochlear nucleus: Wave forms of extracellularly recorded spike potentials, *Science*, 154:667-668.

Rhode, W.S. and Smith, P.H., 1985, Characteristics of tone-pip response patterns in relationship to spontaneous rate in cat auditory nerve fibers, *Hearing Res.*, 18:159-168.

Rhode, W.S. and Smith, P.H., 1986, Encoding timing and intensity in the ventral cochlear nucleus of the cat, *J. Neurophysiol.*, 56:261-286.

Rhode, W.S., Oertel, D., and Smith, P.H., 1983, Physiological response properties of cells labeled intracellularly with horseradish peroxidase in cat ventral cochlear nucleus, *J. Comp. Neurol.*, 213:448-463.

Ritz, L.A., and Brownell, W.E., 1982, Single unit analysis of the posteroventral cochlear nucleus of the decerebrate cat, *Neuroscience*, 7:1995-2010.

Rothman, J.S., 1991, An Electrophysiological Model of Bushy Cells of the Anteroventral Cochlear Nucleus, Masters dissertation, The Johns Hopkins University.

Rouiller, E.M., Cronin-Schreiber, R., Fekete, D.M., and Ryugo, D.K., 1986, The central projections of intracellularly labeled auditory nerve fibers in cats: An analysis of terminal morphology, *J. Comp. Neurol.*, 249:261-278.

Rouiller, E.M. and Ryugo, D.K., 1984, Intracellular marking of physiologically characterized cells in the ventral cochlear nucleus of the cat, *J. Comp. Neurol.*, 225:167-186.

Ryugo, D.K., and Fekete, D.M., 1982, Morphology of primary axosomatic endings in the anteroventral cochlear nucleus of the cat: A study of the endbulbs of Held, *J. Comp. Neurol.*, 210:239-257.

Ryugo, D.K. and Rouiller, E.M., 1988, Central projections of intracellularly labeled auditory nerve fibers in cats: morphological correlations with physiological properties, *J. Comp. Neurol.*, 271:130-142.

Ryugo, D.K. and Sento, S., 1991, Synaptic connections of the auditory nerve in cats: Relationship between endbulbs of Held and spherical bushy cells, *J. Comp. Neurol.*, 305:35-48.

Saint Marie, R.L., Morest, D.K., and Brandon, C.J., 1989, The form and distribution of GABAergic synapses on the principal cell types of the ventral cochlear nucleus of the cat, *Hearing Res.*, 42:97-112.

Smith, P.H. and Rhode, W.S., 1987, Characterization of HRP-labeled globular bushy cells in the cat anteroventral cochlear nucleus, *J. Comp. Neurol.*, 266:360-375.

Smith, P.H. and Rhode, W.S., 1989, Structural and functional properties distinguish two types of multipolar cells in the ventral cochlear nucleus, *J. Comp. Neurol.*, 282:595-616.

Spirou, G.A., Brownell, W.E., and Zidanic, M., 1990, Recordings from cat trapezoid body and HRP labeling of globular bushy cell axons, *J. Neurophysiol.*, 63:1169-1190.

Teich, M.C., Matin, L., and Cantor, B.I., 1978, Refractoriness in the maintained discharge of the cat's retinal ganglion cell, *J. Opt. Soc. Am.*, 68:386-402.

Tolbert, L.P. and Morest, D.K., 1982, The neuronal architecture of the anteroventral cochlear nucleus of the cat in the region of the cochlear nerve root: Electron microscopy, *Neuroscience*, 7:3053-3030.

Tuckwell, H.C. and Richter, W., 1978, Neuronal interspike time distributions and the estimation of neurophysiological and neuroanatomical parameters, *J. Theor. Biol.*, 71:167-183.

Wenthold, R.J., Huie, D., Altschuler, R.A., and Reeks, K.A., 1987, Glycine immunoreactivity localized in the cochlear nucleus and superior olivary complex, *Neuroscience*, 22:897-912.

Westerman, L.A. and Smith, R.L., 1984, Rapid and short-term adaptation in auditory nerve responses, *Hearing Res.*, 15:249-260.

Wu, S.H. and Oertel, D., 1987, Maturation of synapses and electrical properties of cells in the cochlear nuclei, *Hearing Res.*, 30:99-110.

Young, E.D. and Barta, P.E., 1986, Rate responses of auditory nerve fibers to tones in noise near masked threshold, *J. Acoust. Soc. Am.*, 79:426-442.

Young, E.D., Robert, J.M., and Shofner, W.P., 1988a, Regularity and latency of units in ventral cochlear nucleus: implications for unit classification and generation of response properties, *J. Neurophysiol.*, 60:1-29.

Young, E.D., Shofner, W.P., White, J.A., Robert, J.-M., and Voigt, H.F., 1988b, Response properties of cochlear nucleus neurons in relationship to physiological mechanisms, *in*: "Auditory Function: Neurobiological Bases of Hearing", G.M. Edelman, W.E. Gall, and W.M. Cowan, ed., John Wiley & Sons, New York.

# CROSS-CORRELATION ANALYSIS AND PHASE-LOCKING IN A MODEL OF THE VENTRAL COCHLEAR NUCLEUS STELLATE CELL

Murray B. Sachs, Xiaogin Wang and Scott C. Molitor

Department of Biomedical Engineering and Center for Hearing Sciences
Johns Hopkins School of Medicine, Baltimore, Maryland 21205, USA

There is now considerable evidence that stellate cells in the anteroventral cochlear nucleus (AVCN) receive auditory-nerve inputs on or near their cell bodies (Cant, '81, Liberman, '92, Ryugo, this volume). Furthermore, auditory-nerve fibers (ANFs) from all spontaneous rate (SR) groups appear to project to AVCN stellate cells (Ryugo, this volume), although it is not known whether a single stellate cell receives inputs from more than one SR group. This morphological evidence raises a number of issues with regard to our understanding of the physiological properties of stellate cells. For example, results of cross-correlation analysis of spike trains from simultaneously recorded ANFs and AVCN chopper units, which are recorded from stellate cells (Roullier and Ryugo, '84, Smith and Rhode, '89), are consistent with the hypothesis that there are monosynaptic excitatory connections between high SR ANFs and stellate cells, whereas no such evidence exists for connections between low SR ANFs and stellate cells (Young and Sachs, '88). Another issue involves phase-locking in chopper units. It is well known that phase-locking in choppers is degraded with respect to that in ANFs at frequencies above about 500 Hz (Blackburn and Sachs, '89, Bourk, '76, Rhode and Smith, '86). It has been suggested that this reduction in phase-locking results from low-pass filtering of AN inputs by the dendritic tree of the stellate cell (White, et al., '90, Young, et al., '88b). The evidence for ANF synapses on and near the soma of these cells raises the question of whether dendritic filtering is sufficient to produce the observed reductions. In this chapter we will examine both of these issues in terms of a simulation model for the stellate cell (Banks and Sachs, '91).

The model cell (Fig. 1A) is a hypothetical "exemplar cell" that was constructed on the basis of descriptions of CN chopper cells that had been filled with HRP (Rhode, et al., '83, Roullier and Ryugo, '84). The basic electrical parameters chosen are within the range of the standard values (Rall, '77), adjusted to fit intracellular records and data. These parameters yield time constants and input resistances that are close to those recently estimated for stellate cells on the basis of responses to current injection (White, et al., '90). We collapse the dendritic tree of the hypothetical exemplar cell into a single equivalent cylinder (Rall, '77) and separate the cylinder into 10 compartments of equal electrotonic length ($\Delta Z = 0.1$).

The model we use for the dendritic compartments is similar to the one developed by Rall ('77). As illustrated in Fig. 1B we model isopotential sections of the dendritic cylinder as electrical circuits with four branches: a resting branch with battery $E_r = 62$ mv and resting

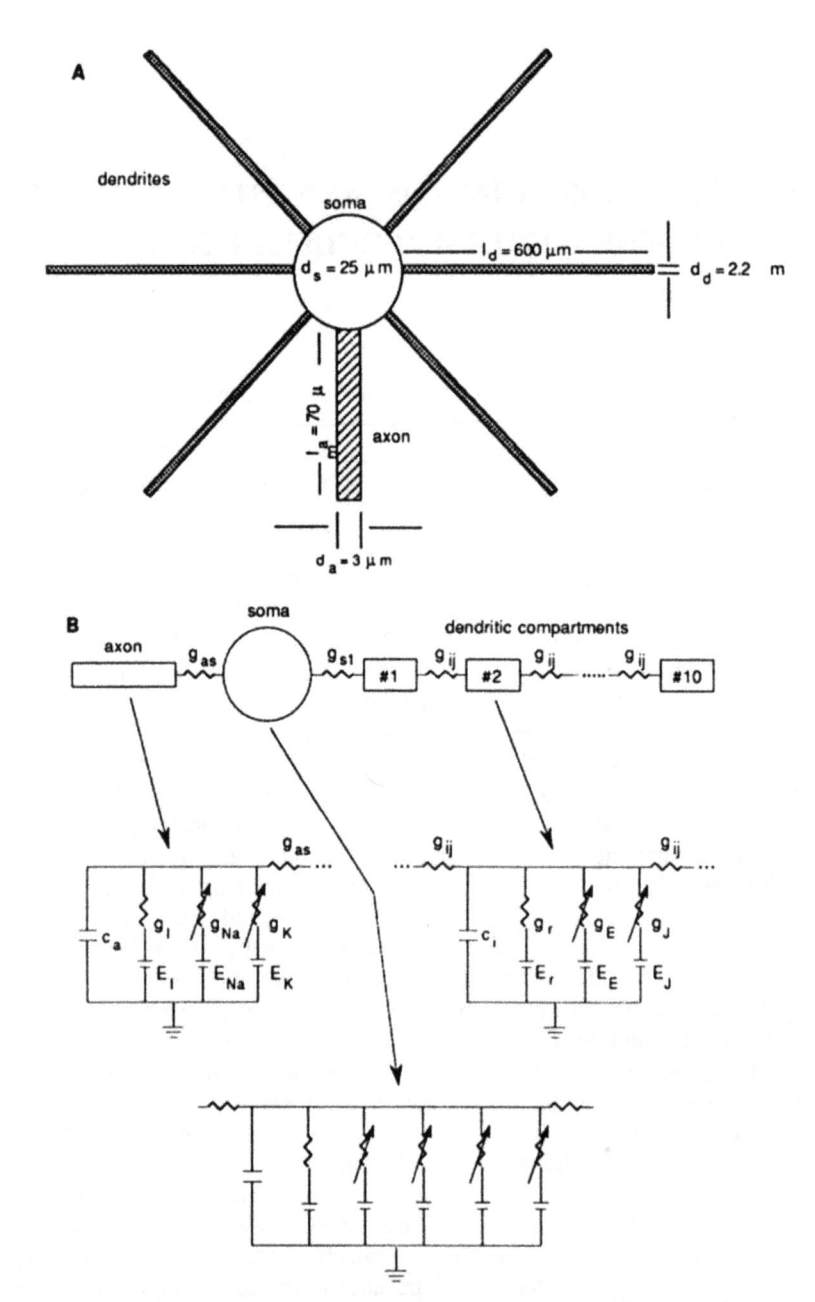

**Figure 1.** A: Schematic diagram of the "exemplar" cell showing the six dendrites, axonal segment and soma, and the anatomic parameters relevant to the model. B: Compartmental model and corresponding electrical circuit representations for the axon compartment, the soma and the ith dendritic compartment. (From Banks and Sachs, 1991).

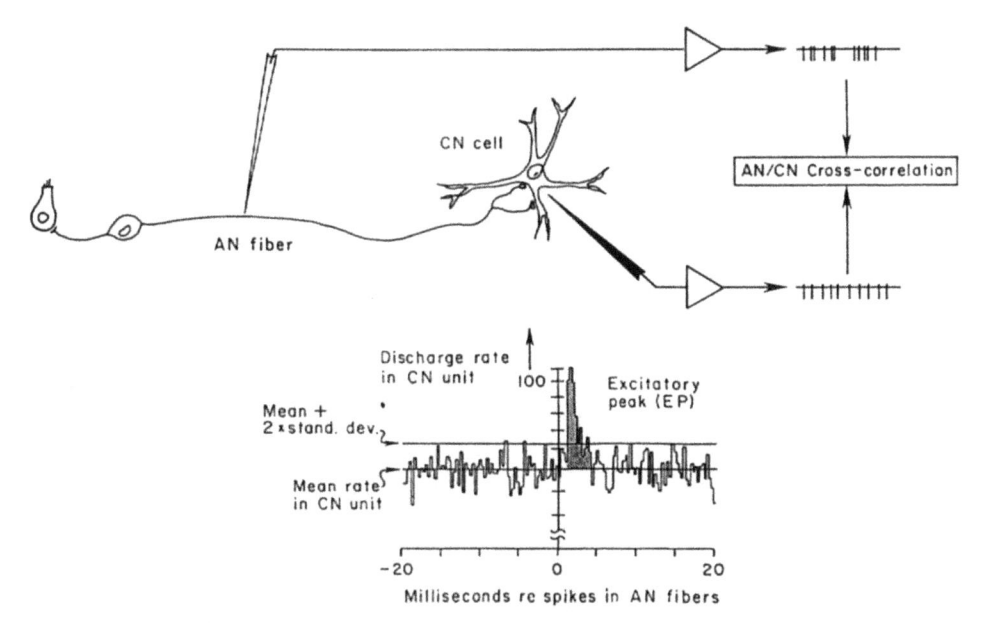

**Figure 2.** Schematic representation of an experiment in which spike trains are simultaneously recorded from an auditory-nerve fiber and cochlear nucleus cell. The cross-correlogram, computed from spike trains from an ANF/AVCN chopper pair, shows discharge rate in the chopper unit as a function of time before and after occurrence of an ANF spike. (Data from (Young and Sachs, 1988)).

conductance $g_r = 0.25 \times 10^{-8}$ S; a capacitive branch with $C_i = 0.75 \times 10^{-4}$ F; an excitatory input branch with battery $E_e = 0$ mv and conductance $g_e(t)$; and an inhibitory input branch with battery $E_j = E_{Cl} = -68$ mv (Wu and Oertel, '86) and conductance $g_j(t)$. The compartments are connected to one another by axial conductances $g_{ij} = 2.48 \times 10^{-7}$S. Excitatory and inhibitory inputs to each compartment are similar to shot noise (Papoulis, '65) with alpha-wave conductance impulse responses. The underlying input processes are either the sequence of spike times measured in single unit recordings in the auditory nerve or simulated nonstationary Poisson processes modified by a deadtime (Young and Barta, '86). We set the time course of the alpha waves to give EPSPs and IPSPs at the soma comparable to those reported in vitro (Oertel, '83, Oertel, et al., '88, Wu and Oertel, '84, Wu and Oertel, '86).

The soma is modeled as an electrical circuit with a leakage branch ($E_l$, $g_l$ a capacitive branch, excitatory and inhibitory input branches (identical to the dendritic input branches) and two branches with voltage and time dependent Na$^+$ and K$^+$ conductances comprising the spike generator. The axonal segment is modeled as a single compartment with circuit branches nearly identical to the somatic compartment but with no input branches. The spike generating mechanism used is a slight modification of the model proposed by Hodgkin and Huxley (Hodgkin and Huxley, '52). It consists of a fast, inactivating sodium conductance $g_{Na}(V,t)$, a slower, delayed rectifier potassium conductance $g_r(V,t)$, and a linear leakage conductance $g_l$.

This model was originally developed to explore the issue of lack of correlation between low SR ANFs and chopper units, mentioned above. Figure 2 illustrates the correlation experiments of interest. Young and Sachs (Young and Sachs, '88) recorded simultaneously from a single ANF and a single VCN chopper unit and computed cross-correlograms between the two recorded spike trains. The correlogram in Fig. 2 shows discharge rate in the chopper unit as a function of time before and after the occurrence of a spike in the ANF. The horizontal line through the mean of the correlogram shows the average

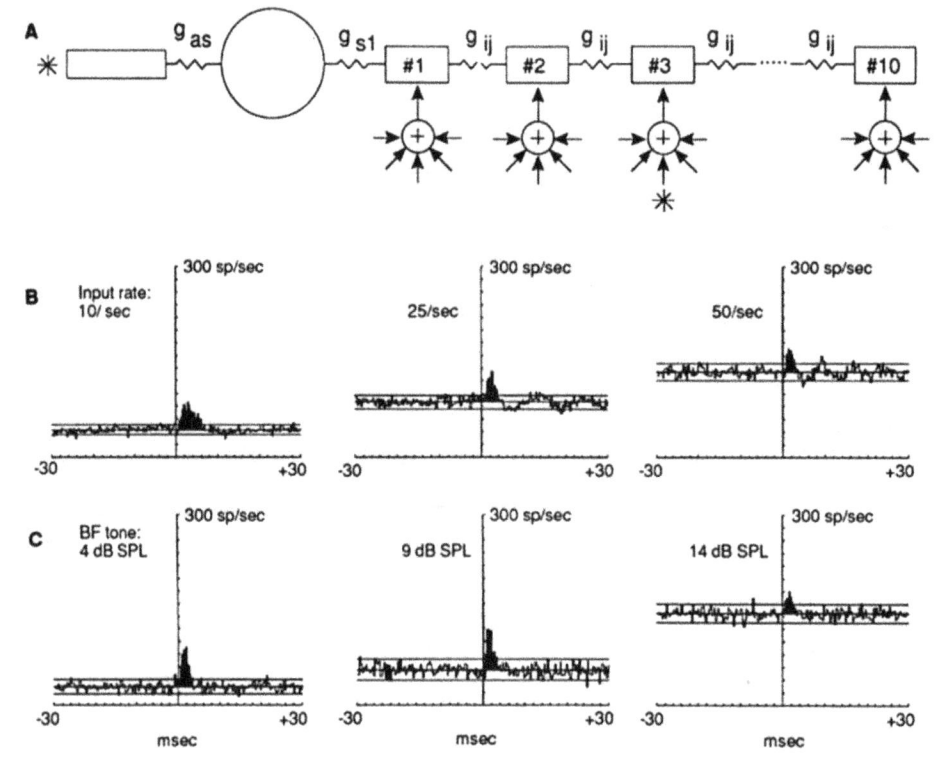

**Figure 3.** A: Model configuration for correlation simulations. Model had five inputs to each compartment. Input spike trains were independent Poisson processes modified by a deadtime. Input rates are the same for each input train and are indicated in the figure. B: Model: Cross-correlograms between input spike train at Compartment 3 (*) and spike train at the axon compartment for the model configuration shown in A. C: Cross-correlograms for an ANF/chopper pair recorded with BF tones as the stimulus (Young and Sachs, 1988); data for three sound levels are shown. 3/25/87.1.1710.

discharge rate of the chopper unit. The peak in discharge rate just to the right of the origin indicates that the likelihood of a spike in the chopper unit increases just after the ANF fires. Such an "excitatory peak" is consistent with the existence of a monosynaptic excitatory connection between the ANF and the stellate cell from which the chopper response was recorded (Moore, et al., '79). Young and Sachs (Young and Sachs, '88) found excitatory peaks in more than 30 correlograms for chopper/high SR ANF pairs in which the BFs of the two units were within 15% of one another. On the other hand, they found only one example suggesting an excitatory connection between a low SR ANF and a chopper.

This result appears at first sight to conflict with the morphological evidence for low SR inputs to stellate cells discussed above (Ryugo, this volume). However, careful analysis of data like those in Fig. 3C suggests an alternative interpretation. Cross-correlograms are shown for an ANF/chopper pair stimulated with best frequency (BF) tones at three sound levels. As sound level increases, or equivalently as average discharge rate of the chopper increases (indicated by the horizontal line through the average value of the correlogram) the salience of the excitatory peak decreases. At higher levels, significant excitatory peaks were not observed for this pair (data not shown). In fact, only one excitatory peak was seen when units were tested at sound levels more than 20 dB above the threshold of the chopper in any pair. Because of the high thresholds of the low SR ANFs (Liberman, '78), it is not usually possible to test chopper/low SR ANF pairs at levels low enough to see excitatory peaks.

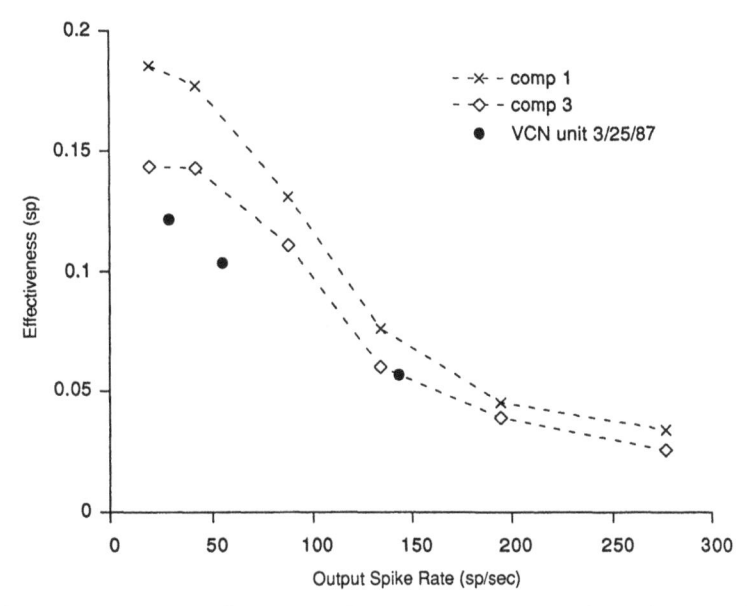

**Figure 4.** Effectiveness plotted as a function of spike rate for the chopper and model data in Figs. 3C and B (filled circles and diamonds respectively). The Xs show effectiveness from correlograms (not shown) computed from the same simulation as the model data in Fig. 3B, but with an input spike train at Compartment 1 as the reference.

Young and Sachs (Young and Sachs, '88) suggest that the decrease in salience of the excitatory peaks with increasing sound level is related to saturation of the spike generating mechanism in the stellate cell. We have explored this suggestion in our computer model of the stellate cell. Figure 3B shows cross-correlograms computed from the model. The input configuration is as illustrated in the Fig. 3A. Each dendritic compartment receives five input conductance trains; the underlying processes are the simulated Poisson processes described above. The amplitude of the conductance waveforms is equal to the model resting conductance ($g_r = 0.25 \times 10^{-8}$ S) in this case. When applied to Compartment 1 (0.1 space constants from the soma) an individual conductance pulse of this amplitude produces an EPSP at the soma of about one millivolt. The correlograms shown are computed between one of the five input spike trains at Compartment 3 and the output spike train measured in the model axonal compartment. As in the real data shown in Figs. 2 and 3C, the model correlograms show an excitatory peak reflecting the monosynaptic excitatory connection between the simulated auditory-nerve fiber and the model stellate cell.

The three correlograms shown in Fig. 3B correspond to three different auditory-nerve input rates. In any one simulation all 50 input spike trains have the same average rate (10, 25, and 50/second, from left to right). The model output rates are 45, 90 and 135 respectively from left to right. As in the data in Fig. 3C, the salience of the excitatory peak decreases as model output rate increases. A quantitative comparison between the data and model correlograms is shown in Figure 4. *Effectiveness* of the synaptic connection is defined to be the area under the excitatory peak (shaded areas in Fig. 3) normalized by the average rate of the chopper. Effectiveness can be interpreted as the number of chopper spikes caused by each spike in the ANF spike train. As shown in Fig. 4 effectiveness for the data in Fig. 3C decreases from about 11% at near threshold sound levels to about 6% with an increase of 10 dB in level. This decrease was a general finding in the Young and Sachs studies (Young and Sachs, '88). Similarly, effectiveness in the model decreases with increasing output rate, as indicated in the Figure 4. Thus, the model with the input parameters and input configura-

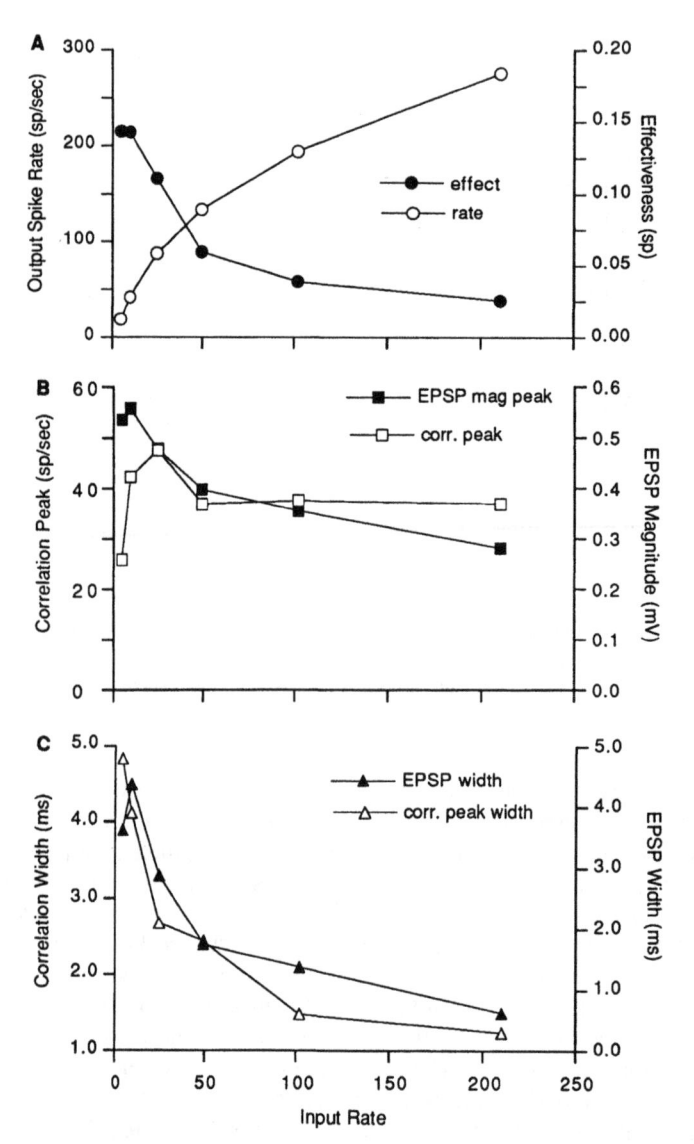

**Figure 5.** A: Effectiveness (filled circles) and output rate (open circles) plotted versus input rate for the model configuration in Fig. 3A. B: Height of the correlogram excitatory peak (measured from the average rate; open squares) and peak height of the EPSP evoked by spikes in the reference train (measured from triggered averages; filled squares). C: Width of the correlogram peak (open triangles) and EPSP (filled triangles) measured at half peak height.

tions as chosen, produces a decrease in effectiveness with chopper rate qualitatively similar to that seen in AVCN choppers. This general result does not seem to depend very strongly on the input configuration although the details do. For example, effectiveness of an input at Compartment 1 as the reference is also shown in Figure 4 for the case of five inputs to the ten dendritic compartments. Although at any output rate the more proximal synapse is more effective than the more distal one, effectiveness decreases with output rate in a similar way in each case.

Figure 5 explores some of the factors involved in this reduction of measured effectiveness with increasing output rate in the model. Figure 5A shows that output rate is

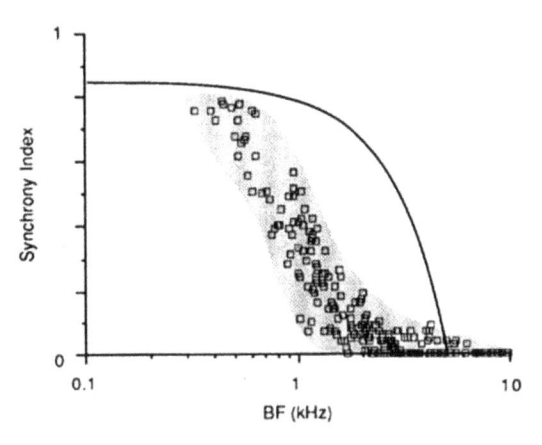

**Figure 6.** Synchronization index versus BF for a population of AVCN choppers. Curve is least squares fit of a quadratic function to the ANF synchrony data of Johnson (1980). Redrawn from Blackburn and Sachs (1989).

a saturating function of input rate over the range where effectiveness decreases rapidly. The DC potential measured at the soma is a similar saturating function of rate over this range (data not shown; somatic potentials were computed with the sodium conductances turned off in order to prevent spike generation). The source of this saturation can be understood in terms of the model diagram in Fig. 1. As input rate increases, the average value of the excitatory dendritic conductances increases. As the excitatory conductance becomes large, the dendritic tree begins to look like a short circuit to the soma and further changes in synaptic conductance will have no effect on somatic potentials.

Figures 5B and C show the effects of this saturation on the EPSPs evoked by dendritic conductance inputs. As input rate (or output rate) increases, there is a decrease in the peak height of the EPSP which is reflected in a decrease in the peak of the correlogram. There is also a distinct shortening in the time course of the EPSPs with increasing input rate, which is also reflected in a narrowing of the correlation peak (Fig. 5C). This shortening is presumably a result of the decrease in effective soma time constant due to loading by an increasing dendritic conductance. Since effectiveness is the area under the correlation peak, both the decrease in height and width contribute to the decrease in effectiveness. The AVCN/AN data of Young and Sachs are inconclusive with respect to the decrease in correlation peak width; more extensive data would provide an important test of the ideas presented here.

It is well known that phase-locking in choppers is much poorer than that in auditory-nerve fibers at frequencies above about 500 Hz as in Fig. 6 (Blackburn and Sachs, '89, Bourk, '76, Rhode and Smith, '86). In terms of models for stellate cells like that in Fig. 1, it has been suggested that this decrease in phase-locking is related to the low-pass filtering of post-synaptic potentials by the dendritic tree (White, et al., '90, Young, et al., '88b). However, as we have pointed out above, there is strong evidence for auditory-nerve inputs on or near the somata of stellate cells (Cant, '81, Ryugo, this volume, Smith and Rhode, '89). On the basis of that evidence, it is important to know whether the model in Fig. 1 provides enough low-pass filtering for inputs on or near the soma to produce phase-locking appropriate to chopper units. In order to explore this issue, we have used phase-locked spike trains recorded from auditory-nerve fibers as the input processes for the model in Fig. 1. Sequences of spike times were used to generate trains of conductance inputs to the model. Figure 7 shows period histograms for the auditory-nerve inputs (Fig. 7C) and for the output spike trains from the model (Fig. 7B), for 0.5, 1.15, and 2.22 kHz stimuli. The model had 10 inputs applied directly to the soma (Fig. 7A); in the case illustrated here the conductance

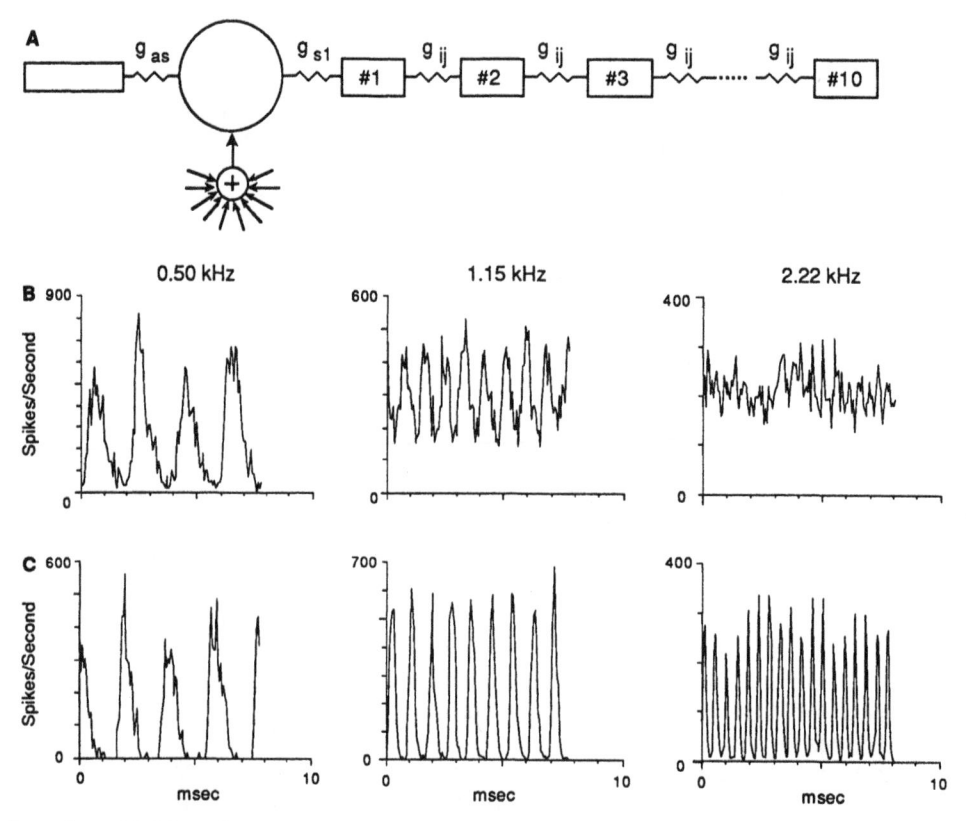

**Figure 7.** A: Model configuration for phase-locking simulations. There are ten inputs to the soma only; the spike times are from phase-locked auditory-nerve data. B: Model Ouput (10 inputs at Soma, conductance=4). Period histograms computed from the model output. C: Period histograms for one of the auditory-nerve inputs.

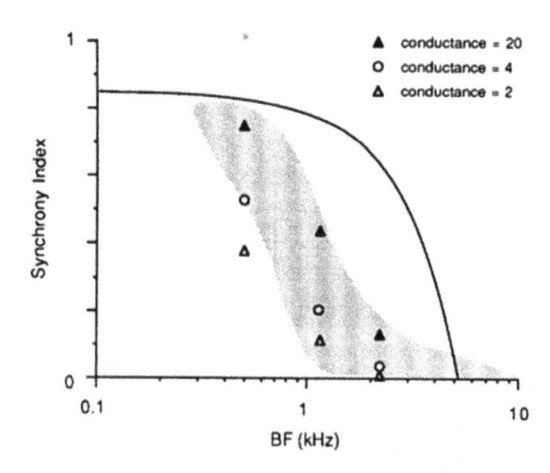

**Figure 8.** Synchrony versus frequency for the model configuration of Fig. 7A. Points are shown for three values of input conductance (2, 4 and 20 x $g_r$). Shaded area and solid line are the same as those in Fig. 6.

418

amplitudes were all $1.0 \times 10^{-8}$ S ($4 \times g_r$). The degradation in phase-locking, even at the soma is evident.

Figure 8 shows synchronization index (Goldberg and Brownell, '73) plotted versus frequency for the model data from Fig. 7 (open circles). The shaded area represents the range of synchrony indices for AVCN choppers from Fig. 6 (Blackburn and Sachs, '89); the solid line represents the average synchrony for ANFs from the data of Johnson (Johnson, '80). The model data fall in the lower half of the range of AVCN data at the three frequencies tested. Phase-locking in the model depends on a number of parameters, including number, rates, and location of the applied inputs (Rothman, '91, Wang, '91). For example, as illustrated in Fig. 8, increasing the conductance amplitude of the inputs increases the phase-locking in the output. With ten inputs on the soma, the model produces the observed range of chopper phase-locking, for a range of conductance amplitudes from 2 to $20 \times g_r$. Figure 8 shows that it is not necessary to postulate distal synaptic inputs in order to account for the degradation in phase-locking in choppers. On the other hand, it is easy to produce appropriate phase-locking with distal inputs if the conductance amplitudes are made large enough (Wang, '91), so that distal inputs are certainly not ruled out.

In summary then, we have shown that a simple computational model of the stellate cell with ANF innervation patterns consistent with those seen in stellate cells in the VCN can produce the correlations between ANFs and AVCN choppers and the degradation in phase-locking observed in these choppers. While it is gratifying that our understanding of the signal processing of stellate cells can be summarized in a simple model, the failures of such a model are often more revealing than its success. We conclude by admitting that the model as described above does not produce the enhancements in amplitude modulation observed in many chopper units (Frisina, et al., '90, Wang, '91), and this failure has led us to examine again some details of the spike-generating mechanism used in the model (Wang, '91).

ACKNOWLEDGEMENTS

This work was supported by grants from the National Institute of Deafness and other Communications Disorders.

REFERENCES

Banks, M.I. and Sachs, M.B., 1991, Regularity analysis in a compartmental model of chopper units in the anteroventral cochlear nucleus, *J. Neurophysiol.*, 65:606-629.
Blackburn, C.C. and Sachs, M.B., 1989, Classification of unit types in the anteroventral cochlear nucleus: post-stimulus time histograms and regularity analysis, *J. Neurophysiol.*, 62:1303-1329.
Bourk, T.R., 1976, Electrical Responses of Neural Units in the Anteroventral Cochlear nucleus of the Cat. Ph.D. Thesis, Massachusetts Institute of Technology.
Cant, N.B., 1981, The fine structure of two types of stellate cells in the anterior division of the anteroventral cochlear nucleus of the cat, *Neuroscience*, 6:2643-2655.
Frisina, R.D., Smith, R.L., and Chamberlain, S.C., 1990, Encoding of amplitude modulation in the gerbil cochlear nucleus: I. A hierarchy of enhancement, *Hearing Res.*, 44:99-122.
Goldberg, J.M. and Brownell, W.E., 1973, Discharge characteristics of neurons in anteroventral and dorsal cochlear nuclei of cat, *Brain Res.*, 64:35-54.
Hodgkin, A.L. and Huxley, A.F., 1952, A Quantitative description of membrane current and its application to conduction and excitation in nerve, *J. Physiol. (Lond.)*, 117:500-544.
Johnson, D.H., 1980, The relationship between spike rate and synchrony in responses of auditory-nerve fibers to single tones, *J. Acoust. Soc. Am.*, 68:1115-1122.
Liberman, M.C., 1978, Auditory-nerve responses from cats raised in a low-noise chamber, *J. Acoust. Soc. Am.*, 63:442-455.
Moore, G.P., Segundo, J.P., Perkel, D.H., and Levitan, H., 1979, Statistical signs of synaptic interaction in neurons, *Biophys. J.*, 10:876-900.
Oertel, D., 1983, Synaptic responses and electrical properties of cells in brain slices of the mouse anteroventral cochlear nucleus, *J. Neuroscience*, 3:2043-2053.

Oertel, D., Wu, S.H., and Hirsch, J.A., 1988, Electrical characteristics of cells and neuronal circuitry in the cochlear nuclei studied with intracellular recording from brain slices, *in*: "Auditory Function", G.M.Edelman, W.E.Gall and W.M.Cowan, eds., John Wiley and Sons, New York, p. 313:336.

Papoulis, A., 1965, "Probability, Random Variables and Stochastic Processes", McGraw Hill, New York

Rall, W., 1977, Core conductor theory and cable properties of neurons, in: "Handbook of Physiology - The Nervous System", E.Kandel and S.Geiger, eds., Am. Physiol. Soc., Washington, D.C., p. 39-97.

Rhode, W.S., Oertel, D., and Smith, P.H., 1983, Physiological response properties of cells labeled intracellularly with horseradish peroxidase in cat ventral cochlear nucleus, *J. Comp. Neurol.*, 213:448-463.

Rhode, W.S. and Smith, P.H., 1986, Encoding timing and intensity in the ventral cochlear nucleus of the cat, *J. Neurophysiol.*, 56:261-286.

Rothman, J., 1991, MS Thesis, Johns Hopkins University.

Roullier, E.M. and Ryugo, D.K., 1984, Intracellular marking of physiologically characterized cells in the ventral cochlear nucleus of the cat, *J. Comp. Neurol.*, 255:167-186.

Smith, P.H. and Rhode, W.S., 1989, Structural and functional properties distinguish two types of multipolar cells in the ventral cochlear nucleus, *J. Comp. Neurol.*, 282:595-616.

Wang, X., 1991, Neural encoding of single-formant stimuli in the auditory nerve and anteroventral cochlear nucleus of the cat, PhD Thesis, Johns Hopkins Univeristy.

White, J.A., Young, E.D., and Manis, P.B., 1990, Application of new electrotonic modeling methods: Results from Type I cells in guinea pig ventral cochlear nucleus, *Soc. Neurosci. Abstr.*, 16:870.

Wu, S.H. and Oertel, D., 1984, Intracellular injection with horseradish peroxidase of physiologically characterized stellate and bushy cells in slices of mouse anteroventral cochlear nucleus, *J. Neuroscience*, 4:1577-1588.

Wu, S.H. and Oertel, D., 1986, Inhibitory circuitry in the ventral cochlear nucleus is probably mediated by glycine, *J. Neuroscience*, 6:2691-2706.

Young, E.D. and Barta, P.E., 1986, Rate responses of auditory-nerve fibers to tones in noise near masked threshold, *J. Acoust. Soc. Am.*, 79:426-442.

Young, E.D. and Sachs, M.B., 1988, Interactions of auditory nerve fibers and cochlear nucleus cells studied with cross-correlation, Soc. Neurosci. Abstr., 14:646.

Young, E.D., Shofner, W.P., White, J.A., Robert, J.M., and Voigt, H.F., 1988b, Response properties of cochlear nucleus neurons in relationship to physiological mechanisms, *in*: "Auditory Function", G.M.Edelman, W.E.Gall and W.M.Cowan, eds., John Wiley and Sons, New York, pp.277-331.

# VIII. APPENDIX: COCHLEAR NUCLEUS PROSTHESES

# THE DEVELOPMENT AND EVALUATION OF COCHLEAR NUCLEUS PROSTHESES

John K. Niparko[1]*, David J. Anderson[1,2], Kensall D. Wise[2] and Josef M. Miller[1]

[1]Kresge Hearing Research Institute, Department of Otolaryngology; and
[2]College of Engineering, The University of Michigan, Ann Arbor, Michigan 48109, USA

The advent of the cochlear prosthesis has had a profound impact on the rehabilitation of profoundly deaf individuals whose hearing impairment is the result of a loss of sensory function within the cochlea. There exists, however, a significant number of deaf individuals who are not candidates for a cochlear prosthesis because they lack viable innervation of the cochlea or have an un-implantable cochlea. The vast majority of these cases are the result of von Recklinghausen's disease (Martuza et al., '88), a neurocutaneous syndrome, also known as neurofibromatosis II, that is characterized by multiple benign tumors involving peripheral nerves. These tumors have a propensity for forming within the vestibuloauditory nerve and their surgical removal often results in nerve ablation, thereby precluding the use of a traditional cochlear implant. In such cases, direct stimulation of the cochlear nucleus (CN) within the pontomedullary brainstem has shown promise in providing an alternative approach to auditory rehabilitation (Eisenberg et al., '87).

This chapter reviews research at the University of Michigan Kresge Hearing Research Institute and the Solid State Electronics Laboratory that address issues of the histocompatibility and physiology of CN stimulation and the development of electrode devices designed to permit reliable, safe, selective stimulation of discrete populations of cells in the CN. In the first section we address issues of biocompatibility and physiology, and in the second, issues of design and manufacture of solid substrate electrodes.

## Biocompatibility

We have investigated the feasibility, biocompatibilty and performance of a penetrating CN prosthesis. The rationale for this strategy is based on the notion that 1) the CN is an obligatory synapse of the auditory tract and represents the first central auditory nucleus along the ascending pathway, 2) there is wide and complete frequency representation in the small space of the CN, 3) while cochlear hair cells and spiral ganglion cells degenerate with

---

[1]*Dr. Niparko is currently in the Department of Otolaryngology-Head and Neck Surgery, Johns Hopkins University, Baltimore, Maryland 21203, USA

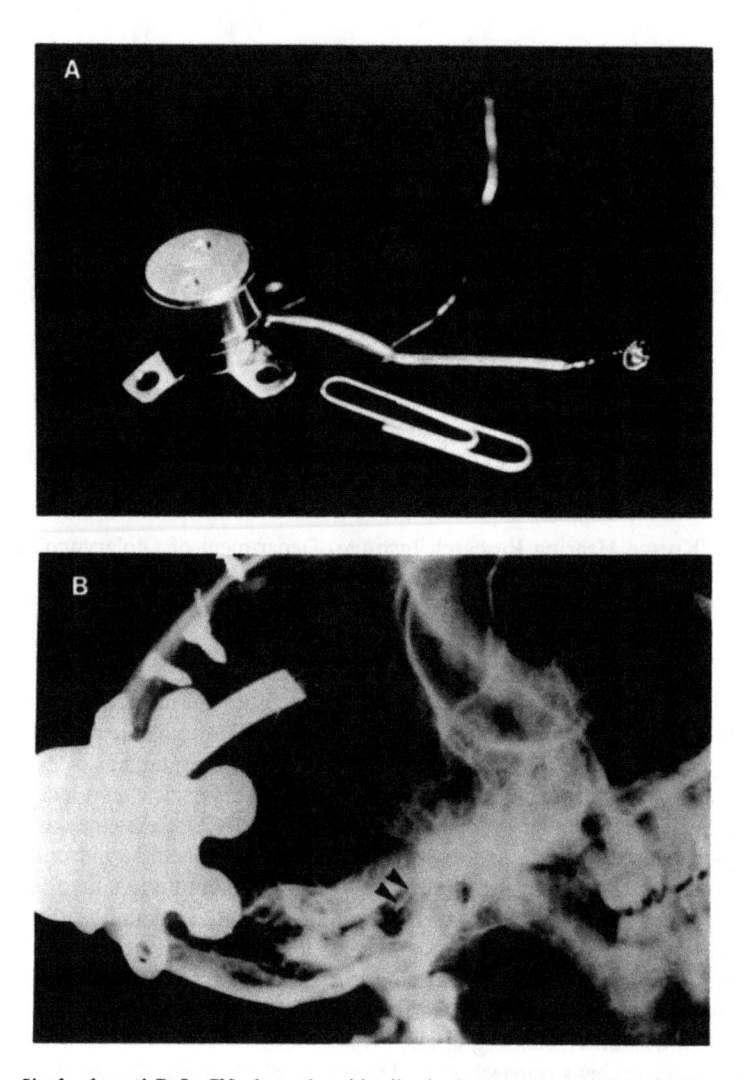

Figure 1. A. Single-channel Pt-Ir CN electrode with silastic flange at electrode hub for stabilization. B. Electrode assembly implanted in a monkey subject with electrode in pontomedullary brain stem (arrow).

sensorineural hearing impairment, CN (second order) neurons tend to survive for long periods, and 4) because of the relative simplicity of the topographic organization of frequency representation within the CN, the stimulating device can be conceptually simple. Direct intranuclear CN stimulation provides other potential advantages, i.e., threshold for activation may be quite low and it may be possible to selectively activate functionally distinct neural subpopulations given the close proximity of stimulating electrodes and nerve cells.

Placing an electrical prothesis within the CN entails a variety of potential intra- and postoperative complications. The CN is located in the ventrolateral aspect of the caudal pons and is flanked by the cerebellum, medulla, vestibular and pontobulbar nuclei, and middle and inferior cerebellar peduncles. For exposure of the CN target, the suboccipital or translabyrinthine approach is routinely performed with intraoperative monitoring to continuously assess regional cranial nerves at risk for compromised function in the course of acoustic neuroma resection (Niparko et al., '89a). Fastidious packing of dural and bone defects is necessary to prevent CSF (cerebrospinal fluid) leakage and potential meningitis.

Stability of the implant is an issue deserving special care at surgery and additional experimental work. The placement of a relatively rigid prothesis within the relatively pliable and pulsating brainstem structure may yield a change in the relative position of the implant (Hitselberger et al., '84). Moreover, exposure of the brainstem will introduce a variable amount of tissue edema, followed by contraction of dura and other soft tissue overlying the craniotomy during recovery, which may also lead to electrode migration.

Our initial efforts to assess the feasibility of a penetrating CN prosthesis were complicated by electrode migration (Niparko et al., '89b). In most cases, placing the electrode within the CN, deep within the cerebellopontine recess, resulted in later migration of the electrode in a lateral direction in both rodent (guinea pig) and monkey subjects. A simple modification (Figures 1A and 1B) placed a flange on the hub of the electrode for stabilization. Once the electrode was engaged in the CN, a soft tissue pack was placed on the lateral aspect of the flange, thereby closing the cerebellopontine angle and temporoccipital defect. Radiographs demonstrated that this method provided electrode stability over intervals as long as eight weeks following implantation in the monkey model. These observations suggest that electrodes placed within the auditory brainstem maintain the potential for migration. To minimize this risk, it appears that the electrode must be secured at the brainstem/electrode interface and the connecting lead cannot be relied upon to maintain positional stability. Rather, this must be achieved at the brainstem interface.

To assess the tolerance and histologic reaction to prolonged, periodic electrical stimulation of the CN, we performed a series of CN implants followed by 20 hours of stimulation in the guinea pig model. Single channel electrodes, fabricated from platinum-iridium alloy, $75\mu m$ in diameter, were stimulated with biphasic, charge-balanced pulses that ranged from 200 to $350\mu s$ per phase. During stimulus trials we monitored auditory tract activation with the middle latency response (MLR). This response is thought to represent activity of higher auditory centers (Kraus et al., '85) and its latency of onset is longer than that of the electrically-evoked auditory brainstem response, thereby reducing interference by the shock artifact. We found that with direct electrical stimulation of the CN, MLR waveform peaks were of greater amplitude and had a more compact configuration than observed with acoustic stimulation. This is consistent with an expected increased synchrony of activation with electrical stimulation. Although the peaks were comparable to those obtained with acoustic stimulation, a latency shift of 2-5 ms for the electrically-evoked response was found. This shift in latency of direct CN stimulation is consistent with activation of the auditory pathway at a site that bypasses the auditory periphery.

The biocompatibility trials demonstrated that the threshold necessary to generate the MLR was typically below 50 $\mu A$ (Niparko et al., '89c). In animals that received stimulation at currents of 50 and 100 $\mu A$, adverse tissue reaction was minimal and glial proliferation along the electrode track never exceeded 25 $\mu m$ in width. Stimulation at intensities of 150 and 200 $\mu A$ (approximately 600 and 800 $\mu Coul/cm^2/phase$) produced significant tissue response at the site of the electrode terminus with neuronal loss, fiber necrosis and reactive cells present. Across all subjects studied the injury threshold was found to exceed the threshold for functional activation by a factor of at least five.

## In-depth stimulation of the CN: Specificity of stimulation

In a subsequent series of subjects we evaluated theoretical advantages of place-specificity of stimulation resulting from stimulation with a penetrating electrode within the CN. We evaluated whether discrete activation of cell groups at two sites in the ventral CN produced differential activation of higher auditory centers.

To evaluate the pattern of activation, we utilized the 2-deoxyglucose (2-DG) technique (Sokoloff, '81). The 2-DG molecule is a radio-labelled analog of glucose that enters the Krebs cycle, but due to stereochemical differences, is trapped within the cycle. This radio-metabolite thereby reflects metabolic and functional activation within the CNS. Autoradiographs are obtained by exposing x-ray film to radio-labelled tissue. Activity is reflected in a higher density of 2-DG accumulation and film exposure. Evoked metabolic activity was evaluated with autoradiographs at the level of the pontine and midbrain auditory nuclei.

Using the 2-DG technique, El-Kashlan et al. evaluated the effects of varying the site of stimulation between the posterior ventral CN (PVCN) and anterior ventral CN (AVCN) (El-Kashlan et al., '90). Autoradiographs from stimulated animals demonstrated qualitative differences between AVCN and PVCN stimulated subgroups, with spatially separated uptake in some higher auditory centers. Activation of the contralateral inferior colliculus was significantly enhanced by a stimulation of both the AVCN and PVCN. Quantified densitometry demonstrated significantly greater inferior colliculus activity in both AVCN- and PVCN-stimulated preparations compared to unstimulated controls. Moreover, at the level of the inferior colliculus, AVCN-stimulated preparations demonstrated a consistent pattern of activation that was spatially distinct from that obtained in PVCN-stimulated subjects. AVCN stimulation produced enhanced activity of the dorsolateral aspect of the inferior colliculus, whereas PVCN stimulation produced higher activity in the ventromedial region of the inferior colliculus. Because isofrequency planes run in a dorsomedial to ventrolateral direction, a pattern of activation that runs orthogonal to that plane would be expected (Oliver et al., '91). The explanation for the differential pattern of activation seen is not entirely clear. Penetrations within the AVCN, however, may have produced a pattern of stimulation that was biased towards activating cells representing low frequencies, and penetrations within the PVCN may have been biased towards stimulating cells representing high frequencies. Alternatively, the effect may have been mediated by stimulating fibers of the acoustic stria or multipolar cells which have a unique tonotopic placement. AVCN multipolar cells are known to represent low frequency information, with PVCN multipolar cells representing high frequencies.

The feasibility of multiple channel electrode stimulation of the CN was assessed in a series of acute experiments by Evans et.al. ('89). Five-channel silicon-substrate electrodes were placed with a micromanipulator in guinea pigs. Each electrode was 2.5mm in length, contained five stimulating pads sputtered with iridium ranging between 1,000 and 5,000 $\mu m2$ with a center spacing of 200 $\mu$m. In the majority of preparations, larger amplitude MLR waveforms resulted from stimulation across broader interelectrode gaps, which would be predicted to produce wider current spread in the CN. The slope of the input/output growth function of the MLR was proportional to the degree of separation between the longitudinally arranged stimulating electrodes: the wider the electrode separation, the greater the slope of the growth function.

The results of the above selectivity experiments indicate that varying the site of stimulation within the CN produces differential activation in higher auditory centers. With the use of multiple channel electrodes, varied stimulation montages produced systematic differences in the evoked response in the majority of preparations. These studies suggest that placement of an electrode within the CN directly adjacent to target second order neurons can provide a strategy for selectively activating functionally distinct neuronal populations within the auditory tract.

## In-depth CN stimulation: Threshold and dynamic range

Another impetus for intranuclear CN stimulation relates to postulated advantages of accessing a larger number of cell bodies and fibers. Implantation within the CN might also provide access to neurons not approached by surface CN electrodes. Theoretically, activation of larger neuronal populations should provide a wider dynamic range of evoked activity. We tested the validity of this notion in a series of trials conducted by El-Kashlan et al. in guinea

Table I. ELECTROPHYSIOLOGICAL RESPONSIVENESS

|  | SURFACE | IN-DEPTH |
|---|---|---|
| Mean Threshold (mA) | 67.5 | 11.4 |
| EMLR Saturation (mA) | 287.5 ± 41.5 | 192.0 ± 49.5 |
| Dynamic Range (dB) | 13.1 ± 2.7 | 24.5 ± 2.6 |

pigs, half of which were implanted with a surface CN electrode and half with an in-depth CN electrode; significant differences in MLR thresholds and dynamic ranges were obtained (El-Kashlan et al., '91). Single-channel platinum-iridium electrodes were matched in their orientation, and the stimulus parameters used were identical. Table I shows the mean threshold and saturation levels of the electrically-evoked MLR (EMLR) input/output functions obtained with surface and in-depth stimulation. In-depth stimulation was associated with significantly reduced threshold ($p < .05$) and a significantly wider dynamic range ($p < .05$).

Autoradiographic analysis of the two stimulation strategies demonstrated no significant differences in activity at the level of the AVCN and PVCN. In-depth CN stimulation, however, produced significantly enhanced 2 DG-uptake of the contralateral inferior colliculus, whereas surface CN stimulation did not, using identical stimulus parameters.

In general, EMLR measures were correlated with the autoradiographic data. Specifically, metabolic activity as reflected in 2-DG uptake correlated with electrophysiologic measures in that both demonstrated saturation at eight times the threshold in most preparations.

Studies comparing surface and in-depth stimulation indicate clear advantages of the penetrating electrode which probably reflect the proximity of a penetrating electrode to target neural populations within the CN, its access to larger neuronal populations, and the elimination of the pia and other surface features of the nucleus that may serve to shunt current spread. In our studies, this was demonstrated by lower thresholds and wider dynamic ranges of activation obtained with in-depth stimulation. By circumventing the pia and reducing CSF contact, in-depth stimulation provides a strategy for producing a reliable pattern of current distribution.

## Direct CN stimulation: Effects of deafferentation

All of the experiments thus far described have been performed in normal hearing animal subjects. Although the survival of the second order neuron is not strictly dependent upon ongoing afferent input, functional and anatomical changes with deafferentation have been described. We assessed the effects of first order neuron deafferentation on the responsiveness of the CN to direct electrical stimulation. An ototoxically-treated guinea pig model was used. This model has been demonstrated to develop a fifty percent reduction in spiral ganglion cells nine weeks following treatment (Webster et al., '81; Jyung et al.,'89). We investigated the central auditory tract response to direct in-depth CN stimulation at different intervals following ototoxic drug administration using electrophysiologic (MLR) recording and autoradiographic analysis.

We saw no significant elevation in EMLR thresholds over a one year period of time following ototoxic deafening. With regard to evoked metabolism, we observed no significant differences in resting metabolism of the CN and inferior colliculus over 4, 9, 16 and 75 weeks after deafening. While substantial differences in CN-stimulation evoked metabolism were observed at the level of the CN, no changes in evoked activity levels were observable in the inferior colliculus. At the level of the CN there was a trend towards early hypersensitivity to stimulation (at four weeks). An insignificant drop-off in response occurred

**Figure 2.** Example probe shapes are shown which reveal the versatility of the micromachining method. The typical probe lengths in this figure are 2 to 3 mm.

at nine and sixteen weeks following deafening. Sixty-five weeks after deafening, a significant difference in CN-evoked metabolism with equal levels of stimulation was observed. At the level of the inferior colliculus, however, no significant differences in evoked metabolism were found. At this level there was an insignificant trend towards early hypersensitivity and only a minimal and insignificant reduction in the metabolism evoked by CN stimulation, even after 65 weeks of deafening.

These results are consistent with those previously demonstrating early denervation hypersensitivity. This may result from an accumulation of transmitter substances at the level of the synapse, released from the inhibition following the elimination of spontaneous activity and perhaps even a reduction in spontaneous-activity-induced masking (Gerken, '79). Within the time limits of this study, however, the central auditory pathway demonstrated preserved responsiveness to CN stimulation despite the loss of first order neuron afferent activity.

## Micromachined solid-state electrodes

The complex anatomy of the cochlear nucleus and its distinct subdivisions and their intrinsic and extrinsic projections raises many fundamental questions for the development of a central auditory prosthesis. The optimum location for prosthetic input to the cochlear nucleus will not be known for some time, but the geometric complexity of the structure will certainly demand a prosthetic device with matching geometric complexity, and this defies manufacture by discrete component methodologies. The life cycle of an individual component must be several tens of years, therefore placing extraordinary demands on reliability. On the other hand, the batch processing techniques of the electronics industry would seem to be a perfect fit for the manufacture of prosthetic devices. Micromachining thin silicon substrates merging such microstructures with appropriate integrated electronics, and making use of thin-film conductors, inlayed iridium stimulating sites and integrated ribbon cables offers a method to accomplish the complex structural shape and processing demands of a cochlear nucleus prosthetic device.

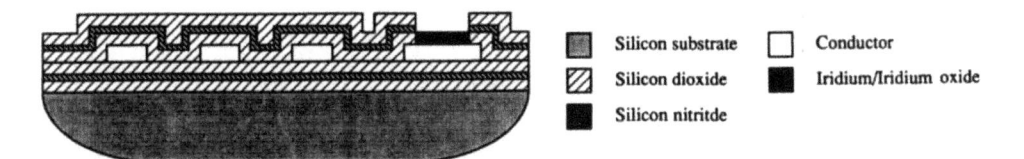

**Figure 3.** The cross-section of a probe shank shows that the silicon substrate is covered by the silicon dioxide/silicon nitride pacification layer, conductors, a stimulation site and another pacification layer. The drawing is not to scale. The substrate thickness is on the order of 15 $\mu$m and the layers are each less than 1$\mu$m thick.

The exploitation by Najafi (Najafi et al., '85) of a 1971 observation (Bohg, '71) that the behavior of certain anisotrophic silicon etchants (specifically ethylene diamine-pyrocatechol) are dramatically changed by boron doping has resulted in the ability to form silicon substrates in virtually any desired planar shape. One needs only to selectively diffuse boron into the surface of a silicon wafer of a type typically used for the manufacture of microelectronic devices. The boundaries of the diffusion will become the boundaries of the etch stop and thus the boundaries of the device. The laws of diffusion and limitations of the photolithographic process which defines the access of the boron to the silicon surface are the only limitations to the achievable mechanical detail.

This process was first used and tested for making recording electrodes (BeMent et al., '86; Drake et al., '88) and was later extended to stimulation electrodes (Anderson et al., '89; Tanghe et al., '90). Figure 2 shows several examples of electrodes that have been designed using these boron etch-stop techniques. Immediate questions arise given the fact that modern photomicrographic processes allow feature sizes in microelectronic devices below one $\mu$m. What is the minimum practical size for an implantable device compatible with the ability to deliver sufficient current to stimulation sites and with sufficient mechanical strength to enter the tissue? Our experience in the cochlear nucleus is that we have had minimal breakage of devices with substrate widths on the order of 30 $\mu$m. Limitations on the substrate widths and thicknesses are imposed by the mechanical forces required to penetrate the target tissue, while other constraints are imposed by the ability to pass current through conductors located on the surfaces of the devices, the pacification layers which protect the conductors, electrical cross-talk, and the size of stimulation sites which are located periodically along the substrate. Najafi has used the equations of statics to show the relationships (Najafi et al., '90). Generally, the buckling load of the devices grows linearly with the width of the probe shanks as does probe function measured by the number of conductors and sites present. Strength is also a third order function of the thickness, so that strength decreases rapidly as the thickness diminishes. If tissue damage is related to total volume of the penetration per probe shank, then shank width is the obvious parameter to reduce while distributing function over many shanks. We have successfully adopted such a strategy for the design of all of our probes.

The passive version of the device is realized using a simple four-mask process that can be completed in most University silicon laboratories. Figure 3 shows the cross-section of a substrate with its thin-film layers. There are three conductors passing a stimulation site on the substrate which is in the cross-section. Alternating layers of silicon oxide and silicon nitride form pacification layers which serve to insulate the conductors from each other and to protect them from the conductive substrate and the conductive environment of the tissue surrounding the probe. The layers also serve to compensate each other to produce a composite whose thermal expansion properties approximately match silicon. The stimulation site is open at the top and therefore allows current to pass into the adjoining environment. This composite has mechanical and materials biocompatibility with the central nervous system and therefore can be inserted and maintained in the brain with minimal short term and long

**STRUCTURE OF AN INTEGRATED
CABLE AND MICROPROBE**

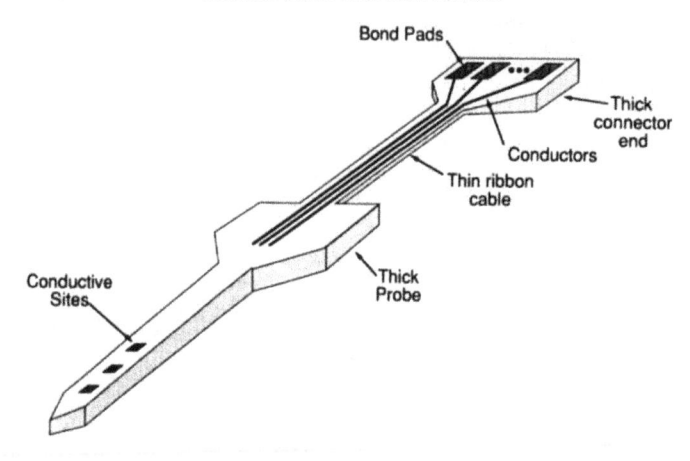

**Figure 4.** The 'long probe' has a continuous substrate which varies in thickness from 5 μm in the cable region to 15 μm in the probe and bonding regions. The fabrication methods are identical to those used for all passive probes.

term consequences. The level of stimulation must be controlled to limit the reaction products which pass into the tissue and any damage due to overstimulation.

Another important characteristic of silicon is that it is the substrate commonly used for all but a small segment of the semiconductor industry. Circuits can be built into the silicon and can be interconnected effectively with the conducting thin-film structures on the probe surface. The importance of this feature is the ability to multiplex signals gathered from many different sites out over very few connecting leads. Minimizing the number of leads traveling to the probe is very important in reducing cable size and stiffness.

## Planar iridium stimulation sites

The site selection, timing and current-driving electronics integrated into these stimulation probe substrates operate at low voltages. The combination of a small stimulation site surface area, short pulse durations, and low voltage demands that the charge delivery capabilities of the material used for the sites be very efficient. We have found that the use of iridium oxide as the stimulation site is not only compatible with the batch processing required for mass production and uniformity of product, but also meets the circuit and stimulation requirements. When operated carefully, activated iridium meets the requirement of not evolving gas, causing pH changes, or corroding.

Recent research with iridium activated by the growth of an iridium oxide layer through electrochemical means has shown great promise for excellent charge delivery. Our technique of sweeping the site voltage between safe gassing limits (-0.9 V and +1.2 V) with respect to a saturated calomel electrode for 15 minutes or more in carbonate-buffered saline yields electrodes with a total charge capacity of over 100 mCoul/cm$^2$. Because of diffusion time-constant limitations combined with short driving pulses, film growth to total charge capacities greater than 100 mCoul/cm$^2$ is not necessary. In fact, the film can become mechanically unstable with excessive growth of oxide. The typical charge delivery (~2400 μCoul/cm$^2$) for a 200 μs pulse is only a small percentage of the total charge capacity. Our

■ Gold
◨ Passification
▨ Iridium
■ Silicon

|← bonding area →|← cable →|← electrode →|

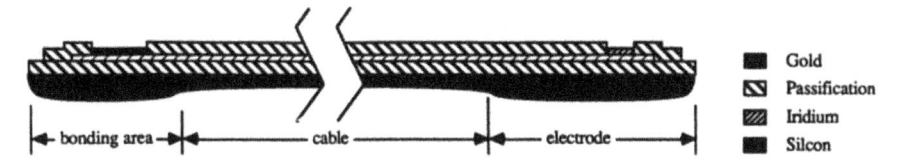

**Figure 5.** A section through a probe with a built-in cable shows conductors passing the length of the probe uninterrupted. This removes the necessity for interconnections at any location other than the bonding area at the percutaneous plug. The thickness of the structure is 15-30 μm at the bonding area but only about 5 μm over the length of the cable portion.

electrodes have been cycled several million times using a current pulse amplitude of 100 μA for 200 μs/phase without physical deterioration of the stimulation sites. Under the same conditions, the charge delivery of gold, stainless steel and platinum sites are 20, 50 and 75 μCoul/cm$^2$ respectively, with accompanying corrosion and metal dissolution problems.

## Silicon cable interconnects

Interconnects are the single biggest problem in the hardware design of neural prosthetic systems. The integrity of the insulation along electrical leads and insulation of junctions among components is most critical. Improvements in the reliability of insulation fabrication or the elimination of junctions can contribute greatly to overall system reliability and lifetime. The same process that has allowed us to form the substrate structure to any desirable planar shape has also contributed to an interconnect solution (Hetke et al., '91). Figure 4 demonstrates the overall features of the 'long' probe which has been fabricated at the University of Michigan. It has a single silicon substrate fabricated with the same techniques described above. In the final system, the conductors are protected from the bonding pads to the stimulation sites. Figure 5 shows a longitudinal cross-section of a single silicon structure combining a bonding area, a cable and a passive stimulation probe. The only difference between this structure and the one previously reported is in its length and the substrate thickness profile. The bonding area and the probe area have a thickness of up to 30 μm compatible with the required strength for these structures, while the cable thickness has been reduced to 5 μm by reduced exposure to the boron diffusion to increase flexibility. A cable 2 cm in length, 75 μm wide, and 5 μm thick is extremely flexible. We have been successful in tightly wrapping the cable around a 1 mm diameter cylinder. To date all *in vivo* and *in vitro* tests have been successful. Cabled probes under 5 volt bias have been tested for over 6 months and have maintained very low (picoampere) leakage levels. Breakage sometimes occurs in surgery but, with improved handling tools and experience, we anticipate increased reliability.

The observations of our biological studies indicate that penetrating stimulation electrodes may be developed as a CNS prosthesis for the profoundly deaf. It is surgically realistic to introduce a stimulation device at the CN level of the central auditory pathways. Devices can be developed which show acceptable levels of biocompatibility and stability, and which will allow stimulation of discrete populations of neural elements at safe levels. The development of such prostheses for humans offers a special technical challenge for the engineer. However, developments over the last few years in the micromachining of silicon substrates indicate that it will be possible to rise to this challenge. Thus, multi-point stimulation systems with appropriate geometry and strength, containing integrated electronics, are at hand. Activated iridium has been shown to be eminently capable of passing the charge densities required for these devices; and recent technical development of silicon cable interconnects is now allowing the development of these devices for tests in animals and eventual application in humans.

ACKNOWLEDGEMENTS

This work was supported in part by NIH grants 5-PO1-DC00274, DC00037-02, N01-NS5-2387, and NIH-NINDS-N01-NS-9-2359.

REFERENCES

Anderson, D., Najafi, K., Tanghe, S., Evans, D., Levy,K., Hetke, J., Xue, X., Zappia, J., and Wise, K., 1989, Batch-fabricated thin-film electrodes for stimulation of the central auditory system, *IEEE Trans. Biomed. Engr.*, 36:693-704.

BeMent, S., Wise, K., Anderson, D., Najafi, K., and Drake, K., 1986, Solid-state electrodes for multichannel multiplexed intracortical neuronal recording, *IEEE Trans. Biomed. Eng.*, 33:230-241.

Bohg, A., 1971, Ethylene Diamine-Pyrocatechol-Water Mixture shows etching anomaly in Boron-doped Silicon, *J. Electrochem. Soc.*, 118:401-402.

Drake, K., Wise, K., Hetke, J., Anderson, D., and BeMent, S., 1988, Performance of planar multisite microprobes in recording extracellular single-unit activity, *IEEE Trans. Biomed. Engr.*, 35:719-732.

Eisenberg, L.S., House, W.F., Mobley, J.P. et al, 1987, The Central Electro-auditory Prosthesis: Clinical Results, in: "Artificial Organs: The WJ Kolft Festschrift", VCH Publishers, New York, pp. 91-101.

El-Kashlan, H., Niparko, J., Kileny, P., and Altschuler, R., 1990, Direct electrical stimulation of the cochlear nucleus: Autoradiographic patterns of central auditory pathway activation, *Otolaryngol. Head Neck Surg.*, 103:189.

El-Kashlan, H., Niparko, J., Altschuler, R., and Miller,J., 1991, Direct electrical stimulation of the cochlear nucleus: Surface versus penetrating stimulation, submitted for publication to *Otolaryngol. Head Neck Surg.*

Evans, D., Niparko, J., Miller, J., Jyung, R., and Anderson, D., 1989, Multiple-channel stimulatuon of the cochlear nucleus, *Otolaryngol. Head Neck Surg.*, 101:651-657.

Gerken, G., 1979, Central denervation hypersensitivity in the auditory system, *JASA*, 66:721-727.

Hetke, J., Najafi, K., and Wise, K., Flexible Silicon interconnects for microelectromechanical systems, Digest IEEE Int. Conf. on Solid-state Sensors and Actuators, June 1991, pp. 764-767.

Hitselberger, W., House, W., Edgerton, B., and Whitaker,S., 1984, Cochlear nucleus implant, *Otolaryngol. Head Neck Surg.*, 92:52-54.

Jyung, R.W., Miller, J.M., and Cannon, S.C., 1989, Evaluation of eighth nerve integrity using the electrically evoked middle latency response, *Otolaryngol. Head Neck Surg.*, 101:670-682.

Kraus, N., Smith, D., and Grossman, J., 1985, Cortical mapping of the auditory middle latency response in the unanesthetized guinea pig, *Electroenceph. Clin. Neurophysiol.*, 62:219-226.

Martuza, R.L. and Eldridge, R., 1988, Neurofibromatosis 2, *N. Engl. J. Med.*, 318:684-688.

Najafi, K., Wise, K., and Mochizuki, T., 1985, A high-yield IC-compatible multichannel recording array, *IEEE Trans. Elect. Devices*, 32:1206-1211.

Najafi, K. and Hetke, J., 1990, Strength Characterization of Silicon Microprobes in Neurophysiological Tissues, *IEEE Trans. Biomed. Engr.*, 37:474-481.

Niparko, J., Kileny, P., Kemink, J., Lee, H., and Graham,M., 1989a, Neurophysiologic intraoperative monitoring: II. Facial nerve function, *Amer. J. Otol.*, 10:55-61.

Niparko, J., Altschuler, R., Xue, X., Wiler, J., and Anderson, D., 1989b, Surgical implantation and biocompatibilty of a CNS Auditory Prosthesis, *Ann. Otol. Rhinol. Laryngol.*, 98:965-970.

Niparko, J., Altschuler, R., Evans, D., Xue, X., Farraye, J., and Anderson, D., 1989c, Auditory brain stem prosthesis: Biocompatibilty of stimulation, *Otolaryngol. Head Neck Surg.*, 101:344-352.

Oliver, D. and Schneiderman, A., 1991, The anatomy of the inferior colliculus: A cellular basis for integration of monaural and binaural information, in: "Neurobiology of Hearing: The Central Auditory System", Altschuler et al., eds., Raven, New York, pp. 195-221.

Sokoloff, L., 1981, Localization of functional activity in the central nervous system by measurement of glucose utilization with radioactive glucose, *J. Cereb. Blood Flow Metab.*, 1:7-36.

Tanghe, S., Najafi, K., and Wise, K., 1990, A planar IrO Multichannel stimulating electrode for use in neural prostheses, *Sensor and Actuators*, B1, pp. 464-467.

Webster, M. and Webster, D.B., 1981, Spiral ganglion neuron loss following organ of Corti loss: A quantitaive study, *Brain Res.*, 212:17-30.

# IX. SPECIAL CONTRIBUTIONS IN HONOUR OF R. LORENTE DE NÓ

# LORENTE DE NÓ'S SCIENTIFIC LIFE

Antonio Gallego

Department of Physiology, Medical School, Complutense University
Madrid, Spain

Rafael Lorente de Nó, born in Zaragoza, Spain, on the 8th of April, 1902, died in Tucson, USA, at the age of 88. His long and intensive scientific activity granted him the recognition of being one of the top neuroanatomists of the first half of the century and the founder of modern neurophysiology. He was the youngest pupil of Cajal, to whom he owed his outstanding skill in nervous system histological techniques and his deep knowledge of the structure of the central nervous system.

Combining his solid structural knowledge with physiological experiments, Lorente de Nó has to be credited with the discovery of basic facts, both in the structure and physiology of the nervous system, which not only allowed the interpretation of central records, but also opened the way for modern neurosciences.

It is a pity that his stubbornness and aggressiveness in the discussions during the battle of the "connective sheath" of the nerve trunks aroused the enemity of many neurophysiologists to the point that he is not quoted by them when referring to several of his fundamental discoveries, a fact that made his name fall into oblivion and consequently remain practically unknown for younger generations of neurophysiologists.

Since several authors in this meeting have widely discussed some of the most relevant contributions of Lorente de Nó to neuroscience, I will restrict myself to briefly expose his scientific life.

Since he was very young, his intellectual precocity combined with ferocious memory gave him an arrogance which, when questioned, would result in a spectacular loss of temper. Later, many of us were witnesses, and on occasion victims, of such "explosions" at scientific meetings or in private conversations.

His interest in science began very early in his life. When he was 15 years old, in 1917, he published his first article, "Termodinámica", in the *Revista del Ateneo Científico Escolar de Zaragoza*, dealing with thermodynamic concepts. He entered the Zaragoza Medical School and very soon, under the guidance of Pedro Ramón y Cajal, brother of Santiago, started research in the nervous system, studying sections of the brain in frogs and lizards. His first studies on the oculomotor system of the frog, brain coccidiosis in rabbits, and spinal cord compression were published in 1919 and 1920.

Santiago Ramón y Cajal, professor of Histology at the Medical School in Madrid, and Director of the Laboratory of Biologocal Research of the University, knowing through his brother Pedro the virtues as a scientist and the passionate style of working of Lorente de Nó, invited him to continue his medical studies in Madrid and at the same time to do research in his laboratory.

Being still a student at the Medical School, he published, at the age of 22, in the *Trabajos del Laboratorio de Investigaciones Biológicas de la Universidad de Madrid*, a paper "La Corteza Cerebral del Ratón", which is a fundamental description of the structure of the brain cortex as well as the primary critical exposition of the principle that the structure of the cortex is the same in the brain of the mouse as in the brain of man. Several studies on this matter were published in the following years, culminating in the chapter he wrote for Fulton's "Physiology of the Nervous System" (Oxford University Press, London, 1949), a systematic summary of the structure of the somatosensory cortex, a classical work on the subject.

Lorente de Nó reached the conclusion that the cortical neuron chains "are in no way different from chains of internuncial neurons in any part of the central nervous system". Quoting from him: "When the chains of neurons are examined closely it will be observed that they are of two types. Some of the chains include short links with cells of a single layer .... Other links vary but little in different mammals, but the short links increase progressively in number from the mouse to man. Thus, in the cortex of the mouse, cells with ascending axons are relatively numerous, while those with short axons are relatively rare. In the human cortex there is an increase in the number of cells with ascending axons, but the increase in the cells having short axons is much more pronounced, so much so that in some cortical regions they outnumber the cells with ascending axons. But the increase in short axon cells is not restricted to any one layer, but takes place in all of them, although in different cortical regions the increase is more pronounced in certain layers, for example, in the area striata in layer IV and the motor area in layer V. Cajal assumed that the large number of cells with short axons was the anatomical expression of the delicacy of function of the brain of man. At present that assumption is almost a statement of fact, for it is known that synaptic transmission demands the summation of impulses under strict conditions, and it is evident that the more heterogeneous is the origin of the synapses on the cells with descending axons, the more rigid become the conditions for threshold stimulation, and the more accurate the selection of the paths through which the impulses may be conducted. The reduction of the number of cells with short axons, without essential modification of the long links in the chain of cortical neurons, makes the cortex of the mouse the 'skeleton' for the human cortex".

Still a medical student, in 1921, he started his fundamental research in physiology based on his knowledge of the vestibulo-ocular pathways. He performed experimental lesions in rabbits, knowing that a longitudinal section in the midline of the medulla and pons must leave intact the connections between nuclei of the abducens nerve and the ipsilateral vestibular nuclei, while the crossed connections, particularly of the reticular nuclei, were interrupted. The fast component of the nystagmus disappeared in the experimental animals and, as a consequence, it was evident that it was not produced directly by the neurons of the vestibular nuclei, but by the activity of neurons of the reticular formation.

In 1923, professor Barany visited Spain and had the opportunity of discussing with Lorente de Nó his experiments on the nystagmus mechanism. Deeply interested and in agreement with Cajal, he invited Lorente de Nó to work with him at Uppsala University.

After obtaining his MD in 1923, he joined Barany's Department in Uppsala, where he continued his histological study on the labyrinth and vestibular centers, performing physiological experiments on labyrinthine reflexes of eye muscles, both in normal and in rabbits with lesions of the central pathways and nuclei. This study continued in the following years and culminated with his work, published in 1933, "Vestibulo Ocular Reflex Arc", which is a fundamental contribution to the knowledge of the functional organization of the nervous system.

Later in his life, Lorente de Nó described his studies in the following way:"Cajal's and Van Gehuchten's studies had demonstrated the existence of two pathways between the vestibular nuclei and the nuclei of the eye muscles, the direct vestibulo-mesencephalic and the longitudinal posterior, and the general opinion then was that these two pathways were the ones which established nystagmus and the tonic reflexes of the eye muscles. I found that after sectioning these two pathways the vestibulo-ocular reflexes still were produced and as a consequence there was no doubt about the existence of other pathways not yet discovered by anatomists. And I had to dedicate many hours to the study of the medulla and pons to find these pathways revealed by the physiological experiments. In fact there are pathways interrupted one or several times in the reticular formation, which connect the vestibular nuclei and the nuclei of the eye muscles". "Instead of explaining the vestibular reflexes on the basis of a chain of three neurons it was necessary to consider complex chains interconnected several times". "To present the reflex arc with chains of internuncial neurons of two types, multiple and closed chains, and to explain that the reflexes are produced by the collaboration of all of them, were ideas very different to the predominant ones among neurophysiologists". "In fact, in 1933 at the same time of the publication of my work on the vestibular ocular reflex, Sherrington and his pupils discussed the spinal cord physiology without even mentioning the existence of internuncial neurons".

In 1927, he returned to Spain after his years in Uppsala. The lack of adequate equipment and financial resources obliged him to work simultaneously at the Cajal Institute and as otorhinolaryngologist with García Tapia, professor of the Medical School in Madrid. In 1928, he was appointed director of the Department of Otorhinolaryngology of the recently founded Hospital Valdecilla in Santander. During the years 1928-29 he improved his clinical and surgical preparation in Berlin, Köningsberg and Frankfurt, and after returning to Santander towards the end of 1929, he was drawn into massive clinical duties which prevented him from doing more than a minimal amount of research work.

After a few months Lorente de Nó decided to leave the clinic to become a full time scientist, a goal reached when with the help of the Rockefeller Foundation he accepted a research position at the Central Institute for the Deaf in St. Louis, where he went in April, 1933.

In St. Louis he continued his experiments on vestibulo-ocular reflexes and published the classical paper to which I have referred. Simultaneously, he continued his research work started in Uppsala on the acoustic system, resulting in the now classical paper on the general plan of structure of the cochlear nuclei.

According to Goodhill, "A very important concept appeared for the first time in that paper, namely that since there are internal paths connecting the neurons of the various parcels of the acoustic nuclei the discharge of nerve impulses by a given acoustic neuron must depend upon the impulses brought in by the cochlear nerve as well as upon the discharges of other acoustic neurons".

In 1936, he was invited by Gasser to join the Rockefeller Institute, becoming a member in 1941 and remaining until his retirement in 1967. There he began working on the histology of the primary acoustic nuclei and on the physiology of nerve conduction, a path which led him to the study of basic neurophysiology until his retirement.

Starting in 1938, the use of microelectrodes in neurophysiological techniques allowed the recording of neuron responses at the level of soma and dendrites, but, according to Lorente de Nó, if obtaining the records was easy, their interpretation was a problem that could not be solved without previous knowledge. In a paper published in Spain on the theoretical basis of the interpretation of ECG, Lorente wrote: "... biological research requires that a scientist must always be prepared to interrupt his specialized work to acquire the auxiliary knowledge needed to interpret his observations". Consistent with these ideas, he went in 1940 to the California Institute of Technology to improve his mathematical background. He developed afterwards the mathematical theory of the potential in a volume conductor and its application to the study of nervous centers.

His knowledge of the eye motor nuclei and internuncials of the reticular formation allowed him to face the problem on a solid base, showing for the first time in the central nervous system, specifically in the neurons of the III cranial nerve, the characteristics of impulse transmission at synapses and the participation of internuncial neurons in the process.

His work forced the abandonment of the classical ideas and the admission by neurophysiologists of the participation of interneurons in the central elaboration of reflexes as a general rule. He demonstrated also that neurons and presynaptic fibers had special properties, not sufficiently known, which were different of the ones found in axons.

This situation led him to interrupt his studies of the central phenomena and devote nearly ten years to the study of nerve fibers in peripheral nerves, culminating in his "Study of Nerve Physiology" published in 1947, in two volumes which produced a commotion in the scientific world.

According to himself, "the results of these studies were sufficient to go back, with probability of success, to the analysis of the records of central phenomena, which looked enigmatic ten years before". "It was not difficult then, to analyse them and reach a clear concept of how the nervous impulses invade the presynaptic fibers and how they were conducted in the dendrites and soma of the neurons." "We know now - he wrote in 1950 - that besides the complexity created by the existence of internuncial chains there is the complexity created by the phenomena produced at the presynaptic fibers and each neuron".

It was a moment, which I witnessed while working in his laboratory, when Lorente de Nó was considered worldwide as the most prestigious neurophysiologist. But his study of nerve physiology, conducted in nerve trunks of frogs, was very soon criticized by the Cambridge School, which worked with single nerve fibers dissected out of nerve trunks.

The experimental study of single nerve fibers was begun in 1934 by Kato and continued by his pupil Tasaki. Both of them supposed that the perineurium was a barrier to the free diffusion of ions. These experiments were assumed by Huxley and Stämpfli, who showed that sodium-free solutions abolished instantly the action potential in desheathed nerves. Lorente de Nó argued that when the nerve was desheathed a transfer of water was produced from the axon to the endoneurium and myelin, and that the isolated nerve fibers were, in fact, an artefact. Intensive work followed, on Lorente's side in collaboration with his last pupil, Honrubia. The controversy, known then as the "battle of the sheath", ended with the victory of the Cambridge School.

After his retirement from the Rockefeller Institute in 1967, he was invited in 1972 to join the Division of Head and Neck Surgery at the UCLA Center for the Health Sciences where he acted as consultant and teacher. There he wrote several papers, culminating with the publication in 1981 of the book, "The Primary Acoustic Nuclei", a master work concerning which, according to Goodhill, "Future investigations in the fields of audiology, auditory neurophysiology, and otology will depend greatly upon this fundamental historical contribution".

In the preface of the book, Lorente de Nó describes its genesis. "In the summer of 1938 I decided to discontinue research on the acoustic nuclei and to write a monograph describing the work already carried out. After the voluminous first draft of the monograph had been written, it became evident that publication of the monograph constituted an exceedingly difficult problem". "Thus, there was nothing else to do but to place the manuscript and the drawings in the drawers of a cabinet, where they remained, virtually forgotten for over 30 years." "My work on the primary acoustic nuclei was rescued from oblivion by Dr. Victor Goodhill. When I showed him the rough manuscript of the monograph and a number of my anatomical drawings, Dr. Goodhill strongly advised me to complete the manuscript and assured me that publication would offer no significant difficulty". "The preparation of the final form of the text has entailed no little time and effort because I had to study the literature on the Subject published during the last 30 years ... Fortunately, the modern authors have worked mainly with experimental or unspecific histological methods. For these reasons only a few, very few of my "old" observations, have been duplicated in the "modern" literature. In revising the original manuscript I also have taken due notice of important papers on the physiology of the acoustic nerve and its primary nuclei. The revision was closed early in 1979".

There were numerous neurophysiologists and postdoctoral fellows coming from several countries who worked under Lorente de Nó's guidance at the Rockefeller Institute, among them Feng, Marrazzi and Renshaw. In 1946, when I joined his laboratory, already there were Laporte from France and Chang from China. In 1947, joining the laboratory were Cazullo from Italy, Soriano from Uruguay, García Ramos from Mexico, Lundberg from Sweden, and, for a short period of time, Fernando de Castro, his friend since both of them worked in Madrid in Cajal's Laboratory.

Gasser and Lloyd, whose laboratories were close to Lorente de Nó's, very often went to his office to discuss electrical records or consult about structural details of the nervous system. According to Lloyd, it was Lorente de Nó who called his attention, in 1942, to the functional schemes of the spinal cord published by Cajal thirty years before.

Lorente's disgrace was to stick twice, with passion and stubbornness, to scientific theories which at long range were untenable. The first time was his defense of the electrical transmission at synapses, a position shared with many physiologists, among them for a certain time Eccles, against the chemical theory whose champion was Dale. The evidence in favour of chemical transmission increased, and was so convincing that finally, in 1951, in a meeting held in Stockholm on the occasion of the inauguration of the laboratories of Physiology and Pharmacology of the Karolinska Institutet, Lorente de Nó finally gave up. This controversy finished, and without pause, he got involved in the "Battle of the sheath" to which I have already referred.

Up to this meeting, few homages have been paid to Lorente de Nó. Perhaps the most relevant was the symposium on Integration in the Nervous System held at the Rockefeller University in 1978, in his and Lloyd's honor, where Ann Graybiel pointed out his enormous contributions in many areas of the nervous system and underlined that he had anticipated by some 30 or 40 years modern ideas related to the reticular formation. In Spain, only the Royal Academy of Medicine held a conference on his contributions to Neuroscience, and the College of Physicians of Madrid organized a symposium where Spanish neuroscientists emphasized his scientific achievements.

Lorente de Nó was the last of the great scientists who worked individually without the help of a team of specialists. In the history of Neuroscience, he has been unique, mastering both structural and physiological techniques and combining them to solve basic functional problems of the nervous system. He was one of the giants in the development of neuroscience during this century and deserves an outstanding place in the history of Science.

# SENSORITOPIC AND TOPOLOGIC ORGANIZATION OF THE VESTIBULAR NERVE

Vicente Honrubia, Larry F. Hoffman, Anita Newman,
Eri Naito, Yasushi Naito and Karl Beykirch

Division of Head and Neck Surgery, UCLA School of Medicine
University of California, Los Angeles, USA

Anatomical studies in the vestibular system represent one of the earlier steps of Professor Rafael Lorente de Nó's search for understanding of brain function. His investigations into the anatomy and physiology of the vestibular system brought the attention of the neuroscience community to this very young (early 20's), curious, creative and tenacious worker, one of the last students of Nobel Laureate Santiago Ramón y Cajal. Lorente de Nó made many contributions which remain the foundation of vestibular science, such as the well-known descriptions of the anatomy and physiology of the three neuron vestibulo-ocular reflex arc and interneurons in vestibular reflex function (Lorente de Nó, '33b; Lorente de Nó, '38).

Owing to the originality of Dr. Lorente's work, he was invited to study with Bárány at Uppsala (Lorente de Nó, '87), and visited the preeminent laboratories in Europe, including Magnus in Utrecht where he performed experimental demonstrations during a short visit to Holland (personal communication). Additionally, he was invited to the US to work in the world's premier laboratories, the Central Institute for the Deaf and the Rockefeller Institute. In the course of his earlier anatomical studies (Lorente de Nó, '26; Lorente de Nó, '33a) he raised two fundamental issues regarding the vestibular nerve: 1) peripheral organ innervation and physiology, and 2) central nervous system pathways. These issues, which are fundamental to the understanding of the physiology of the vestibular nerve, have remained unanswered until recently. This presentation summarizes some of the work conducted with a number of collaborators regarding both issues of the anatomy and physiology of the vestibular nerve.

## I. Peripheral vestibular innervation and physiology

Lorente de Nó expressed his view with regard to the innervation and physiology of the semicircular canal crista as follows: "The anatomical study of the termination of the vestibular nerve is of capital interest for the physiologist... The crista receives different innervation at the summit than at the base." (Lorente de Nó, '26). On the basis of such a differential innervation pattern, he postulated that the center of the crista and the periphery must have different physiological characteristics (Lorente de Nó, '26). This anatomical observation, made in the vestibular nerve of mice, must be put in the perspective of the time.

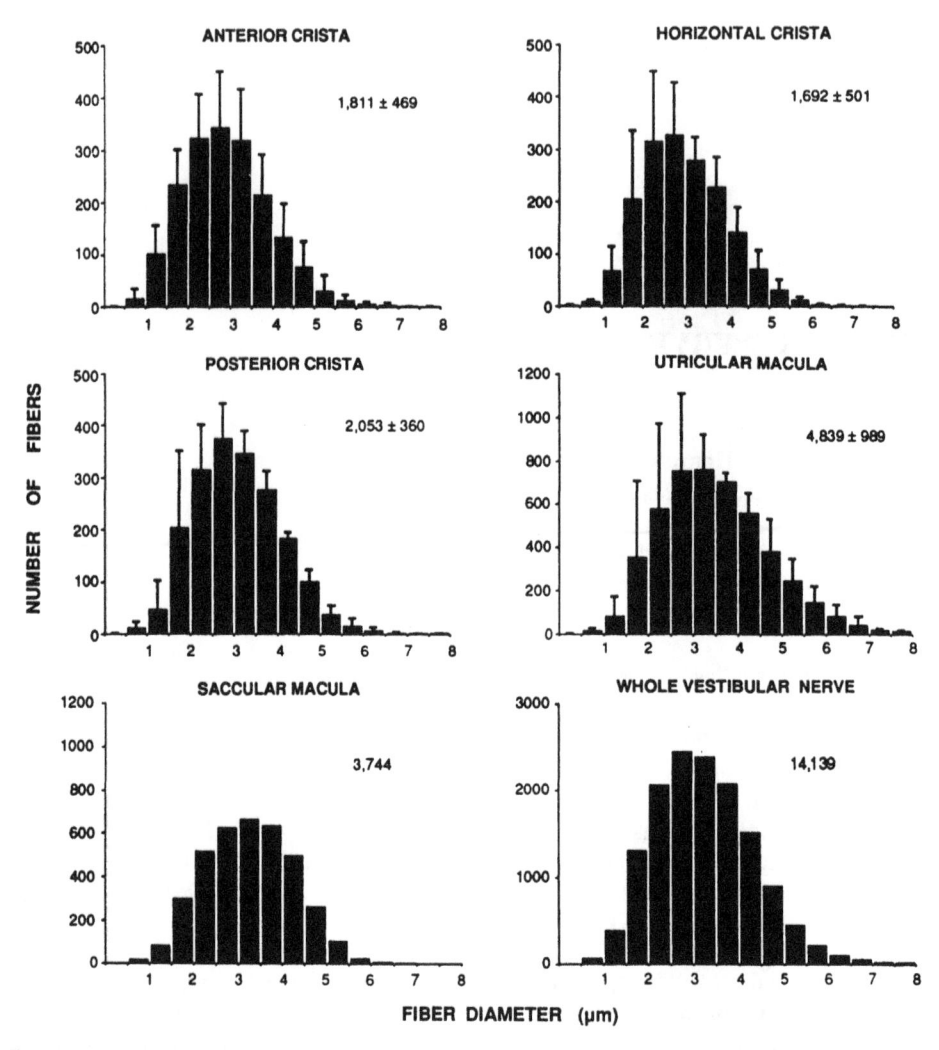

**Figure 1.** Histograms showing the number of fibers of different diameters innervating each one of the vestibular end organ receptors of man.

It was made two years before the publication of Matthews' oscillograph (Matthews, '28) which made possible the documentation of single nerve fiber action potentials.

### Anatomy of the vestibular nerve

The gross anatomy of the vestibular receptors and of their innervation patterns has experienced a progressive evolution through the phylogenetic scale (Baloh and Honrubia, '90). In mammals, it has reached a degree of consistency in which the number of receptors and of nerve bundles from the vestibular nerve have remained constant in a great number of animal species, including man. Still, information about number and size of fibers in the vestibular nerve of some mammals has only recently been elucidated through the use of computer-based graphic methods that greatly simplify the acquisition of quantitative information.

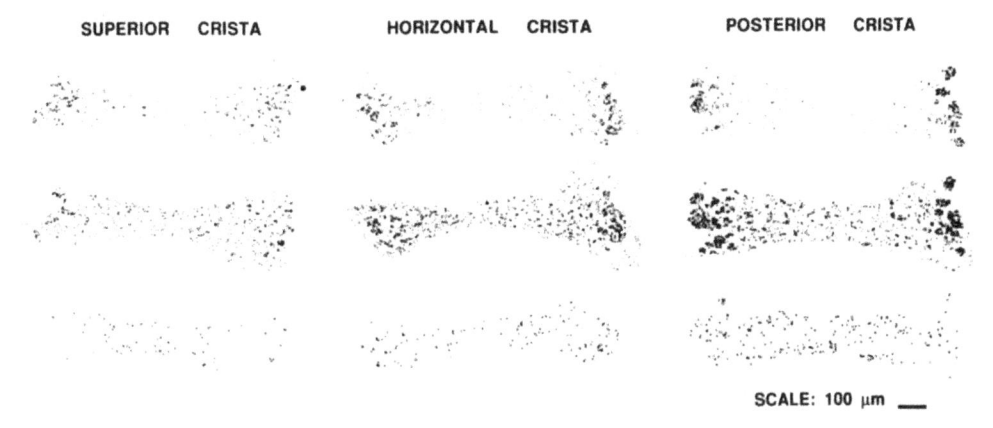

**Figure 2.** Computer reconstruction of innervating fibers according to diameter in anterior, horizontal and posterior crista nerves. Each dot represents the exact site of the individual fiber as it was located in the nerve. Thin fibers (less than 2.5 μm, top row) were distributed predominantly in the ends and thick fibers (greater than 4.5 μm, bottom row) in the central and intermediate areas. Medium-diameter fibers (between 2.5 μm and 4.5 μm, middle row) innervated the entire surface, but their population was highest in the ends.

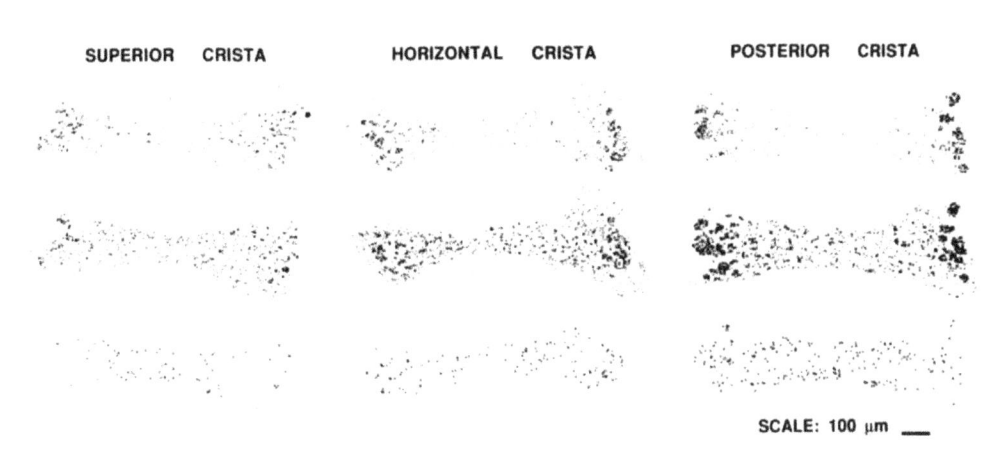

**Figure 3.** Computer reconstructions showing the location of fibers according to diameter in the anterior cristae of the indicated species. Thin fibers (top row) are distributed predominantly in the ends, medium-size fibers (middle row) over the entire surface, but with a higher density in the ends, and thick fibers (bottom row) in the central and intermediate areas.

The broad diameter distribution of fibers in the vestibular nerve, which includes some of the largest in the nervous system, was noted early in the century by Cajal (Ramon y Cajal, '08). Quantitative studies, including those completed in our laboratory of the number of fibers and the diameter distribution in the vestibular nerve have been conducted in man (Lee et al., '90), squirrel monkey (Honrubia et al., '87), chinchilla (Boord and Rasmussen, '58) and frogs (Dunn, '78). In regard to the nerves to the cristae organs, it has been found that all species studied have the same features. There is a broad representation of fiber diameters from $<1\mu m$ to $>7\mu m$, with fibers having a diameter of less than $3\mu m$ representing more than 50% of the nerve fiber population. An example of the distribution of nerve fibers of different diameters in nerve branches to individual vestibular receptors of man is shown in Figure 1. The number of fibers varies among species, with monkeys having approximately

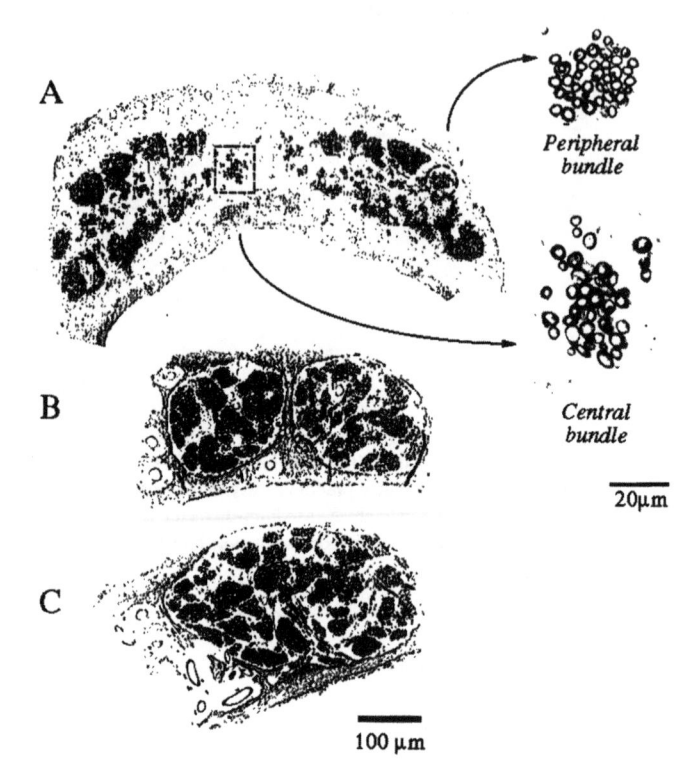

**Figure 4.** Computer reproduction of microphotographs illustrating the sequence of nerve bundle subdivision in the anterior semicircular canal nerve of the chinchilla. The sections were obtained within one millimeter of the crista epithelium at three different levels, immediately below the hair cells (top) and after the nerve separates from the vestibular superior branch. Inset shows two bundles of fibers selected for reconstruction of their trajectory in the crista.

the same number of myelinated fibers as man (squirrel monkey 16,000 vs. 14,000 in man), while chinchillas (7,800) have more than frogs (5,500).

Further details about the innervation of the crista in man were obtained with a study on the distribution of fibers immediately before the receptor area as the fibers approach the hair cells (Fig. 2). A cross-section of the nerve resembles the projection of the dumbbell topography of the crista, whereby it is easy to recognize the pattern of distribution of fibers as they approach hair cells. The figure shows a computer reconstruction of the location of fibers in each of the three crista, separated according to three different diameter categories. In the top row is shown the location of thin fibers, in the middle row medium-sized fibers, and in the bottom row the thickest fibers. In all the cristae, the proportion of thin fibers, is greater in the periphery ($\approx 75\%$) and the opposite is true for the larger fibers, which predominate in the center and intermediate areas.

The anatomic similarities between the human and animal cristae nerves is illustrated in Figure 3, which shows data from monkeys, chinchillas and frogs, together with that of man. The distribution of fibers is similar, with a preponderance of small fibers in the periphery and of the largest in the center of the crista, confirming the generality of Lorente de Nó's original observation about the differential distribution of fibers in the center and the periphery of the crista.

More details about the pattern of innervation of the cristae receptors have been obtained in the chinchilla as shown in Figures 4 to 6. The canalicular nerve to the anterior

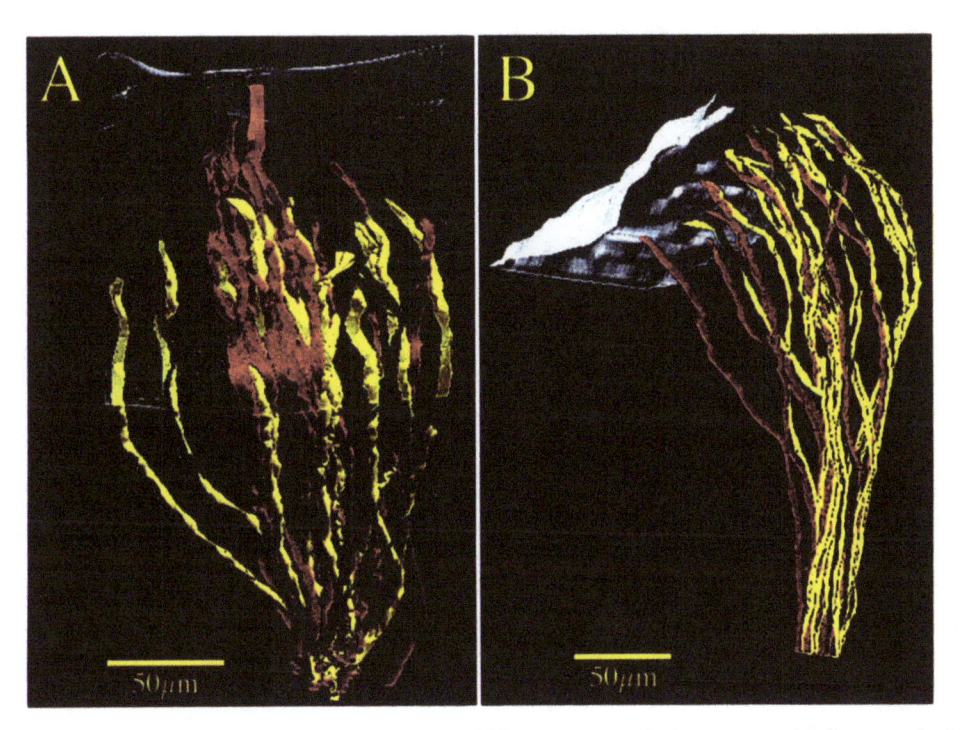

**Figure 5.** Computer reconstruction of two bundles of fibers as they reach the sensory epithelium, one in the center of the crista (A) and another from the periphery (B). There are forty fibers in each bundle and their location in the canalicular nerve is indicated in Fig. 4. Fibers of diameter $\leq 3$ $\mu$m are rendered in green, fibers between 3 and 4 $\mu$m in yellow and fibers $\geq 4$ $\mu$m in red. The area of the crista innervated by the bundles is shown as a transparent surface. The bundle in A is viewed from outside the crista surface, while the bundle in B is viewed from within the crista from a downward-lateral position.

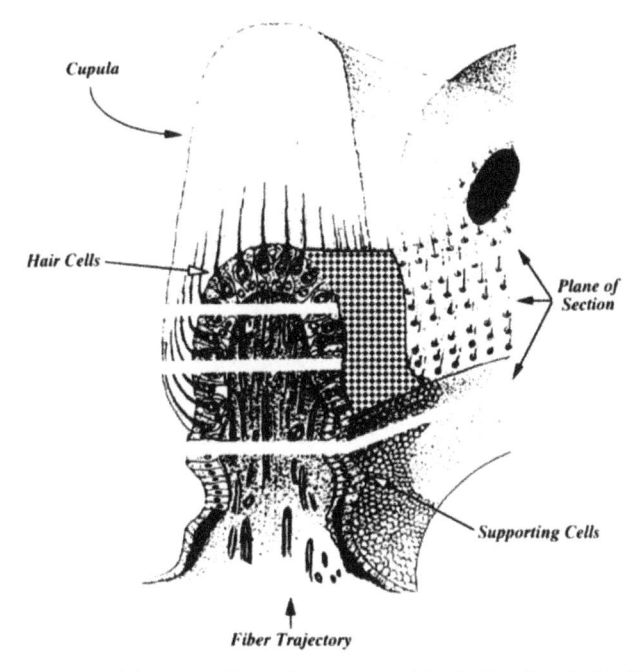

**Figure 6.** Schematic diagram of the areas of the crista innervated by the bundles depicted in Fig. 5. Discrete areas of the crista center, equivalent to half-sided slabs of a crista cross-section, are innervated by a discrete bundle of fibers. The arriving fibers have a straight trajectory toward the receptor surface without intermingling with the fibers from neighboring bundles. Likewise, the bundle to the shallow part of the crista periphery innervates a unique assigned area.

crista, consisting of a single bundle of fibers at a distance of about 1- 2 mm below the crista, divides in two bundles one for each half of the crista (lateral-medial). Further in the periphery at approximately 100 $\mu$m, below the receptor area, each of the the nerves divide in two parallel rows of smaller bundles along the longest crista axis. Each row innervates one of the two slopes of the crista, one row to the utriculopetal and another to the utriculofugal side (Fig. 4). To elucidate the detailed trajectory of the individual fibers, we used computerized three-dimensional reconstruction methods. Two bundles of 40 fibers each were selected, one from the center and another from the periphery of the nerve. The bundles are indicated in Figure 4. A view of the fibers' trajectory for each bundle is shown in Figure 5. In each of the reconstructions, fibers $< 3$ $\mu$m are shown in green, fibers between 3 and 4 $\mu$m in yellow and fibers $> 5$ $\mu$m in red. The region of the crista innervated by the bundles is shown as a white surface. It is viewed from the outside for the central bundle (Fig. 5A) and from the inside for the peripheral bundle (Fig. 5B). It can be appreciated that in the central bundle, the red (largest) fibers are directed toward the top-most part of the crista, while the green fibers are placed in the lowest part of the crista slope. In the peripheral bundle, fibers of different diameters are intermingled. A statistical analysis of fiber composition from similar bundles from the three crista (5 superior and anterior and 3 posterior) showed as expected, that the percentage of large fibers greater than 4$\mu$m was greater in the central bundle than in the peripheral bundles ($p < 0.05$) while fibers of less than 2$\mu$m were more abundant in the peripheral bundles (p, 0.05). In summary, there is a differential innervation between the center and periphery, shown in Figure 3, and a differentiation between the top and the lower part of the center of the crista, with larger fibers innervating the summit and smaller fibers the bottom of the slope, as shown in Figure 4.

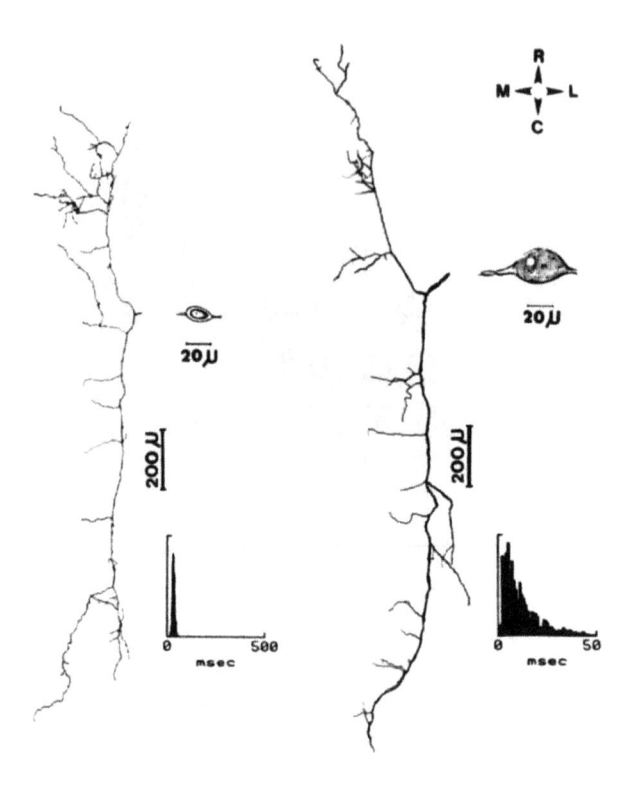

**Figure 7.** Camera lucida reconstruction of the soma and central arborization of two neurons from the vestibular nerve of the bullfrog after visualization with intracellular injection of HRP. The insets show the histograms of the interspike intervals of spontaneous activity, indicating that one - the largest - was highly irregular, but the other was more regular.

Whether at the center or the periphery, the small bundles of fibers innervate restricted areas of the crista surfaces as illustrated in the summary diagram of Figure 6. Of greater significance, is the selective innervation of the crista in the form of small columns in the periphery as in the center in the form of small cross-sectional slabs with separation of fibers from the two sides, the canal vs. the vestibular sides. It can be said that information from the crista is conveyed to the vestibular centers by discrete nerve bundles which create the opportunity to the CNS of receiving topographic (or place) information about the receptor organ.

### The physiology of the vestibular nerve

The semicircular canal cristae and vestibular nerve encode and transmit angular head motion information, which, in combination with that from other sensory systems, contributes to animal orientation. The sites for this mechano-bio-electrical transduction are the hair cells of the crista and the nerve endings of the vestibular nerve synapsing upon them. The functional significance of the particular topography of the crista and the innervation patterns within the receptor organ is, however, yet unclear. The initial studies in the 1940's and 1950's of the physiological properties of the responses from primary vestibular afferents from the semicircular canal (Lowenstein and Sand, '40; Groen et al., '52) suggested that neurons in the vestibular nerve were similar and reflected the dynamics of the motion of the cupula and endolymphatic fluid in the ampula; the behavior of the neurons was consistent with that predicted by the pendulum model of vestibular function (Steinhausen, '27; Dohlman, '35).

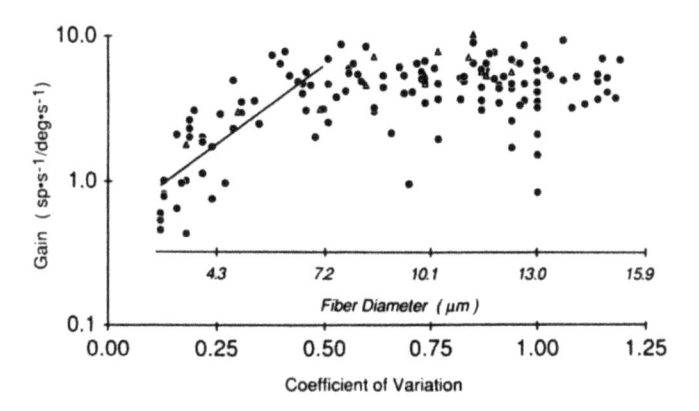

**Figure 8.** Scatter plot showing relationship between gain (relative to velocity of a 0.05 Hz stimulus) and CV for individual HRP-labeled (open triangles) and unlabeled (filled circles) afferents. Line represents the least-square best-fitted linear regression description of the dependence of gain on the value of CV for the neurons with CV ≤0.5. The inserted fiber diameter axis was obtained from analysis of individually labelled fibers and of their recorded CV. Thus, the figure illustrates the relationship between gain, CV and <u>diameter</u> of vestibular afferent fibers.

Fibers innervating the crista could be thought of as representing parallel channels of information, all carrying similar data about head acceleration. This conclusion was, however, inconsistent with the predictions of Lorente de Nó's anatomical observations. More than two decades later, in collaboration with O'Leary and Dunn (O'Leary et al., '74) working in the isolated preparation of the guitarfish, *Rhinobatos productus*, it was found that the response of individual neurons from bundles that innervated the center of the crista had a higher sensitivity and faster dynamics than fibers from bundles innervating the peripheral end of the crista. Other fibers, from bundles intermediate between the center and the periphery of the crista, showed a gradation of sensitivity and dynamics. The data suggested that there must be an anatomically dependent excitation mechanism reflecting a more complex organization in the receptor than was anticipated from earlier physiological recordings. Unknown at that time was whether or not individual neurons with different characteristics also differ in their anatomical properties, such as the diameter of the fibers, the precise locus of innervation in the crista, the type of hair cells they contact or their nerve endings. One decade later, such an opportunity developed through application of intraaxonal recording and staining techniques in the bullfrog because of the relative accessibility of the VIIIth nerve in this animal (Honrubia et al., '81; Honrubia et al., '84). Intraaxonal injections of a solution of horseradish peroxidase (HRP) allowed the visualization of the trajectory of the neuron terminals in the periphery, as well as the ganglion cells, demonstrating that large neurons of the horizontal and anterior canal nerves were the most irregular and sensitive. In a subsequent study of 126 neurons recorded in the nerve to the anterior semicircular canal nerve, detailed information was obtained about the anatomo-physiological characteristics (Honrubia et al., '89). Their spontaneous activity was found to range between 2 and 95 spikes per second. There was a close relationship between the degree of regularity of the firing of action potentials as expressed by the the coefficient of variation (C.V.) and the mean inter-spike interval. Neurons in this example had a coefficient of variation as small as 0.12, and greater than 1.90. Among these neurons, a group of 37 were individually injected with HRP, and it was found that the axon diameter and the coefficient of variation of the neuron were statistically correlated. Based on this correlation analysis, the axon diameter corresponding to neurons with C.V. of 0.5 was calculated to be 7$\mu$m. Neurons with 1$\mu$m diameter have a predicted C.V. of 0.01.

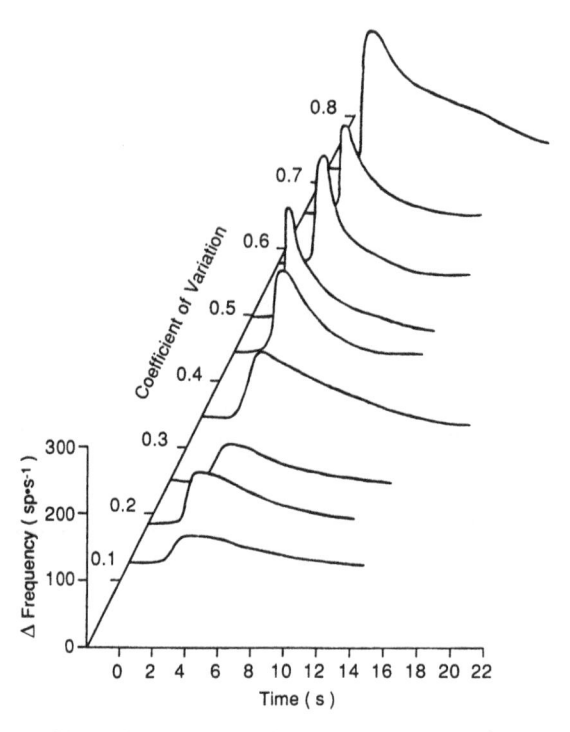

**Figure 9.** Plot of responses of 9 anterior canalicular afferents to acceleration impulses in which change in firing frequency is plotted against time (in seconds) following onset of impulse stimulus (i.e., impulse onset at 0 s). Units' responses are specially distributed along 3rd axis according to their CV. These neurons were selected to illustrate variation in impulse response dynamics characteristics with each neuron's spontaneous firing regularity (i.e., CV).

An example of some of the anatomical and physiological characteristics of two representative neurons is shown in Figure 7. The traces are camera lucida drawings of the soma and of the trajectories in the central nervous system of one very irregular (right) and of a more regular neuron (left). In general, large neurons innervated the center of the crista making contact with few hair cells. Thin neurons innervated the periphery of the crista and had synapses with multiple hair cells (Honrubia et al., '89). An interesting finding of this study was the relationship between the degree of regularity and the sensitivity of the neuron to rotatory stimulus at .05 Hz. The results are shown in Figure 8. The data illustrates that neurons with low C.V. are less sensitive than those with high C.V. The gain of the more irregular neurons (C.V. > 0.5) was independent of the C.V. but, for more regular neurons (C.V. < 0.5), there was a positive correlation between the logarithm of the gain and the C.V. of the spontaneous firing. The estimated slope of this correlation was 2.5, which corresponds to a 5 dB increase in gain per increase of 1 C.V. unit (N=42, R=0.78, P<.001). To further facilitate interpretation of the data in terms of sensitivity and diameter of the axons, an upper abscissa is included in the scatter plot, indicating the diameters of fibers corresponding to the values of the coefficient of variation as shown in the lower abscissa. Phase measurements from these neurons demonstrated the differences which were related to the degree of regularity, again suggesting differences in the dynamics of the excitation mechanisms. Such differences can be appreciated by inspection of impulse response of a group of neurons shown in Figure 9. The coordinates of the three dimensional plot indicate the time immediately before and after the impulse stimulus, the magnitude of the response, and the value of the coefficient of variation of the spontaneous activity. As shown,

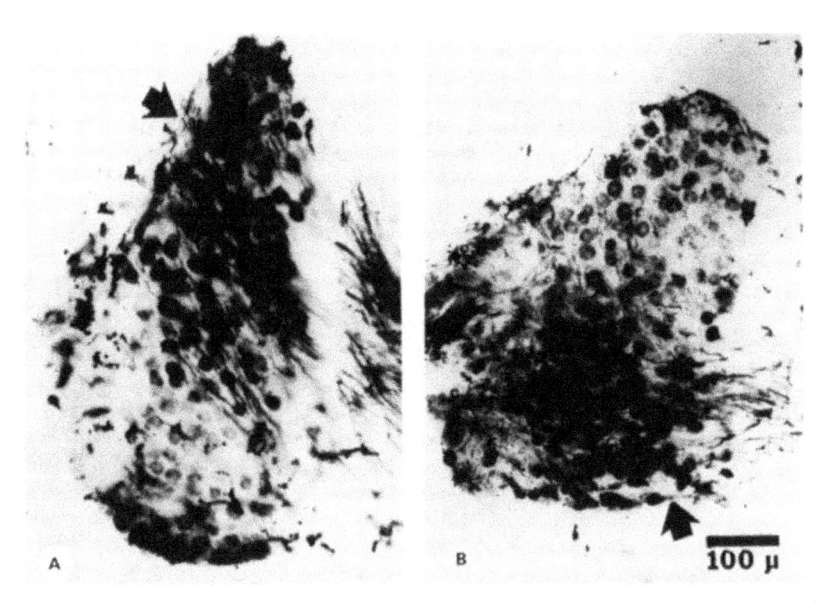

**Figure 10.** Cross section of the inferior part of the Scarpa ganglion innervating the posterior vestibular nerve. The photographs illustrate the different locations of ganglion cells from the posterior semicircular canal (A) and the saccular macula (B). Neurons were labelled with HRP following the scratching method.

neurons with diameter smaller than 7.1 $\mu$m (C.V. <0.5) which represent the majority of the nerve population (>90%), have physiological sensitivities and dynamics which are related to their diameter. It is noteworthy that the thinnest neuron recorded had an estimated diameter of 2.3 $\mu$m; smaller neurons have yet to be recorded. Similar results have been obtained by Goldberg and collaborators in their study of chinchilla primary afferents (Fernandez et al., '88; Baird et al., '88). Although some differences exist in the quantitative relationships, their data is also consistent with our findings and, above all, Lorente de Nó's hypothesis.

In summary, the afferent fibers of the semicircular canal could be conceived as sensory-specific band-pass filters, whereby individual afferents could be considered to be tuned to a particular range of head accelerations. Together with the topographical information described earlier, it is justified to postulate that the crista neuroepithelium is organized along sensoritopic guidelines - very similar to those used by auditory neurons to convey auditory information to the CNS. The small bundles from the center because of the high frequency dynamics of the thick fibers in the summit, together with the small percentage of thin fibers from the slopes, could represent a physiological unit capable of decoding all types of natural head movements. The predominance of thin fibers in bundles from areas of the periphery makes this part of the receptor less suitable for decoding quick head movements. Instead, they may have other physiologic functions, including those unrelated to head movements, but related to producing tonus and posture (Baloh and Honrubia, '90).

## II. Central vestibular innervation

The second fundamental issue raised by Lorente de Nó was whether there are anatomical and physiological differences in the central division of the vestibular nerve, just as in the peripheral organ. The issue as he presented it was as follows: "This is a fundamental problem of considerable importance for the understanding of the physiology of the labyrinth.

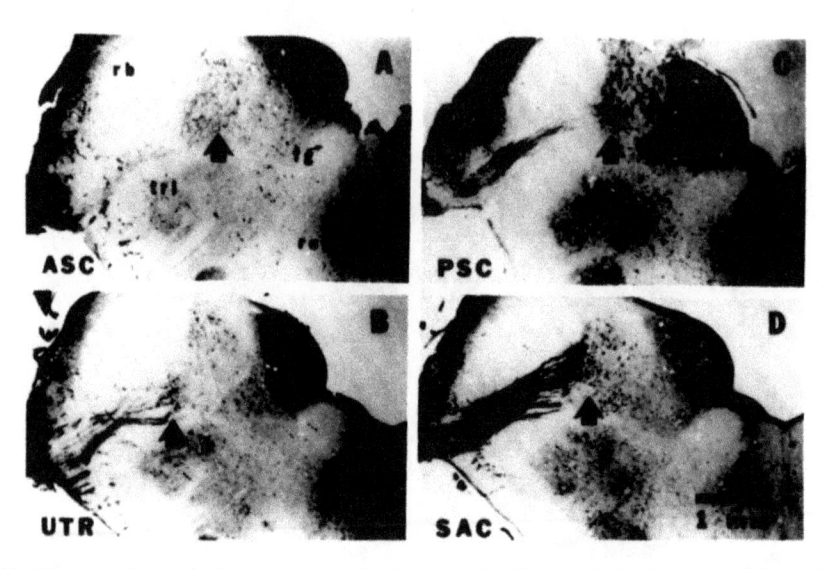

Figure 11. Photographs obtained at approximately the same location in the brain stem of four chinchillas. Microphotographs A and B illustrate the location of HRP-labelled fibers innervating two receptors from the superior vestibular nerve, the anterior semicircular crista (ASC) and the utricle (UTR). Microphotographs C and D illustrate the location of the fibers innervating the posterior semicircular crista (PSC) and the saccular macula (SAC).

If all the end organs of the labyrinth were connected with the same nerve cells, they should have very similar functions, the difference being conditioned chiefly by the mechanics of peripheral stimulation; but if each end organ should have a special function, the central representation ought to be different." (Lorente de Nó, '33a). It is a sobering thought that only fifty years ago, so little was known about the physiology of the organs for equilibrium!! A particularly bothersome question to him was whether or not the macula and the sacculus played a role in hearing. He found that the fibers innervating the vestibular receptors "had nothing in common with the primary cochlear centers" as he followed them in Golgi stained preparations to innervate the vestibular centers.

Again, the use of new techniques for labeling of individual or small groups of neurons which innervate restricted parts of the inner ear (e.g., one peripheral receptor) has provided new opportunities to elucidate the question of the specificity of central representation of vestibular receptors with much more accuracy than was possible earlier, and beyond the expectations that Dr. Lorente could have had. He was particularly pleased with the possibilities of these techniques of cell labeling as he studied some of my preparations and read about the contributions of other investigators such as C. Fernández and J. Goldberg (personal communication). At UCLA, a method for labeling fibers from individual receptor organs was developed in collaboration with Dr. W.S. Lee (Yonse University, Seoul, South Korea) to study the location of fibers in each of the branches of the peripheral vestibular nerve, in the Scarpa ganglion, and in the CNS (Lee et al., '89; Lee et al., '91). After introducing a solution of HRP inside the endolymphatic spaces of the vestibule, the only labeled neurons were those from receptor organs which were very carefully injured by scratching the surface of the sensory epithelia with an electrically polished needle mounted in a micromanipulator to properly control the extent of damage.

One aspect of the histological verification of the process of selective labeling of fibers from different receptors in the chinchilla vestibular nuclei is shown in Figures 10 and 11. Figure 10 shows sections at the same level in ganglion cells from the posterior vestibular nerve, but from two different animals. It can be appreciated that ganglion cells from the

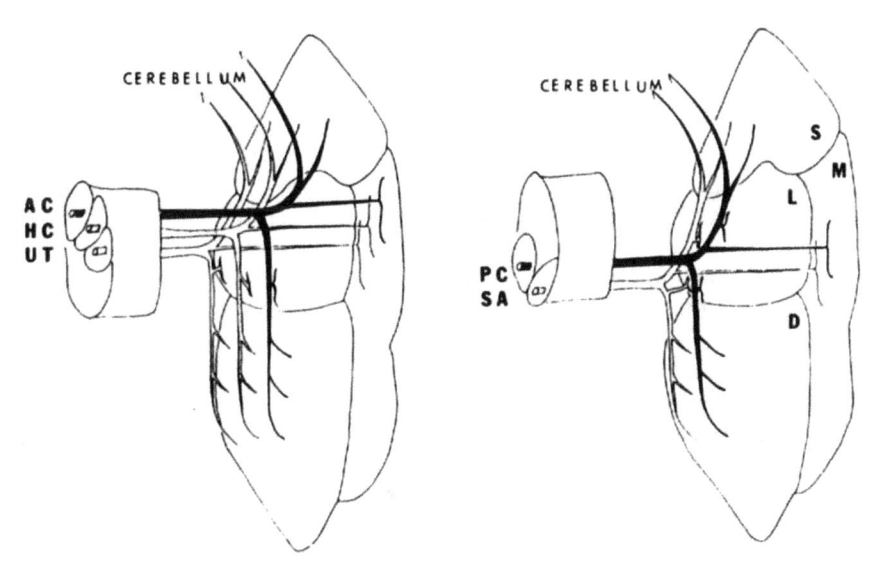

Figure 12. Diagram illustrating the relative location of fibers from each vestibular receptor in the vestibular root, the vestibular tract and the vestibular nuclei of the chinchilla. Abbreviations are as follows: AC, anterior semicircular crista; HC, horizontal semicircular crista; UT, utricle; PC, posterior semicircular crista; SA, sacculis; S, M, L and D refer to the superior, medial, lateral, and descending vestibular nuclei, respectively.

animal whose sacculus was labeled are located in the opposite side of the nerve from the animal whose posterior semicircular canal was labelled. Likewise, in the brain stem, it could be demonstrated that fibers from each of the receptor organs have different trajectories into the vestibular tract as shown in Figure 11. All the sections in the illustration were obtained at the same level in the brain stem, corresponding to the location of the genu of facial nerve and the abducens nucleus. As the arrows indicate in each one of the sections, the fibers in the vestibular tract are located at different positions depending on the receptor organ that was labeled. Fibers from two organs from the superior and two from the posterior nerves are shown for illustration. Fibers from the anterior semicircular canal are seen to enter the vestibular root and bifurcate at a level slightly dorsal and lateral to the genu of the facial nerve, in the lateral part of Deiters nuclei. The fibers of the utricle are seen to be located more dorsolateral than those of the anterior semicircular canal. From the posterior vestibular nerve, the ascending fibers of the sacculus are in a dorsolateral position to the fibers from the posterior semicircular canal. These posterior canal nerve fibers are also dorsal to the anterior canal and utricular fibers.

All the receptors send projections to the cerebellum, mainly the maculae organs, but not all areas of the vestibular nuclei are innervated by fibers from the vestibular periphery. A summary of the relative distribution of the fibers in the root and the distribution into the vestibular nuclei are shown in Figure 12. The diagram indicates that each vestibular receptor innervates all the vestibular nuclei and there is, together with a degree of differentiation, some overlap among the various receptors. In the superior nucleus (SN), fibers projected in the dorsolateral portion of the nucleus throughout its rostral caudal extent. Fibers from the semicircular canals were medial to those of the maculae. The most medial portion of the SN did not receive afferent projections from any of the vestibular receptors. In the MN, fiber projections were observed from the most rostral part of the nucleus to the rostral portion of the DN. In the caudal portion of the medial nucleus (MN), no afferent projections from any vestibular receptors were seen. The descending nucleus (DN) received afferent projections laterally in the rostral and middle areas. The rostromedial portion received a few fibers. In the most caudal DN, afferent projections from the vestibular receptors were not seen. The afferent projections to the lateral nucleus (LN) are located laterally in the rostral and middle areas and ventrolaterally in the caudal area. Laterally in the nucleus, the fibers overlapped extensively. Despite this, individual areas of projection were identified for each receptor. The medial area of the upper half and the dorsomedial part of the caudal LN do not receive afferent projections. Similar findings have been obtained earlier in the frog, studying the representation of the vestibular receptors in the vestibular nuclei (Kuruvilla et al., '85; Suarez et al., '85; Honrubia et al., '85b). In this animal, it was argued that the overlap is designed to support the anatomo-physiological organization as represented in the projections of vestibular receptors via the vestibular nucleus to the various oculomotor nuclei (Honrubia et al., '85a).

Thus, the present data contribute to the elucidation of the issue raised by Lorente de Nó regarding whether each organ could act differently on the basis of central connections. The question still remains to be answered as to whether or not fibers that have different topographical origins in a particular organ and differ in their sensitivities have exclusive representation into the nuclei. The present data strongly suggest that a sensoritopic and topological organization exist in the vestibular system reminiscent of Lorente de Nó's findings of the central projection of the cochlear nerve (Lorente de Nó', '33a). These questions are now being investigated and offer new challenges to the biologists of vestibular science. More refined techniques for anatomical labelling and physiological characterization need to be developed to completely answer Lorente de Nó's question at the cellular level.

REFERENCES

Baird, R.A., Desmadryl, G., Fernandez, C., and Goldberg, J.M., 1988, The vestibular nerve in the chinchilla. II. Relation between afferent response properties and peripheral innervation patterns in the semicircular canals., *J. Neurophysiol.*, 60:182-203.

Baloh, R.W. and Honrubia, V., 1990, "Clinical Neurophysiology of the Vestibular System.," F.A. Davis Co., Philadelphia.

Boord, R.L. and Rasmussen, G.L., 1958, Analysis of the myelinated fibers of the acoustic nerve of the chinchilla, *Anatomical Record*, 130:394.

Dohlman, G., 1935, Some practical and theoretical points in labyrinthology, *Proceedings of the Royal Society of Medicine*, 28:1371-1380.

Dunn, R.F., 1978, Nerve fibers of the eighth nerve and their distribution to the sensory nerves of the inner ear in the bullfrog, *J. Comp. Neurol.*, 182:621-636.

Fernandez, C., Baird, R.A., and Goldberg, J.M., 1988, The vestibular nerve of the chinchilla. I. Peripheral innervation patterns in the horizontal and superior semicircular canals., *J. Neurophysiol.*, 60:167-181.

Groen, J.J., Lowenstein, O., and Vendrik, A.J.H., 1952, The mechanical analysis of the responses from the end organs of the horizontal semicircular canal in the isolated elasmobranch labyrinth, *Journal of Physiology* (London), 117:329-346.

Honrubia, V., Sitko, S., Lee, R., Kuruvilla, A., and Schwartz, I.R., 1984, Anatomical characteristics of the anterior vestibular nerve of the bullfrog, *Laryngoscope*, 94:464-474.

Honrubia, V., Kuruvilla, A., Suarez, C., and Sitko, S., 1985a, Individual projections of the semicircular canal cristae in the vestibular nuclei of the frog, *in*: "New Dimensions in Otorhinolaryngology - Head and Neck Surgery, Proceedings of the XIII World Congress of Otorhinolaryngology," E.N. Myers, ed., Elsevier, Amsterdam.

Honrubia, V., Suarez, C., Kuruvilla, A., and Sitko, S., 1985b, Central projections of primary vestibular fibers in the bullfrog. III. The anterior semicircular canal afferents, *Laryngoscope*, 95:1526-1535.

Honrubia, V., Kuruvilla, A., Mamikunian, D., and Eichel, J.E., 1987, Morphological aspects of the vestibular nerve of the squirrel monkey, *Laryngoscope*, 97:228-238.

Honrubia, V., Hoffman, L.F., Sitko, S., and Schwartz, I.R., 1989, Anatomic and physiological correlates in bullfrog vestibular nerve., *J. Neurophysiol.*, 61:688-701.

Kuruvilla, A., Sitko, S., Schwartz, I.R., and Honrubia, V., 1985, Central Projections of primary vestibular fibers in the bullfrog. I. The vestibular nuclei, *Laryngoscope*, 95:692-707.

Lee, W.-S., Newman, A.N., and Honrubia, V., 1989, HRP labelling of afferent vestibular nerve fibers in mammals, *Cellular Biology International Reports*, 13:635-636.

Lee, W.-S., Suarez, C., Honrubia, V., and Gomez, J., 1990, Morphological aspects of the human vestibular nerve, *Laryngoscope*, 100:756-764.

Lee, W.-S., Suarez, C., Newman, A.N., and Honrubia, V., 1991, Central projections of the individual vestibular sensory end-organs of the chinchilla, in: "Otorhinolaryngology Head and Neck Surgery (Proceeding of the XIV World Congress - Madrid, Spain, September 10-15, 1989). Vol. I," T. Sacristan, J.J. Alvarez-Vicent, J. Bartual, F. Antoli-Candela and L. Rubio, ed., Kugler and Ghedini Publications, Amsterdam.

Lorente de Nó, R., 1926, Etudes sur l'anatomie et la physiologie du labyrinthe de l'oreille et du VIIIe nerf. Deuxieme partie., *Trav. Lab. Rech. Biol. Univ. Madrid*, 24:53-153.

Lorente de Nó, R., 1933a, Anatomy of the eighth nerve. The central projection of the nerve endings of the internal ear., *Laryngoscope*, 43:1-38.

Lorente de Nó, R., 1933b, Vestibulo-ocular reflex arc, *Arch. Neurol. Psych.*, 30:245.

Lorente de Nó, R., 1938, Analysis of the activity of the chains of internuncial neurons, *J. Neurophysiol.*, 1:207-244.

Lorente de Nó, R., 1987, Facets of the Life and Work of Professor Robert Barany (1886-1936), in: "The Vestibular System: Neurophysiologic and Clinical Research," M.D. Graham and J.L. Kemink, ed., Raven Press, New York.

Lowenstein, O. and Sand, A., 1940, The individual and integrated activity of the semicircular canals of the elasmobranch labyrinth, *Journal of Physiology* (London), 99:89-101.

Matthews, B.H.C., 1928, A new electrical recording system, *J. Physiol.*, 71:64-72.

O'Leary, D.P., Dunn, R.F., and Honrubia, V., 1974, Functional and anatomical correlation of afferent responses from the isolated semicircular canal, *Nature*, 251:225-227.

Honrubia, V., Sitko, S., Kimm, J., Betts, W., and Schwartz, I.R., 1981, Physiological and anatomical characteristics of primary vestibular afferent neurons in the bullfrog., *Int. J. Neuroscience*, 15:197-206.

Ramon y Cajal, S., 1908, Sur un ganglion special du nerf vesibulaire des poissons et de oiseaux, *Trav du Lab de Recherches biol.*

Steinhausen, W., 1927, Uber Sichtbarmachung and Funktionsprufung der Cupula terminalis in den Bogengangs-ampullen der Labyrinths., *Arch. Ges. Physiol.*, 217:747.

Suarez, C., Kuruvilla, A., Sitko, S., Schwartz, I.R., and Honrubia, V., 1985, Central projections of primary vestibular fibers in the bullfrog. II Nerve branches from individual receptors, *Laryngoscope*, 95:1238-1250.

# LORENTE DE NÓ AND THE HIPPOCAMPUS: NEURAL MODELING IN THE 1930s

Larry W. Swanson

Department of Biological Sciences, University of Southern California
Los Angeles, California, 90089-2520, USA

Lorente de Nó's substantial contributions to our understanding of the structure and function of the hippocampus are largely contained in a single, extensive paper that was published in 1934 in the *Journal für Psychologie und Neurologie* (Lorente de Nó, '34) while he was working at the Central Institute for the Deaf in St. Louis, Missouri (USA), although most of the drawings in the paper were dated 1927 or 1928, during a period when he was in Spain. The experimental work underlying this paper dealt with the cellular organization (cytoarchitecture) and connections of the hippocampus in the mouse, monkey, and human brain, and contributions emerging from this analysis may be divided in factual and theoretical components. As we shall see, Lorente de Nó advanced our knowledge of hippocampal circuitry in several important ways. However, his real genius emerged from theoretical considerations of the physiology of individual neurons and then of interconnected groups of neurons. From this he made the remarkable prediction that conduction in axons and dendrites must be fundamentally different: while conduction in axons is by way of all-or-none impulses that are followed by a refractory period, conduction at dendrites was predicted to be subthreshold, additive, and nonrefractory. In short, he proposed that the *neuron is a summation apparatus* that generates a pattern of impulses in the axon. And this was only the highlight of the paper; many other profound issues were raised and dealt with in a more or less complete way.

It is worth pointing out early on that Lorente de Nó dealt with general principles in neurobiology, and was helped immeasurably in his thinking by the comparative approach: he tried not only to determine whether a common, general structural plan applies to what appears to be the same area in the brains of different mammals, as well as the extent to which similar organizing principles apply to circuits in different parts of the brain in the same species. Thus, he examined three different species in this paper, and compared the results with similar work he had carried out in other parts of the cerebral cortex, in the cochlear nuclei, and in circuits underlying the vestibulo-ocular reflex.

## Structural plan of the ammonic system

Lorente de Nó was deeply concerned about formal ways to establish the basic subdivisions of the central nervous system, whether they be cortical fields or subcortical nuclei. By the time this paper was published he had come to the conclusion that major inputs defined *cortical systems* and *subcortical nuclear complexes*, while other, partially overlapping inputs serve to divide each cortical system into fields or each subcortical complex into nuclei. For example, the olfactory cortical system defined by the lateral olfactory tract is divided into a number of fields, just as the cochlear nuclear complex defined by the auditory nerve is divided into a number of nuclei.

Based on this reasoning, and on current knowledge of the longer intracortical pathways, Lorente de Nó defined an "Ammonic system" consisting of the hippocampus (Ammon's horn or cornu ammonis and the fascia dentata) and three other cortical regions that project in partially overlapping ways to it. According to Lorente de Nó, these regions included the entorhinal area, which projects to the ventral two-thirds of Ammon's horn, the cingulate gyrus, which projects to the middle third or so, and the induseum griseum, which projects to dorsal regions. We now know that this scheme is fundamentally incorrect (Swanson et al., '87), but fortunately this is of little consequence for the work Lorente de Nó carried out on the hippocampus itself.

## Cytoarchitecture of the hippocampus

As we shall see, Lorente de Nó firmly believed that cytoarchitectonics is a necessary approach, but only as a prelude to the application of other methods, specifically those dealing with the connections of the neurons under consideration. In any event, he advanced a new terminology for the cortical fields of Ammon's horn, and the fact that it is still widely used today testifies to the possibility that it reflects accurately the basic structure of this cortical region. Briefly, his "CA" (cornu ammonis) fields are as follows: field $CA_1$ roughly corresponds to the lower or subicular blade of the early anatomists; field $CA_2$ corresponds to the narrow strip containing large pyramidal cells but no mossy fibers, identified by Doinikow ('08); field $CA_3$ roughly corresponds to the upper or ventricular blade of the older workers, and field $CA_4$ roughly corresponds to the endblade of the older workers (and deeper parts of the dentate polymorph layer of Cajal, and a number of modern workers) (Swanson et al., '87). There is no easy way to correlate this CA fields with the "H" fields of M. Rose that some neuropathologists still insist on using today.

Lorente de Nó also divided the $CA_1$ and $CA_3$ fields into a, b, and c subfields (progressing in a traverse, subiculum to dentate gyrus direction). While these subdivisions have occasionally proven useful for descriptive purposes, their borders cannot be defined (as Lorente admitted), and very few clear differences in their connections have been established.

## Connections of the hippocampus

Lorente de Nó's analysis of hippocampal connections was limited to what he could see in Golgi preparations. His description of cell types with this method supplemented Cajal's ('09, '11) classical overview, and together they still constitute most of what we know about the form of neurons in Ammon's horn and the dentate gyrus.

In addition, he identified two association pathways that added significantly to our understanding of the trisynaptic circuit so elegantly presented by Cajal. As is well known, the trisynaptic circuit consists of a series of pathways that is oriented perpendicular to the long axis of the hippocampal formation; the first link in this circuit is the perforant path from superficial neurons of the entorhinal area to the molecular layer of the dentate gyrus, the second is the mossy fiber system from granule cells in the dentate gyrus to pyramidal cells in field $CA_3$, and the third is the Schaffer collateral system from large pyramidal cells in field $CA_3$ to small pyramidal cells in field $CA_1$.

The first pathway discovered by Lorente de Nó was a projection from field $CA_1$ back to the subiculum and entorhinal area, thus in a sense "closing" the trisynaptic circuit, although the fibers end in layer 4 of the entorhinal area, not in more superficial layers. This pathway was finally confirmed experimentally in 1978 (Swanson et al., '78), and later studies showed that deeper layers of the entorhinal area project widely to virtually the entire cortical mantle (Swanson et al., '86). Thus, information from a number of isocortical areas converges on the entorhinal area, is processed through the trisynaptic circuit, and appears to be relayed back out into widespread parts of the cortex.

The second connection revealed in this study was what Lorente de Nó called the "longitudinal association path". He showed that field $CA_3$ gives rise to a longitudinal association path to other parts of Ammon's horn, in addition to the essentially transverse system of Schaffer collaterals. This pathway was also confirmed many years later with the autoradiographic method (Swanson et al., '78), and was shown to arise in temporal parts of field $CA_3$, and to course primarily in a "dorsal" (septal) direction.

These are the major anatomical contributions that emerged from Lorente de Nó's paper, at least from our current perspective. We shall now go on to consider the important conceptual issues that this information raised in his mind.

## Comparative cytoarchitectonics and the general structural plan

As mentioned earlier, one of Lorente de Nó's longstanding concerns (going back to his early paper on the cerebral cortex in 1922) (Lorente de Nó, '22) was the extent to which particular regions of the brain differ between species, and as a correlate to this, how best to establish such differences or similarities. In the paper under consideration here, he pointed out that no formal procedure for defining distinct cortical fields or subcordical nuclei had gained wide acceptance, and went on to suggest at least some guidelines. His basic conclusion was that a cell group can only be defined on the basis of its cytoarchitectonics, combined with a knowledge of the morphology of its neurons gained from the Golgi method and an understanding of its longer inputs and outputs.

He clearly appreciated the cytoarchitectonic approach, as illustrated in the following passage: "Architectonics is of enormous value however in making a first analysis and determining what has to be studied with other methods" (p. 166). However, he noted that in comparative studies it is important to use the same criteria for drawing boundaries in different species, and to avoid establishing arbitrary boundaries. These may seem like obvious points, but he used the work of M. Rose - a highly respected practitioner of the art - on the parcellation of the entorhinal area, as an example of problems that may arise. Thus, Rose identified five subdivisions of this area in the mouse, and 21 subdivisions in the human. According to Lorente de Nó, Rose used different criteria in different species, and within the same species drew boundaries in some places where they are obvious, but failed to draw them in other places where they were equally obvious.

From his own work (Lorente de Nó, '33), he concluded instead that the entorhinal area in all mammals has the same *basic structural plan*. Exactly what he meant by this is unclear, but he did state that this cortical area has the same basic subdivisions in all mammals, based on cytoarchitecture and connections, not the classes involved. In short, he seems to have concluded that *mammalian brains differ only in quantitative rather than qualitative ways*, since his arguments extended to all parts of the brain that he had studied. Expressed in yet another way, he believed that *the mammalian brain shares a general structural plan*. The problem of defining its basic cell groups (systems or complexes) and their corresponding subdivisions (fields and nuclei), as discussed above, remains.

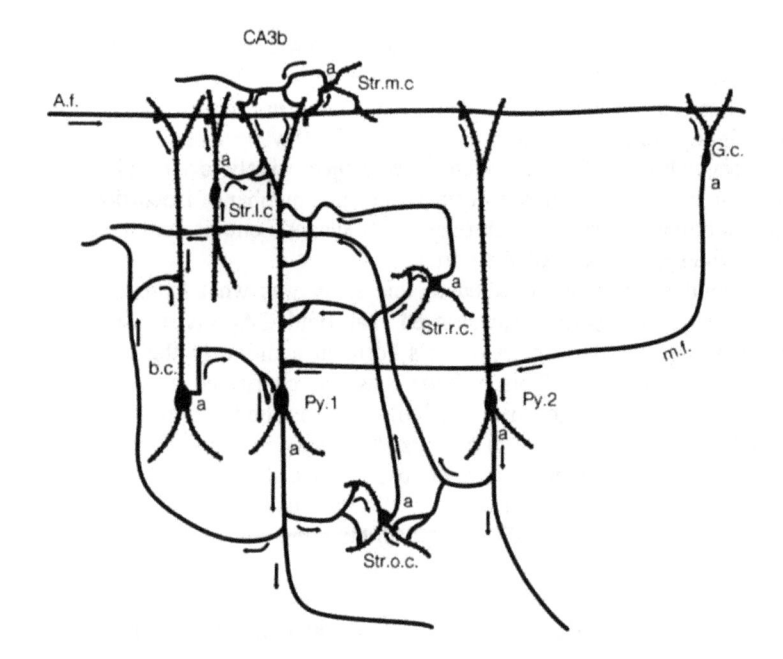

**Figure 1.** Lorente de Nó's legend to this figure (Redrawn from Fig. 35 in Ref. 1) reads as follows: "Several of the many possible paths between the afferent fibres of the perforant path and the effector pyramid, "Py. 1"; field $CA_3b$. The afferent fibre "A.f." establishes contacts with the pyramids "Py. 1" and "Py. 2", with the cells with short axis cylinder of the Stratum moleculare ("Str.m.c."), lacunosum ("Str.l.c.") and pyramidale (basket cell, "B.c.") and with the granules of the Fascia dentata ("g.c."). When these cells discharge, their impulses are transmitted to cell "Py. 1". Besides when cells "Py." discharge other cells with short axis cylinder, of the Stratum radiatum ("Str.r.c.") and oriens ("Str.o.c.") which were not affected by the afferent impulse are brought into activity; their impulses are again transmitted to cell "Py.1". The axons are marked with a; the arrows indicate the direction of transmission of the impulses according to Cajal's law of axonal polarisation. If this law is not accomplished, i.e. if the synapse is not irreversible the interpretation of the diagram would be quite different from that proposed in the text."

## Law of partly shifted, overlapping inputs

While all of the major systems or complexes in the central nervous system may not be known, Lorente de Nó believed that he had discovered a fundamental law governing the establishment of such units, as well as their subdivision. He believed that the basic units of the nervous system are defined on the basis of a major, dominant input, and equally importantly, that multiple inputs to a region never overlap completely, whether they arise in functionally similar or in different areas. Stated somewhat more formally he proposed that, in general, *multiple inputs to a region are never uniformly distributed, and there are always association pathways between the subdivisions thus formed.*

## The physiology of the neuron

In addition to the physiological unit of the nervous system, Lorente de Nó turned his attention to the physiology of the individual neuron, and based on what he knew about hippocampal pyramidal cells and other cell types, arrived at the brilliant deduction that conduction in axons and dendrites must be fundamentally different.

He began by summarizing the evidence that different classes of inputs to pyramidal cells are segregated on different parts of the dendritic tree, cell body, and initial segment of the axon, just as Cajal had originally shown for the Purkinje cell. For Lorente de Nó, this fact alone was enough to show that central synapses must differ functionally, unlike the synapses of motor neurons at the neuromuscular junction.

Lorente de Nó then placed a $CA_3$ pyramidal cell within the context of its local circuitry (Fig. 1). The key feature of his model was that a single perforant path fiber from the entorhinal area synapses on the distal dendrites of the pyramidal cell, as well as on the distal dendrites of other cell types including adjacent basket cells and dentate granule cells. He developed his argument by assuming first that there is a 1:1 correspondence between incoming impulses and impulse initiation in the postsynaptic neuron. If this were true, activity in the perforant path would set up a volley in the pyramidal cell, as well as in the basket cell and granule cell, both of which also synapse on the pyramidal cell. Since distances are so short in this circuitry, volleys to the pyramidal cell from the basket and granule cells would arrive while it was still in its refractory period from the perforant path stimulation, and would thus be "wasted". And since the circuitry is much more complex than this, a large series of volleys would arrive at the pyramidal cell while it was still at some stage of its refractory period.

Clearly, Lorente de Nó argued, nature could not operate under such inefficient conditions. And while the meaning of circuitry like this (which is typical of that found throughout the brain) is not known, this arrangement suggests that dendrites conduct differently than axons. In fact, if impulses are conducted 1:1 at synapses the absurd conclusion follows that only the first impulse arriving at the pyramidal cell is effective.

This led Lorente de Nó to speculate that dendrites only generate subthreshold responses that summate and have no refractory period. When the summed response is large enough, it is detected by the axon, which fires an all-or-none impulse with a refractory period; meanwhile, summation goes on uninterrupted in the dendrites and cell body. He noted that his view was similar in many ways to Sherrington's concept of the "central excitatory state" of neurons, except that Sherrington though the refractory period involved the entire neuron (Creed et al., '32).

This prediction is remarkably close to our modern view of central neuron physiology, which awaited the impetus of the microelectrode (Brock et al., '52). Nevertheless, as early as 1934, Lorente de Nó stated that the greatest question in neurophysiology was how dendritic changes are summated, even though informed opinion at the time was skeptical of subthreshold summation in central neurons (Gasser and Graham, '32).

## The neuron as a summation apparatus

Lorente de Nó closed this remarkable paper by applying the results of his speculations about the physiology of the neuron to a classical problem: how can a point in visual space be represented at a corresponding point in the visual cortex when there is so much divergence in the pathway from the retina? As shown in Fig. 2, a point in visual space may excite a number of retinal ganglion cell (A1-A5). However, if the neuron is regarded as a summation apparatus, then these five ganglion cells may only deliver enough impulses to excite the central neurons $A'_2$-$A'_4$ in the lateral geniculate nucleus, which in turn may only activate neuron $A_3$ (Swanson et al., '87) in the visual cortex, thus providing a physiological "focusing" of activity in an anatomically divergent system. It is interesting to note in passing that the physiological focusing in this example takes place without involving inhibitory mechanisms, which Lorente de Nó mentioned only briefly.

In closing, he asked rhetorically why nature would design such a divergent system, if point-to-point transfer of information were the desired end product, and the answer was clear: the anatomical divergence allowed information to influence other system while the

Figure 2. Lorente de Nó's legend to this figure (redrawn from Fig. 36 Lorente '34) reads as follows: "Diagram explaining how point P of the retina, in spite of overlapping of the dendritic and axonal arborisations may be physiologically projected in a point P" of the cerebral cortex. The arborisations of the dendrites of the ganglionic cells of the retina ($A_1$-$A_5$) have been drawn according to Cajal's description, but only one type of cells has been considered. The arborisations in the geniculate body and cerebral cortex have been drawn according to my own (unpublished) observations".

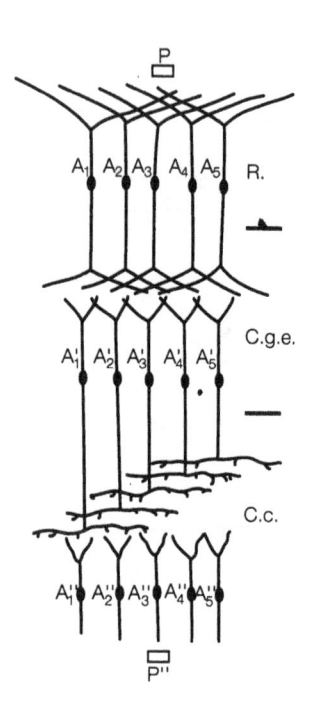

physiological properties of dendritic summation allowed functional convergence. Thus, Lorente de Nó pointed out, the so-called avalanche of conduction predicted by Cajal and Herrick strictly on the basis of anatomical divergence does not necessarily apply to the flow of information through this circuitry.

Lorente de Nó's work on the hippocampus provides one of the very best examples of how detailed neuroanatomical information may be reported, and at the same time how the possible functional significance of this information may be used to generate models that yield testable predictions about the fundamental organizing principles of neural circuitry.

REFERENCES

Brock, L.G., Coombs, J.S., and Eccles, J.C., 1952, The recording of potentials from motoneurones with an intracellular electrode, *J. Physiol.*, 117:431-460.
Cajal, S. Ramón y, 1909, 1911, "Histologie du Système nerveux de l'homme et des vertébrés, Norbert Maloine, 2 vols.", Paris.
Creed, R.S., Denny-Brown, D., Eccles, J.C., Liddell, E.G.T. and Sherrington, C.S., 1932, "Reflex activity of the spinal cord", Clarendon Press, Oxford.
Doinikow, B., 1908, Betrag zur Histologie des Ammonshorns, *J. Psychol. Neurol.*, 13.
Gasser, H.S. and Graham, H.T., 1932, Potential produced in the spinal cord by stimulation of dorsal roots, *Amer. J. Physiol.*, 103:303-320.
Lorente de Nó, R., 1922, La corteza cerebral del ratón, *Trab. Labor. Invest. Biol. Univ. Madrid*, 20.
Lorente de Nó, R., 1933, Studies on the structure of the cerebral cortex, I., The area entorhinalis, *J. Psychol. Neurol.*, 45:381-438.
Lorente de Nó, R., 1934, Studies on the structure of the cerebral cortex. II. Continuation of the study of the ammonic system, *J. Psychol. Neurol.*, 46:113-177.
Swanson, L.W., Wyss, J.M., and Cowan, W.M., 1978, An autoradiographic study of the organization of intrahippocampal association pathways in the rat, *J. Comp. Neurol.*, 181:681-716.
Swanson, L.W., Kohler, C., and Bjorklund, A., 1987, The limbic region. I: The septohippocampal system, *in*: "Handbook of Chemical Neuroanatomy, Vol. 5, Integrated Systems of the CNS, Part II, T.Hokfelt, A.Bjorklund and L.W. Swanson, eds., Elsevier, Amsterdam, pp. 125-277.
Swanson, L.W. and Kohler, C., 1986, Anatomical evidence for direct projections from the entorhinal area to the entire cortical mantle in the rat, *J. Neurosci.*, 6:3010-3023.

# THE RAT ENTORHINAL CORTEX. LIMITED CORTICAL INPUT, EXTENDED CORTICAL OUTPUT

Ricardo Insausti

Department of Anatomy, University of Navarra, Apdo. 273
31080 Pamplona, Spain

There is little doubt that one of the most outstanding contributions of Rafael Lorente de Nó to the neurosciences is the accuracy of his description of the rat entorhinal cortex almost sixty years ago. A measure of his modernity (and probably one of the reasons by which his work is almost systematically cited in every paper dealing with the entorhinal cortex or the hippocampus) is the fact that he organized the entorhinal cortex according not only to cytoarchitectonic criteria (i.e. as in Brodmann , '09) but to his observations on its projections to the hippocampus as well. In his classic paper on the ammonic system (Lorente de Nó, '33) he states "At the beginning of the present study, it was found that few problems could be solved without studying at the same time the other cortical areas tributary to the Ammon's horn, i.e. those areas which send afferent fibers to it". This statement makes a clear assertion on the importance of the knowledge of the exact sources of afferents to the hippocampal formation[1] before a comprehensive picture of its functional implications arises.

Modern tracing techniques were not available at the time Lorente de Nó made his work, so he had to rely mostly on Golgi preparations. More recent techniques, beginning with the degeneration studies of Whitlock and Nauta ('56) and other subsequent studies (Jones and Powell, '70; Leichnetz and Astruc, '75; Leichnetz and Astruc, '76; Van Hoesen and Pandya, '75; Van Hoesen et al., '75) revealed that the entorhinal cortex was not devoid of interconnections with the neocortex. Further anatomical studies were adding more information on various sources of afferents to the entorhinal cortex (Amaral et al., '83; Goldman-Rakic et al., '84; Insausti et al., '87; Mesulam, '76; Room and Groenewegen, '86; Turner et al., '80). However, most of the information comes from studies in primates and cat, and fewer reports exist on the cortical inputs to the entorhinal cortex in the rodent. It was not until 1979 that Beckstead ('79) reported in a series of experiments using autoradiographic experiments that an indication of direct projections from neocortex to the rat entorhinal cortex could be demonstrated. Curiously, one year earlier, the same author (Beckstead, '78) reported no retrogradely labeled cells in any portion of the neocortex after HRP injections into the rat entorhinal cortex. Only the olfactory projection to the entorhinal cortex originating in the olfactory bulb and piriform cortex was well documented (Beckstead, '78; Kosel et al., '81).

---

[1] The term hippocampal formation applied here includes the entorhinal cortex, the dentate gyrus, the CA fields and the subicular complex.

Ancillary information on the projection from neocortex to the entorhinal cortex can be obtained from various reports on which different cortical areas were injected with anterograde tracers. Among these regions it is better documented the projection from the anterior insular cortex (Saper, '82; Yasui et al., '91). A similar projection has also been described in the hamster (Reep and Winans, '82). The frontal cortex also projects to the entorhinal cortex in the rat. The projections originate in both prelimbic (Beckstead, '79) and infralimbic (Swanson et al., '87), as well as in anterior cingulate and orbital frontal cortices (Beckstead, '79; Deacon et al., '83). Some of these projections have also been observed in the guinea pig (Sorensen, '85). More posterior parts of the frontal cortex also have projections to the entorhinal cortex such as medial and lateral agranular cortices, the former being more abundant than the latter (Deacon et al., '83; Reep et al., '87). Projections from perirhinal cortex in the temporal cortex seem to be the most abundant and they have been reported in the rat (Deacon et al., '83; Kohler, '83).

The entorhinal cortex is in a position of privilege to rely to the hippocampus the neocortical information. On the other hand, after receiving the hippocampal output the entorhinal cortex returns the information to the cerebral cortex and other subcortical structures. The limited direct input from cortical areas stands in contrast to the extended projections reported in the rat (Kosel et al., '81; Sarter and Markowitsch, '85; Swanson and Kohler, '86; Vaudano et al., '91). The functional significance of such an uneven reciprocal relationship between the entorhinal cortex and the neocortex is not known. It is generally accepted that the hippocampal formation, by way of its cortical input mediated through the entorhinal cortex, receives a complex information. Any disruption of the connectional links in the hippocampal formation leads to amnesia although the mechanisms are far from being understood (Zola-Morgan et al., '86).

In this context, it thus seemed necessary to evaluate with a sensitive tracing method the cortical input of the rat entorhinal cortex keeping in mind topographical differences and the extended cortical output.

The rat entorhinal cortex lies along the ventrolateral rim of the cerebral hemisphere, extending medially along the ventral surface of the brain, as far as the hippocampal fissure. Therefore, a conventional stereotaxic approach has to pass through the whole dorsoventral extent of the hemisphere before reaching the entorhinal cortex. To circumvent this shortcoming, the experiments described below were performed by exposing directly the cortical surface of the entorhinal cortex using an adaptation of the technique of Powell et al. ('65). Briefly, male young adult Wistar rats were anesthetized with sodium pentobarbital (32 mg/kg) and placed lying on its right side. The temporal muscle was detached, a burr hole drilled in the temporal bone and the cortical surface above and below the rhinal fissure exposed. Deposits of 45 nl of a 1% solution of wheat germ-agglutinin conjugated horseradish peroxidase were made with air pulses through glass micropipettes (o.d. 30 $\mu$m). In addition, small deposits were made by iontophoresis with an intensity of 8-10 $\mu$A for 20 minutes at intervals of 7 seconds. After a survival of 20-24 h the animals were deeply anesthetized with sodium thiopental and perfused transcardially with a solution of 1% paraformaldehyde and 1.25% glutaraldehyde in 0.1M phosphate buffer at pH 7.4. The brain was stored overnight in a 30% sucrose solution in 0.1M phosphate buffer, and sectioned coronally at 50 $\mu$m in a freezing microtome. Three series of one-in-five sections were immediately mounted. Two adjacent series were reacted according to Mesulam's protocol ('76). One of the reacted series was counterstained with Richardson's stain. The third series was stained with thionin to determine the cytoarchitectonic fields. The series were analyzed with bright- and dark-field microscopy and the labeled cells charted with an X-Y recording system coupled to the microscope stage.

Our previous cytoarchitectonic analysis of the rat entorhinal cortex (Insausti, Witter and Herrero, in preparation) resulted in the six separate divisions represented in the two-dimensional reconstruction (Van Essen and Maunsell, '80) (Fig. 1). Each one of these divi-

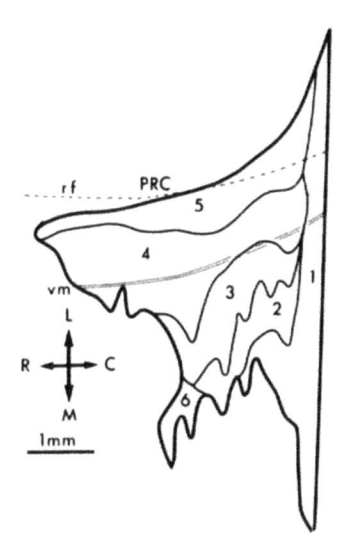

**Figure 1.** Two-dimensional reconstruction of the rat's entorhinal cortex. The arrows indicate the general direction of the cortical surface. Abbreviations: C, caudal; L, lateral, M, medial, PRC, perirhinal cortex; R, rostral; rf, rhinal fissure; vm, ventromedial border of the cerebral hemisphere.

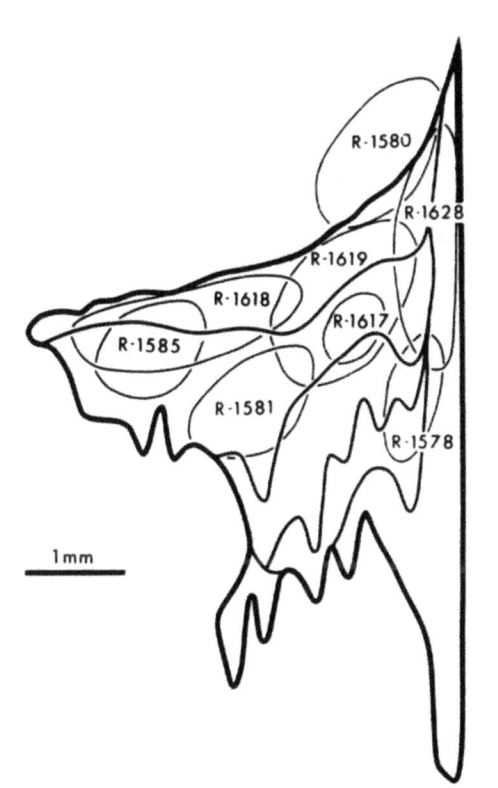

**Figure 2.** Representation of the WGA-HRP deposit extent of some selected cases used in this study. The numbers of each experiment are given, and they were plotted onto the two-dimensional map of Fig. 1. Cases R-1618, R-1628 and R-1578 are represented in Figs. 3, 5 and 7.

sions has a definite projection to the molecular layer of the dentate gyrus: the division closest to the rhinal fissure and the caudalmost pole of the entorhinal cortex (5 and 1 in Fig. 1) project to the septal pole of the dentate gyrus, while the medialmost divisions (3 and 6 in Fig. 1) project to the temporal extreme of the dentate gyrus. The remainder division (4 in Fig. 1) projects to intermediate portions of the dentate gyrus.

The extent of the injection sites of some of the cases is depicted in Figure 2. Some of the deposits were aimed at the lateral part of the entorhinal cortex (that is, closer to the

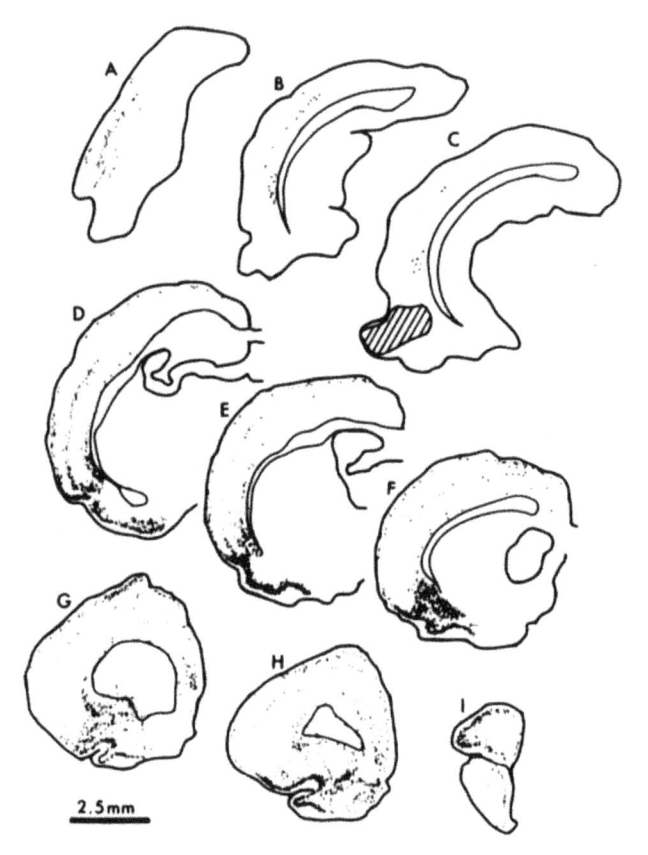

**Figure 3.** Line drawings of a series of sections from caudal (A) to rostral (I) on which the location of retrogradely labeled neurons has been plotted after a deposit of WGA-HRP into the lateral portion of the entorhinal cortex. Notice the abundant labeling in temporal and frontal cortices.

rhinal fissure). Of these, case R-1618 is the most extensive and will be presented in more detail. The posterior limit of the entorhinal cortex was injected in another group, from which case R-1628 has been selected as representative. Finally, other deposits were aimed at the ventromedial part of the entorhinal cortex, from which case R-1578 is the most representative. Additional information was available from other experiments with retrograde tracers injected in perirhinal cortex that we define as the band of cortex adjacent to the rhinal fissure that does not project to the dentate gyrus.

The analysis of the retrogradely labeled cells revealed a number of cortical areas that seem to project directly to the entorhinal cortex. The distribution of the cortical innervation was not, however, homogeneous throughout the entorhinal cortex, as for instance, the projections from the olfactory bulb (Kosel et al., '81). The lateral part of the entorhinal cortex (division 5 in Fig. 1, that sends its axons to the septal part of the dentate gyrus) received the most abundant cortical innervation. The caudal part of the entorhinal cortex (division 1 in Fig. 1, that also sends its axons to the septal part of the dentate gyrus) received a somewhat more restricted cortical input, only partially overlapping with that of division 5. The medial part of the entorhinal cortex (divisions 2 and 3 in Fig. 1, that innervate the temporal extreme of the dentate gyrus) was the recipient of a rather minor cortical input. Finally, the intermediate part had a cortical innervation from a few areas, all of which projected to division 5.

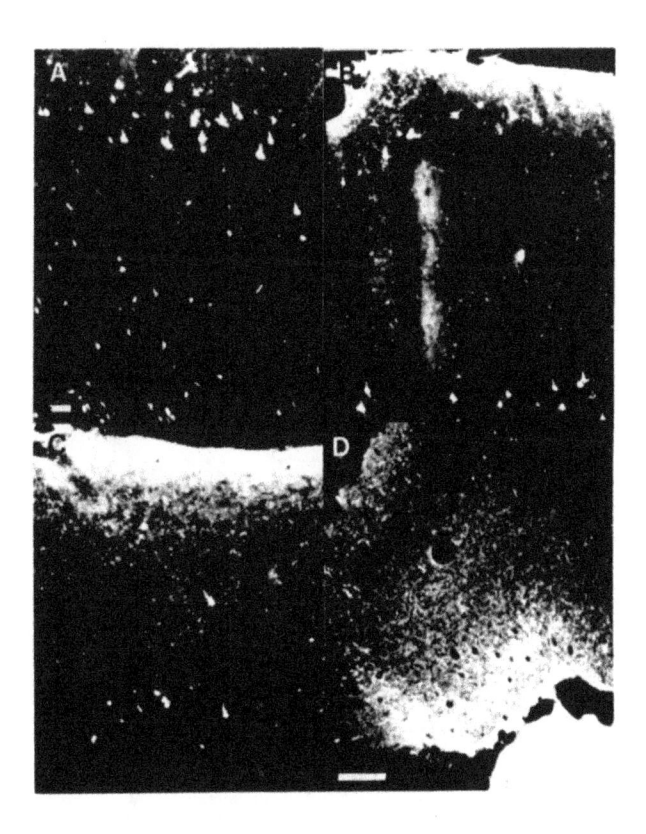

**Figure 4.** Darkfield photomicrographs of the labeling obtained in case R-1628 with a WGA-HRP deposit in the lateral part of the entorhinal cortex. In A, labeled neurons in perirhinal cortex (layers V and VI) are shown. In B, an upper row of labeled neurons in layers II and upper III, and a lower row of layer V neurons can be appreciated. In C, a group of labeled neurons in layer V of the somatosensory cortex is presented. In D, a low power photomicrograph of the anterior orbital cortex with the labeling present is shown. Also, anterograde transport in layer I can be observed in B, C and D. Scale bar in A, B and C is 50 μm, in D is 250 μm.

The distribution of retrogradely labeled neurons in case R-1618 (lateral part of the entorhinal cortex) is represented in Figure 3 in a series of sections arranged from caudal to rostral.

The majority of the projections were located in the temporal cortex adjacent to the rhinal fissure and slightly rostral to the injection site. Most of the retrogradely labeled neurons fell in the territory of perirhinal cortex (Fig. 4A), but they were also very abundant in the adjacent temporal cortex including areas Te2 (Zilles and Wree, '85) (Fig. 4B) and primary auditory cortex. This band of labeled cells was continuous rostrally with the cortex along the rhinal fissure, mostly the agranular insular cortex. Fewer cells were present in more dorsal parts of the cerebral cortex as parietal cortex (somatosensory, Fig. 4C) and lateral agranular cortex (motor). A somewhat higher concentration of labeled cells was present in medial agranular cortex. Only occasional labeled cells were seen in the cingulate cortex. The number of retrogradely labeled cells increased in the anterior frontal cortex where they were present in both medial and lateral orbital cortices and also in the frontal tip (Fig. 4D). Both prelimbic and infralimbic cortices contained a moderate amount of labeled neurons. Caudal to the injection site labeled neurons were present in caudal temporal cortex Te3 and visual cortex (area 18). No labeled cells were observed in retrosplenial cortex. A different pattern of cortical labeling was noticed between the regions located caudally compared to those rostral to the injection site. The former had most of the labeled cells in layers V and VI while fewer were in layer II and upper layer III. In contrast, the temporal cortex contained labeled

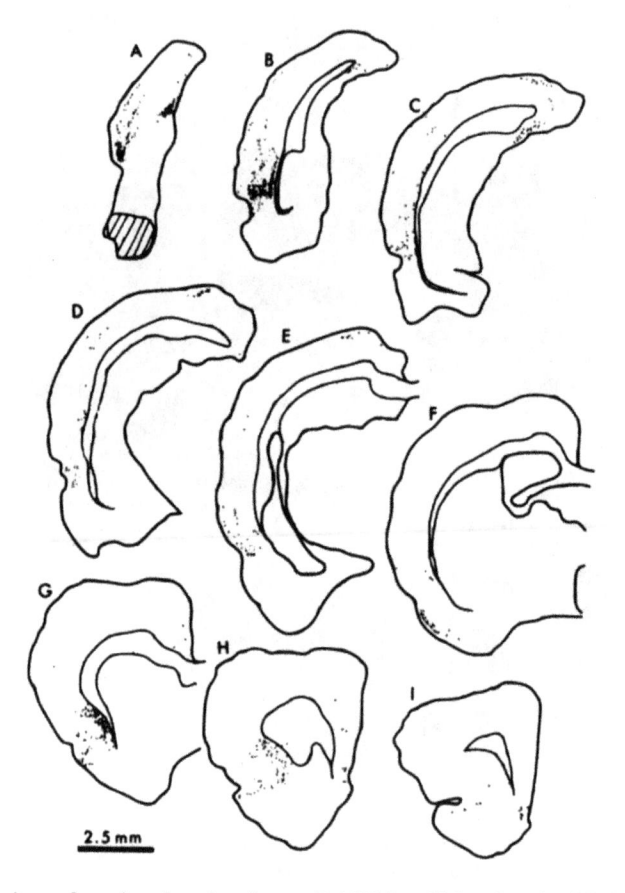

**Figure 5.** Line drawings of a series of sections in case R-1628 in which a deposit of WGA-HRP was made at the caudal extreme of the entorhinal cortex. The sections are arranged from caudal (A) to rostral (I). Notice the less dense labeling, especially in the anterior half of the brain.

neurons in layer II and upper layer III that were as numerous as those in layers V and VI. Deep labeled neurons (layers V and VI) outnumbered the superficial ones (layers II and III) in somatosensory, motor and anterior frontal areas. In agranular insular, infralimbic and prelimbic cortices they were distributed throughout the cortical thickness without any predominant layer.

The distribution of retrogradely labeled neurons after deposits in the caudalmost tip of the entorhinal cortex (division 1 in Fig. 1) is represented in case R-1628 (Figs. 2 and 5). The injection site is shown in Fig. 6A. The pattern of labeling was substantially different to that seen after deposits of the lateral part of the entorhinal cortex. By far, perirhinal cortex (Fig. 6B) contained the vast majority of retrogradely labeled cells. As in the previous case, the highest density of cells was located slightly rostral to the injection site (Fig. 5B). Most of the labeled neurons were located in layers V and VI, while a substantially lower density was present in layers II and upper III. Further rostrally the overall amount of labeled neurons in perirhinal cortex decreased dramatically. Other cortical regions coincident with the previous case were the anterior insular and anterior cingulate cortices. Both prelimbic and infralimbic cortices had only occasional neurons. No labeled neurons were seen in anterior frontal cortex (orbital and dorsolateral cortex). Far fewer labeled neurons were observed in temporal association cortex (Fig. 6D) or in parietal and lateral frontal cortices (almost exclusively in layer V). In contrast, groups of neurons were labeled in the ventral part of

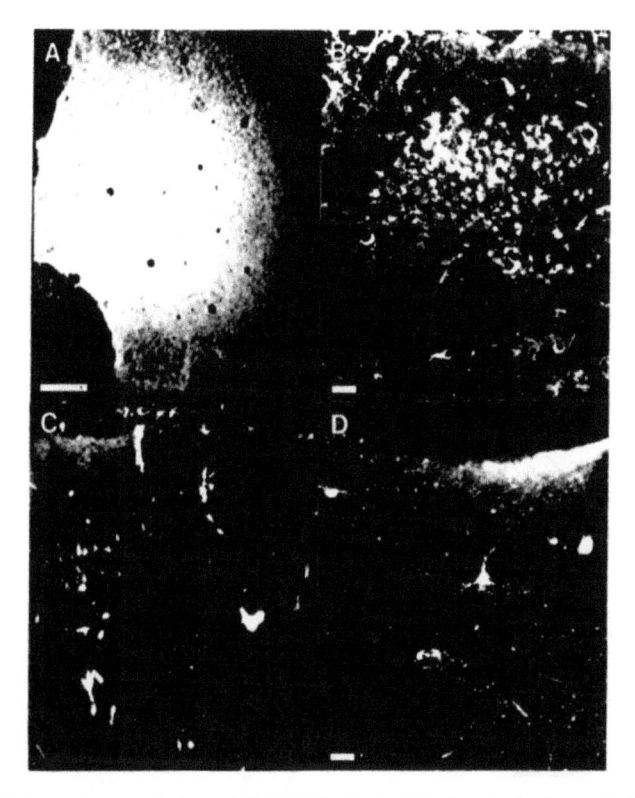

**Figure 6.** Darkfield photomicrographs in case R-1628. In A, the injection site is shown. In B, a group of labeled neurons in layers III through V of perirhinal cortex can be seen. In C, labeled neurons in layer III of medial agranular cortex can be noticed. In D, a detail of a labeled neuron in layer II of temporal neocortex is presented. Scale bar in A is 250 µm, in B, C and D is 50 µm.

retrosplenial cortex (Fig. 5B and C) and in the posterior part of medial agranular field, where they extended cranially for some distance. The cells were located exclusively in layers II and III (Fig. 6C). The previous group of experiments in which the lateral band (division 5 in Fig. 1) of the entorhinal cortex was injected, did not present retrogradely labeled cells neither in retrosplenial cortex nor in the posterior part of the medial agranular field. Therefore these two cortical regions seem to be sending their output exclusively to the caudal extreme of the entorhinal cortex.

The third differential pattern of labeling was observed after deposits in medial portions of the entorhinal cortex. One of these cases is represented in Figure 7 (case R-1578, Fig.2). The first difference compared to the other two previous patterns described above was the paucity of labeling in the cortex. However, it was unquestionably present, albeit in reduced numbers, in temporal cortex (perirhinal cortex and temporal association cortex Te3, Fig. 7) slightly rostral to the injection site. Only occasional labeled neurons were observed in visual association cortex. In clear contrast to the two previous groups, the rostral half of the cerebral cortex, including agranular insular cortex, infralimbic, prelimbic and frontal orbital cortices had no labeled neurons.

It is now generally accepted that the entorhinal cortex is an interface between neocortex and the hippocampus proper, as well as that the hippocampal formation is necessary for the sensorial impressions to be permanently stored in a repository site, probably in the neocortex. This function is of an extreme importance for learning strategies of all organisms, and represents the substrate of the flexibility of adaptation to the changing and challenging

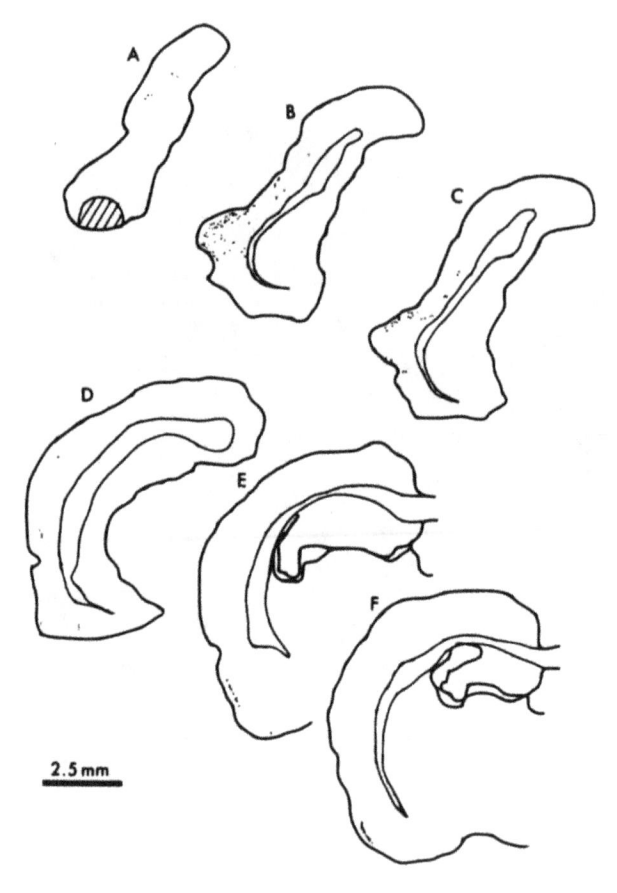

**Figure 7.** Series of line drawings arranged from caudal (A) to rostral (F) in case R-1578 in which the medial part of the entorhinal cortex was injected. It is striking the paucity of retrogradely labeled neurons, most of them located in perirhinal cortex slightly rostral to the injection site.

conditions of the environment. How does the brain perform this process is barely grasped. The substrate of the anatomical relationships of the entorhinal cortex with the neocortex is presented here as an attempt to better define the structure of the entorhinal cortex (cytoarchitectonic parcellation, that is just a supplement to the divisions outlined by Lorente de Nó almost sixty years ago) as well as to understand better the functional role of the entorhinal cortex.

From the results stated above, it seems to be well established that the entorhinal cortex distributes fibers throughout the cerebral cortex (Kosel et al., '82; Sarter and Markowitsch, '85; Swanson and Kohler, '86; Vaudano et al., '91), while it receives projections from the cortex in a much more restricted fashion. This characteristic of the entorhinal cortex applies for much of its extent. However, the lateral band of the entorhinal cortex, closer to the fundus of the rhinal fissure, receives a substantially denser cortical innervation. This topographical trend in the density of cortical projections is also present in cat (Room and Groenewegen, '86) and monkey (Insausti et al., '87; Insausti and Amaral, '88). Previous work in rats support our analysis as they show different projections to the entorhinal cortex originating in temporal cortex (Deacon et al., '83; Kohler et al., '86), anterior agranular insular cortex (Saper, '82; Yasui et al., '91), medial frontal cortex (Beckstead, '79; Deacon et al., '83; Swanson et al., '87) and medial and lateral agranular cortices (Deacon et al., '83; Reep et al., '87). It is worth noting that all these data are based in experiments using anterograde tracers, and therefore, while they offer some information about the topographical and laminar distribution of the cortical afferents to the rat entorhinal cortex, they do not cast any light on the origin and areal distribution of such projection. The use of restricted injections of the sensitive retrograde tracer WGA-HRP helps to disclose a topographical pattern that can be summarized in the statement that the neocortical input can reach the septal portion of the dentate gyrus through a relay in the entorhinal cortex. In fact, our own unpublished results and others (Pohle and Ott, '84; Ruth et al., '88) indicate that the entorhinal strip recipient of the most abundant cortical input is projecting to the septal portion of the hippocampus.

It is well beyond this study the disclosure of what the functional significance of such projections might be, but it seems safe to assume that whatever information is relayed through the projections of the neocortex to the hippocampus, it has to be related to the functions on which the hippocampus itself is implicated. In the light of present evidence memory processing seems to be if not the most important, at least the best known.

ACKNOWLEDGEMENTS

This work has been supported by a grant from the Department of Health of the Government of Navarra (Spain). The technical support of P. Botín, M.A. Erdozáin, C. García Gortari and J. García is appreciated.

REFERENCES

Amaral, D.G., Insausti, R. and Cowan, W.M., 1983, Evidence for a direct projection from the superior temporal gyrus to the entorhinal cortex in the monkey, *Brain Res.*, 275:263.

Brodmann, K., 1909, Verchleichende Lokalisationslehre der Grosshirnrinde. Verlag von Johann Ambrosius Barth, Leipzig.

Beckstead, R.M., 1978, Afferent connections of the entorhinal area in the rat as demonstrated by retrograde cell-labeling with horseradish peroxidase, *Brain Res.*, 152:249.

Beckstead, R.M., 1979, An autoradiographic examination of corticocortical and subcortical projections of the mediodorsal-projection (Prefrontal) cortex in the rat, *J. Comp. Neurol.*, 184:43.

Deacon, T.W., Eichenbaum, H., Rosenberg, P. and Eckmann, K.W., 1983, Afferent connections of the perirhinal cortex in the rat, *J. Comp. Neurol.*, 220:168.

Goldman-Rakic, P.S., Selemon, L.D., and Schwartz, M.L., 1984, Dual pathways connecting the dorsolateral prefrontal cortex with the hippocampal formation and parahippocampal cortex in the rhesus monkey, *Neuroscience*, 12:719.

Insausti, R. Amaral, D.G. and Cowan, W.M., 1987, The entorhinal cortex of the monkey.II. Cortical afferents, *J. Comp. Neurol.*, 264:356.

Insausti, R. and Amaral, D.G., 1988, Distribution of cortical projections to the monkey entorhinal cortex: an autoradiographic study, *Soc. Neurosci. Abstr.*, 14: 858.

Jones, E.G. and Powell, T.P.S., 1970, An anatomical study of converging sensory pathways within the cerebral cortex of the monkey, *Brain*, 93:793.

Kohler, C., Smialowska, M., Eriksson, L.G., Chan-Palay, V., and Davies, S., 1986, Origin of the neuropeptide Y innervation of the rat retrohippocampal region, *Neurosci. Lett.*, 65:287.

Kosel, K.C., Van Hoesen, G.W. and West, J.R., 1981, Olfactory bulb projections to the parahippocampal area of the rat. *J. Comp. Neurol.*, 198:467.

Kosel, K.C., Van Hoesen, G.W. and Rosene, D.L., 1982, Non-hippocampal cortical projections from the entorhinal cortex in the rat and rhesus monkey, *Brain Res.*, 244:201.

Leichnetz, G.R. and Astruc, J., 1975, Efferent connections of the orbitofrontal cortex in the marmoset (Saquinus oedipus), *Brain Res.*, 84:169.

Leichnetz, G.R. and Astruc, J., 1976, The squirrel monkey entorhinal cortex: architecture and medial frontal afferents, *Brain Res. Bull.*, 1:351.

Lorente de Nó, R., 1933, Studies on the structure of the cerebral cortex. I. The area entorhinalis, *J. Psychol. Neurol.*, 45:381.

Lorente de Nó, R., 1934, Studies on the structure of the cerebral cortex. II. Continuation of the study of the ammonic system, *J. Psychol. Neurol.*, 46:113.

Mesulam, M.M., 1976, The blue reaction product in horseradish peroxidase histochemistry. Incubation parameters and visibility, *J. Histochem. Cytochem.*, 24:1273.

Mesulam, M.M. and Mufson, E.J., 1982, Insula of the Old World monkey. III. Efferent cortical output and comments on function, *J. Comp. Neurol.*, 2112:38.

Pohle, W. and Ott, T., 1984, Localization of entorhinal cortex neurons projecting to the dorsal hippocampal formation. A stereotaxic tool in three dimensions, *J. Hirnforsch.*, 25:661.

Powell, T.P.S., Cowan, W.M. and Raisman, G., 1965, The central olfactory connections, *J. Anat. (Lond.)*, 99:791.

Reep, R.L. and Winans, S.S., 1982, Efferent connections of dorsal and ventral agranular insular cortex in the hamster, Mesocricetus auratus, *Neuroscience*, 11:2609.

Reep, R.L., Corwin, J.V., Hashimoto, A., and Warson, R.T., 1987, Efferent connections of the rostral portion of medial agranular cortex in rats, *Brain Res. Bull.*, 19:203.

Room, P. and Groenewegen, H.J., 1986, The connections of the parahippocampal cortex in the cat. I. Cortical afferents, *J. Comp. Neurol.*, 251:415.

Ruth, R.E., Collier, T.J., and Routtenberg, A., 1988, Topographical relationship between the entorhinal cortex and the septotemporal axis of the dentate gyrus in rats. II. Cells projecting from lateral entorhinal cortex, *J. Comp. Neurol.*, 270: 506.

Saper, C.B., 1982, Convergence of autonomic and limbic connections in the insular cortex of the rat, *J. Comp. Neurol.*, 210:163.

Sarter, M. and Markowitsch, H.J., 1985, Convergence of intra- and interhemispheric cortical afferents: Lack of collateralization and evidence for a subrhinal cell group projecting heterotopically, *J. Comp. Neurol.*, 236:283.

Sorensen, K.E., 1985, Projections of the entorhinal area to the striatum, nucleus accumbens, and cerebral cortex in the guinea pig, *J. Comp. Neurol.*, 238:308.

Swanson, L.W., Kohler, C. and Bjorklund, A., 1987, The limbic region. I. The septohippocampal system, *in*: "Handbook of Chemical Neuroanatomy vol. 5: Integrated systems of the CNS, Part I, Hypothalamus, Hippocampus, Amygdala and Retina", A. Bjorklund, T. Hokfelt and L.W.Swanson, eds., Elsevier, Amsterdam.

Swanson, L.W. and Kohler, C., 1986, Anatomical evidence for direct projections from the entorhinal area to the entire cortical mantle in the rat, *J. Neurosci.*, 6:3010.

Turner, B.H., Mishkin, M., and Knapp, M., 1980, Organization of the amygdalopetal projections from modality-specific cortical association areas in the monkey, *J. Comp. Neurol.*, 191:515.

Van Essen, D.C. and Maunsell, J.H.R., 1980, Two-dimensional maps of the cerebral cortex, *J. Comp. Neurol.*, 191:255.

Van Hoesen, G.W. and Pandya, D.N., 1975, Some connections of the entorhinal (area 28) and perirhinal (area 35) cortices of the rhesus monkey. I. Temporal lobe afferents, *Brain Res.*, 95:1.

Van Hoesen, G.W., Pandya, D.N. and Butters, N., 1975, Some connections of the entorhinal (area 28) and perirhinal (area 35) cortices of the rhesus monkey. II. Frontal lobe afferents, *Brain Res.*, 95:25.

Vaudano, E., Legg, C.R., and Glickstein, M., 1991, Afferent and efferent connections of temporal association cortex in the rat: a horseradish peroxidase study, *Eur. J. Neurosci.*, 3:317-330.

Whitlock, D.G. and Nauta, W.J.H., 1956, Subcortical projections from the temporal neocortex in Macaca mulatta, *J. Comp. Neurol.*, 106:183.

Yasui, Y., Breder, C.D., Saper, C.B., and Cechetto, D.F., 1991, Autonomic responses and efferent pathways from the insular cortex in the rat, *J. Comp. Neurol.*, 303:355.

Zilles, K. and Wree, A., 1985, Cortex: areal and laminar structure, *in*: "The Rat Nervous System". Vol. 1, G. Paxinos, ed., Academic Press, San Diego.

Zola-Morgan, S., Squire, L.R., and Amaral, D.G., 1986, Human amnesia and the medial temporal region: Enduring memory impairment following a bilateral lesion limiterd to field CA1 of the hippocampus, *J. Neurosci.*, 6:2950.

# AXONAL PATTERNS OF INTERNEURONS IN THE CEREBRAL CORTEX: IN MEMORY OF RAFAEL LORENTE DE NÓ

Alfonso Fairén

Instituto Cajal, CSIC, Madrid, Spain

It is not a futile exercise to analyze the scientific accomplishments of our predecessors; it may serve to better understand our own intellectual attitudes towards science. We did so for Rafael Lorente de Nó when we were in the classroom of Francisco de Salinas, a professor of music of the Spanish *Edad de Oro*, celebrated by Fray Luis de León:

> El aire se serena
> y viste de hermosura y luz no usada,
> Salinas, quando suena
> la música estremada
> por vuestra sabia mano governada[1]

In retrospect, after having witnessed throughout the workshop the pervading influence of Lorente in the field of auditory neurobiology, we feel that there would have not been a better place for our homage than such a classroom.

Rafael Lorente de Nó was one of the first modern neurobiologists; he was also one of a handful of scientists with a pluridisciplinary personal inclination. Lorente, the last distinguished disciple of Santiago Ramón y Cajal, developed most of his scientific career beyond the frontiers of his native Spain. It was pertinent, though, that this international homage were celebrated in Spain, for we Spaniards are in debt to Lorente de Nó. This is partly because the science authorities in postwar Spain refused to attract Lorente (Gallego, '90) - one of the many political decisions of those years that obviously had enormous negative influence on the development of science in this land. We are also in debt to Lorente because the latest generations of Spanish neuroscientists did not search, and did not benefit from, the contact with a man who would have spread his scientific inspiration and hindsight. Doubtless, his sense of critical discussions and his willingness to maintain his ideas were at times to the point of being obstinate.

The objective analysis of his scientific contributions, as well as the analysis of the influence of these contributions on the science of others, is the best way to render justice to the labor of Lorente de Nó. We analyzed his work on auditory (Dr. Osen) and on vestibular

---

[1] Fray Luis de León, A Francisco de Salinas, 1990, *in*: "Poesías Completas", J.M.Blecua, ed., Gredos, Madrid.

neuroanatomy (Dr. Honrubia), and then moved to other issues which had been a matter of Lorente's concern. From his studies on the cerebral cortex, which have given Lorente a perennial reputation as a giant of neuroanatomy, we went to the issues which involved him in hard and enduring controversies that, unfortunately, eclipsed most of his positive findings (see Gallego, Kruger, this Volume). We hope that the results of this endeavor would have pleased Lorente.

The impressive advances in the understanding of the neuronal organization of the cortex during the last decade places us in a vantage point to better recognize the relevance of the pioneering efforts by Lorente de Nó. In this essay, which will be a reflection on the cerebral cortex, I focus on Lorente's thoughts on the local cortical circuits; in separate chapters, Woolsey, Insausti and Swanson review other aspects of the concepts of Lorente on the cerebral cortex.

Lorente, while still a medical student, joined Cajal in his studies on the fine structure of the brain, using the Golgi method extensively. This style of research was not the most common one in Cajal's laboratory at that time, with the exception of the studies of Cajal on the retina of invertebrates.

"The enduring work by Professor Cajal with the Golgi method ceased in 1903, when he set about to profit from the enormous resolving power of his own reduced silver nitrate method. But, nowadays, he has retraced his own steps and he is exploring again the fine structure of the nerve centres; we have followed him along this line, and the present work is, rather than a sample of our labors, a breath of his direction and example" (Lorente de Nó, '22).

Along this line, Lorente published several papers on brain stem organization and started his cortical studies at the same time as Cajal (Cajal, '22bis); we may suppose there was a mutual influence that favored such accomplishments. The first article of Lorente on cortical organization, a formidable account that we shall analyze in detail (see also Woolsey, this volume), was published when he was only 20 years old (Lorente de Nó, '22).

This excerpt from Pio del Río Hortega ('86) gives a glance at young Lorente's personality:

"Frequently, the unfortunate people who had a poor memory (as was my case) were in a disadvantageous position [in the laboratory] in those times; this was made even worse not only because Villaverde possessed a good memory, but so did an Aragonese fellow just incorporated into the Cajal school. His name was Lorente de Nó; he was a third year medical student who had just moved from Zaragoza to Madrid after a dispute with I do not know which professor in Zaragoza. He was endowed with both the affection and the aptitude for histology, and had already published a paper on work done with such a model of a fatuous professor as was Luis del Río. He spoke in a mood of superiority, standing on his toes to speak to everybody from over his shoulder, a characteristic that alienated a great deal of sympathy from him. He had, however, quite a good reason to be proud, for nobody else knew in such minute detail the works by Cajal and, above all, the monumental 'Textura del sistema nervioso'."

"A true child prodigy, gifted with such a good memory, that intelligent youth held a lot of promise; because of that, and also because of being Aragonese (a not negligible circumstance) and having credentials from D[on] Pedro Ramón y Cajal, he had a hearty welcome from Tello and D[on] Santiago. [Don Santiago] was delighted listening to him speaking about nuclei and nerve pathways and supporting his opinions, like a recognized scholar, about any question under dispute. He worked in the Faculty of Medicine but, frequently, he came to me for consultations, which I attended with my best wishes."

These words, indeed not very kind to Lorente, must be understood in the context in which they were written: the scientific and human conflict between Cajal and del Rio Hortega that arose precisely at about that time. The comments of del Rio Hortega are interesting for their freshness; on their basis, we may better imagine the argumentative character of Lorente's discussions in his first paper on the cerebral cortex, or the apparent paradox of the positive interaction between Cajal and Lorente, freed of any hierarchical prejudices (Kruger and Woolsey, '90).

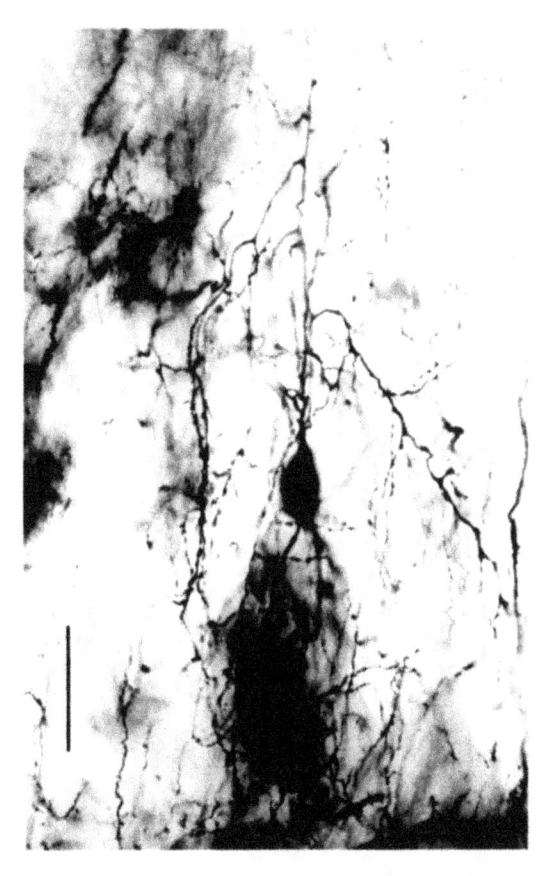

**Figure 1.** A short-axon cell in the primary somatosensory cortex of a 12 days-old mouse. Golgi impregnation by R. Lorente de Nó. Scale bar, 50 μm.

Lorente's neuroanatomical studies were based strongly on the Golgi method. His own Golgi preparations have not been available, but I have recently been fortunate enough to identify one single slide from Lorente in the collection of Cajal preparations of the Museum. The preparation is well preserved, and its quality is excellent. One interneuron from such a slide is shown in Figure 1; its axonal arborization can be compared to those illustrated by Lorente himself (Lorente de Nó, '22) (see Fig. 4).

The Golgi method is excellent for dealing with the organization of short-axon neurons. It provides us with sufficient morphological information as a base for a systematic classification of such cells. Classification of neurons is a complex task (for discussion, see Fairén et al., '84; Fairén and Smith-Fernández, '92) which requires the adoption of certain preconceived criteria. One such criterion, which we now know bears functional significance, may be the distribution of the axon, and then either its general spatial orientation, or the detailed distribution of its terminal synaptic elements among the diverse postsynaptic partners. In classifying cortical nerve cells, Lorente chose this first criterion in his 1938 work (Lorente de Nó, '38b). Previously, in 1922, he limited himself to the enumeration of the different neurons and to an exhaustive description of their characteristics. Thus, some statements in the literature, suggesting that Lorente distinguished an exaggerated number of different neuronal types in his 1922 paper (Lorente de Nó, '22b), are probably unfounded.

Today's classifications of cortical neurons dwell strongly on the distribution of the terminal axonal branches (Fairén et al., '84; De Felipe and Fariñas, '92; Fairén and Smith-Fernández, '92). Unlike, for instance, the cerebellum, neurons in the cerebral cortex seldom

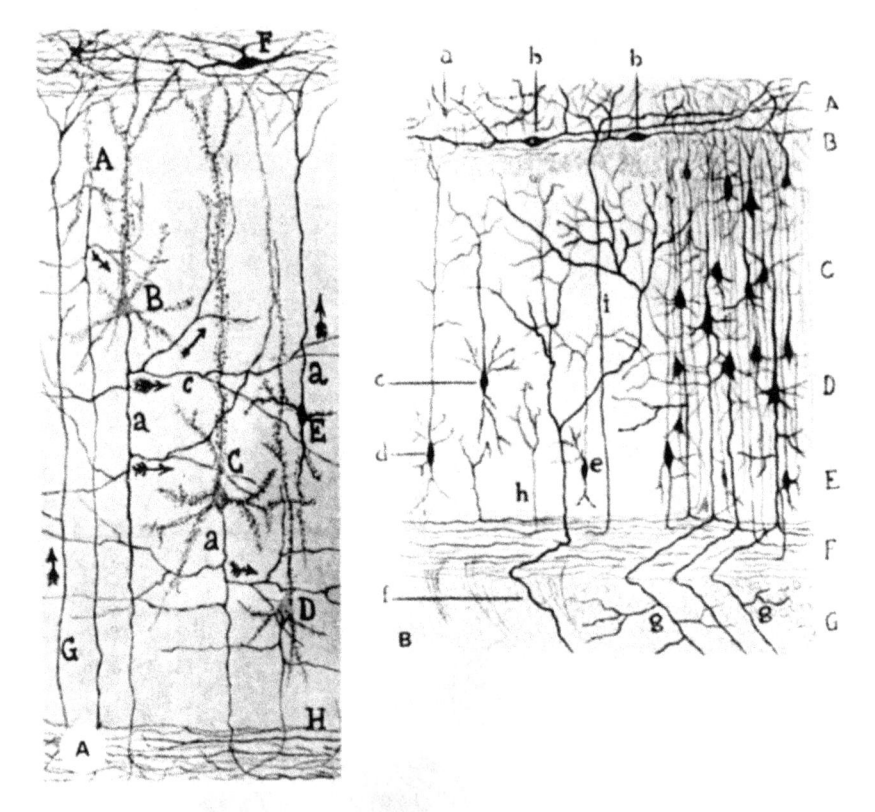

**Figure 2.** The rodent cerebral cortex in Cajal's drawings. A. A-D, pyramidal cells; E, cell with ascending axon; F, "special cells of the first layer of the cerebral cortex"; G, afferent fiber; H, white matter; a, axons; b, "nervous collaterals which appear to cross and touch the dendrites and the trunks [apical shafts] of the pyramids". "The arrows mark the supposed direction of the nervous current." (Cajal, '54). B. "In this figure, I have put together some of my findings in 1890 and 1891." a, small stellate cells of layer I; b, horizontal, fusiform cells of layer I; c - e, cells with ascending axon; f, "a terminal ramification of sensitive fibers"; h, [i], "collaterals from white matter"; g, "collaterals from pyramidal cell axons destined for the corpus striatum." A - G, cortical layers. (Cajal, '23).

have conspicuous terminal formations, so that synaptic distribution needs to be assessed using complementary techniques. Golgi-electron microscopy (Blackstad, '65; Blackstad, '75; Fairén et al., '77), immunocytochemistry and electro-physiological analysis have led to quite a detailed knowledge of cortical circuits (see Peters and Jones, '84; White, '89; De Felipe and Fariñas, '92; Fairén and Smith-Fernández, '92 for review).

How can one appreciate the mutual connections established within an ensemble of neurons, some of them visualized with the Golgi method? Both Cajal ('54) and Lorente ('38b) ventured some interpretations on synaptic connectivity within the cerebral cortex, but this problem may have been elusive in the absence of direct evidence. The Golgi method, however, allows us "...to see, however dimly, the general model of interconnections, and the putative consequences of such linkage patterns" (Scheibel and Scheibel, '78). An example from Cajal's last work ('54) is shown in Figure 2A. This figure refers to the cerebral cortex of the mouse, where Cajal did not accept the existence of sophisticated types of short-axon cells. This is further illustrated in Figure 2B (taken from Ramón y Cajal, '23). The two types of neurons with intracortical axons that Cajal illustrated are the horizontal cells in layer I and the cells with ascending axons. Although the details of connectivity suggested by Cajal have proven to be true by modern studies, Lorente suggested in 1938 the existence of many more different populations of short-axon cells even in the rodent cortex, and he postulated a much more complicated pattern of intracortical relationships (Lorente de Nó, '38b) that is designed as a wiring diagram (Fig. 6). This is clearly an anticipation of recent circuit models (White, '89; Shepherd, '90). I shall discuss these aspects later.

The Scheibels ('78) have pointed out very pertinently that the Golgi is a methodology that lends itself to functional interpretations. The Golgi method provided Lorente with a vision of the circuit diagrams in the brain. This led him to envisage the brain as an ensemble of neuronal chains, modeled after the reflex arcs (Lorente de Nó, '33a; Lorente de Nó, '38a,b). These chains were not linear (serial) chains of neurons. To the complexity generally accepted at his time, he added the chains of internuncial neurons superimposed upon the simple reflex arc (Lorente de Nó, 38a). For Lorente, internuncial neurons were not only the diverse varieties of short-axon cells but also, quite adequately, the projecting neurons which may have an extensive, local axonal arborization. He understood (Lorente de Nó, '33; Lorente de Nó, '33a,b) that neuronal chains were not static entities; the functional state of the other neurons govern the possibilities for certain connections to be functionally effective in a given instant. It is worth noting his early understanding of the importance of time for interpreting circuit diagrams based on anatomical data.

Lorente became progressively interested in function. This interest already pervaded his early anatomical studies. A concluding note in Lorente's paper of 1922 is somewhat of a declaration of intent: "This material is inadequate for providing an outline of the trajectory of nerve currents. In order to achieve this, it would first be necessary to know what parts of the axon are capable of transmitting impulses..." (Lorente de Nó, '22b). His intellectual demand for functional interpretation was not fully satisfied with reflections on morphological observations, as alluded to by the Scheibels ('78). Neurophysiology was behind the times, both conceptually and technically, to analyze complex neuronal circuits. Let us venture the hypothesis that it was precisely because of this situation that Lorente took his chance and became a neurophysiologist.

A logical development of Cajal's heritage was that the best Spanish neuroscience that immediately followed him (in the hands of Tello, del Rio Hortega and de Castro) grew on the understanding that research in morphology is useless if done disregarding function. Lorente, with more advanced facilities in the United States, was able to cross over this disciplinary border; he became the first neurophysiologist of Spanish lineage that felt a devotion to morphology as a foundation for neurophysiology. This attitude continued with Gallego, himself a pupil of Lorente (see Gallego, this volume). This is perhaps one of the bases of the contradictions in Lorente's own perception as a neuroscientist: contrary to his own wishes, he has been credited much more as a neuroanatomist rather than as a physiologist (Kruger and Wooolsey, '90).

## The first contribution of Lorente to cortical organization: the "acoustic" cortex (Lorente de Nó, '22b)

This is an exhaustive description of neuronal morphology in the cortical area known today to contain the representation of individual vibrissae (Woolsey and Van der Loos, '70). Contrary to what has been frequently expressed, Lorente cast a shadow of doubt on the acoustic nature of the cortical area he was studying: "[...] tentatively [we] shall continue to call this region 'acoustic.'" (Lorente de Nó, '22b). This is because the evidence available at that time was based on cytoarchitectonics rather than on a connectional criterion, which he recognized correctly as the definitive one. He pointed out that this cortex, like any other sensory cortex, possesses a prominent terminal plexus in layer IV and, therefore, he stated: "...although we cannot positively affirm that it is acoustic, at least we are probably safe in assigning a sensory function to it." (Lorente de Nó, '22b).

When Lorente initiated this study, he had several fundamental questions to address. One of his first questions was how "*sensory excitation*" could be localized in the brain in a very precise manner, in spite of the "numerous relay stations of sensory excitation" and of its dispersion in the cerebral cortex, where "...it is unlikely that one single cell is not influenced by a part of such excitation." He added the comment, predicting his interpretation of internuncial neurons, that such dispersion is exerted by the short (local) axons and by the axon collaterals of projecting neurons. Years later, he elaborated a theory to explain localization of impulses in the brain (Lorente de Nó, '34) (See Swanson, this volume).

How is phylogeny related to the neuronal organization of the cerebral cortex? - this was another problem approached by Lorente. The basic question was whether the plan of organization of the rodent cortex was similar to that of higher species. In 1922, Lorente was very impressed with the structural complication of the rodent cortex. He accepted that the organization of this cortex may be simple with respect to that of man only with regard to the number of cortical areas, but not "when dealing with the intimate details of [cortical] structure" revealed by the Golgi method.

This point of view became more elaborated in his 1938 paper (Lorente de Nó, '38b), when he claimed to be completely in agreement with Cajal's ideas. According to Cajal, the complexity of the cerebral cortex reaches a peak in humans; "Cajal assumed that the large number of cells with short axons was the anatomical expression of the delicacy of function of the brain of man" (Lorente de Nó, '38b). Lorente admitted that the cortex of the mouse contains a lower number of short-axon cells, "without essential modification of the long links in the chains of cortical neurons." Long links were those connecting cells located in different cortical layers; these connections, formed by the cells with ascending axons, would vary very little during evolution. Short-axon cells would be responsible for short intracortical links; their number would be limited in rodents. Lower species would contain, therefore, an *elementary pattern* of cortical organization, essentially constant during mammalian evolution. In recent times, there have been opposing points of view about the relative abundance of cortical interneurons in different species of mammals (Fairén et al., '84), and the question cannot be considered as definitively settled. Lorente ('38b) commented in passing that, in higher species, there is also an elaboration of the dendritic and axonal plexuses; this was indicated already by Cajal ('90).

When Lorente initiated his cortical studies, cytoarchitectonics was in the mainstream of cortical anatomy. Already in 1922, Lorente had expressed a lively opposition to what he believed to be an oversimplification in this approach; he elaborated this idea further in 1933 and 1938. Lorente did not oppose the functional importance of parcellation. Instead, he opposed the fact that cytoarchitectonics ignored the type of information that the Golgi method provides about the distribution of axonal and dendritic arborizations. This criticism led him to a certain underestimation of the meaning of layering in cortical organization: contrary to his arguments, while each cortical layer can be considered as an effector, the distant projection of each cortical layer is specific (Jones, '84). Lorente's attempt to deal with cytoarchitectonics was certainly ambitious: in line with the studies of Cajal ('21; '22), Lorente wished to find in the fine structure, such as revealed by the Golgi method, a substrate for cortical parcellation.

A most remarkable achievement of the 1922 paper is the description of "*homodynamic groups of neurons*" that Lorente called *glomérulos*. What does it mean? It means that neurons are not described in isolation, but with regard to the position they happen to occupy within the cortex. This is, clearly, a novel finding, in line with the detailed analysis of cortical parcellation he proposed. The somatosensory cortex of rodents possesses, in its layer IV, multicellular aggregates that Woolsey and Van der Loos (Woolsey and Van der Loos, '70) called *barrels* (see Woolsey, this volume). Lorente made an exhaustive description of these cellular aggregates and of the neuronal elements participating in their formation. The presence of these multineuronal units exerts a certain constraint on the expression of the morphology of the neurons that happen to be under their sphere of influence; this makes them extremely

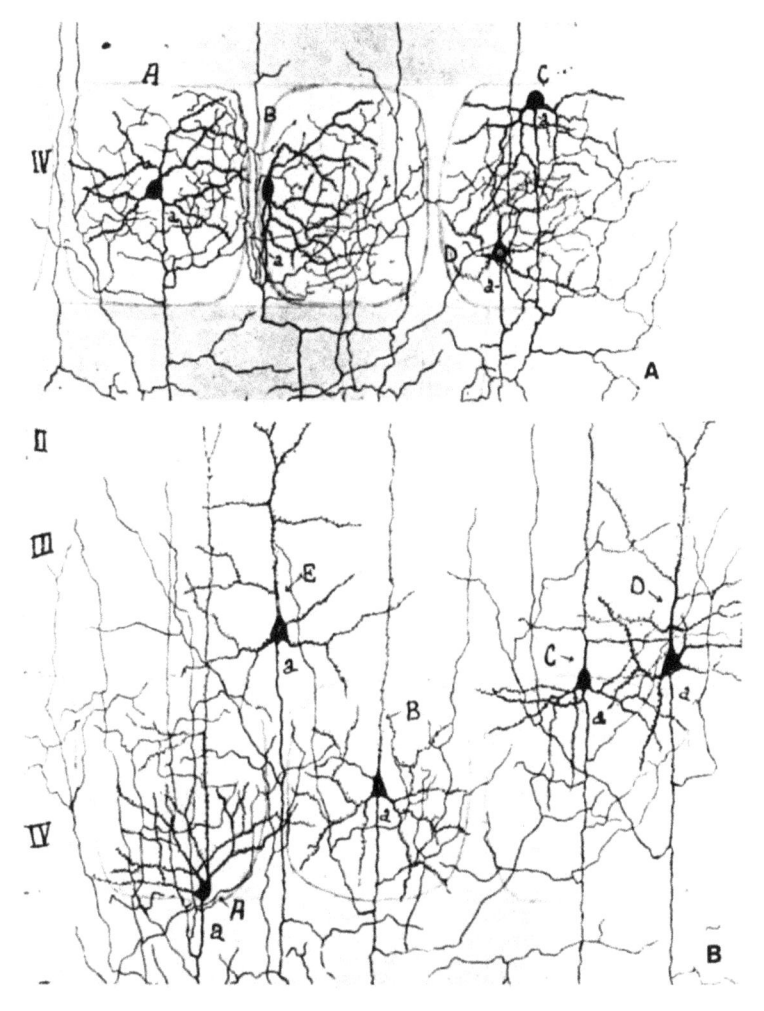

**Figure 3.** Spiny cells in mouse primary somatosensory cortex (Lorente de Nó, '22b). Glomeruli were outlined by Lorente. II-IV, cortical layers; a, axons. A. Spiny stellate cells or star cells (A-D). B. Star-pyramids (A-C) and conventional pyramidal cells (D, E).

interesting for developmental studies. Woolsey et al. ('75) made the case for "context-dependent characteristics", which are those characteristics of a neuronal set that are dictated by influences of the local environment.

Did Lorente really appreciate such influences? This is evident in the cases of spiny cells - the "granule" or "star" cells (Fig. 3A) and the "star" pyramids (Fig. 3B). Lorente represented star cells as having their dendrites confined to one barrel, and described their changing morphology depending upon the position they occupied within the barrel (cells A - C in Fig. 3A). If cells that qualify as pyramidal cells had basal dendrites with an identical configuration to that of star cells, and related to the presence of barrels, then they were the "star pyramids" (cells A - C in Fig. 3B). Typical pyramidal cells (cells E and F in Fig. 3B) were quite similar to these star pyramids, but their basal dendrites did not form part of the barrels.

Neurons with spiny dendrites were not the only ones described by Lorente. Figure 4 shows several examples of spine-free or sparsely-spined cells. All of these cells have axonal arborizations which are local in nature; some have a relationship with the barrels, while

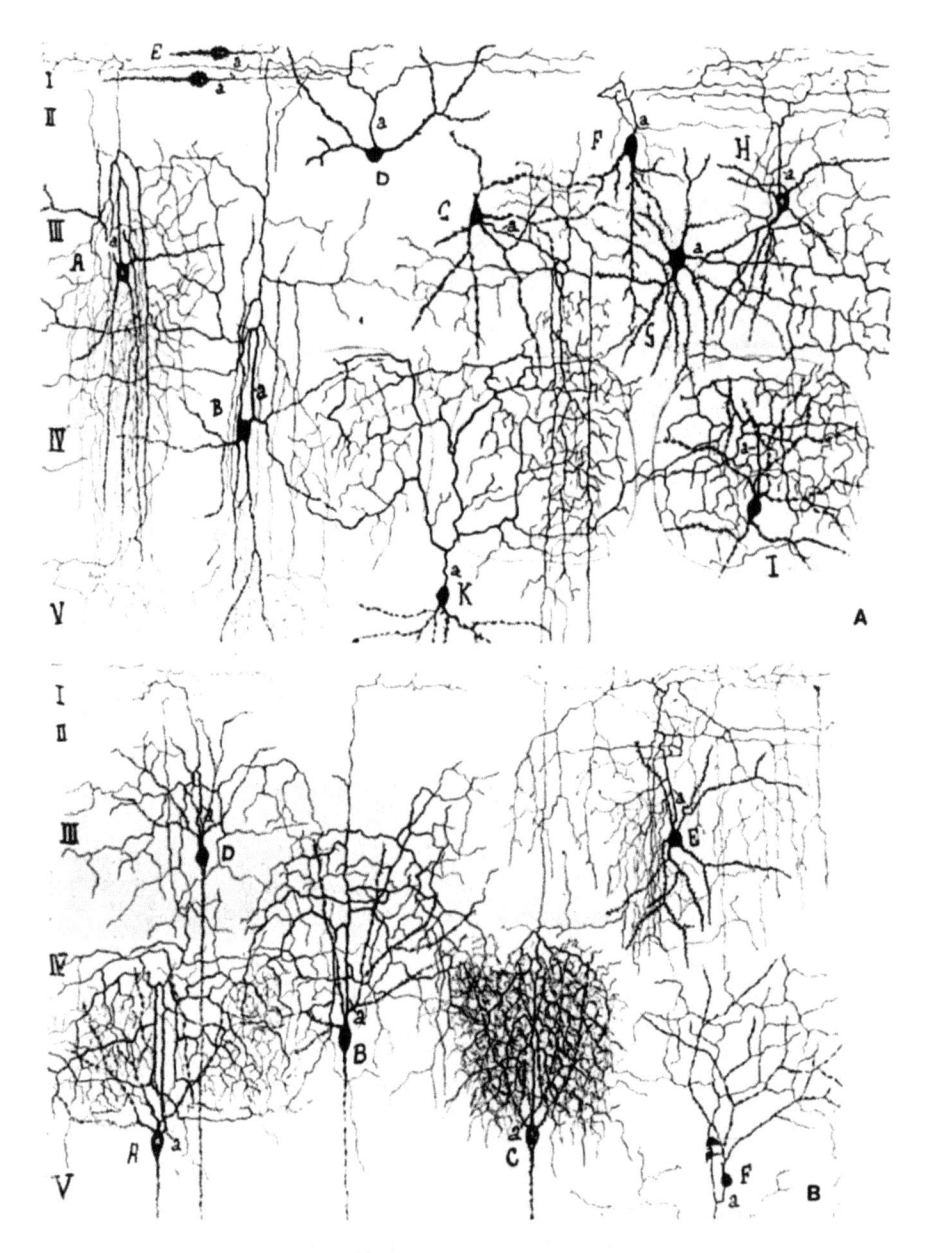

**Figure 4.** Neurons with intracortical axons (Lorente de Nó, '22b). I-V, cortical layers; a, axons.

others are apparently independent. Briefly, following criteria detailed elsewhere (Fairén et al., '84), some of these cells could be grouped with the generalized cells with axonal arcades (Fig. 4A, cell A; Fig. 4B, cell E). Some are diverse types of neurons with ascending axons (Fairén et al., '84; Fairén and Valverde, '79; Fairén and Smith-Fernández, '92) (Fig. 4A, cells B, H; Fig. 6B, cell D). In Figure 4A, cell G has horizontal axonal collaterals predominating (Fairén and Smith-Fernández, '92) while cell C is a cell with a vertical, bi-tufted axon (Lorente de Nó, '22b), i.e., a double bouquet cell. Another variety is formed by cells showing a single, descending dendrite and a bouquet of ascending dendrites; their axons are also ascending and distribute within a relatively small area, corresponding to one or two barrels. These are the cells that were named (Cobas et al., '87) "Lorente de Nó cells." This

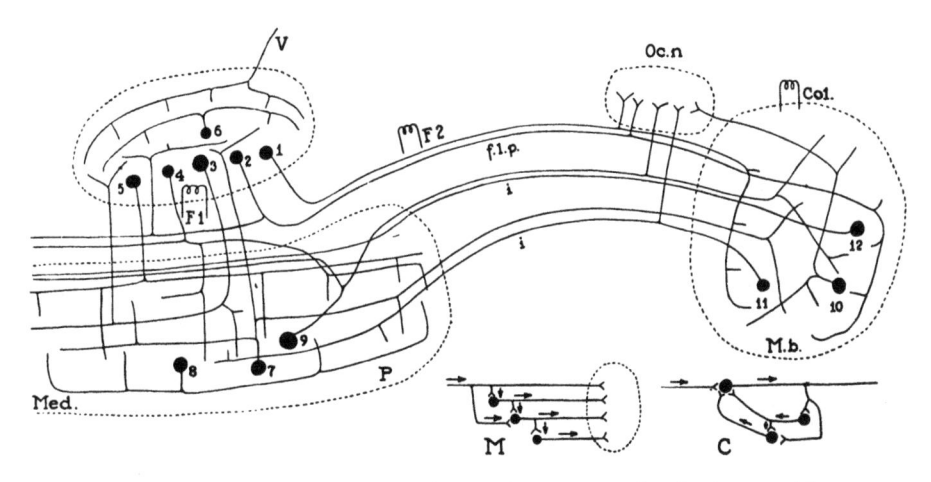

**Figure 5.** Lorente's legend ('38a): "Diagram of the pathways connecting the internuncial neurons among themselves and with the ocular motoneurons. V, vestibular nerve; 1 to 6, cells in the primary vestibular nuclei; 7, 8, 9, cells in the reticular formation in the medulla (Med.) and pons (P.) [sic]; 10, 11, 12, cells in the reticular nuclei in the midbrain (M.b.); Oc. n., oculomotor nuclei; f.l.p., fasciculus longitudinalis posterior and similar pathways; i, internuncial pathways; F1, F2 and Col., position of the stimulating electrodes. The diagrams below indicate the two types of chains formed by internuncial cells; M, multiple and C, closed chain."

is one of the examples in which context seems to influence neuronal morphology (See Woolsey, this Volume, for a functional interpretation of cell K in Fig. 4A). In other cases (cell I in Fig. 4A), all of the dendritic and axonal arborizations are restricted to a single barrel, as in the case of star cells. Looking at this richness of different dendritic and axonal plexuses, one can easily understand Lorente's reservations about cytoarchitectonics and his long-standing emphasis on vertical connectivity. To conclude these comments, I hasten to add that this description of neurons in the mouse somatosensory cortex has not yet been surpassed.

Lorente ignored the possible controversy on the peculiarity of the rodent "*sensory*" cortex when dealing in 1938 with the cortex in general (Lorente de Nó, '38b). This final account was based upon his observations in the rodent cortex, which he considered a good model of cortical organization. It is probable that he had explored, without publishing, more cortical areas. Perhaps we have lost the opportunity to see a compilation of his complete works on the cerebral cortex similar to his great *The primary acoustic nuclei* (Lorente de Nó, '81).

## The modular cortex

For Lorente, research in neurophysiology was greatly facilitated by the use of realistic circuit diagrams derived, as discussed above, from Golgi observations: "The interest of the analysis consists in that it is possible to reduce the actual anatomical complexity of the nerve centres to simple diagrams suitable for theoretical arguments..." (Lorente de Nó, '38) and, evidently, for experimental testing. An example is illustrated in Figure 5, which shows the relationships of internuncial neurons with a group of motoneurons (Oc. n.). The two small schemes represent the two possible modalities that can be distinguished in neuronal chains: M, multiple (obeying the principle of plurality of connections), and C, closed (obeying the principle of reciprocity of connections) (Lorente de Nó, '33a).

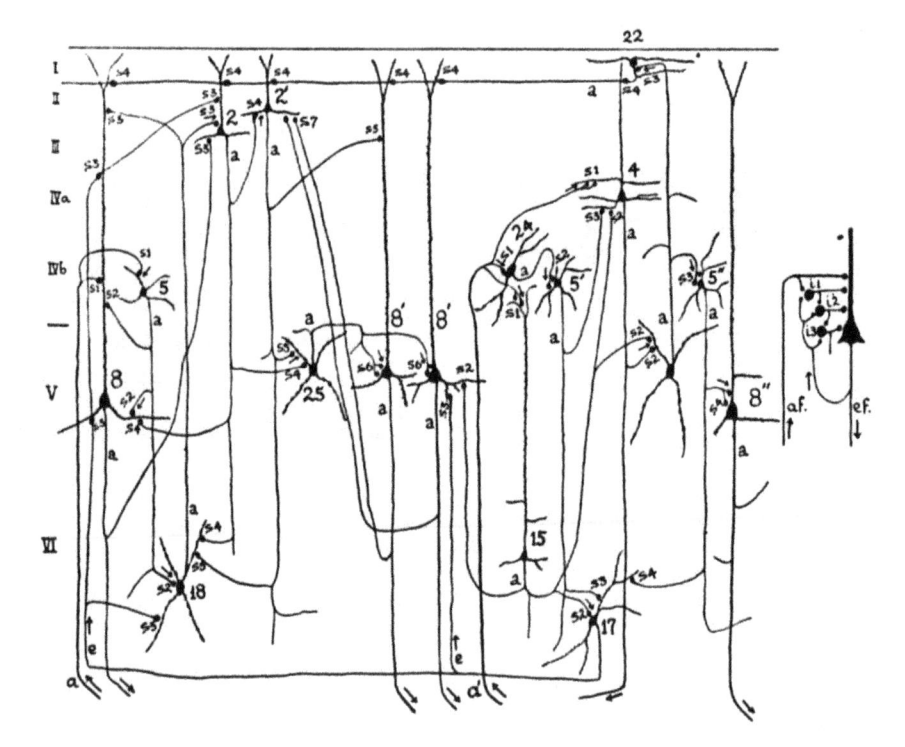

**Figure 6.** Lorente's wiring diagram of the cerebral cortex (Lorente de Nó, '38b). I-VI, cortical layers; a, axons, si, synapses (also "indicated with a thickening of the axon"). "It is assumed that the synapses marked with an arrow are passed by the impulses." Arabic numerals refer to the neurons described in the text. In the inferior part of the diagram, a, a' and e are afferent fibres. "The small diagram at the right is a simplification of the diagram at the left", and "summarizes the plan upon which the central nervous system is built." ii, internuncial neurons, af., afferent fiber; ef., efferent fiber.

The cerebral cortex can only be modeled after one such circuit diagram (Fig. 6) on the condition that "after important simplification the cortex is considered as a chain of neurons built on the plan of reflex arcs" (Lorente de Nó, '38b). A simplification is needed, naturally, because each cortical region has specific structural traits, but "*what remains constant is the arrangement of the plexuses of dendritic and axonal branches, i.e., of the synaptic articulations through which nerve impulses are transmitted*" (Lorente de Nó, '38b) (italics Lorente's).

In this definitive report (Lorente de Nó, '38b), Lorente distinguished among the cells with intracortical axons: *(i) short-axon cells* (cells 24 and 25 in Fig. 6 - he explicitly included basket cells within this group); *(ii) cells with ascending axons* (cell *18* in Fig. 6); *(iii) cells with horizontal axons*: cell 22 in Fig 6. The examples chosen by Lorente to construct his diagram correspond clearly to the distinct types of nonpyramidal cells described in his previous papers (Lorente de Nó, '22b; Lorente de Nó, '33b), but he did not choose to show some of the neurons which had axonal arborizations related to barrels (e.g., Lorente de Nó cells, see above). Additionally, he considered one ample class of cells with descending axons, corresponding to the diverse types of pyramidal cells (*2, 2', 4, 8, 8', 8''* in Fig. 6) and to spiny stellate cells (*5, 5', 5''* in Fig. 6). In Figure 6, cell *15* is a particular class of pyramidal cell: it is in layer VIa, its apical dendrite forms a tuft in layer IV, and its axon has recurrent colaterals ending in layer IV (Lorente de Nó, '22b; McGuire et al., '84). Cell number *17* is in layer VIb and gives rise to an association axon (Lorente de Nó, '22b). Lastly, a and e point to two of the types of cortical afferent fibers described in his 1922 paper. Lorente was searching for the elementary unit of the cortex, in which "the whole process of the transmission of impulses from the afferent fibre to the efferent axon may be accomplished". He conceived such an elementary unit as a vertical column, or module: "[the cortex] is a unitary system composed of vertical chains of neurons." Within the vertical chains, virtually all classes of neurons are included. One important aspect of this concept of the elementary unit is that the unit is not an entity rigidly defined by its anatomy; the extent of the neuron chain depends on the functional state and may change over time. Therefore, one may well imagine a multitude of possible configurations for the cortical modules. This is explicitly formulated in his paper on the entorhinal cortex (Lorente de Nó, '33b): any cortical element (be it an afferent fiber, a pyramidal cell or a short-axon cell) can be considered the axis of a functional unit. Lorente set the basis for more recent theoretical models of cortical organization (Szentágothai, '75). One of the limitations of Lorente's concept of cortical organization lies in the fact that, at the time he wrote the chapter for Fulton's Physiology, he did not recognize a possible role for inhibition in the functioning of the cortex. This is unfortunate, since he had previously postulated the existence of inhibitory neurons to explain the mechanism of the nystagmus (Lorente de Nó, '33a; Lorente de Nó, '38a), and since he had written in his paper ('33b) on the entorhinal cortex: "...it is important to note that internuncial neurons may be the site of inhibition". Probably, he failed to attend to inhibition because he was supporting the notion of closed, re-exciting neuronal chains (Fig. 6, small diagram), in which cells such as the basket cells would play an important role in maintaining cortical activity (Lorente de Nó, '38b,b). The nature of inhibition, however, was always a controversial issue at those times, before the discovery of the synaptic mechanisms involved (Brook et al., '52). Even with this reservation, perusing Lorente's circuit diagram (Fig. 6) is still exciting, since not all the proposed links in the synaptic chains proposed there have yet yielded to experimental test. Interestingly, some have been analyzed only recently (e.g., the relationships between the afferent fiber and cells number 15 and number 24 in Figure 6 (McGuire et al., '84).

We are now at the close of a decade where advances in our understanding of the organization of the cerebral cortex have been unprecedented. In this situation, it gives a stimulating sense of permanence to realize how the concepts set forth by Lorente more than fifty years ago still pervade our understanding of the cortex.

ACKNOWLEDGEMENTS

Supported by DGICYT Grant PB 87-0223. Discussions with Steven Jones on English usage are acknowledged with thanks.

REFERENCES

Blackstad, T.W., 1965, Mapping of experimental axon degeneration by electron microscopy of Golgi preparations, *Z. Zellforsch.*, 67:819.
Blackstad, T.W., 1975, Electron microscopy of experimental axonal degeneration in photochemically modified Golgi preparations: A procedure for precise mapping of nervous connections, *Brain Res.*, 95:191.
Brook, L.G., Coombs, J.S., and Eccles, J.C., 1952, The recording of potentials from motoneurones with an intracellular electrode, *J. Physiol.*, 115:320.
Cobas, A., Welker, E., Fairén, A., Kraftsik, R., and Van der Loos, H., 1987, The GABAergic neurons of the barrel cortex of the mouse. A study using neuronal archetypes, *J. Neurocytol.*, 16:843.
del Rio Hortega, P., 1986, "El Maestro y Yo", A. Sánchez Alvarez-Insúa, ed., Consejo Superior de Investigaciones Científicas, Madrid.
DeFelipe, J. and Fariñas, I., 1992, The pyramidal neuron of the cerebral cortex: morphological and chemical characteristics of the synaptic inputs, *Progress Neurobiol.*, in press.
Fairén, A., Peters, A., and J. Saldanha, 1977, A new procedure for examining Golgi impregnated neurons by light and electron microscopy, *J. Neurocytol.*, 6:311.

Fairén, A. and F. Valverde, F., 1979, Specific thalamo-cortical afferents and their presumptive targets in the visual cortex. A Golgi study, *in*: "Development and Chemical Specificity of Neurons" Progress in Brain Research, Vol. 51. M. Cuénod, G.W. Kreutzberg, and F.E. Bloom, eds., Elsevier/North-Holland Biomedical Press, Amsterdam.

Fairén, A., DeFelipe, J., and Regidor, J., 1984, Nonpyramidal neurons. General Account, *in*: "Cerebral Cortex." Vol. 1. A. Peters, and E.G. Jones, eds. Plenum, New York.

Fairén, A., and Smith-Fernández, A., 1992, Electron microscopy of Golgi-impregnated interneurons: notes on the intrinsic connectivity of the cerebral cortex, *Microsc. Res. Techn.*, in press.

Gallego, A., 1990, La obra científica del doctor Lorente de Nó, *Anales de la Real Academia de Medicina (Madrid)*, 107:467.

Jones, E.G., 1984, Laminar distribution of cortical efferent cells, *in:* "Cerebral Cortex," Vol. 1. A. Peters, and E.G. Jones, eds., Plenum, New York.

Kruger, L., and Woolsey, T.A., 1990, Rafael Lorente de Nó: 1902-1990, *J. Comp. Neurol.*, 300:1.

Lorente de Nó, R., 1922a, Contribución al conocimiento del nervio trigémino, *in:* "Libro en Honor de D. S. Ramón y Cajal. Trabajos originales de sus admiradores y discípulos, extranjeros y nacionales." Junta para el Homenaje a Cajal, Vol. 2, Jiménez y Molina, ed., Madrid.

Lorente de Nó, R., 1922b, La corteza cerebral del ratón. (Primera contribución.- La corteza acústica), *Trab. Lab. Invest. Biol. Univ. Madrid*, 20:41. Translated into English, as: The cerebral cortex of the mouse. (A first contribution.- The acoustic cortex), by A. Fairén, J.Ragidor and L. Kruger, *Somatosensory Motor Res.*, in press.

Lorente de Nó, R., 1933a, Vestibulo-ocular reflex arc, *Arch. Neurol. Psychiat.*, 30:245.

Lorente de Nó, R., 1933b, Studies on the structure of the cerebral cortex, *J. Psychol. Neurol.*, 45:381.

Lorente de Nó, R., 1934, Studies on the structure of the cerebral cortex. II. Continuation of the study of the ammonic system, *J. Psychol. Neurol.*, 46:113.

Lorente de Nó, R., 1938a, Analysis of the activity of the chains of internuncial neurons, *J. Neurophysiol.*, 1:207.

Lorente de Nó, R., 1938b, Architectonics and structure of the cerebral cortex, *in*: "Physiology of the Nervous System", Chapter XV., J.F. Fulton, ed., Oxford University Press, New York.

Lorente de Nó, R., 1981, "The Primary Acustic Nuclei." Raven, New York.

McGuire, B.A., Hornung, J.-P., Gilbert, C.D., and Wiesel, T.N., 1984, Patterns of synaptic input to layer 4 of cat striate cortex, *J. Neurosci.*, 4:3021.

Peters, A. and Jones, E.G., 1984, "Cerebral Cortex," Vol. 1. Plenum, New York.

Ramón y Cajal, S., 1921, Textura de la corteza visual del gato, *Archivos de Neurobiología*, 2:338. Translated into English in "Cajal on the Cerebral Cortex," J. DeFelipe and E.G. Jones, eds., Oxford University Press, New York 1988.

Ramón y Cajal, S., 1922 Estudios sobre la fina estructura de la corteza regional de los roedores. I. Corteza suboccipital (retroesplenial de Brodmann), *Trab. Lab. Invest. Biol. Madrid*, 20:1. Translated into English in "Cajal on the Cerebral Cortex," J. DeFelipe and E.G. Jones, eds., Oxford University Press, New York, 1988.

Ramón y Cajal, S., 1923, "Recuerdos de mi Vida." Juan Pueyo, Madrid.

Ramón y Cajal, S., 1954, "Neuron Theory or Reticular Theory? Objective Evidence of the Anatomical Unity of Nerve Cells." English translation by M. Ubeda Purkiss and C.A. Fox, Consejo Superior de Investigaciones Científicas, Madrid.

Ramón y Cajal, S., 1990, "New Ideas on the Structure of the Nervous System in Man and Vertebrates." English translation by N. Swanson and L.W. Swanson, MIT Press, Cambridge.

Scheibel, M.E. and Scheibel, A.B., 1978, The methods of Golgi, *in*: "Neuroanatomical Research Techniques," R.T. Robertson, ed. Academic Press, New York.

Shepherd, G.M., 1990, "The Synaptic Organization of the Brain," 3rd. edition. Oxford University Press, New York.

Szentágothai, J., 1975, The "module-concept" in cerebral cortex architecture, *Brain Res.*, 95:475.

White, E.L., with A. Keller, 1989, "Cortical Circuits: Synaptic Organization of the Cerebral Cortex. Structure, Function and Theory," Birkhauser, Boston.

Woolsey, T.A. and Van der Loos, H., 1970, The structural organization of layer IV in the somatosensory region (S I) of mouse cerebral cortex, *Brain Res.*, 17:205.

Woolsey, T.A., Dierker, M.L., and Wann, D.F., 1975, Mouse SmI cortex: Qualitative and quantitative classification of Golgi-impregnated barrel neurons, *Proc. Natl. Acad. Sci. USA*, 72:2165.

# *GLOMÉRULOS*, BARRELS, COLUMNS AND MAPS IN CORTEX: AN HOMAGE TO DR. RAFAEL LORENTE DE NÓ

Thomas A. Woolsey

James L. O'Leary Division of Experimental Neurology and Neurological Surgery, Department of Neurology and Neurological Surgery, Washington University School of Medicine, St. Louis, Missouri 63110, USA

## Prologue

My first contact with Dr. Rafael Lorente de Nó was by mail. I was working on the anatomy of the mouse cerebral cortex in the summer of 1966 and found a reference to his classic paper on the cerebral cortex of the mouse (Lorente de Nó, '22) (*See note added in proof, p.498). The Medical Library at the University of Wisconsin did not have the *Trabajos* in its serials collection and I wrote Dr. Lorente de Nó at the Rockefeller to request a reprint. A short time later, I received a polite, slightly bemused and, I fancy, flattered response written in a hand I later would recognize well. The supply of reprints was exhausted decades earlier, I was told, but the author appreciated my request.

Several months later, I found the paper in a library in Baltimore. I could see the beauty and thoroughness of the work. Later I understood its importance. The paper lays out the anatomical concepts that anticipated the now generally accepted idea that the cerebral cortex is arranged in vertical, functional, columnar arrays of neurons that process information. Much later I was to meet Dr. Rafael Lorente de Nó, to share personal and scientific thoughts with him and, quite incredibly, to crisscross paths and acquaintances. Today I hold a position that is named for one of Lorente's early physiological colleagues, George H. Bishop, and direct a Division named for one of his anatomical collaborators, James L. O'Leary.

Others at this symposium have covered in detail the fine anatomy of cortical neurons. I shall here focus on several consequences of his illustration and conceptualization of the *glomérulos* of the *corteza acústica* of the mouse and of the vertical chains of cortical neurons that make up the columns. I confine my discourse to the somatic sensory cortex of the rodent. I shall end with an account of the unlikely coincidences that brought me into much closer contact with this remarkable man than I ever imagined possible when I wrote for that reprint in 1966.

## Recording from the mouse brain

My father suggested, of course, that I make a functional "map" of the mouse cortex for my project as a summer student in his Laboratory of Neurophysiology at the Univer-

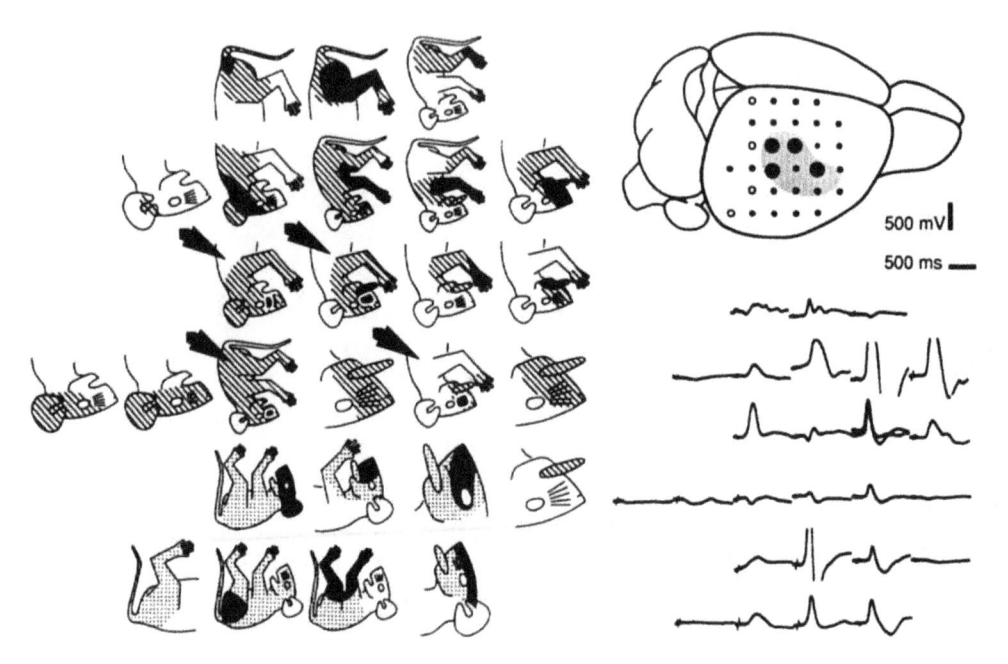

**Figure 1.** Evoked potential map defining the mouse somatosensory cortex. Shading on the "figurines" (upper left) indicates the peripheral stimulation sites producing potential changes (lower right) with the black indicating the larger amplitudes. Figurines with extremities pointing rightward are at sites assigned to SmI; those with the feet pointing upward to SmII on the basis of latency and lateralization. Sites with responses to stimulation for the whiskers are marked with arrows. The drawing of the brain shows the location of sites responding to tactile stimulation (•) and several not responding (°). The four sites at which the whiskers specifically activated the cortex are marked by larger dots. The stippling outlines approximate extent of the cytoarchitecture shown in Figure 2 and outlined in Figure 6 (modified from reference Woolsey, '67).

sity of Wisconsin in Madison. Surface evoked potentials were the standard method to map and it was fairly straight-forward to collect the data. The somatosensory map from one experiment is shown in Figure 1. The plan and most of the details were as expected from the well known maps of the rat that my father and his colleagues had generated two decades earlier (Woolsey, '52). In particular, the bulk of the representation in the first and second somatic cortex (SmI and SmII) was devoted to the head. The whiskers strongly activated several points in this map (see arrows in Fig. 1). The solid tradition of correlating anatomy with function (Woolsey, '83) required that a brain studied functionally, be sectioned and evaluated with the microscope.

Experimental brains, cut in the cardinal planes, were prepared histologically while I studied medicine and awaited inspection when I returned the following summer somewhat wiser from courses in neuroanatomy and physiology. Standard procedures at Wisconsin made it easy to match the physiology to the anatomy. In the sections, the unusual character of neocortical layer IV quickly caught my eye. The pattern is shown in Figure 2 along with an adjacent section stained for the mitochondrial enzyme cytochrome oxidase (CO). The latter is one of many stains that now have been shown to mark this unusual cortex. Jerzy Rose directed me to the paper by his uncle which referenced the paper by Dr. Lorente de Nó for which I wrote (Rose, '29). There were four other cytoarchitectonic studies as well (Rose, '29; Isenschmid, '11; de Veries, '12; Rose, '12; Broogleever Fortuyn, '14). All illustrated the peculiar architecture of this layer of the mouse cortex. What was new, after construction of a model from serial sections, was that the architectonic field which the *glomérulos* defined

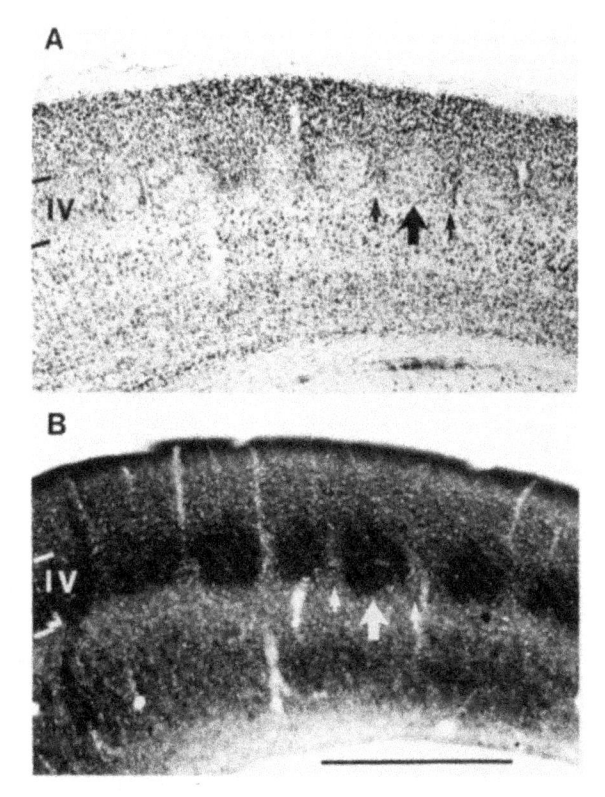

**Figure 2.** Sections through the whisker representation of somatosensory cortex showing the layer IV glomérulos/barrels in Nissl (A) and cytochrome oxidase (CO) (B) (Wong-Riley, '79). The pia is up and layer IV is identified to the left. Reconstructions from serial coronal sections were used to outline the extent of this anatomy in relation to surface evoked potential recordings as in Figure 1. Numerous staining methods show segregation of elements in the barrels such as CO which stains mitochondria (e.g. Nachlas et al., '56; Cooper and Steindler, '86). Small arrows mark septa between barrels; the large arrows mark the barrel hollows. Bar = 500 μm.

coincided with the head representation of the SmI cortex, especially the area activated by the whiskers (vibrissae) (Woolsey, '67; Woolsey and Van der Loos, '70).

I advanced the notion that there was a direct relationship between the unusual layer IV and the whiskers (Woolsey, '67). It took 2 years to get back to this problem. Working in Van der Loos' laboratory at Hopkins, I made a direct approach to the architecture of layer IV by cutting sections parallel to the cortical layers. At the time, this was a novel approach for cytoarchitecture. In this plane of section it was immediately clear that layer IV mimicked the pattern of the whiskers on the face. From cytoarchitecture it seemed that each whisker had a little cask-like group of cells which for a variety of reasons we chose to re-christen the barrels (Woolsey and Van der Loos, '70). (It is obvious that Dr. Lorente understood well the arrangement of the neuropil which he called the *glomérulos*. See below.) Figure 3 shows the pattern for layer IV stained with a mitochondrial enzyme (Nachlas et al., '56; Wong-Riley, '79). In addition to the representation of the whiskers other neuronal groupings relate to the digits, and, in the rat, the whole body form can be seen in layer IV (Welker, '76; Dawson and Killackey, '87).

The relationship between the whiskers and the barrels is schematically represented in Figure 4 which shows the similarities in the patterns of the two surfaces. Experiments later

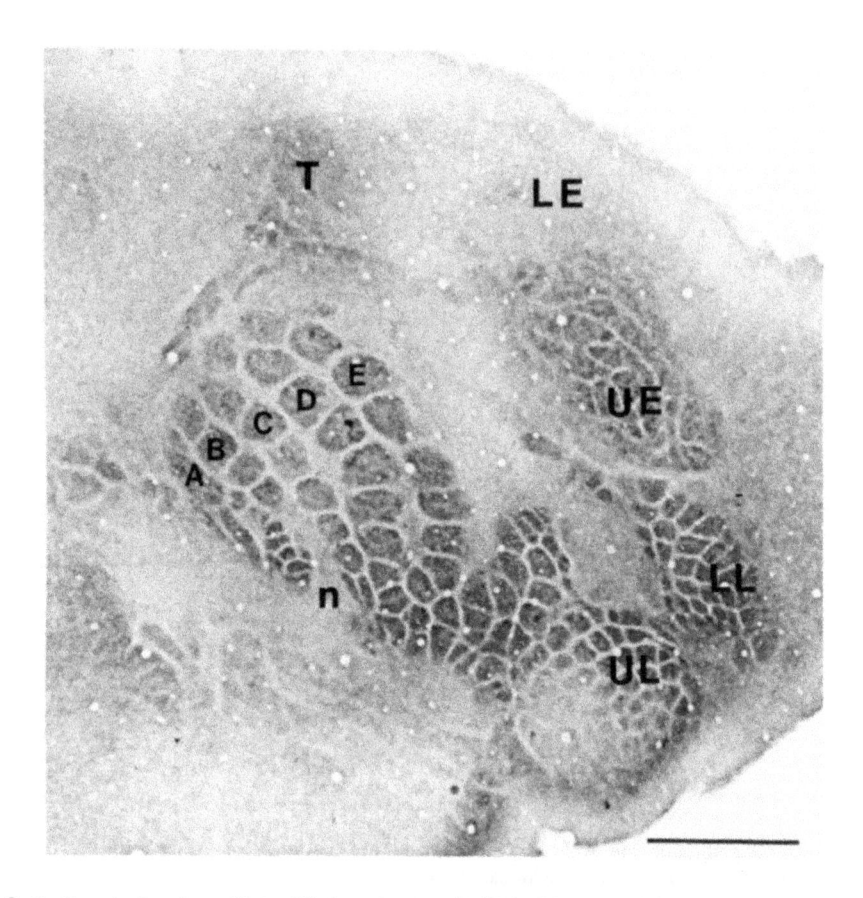

**Figure 3.** Sections in the plane of layer IV show the organization of the layer. The patterns of cells and other elements make a map that is highly consistent from one animal to the next. The larger barrels in five rows A-E mimic the pattern of the whiskers on the contralateral face. Other features are related to the rest of the body map: T = trunk; LE = lower extremity; UE = upper extremity; LL = lower lip; UL = upper lip; n = naris. This section, stained for the mitochondrial enzyme CO, shows this pattern for the right hemisphere of juvenile rat. Medial is up and anterior to the right. Bar = 1mm.

confirmed functionally what, in 1970, was an hypothesis based on anatomical correlation. One area that the work stimulated was a reinvestigation of the organization of the sensory periphery and the central somatic pathways associated with the whiskers (Rice and Munger, '86; Jacquin et al., '84; Harris, '86). Figure 5 shows, with the modern Cajalian "unexpected spectacle" (Loewy, '71), the fluorescent lipophilic dyes (Godement et al., '87), the organization of the periphery. Each whisker is an isolated tactile organ, has a separate innervation, and projects to the brainstem and thalamus in a pattern that replicates the periphery. The behavioral importance of the whiskers to the rat is great and the tactile capabilities of a rat with whiskers is as good as a human with active touch (Welker, '64; Carvell and Simons, '90; Huston and Masterson, '86). These are all convenient features that have been exploited to better understand connectivity, processing, development, genetics, cerebrovascular control and the activity of numerous molecules in the mammalian brain (Jhaveri et al., '91; Crandall et al., '90; Steindler et al., '90; Woolsey and Rovainen, '91; Rhoades et al., '90).

However this gets ahead of the story. In the 6 different studies of the mouse cortical cytoarchitecture that I found in the literature, almost none agreed on boundaries *except* in the area that contained the barrels. Figure 6 reproduces some of these. The major insight we had that the earlier workers lacked was the somatic function of this cortex. Lorente followed

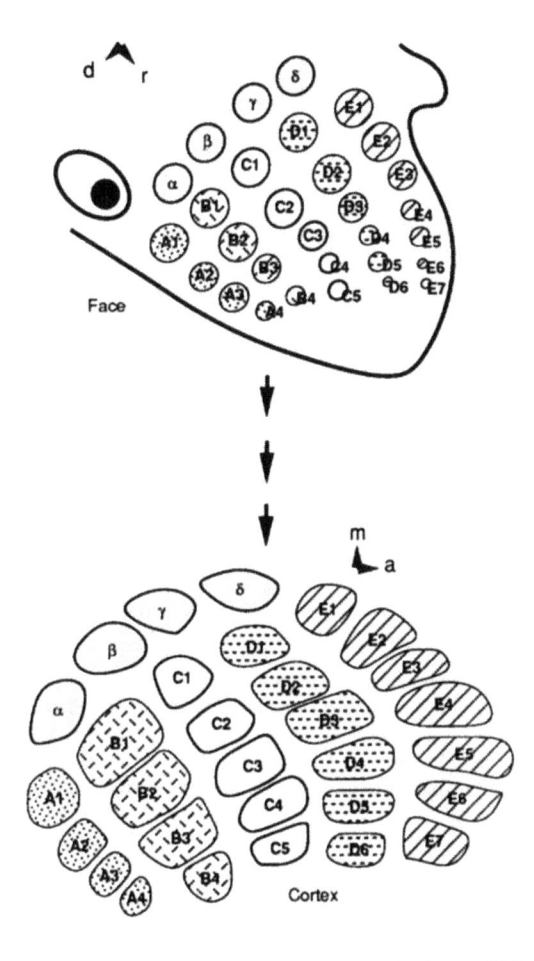

**Figure 4.** A schematic diagram showing the correspondence between the large whiskers (of the right face) and the barrels (of the left cortex) of the mouse, which is slightly different from the rat (compare Figure 3 to Figure 16), indicates the nomenclature. Five rows of whiskers are labeled A-E dorsal to ventral and are numbered forward (d = dorsal; r = rostral). "Straddler" whiskers are labeled α-σ. Arrows indicate the synapses between the periphery and the contralateral cortex. The nomenclature in the cortex is the same as emphasized by the shading patterns (m = medial; a = anterior).

Rose's 1912 map (Rose, '12), and, skeptically, his assignment of auditory function to this cortex. That is why the subtitle to "La corteza cerebral del ratón" reads "Primera contribución - La corteza acústica 1". Lorente could not have been certain of the function of this cortex, but on its detailed cellular organization he was absolutely correct.

## Dr. Lorente de Nó's anatomy of the mouse cortex

We have described elsewhere the unique circumstances that lead Lorente de Nó to study the mouse cortex (Kruger and Woolsey, '90). Although he surveyed the whole of the mantle, as his subtitle suggests, he was first attracted by the *corteza acústica*. Happily, Professor Fairén has translated this paper into English for many modern readers. The plates

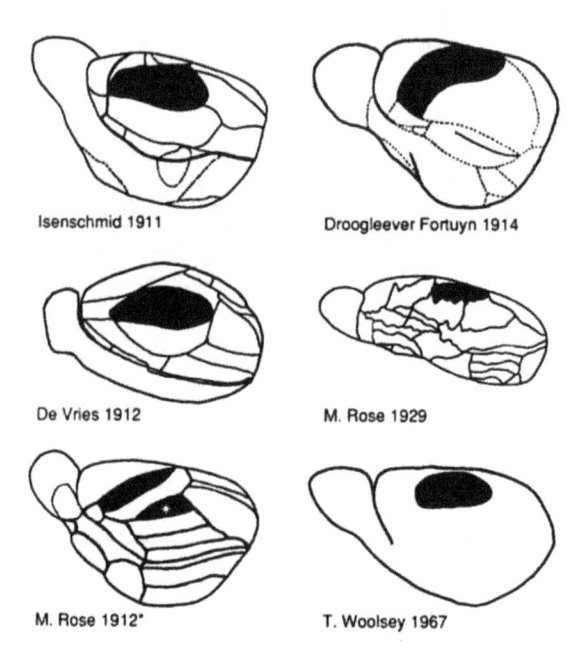

**Figure 6.** Summary of several cytoarchitectonic studies of the mouse brain. The area described or illustrated by several authors having cytoarchitecture such as shown in Figure 2 is blackened (Rose, '29; Isenschmid, '11; de Veries, '12; Rose, '12; Droogleever Fortuyn, '14; Woolsey, '67). There is good agreement on the location and extent of the glomerular/barrel cortex but less for other architectonic divisions. The * marks "La corteza acústica" (area 22) identified by Rose in 1912. This was the map used by Lorente de Nó for his 1922 paper (Lorente de Nó, '22).

in the original are all stunning works of art. Lorente Illustrated the *glomérulos* in 8 of his 26 Figures. I have reproduced parts of them here, highlighting the delicate outlines he drew in layer IV. In Figure 7 A & B, the location of the cortex is depicted as is its cytoarchitecture. He described two kinds of cells with spines in layer IV - the stellate cells (Fig. 7 D) and for the first time - the star-pyramids (Fig. 7C). Our quantitative studies of Golgi-Cox material in the mouse and the rat (Woolsey et al., '75; Simons and Woolsey, '84) show that the topology and measures of dendrites of these cells excepting the apical dendrite of the star pyramids are identical. Recently, it has been shown that the stellate cells have an apical dendrite which they lose in normal development (Peinado and Katz, '90).

The short axon cells shown in Figure 7 E and F are interesting in several respects. They are most certainly GABAergic inhibitory interneurons and have been carefully studied and classified by several groups (Cobas et al., '87; Lin et al., '85; Keller and White, '87). We have recently become especially interested in the cell type labeled K in Figure 7 F for reasons outlined below. This cell type is known to receive relatively few inputs from the somatosensory thalamus (the venrobasal complex, VB) (White, '78). From these pictures it is obvious that Lorente had a context for grouping these cells and thinking about the very difficult but central problem of local connectivity. The small size of the mouse brain made it possible for Rafael to be confident, and surprisingly accurate, of the distant targets of many of the cells and the sources of axons he drew.

One of Dr. Lorente's Figures is an exceptionally clear depiction of the organization of the processes of the layer IV neurons in layer IV. It is easy to see why he called these groupings *glomérulos*; he was struck by the similarity to the glomeruli of the olfactory bulb.

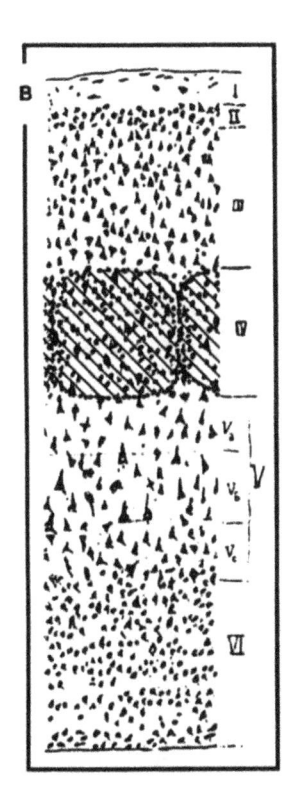

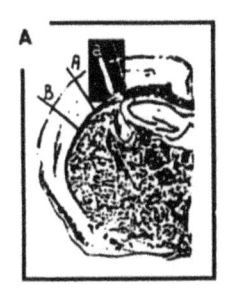

Figure 7. Figures or parts of Figures from Lorente de Nó ('22) illustrating his delineation of the glomérulos and the neurons that comprise them. A. Low power drawing of a coronal section (Fig. 1 from Lorente de Nó, '22) showing the location, "a", which is blackened, of the higher power view in B. B. Lorente outlined the glomérulos in this and other sections which I have emphasized by hatching. This pattern of glomérulos was similar to the published pictures of M. Rose for his area 22 (Rose, '12). C. Stellate pyramids, some with callosal axons, (Fig. 5 from reference Lorente de Nó, '22) which have basal dendrites confined to a glomerulus and some recurrent collaterals to adjacent glomeruli. D. Spiny stellate cells (Fig. 9 from reference Lorente de Nó, '22) lack an apical dendrite but in many other respects resemble the cells shown in C. E. Short axon cell with coextensive axon and dendrites confined to a single glomerulus (Fig. 6 from reference Lorente de Nó, '22). F. A short axon cell (labeled I) which is similar to that show in E and a large cell in upper layer V (labeled K) which has axons ramifying in two adjacent glomeruli (Fig. 9 from reference Lorente de Nó, '22).

I juxtapose the Figure to a sketch of a barrel in Figure 8. The truncated appearance of the barrel to the right suggests that the barrel in layer IV corresponds to a cortical column which, as Lorente later postulated, ought to extend through the full thickness of the cortex.

The 1922 (Lorente de Nó, '22) paper also illustrated the extrinsic afferents to the cortex. As with the targets of the cortical neurons, Dr. Lorente was remarkably accurate in his identification of these fibers. Figure 9 shows these drawings. With the lipophilic dye, DiA, it has been possible to label these fibers brilliantly and Figure 10 illustrates how the dense fibers from VB cluster in the center or hollow of the barrels around which the cell bodies are organized as barrels as can be seen directly with a kind of Nissl stain (Senft and Woolsey, '92).

Lorente later summarized his thoughts on the organization of the cerebral cortex in Fulton's book (Lorente de Nó, '38):

**Figure 5.** The central whisker pattern is established by the sensory periphery. A. Axons to two adjacent vibrissae, which have a red autofluorescence, from the follicle nerves are labeled with two different lipophilic dyes showing the segregation of the inputs to the whisker system. Arrows point to yellow colored axons in the upper follicle stained with DiI and to deep red axons in the lower follicle stained with DiQ. The deep aspect of the follicles is to the lower right and the skin surface is to the upper left. Bar = 100 μm. (Preparation by David T, Miller.) B. Axons from the infraorbital nerve innervate vibrissae and hairy skin (s) between the vibrissal follicles (e.g., f). These conus nerves (smaller arrows) are between the stout follicle nerves (larger arrows) and were labeled from the infraorbital nerve with DiI. C. Neurons in the trigeminal ganglion are labeled and lack a whisker like organization. The arrow pints to the "t" segment of an axon from a trigeminal ganglion cell. Bar = 50 μm. D. The terminal arbors of the trigeminal ganglion cells end all along the length of the trigeminal complex in the brainstem. The pattern of these afferents and the cell bodies they innervate is whisker-like in three brainstem nuclei. The axons enter the nuclei from the thick tract of the Vth nerve (tr V) which labels brightly. The rows of terminal patches associated with the different rows of whiskers in the spinal subnucleus interpolaris are labeled A-E. Bar = 500 μm is the scale for B and D. (B-D prepared by Franz Paul.)

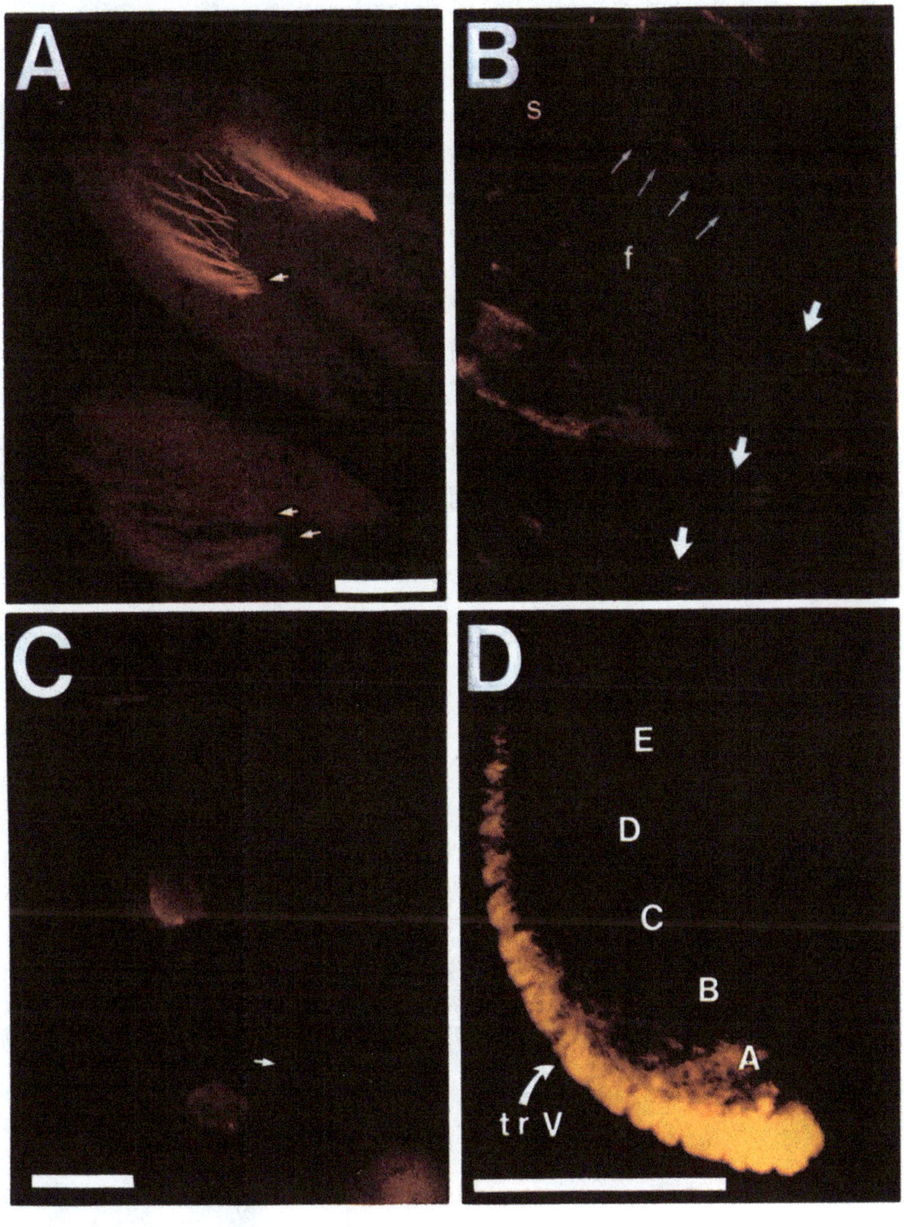
A
B
s
f
C
D
E
D
C
B
A
tr V

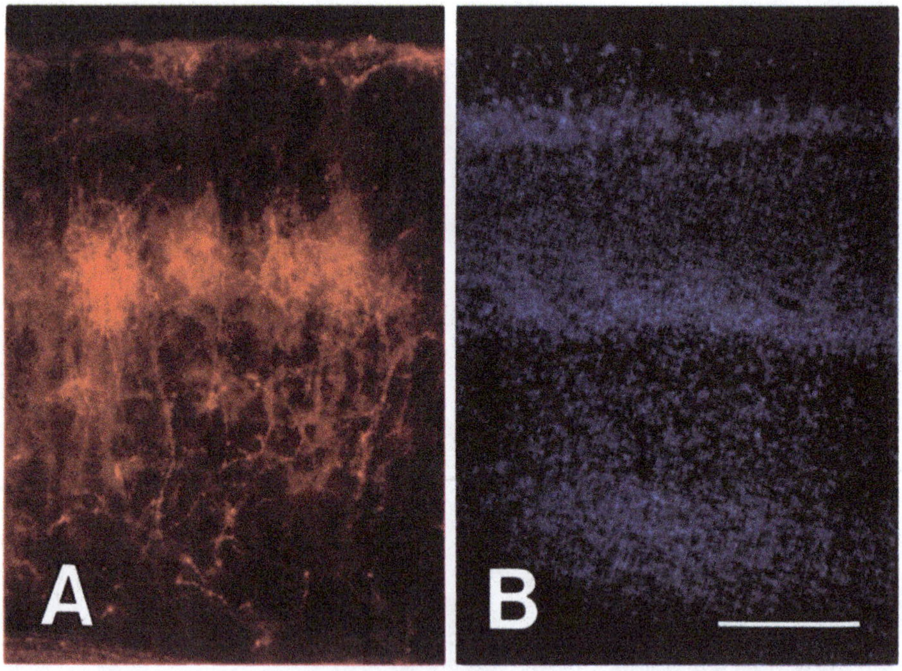
A
B

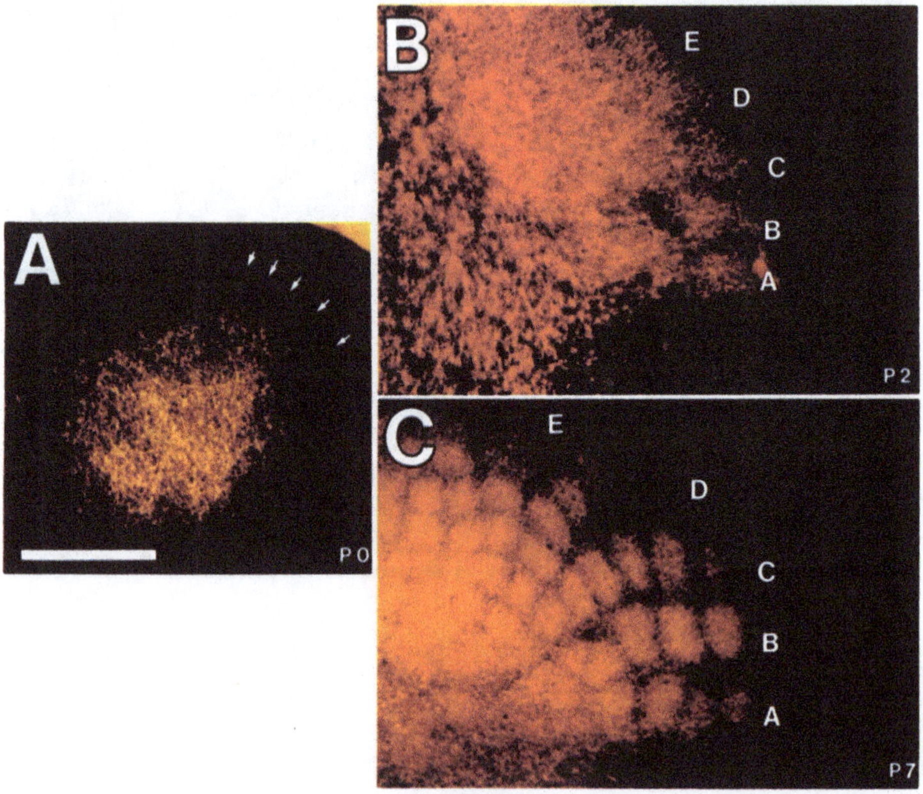
A
P0
B
E
D
C
B
A
P2
C
E
D
C
B
A
P7

**Figure 10.** Fibers from the ventrobasal thalamus labeled directly with DiI from VB of a P7 mouse. A thick coronal section illuminated with different filters shows several clusters of afferent terminals in "A" which are in register with the layer IV barrels shown with the Bisbenzimide counterstain for nuclei in "B". The Figure corresponds well to the earlier drawings of Lorente shown in Figure 9. At this age the cortex is still not fully differentiated and the supragranular layers have not reached their full thickness. Bar = 200 μm, pia is up.

**Figure 15.** The thalamocortical axons of the mouse have an explosive postnatal development. Here they are selectively labeled in tangential sections of mice of different ages with DiA or DiI placed in VB. A. Deep layers of somatosensory cortex at birth (P0). The fibers outline roughly the map of the skin and a region destined to represent the whiskers is marked with arrows. B. In the deeper layers the whisker map begins to emerge first on P2 where rows of TCAs, but not groups of barrel-like TCAs, are seen. C. Over the next several days the barrels within the rows are defined and the map is sharp at the end of the first week of life (P7). Scale in A = 500 μm and applies to all parts of the Figure. The orientation of all panels is the same with medial up and anterior to the left.

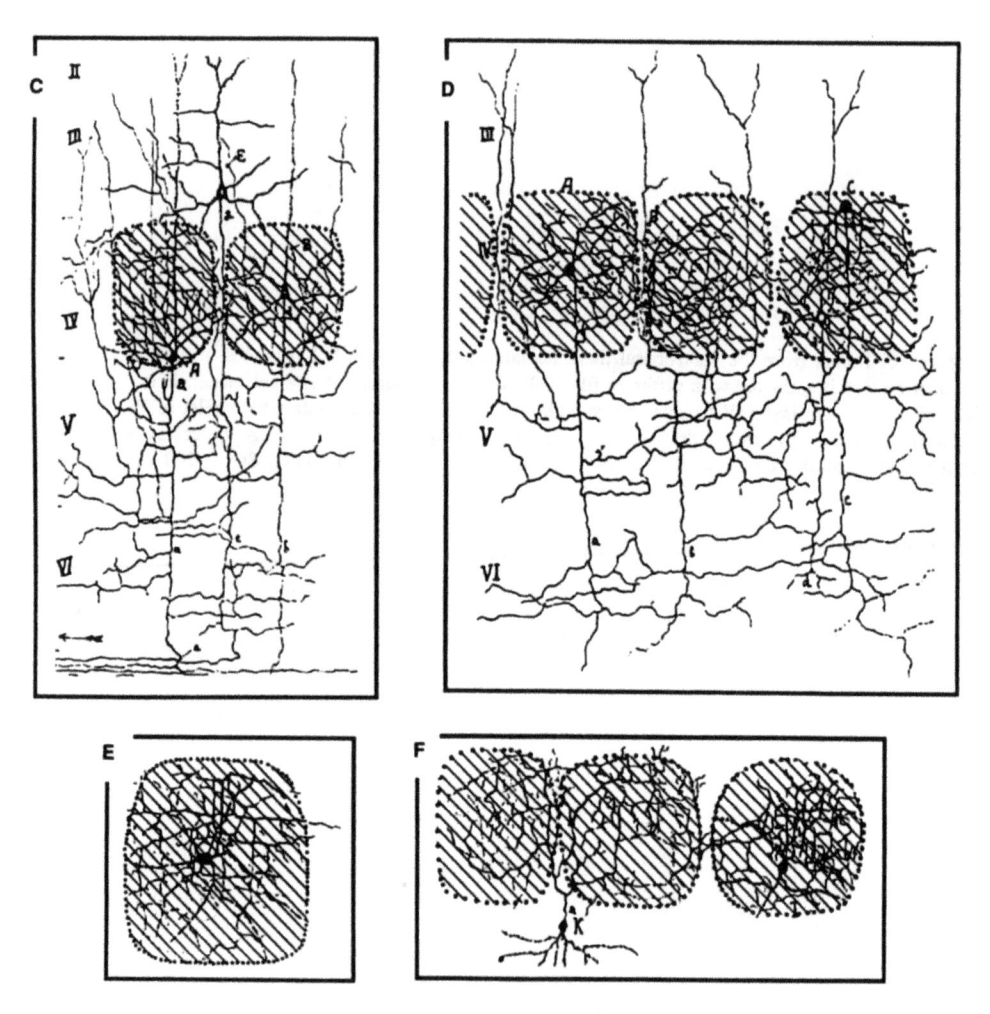

*"The small strip ... [of cortex]...is the vertical section of a cylinder having a specific afferent fibre like a [see our Fig. 9] as axis. All the elements of the cortex are represented in it, and therefore it may be called an elementary unit, in which, theoretically, the whole process of the transmission of impulses from the afferent fibre to the efferent axon may be accomplished. Within the elementary unit there are cells which make no such connections; the latter cells, of course, will be stimulated only as a result of cortical activity. "*

and,

*"In these circumstances it is obvious that there is no basis for considering the cortex as composed of several layers with specific primordial functions: reception, association and projection. From the functional point of view it is a unitary system composed of vertical chains of neurons, among which anatomically the most important are those starting at the articulation of the specific afferents and the cells of the external lamina. "*

He returned to the mouse cortex before writing this chapter and this chapter influenced Mountcastle's functional columnar hypothesis (Mountcastle, '57) that was quickly extended by Hubel and Wiesel and may others (Hubel and Wiesel, '63; Jones et al., '75). One of our early interests was to determine how a barrel is related to a functional column.

**Figure 8.** Lorente de Nó's illustration of two *glomérulos* (Fig. 8 from reference Lorente de Nó, '22) shows the highly oriented dendrites of spiny stellate cells and short axon cells lacking spines in layer IV. On the bottom is a cartoon illustrating the cask like three-dimensional organization of the cell bodies in the Nissel dye (Fig. 8, from reference Woolsey and Van der Loos, '70).

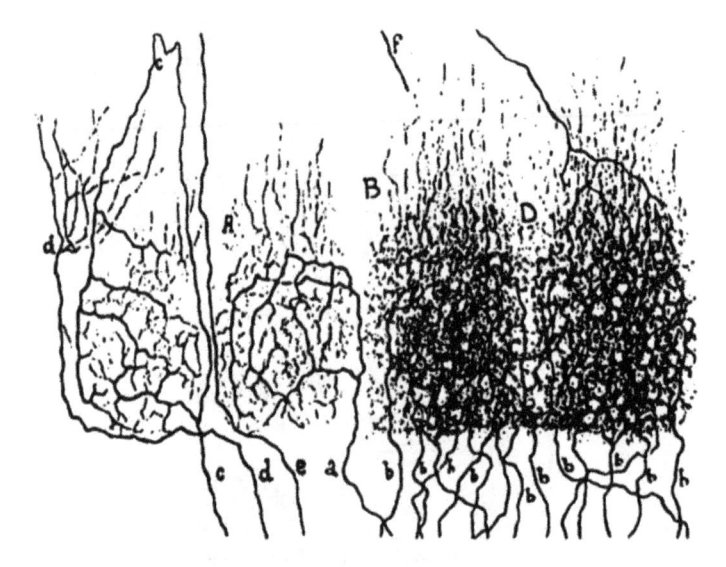

**Figure 9.** A portion of Lorente's Figure 26 in which he summarized the afferents to the *glomérulos*. Fibers labeled a, b and c are projection fibers which he identified as from the thalamus. They come from VB. The fiber c hooks back from supragranular layers. Fiber d likely comes from other cortex. The fibers labeled f (upper center-right) could arise in the brainstem. A and B identify the dense plexes of fibers to the glomeruli; D the interglomerular space includes what we later termed the septum.

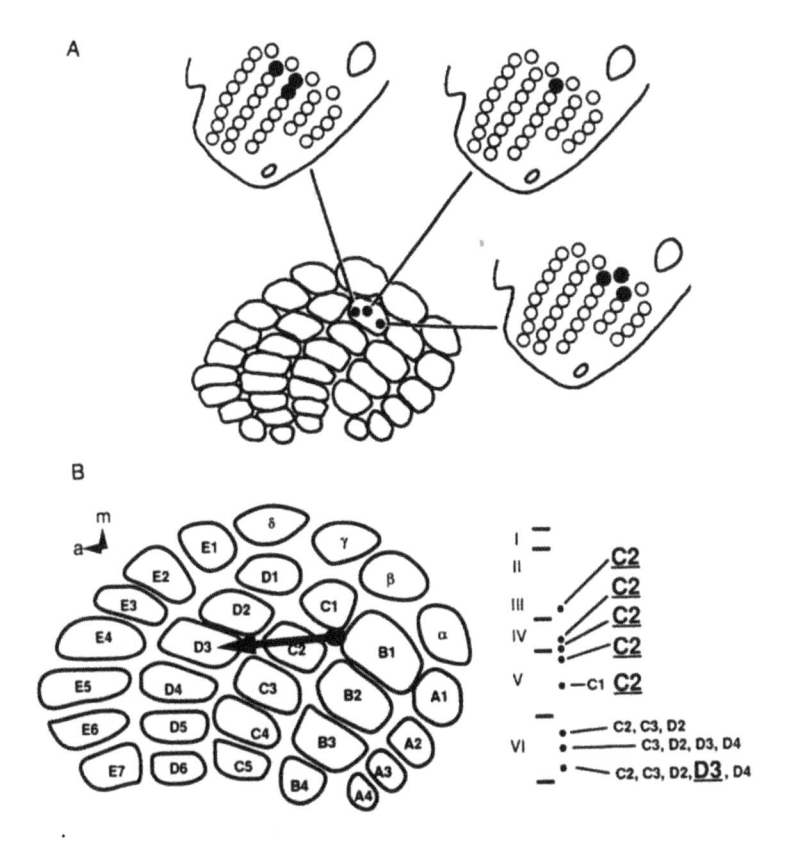

Figure 11. The glomérulos are in vertical arrays. A. Cells in a particular barrel are driven by the appropriate contralateral whisker. Three cells in the barrel C1 column are all strongly activated by whisker C1. They are also driven by adjacent whiskers in a way that depends on the layer of the cortex. B. A slightly oblique microelectrode penetration from the pia enters the brain just caudal to the C2 barrel and ends deep to the D3 barrel. To the right are the depths of eight units isolated in the penetration and their relation to the cortical layers. Cells near layer IV are activated mainly by the C2 whisker. In deeper layers and, in other penetrations, in more superficial layers the receptive fields are larger. This is a property that emerges because of the intracortical connections which Dr. Lorente described (modified from Simons and Woolsey, '79).

## Each barrel is in a cortical column

Surface recording of evoked potentials is a good way to survey the cortical landscape but lacks the spatial and temporal precision to define groupings of cells with dimensions on the order of a millimeter or less in diameter which was the known dimension of the columns (Hubel and Wiesel, '63; Jones et al., '75) and of the barrels (Woolsey and Van der Loos, '70). Carol Welker first recorded multiunit activity (neural hash) from the rat whisker cortex with low impedance microelectrodes (Welker, '76). Under deep anesthesia she observed responses to single vibrisssae in layer IV and marked the sites with small lesions to establish the map at higher spatial resolution. Her work demonstrated that the anatomical map represented a functional map, in detail, and it was she who first showed that the granule cells of layer IV outline the full body form of the rat (Welker, '76). A short while later in St. Louis, Dan Simons completed an elegant thesis which characterized information processing in the barrel cortex ofthe rat (Simons, '78). When he joined our laboratory, Dan repeated the

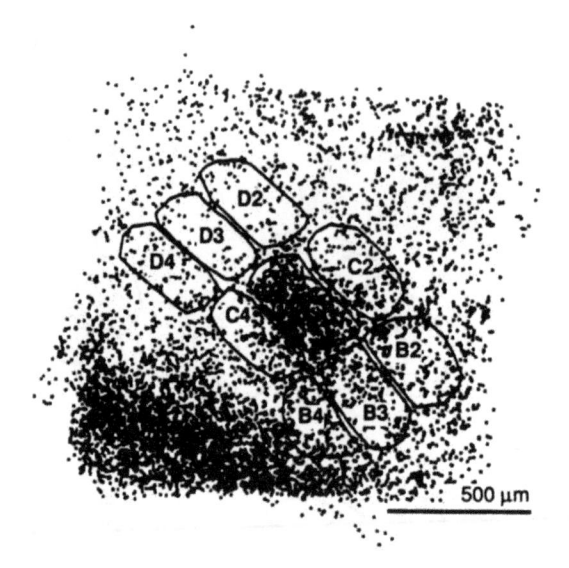

**Figure 12.** Plots of neurons labeled with tritiated 2-DG with stimulation of the right C3 whisker. All the larger whiskers except C3 were clipped from the right face. The animal was given radioactive 2-DG IP and allowed to roam for 45 minutes. Emulsion autoradiography of serial tangential sections, which were counterstained for cell bodies, was used to identify labeled neurons. This is a map of all labeled neurons (·) from 12 aligned serial sections. The heavily labeled neurons are largely confined to the column of tissue outlined by the C3 barrel in layer IV. The labeled cells to the lower left reflect activity driven from the small whiskers on the upper lip which were not clipped and which the animal evidently used in exploration. The number of neurons in a whisker column is estimated to be about 10,000; in this experiment about half of them were heavily labeled. For further details see text (modified from McCasland and Woolsey, '88b).

analysis on the mouse cortex (Simons and Woolsey, '79). With the higher impedance electrodes and a different anesthetic protocol, the localization was precise and the first glimmerings of the richness of the intracortical processing became clear. For instance, as shown in Figure 11, three electrode tracks localized to the left C1 barrel all responded most vigorously to stimulation of the right C1 vibrissa.

Electrodes advanced through the cortex always detected cells that responded best to the whisker of the appropriate barrel but receptive fields changed systematically according to depth in the cortex in a way that was lamina specific. For instance, as shown in Figure 11, cells outside layer IV respond to other whiskers; generally those in the supragranular layers respond to other adjacent whiskers mainly in the same row while those in the infragranular layers respond to more whiskers still, many in other rows. This feature is a clue to the integration of information from different barrels (see below). The critical point here is that, as Simons had shown for the rat, the cells in a vertical array spanning the cortical depth all were related to the same part of the periphery. This was a functional confirmation in the mouse barrel cortex of the hypothesis that Lorente had advanced from his anatomical studies of the mouse barrel cortex.

Even the fine microelectrode has limitations. The spatial resolution is only secure to the single cell level if the recorded neurons are marked; the number of units that can be recorded in a tissue the size of a cortical column is relatively small because the procedure is invasive. We began studies with Professor Kruger which ultimately allowed us to circumvent these limitations, with a loss of temporal resolution which is tolerable for certain problems (Durham et al., '81). When Jim McCasland joined the laboratory, one of the issues he was interested in was how many cells are active in a cortical column and what is their spatial localization. With high-resolution autoradiography and low energy $^3$H isotopes of 2 deoxy-

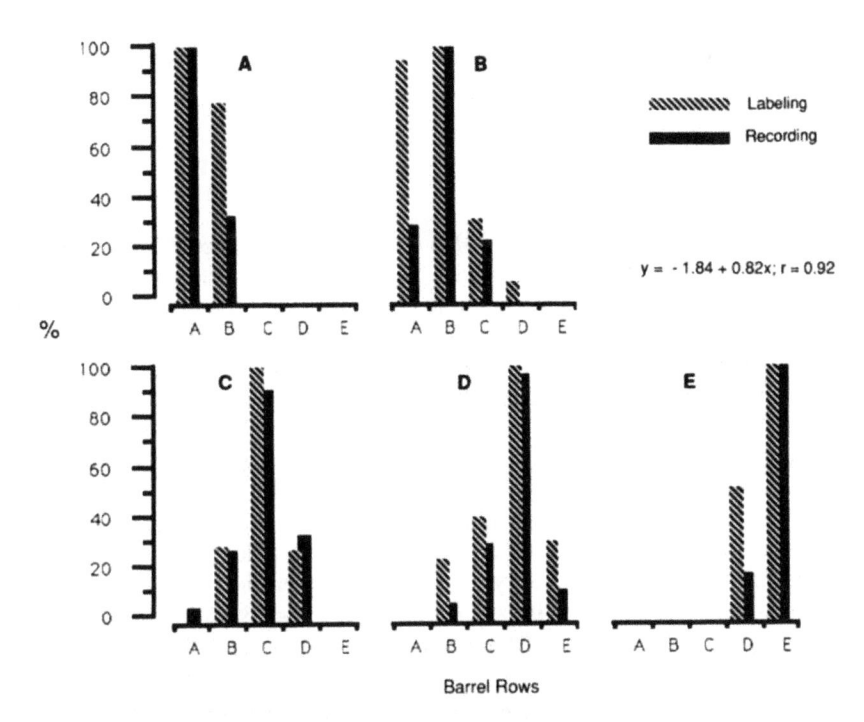

**Figure 13.** Correlation of intracortical interbarrel connectivity and functional integration from different whiskers. Injections of horseradish peroxidase (HRP) were made into the barrel cortex and the positions of labeled cells quantified. Given a labeled cell in one barrel the probability that a cell in any other barrel would be labeled was computed. These probabilities estimate the strength of connectivity within a barrel row (e.g., A1 and A2 labeled) and the strength of connectivity between whisker rows (e.g., A2 and B3 labeled). The results are presented as five histograms, each representing a barrel row. A value of 100% means a barrel is always interconnected with other barrels in a barrel row and 0% never. The middle histogram indicates that label in a C barrel always is associated with label in another C barrel, and label in a row B and/or a row D barrel about 25% of the time. All histograms indicate stronger connections between barrels in the same row than between barrels in different rows. These data are compared to normalized data from recordings of single units in the mouse barrel field. The plots indicate the observed frequency of units in a particular barrel's column that are driven by whiskers in the same row or by whiskers in different rows. The bulk of the multi-whisker activity is due to intracortical connectivity. The high correlation between the two histograms indicates that this property is likely carried by the asymmetrical intracortical projections (modified from Bernardo et al., '90a and Simons et al., '84).

D-glucose we mapped the location of neurons activated by the C3 vibrissa (McCasland and Woolsey, '88a; McCasland and Woolsey, '88b). As shown in Figure 12, the cells labeled after 45 min of stimulation (the effective stimulus period is less than 10 min) are mostly confined to the vertical column of cells outlined by the projection of the C3 barrel outline through the cortical thickness. Jim estimates that over the 10 min period of the effective stimulus over 50% of all the neurons in the C3 column were labeled, and, this must be a minimal estimate of the percentage of cells in this column activated by the simulation of this whisker. This pattern of active cells is consistent with the hypothesis advanced by Lorente de Nó in 1938 (Lorente de Nó, '38).

Not shown, and relevant in this context, is the processing that Simons and his colleagues have demonstrated in the barrel cortex. The neurons are tuned to stimulus frequency and deflection direction, monitor the duration and amplitude of the whisker deflections and integrate information from stimuli crossing the whiskers. It is still not clear how "pure" the inputs from the TCAs to a single barrel are (Simons, '83; Simons, '85; Simons and Carvell, '89). Nonetheless, much of the integration of information from different whiskers is achieved by interactions and therefore connections between adjacent cortical columns.

## A cortical map is made of columns

The *glomérulos* gave Lorente de Nó a unit to which he could relate observations of different neuron types. The barrels provide a context in which to study the functional and anatomical interactions between the different columns. These interactions are implied by the convergence of activity from different whiskers and, therefore different columns; the route of information flow is largely as inferred by Lorente. With carefully and independently controlled stimuli to different whiskers, Simons has shown that neurons likely in layer V are highly sensitive to the direction of movements across the whiskers (Simons, '85). For instance, a neuron can be strongly excited by sequences of stimuli from front to back and strongly inhibited by a converse sequence. More global interactions are evident in McCasland's data where the number of labeled neurons in a whisker column is less when all the whiskers are stimulated than when only one is active (McCasland and Woolsey, '88b). This result, in part, implies inhibitory interactions between the columns.

The anatomical bases for these interactions are suggested from the careful drawings of Lorente de Nó (see Fig. 6 F). But given the nature of the spatial interactions between the columns that create the tile-like map (Senft and Woolsey, '91a), the best way to see these relationships is as if looking down from pia. Kerry Bernardo and Jim McCasland investigated this issue quantitatively and systematically by placing small amounts of the tracer horseradish peroxidase (HRP) into different laminae in the barrel field. The positions of retrogradely labeled somata were highly lamina specific both in terms of the somata that were labeled and the location of their axons in different cortical layers (Bernardo et al., '90a; Bernardo et al., '90b). Figure 13 shows an aspect of these asymmetrical and highly ordered intracortical projections. The strongest intercolumnar connections of the supragranular layers link barrels and information from whiskers in the same row while those in the infragranular layers also link information from whiskers in different rows. The pattern was partially anticipated by Rafael (Lorente de Nó, '38) and is consistent with the interactions described by Simons ('78; '85). Obviously in the mouse cortex, as in the intensively studied visual cortex (Lorente was especially impressed by the work of Jennifer Lund (Lund and Booth, '75), the columns are packed in a map of the periphery and linked to extract features relevant to what is occurring in that periphery.

These connectional asymmetries are especially interesting in relation to the inhibitory interactions in layer IV of the barrel cortex which result from prior movements of the whiskers next to the one directly associated with a particular barrel. Dan Simons and George Carvell noticed that all of the adjacent whiskers inhibit the principal whisker response if they are stimulated first. The pattern of this inhibition is asymmetrical in that the caudal whisker inhibits the principal whisker more strongly than the rostral one (Simons and Carvell, '89). This is shown schematically in Figure 14. To test whether this asymmetry was a feature general to the whole barrel field, we calculated the propagation of the asymmetry across a grid of nine barrels and used the relative activity in these same barrels as measured with 2 DG as the experimental test of the model (McCasland et al., '91). The model accurately predicted the 2DG results. This suggests that there is strong asymmetrical inhibitory influence across the barrel field which for a variety of reasons we believe develops in the postnatal period probably as a result of experience. Further, studies in which we combined the 2 DG method with immunocytochemistry for GAD immunoreactivity suggest that cells described by Lorente de Nó (see Fig. 7F- cell K) are responsible for this pattern of asymmetrical intracortical inhibition.

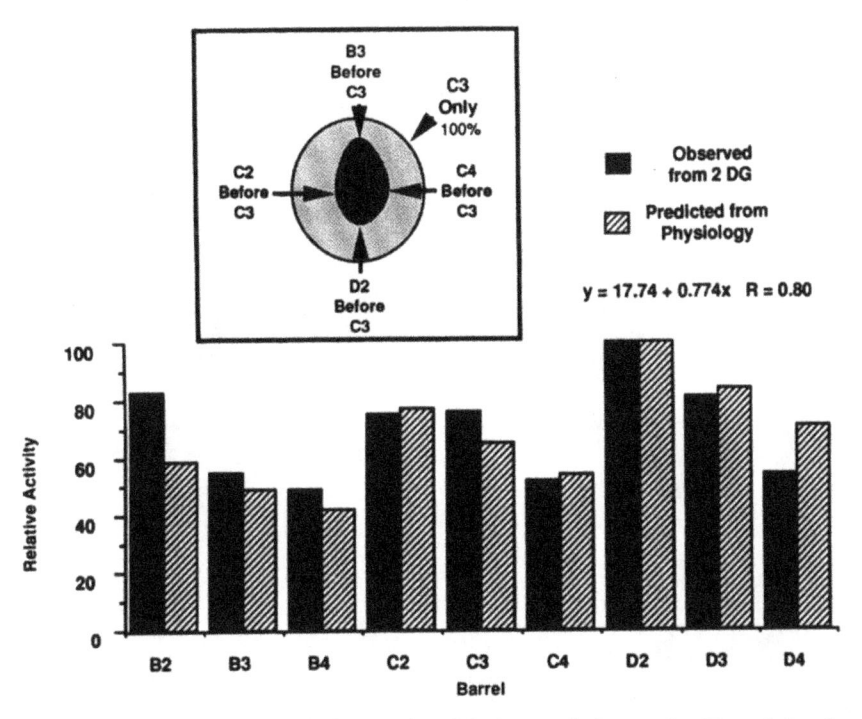

**Figure 14.** Another test of the asymmetrical properties of the intracortical networks. The activity of a cell in a particular barrel when it is driven by the homologous whisker is inhibited when an adjacent whisker is stimulated 20 ms earlier (Simons and Carvell, '89). The box shows this schematically for the C3 whisker. The pattern of the inhibition in rats is asymmetrical such that caudal (C2) and ventral (D3) whiskers are stronger at suppressing the subsequent firing of a barrel neuron than a rostral (C4) or dorsal (B3) whiskers. If the effect is general, then the activity in the barrel columns can be calculated or "predicted". The effect is evident in the asymmetrical pattern of 2DG labeling observed in whisker columns when a behaving mouse is using all its whiskers. The high correlation suggests that inhibitory mechanisms, possibly mediated by cells of the kind described by Dr. Lorente (see Fig 7, F) link adjacent cortical columns in behaving animals (modified from McCasland et al., '91).

## Development depends on integrity of the periphery

Certain manipulations can change the function and structure of the barrels. Results from many laboratories now show that the mouse and rat whisker systems are outstanding models for brain development and plasticity (Killackey et al., '90; Rhoades et al., '90; Woolsey, '90). The central feature of the system is that it permits results to be interpreted in a context that extends along the axis of the whisker system from the skin to the cortex. At birth neural differentiation is still in progress (Senft and Woolsey, '91c; Rice and Van der Loos, '77; Belford and Killackey, '79; Killackey and Belford, '79). For instance, the thalamocortical fibers which are in the cortex at birth grow and sort over the first week of life. Recently Steve Senft analyzed the way in which this happens in detail (Senft and Woolsey, '91a; Senft and Woolsey; '91b; Senft and Woolsey, '91c). The fibers at first outline homogeneously the somatic cortex (see fig 15 A). Three days after birth, the pattern formed is predominantly row like (Fig. 15 B) and only two days later is segregated barrel like in layer IV and upper layer VI which are the thalamic recipient layers in all sensory corticies (Land and Simons, '85). Senft has demonstrated that the progression is due to a focal increase in branching and a selective pruning of branches that initially extend too widely (Senft and Woolsey, '91b). With many of new probes generated by molecular biologists (Crandall et al.,

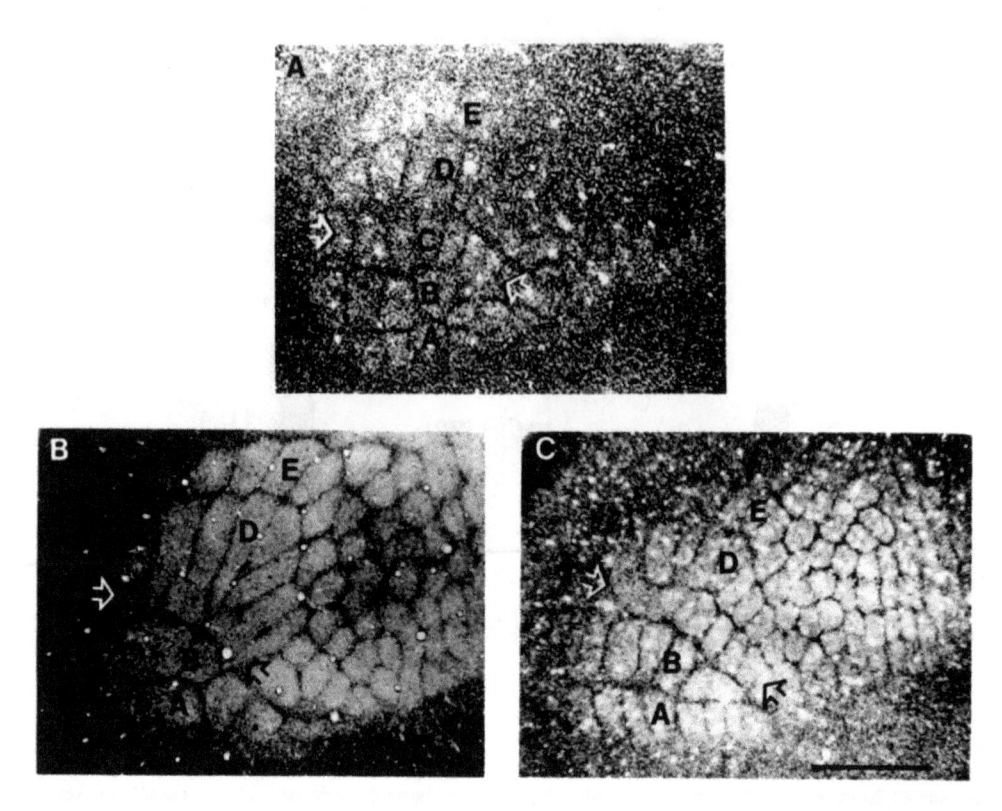

Figure 16. A lectin from the peanut plant transiently labels extracellular matrix molecules in young rodents (Cooper and Steindler, '86). The tile-like patterns of the whisker barrels is exceptionally clear on P7 as the densest staining is between the barrels. A. The normal pattern of representation of the whiskers is reminiscent of the vibrissae which supply them. If the middle row of whiskers, row C, is destroyed in early life, the appropriate barrels do not form normally. B. Whisker damage at birth obliterates the barrels in row C (arrows); the adjacent and anterior row D enlarges. C. Whisker damage 3 days after birth prevents the formation of separate barrels but the row-like arrangement for C is pronounced (arrows). Bar = 500 $\mu$m; medial up and anterior right for all sections. (Prepared by Jon Christensen.)

'90; Crossin et al., '89; Cobas et al., '91) and some older ones from their biochemical forefathers (Steindler et al., '90) it is very clear that numerous different molecules, which are related to the extracellular matrix, cytoskeleton, metabolism, supporting cells and transmitters each have different timetables of expression. It is probably no coincidence that many of these molecules are expressed only in this developmental period, although the functional meaning of these expression patterns is not yet obvious.

In 1973 Van der Loos and I showed that simple whisker damage in early life could change the whisker maps in the cerebral cortex (Van der Loss and Woolsey, '73). If we had read Lorente's favorite, the Don Quixote of Cervantes, the experimental idea might have occurred sooner. As the priest suggested "*I ... pulled out one of my moustaches*" (Cervantes Saavedra, '65). When we lesioned the whiskers in the postnatal period, the appropriate part of the map changed always and the changes that could be provoked were graded according to the time they were induced. There is a critical period after which there are no detectable anatomical changes. This period coincides with the periods of normal development and the changes provoked by whisker damage or section of the infraorbital nerve which supplies them are consistent with an arrest of development at the time the peripheral lesion was made. I have elsewhere reviewed this problem in some detail as have others (Killackey et al., '90; Rhoades et al., '90; Woolsey, '90) and a full discussion of the issues would take us well beyond the relationship of our work to the early observations of Lorente de Nó. The point here is that the thalamocortical axons that he identified as the central element in a whisker column are responsible for setting up the initial map and engaging the neurons in a column. The patterns shown in Figure 16 illustrate these altered patterns and show the time dependence of the changes. Recently Schlaggar and O'Leary (Schlaggar and O'Leary, '91) showed that occipital cortex transplanted to the parietal area develops barrels while parietal cortex transplanted to an occipital location developed as visual cortex. While it has yet to be demonstrated that the transplants function as normal cortex does, it is intriguing that cortex in the rodent is interchangeable and its structure evidently can be determined by the extrinsic connections it receives.

## Postlogue - a personal perspective

Dr. Kruger and I have told elsewhere of the motivation for and the remarkable circumstances under which Lorente de Nó undertook his study on the mouse cerebral cortex in Cajal's laboratory (Kruger and Woolsey, '90; Gallego, '90). He extended his training and his intellectual horizons remarkably in the 15 or so years from the time the first paper was published until he wrote his last article on the subject for John Fulton (Lorente de Nó, '38). Rafael told me he found the *glomérulos* the same way I did. They are so obvious in the mouse that they cannot be easily missed. In retrospect it seems that Rafael took the anatomy he had discovered and incorporated subsequent information and thinking from his experience with brainstem polysynaptic reflexes and from his work on the pyriform cortex and the hippocampus for the chapter in Fulton's book. That book, in turn, the benchmark text of its time, was widely read by many young neurophysiologists who doubtless were persuaded by Lorente's clear exposition (Mountcastle, '57; Hubel and Wiesel, '63). I think that his work had a profound impact on the development of the columnar hypothesis which has been widely confirmed in many cortices including the barrel cortex of the mouse.

With regard to our own work, much of the detail from our studies of the Golgi architecture, confirms Lorente's initial clear understanding of how cells in this part of the brain are organized for information processing. It is on the interactions of all of these cells that we and mice depend. I have tried to show here how the simple and correct idea of a grouping of cells in relation to afferents from the thalamus to the cortex has lead us and others to explore the functional organization of this columnar system, the grouping of many columns into a map where they interact and the specification of the map by the columns in development. It is remotely possible that we could have figured things out in the absence of Lorente's work, but it is more likely that without his work our attention might have been directed to other, less fundamental things. I think the burgeoning literature on many aspects of the barrel system, as a model system for the brain (O'Leary, '89; White, '89), indeed as a brain system to be modeled (Senft and Woolsey, '91a; Montague et al., '91), will lead us and others in new and interesting directions. Because mice are the mammal for genetic manipulations, it is likely that the system will yield to inquiries on the relation of the brain to the genome (Marx, '82; Hogan et al., '86; Van der Loos et al., '86).

There is a different, more personal, side to my interactions with Rafael Lorente de Nó. I am reminded of an improbable situation in his beloved Don Quixote where two brothers, one a judge the other a captain in the army, who had been separated by hundreds of miles for decades meet by chance in a small Manchegian Inn (Cervantes Saavedra, '65). In 1978 I invited Rafael Lorente de Nó to St. Louis to give 4 lectures (see Fig. 17). On the way to our house from the airport, I asked him whether he had any friends from his days in St. Louis in the mid 30s when Rafael served as director for research at the Central Institute for the Deaf. He doubted that they were around but he recalled one person whom he recalled as ".... the most intelligent and interesting woman I ever met." From his description I had

# XV

## CEREBRAL CORTEX: ARCHITECTURE, INTRACORTICAL CONNECTIONS, MOTOR PROJECTIONS *

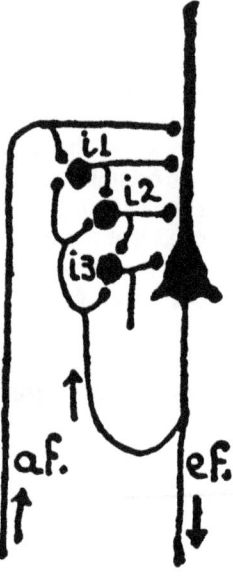

**Figure 17.** The title of Lorente de Nó's chapter in Fulton's textbook (Lorente de Nó, '38) which he endorsed this during his 1978 visit to St. Louis. The small schema to the right of cortical cortical connectivity summarizes Lorente's ideas about "some of intracortical chains of neurons" ('38). Dr. Lorente de Nó described the schema as follows: "This diagram exemplifies the broad plan upon which the central nervous system is organized."

no trouble identifying this person as Mrs. Samuel B. Grant, who lived with Dr. Grant in a house directly across the street from ours. The unit in which I work commemorates Dr. O'Leary to whom Lorente taught the Golgi techniques (O'Leary and Bishop, '38). In the 30's, in St. Louis, Lorente introduced Dr. O'Leary to Mrs. O'Leary, who became colleagues and friends.

It is clear that Lorente's work on the neocortex was a landmark which has profoundly influenced the work and thinking of many. His decision to change his scientific direction in the mid-thirties was not odd. He could not easily answer the functional questions that his final summary diagram directly led him to (see Fig. 17) ; the methods were not available. Nevertheless, I presume that in paying homage to Rafael Lorente de Nó now as we did when he was living we follow the suggestion of Don Quixote:

"'One of the things,' said Don Quixote ..., 'which must give the greatest pleasure to a virtuous and eminent man is to see himself, in his life-time, printed in the Press, and with a good name on people's tongues.'" (Cervantes Saavedra, '65)

Were Dr. Rafael Lorente de Nó to have done nothing more than to study the mouse somatic cortex and from that develop a general view of the functional organization of cortical neurons, we would still recognize his work in the press and speak well of him. In these two works, he left his mark for all students of the structure and function of the cerebral cortex for all time.

**Note added in proof:**
* See the recently published excellent English translation of Lorente de Nó 1922 (Lorente de Nó, '92) by A.Fairén, J.Regidor and L.Kruger.

ACKNOWLEDGEMENTS

I thank Drs. James S. McCasland, Stephen L. Senft and Kerry L. Bernardo for their expertise and innovation and Jon Christensen and Sandra Kalmbach for excellent technical assistance. Supported by NIH grants NS 17763 and DE 07734, the McDonnell Center of Higher Brain Function, and the Spastic Paralysis Foundation of the Illinois-Eastern Iowa District of Kiwanis International.

REFERENCES

Belford, G.R. and Killackey, H.P., 1979, The development of vibrissae representation in subcortical trigeminal centers of the neonatal rat, *J. Comp. Neurol.*, 188:63.

Bernardo, K.L., McCasland, J.S. and Woolsey, T.A., 1990, Local axonal trajectories in mouse barrel cortex, *Exp. Brain Res.*, 82:247.

Bernardo, K.L., McCasland, J.S., Woolsey, T.A. and Strominger, R.N., 1990, Local intra- and interlaminar connections in mouse barrel cortex, *J. Comp. Neurol.*, 291:231.

Carvell, G. and Simons, D.J., 1990, Biometric analyses of vibrissal tactile discrimination in the rat, *J. Neurosci.*, 10:2638.

Cervantes Saavedra, M., 1965, "The Adventures of Don Quixote," J.M. Cohen (trans.), Penguin, Baltimore.

Cobas, A., E. Welker, Fairén, A., Kraftstik, R. and Van der Loos, H., 1987, The GABAergic neurons in the barrel cortex of the mouse: an analysis using neuronal archetypes, *J. Neurocytol.*, 16:843.

Cobas, A., Fairén, A., Alverez-Bolado, G., and Sánchez, M.P., 1991, Prenatal development of the intrinsic neurons of the rat neocortex: a comparative study of the distribution of GABA-immunoreactive cells and the GABAa receptor, *Neurosci.*, 40:375.

Cooper, N.G.F. and Steindler, D.A., 1986, Lectins demarcate the barrel subfield in the somatosensory cortex of the early postnatal mouse, *J. Comp. Neurol.*, 249:157.

Crandall, J.E., Misson, J.-P., and Butler, D., 1990, The development of radial glia and radial dendrites during barrel formation in mouse somatosensory cortex, *Dev. Brain Res.*, 55:87.

Crossin, K.L., Hoffman, S., Tan, S.-S., and Edelman, G.M., 1989, Cytotactin and its proteoglycan ligand mark structural and functional boundaries in somatosensory cortex of the early postnatal mouse, *Dev. Biol.*, 136:381.

Dawson, D.R. and Killackey, H.P., 1987, The organization and mutability of the forepaw and hindpaw representations in the somatosensory cortex of the neonatal rat, *J. Comp. Neurol.*, 256:246.

de Veries, I., 1912, Über die Zytoarchitectonik der Grosshirnrinde der Maus und über die Beziehungen der einzelnen Zellschichten zum Corpus Callosum auf Grund von experimentellen Läsionen, *Folia Neurobiol.* (Lpz.), 6:288.

Droogleever Fortuyn, A.B., 1914, Cortical cell-lamination of the hemispheres of some rodents, *Arch. Neurol. Psychiat.* (Lond.), 6:221.

Durham, D., Woolsey, T.A. and Kruger, L., 1981, Cellular localization of $^3$H-2-deoxy-d-glucose (2DG) from paraffin embedded brains, *J. Neurosci.*, 1:519.

Friede, R.L., 1966, "Topographic Brain Chemistry," Academic, New York.

Gallego, A., 1990, La obra científica del doctor Lorente de Nó, *An. Real Acad. Nacional Med.*, 107:467.

Godement, P., Vaneslow, J., Thanos, S., and Bonhoeffer, F., 1987, A study in developing visual system with a new method of staining neurons and their processes in fixed tissue, *Development*, 101:697.

Harris, R.M., 1986, Morphology of physiologically identified thalamocortical relay neurons in the rat ventrobasal thalamus, *J. Comp. Neurol.*, 251:491.

Hogan, B., Costantini, F., and Lacy, E., 1986, "Manipulating the Mouse Embryo: A Labororatory Manual," Cold Spring Harbor, Cold Spring Harbor.

Hubel, D.H. and Wiesen, T.N., 1963, Shape and arrangement of columns in cat's striate cortex, *J. Physiol.*, (Lond.), 165:559.

Huston, K.A. and Masterson, R.B., 1986, The sensory contribution of a single vibrissa's cortical barrel, *J. Neurophysiol.*, 56:1196.

Isenschmid, R., 1911, Zur Kenntnis der Grosshirnrinde der Maus, *Abh. kön. preuss. Akad. Wiss., Anhang.*, 3:1.

Jacquin, M.F., Mooney, R.D., and Rhoades, R.W., 1984, Axon arbors of functionally distinct whisker afferents are similar in medullary dorsal horn, *Brain Res.*, 298:175.

Jhaveri, S., Ersurumlu, R.S. and Crossin, K., 1991, Barrel construction in rodent neocortex: role of thalamic afferents versus extracellular matrix molecules, *Proc. Nat. Acad. Sci. USA*, 88:4489.

Jones, E.G., Burton, H. and Porter, R., 1975, Commisural and cortico-cortical "columns" in the somatosensory cortex of primates, *Science* 190:572.

Keller, A. and White, E.L., 1987, Synaptic organization of GABAergic neurons in the mouse SmI cortex, *J. Comp. Neurol.*, 262:1.
Killackey, H.P. and Belford, G.R., 1979, The formation of afferent patterns in the somatosensory cortex of the neonatal rat, *J. Comp. Neurol.*, 183:285.
Killackey, H.P., Jacquin, M.F. and Rhoades, R.W., 1990, Development of somatosensory system structures, in: "Development of Sensory Systems in Mammals", J. Coleman, ed., Wiley, New York.
Kruger, L. and Woolsey, T.A., 1990, Rafael Lorente de Nó: 1902-1990, *J. Comp. Neurol.*, 300:1.
Land, P.W. and Simons, D.J., 1985, Cytochrome oxidase staining in the rat SmI barrel cortex, *J. Comp. Neurol.*, 238:225.
Lin, C.-S., Lu, S.M., and Schmechel, D.E., 1985, Glutamic acid decarboxylase immunoreactivity in layer IV of barrel cortex of rat and mouse, *J. Neurosci.*, 7:1934.
Loewy, A.D., 1971, Ramón y Cajal and methods of neuroanatomical research, *Persp. Biol. Med.*, 15:7.
Lorente de Nó, R., 1922, La corteza cerebral del ratón, *Trab. Lab. Invest. Biol. Univ. Madrid*, 20:41.
Lorente de Nó, R., 1938, Architectonics and structure of the cerebral cortex, in: "Physiology of the Nervous System," J.F. Fulton, ed., Oxford, London.
Lorente de Nó, R., 1992, The cerebral cortex of the mouse (A first contribution - the "acoustic" cortex), *Somatosensory and Motor Res.*, 9:3.
Lund, J.S. and Booth, R.G., 1975, Interlaminar connections and pyramidal neuron organization in the visual cortex, area 17, of the macaque monkey, *J. Comp. Neurol.*, 159:305.
Marx, J.L., 1982, Tracking genes in developing mice, *Science*, 215:44.
McCasland, J.S. and Woolsey, T.A., 1988, A new high resolution 2-deoxyglucose method for featuring double labeling and automated data collection, *J. Comp. Neurol.*, 278:543.
McCasland, J.S. and Woolsey, T.A., 1988, High resolution 2DG mapping of functional cortical columns in mouse barrel cortex, *J. Comp. Neurol.*, 278:555.
McCasland, J.S., Carvell, G.E., Simons, D.J. and Woolsey, T.A., 1991, Functional asymmetries in the rodent barrel cortex, *Somatosensory and Motor Res.*, 8:111.
Montague, P.R., Gally, J.A., and Edelman, G.M., 1991, Spatial signaling in the development and function of neural connections, *Cerebral Cortex*, 1:199.
Mountcastle, V.B., 1957, Modality and topographic projection of single neurons of cat's somatic sensory cortex. *J. Neurophysiol.*, 20:408.
Nachlas, M.M., Tsou, K.-C., de Souza, E., Cheng C.-S., and Seligman, A.M., 1956, Cytochemical demonstration of succinic dehydrogenase by the use of a new p-nitrophenyl substituted ditetrazole, *J. Histochem. Cytochem.*, 5:420.
O'Leary, D.D.M., 1989, Do cortical areas emerge from a protocortex? *TINS*, 12:400.
O'Leary, J.L. and Bishop, G.H., 1938, The optically excitable cortex of the rabbit, *J. Comp. Neurol.*, 68:423.
Peinado, A. and Katz, L.C., 1990, Development of cortical spiny stellate cells: retraction of a transient apical dendrite, *Soc. Neurosci. Abstr.*, 16:1127.
Rhoades, R.W., Bennett-Clarke, C.A., Chiaia, N.L., White, F.A., Macdonald, G.J., Haring, J.H., and Jacquin, M.F., 1990, Development and lesion induced reorganization of the cortical representation of the rat's body surface as revealed by immunocytochemistry for serotonin, *J. Comp. Neurol.*, 293:190.
Rhoades, R.W., Killackey, H.P., Chiaia, N.L. and Jacquin, M.F., 1990, Physiological development and plasticity of somatosensory neurons, in: "Development of Sensory Systems in Mammals", J. Coleman, ed., Wiley, New York.
Rice, F.L. and Van der Loos, H., 1977, Development of the barrels and barrel field in the somatosensory cortex of the mouse, *J. Comp. Neurol.*, 171:545.
Rice, F.L. and Munger, B.L. 1986, A comparative light microscopic analysis of the sensory innervation of the mystacial pad. I. Innervation of vibrissal follicle-sinus complexes, *J. Comp. Neurol.*, 252:154.
Rose, M., 1929, Cytoarchitektonischer Atlas der Grosshirnrinde der Maus, *J. Psychol. Neurol.*, (Lpz.), 40:1.
Rose, M., 1912, Histologische Lokalisation der Grosshirnrinde der kleinen Säugetiere (Rodsentia, Insectivora, Chiroptera), *J. Psychol. Neruol.* (Lpz.), 19:389.
Schlaggar, B.L. and O'Leary, D.D.M., 1991, Potential of visual cortex to develop an array of functional units unique to somatosensory cortex, *Science*, 252:1556.
Senft, S.L. and Woolsey, T.A., 1991a, Mouse barrel cortex viewed as Dirichlet domains, *Cerebral Cortex*, 1:348.
Senft, S.L. and Woolsey, T.A., 1991b, Computer-aided analysis of thalamocortical afferent ingrowth, *Cerebral Cortex*, 1:336.
Senft, S.L. and Woolsey, T.A., 1991c, Growth of thalamic afferents into mouse barrel cortex, *Cerebral Cortex*, 1:308.
Simons, D.J., 1978, Response properties of vibrissa units in rat SI somatosensory neocortex, *J. Neurophysiol.*, 41:798.
Simons, D.J. and Woolsey, T.A., 1979, Functional organization in mouse barrel cortex, *Brain Res.*, 165:327.

Simons, D.J., 1983, Multi-whisker stimulation and its effects on vibrissa units in rat SmI barrel cortex, *Brain Res.*, 276:178.
Simons, D.J. and Woolsey, T.A., 1984, Morphology of Golgi-Cox impregnated barrel neurons in rat SmI cortex, *J. Comp. Neurol.*, 230:119.
Simons, D.J., Durham, D. and Woolsey, T.A., 1984, Functional organization of mouse and rat SmI barrel cortex following vibrissal damage on different postnatal days, *Somatosensory Res.*, 1:207.
Simons, D.J., 1985, Temporal and spatial integration in the rat SI vibrissa cortex, *J. Neurophysiol.*, 54:615.
Simons, D.J. and Carvell, G.E., 1989, Thalamocortical response transformation in the rat vibrissa/barrel system, *J. Neurophysiol.*, 61:311.
Steindler, D.A., O'Brien, T.F., Laywell, E., Harrington, K., Fassner, A., and Schachner, M., 1990, Boundaries during normal development: in vivo and in vitro studies of glia and glycoconjugates, *Exp. Neurol.*, 109:35.
Van der Loos, H. and Woolsey, T.A., 1973, Somatosensory cortex: structural alterations following early injury to sense organs, *Science.*, 179:395.
Van der Loos, H., Welker, E., Dörfl, J., and Rumo, G., 1986, Selective breeding for variation in patterns of mystacial vibrissae of mice, *J. Hered.*, 77:66.
Welker, C., Receptive fields of barrels in the somatosensory neocortex of the rat, *J. Comp. Neurol.*, 166:173.
Welker, W.I., 1964, Analysis of sniffing of the albino rat, *Behaviour*, 22:223.
Welker, C., 1976, Receptive fields of barrels in the somatosensory neocortex of the rat, *J. Comp. Neurol.*, 166:173.
White, E.L., 1989, "Cortical Circuits: Synaptic Organization of the Cerebral Cortex - Structure, Function, and Theory," Birkhäuser, Boston.
White, E.L., 1991, Identified neurons in mouse SmI cortex which are post-synaptic to thalamocortical axon terminals: a combined Golgi-electron microscopic and degeneration study, *J. Comp. Neurol.*, 181:627.
Wong-Riley, M., 1979, Changes in the visual system of monocularly sutured or enucleated cats demonstrable with cytochrome oxidase histochemistry, *Brain Res.*, 171:11.
Woolsey, C.N., 1952, Patterns of localization in sensory and motor areas of the cerebral cortex, in: "The Biology of Mental Health and Disease," Hoeber, New York.
Woolsey, T.A., 1967, Somatosensory, auditory and visual cortical areas in the mouse, *Johns Hopk. Med. J.*, 121:91.
Woolsey, T.A. and Van der Loos, H., 1970, The structural organization of layer IV in the somatosensory region (SI) of mouse cerebral cortex: the description of a cortical field composed of discrete cytoarchitectonic units, *Brain Res.*, 17:205.
Woolsey, T.A., Dierker, M.L. and Wann, D.F., 1975, Mouse SmI cortex: qualitative and quantitative classification of Golgi-impregnated barrel neurons, *Proc. Natl. Acad. Sci. USA*, 72:2165.
Woolsey, T.A., 1983, Gerard award presentation, *Neurosci. Newsletter*, 14(4):5.
Woolsey, T.A., 1990, Peripheral alteration and somatosensory development, in: "Development of Sensory Systems in Mammals", J. Coleman, ed., Wiley, New York.
Woolsey, T.A. and Rovainen, C.M., 1991, Whisker barrels: a model for direct observation of changes in the cerebral microcirculation with neuronal activity, in: "Alfred Benzon Symposium No. 31, Brain Work II," D.H. Ingvar, N.A. Lassen, M.E. Raichle and L. Friberg, eds., Munksgaard: Copenhagen.

# LORENTE DE NÓ: THE ELECTROPHYSIOLOGICAL EXPERIMENTS OF THE LATTER YEARS

Lawrence Kruger

Departments of Anatomy and Cell Biology, Anesthesiology and the Brain Research Institute, UCLA Medical Center, Los Angeles, CA 90024, USA

*"Cosas veredes del Cid que las piedras hablaran".*
*(M. Cervantes)*

Among the truly prominent figures of 20th century neuroscience it is difficult to identify anyone who can approach the breadth and versatility of Rafael Lorente de Nó. In recent years there has been a strong revival of interest in the anatomical studies of his early youth, especially the Golgi stain analyses of the cerebral cortex. He remained interested in this for the remainder of his extraordinary career, but shortly after his arrival in the United States, working at the Central Institute for the Deaf in St. Louis, he was inevitably caught up in the excitement generated by the newly emerging electrophysiological methods used with such impressive success by Erlanger, Bishop, Gasser, Heinbecker, O'Leary and others in that vibrant era at Washington University. Expressing a life long penchant for originality, Lorente soon seized upon the application of electrophysiology to the problem of synaptic integration in the central nervous system. He was particularly enamored of the idea of reverberating circuits and was certain that he would find a role for the ubiquitous interneurons in the production of prolonged electrical events that must underlie the enhancement of reflexes.

The years at the Rockefeller Institute in New York placed Lorente in an extraordinarily fertile intellectual climate central to the development of electrophysiology in the United States. Gasser and Erlanger had already been awarded a Nobel prize for elucidating the fundamental functional correlates of different types of axonal impulse conduction, and Herbert Gasser, Lorente's laboratory neighbor, was turning his attention to conduction in unmyelinated axons, the principle of decremental conduction, and the meaning of after-potentials. For Lorente, decremental conduction was the general principle and a key component in the larger problem of integration involving synaptic interaction and dendritic events. It is remarkable that he attempted to cover the whole field, including an amazingly industrious survey of the basic experiments on myelinated axon impulse conduction. His published recordings were not merely numerous, they set a high standard of technical excellence. He also wrote in precise, fluent and rather elegant English. This was a man with a flair for language and a passion for literature - especially his favorite, *Don Quixote*. In his latter years, he identified with the noble, aging fictional knight and, indeed, the similitude and a moment's reflection provide an endearing image of the latter years of Don Rafael's scientific career. In the following account, an affectionate tribute to a great man no longer with us, I shall refer to him as his colleagues and friends did - Lorente in scientific matters, and Don Rafael in intimate, yet deferential terms.

A crucial feature of Lorente's intellectual thrust and energy derived from his preoccupation with understanding the role of the connective tissue sheaths of nerves and especially the prevailing belief among electrophysiologists that the epineurium acts as an "effective diffusion barrier which prevents the exchange of solutes between a test solution applied to its surface". Moreover, he was concerned about the general view "that the epineurium has a large electric impedance, for which reason it acts as the external sheath of a double core conductor and causes a serious distortion of currents which are applied to the nerve through electrodes on its surface". This was a profoundly emotional issue because if correct, as he put it, "Work done with intact nerve trunks would be of little if any value: valid experiments could be conducted only with nerve deprived of epineurium (denuded nerve) or with single isolated nerve fibers" (Lorente de Nó, '47). In essence, he was defending the enormous body of his work published in the two volume *Studies in Nerve Physiology* (Lorente de Nó, '47) and subsequent papers against a new body of evidence that dismissed much of what had been learned from intact nerve.

The controversy may seem an esoteric episode of only historical interest, but an examination of Lorente's reasoning on this subject is not only impressive, but provides the modern axonologists with some interesting observations for contemplation. Lorente believed that the epineurium is freely permeable to solutes, and although this notion required modification when desheathed and resheathed nerve experiments indicated that it was untenable, the nature of the epineurium as a diffusion barrier remains a subject of interest. He had the peculiar notion that "the epineurium is only a relatively small part of the connective tissue sheath of nerve. The major part of the sheath is the endoneurium, etc...". Yet he clearly understood that removal of the epineurium had profound consequences and further that it was quite different from loose collagen and elastin; indeed, he thought it had significant mechanical properties because a slit in the epineurium results in rapid extrusion of fluid and nerve fibers. He did not understand that the "epineural sheath" includes an epithelial barrier, and he was mistaken in thinking that endoneural collagen contributed to the "protein content of the interstitial fluid of the nerve," but the morphology had not been adequately explored at that time, and Lorente was quite rightly struck by the profound changes in apparent swelling and in electrical properties when the epineurium was opened. Modern electron microscopy provides micrographs of the epithelial sheath, now commonly called the "perineurial epithelium" and containing cells coated with a distinct basal lamina, but the explanation of the pressure and fluid content differences on both sides of this sheath of cells remains inadequately explored.

There is something to be learned from this torrid episode in the history of neurophysiology in which powerful intellects were in combat over a number of related problems. With hindsight we can detect some of their slips, most of which might be appealing as gossip but of far less intellectual interest than the remarkably insightful observations. From Lorente's point of view it is essential to understand why this is so important, and I think there is much to be learned from this that exceeds mere anecdotal interest.

The context derives from the various solutions applied to whole nerve trunks, much of which is described in the two volume *Study of Nerve Physiology* in 1947 that was sometimes malphemistically called "the telephone book" and often treated as such, i.e. acknowledged to exist and occasionally opened for some detail but rarely read or referred to according to Lorente. Among the key observations was the use of high potassium solutions demonstrating that the resting membrane potential of frog nerve is not proportional to the logarithm of the ratio of internal to external K+ concentration. The other was the demonstration that, in the absence of external sodium ions, it is still possible to elicit many thousands of impulses in frog nerve fibers. Taken at face value, these observations placed the applicability of the Nernst equation in some jeopardy, but this was happening in the heyday of the Physiological laboratories in Cambridge, and the role of sodium and potassium ions in accounting for the form of the action potential of nerve was the focus of a brilliant advance, principally by Hodgkin and Huxley, that certainly requires no subjective comment of approval today.

The controversy extended to other related issues, all of which revolved around the functional significance of the ensheathments of nerve. As already noted, Lorente viewed the epineurium as "only a relatively small part of the connective tissue sheath of nerve. The major part of the sheath is the endoneurium, which consists of a dense network of collagen and elastic fibrils ...". The longitudinal distribution of currents was crucial to understanding conduction, and Lorente had concluded that the epineurium had no significant role in either fast or slow electrotonus, which he believed is established across the myelin of large axons. If the epineurium is not a polarizable membrane, this would fit his belief that it must be freely permeable to ions. Despite his beautiful records supporting his view, he found opposition from Cambridge and an exuberant combatant in William Rushton who, together with Rashbass (Rashbass and Rushton, '49) and armed by the findings of Anders Lundberg (Lundberg, '51), argued that the largest fraction of the fast electrotonus is established across the epineurium. Indeed, Lorente himself had shown that after a nerve was denuded and kept in Ringer's solution for several hours, the fast electrotonus was rather small compared to that observed in intact nerve although the decrease progressed rather rapidly - in a matter of minutes. He felt that the conclusion of the Cambridge workers, that the largest fraction of the fast electrotonus is established across the epineurium, was erroneous and directed at discrediting his work on intact nerve. His ire was not fully expressed in print, but he did write that the Rashbass and Rushton paper "has features that make it unique in the history of nerve physiology" and, in repeating Lundberg's microelectrode experiments, concluded "that Lundberg recorded practically nothing but artifacts" (Lorente de Nó, '52). The considerable energy and intellect Lorente devoted to this issue is impressive, and he would not allow the dispute to be resolved as a mere argument about the magnitude of the epineural contribution because he had much at stake in arguing that even inadvertent damage to the epineurium in dissecting nerve fibers would result in grossly altered functional properties. The most spectacular change, of course, was an increase in sensitivity to the lack of external sodium such that he was able to observe "denuded nerve usually becomes inexcitable in a sodium-free medium within six to eight minutes" - a clear indication to Lorente that an intact epineurium was of profound importance in axonal conduction. He also noted that certain quaternary ammonium ions restore impulse conduction in the thin, mostly unmyelinated axons in a sodium-free medium, but that this does not restore the excitability of the large myelinated (A) fibers (Lorente de Nó, 1949). In contrast, the guanidium ion restores conduction in sodium-deficient A fibers but fails in recovery of small axon conduction (Larramendi et al., '56). These observations were made in both desheathed nerve and in spinal roots.

With the hindsight of subsequent developments, it may be difficult to recognize that the clash between Lorente and the more benign Andrew Huxley and Alan Hodgkin was a disagreement among intellectual giants, and for various reasons, mercifully these workers rarely referred to each others papers subsequently. It would be a sad mistake to think of winners and losers or to trivialize the enormous achievements of Lorente. Because his reasoning remains instructive, rather than dwell upon the "sodium hypothesis", it may be worth recognizing the problems that were disturbing to his great man, some of which remain poorly understood.

Much of the early folklore in understanding the nature of the nerve impulse, especially the "all-or-none response" concept, derived from work in the first decade of this century by Keith Lucas (Lucas, '17) and Adrian (Adrian, '30) in Cambridge. They introduced the idea of measuring the impulse by its capacity to extend into adjacent segments of nerve or by its ability to traverse a narcotized stretch of nerve. Tasaki (Tasaki, '59) described the logic of these ingenious experiments as "being analogous to measuring the power of a man by his ability to cross a desert". The mechanism of narcotic action turned out to be a complicated matter that naturally attracted Lorente (Honrubia and Lorente de Nó, '62) and continues to attract considerable interest. But despite the disputes and some erroneous assumption *en route*, Lucas and Adrian turned out to be essentially correct in concluding that no matter how an impulse is generated, it obeys the all-or-none law. By the time Lorente de Nó entered the foray, attention was centered on the 'local' or 'subthreshold' response from Rushton, Hodgkin, and Huxley in Cambridge, a group interacting with the Americans, Cole and Curtis and their colleagues, who were exploiting the study of measurement of membrane properties using the giant axon of the squid. The outbreak of World War II interfered with collaboration and even interaction, but it accelerated interest in some practical studies of agents that might alter nerve conduction, and from that time onward, the problem of synaptic integration and of impulse transmission became the overwhelming thrust of the remainder of Lorente's life.

The permeability of the "epineurium" was apparent to him for several reasons, and it is noteworthy that observations made in the literature were almost invariably repeated by Lorente and usually more elegantly documented with illustrations revealing technical excellence and, parenthetically, the date and time indications reveal the extraordinarily late hours he spent in the laboratory. Noting the rapidity with which the respiratory gases ($O_2$ and $CO_2$) acted upon conduction in nerve, indicated unobstructed rapid diffusion reaching equilibrium in 6-10 *minutes*. He was further struck by the action of sodium ions and cocaine on sodium-deficient nerve in a few *seconds*. In some respects, the most persuasive argument for epineural permeability derived from the observation of how rapidly a fairly large ionized molecule such as methylene blue (M.W.320) not only "penetrated across the epineurium" but also stained the axons intensely and revealed nodes of Ranvier only by the slight constrictions of the axis cylinder, while the myelin sheath was unstained, and all of this could be visualized through the transparent, unstained epineurium. The axonal staining was indeed beautiful, but the methylene blue was poured into the abdominal cavity, not on the epineurium. In later years, Lorente acknowledged his awareness of Paul Ehrlich's 19th century observation of very rapid vital staining of nerve by intravenous injection of methylene blue, but admittedly he did not discuss this in the context of these arguments about the epineurium. Some modern experiments with the fluorescent dye fast blue continue to mislead neuroscientists about routes of diffusion and axonal transport.

In all fairness, it should be acknowledged that Lorente was struck by the eruption of fluid and swelling of the diameter when the nerve was desheathed. This, plus the change in properties to applied solutions of ions, anesthetics, etc., were indications of the functional importance of the sheath that we tend to ignore today. He also performed many of the critical experiments on the spinal nerve roots because they lack an epineural sheath and tried to emphasize the differences between this 'natural' condition from the profound changes he observed in opening the epineurium. There are some traces of naiveté in his interpretations, but it is difficult to be unimpressed with the ingenuity of his arguments especially the explanation of resheathed nerve properties shown by Feng and Liu (Feng and Liu, '49) and Crescitelli (Crescitelli, 51), who countered Lorente's argument that slitting the epineurium altered the properties of the encased axons. The replacement of the "epineurium", by pulling it back over the nerve like a sausage casing, did not wholly restore the nerve to its original condition, but it was close enough to render difficult the argument that it could not act as a diffusion barrier to certain substances. But this takes us too far afield from the valuable insights still to be learned from the master.

The arguments about the methylene blue staining of axons extended further (Lorente de Nó, '53) as part of another related battle concerning the properties of the nodes of Ranvier and the concept of 'saltatory' conduction between nodes championed by the now classic contribution of Huxley and Stämpfli (Huxley and Stämpfli, '49; Huxley and Stämpfli, '51). The arguments are complex and interesting in the context of the history of morphological knowledge and interpretation.

Ranvier (1878) believed that penetration of substances, notably methylene blue, takes place preferentially at the nodes rather than through myelin, and the greater intensity of staining at the nodal and, to some extent, the paranodal portion of the axon was supported by Cajal (Cajal, 1896; Cajal, '09). The lone dissenter in the post-war years of controversy was Lorente de Nó who berated Hodgkin as well as Huxley and Stämpfli (Huxley and Stämpfli, '51) for misunderstanding Ranvier's description of an endothelial lining in the lamellated myelin sheath, which they interpreted as a barrier to ion diffusion. The argument is now forgotten because electron microscopy subsequently revealed that all of these giants of neuroscience were mistaken about the existence of an endothelial lining in myelin when it became evident that myelin was actually the plasmalemma of Schwann cells. But there were other flies in the ointment, and Lorente's other arguments were not totally ignored by others. Much to his credit, Huxley (see Huxley and Stämpfli, '49 and '51) considered both sides of the issue and presented observations that were not easily reconciled with the notion of saltatory conduction, although it must have irritated Lorente immensely that his work was uncited . The problems included Sanders and Whitteridge's (Sanders and Whitteridge, '46) observation that a large decrease in node spacing can occur without a drop in conduction velocity. This was explained rather ingeniously by assuming a maximum conduction velocity at a particular "normal" node spacing wherein even "considerable deviations from the normal spacing would cause only small changes in velocity" (Huxley and Stämpfli, '49). Another problem was the apparent rarity of nodes of Ranvier in myelinated axons of the central nervous system except at branch points. Huxley believed it unlikely that the basic mechanism of conduction is different in central and peripheral nervous systems but rather ingenuously referred to Bielschowsky's opinion that the interruptions in myelin exist despite the deficient direct evidence.

Another feature that convinced Lorente that conduction can only be understood in intact nerve was based on the weak interaction between nerve fibers in a common nerve trunk. The barely detectable alteration in the excitability threshold produced by a passing impulse, at best a 10% reduction, was most notably analyzed in a paper by Marrazzi and Lorente de Nó (Marrazzi and Lorente de Nó, '44) according to Tasaki's handbook review (Tasaki, '59). Despite the technical elegance in such measurements and a flurry of brief interest in "ephaptic" interaction, Katz and Schmitt (Katz and Schmitt, '40) provided persuasive evidence that the result was electrical in nature by demonstrating that the interaction is enhanced by reducing the shunting effect of fluid around the nerve fiber.

Much of the remainder of Lorente's career as an electrophysiologist dwells on the flow of ionic currents and the principle of continuous decremental conduction in myelinated axons (Lorente de Nó and Condouris, '59), culminating in a long series of papers based on ingeniously contrived experiments with Vicente Honrubia, his compatriot and last disciple in the laboratory. It is argued in these papers, in keeping with his previous belief, that "desheathed nerves are partly demyelinated fibers, and that the isolated myelinated fiber actually is a demyelinated fiber (Lorente de Nó and Honrubia, '64) and later stated that in this condition "electrotonus spreads much as it does in normal unmyelinated fibers" (Lorente de Nó and Honrubia, '65). They also observed that in isolated myelinated fibers "propagation may occur continuously forward or may involve what in figurative language may be called jumps of the nerve impulse from one zone of the internode to another zone of the same or of another internode" (Lorente de Nó and Honrubia, '65a). The "jumps" should not be confused with the unmentioned neologism "saltatory" devised by Lorente's Cambridge adversaries who believed that the action potential is produced at the nodes rather than at the internodes. But Lorente (Lorente de Nó and Honrubia, '65b) believed that the distribution of membrane excitability and the depolarization waveform indicated more than a single (nodal) zone of axonal excitability, and indeed, modern patch clamp studies have revealed paranodal ion channels on both axonal and adjacent Schwann cell membrane patches (Wilson and Chiu, '90). The methods used by Lorente and Honrubia did not provide sufficient spatial resolution to convincingly demonstrate the existence of internodal currents in the paranodal regions, but the experiments provide early clues to the modern patch clamp studies of the three types of $K^+$ channels, which have a density 20 times higher at the node, and the single type of $Na^+$

channel, which is 500 times higher in density at the node than the internode (Grissmer, '86; Jonas et al., '89).

Later experiments based on an enormous number of observations revealed the likelihood of more than a single zone of increased axonal excitability and current flow extending beyond the node (Lorente de Nó and Honrubia, '65a; Lorente de Nó and Honrubia, '65b) and indeed, years later, studies of demyelinating disease models (Bostock and Sears, '78) provided persuasive evidence for continuous decremental, i.e. non-saltatory, conduction in demyelinated axons. These experiments on single axons in what Lorente clearly believed were abnormal conditions, including drying and damage, must have been frustrating, for they produced what he called "a seemingly kaleidoscopic variety of experimental observation" (Lorente de Nó and Honrubia, '65a). He and Honrubia must have worked feverishly; in one spurt from May 25 to July 7 they performed 125 successful experiments on isolated fibers! I cannot pretend to fully appreciate and explain the findings reported in this long series of papers, but it is crucially important to understand the thrust of Lorente's thought.

There was some logic and an important philosophical impetus, deriving from his fundamental original interest in integration at central nervous system synapses, for Lorente to support his arguments about the presence and significance of graded potentials conducted in excitable tissue. He clearly believed in and demonstrated the existence of propagated, decremental potentials in desheathed frog sciatic nerve in a sodium-free medium (Lorente de Nó, '59; Lorente de Nó, '61), yet believed that the all-or-none law was valid in "untreated nerve" (Lorente de Nó, '61). However, his lifelong interest in the cerebral cortex containing neurons with numerous synapses on long dendritic processes required "the existence of graded responses capable of spreading without leaving a refractory period, and capable of summating", something that could not be achieved with a discontinuous, all-or-none nerve impulse (Lorente de Nó, '59). He postulated this concept in a paper as early as 1934 (Lorente de Nó, '34) and developed it more fully in his pioneer papers on the graded potentials he saw recording from motoneurons (Lorente de Nó, '39; Lorente de Nó, '47), thereby presaging the later interpretation of intracellularly recorded synaptic potentials and the cable properties of dendritic conduction. Thus, decremental conduction was the *general* principle and essential in the larger problem of integration. He did not believe that integration occurred in a strict mathematical sense because he was also enamored of the idea of "reverberating circuits" - a popular topic of intellectual ferment at the Rockefeller Institute during Lorente's many productive post-war years there. His passion for the interpretation of electrical fields in nervous tissue was a source of mixed delight and frustration among his colleagues, notably in heated discourses with the stalwarts David P.-C.Lloyd, Alexander Mauro and others, but most notoriously with Herbert Gasser in the corner of the Institute dining room where the gradually rising volume of their arguments over the lunch table often sounded scandalous. The Spanish disposition and spirit rendered all of this permissible from Don Rafael's point of view, and when he retired from the Rockefeller and accepted the invitation of Vicente Honrubia to join him at the University of California, Los Angeles (UCLA) and assume a position as an Emeritus Professor, he found a coterie of young colleagues with whom he could debate about the interpretation of electrical potentials in Castillian *bel voce*. It was not without interest to Lorente that the UCLA campus was created by the development of the last of the large original Spanish land grants; a piece of Spain resembling the terrain of his native land in a city where Spanish was becoming the dominant language. The geography and intellectual climate proved felicitously *simpatico* but for periods of air pollution. In this last period of his scientific career his passionate outbursts, and sometimes brooding anger, seemed reserved for his *Españolaphonic* friends Honrubia, Emilio Decima, José (Pepe) Segundo and Jorge Larriva-Sahd, for he remained courteous, controlled and almost *sotto voce* with the rest of us. With women he was frankly courtly as well as charming in much the chivalrous manner of Don Quixote and usually energetically conscious of trying to entertain non-scientist companions - especially if they were beautiful!

Because this essay constitutes an historical memoir, it would be remiss to fail noting Don Rafael's keen pursuits and depth of scholarship in the history of his several fields of interest. He understood and indeed fully credited the monumental advances in the knowledge of nerve conduction made by Keith Lucas (Lucas, '17) and later by Kato (Kato, '34) in establishing the all-or-none conduction principle in *intact* peripheral myelinated axons, and his knowledge of 19th century electrophysiology was astonishing in its scope, covering the significant German, as well as English, literature extending to the more esoteric, at times, as in his discussion of Lord Kelvin's treatise "On peristaltic induction of electric currents in submarine telegraph wires " (Marrazzi and Lorente de Nó, '44). He erred in some of his attacks on Hodgkin, Huxley, Katz and Rushton and clearly put his money on the wrong horse in supporting Nachmansohn's belief in the role of acetylcholine in axonal impulse conduction before its synaptic role became evident, but we must honor him for the prescience of so many of his insights and for the amazing enormity of his original personal observations at the lab bench. You may have noted the gradual *glissando* of referring to the maestro as Lorente and then as Don Rafael, deliberately reflecting the austere and the intimate relationship he maintained with his friends. For those of us who experienced the vociferous, heated and sometimes truly passionate arguments that constituted a considerable component of the joy in having Rafael's friendship, he is remembered with a warmth that invariably extends beyond the descriptor of mere fondness. His encounters with the new personal computers of the seventies was not unlike that of a child with his first set of electric trains, except that a PC was more socially acceptable.

His last contributions to the electrophysiological literature contained a formidable, sophisticated mathematical analysis of the theory underlying voltage clamp of nerve membranes, which he unabashedly pronounced was "radically and essentially different from what has been offered in the literature" (Lorente de Nó, '71) maintaining that "membrane conductance... remains practically constant during the production of the nerve impulse". He recognized that the spatial resolution required for visualizing the spread of current from an action potential of long time constant within the largest internode lengths of 3 mm in frog nerve provided a serious obstacle (Lorente de Nó and Honrubia, '64). I cannot pretend to interpret or offer opinions about the final period of his scientific activities when he transformed into a mathematician and was seduced by the new technology of the computer.

If there were errors in interpretation, and there were many, it is not a rare occurrence in fertile, imaginative minds and in a way reflects on the excellence, integrity and energy of the remarkable observations of this great scientist. Like his master, Santiago Ramón y Cajal, the sheer magnitude of his original observations was so great that even a large number of erroneous interpretations must be discounted as almost trivial on a proportional basis. As Cajal's last surviving protegé, he left a profound mark on the history of 20th century neuroscience and sustained the great tradition of Spanish influence on the development of this field. In the latter years of his life, my friend Rafael remained passionately interested in neuroscience and in new ideas and discoveries, but he retained a consuming desire to have the world acknowledge that it misunderstood the Hodgkin-Huxley formulation, and he pursued this with vigor and anger, but not despair, Fortunately, he also was able to look further back into an intellectually rich life and reinvigorated a life-long interest in Cervantes and his beloved book Don Quixote de la Mancha. He quoted passages from memory in both pure Castillian Spanish and in a charmingly accented English, and saw in himself the trials of Don Quixote, including the recognition of his personal sense of tilting at windmills. His legacy lies in the insights he provided in every field he touched, and his scope was amazingly vast; perhaps greater than any other neuroscientist of this century. There were few topics in neuroanatomy or in electrophysiology in which he could not offer fresh, and often brilliant, insights. Although his health gradually decayed with pulmonary disfunction in his latter years, ultimately forcing him to leave Los Angeles for the dry air of Arizona, he showed no signs of senility, and indeed his last major publication in 1981 was an up-to-date, sophisticated monograph on the Acoustic Nuclei (Lorente de Nó, '81) based on the elegant Golgi preparation drawings that he had prepared almost 50 years earlier. For those few of us who had the good fortune of sharing his friendship in his latter years, we preserve the memory of a gentle, witty and kind man whose zest and passion for science and for literature provided inspiration of lasting value. His death prompted turning to his beloved Cervantes's account of that imaginative gentleman, Don Quixote, that prophetically ends with an epitaph Rafael recited in elegant style and which reads (Cervantes Saavedra, '57):

> *Here lies the noble fearless knight,*
> *whose valour rose to great height;*
> *When Death at last had struck him down,*
> *His was the victory and renown.*
> *He deemed the world of little prize,*
> *And was a dreamer in men's eyes;*
> *But had the fortune in his age*
> *To live a fool and die a sage.*

## REFERENCES

Adrian, E.D., 1930, The effects of injury on mammalian nerve fibres, *Proc. roy. Soc.*, B106:596.
Bostock, H. and Sears, T.A., 1978, The internodal axon membrane: Electrical excitability and continuous conduction in segmental demyelination, *J. Physiol.*, 280:273-301.
Cajal, S.R., 1896, El azul de metileno en los centros nerviosos, *Rev. trim. microgr.*, 1:151.
Cajal, S.R., 1909, "Histologie dy système nerveux". vol. 1. A.Maloine, Paris.
Crescitelli, F., 1951, Nerve sheath as a barrier to the diffussion of certain substances, *Am. J. Physiol.*, 166:229.
de Cervantes Saavedra, M., 1957, "Don Quixote of La Mancha", New American Library, New York. Modified. from the W.Starkie translation.
Feng, T.P., and Liu, Y.M., 1949, The connective tissue sheath of nerve as effective diffusion barrier, *J. Cell. Comp. Physiol.*, 34:33.
Grissmer, S., 1986, Properties of potassium and sodium channels in frog internode, *J. Physiol.*, 381:119.
Honrubia, V. and Lorente de Nó, R., 1962, On the effect of anesthetics upon isolated, single frog nerve fibers, *Proc. Natl. Acad. Sci.*, 48:2065.
Huxley, A.F. and Stämpfli, R., 1951a, Evidence for saltatory conduction in peripheral myelinated nerve fibers, *J. Physiol.*, 108:315.
Huxley, A.F. and Stämpfli, R., 1951b, Effects of potassium and sodium on resting and action potentials of single myelinated fibers, *J. Physiol.*, 112:496.
Jonas, P., Bräu, M.E., Hermsteiner, K., and Vogel, W., 1989, Single-channel recording in myelinated nerve fibers reveals one type of Na channel but different K channels, *Proc. Natl. Acad. Sci. USA*, 86:7238.
Kato, G., 1943, "The Microphysiology of Nerve", Maruzen, Tokyo.
Katz, B. and Schmitt, O.H., 1940, Electric interaction between two adjacent nerve fibres, *J. Physiol.*, 97:471.
Larramendi, L.M.H., Lorente de Nó, R., and Vidal, F., 1956, Restoration of sodium-deficient frog nerve fibres by an isotonic solution of guanidinium chloride, *Nature*, 178:316.
Lorente de Nó, 1934, Studies on the structure of the cerebral cortex.II. continuation of the study of ammonic system, *J. Physiol. Neurol.*, 46:113.
Lorente de Nó, R., 1939, Transmission of impulses through cranial motor nuclei, *J. Neurophysiol.*, 2:402.
Lorente de No, R., 1947a, Action potential of the motoneurons of the hypoglossus nucleus, *J. Cell Comp. Physiol.*, 29:207.
Lorente de Nó, R., 1947b, A study of nerve physiology. Studies from the Rockefeller Institute, New York, vols. 131 and 132.
Lorente de Nó, R., 1949, On the effect of certain quaternary ammonium ions upon frog nerve, *J. Cell Comp. Physiol.*, 33(suppl.1):3.
Lorente de Nó, R., 1952, Observations on the properties of the epineurium of frog nerve, *Cold Spr. Harb. Symp. quant. Biol.*, XVII:299.
Lorente de Nó, 1953, Observations on vital staining of the axons of myelinated fibers, *Folia Psychiatr. Neurol. Neurochir. Neerl.*, 56:3.

Lorente de Nó, R., 1959, Decremental conduction and summation of stimuli delivered to neurons at distant synapses, *in*: "Structure and Function of the Cerebral Cortex", Proc. 2nd int. Mtg. Neurobiologist, D.B.Tower and J.P.Schadé, eds., Elsevier, Amsterdam.

Lorente de Nó, R., 1961, Decremental conduction in peripheral nerve. Nature of the nerve impulse, "Bioelectrogenesis", Proc. Symp. Comp. Bioelectrogen., C.Chagas and A.Paes de Carvalho, eds., Elsevier, Amsterdam.

Lorente de Nó, R., 1971, Theory of the voltage clamp of the nerve membrane, I. Ideal clamp, *Proc. Natl. Acad. Sci.*, 68:192.

Lorente de Nó, R., 1981, "The Primary Acoustic Nuclei", Raven Press, New York, pp. 139.

Lorente de Nó, R. and Condouris, G.A., 1959, Decremental conduction in peripheral nerve. Integration of stimuli in the neuron, *Proc. Natl. Acad. Sci.*, 45:592.

Lorente de Nó, R. and Honrubia, V., 1964a, Electrical stimulation of the internodes of single fibers of desheathed nerves, *Proc. Natl. Acad. Sci.*, 52:1142.

Lorente de Nó, R. and Honrubia, V., 1964b, Electrical stimulation of the internodes of single fibers of nerves with intact sheath, *Proc. Natl. Acad. Sci.*, 52:783.

Lorente de Nó, R., and Honrubia, V., 1965a, Production of action potentials by the internodes of isolated myelinated nerve fibers, *Proc. Natl. Acad. Sci.*, 53:757.

Lorente de Nó, R. and Honrubia, V., 1965b, Theory of the flow of action currents in isolated myelinated nerve fibers, I., *Proc. Natl. Acad. Sci.*, 53:938.

Lucas, K., 1917, "Conduction of the nerve impulse", Longmans, London, pp. 102.

Lundberg, A., 1951, Electrotonus in frog spinal roots and sciatic trunk, *Acta Physiol. Scand.*, 23:234.

Marrazzi, A.S. and Lorente de Nó, R., 1944, Interaction of neighboring fibres in myelinated nerve, *J. Neurophysiol.*, 7:83.

Rashbass, C. and Rushton, W.A.H., 1949, The relation of structure to the spread of excitation in the frog's sciatic nerve. *J. Physiol.*, 110:110.

Sanders, F.K. and Whitteridge, D., 1946, Conduction velocity and myelin thickness in regenerating nerve fibres, *J. Physiol.*, 105:152.

Tasaki, I., 1959, Conduction of the nerve impulse, in: "Handbook of Physiology. Neurophysiology, Vol.I", J. Field, ed., American Physiological Society, Washington, D.C., pp. 75-121.

Wilson, G.F. and Chiu, S.Y., 1990, Ion channels in axon and Schwann cell membranes at paranodes of mammalian myelinated fibers studies with patch clamp, *J. Neurosci.*, 10:3263.

# INDEX

# Index

ACh receptors, 280
Amino acids, 155, 159, 179, 181, 187, 196, 205, 211, 212, 213, 226, 229, 240, 273, 279, 335
AMPA receptor, 180
Ascending, 1, 3, 22, 25, 45, 46, 65-67, 78, 87, 92, 143-145, 147-149, 153, 159-161, 169, 183, 208, 211, 276, 287, 298, 305, 321, 322, 323, 335, 344, 421, 432, 447
Auditory nerve, 29-31, 35-40, 43, 45, 48, 50, 51, 55, 56, 65, 66, 68-70, 72, 75-79, 83, 86, 87, 92, 100, 101, 103, 116, 167, 181-183, 191, 211, 212, 395, 405, 406
AVCN, 3, 11, 12, 20, 22, 30, 31, 34, 35, 44-47, 55-59, 61, 66, 75-79, 92-94, 96, 97, 121-126, 146-149, 156, 170-172, 183, 186, 187, 191, 196-198, 201, 204, 205, 212, 215, 219, 227, 233, 239-243, 246-249, 267, 275, 303-307, 310, 312-314, 323, 329- 331, 335-340, 346, 354, 356, 358, 389, 395, 406, 411, 416, 417, 419, 424, 425, 406
Barrels, 242, 254, 472-474, 476, 479, 481, 482, 488, 491, 492, 494, 495, 497
Bat, 30, 36, 75, 145, 147-149, 155, 161, 233, 303, 305-307, 312, 315, 316, 322-326, 330, 331
Bushy cells, 6, 7, 9, 10, 12, 16, 22, 68, 70, 79, 93, 122, 125, 212-215, 218, 219, 231, 234, 239-242, 244, 248, 249, 287, 305, 310, 312, 313, 323, 330, 331, 335, 337, 338, 350, 354, 389, 395-398, 400, 402, 406, 407

C-fos, 23
Cartwheel cells, 110, 114, 116, 183, 186, 189, 191, 225, 226, 228, 230, 234, 262, 361, 362, 364, 365, 366, 368
Cat, 1, 8, 30, 31, 46, 58, 59, 65, 66, 70, 75, 79, 95, 101, 102, 125, 126, 133, 136, 137, 145, 148, 155, 157, 159, 160, 162, 174, 182, 183, 189, 253, 268, 280, 287, 306, 322-325, 329-331, 338, 339, 342, 346, 349, 350, 354, 356, 358, 377, 457, 464
Cerebellum, 83, 107, 109, 111-116, 173, 269, 361, 366, 367, 422, 447, 469
Cochlear root, 273, 275, 276, 291

Coding, 12, 13, 56, 65, 81, 83, 86, 87, 143, 180, 242, 373, 386
Columns, 323, 442, 479, 491, 494, 497
Colloidal gold, 169-172
Complex sounds, 373
Cortex, 15, 107, 108, 112, 114, 137, 153, 156, 187, 226, 287, 338, 344, 361, 432, 451, 453, 455-472, 475-477, 479-497, 503, 508
Characteristic frequency, 66, 233, 239, 350, 388
ChAT, 136, 267-276, 279-288
Choppers, 11, 83, 324, 328, 330-332, 354-356, 381, 389, 397, 411, 416, 417, 419

Descending projections, 126, 143, 149, 155, 160, 161
Development, 8, 12, 15, 19, 21, 22, 25, 29, 30, 36, 38-40, 59, 180, 182, 226, 293, 294, 322, 421, 426, 429, 430, 435, 467, 471, 482, 484, 487, 495, 496, 497, 503, 505, 508, 509
Dextran, 335-336, 338, 346
Discharge rate, 55, 56, 66, 241-242, 244, 247-249, 258, 259, 263, 362, 396, 402, 403, 405-407, 413-414
Dorsal Cochlear Nucleus, 14, 20, 30, 31, 44, 45, 91, 103, 107, 125, 154, 160, 182, 225, 231, 240, 253, 254, 258, 259, 305, 329, 335, 361, 391

Echolocation, 303, 307, 316
Electrophysiology, 16, 503, 509
Excitation, 23, 78, 87, 168, 259, 263, 265, 443, 444, 472

Fluorescent, 145, 155, 310, 336, 482, 506
Frog, 7, 181, 431, 431, 434, 439, 440, 443, 448, 504, 508, 509
Fundamental frequency, 373
Fusiform cells, 79, 160-162, 182, 183, 186, 189, 230, 231, 233, 234, 335, 392

GABA, 12, 116, 122, 124-128, 133-136, 162, 168, 169, 195-199, 201, 205-207, 211-222, 225, 229, 231, 233, 239-242, 244, 247, 248, 254, 273, 279, 287, 296, 340
$GABA_A$ receptor, 212, 213, 215, 217, 231, 233, 242, 244, 247, 248, 254
$GABA_B$ receptor, 212, 214, 216, 244, 248, 254

Gerbil, 45, 75, 145, 147, 148, 233, 291, 293, 296,298, 299, 313, 322
Glomerulos, 472, 479, 480, 481, 484, 494, 497
Glutamate, 7-9, 77, 122, 133, 162, 169, 170, 179, 181, 182, 211, 213, 219, 279, 353, 368
Glutamate receptors, 7, 8, 211
Glycine receptor, 122, 214, 229, 231, 233, 244, 254
Golgi, 1, 22, 30, 44, 68, 75, 93-97, 114, 116, 122, 125, 148, 154, 159, 182, 189, 275, 291, 292, 446, 452, 453, 457, 468-475, 484, 497, 498, 503, 509
Guinea Pig, 47, 48, 75, 79, 114, 116, 125, 135, 145, 155, 157, 158, 162, 169, 201, 205-207, 212, 213, 220, 233, 253, 361, 374, 377, 382, 402, 423-425, 458

Harmonic complexes, 382
Hippocampus, 15, 187, 368, 451, 456-458, 463, 465, 497
HRP, 30, 31, 34, 36-8, 44, 51, 55-57, 61, 67, 72, 75-79, 122, 125, 126, 155, 158, 159, 161, 162, 220, 226, 240, 241, 243, 269, 291, 295-297, 299, 305, 315, 324, 329-331, 336, 342, 343, 349, 350, 411, 443, 446, 457, 465, 494
Hybridization, in situ, 8. 179, 180, 183, 184, 186, 187, 190, 191

Immunohistochemistry, 268
Implant, 21, 423
Inferior colliculus, 153
Inhibition, 9, 78-81, 83, 86, 128, 168, 174, 214, 222, 225, 228-234, 239, 241, 242, 244, 253-265, 363, 374, 389, 392, 407, 426, 477, 494
Interneurons, 233, 467
Iontophoresis, 458
Isointensity contours, 243, 244

L7, 108, 112, 114
Lateral lemniscus, 7, 21, 87, 153, 155, 157, 160, 205, 296, 297, 298, 321-324, 330, 332, 335-338, 346, 352, 354, 355
Lateral superior olive (see also LSO), 145, 303, 335
LSO, 7, 9, 145, 146, 148, 220, 274, 303, 305-307, 309, 312, 313, 316, 335, 337, 340-344, 346, 352, 354, 355

Maps, 29, 95, 259, 265, 354, 479, 480, 496
Medial superior olive (see also MSO), 22, 137, 145, 303, 354
Medial nucleus of the trapezoid body (see also MNTB), 22, 137, 145, 303, 354
Midbrain, 30, 300, 314, 321, 335, 344, 424
MNTB, 7, 137, 145, 146, 206, 303, 305, 306, 312, 313, 329, 330, 331, 337, 353, 356
Model, 4, 12, 14, 16, 29, 56, 70, 107, 116, 179, 354, 386, 388-390, 392, 395-408, 411, 413, 415, 416, 417, 419, 423, 425, 442, 455, 456, 468, 470, 471, 475, 477, 480, 494, 459, 497, 508
Modelling, 40, 385-389, 392, 408, 451
Molecular layer, 22, 107, 109, 110, 112, 114, 159, 160, 170, 173, 271, 281, 284, 287, 361, 366, 367, 368, 452, 459
Monodelphis, 29-31, 36, 37, 39, 40
Mouse, 8, 30, 36-39, 45, 75, 79, 101, 108-110, 145, 147, 155, 159, 174, 182, 219, 226, 291, 296, 298, 299, 313, 350, 353, 358, 432, 451, 453, 470, 472, 475, 479, 480, 482-484, 489, 492, 493, 494, 495, 497, 498
MSO, 7, 145, 146, 148, 303, 305-307, 309, 310, 312-316, 335, 337, 340, 341, 343, 344, 346, 354, 356, 358
Mutant mice, 109, 110

Neurotransmitters, 168, 179, 181, 182, 212, 214, 222, 240, 267, 279, 340
NMDA receptor, 7, 180, 182, 368
Non-primary, 133, 136, 137, 140, 269

Octopus cells, 13, 14, 31, 79, 87, 136, 162, 183, 189, 212, 215, 218, 219, 230, 231, 286, 287, 298, 305, 331, 338, 356, 406
Olivo-cochlear neurons, no hay
Onset, 6, 10, 13, 14, 19, 22, 25, 68, 81, 83, 222, 258, 265, 309, 316, 321, 324, 325, 328, 331, 332, 350, 356-358, 373-382, 389, 392, 407, 423

Phase-locking, 13, 350, 353, 357
Primary-like, 6, 79, 83, 231, 239, 249, 325, 353, 389
Prostheses, 421, 429
Purkinje cells, 107-116, 362, 364-366, 368
PVCN, 12, 20, 30, 31, 34, 35, 38, 44, 45, 75, 76, 83, 122, 124-126, 137, 148, 156, 183, 186, 201, 215, 219, 233, 275, 305, 329-331, 336-338, 342, 344, 354-358, 395, 406, 424, 425

Rat, 19-25, 30, 39, 40, 83, 136, 155-162, 169, 174, 179, 181, 183, 187, 189, 190, 231, 267, 268, 269, 271, 273-276, 279-284, 287-299, 324, 330, 331, 336, 338, 346, 350, 352, 353, 356, 366, 367, 457, 458, 464, 465, 480-482, 484, 491, 495
Receptors, 7, 8, 13, 29, 65, 122, 168, 179-182, 190, 191, 195, 211-217, 222, 229, 231, 233, 240, 244, 247, 248, 254, 262, 265, 267, 280, 300, 367, 368, 438-440, 446-448
Response patterns, 79, 244, 249, 279, 325, 363, 389, 392, 404, 408

Small cells, 46, 68-70, 91, 94, 99, 100, 103, 161, 186, 215, 218, 281
Small cell cap, 12, 56, 67, 68, 91-97, 99, 101, 103, 148, 149, 217, 226
Speech, 222, 373, 376, 381, 282

Spherical cells, 148, 183, 186, 189, 287, 305, 395
Spiral ganglion, 29, 36, 37, 39, 40, 43, 44, 47-51, 55-59, 61, 65, 173, 174, 191, 294-296, 425
Spontaneous discharge rate, 55, 56, 66, 244, 406, 407
Stellate cells, 6, 8, 12, 14, 15, 31, 68, 70, 79, 93, 96, 124, 125, 136, 137, 139, 186, 215, 217, 225, 226, 228, 230, 234, 310, 330, 331, 335, 338, 354, 411, 414, 417, 419, 476, 484
Superior olive, 22, 126, 128, 137, 144-147, 268, 288, 303, 306, 313, 314, 323, 335, 339, 341, 346, 354
Synapses, 4, 8, 9, 10, 13, 15, 21, 22, 29, 51, 55-57, 59-61, 68, 70, 71, 86, 97, 99, 101, 113, 121, 122, 125, 128, 153, 162, 167-169, 173, 174, 179, 181, 182, 191, 195, 211, 212, 214, 215, 229, 267, 268, 279, 288, 340, 341, 361, 366-368, 403, 411, 432, 433, 435, 444, 455, 508, 509
Synaptic endings, 1, 9, 16, 68, 71, 122, 124-126, 168-174, 195, 197-199, 205, 341

Tonotopic, 2, 3, 19, 22-24, 29, 46, 55, 78, 155, 158, 160, 161, 240, 242, 305, 312, 323, 324, 331, 335, 424
Tuberculoventral neurons, 75-79, 81, 83, 86, 87, 225-228, 233, 234
Type I cells (*see also* type I neurons), 259
Type II cells (*see also* type II neurons), 79, 258
Type III cells, 265
Type IV cells, 258, 259, 265
Type I neurons, 43, 45, 46, 49, 51, 65
Type II neurons, 43, 44, 46, 47, 51, 65, 79

Ventral cochlear nucleus, 20, 31, 45, 46, 55, 56, 66, 70, 75, 91-97, 101-103, 121, 148, 183, 212, 215, 225, 226, 230, 231, 239, 296, 298, 303, 305, 323, 328-332, 335-337, 374, 381, 382, 395, 411
Vertical cells (*see also* tuberculoventral neurons), 79, 125, 240, 249, 259, 265
Vestibular nerve, 36, 50, 437-439, 442, 445-447

Printed in the USA
CPSIA information can be obtained
at www.ICGtesting.com
CBHW082127011224
18288CB00005B/146

9 781461 362739